Advanced Technologies for Rechargeable Batteries

This volume covers recent advanced battery systems such as metal ion, hybrid, and metal-air batteries. It includes an introduction to fluoride, potassium, zinc, chloride, aluminium, and iron-ion batteries; special or hybrid batteries are included, with calcium, nuclear, thermal, and lithium-magnesium hybrid batteries also explained. It summarizes the recent progress and chemistry behind the popular metal-air batteries, including a systematic overview of the components, design, and integration of these new battery technologies.

Features:

- Covers recent battery technologies in detail, from the chemistry to advances in post-lithium-ion batteries.
- Various post-lithium-ion batteries are discussed in detail.
- Includes a section on ion batteries, exploring new types of metal ion batteries.
- Focuses in each chapter on a particular battery type, including different metal ion batteries such as zinc, potassium, aluminium, and their air version batteries.
- Provides authoritative coverage of scientific content via global contributing experts.

This book is aimed at graduate students, researchers, and professionals in materials science, chemical and electrical engineering, and electrochemistry.

Advanced Technologies for Rechargeable Batteries

Metal Ion, Hybrid, and Metal-Air Batteries

Volume 2

Edited by

Prasanth Raghavan, Akhila Das, and Jabeen Fatima M. J.

CRC Press
Taylor & Francis Group
Boca Raton London New York

CRC Press is an imprint of the
Taylor & Francis Group, an **informa** business

Designed cover image: © Shutterstock Images

First edition published 2025
by CRC Press
2385 NW Executive Center Drive, Suite 320, Boca Raton FL 33431

and by CRC Press
4 Park Square, Milton Park, Abingdon, Oxon, OX14 4RN

CRC Press is an imprint of Taylor & Francis Group, LLC

ISBN: 9781032315362 (hbk)
ISBN: 9781032315379 (pbk)
ISBN: 9781003310174 (ebk)

DOI: 10.1201/9781003310174

Typeset in Times
by codeMantra

Dedication

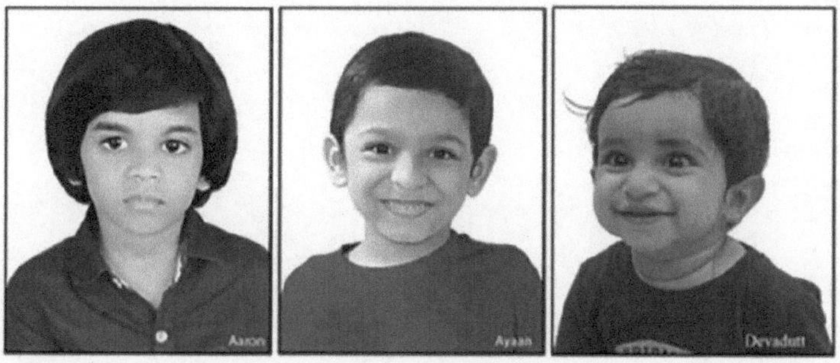

Dedicated to the Little Lanterns, who gave the best support for the successful completion of this book, but if they are understanding, then the book must have been completed much earlier than expected.

Contents

Preface

The realization of the depletion of fossil fuels by the exploitation of proliferating populations has impacted negatively on the environment and mankind. Increasing demand for energy led to the misuse of non-renewable energy sources, which resulted in global warming, climatic changes, and pollution of the environment. Addressing the booming demand for energy following the green protocol led to the search for energy sources that are sustainable and efficient. Such inquisitive nature of humans found the presence of green energy in renewable energy sources such as solar, wind, and geothermal sources, which are environmentally benign. Extraction of energy from these sources has less carbon footprint and contributes a decent amount of energy towards the grid. Since renewable sources are intermittent, the consistency of energy extraction from and reliance on them are being questioned. Therefore, developments in energy storage technology are instigated, where supercapacitors and batteries were shortlisted as an effective solution. Such devices store energy and can power portable electronic devices and contribute towards the realization of powering electric vehicles. For such realization, high-energy-density storage devices can unequivocally outperform conventional energy storage devices. The developments in battery technology have been initiated in the past century, which trace back to the Parthian Period. Rechargeable batteries were popularized through lead acid and nickel–cadmium batteries. Greater developments were recorded in the 20th century, when alkaline batteries and nickel-metal-hydride batteries occupied a greater share of the battery market. It can be ubiquitously said that revolutionary developments were reported after the commercialization of lithium-ion batteries (LIBs) by Sony Corporation in 1991.

The past few decades have seen exponential growth in the optimization and enhancement of LIB technology. LIB has become a benign technology for powering portable devices, electronic gadgets, smartwatches, and electrifying electric vehicles. But, consistency regarding the raw material prices and availability of materials has always been interrogated. As greater advancements in the electronics sector and electric vehicle technology arose, demand for high-energy-density batteries increased. Many battery technologies have been introduced in the past few decades, of which other batteries were popular in the post-LIB technology. The book titled *Advanced Technologies for Rechargeable Batteries* highlights developments in battery technology. The book introduces general considerations and then dives deep into the mechanism and electrochemistry of each battery technology, unveiling the complexities of battery chemistry.

In a nutshell, the book sheds light on the topics of different types of batteries, which ubiquitously enhance knowledge on the future of post-LIB era batteries. The book will serve as a reference material for those who work in the areas of energy storage technology and materials for battery technology.

About the Editors

Prasanth Raghavan, Professor, at Materials Science and NanoEngineering Lab (MSNE-Lab), Department of Polymer Science and Rubber Technology, Cochin University of Science and Technology (CUSAT); Visiting Professor at the Department of Materials Engineering and Convergence Technology, Gyeongsang National University, Republic of Korea, International and Inter University Centre for Nanoscience and Nanotechnology (IIUCNN), Mahatma Gandhi University, Kottayam, Kerala, India, and at Biorefining and Advanced Materials Research Center, SRUC (Scotland's Rural College), Edinburgh, UK. He received his PhD in Engineering under the guidance of Prof. Jou-Hyeon Ahn, from the Department of Chemical and Biological Engineering, Gyeongsang National University, Republic of Korea, in 2009, under the prestigious Brain Korea (BK21) Fellowship. He completed his BTech and MTech from CUSAT, India. After a couple of years of attachment stint as Project Scientist at the Indian Institute of Technology (IIT-D), New Delhi, he moved abroad for his PhD studies in 2007. His PhD research was focused on the fabrication and investigation of nanoscale fibrous electrolytes for high-performance energy storage devices. He completed his Engineering doctoral degree in less than 3 years and still it's an unbroken record in the Republic of Korea. After PhD, Dr. Prasanth joined as Research Scientist at Nanyang Technological University (NTU), Singapore, in collaboration with Energy Research Institute at NTU (ERI@N) and TUM CREATE, a joint electromobility research centre with Germany's Technische Universität München (TUM) and NTU, where he was working with Prof. Rachid Yazami, one who successfully introduced graphitic carbon as an anode for commercial lithium-ion batteries and received Draper Prize, along with the Nobel Laurates Prof. J. B. Goodenough and Prof. Akira Yoshino. After 4 years in Singapore, Dr. Prasanth moved to Rice University, USA, as Research Scientist where he worked with Prof. Pulickal M Ajayan, the co-inventor of carbon nanotubes, and was lucky to work with 2019 Chemistry Nobel Laureate Prof. J. B. Goodenough. Dr. Prasanth was selected for the Brain Korea Fellowship (2007); SAGE Research Foundation Fellowship, Brazil (2009); Estonian Science Foundation Fellowship and European Science Foundation Fellowship (2010); Faculty Recharge, UGC (2015); etc. He received many international awards including the Young Scientist Award, Korean Electrochemical Society (2009), and was selected for the Bharat Vikas Yuva Ratna Award (2016). He developed many products such as high-performance braking parachutes, flex wheels for space shuttles, and high-performance lithium-ion batteries for leading portable electronic devices and automobile industries. He has a general research interest in polymer synthesis and processing, nanomaterials, green/nano-composites, and electrospinning. His current research focused on nanoscale materials and polymer composites for printed and lightweight charge storage solutions including high-temperature supercapacitors and batteries. He has published a good number of research papers and book/book chapters in high-impact factor journals and has more than 5,000 citations and an h-index of 45 plus. Apart from science and technology, Dr. Prasanth is a poet, activist, and columnist in online portals and printed media.

Akhila Das is a graduate student pursuing a PhD at Materials Science and NanoEngineering Lab (MSNE-Lab), Department of Polymer Science and Rubber Technology (PSRT), Cochin University of Science and Technology (CUSAT), India. She completed her master's degree in Chemistry from St. Thomas College Thrissur after receiving her Bachelor's degree in Chemistry from the University of Calicut (Christ College, Irinjalakuda). She has qualified in the IIT JAM exam. She has expertise in the extraction of natural oils and medicinal components from medicinal plants. She was an intern/project student in Dr. Ajay Ghosh's group (Winner of the Shanti Swarup Bhatnagar prize). Her MS project was entitled "An Efficient Supramolecular Approach for CO_2 Detection". Her research area focused on the "Development of Alternative Battery Technology Other Than Li Ion Technology". More than 25 peer-reviewed international publications and a couple of international conference publications are in her credits. Her current area of interest includes batteries, supercapacitors, and e-waste management for a sustainable future and a green environment.

Jabeen Fatima M. J. has recently joined Carborundum Universal Limited as Manager of Technology, Kakkanad, Cochin, Kerala, India. She was previously engaged as a research scientist at the Materials Science and Nano Engineering Lab (MSNE-Lab), Department of Polymer Science and Rubber Technology (PSRT), Cochin University of Science and Technology (CUSAT), India. Before joining MSNE Lab, she was working as a tentative Assistant Professor at the Department of NanoScience and Technology, University of Calicut, India. She received her PhD in Nanoscience and Technology from the University of Calicut in 2016, India, with a prestigious National Fellowship JRF/SRF from the Council of Scientific and Industrial Research (CSIR), under the Ministry of Science and Technology, Government of India. Her research area was focused on the synthesis of nanostructures for photoelectrodes for photovoltaic applications, energy storage devices, photoelectrochemical water splitting, catalysis, etc. She received her MS degree in Applied Chemistry (University First Rank) after receiving her BSc degree in Chemistry from Mahatma Gandhi University (MGU), Kottayam, India. She received many prestigious fellowships including a Junior/Senior Research Fellowship (JRF/SRF) from the Centre for Science and Research, Department of Science and Technology, Ministry of India; Post-Doctoral/Research Scientist Fellowship from Kerala State Council for Science, Technology and Environment (KSCSTE), and InSc Research Excellence Award. She has published a good number of full-length research articles in peer-reviewed international journals and book chapters with international publishers. She is serving as a reviewer for many STM journals published by Wiley International, Elsevier, Springer Nature, etc. Her current area of interest includes the development of flexible and free-standing electrodes for printable and stretchable energy storage solutions and the development of novel nanostructured materials and ternary composite electrodes and electrolytes for sustainable energy applications like supercapacitors, fuel cells, and lithium-ion batteries.

Contributors

Jou-Hyeon Ahn
Department of Material Engineering and Convergence Technology
Gyeongsang National University
Jinju, Korea

Ajish Babu
Department of Metallurgical and Materials Engineering
Indian Institute of Technology Patna
Bihta, India

Neethu T. M. Balakrishnan
Materials Science and NanoEngineering Lab (MSNE-Lab), Department of Polymer Science and Rubber Technology
Cochin University of Science and Technology
Cochin, India

Akhila Das
Materials Science and NanoEngineering Lab (MSNE-Lab), Department of Polymer Science and Rubber Technology
Cochin University of Science and Technology
Cochin, India

Pritam V. Dhawale
Department of Polymer and Surface Engineering
Institute of Chemical Technology
Mumbai, India

Nutan Gupta
School of Materials Science and Engineering
NTU
Singapore, Singapore

Jabeen Fatima M. J.
Materials Science and NanoEngineering Lab (MSNE-Lab), Department of Polymer Science and Rubber Technology
Cochin University of Science and Technology
Cochin, India

N. S. Jishnu
Institute of Materials Science
Technical University of Dresden
Dresden, Germany

Roshny Joy
Materials Science and NanoEngineering Lab (MSNE-Lab), Department of Polymer Science and Rubber Technology
Cochin University of Science and Technology
Cochin, India

R. Kiruthiga
Institute of Scientific and Technological Research
University of Atacama
Copiapo, Chile

Devika S. Lal
School of Energy Materials
Mahatma Gandhi University
Kottayam, India

Kee Wah Leong
Department of Mechanical Engineering
The University of Hong Kong
Hong Kong, China

Dennis Y. C. Leung
Department of Mechanical Engineering
The University of Hong Kong
Hong Kong, China

Nikhil Medhavi
Materials Science and NanoEngineering Lab (MSNE-Lab), Department of Polymer Science and Rubber Technology
Cochin University of Science and Technology
Cochin, India

Shakir Bin Mujib
Alan Levin Department of Mechanical and Nuclear Engineering
Kansas State University
Manhattan, Kansas

Santanu Mukherjee
Alan Levin Department of Mechanical and Nuclear Engineering
Kansas State University
Manhattan, Kansas

Anandu M. Nair
School of Energy Materials
Mahatma Gandhi University
Kottayam, India

C. Nithya
Department of Chemistry
PSGR Krishnammal College for Women
Coimbatore, India

L. K. Nivedha
Department of Chemistry
Indian Institute of Technology Madras
Chennai, India

Vijaymohanan Pillai
Department of Chemistry
Indian Institute of Science Education and Research (IISER)
Tirupati, India

Abhilash Pullanchiyodan
Materials Science and NanoEngineering Lab (MSNE-Lab), Department of Polymer Science and Rubber Technology
Cochin University of Science and Technology
Cochin, India

Prasanth Raghavan
Materials Science and NanoEngineering Lab (MSNE-Lab), Department of Polymer Science and Rubber Technology
Cochin University of Science and Technology
Cochin, India

Department of Material Engineering and Convergence Technology
Gyeongsang National University
Jinju, Korea

International and Inter University Centre for Nanoscience and Nanotechnology (IIUCNN), Mahatma Gandhi University
Kerala, India;
and
Biorefining and Advanced Materials Research Center
SRUC (Scotland's Rural College)
Edinburgh, United Kingdom

Kothandaraman Ramanujam
Department of Chemistry
Indian Institute of Technology Madras
Chennai, India
and
DST-IITM Solar Energy Harnessing Centre
Indian Institute of Technology Madras
Chennai, India

Janraj Naik Ramavath
Department of Chemistry
Indian Institute of Technology Madras
Chennai, India

Leya Rose Raphael
Materials Science and NanoEngineering Lab (MSNE-Lab), Department of Polymer Science and Rubber Technology
Cochin University of Science and Technology
Cochin, India

Carla Giselle Martins Real
Alan Levin Department of Mechanical and Nuclear Engineering
Kansas State University
Manhattan, Kansas

M. V. Reddy
Nouveau Monde Graphite (NMG)
Saint-Michel-des-Saints
Quebec, Canada

Zhongkan Ren
Alan Levin Department of Mechanical and Nuclear Engineering
Kansas State University
Manhattan, Kansas

Aswith R. Shenoy
Department of Mechanical Engineering
Albertian Institute of Science and Technology
Kochi, India

Gurpreet Singh
Alan Levin Department of Mechanical and Nuclear Engineering
Kansas State University
Manhattan, Kansas

Davi Marcelo Soares
Alan Levin Department of Mechanical and Nuclear Engineering
Kansas State University
Manhattan, Kansas

Vijay Kumar Thakur
Biorefining and Advanced Materials Research Center
SRUC (Scotland's Rural College)
Edinburgh, United Kingdom

Anusree Thilak
Materials Science and NanoEngineering Lab (MSNE-Lab), Department of Polymer Science and Rubber Technology
Cochin University of Science and Technology
Cochin, India

Anjumole P. Thomas
Materials Science and NanoEngineering Lab (MSNE-Lab), Department of Polymer Science and Rubber Technology
Cochin University of Science and Technology
Cochin, India

S. K. Vineeth
Department of Polymer and Surface Engineering
Institute of Chemical Technology
Mumbai, India
and
Materials Science and NanoEngineering Lab (MSNE-Lab), Department of Polymer Science and Rubber Technology
Cochin University of Science and Technology
Cochin, India

Alexandru Vlad
Institute of Condensed Matter and Nanosciences
Université catholique de Louvain
Ottignies-Louvain-la-Neuve, Belgium

Yifei Wang
Department of Mechanical Engineering
The University of Hong Kong
Hong Kong, China

Hudson Zanin
Center for Innovation on New Energies
University of Campinas
Campinas, Brazil

Abbreviations

AA	acrylic acid
AAB	aluminium-air battery
AAFB	aluminium air flow battery
AFM	atomic force microscopy
AIB	aluminum-ion battery
ALD	atomic layer deposition
APC	all phenyl Complexes
ASSB	all-solid-state battery
ASS-RCIB	all solid-state rechargeable chloride ion batteries
BEV	battery electric vehicle
BMIM|OTF	1-butyl-3-methylmidazolium trifluoromethanesulphonate
[BMIM][BF_4]	1-butyl-3-methylimidazolium tetrafluoroborate
BP	biphenyl
BpyCl	1-butyl-1-methylpiperidinium chloride
BTT	butantetraol
C4Q	calix[4]quinone
CB	carbon black
CC	carbon cloth
CE	coulombic efficiency
CG	coarse grained
CIBs	chloride ion batteries
CMC	carboxy methyl cellulose
CMK	ordered mesoporous carbon
CNT	carbon nanotube
CPL	composite protective layer
CTAB	cetyl trimethyl ammonium bromide
CV	cyclic voltametry
DDAB	dihexadecyl dimethyl ammonium bromide
DDBAB	dodecyl dimethyl benzyl ammonium bromide
DEC	diethyl carbonate
DEGDME	diethylene glycol dimethyl ether
DMC	dimethyl carbonate
DME	dimethoxyethane
DMSO	dimethylsulphoxide
DOL	dioxolane
DTT	dithiothreitol
3D-NSCNs	three dimensional nitrogen and sulphur co-doped porous carbon nanosheets
EC	ethylene carbonate
ECAP	equal channel angular pressing
EDX	energy dispersive X-ray analysis
EEES	electrochemical energy storage systems
EG	ethylene glycol
EIS	electrochemical impedance spectroscopy

EMF	electromotoric force
EMI-DCA	1-ethyl-3-methylimidazolium dicyanamide
EV	electric vehicles
FCEV	fuel cell electric vehicle
FEC	fluoroethylene carbonate
FEC	fluroethylene carbonate
FESEM	field emission scanning electron microscopy
FTIR	Fourier transform infrared spectroscopy
GA	graphene aerogels
GCD	galvanostatic charge-discharge
GDL	gas-diffusion layer
GIXRD	grazing incidence X-ray diffraction
GN	graphene
GNS	graphene nanosheets
GO	graphene oxide
GPE	gel polymer electrolyte
HEV	hybrid electric vehicle
HOMO	highest occupied molecular orbital
1H NMR	proton nuclear magnetic resonance spectroscopy
HRTEM	high-resolution transmission electron microscopy
IL	ionic liquid
K[FSA]	potassium bis(fluorosulphonyl)amide
K[FTA]	potassium (fluorosulphonyl)(trifluoromethylsulphonyl) amide
KAB	potassium-air battery
$KClO_4$	potassium perchlorate
KFSI	potassium bis(fluorosulphonyl)imide
KIB	potassium-ion battery
KPF_6	potassium hexafluorophosphate
KTFSI	potassium bis(tri-fuoromethylsulphonyl)imide
LAB	lithium-air battery
LDHs	layered double hydroxides
Li	lithium
Li_2O_2	lithium oxide
LIB	lithium-ion battery
LiClO4	lithium perchlorate
$LiCoO_2$	lithium cobalt oxide
$LiMnO_2$	lithium manganeseoxide
$LiNO_3$	lithium nitrate
LiOH	lithium hydroxide
LISICON	LIithium Super Ionic CONductor
LiTFSI	lithium bis(tri-fuoromethylsulphonyl)imide
LUMO	lowest unoccupied molecular orbital
MAB	metal-air battery
MFCs	carbonization of melamine foams
MLD	molecular layer deposition
MWCNT	multiwalled carbon nanotube

n	no of electrons involved in the cell reactions
$NaClO_4$	sodium perchlorate
NASICON	NAtrium Super Ionic CONductor
^{23}Na-NMR	sodium nuclear magnetic resonance spectroscopy
N-CNT	nitrogen-doped carbon nanotube
NDE	negative difference effect
NHE	normal hydrogen electrode
Ni	nickel
NIBs	sodium ion batteries
Ni–Cd	nickel–cadmium
Ni–MH	nickel–metal halide
NO-WCTs	nitrogen/oxygen dual-doped highly wrinkled carbon tubes
NRs	nanorods
NVP	natrium superionic conductor
OCV	open circuit voltage
OER	oxygen evolution reaction
ORR	sluggish oxygen reduction reaction
[OMIM][Cl]	1-methyl-3-octylimidazolium chloride
PAN	polyacrylonitrile
PANa	sodium polyacrylate hydrogel
PANI	polyaniline
PC	propylene carbonate
PEALD	plasma-enhanced atomic layer deposition
PEG	polyethylene glycol
PEO	poly(ethylene oxide)
PHEV	plug-in hybrid electric vehicle
PMMA	poly(methyl methacrylate)
$PP_{14}Cl$	1-butyl-1-methylpiperidinium chloride
$PP_{14}TFS$	1-butyl-1-methylpiperidinium bis(trifluoromethylsulphonyl)imide
PPy	polypyrrole
PBA	Prussian blue analogues
PS	poly(styrene)
Pt	platinum
PT	5,7,12,14-tetraone
PVA	polyvinyl alcohol
PVC	polyvinylchloride
PVDF	polyvinylidene fluoride
PVdF-co-HFP	polyvinyldifluoride-hexafluoropolymer
PVDF-HFP	polyvinylidene fluoride-co-hexafluoropropylene
PVP	polyvinyl pyrrolidone
RAMAN	Raman spectroscopy
rGO	reduced graphene oxide
RT	room temperature
RTIL	room temperature ionic liquid
SAB	sodium-air battery
SAED	selected area electron diffraction

SDS	sodium dodecyl sulphate
SEI	solid electrolyte interface
SEM	scanning electron microscopy
SHE	standard hydrogen electrode
SIB	sodium-ion battery
SN	succinonitrile
SPE	solid polymer electrolyte
SS	stainless steel
SSE	solid-state electrolyte
STEM	scanning transmission electron microscopy
SWCNT	single-wall CNT
TBACl	tetrabutyl ammonium chloride
TBMACl	tributylmethylammonium chloride
TEACl	tetraethyl ammonium chloride
TEGDME	tetraethylene glycol dimethyl ether
TEM	transmission electron microscopy
TEMPO	2,2,6,6,-tetramethylpiperidinyl-1-oxyl
TMA	trimethyl aluminium
TMCs	transition/non-transition metal chalcogenides
TMD	transition metal dichalcogenide
TOACl	octyltrimethyl ammonium chloride
UFG	ultrafine grained
VN/CNF	vanadium nitride based carbon nanofibres
XPS	X-ray photoelectron spectroscopy
XRD	X-ray diffraction
ZIB	zinc ion battery
ΔrG	Gibbs free energy

Symbols

Ag	silver
Al	aluminium
Bi	bismuth
$BiCl_3$	bismuth chloride
BiOCl	bismuth oxychloride
Ca	calcium
$CaCl_2$	calcium chloride
Ce	cerium
$CeCl_3$	cerium chloride
Cl	chlorine
Co	cobalt
$CoCl_2$	cobalt chloride
CuCl	cuprous chloride
$CuCl_2$	cupric chloride
Fe	iron
Fe_2O_3	iron oxide
$FeCl_2$	ferrous chloride
$FeCl_3$	ferric chloride
FeOCl	orthorhombic iron oxychloride
Li	lithium
LiCl	lithium chloride
Mg	magnesium
$MgCl_2$	magnesium chloride
Mn	manganese
$MnCl_2$	manganese chloride
Na	sodium
NaCl	sodium chloride
Ni	nickel
$NiCl_2$	nickel chloride
$Sb_4O_5Cl_2$	antimony oxychloride
V	vanadium
VCl_3	vanadium chloride
VOCl	vanadium oxychloride

Units

Ah kg^{-1}	ampere hour per kilogram
Ah L^{-1}	ampere hour per litre
K	Kelvin
kJ mol^{-1}	kilojoules per mole
mA g^{-1}	milliampere per gram
mAh cm^{-3}	milliampere hour per cubic centimetre
mAh g^{-1}	milliampere hour per gram
°C	degree Celsius
S cm^{-1}	Siemens per centimetre
V	volt
Wh kg^{-1}	watt-hour per kilogram
Wh L^{-1}	watt-hour per litre
%	percentage

1 Chloride Ion Batteries

*Leya Rose Raphael, Akhila Das, Devika S. Lal,
Anandu M. Nair, Jou-Hyeon Ahn,
Alexandru Vlad, and Prasanth Raghavan*

1.1 INTRODUCTION

The requirement for novel batteries with high gravimetric and volumetric energy densities has paved the way for exploring several cations and anions for charge transfer in electrolytes. The initial interest for such electrolyte systems was focused on the conduction of alkali metal ions, and hence they are continuously explored and expanded. The anion conduction in solid polymer electrolytes (SPEs) was initially reported by Hardy et al. [1] with conductivity in the same order as that of the conventional polymer-alkali metal salt systems. A novel battery concept involving the fluoride shuttle, followed by the introduction of the chloride ion batteries (CIBs), was reported for the first time by Fichtner et al. [2,3]. In CIBs, the chloride ion shuttles between a metal/metal chloride pair and yields a high electromotive force (EMF) in virtue of its significant Gibbs free energy change. The battery reactions can be expressed as follows:

$$\text{At cathode, } M_cCl_n + ne^- \leftrightarrow M_c + nCl^- \tag{1.1}$$

$$\text{At anode, } M_a + mCl^{-1} \leftrightarrow M_aCl_m + me^- \tag{1.2}$$

$$\text{Overall cell reaction, } mM_cCl_n + nM_a \leftrightarrow mM_c + nM_aCl_m \tag{1.3}$$

where M_c is the cathode metal, M_a is the anode metal, and m or n is the number of chloride ions. Compared with the conventional lithium ion battery (LIB), a higher theoretical energy density is evident for the CIBs. The parameters for several electrochemical couples for CIBs are listed in Table 1.1. Thus, the first-ever demonstration of these systems was carried out with several electrochemical metals/metal chlorides, coupled with binary ionic liquid (IL) electrolytes.

1.2 CATHODES FOR CIBS

The choice of electrolytes and electrodes is the key challenge for the development of CIBs. Table 1.2 encompasses the research on CIBs in a nutshell with their performance using different electrodes and electrolyte components. The initial rechargeable CIB concept was demonstrated using an IL as the electrolyte, lithium foil as the anode, and $CoCl_2$, VCl_3, or $BiCl_3$ as the cathode, in which good stability for $BiCl_3$/Li system was displayed by its operation at room temperature (RT), in addition to comparably low volume change during cycling. Owing to the high safety feature, good chloride ion conductivity, and wide electrochemical stability window, a mixture of imidazolium-based ILs, namely, 1-methyl-3-octylimidazolium chloride ([OMIM][Cl]) and 1-butyl-3-methylimidazolium tetrafluoroborate ([BMIM][BF_4]) in the ratio of 3:1, were used. In addition to the electrolyte choice, the cathode dissolution with regard to its Lewis acid nature is to be considered and suppressed.

DOI: 10.1201/9781003310174-1

TABLE 1.1
Battery and Electrode Performance of Some Electrochemical Couples for Chloride Ion Batteries[a]

Cell Reactions	Δ_rG^b (kJ mol^{-1})	*n*	EMF[c] (V)	Specific Capacity of Cell (Theoretical) Ah kg^{-1}	Ah L^{-1}	Energy Density (Theoretical) Wh kg^{-1}	Wh L^{-1}
$CoCl_2$ (c)+2Li (a) → 2LiCl+Co	−499.0	2	2.58	372.9	827.9	962.1	2,136.0
VCl_3 (c)+3Li (a) → 3LiCl+V	−642.0	3	2.21	541.3	891.4	997.3	1,970.0
$BiCl_3$ (c) +3Li (a) → 3LiCl+Bi	−838.2	3	2.89	239.2	761.0	691.3	2,199.3
2$BiCl_3$ (c)+3 Mg (a) → 3$MgCl_2$+2Bi	−1,145.4	6	1.98	228.6	920.8	452.6	1,823.2
$BiCl_3$ (c)+Ce (a) → $CeCl_3$+Bi	−669.8	3	2.31	176.5	923.4	407.7	2,133.1
$CuCl_2$ (c)+Ca (a) → $CaCl_2$+Cu	−573.1	2	2.97	307.2	818.1	912.5	2,429.8
$CuCl_2$ (c)+Mg (a) → $MgCl_2$+Cu	−416.1	2	2.15	337.6	998.6	725.9	2,147.0
$CuCl_2$ (c)+2Na (a) → 2NaCl+Cu	−592.5	2	3.07	297.0	614.5	911.9	1,886.5
$CuCl_2$ (c)+2Li (a) → 2LiCl+Cu	−593.1	2	3.07	361.3	813.3	1,109.0	2,496.8
CuCl (c)+Li (a) → LiCl+Cu	−264.5	1	2.74	252.9	724.7	693.1	1,985.7
$FeCl_2$ (c)+2Na (a) → 2NaCl+Fe	−465.9	2	2.41	311.9	611.7	751.6	1,474.2
$NiCl_2$ (c)+2Na (a) → 2NaCl+Ni	−509.2	2	2.64	305.2	637.9	805.7	1,684.0
$FeCl_3$ (c)+Ce (a) → $CeCl_3$+Fe	−650.8	3	2.25	265.9	1,048.7	598.2	2,359.8
2$FeCl_3$ (c)+3 Mg (a) → 3$MgCl_2$+2Fe	−1,107.4	6	1.91	404.7	1,045.3	773.0	1,996.5
$MnCl_2$ (c)+2Li (a) → 2LiCl+Mn	−328.3	2	1.70	383.6	782.9	652.1	1,330.9

Source: Adapted and reproduced from Ref. [3]. Copyright © 2013 Elsevier.

[a] The calculation of the specific capacity and the energy density is based on the active materials values of the cathode and anode, excluding typical inert components of a cell such as a collector, electrolyte, and separator.

[b] The Gibbs free energy data of metal chlorides are derived from the literature entitled "Standard thermodynamic properties of chemical substances".

[c] EMF is the electromotoric force of the battery reaction.

TABLE 1.2
Existing Literature on CIBs

Cathode	Anode	Electrolyte	Separator	Discharge Capacity (mAh g^{-1})	Coulombic Efficiency (%)	References
$CoCl_2$	Li	[BMIM][BF_4] & [OMIM] [Cl]	Glass fibre	67.1		[3]
VCl_3	Li	[BMIM][BF_4] & [OMIM] [Cl]	Glass fibre	176.6		[3]
$BiCl_3$	Li	[BMIM][BF_4] & [OMIM] [Cl]	Glass fibre	142.9		[3]
FeOCl	Li	0.5 M $PP_{14}Cl$ in $PP_{14}TFSI$	Glass fibre	60		[4]
BiOCl	Li	0.5 M $PP_{14}Cl$ in $PP_{14}TFSI$	Glass fibre	60		[4]
FeOCl	Li	0.5 M $PP_{14}Cl$ in $PP_{14}TFSI$	Celgard 2,400	44		[5]
FeOCl/GN	Li	0.5 M $PP_{14}Cl$ in $PP_{14}TFSI$	Celgard 2,400	184	99	[6]

(*Continued*)

TABLE 1.2 (*Continued*)
Existing Literature on CIBs

Cathode	Anode	Electrolyte	Separator	Discharge Capacity ($mAh\ g^{-1}$)	Coulombic Efficiency (%)	References
FeOCl/CN-250	Li	0.5 M $PP_{14}Cl$ in $PP_{14}TFSI$	Celgard 2,400	103		[6]
FeOCl/CN-450	Li	0.5 M $PP_{14}Cl$ in $PP_{14}TFSI$	Celgard 2,400	165		[6]
FeOCl/CB	Li	0.5 M $PP_{14}Cl$ in $PP_{14}TFSI$	Celgard 2,400	73	46	[6]
FeOCl/CMK-3	Li	0.5 M $PP_{14}Cl$ in Acetonitrile (ACN) and $PP_{14}TFSI$		202		[7]
$Sb_4O_5Cl_2$/ graphene aerogel	Li	0.5 M $PP_{14}Cl$ in PC		~65		[8]
CoFe-Cl LDH	Li	0.5 M BpyCl in PC and $PP_{14}TFSI$		~160	99	[9]
$Ni_2V_{0.9}Al_{0.1}$-Cl LDH	Li	0.5 M BpyCl in PC and $PP_{14}TFSI$	Glass fibre filter	113.8	92	[10]
NiFe-Cl LDH	Li	0.5 M BpyCl in PC	Glass fibre discs	101.1	90.3	[11]
NiMn-Cl LDH/ CNT	Li	0.5 M BpyCl in PC	Glass fibre discs	130	99	[12]
BiOCl	Mg/C	0.5 M $PP_{14}Cl$ in $PP_{14}TFSI$		69	68	[13]
BiOCl	MgH_2/C	0.5 M $PP_{14}Cl$ in $PP_{14}TFSI$			78	[13]
FeOCl	Mg/C	0.5 M $PP_{14}Cl$ in $PP_{14}TFSI$		111		[13]
FeOCl	MgH_2/C	0.5 M $PP_{14}Cl$ in $PP_{14}TFSI$		130	95	[13]
VOCl	$MgCl_2$/ Mg/C	0.5 M $PP_{14}Cl$ in $PP_{14}TFSI$	Glass fibre filter	60	60	[14]
Ag	$Sb_4O_5Cl_2$	1 M NaCl		34.6		[15]
AgCl	Sb@rGO	1 M NaCl		51.6		[16]
BiOCl	Ag	1 M NaCl	GB100R (glass fibre membrane)	92.1	~100	[17]
VOCl	Li	0.5 M $PP_{14}Cl$ in PC		189 (0.5 C) 113 (2 C)	98	[18]
PANI	Zn	TBACl/PVdF-*co*-HFP		150		[19]
FeOCl	Li	PEO-TBMACl		18 (298 K) 40 (313 K)		[20]
FeOCl	Li	PEO-TBMACl-SN		73 (298 K) 87 (313 K)		[20]
FeOCl	Li	PMMA/0.5 M $PP_{14}Cl$ in $PP_{14}TFSI$		122 (C/25)		[21]
PPyCl/CNTs	Li	0.5 M $PP_{14}Cl$ in $PP_{14}TFSI$		118	>95	[22]
PANICl/CNTs	Li	0.5 M $PP_{14}Cl$ in $PP_{14}TFSI$	Celgard 2,400	80	99	[23]
PANI/FeOCl	Li	0.5 M $PP_{14}Cl$ in $PP_{14}TFSI$	Celgard 2,400	120	97	[24]
PPy/FeOCl	Li	0.5 M $PP_{14}Cl$ in $PP_{14}TFSI$	Celgard 2,400	187		[25]
β-FeOOH(Cl)	Li			122		[26]

Considering these factors, a new type of chlorine compound, the metal oxychloride, was reported as the potential cathode material for CIBs [4]. The overall reaction of the CIB involving these metal oxychlorides as cathodes can be expressed as follows:

$$m\mathrm{M}_c\mathrm{OCl}_n + n\mathrm{M}_a \leftrightarrow m\mathrm{M}_c\mathrm{O} + n\mathrm{M}_a\mathrm{Cl}_m \quad (1.4)$$

where M_c is the metal element in the cathode, M_a is the metal used in the anode, and *m* and *n* are the numbers of chloride ions. Thus, upon investigating monoclinic bismuth oxychloride (BiOCl) and orthorhombic iron oxychloride (FeOCl) as cathode components, reversible initial discharge capacities of 60 and 158 mAh g^{-1} (58% and 63% of the theoretical capacity) were obtained, respectively. FeOCl, the layered metal oxychloride salt when used in CIBs as cathode material, has proven to show better intercalation with organic components and good electrochemical behaviour [5].

1.2.1 Carbon-Based Cathodes

Incorporating carbon materials like graphene, carbon nanotubes (CNTs), carbon black (CB), porous carbon, activated carbon, or their hybrids has provided a conductive matrix for the electrodes in rechargeable batteries. Thus this method can also enhance the conductivity and the structural stability of the metal oxychlorides, which otherwise exhibit low experimental discharge capacity compared to their corresponding theoretical capacity. Zhao et al. [6] prepared FeOCl/carbon composites as cathodes (Figure 1.1) using CNTs, CB, and graphene by mechanical milling and proved its worth by retraining the formation of a part of the decomposition products Fe_2O_3 and $FeCl_3$. The graphene-incorporated FeOCl cathode showed a better electrochemical kinetic performance with a high reversible capacity of 184 mAh g^{-1}, 73% of the theoretical capacity. This study also showed new insights into the reaction mechanism of the dissociation of chloride ions at the cathode using density functional theory. A nanoconfined FeOCl/CMK-3 (Figure 1.2) composite cathode was prepared via vacuum impregnation, and subsequent thermal decomposition at mild conditions showed a high discharge capacity of 202 mAh g^{-1} [7].

For the first time, antimony oxychloride ($Sb_4O_5Cl_2$) has been explored as cathode material for aqueous CIB. The microstructures were embedded in a graphene aerogel matrix, which showed

FIGURE 1.1 Scanning electron microscopy images of the as-prepared FeOCl and FeOCl/carbon composites: (a) FeOCl; (c) FeOCl/CN-450; (d) FeOCl/CB-450; and (e) FeOCl/GN-450. Adapted and reproduced from Ref. [6]. Copyright © 2016 Springer Nature.

FIGURE 1.2 Scanning electron microscopy images of FeOCl/CMK-3 nanocomposite (a) at lower and (b) higher magnifications, and (c) transmission electron microscopy image and the corresponding selected area electron diffraction (SAED) and elemental mapping images of the FeOCl@CMK-3 nanocomposite. Adapted and reproduced from Ref. [7]. Copyright © 2017 American Chemical Society.

improved electrochemical performance with a capacity of around 65 mAh g^{-1} after 100 cycles. The reduction in volume expansion and the improved ionic conductivity can be ascribed to the synergistic effect of graphene and $Sb_4O_5Cl_2$ [8]. The carbon materials mentioned earlier play a pivotal role in the enhancement of the conductivity and imparting structural stability to the cathodes. Thus carbon-based cathodes find potential applications in CIB systems supported by their property enhancement features. The role of carbon materials is not limited to the cathodes alone but can buffer the volume change in the electrodes, thereby the CIB.

1.2.2 Layered Double Hydroxides as Cathodes

A new class of cathode materials, layered double hydroxides (LDHs) with chloride ions in their interlayers, was developed, namely CoFe-Cl LDH. The structural and component characteristics of these 2D materials showed superior energy storage performance as cathodes in anion-shuttling CIBs [9]. Yin et al. [10,11] studied nickel-based trimetallic and dimetallic LDHs ($Ni_2V_{0.9}Al_{0.1}$–Cl and NiFe–Cl LDHs), which showed a significant improvement in the electrochemical performances. By applying a current density of 200 and 100 mA g^{-1} to the trimetallic and dimetallic LDHs, respectively, a stable reversible capacity of ~113.8 mAh g^{-1} (1,000 cycles) and 101.1 mAh g^{-1} (800 cycles) was attained (Figure 1.3). A hybrid material composed of NiMn–Cl LDH nanoplate arrays anchored on CNTs prepared via an in situ co-precipitation method has displayed a cross-linked network structure with a well-defined core-shell configuration. Their enlarged surface area and enhanced electrical conductivity have helped them achieve improved rate performance, high structural stability, suitable reversible reaction mechanism, and a long cycle life [12]. Thus LDHs and their hybrid composites hold great promise for new cathode materials for CIBs.

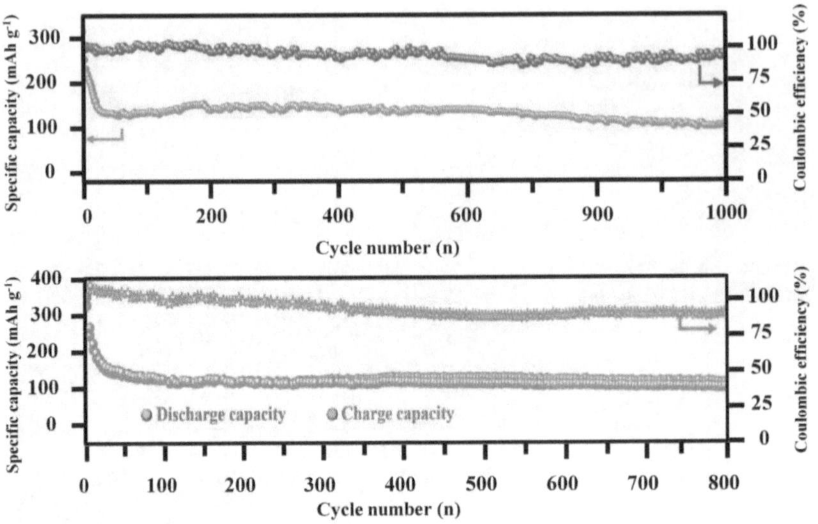

FIGURE 1.3 Cycle performances and corresponding coulombic efficiency of the $Ni_2V_{0.9}Al_{0.1}$-Cl LDH at a current density of (a) 200 mA g^{-1} within 1,000 cycles and (b) 100 mA g^{-1} within 800 cycles. Adapted and reproduced from Ref. [10,11], respectively. Copyright © 2019 John Wiley and Sons. Copyright © 2020 The Royal Society of Chemistry.

1.3 ANODES FOR CIBS

Intending to develop battery alternatives to cation-based conventional systems, focusing on battery chemistries utilizing abundant metals as anodes, has gained attention. This advantageous feature of CIBs has triggered more possibilities for formulating better electrochemical systems. With respect to the previous reports, FeOCl/Mg and BiOCl/Mg electrochemical couples were investigated and they showed high coulombic efficiency, which proved the feasibility of Mg anodes in CIBs. Electrolyte instability to a small extent with large volume change resulted in decay in capacity during cycling [13]. Another metal oxychloride, vanadium oxychloride (VOCl), was used as a cathode with $MgCl_2$/Mg anode and electrolyte composed of 1-butyl-1-methylpiperidinium chloride ($PP_{14}Cl$) in 1-butyl-1-methylpiperidinium bis(trifluoromethylsulfonyl)imide ($PP_{14}TFSI$) IL. Its charge and discharge reactions are shown in Equations 1.5–1.7. This choice of electrodes and electrolytes for CIB operation delivered a discharge capacity of 101 mAh g^{-1} and retained 60 mAh g^{-1} after 53 cycles [14]. Thus magnesium is a good choice as electrodes in CIBs owing to its higher theoretical specific volumetric capacity of 3,833 mAh cm^{-3}, which is much higher than lithium (2,046 mAh cm^{-3}).

$$\text{Charge, } 2VOCl + MgCl_2 \rightarrow 2VOCl_2 + Mg \tag{1.5}$$

$$\text{Discharge, } 2VOCl_2 + Mg \rightarrow 2VOCl + MgCl_2 \tag{1.6}$$

$$2VOCl + Mg \rightarrow 2VO + MgCl_2 \tag{1.7}$$

Another material of interest explored as anode material for CIBs was $Sb_4O_5Cl_2$ by Hu et al. [15]. This material was also used for the first time as a cathode and has been mentioned in the preceding paragraph. Unlike the earlier mention, a carbon-based cathode material, in this study, $Sb_4O_5Cl_2$, was used with salt water electrolyte and silver cathode to form a water-based chloride shuttle battery. On applying a current density of 600 mA g^{-1}, a discharge capacity of 34.6 mAh g^{-1} was maintained after 50 cycles. Sb@rGO composites formed by the dispersion of Sb nanoparticles on reduced graphene oxide (rGO) via hydrothermal method have been effectively studied as cathode for water-based CIBs. A specific discharge capacity of 51.6 mAh g^{-1} can be obtained after 200 cycles at a current density of 400 mA g^{-1} [16]. These water-based rechargeable CIBs can achieve excellent electrochemical performances concerning the synergy between the salt water electrolyte and the electrodes, which can contribute largely to salty water batteries and seawater desalination.

1.4 ELECTROLYTES FOR CIBS

The compatibility between the electrodes and the electrolytes has always been the driving force for expanding the research on CIBs. Conventionally ILs have been explored as potential electrolytes for CIBs, whereas, other electrolyte materials in different states such as aqueous, plasticized (gel), and solid have been explored. A new aqueous battery system with a BiOCl anode and a silver cathode in aqueous sodium chloride (NaCl) electrolyte solution was developed for the first time. A stable discharge capacity was obtained at a current density of 400 mA g^{-1} with a coulombic efficiency comparable to or even better than the existing aqueous batteries [17]. Fichtner and co-workers [18] reported a CIB with high stability over 100 cycles using $PP_{14}Cl$ in propylene carbonate (PC) electrolyte and VOCl cathode. They have proven that the ion transport mechanism in the developed CIB has resulted from multistep reactions involving partial reduction, oxidation, intercalation, and expansion of VOCl interlayers. The development of solid-state electrolytes in CIBs has been explored by Xia et al. [27], using a RT inorganic halide perovskite of $CsSnCl_3$. The as-prepared cubic electrolyte achieves excellent electrochemical performance with the highest ionic conductivity of 3.6×10^{-4} S cm^{-1}. When coupled with $SnCl_2$/Sn and $BiCl_3$/Bi electrochemical couples, these electrolytes portrayed a wide electrochemical window of 6.1 V at 298 K.

1.5 POLYMER-BASED CHLORIDE ION BATTERIES

Polymer materials have found themselves familiar in almost all aspects of rechargeable batteries, which have been already established. The leakage, volatility, flammability, and unsafe aspects of the initially investigated electrolytes have paved the way for redox-active polymers. The solid dry and gel-based polymer systems also prevent electrode dissolution, which is an essential requirement as it is a challenging aspect of CIBs. Thus their application in developing systems can be all the more attractive due to their inherent high-performance, stability, and safety properties. The electrodes and electrolytes have exploited polymers for their individual requirements. The following sections thus deal with this prospect.

1.5.1 Polymer Electrolytes

The concept of solid-state batteries is desirable in terms of safety requirements and has been explored in the existing technologies. Thus, in the research of CIBs, a novel RT solid-state CIB was initiated with polymer-based electrolytes. A pair of commonly used polymers in other rechargeable technologies like LIBs and sodium ion batteries such as polyvinylchloride (PVC) and polyvinyl difluoride-hexafluoropolymer (PVdF-*co*-HFP) as well as commercial gelatin were evaluated. Thus by combining suitable chloride salts with larger cations, octyltrimethyl ammonium chloride (TOACl) for PVC, tetrabutyl ammonium chloride (TBACl) for PVdF-*co*-HFP, and tetraethyl ammonium chloride (TBACl) for gelatin, free-standing gel electrolyte membranes were investigated. These membranes (Figure 1.4) are feasible for CIB electrolytes that otherwise use ILs, which promote electrode dissolution [19]. An all solid-state rechargeable CIB (ASS-RCIB) was fabricated using polyethylene oxide (PEO), tributylmethylammonium chloride (TBMACl), and succinonitrile (SN)–based ternary SPE, FeOCl cathode, and lithium anode (FeOCl/PEO-TBMACl-SN/Li) (Figure 1.5). These SPEs showed conductivities of 10^{-5} to 10^{-4} S cm^{-1} in the temperature range of 298–343 K. They also portrayed electrochemical stability of 4.2 V vs Li, making the SPEs in CIBs a promising component [20].

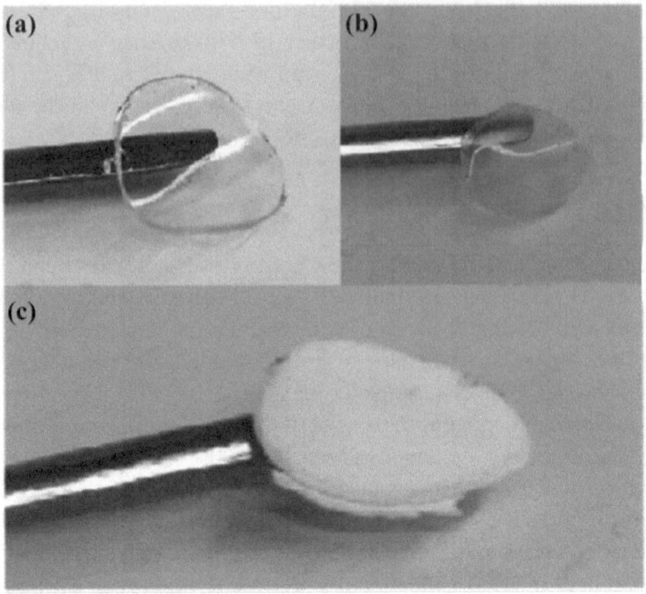

FIGURE 1.4 Digital photographs of free-standing (a) gelatin membrane, (b) PVC membrane, and (c) PVdF-*co*-HFP membrane. Adapted and reproduced from Ref. [19]. Copyright © 2016 John Wiley and Sons.

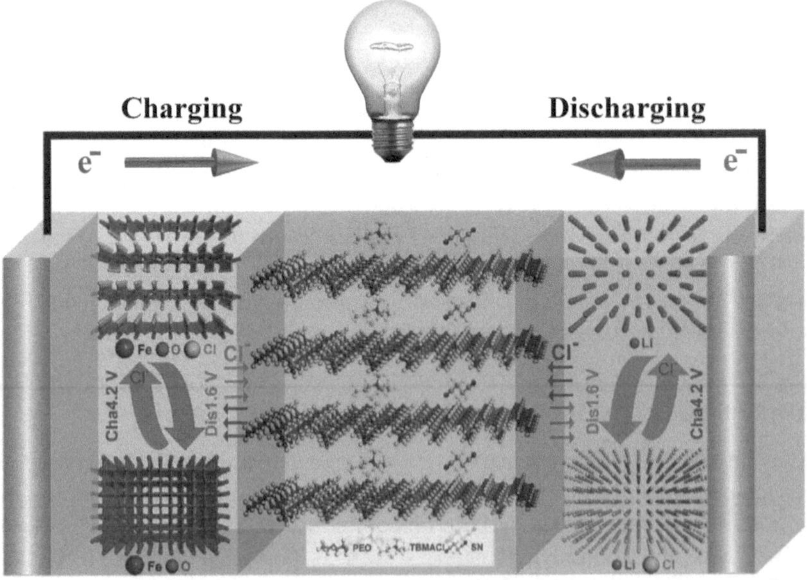

FIGURE 1.5 Schematic diagram of the ASS-RCIB with the PEO_1-$TBMACl_1$-SN_3 SPE and the FeOCl/Li electrode couple. Adapted and reproduced from Ref. [20]. Copyright © 2019 John Wiley and Sons.

Chloride ion-conducting polymer electrolyte-based ion gel membranes were prepared using polymethylmethacrylate (PMMA) with ILs to obtain gel polymer electrolytes (GPEs) suitable for CIBs. These flexible, self-standing, transparent GPEs exhibited superior thermal stability of up to 340°C, electrochemical stability of 5 V vs SS/SS, and ionic conductivity of 0.9×10^{-4} S cm^{-1}. The Li/IL-GPE/FeOCl cell showed a capacity above 122 mAh g^{-1} at a C/25 rate [21]. The investigations through solid, quasi-solid, and IL-based GPEs have shown the possibilities of polymer-based CIBs. The role of polymer constituents in rechargeable batteries has been proven to be a good choice, evident from their applicability. The establishment of the synergy between the polymer-based electrolytes and the CIBs' other components can lead to the development of safe, flexible, and stable systems.

1.5.2 Polymer-Based Cathodes

The use of polymers as potential cathode materials for CIBs has been explored too, similar to the existing technologies. In addition to this, the incorporation of CNTs can contribute to the improvement of electrochemical performance. A chloride ion-doped polymer, polypyrrole chloride with multiwalled CNTs to form nanostructured composite material (PPyCl/CNTs), was synthesized using chemical oxidation polymerization (Figure 1.6). The as-prepared cathode showed a high reversible capacity of 118 mAh g^{-1} with good cycling stability [22]. The nanostructured chlorine-doped polyaniline with CNTs (PANICl/CNTs) prepared by a similar method also portrayed good cycling stability with a coulombic efficiency of 99% [23]. The cathode made of PANI-intercalated FeOCl possesses a uniform expanded laminated structure (Figure 1.7), contributing to a significant improvement in electrochemical performance [24]. By coating polypyrrole

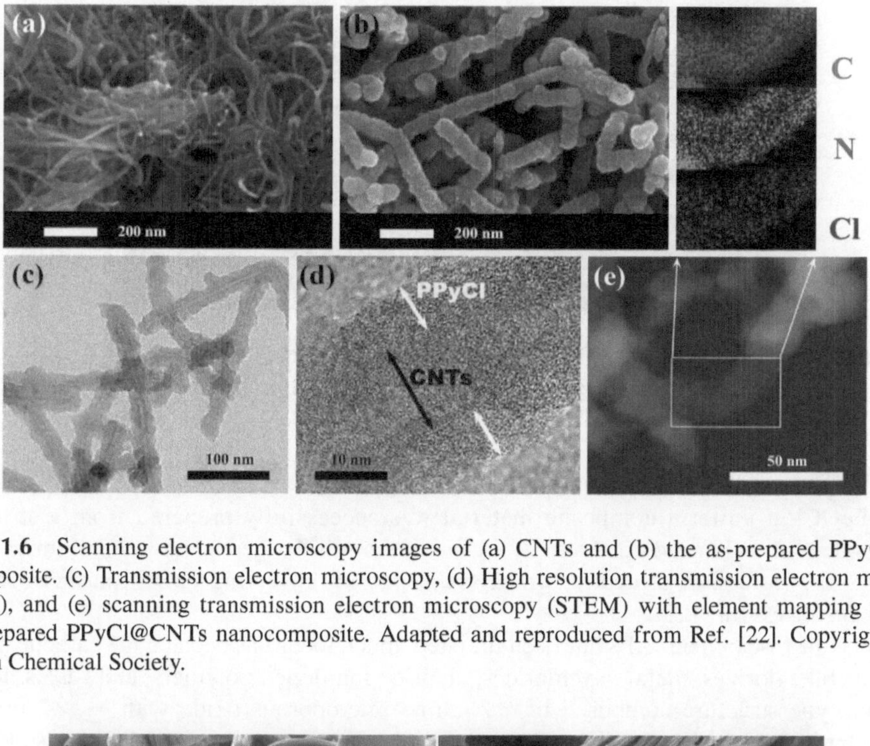

FIGURE 1.6 Scanning electron microscopy images of (a) CNTs and (b) the as-prepared PPyCl@CNTs nanocomposite. (c) Transmission electron microscopy, (d) High resolution transmission electron microscopy (HRTEM), and (e) scanning transmission electron microscopy (STEM) with element mapping images of the as-prepared PPyCl@CNTs nanocomposite. Adapted and reproduced from Ref. [22]. Copyright © 2017 American Chemical Society.

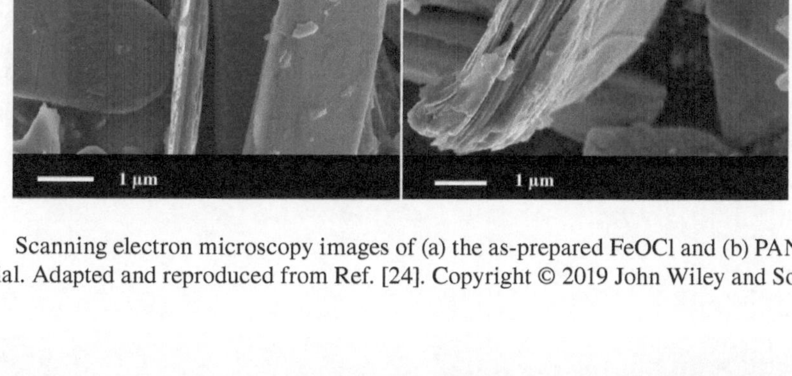

FIGURE 1.7 Scanning electron microscopy images of (a) the as-prepared FeOCl and (b) PANI-intercalated FeOCl material. Adapted and reproduced from Ref. [24]. Copyright © 2019 John Wiley and Sons.

FIGURE 1.8 The morphological characteristics of β-FeOOH NRs. (a) The model structure of a hollandite-type β-FeOOH and (b) scanning electron microscopy image of β-FeOOH NRs. Adapted and reproduced from Ref. [26]. Copyright © 2019 John Wiley and Sons.

(PPy) on FeOCl, a uniform composite material was successfully prepared from a structurally stable cathode with enhanced conductivity. A maximum discharge capacity of 187 mAh g^{-1} was obtained for this core-shell structure, which succeeded in preventing FeOCl particles from pulverization during cycling [25].

Ever since the research on CIBs has been initiated, much focus has been on the cathode materials involving metal chlorides, metal oxychlorides, chloride ion-doped polymers, and LDHs. Recently, a hollandite-type structure composed of crystalline microporous oxide with a 2×2 tunnel has attracted interest as cathodes for CIBs. These materials have already been explored as supercapacitor electrodes and as anodes in potassium-ion batteries. A typical hollandite-type structure with chloride ions in the tunnel space (β-FeOOH(Cl)) as nanorods (NRs) (Figure 1.8) has been shown to exhibit superior rate capability and high reversible capacity over 100 cycles [26]. Several studies have been carried out to find a compatible combination of the electrodes in terms of theoretical capacity, volumetric energy density, and gravimetric energy density. Table 1.3 displays the cell combinations produced from theoretically calculated data to investigate how realistic they would be in an actual battery [28].

TABLE 1.3
Cell Combinations Yielding the 20 Highest Theoretical Capacities and Volumetric and Gravimetric Energy Densities

+	–	mAh g^{-1}	+	–	Wh L^{-1}	+	–	Wh kg^{-1}
CCl_4	Al	564	$SeCl_4$	Sr	2,219	CCl_4	Ca	1,728
CCl_4	Ti	531	CCl_4	Sr	2,189	CCl_4	Na	1,706
CCl_4	Mg	529	$SeCl_4$	Yb	2,146	CCl_4	Mg	1,574
CCl_4	V	523	$SeCl_4$	Na	2,130	CCl_4	K	1,442
CCl_4	Sc	501	CCl_4	Yb	2,120	CCl_4	Sc	1,415
CCl_4	Ti	492	$SeCl_4$	Ca	2,106	$SeCl_4$	Ca	1,294
CCl_4	V	483	CCl_4	Na	2,105	$SeCl_4$	Na	1,293
CCl_4	Cr	480	CCl_4	Ca	2,082	CCl_4	Sr	1,285
$TiCl_4$	Al	475	$SeCl_4$	La	2,060	CCl_4	Y	1,221
$AlCl_3$	Mg	473	$SeCl_4$	Ce	2,056	CCl_4	Al	1,178
CCl_4	Nb	469	$SeCl_4$	Nd	2,049	$SeCl_4$	K	1,147
CCl_4	Fe	469	$SeCl_4$	Sm	2,049	$SeCl_4$	Mg	1,128
VCl_4	Al	468	CCl_4	La	2,036	CCl_4	La	1,058
CCl_4	Se	460	CCl_4	Ce	2,032	CCl_4	Ti	1,055
CCl_4	Ca	458	CCl_4	Nd	2,025	VCl_4	Na	1,049
$AlCl_3$	Sc	450	CCl_4	Sm	2,024	CCl_4	Ce	1,040
$TiCl_4$	Mg	449	$SeCl_4$	Gd	2,016	VCl_4	Ca	1,039
$MgCl_2$	Mg	448	WCl_4	Sr	2,002	$SeCl_4$	Sr	1,030
VCl_4	Ti	445	CCl_4	Gd	1,993	$SeCl_4$	Sc	1,025
VCl_4	Mg	444	$SeCl_4$	Ba	1,983	CCl_4	Ba	1,025

Source: Adapted and reproduced from Ref. [28]. Copyright © 2017 John Wiley and Sons.
Notes: "+" denotes the cathode, while "–" denotes the anode.

1.6 CONCLUSION

The world of battery chemistries has been involved with conventional cation-based systems. Several anion-based rechargeable batteries were explored, which also include halide ion batteries. Thus batteries involving the chloride ion conversion mechanism have been developed lately as an alternative to the typical LIB. In CIBs, metal chloride salts and metal oxychlorides have been chiefly used as cathodes with conventional anode materials like Li, Mg, and Ag. As electrolytes, ILs have been implemented mostly, but certain studies have explored polymer electrolytes. Thus the surge for the exploration of chloride ion-based rechargeable systems has been initiated in these recent years.

Initially, metal chloride cathodes were used as they could provide more than one electron per redox reaction. Their dissolution with electrolytes like ILs shifted towards the use of metal oxychloride salts, which can withstand the electrolyte apparently better. Then several carbon-based cathodes were studied in addition to the LDH-structured cathodes. The traditional Li anode employed contributed to the issue of dissolution, which prompted the CIB with Li-free anode such as Mg. Several other anodes such as silver- and antimony-based chlorides were also scrutinized for CIB research. Polymer-based organic cathodes and electrolytes were also subjected to preliminary studies in the run for the CIB research.

The main challenge of CIBs is the cathode dissolution into the electrolyte employed, which still needs to be addressed whenever a new or better cathode material is investigated. The significant volume change that accompanies the transformation of metals to metal chlorides and vice versa also contributes to the drawback of establishing this system. These issues can lead to the disruption of the ionic conductivity path and thus lower the battery performance. CIBs are not capable of achieving high capacities or voltages like the conventionally used Nobel prize–winning LIB technology. Nevertheless, they remain under the spotlight among a milieu of new battery chemistries because of the availability of chloride materials. They are considered promising, sustainable, and safe energy storage systems owing to these characteristics. Thus this chapter has extensively reviewed the research advances on the benign and tenable anion-based systems, providing good reading material for further investigations on CIBs.

REFERENCES

1. Hardy LC, Shriver DF (1984) Chloride ion conductivity in a plasticized quaternary ammonium polymer. *Macromolecules* 17:975–977. https://doi.org/10.1021/ma00134a077
2. Anji Reddy M, Fichtner M (2011) Batteries based on fluoride shuttle. *J Mater Chem* 21:17059–17062. https://doi.org/10.1039/c1jm13535j
3. Zhao X, Ren S, Bruns M, Fichtner M (2014) Chloride ion battery: a new member in the rechargeable battery family. *J Power Sources* 245:706–711. https://doi.org/10.1016/j.jpowsour.2013.07.001
4. Zhao X, Zhao-karger Z, Wang D, Fichtner M (2013) Metal oxychlorides as cathode materials for chloride ion batteries. *Angew Chemie – Int Ed* 125:13866–13869. https://doi.org/10.1002/ange.201307314
5. Yu T, Zhao X, Ma L, Shen X (2017) Intercalation and electrochemical behaviors of layered FeOCl cathode material in chloride ion battery. *Mater Res Bull* 96:485–490. https://doi.org/10.1016/j.materresbull.2017.03.070
6. Zhao X, Li Q, Yu T, et al (2016) Carbon incorporation effects and reaction mechanism of FeOCl cathode materials for chloride ion batteries. *Sci Rep* 6:1–8. https://doi.org/10.1038/srep19448
7. Yu T, Li Q, Zhao X, et al (2017) Nanoconfined iron oxychloride material as a high-performance cathode for rechargeable chloride ion batteries. *ACS Energy Lett* 2:2341–2348. https://doi.org/10.1021/acsenergylett.7b00699
8. Lakshmi KP, Janas KJ, Shaijumon MM (2019) Antimony oxychloride embedded graphene nanocomposite as efficient cathode material for chloride ion batteries. *J Power Sources* 433:126685. https://doi.org/10.1016/j.jpowsour.2019.05.091
9. Yin Q, Rao D, Zhang G, et al (2019) CoFe-Cl layered double hydroxide: a new cathode material for high-performance chloride ion batteries. *Adv Funct Mater* 29:1900983. https://doi.org/10.1002/adfm.201900983
10. Yin Q, Luo J, Zhang J, et al (2020) Ultralong-life chloride ion batteries achieved by the synergistic contribution of intralayer metals in layered double hydroxides. *Adv Funct Mater* 30:1–8. https://doi.org/10.1002/adfm.201907448
11. Yin Q, Luo J, Zhang J, et al (2020) High-performance, long lifetime chloride ion battery using a NiFe-Cl layered double hydroxide cathode. *J Mater Chem A* 8:12548–12555. https://doi.org/10.1039/d0ta04290k
12. Luo J, Yin Q, Zhang J, et al (2020) NiMn-Cl layered double hydroxide/carbon nanotube networks for high-performance chloride ion batteries. *ACS Appl Mater Interfaces* 3:4559–4568. https://doi.org/10.1021/acsaem.0c00224
13. Zhao X, Li Q, Zhao-karger Z, et al (2014) Magnesium anode for chloride ion batteries. *ACS Appl Mater Interfaces* 6:10997–11000
14. Gao P, Zhao X, Zhao-Karger Z, et al (2014) Vanadium oxychloride/magnesium electrode systems for chloride ion batteries. *ACS Appl Mater Interfaces* 6:22430–22435
15. Hu X, Chen F, Wang S, et al (2019) Electrochemical performance of Sb_4 O_5Cl_2 as a new anode material in aqueous chloride-ion battery. *ACS Appl Mater Interfaces* 11:9144–9148. https://doi.org/10.1021/acsami.8b21652
16. Zhang Q, Karthick R, Zhao X, et al (2020) Sb nanoparticles decorated with rGO as a new anode material in aqueous chloride ion battery. *Nanoscale* 12:12268–12274. https://doi.org/10.1039/D0NR00862A
17. Chen F, Leong ZY, Yang HY (2017) An aqueous rechargeable chloride ion battery. *Energy Storage Mater* 7:189–194. https://doi.org/10.1016/j.ensm.2017.02.001

18. Gao P, Reddy MA, Mu X, et al (2016) VOCl as a cathode for rechargeable chloride ion batteries. *Angew Chemie – Int Ed* 128:4357–4362. https://doi.org/10.1002/ange.201509564
19. Gschwind F, Steinle D, Sandbeck D, et al (2016) Facile preparation of chloride-conducting membranes: first step towards a room-temperature solid-state chloride-ion battery. *Chem Open* 5:525–530. https://doi.org/10.1002/open.201600109
20. Chen C, Yu T, Yang M, et al (2019) An all-solid-state rechargeable chloride ion battery. *Adv Sci* 6:1802130. https://doi.org/10.1002/advs.201802130
21. Wu L, Chen Z, Ma X (2019) Chloride ion conducting polymer electrolytes based on cross-linked PMMA-PP_{14}Cl-PP_{14}TFSI ion gels for chloride ion batteries. *Int J Electrochem Sci* 14:2414–2421. https://doi.org/10.20964/2019.03.49
22. Zhao X, Zhao Z, Yang M, et al (2017) Developing polymer cathode material for the chloride ion battery. *ACS Appl Mater Interfaces* 9:2535–2540. https://doi.org/10.1021/acsami.6b14755
23. Zhao Z, Yu T, Miao Y, Zhao X (2018) Chloride ion-doped polyaniline/carbon nanotube nanocomposite materials as new cathodes for chloride ion battery. *Electrochim Acta* 270:30–36. https://doi.org/10.1016/j.electacta.2018.03.077
24. Yu T, Yang R, Zhao X, Shen X (2019) Polyaniline-intercalated FeOCl cathode material for chloride-ion batteries. *ChemElectroChem* 6:1761–1767. https://doi.org/10.1002/celc.201801803
25. Yang R, Yu T, Zhao X (2019) Polypyrrole-coated iron oxychloride cathode material with improved cycling stability for chloride ion batteries. *J Alloys Compd* 788:407–412. https://doi.org/10.1016/j.jallcom.2019.02.234
26. Zhao G, Wan G, Tang Y, et al (2020) The hollandite-type β-FeOOH(Cl) as a new cathode material for chloride ion batteries. *Chem Commun* 56:12435–12438. https://doi.org/10.1039/D0CC04762G
27. Xia T, Li Y, Huang L, et al (2020) Room-temperature stable inorganic halide perovskite as potential solid electrolyte for chloride ion batteries. *ACS Appl Mater Interfaces* 12:18634–18641. https://doi.org/10.1021/acsami.0c03982
28. Gschwind F, Euchner H, Rodriguez-Garcia G (2017) Chloride ion battery review: theoretical calculations, state of the art, safety, toxicity and an outlook towards future developments. *Eur J Inorg Chem* 2017:2784–2799

2 Potassium-Ion Batteries
Recent Trends and Challenges

C. Nithya and R. Kiruthiga

2.1 INTRODUCTION

Lithium-ion batteries (LIBs) are widely used as a power source not only for portable devices such as laptops, mobile phones, and torch lights but also for electric vehicles (hybrid electric vehicles and plug-in hybrid electric vehicles) and play a vital role in our day-to-day lives. Unfortunately, LIBs do not satisfy the demand for energy storage devices because of lithium scarcity (0.0017 weight %) and their high cost [1]. Sodium-ion batteries (NIBs) are considered a promising alternative to LIBs [2] owing to their high abundance and low cost. The standard reduction potential of sodium is -2.71V *vs* standard hydrogen electrode (SHE) inherently determines the Na-ion chemistry. This high potential ultimately decreases the energy density of NIBs, making them unsuitable for practical applications [3].

Recently, potassium-ion batteries (KIBs) have obtained a lot of attention among researchers due to various factors like the uneven distribution of lithium resources and today's energy demand. The abundant resources of potassium (the seventh most abundant element in Earth's crust) and relatively low standard reduction potential [–2.931 V vs SHE, which is close to that of Li (–3.04 V vs SHE)] make KIB a promising energy storage technology for electric vehicles. The KIB technology has many advantages over LIBs in terms of cheap raw material cost and the use of aluminium foil, which can be used as a current collector for manufacturing anodes, reducing not only the cost of the battery but also its weight [4]. Another notable advantage of KIBs is that their low standard reduction potential produces lower cut-off voltages for the negative electrode materials available for LIBs and NIBs without any potassium deposition. This factor is further helpful to make KIBs operate over a wide potential window compared to LIBs and NIBs [5]. The working principle of KIB is similar to LIB and NIB technology, which follows "rocking chair" chemistry. This is based on the shuttling of K^+ ion that takes place between the electrodes during the charge/discharge process. The redox reaction in KIBs constitutes "intercalation/deintercalation" at the cathode (positive electrode) and the anode (negative electrode). K atoms deintercalate from the cathode during charging by releasing the electrons (travel through an external circuit) and form K^+ ions, which travel through the separator and the electrolyte. K^+ ions reach the anode and associate with electrons coming from the external circuit. During this process, the electrical energy from the external power source is converted into chemical energy in the battery, and, during discharge, the reverse phenomenon happens. Due to the similar electrochemistry of LIB technology, it is expected that the electrode materials available for LIBs work out well for KIBs. When KIBs are assembled with graphite anode against potassium [6], they form an intercalation compound KC_8, leading to the theoretical capacity of 279 mAh g^{-1}. The intercalation potential of K^+ ions for the majority of anode systems is 0.2 V vs K/K^+ (it is 0.05 V vs Na/Na^+ for NIBs), thus reducing the possibility of plating during cycling [7]. KIB technology is still in an infant stage; hence various scientific and engineering advancements are required to overcome the current challenges. For example, the larger K^+ ions (1.38 Å), compared to Li^+ (0.68 Å) and Na^+ (0.97 Å), alter the structure of common cathode and anode materials available for LIBs and NIBs, resulting in poor capacity, cyclability, and rate capability. Therefore, there is an immediate need to understand the structural and electrochemical activity of various electrodes to design fully functional KIB technology. Various challenges associated with KIBs are low ion

 DOI: 10.1201/9781003310174-2

diffusivity, large volume variation during cycling, severe side reactions, and dendrite growth, resulting in limited energy/power density [8]. To address these issues, research and development of various electrode/electrolyte materials have started since 2018. In this chapter, we briefly summarize the recent developments of electrode/electrolyte materials in the nanoscale for KIBs over the past 1 year. In the end, we address the practical challenges of KIB technology and shared our views on new strategies and future perspectives of potassium-ion-based storage technology.

2.2 ANODE MATERIALS FOR KIBS

Potassium is the first choice of anode materials for KIB applications owing to its high theoretical capacity of 685 mAh g^{-1}. Nevertheless, this K anode leads to undesirable side reactions because of the irregular distribution of electrons, resulting in the formation of dendrite growth and the breaking of solid electrolyte interphase (SEI) layers. This ruptured the SEI layer (passivation layer formed on the surface of the anode due to the decomposition of electrolyte) further because of the formation of a new SEI layer on the surface of the electrode, which consumes a lot of electrolytes, and increased electrode polarization leads to severe capacity decay [7]. Hence, excessive side reactions, irregular formation of the SEI layer, and dendrite growth ultimately result in poor cycling life and failure of the battery. This makes a great challenge for the use of potassium metal as an anode for full-cell KIB technologies. Therefore, to develop a fully functional potassium-ion cell, a suitable ideal host material is required. The present set of reported negative anode materials for KIBs are classified into carbonaceous anode materials, alloy-based materials, metal chalcogenides, polyanionic materials, and organic anode materials. In the following section we discuss the recent advances in various anode materials for KIBs.

2.2.1 Carbonaceous Anode Materials

Carbon-based anode materials exhibit several advantages such as they are easy to synthesize, inexpensive, and environmentally friendly and have a flexible interlayer arrangement, which make them extremely worthy negative electrode materials for KIBs. Among the carbonaceous materials, graphite is the first reported [9] intercalated electrode material that forms KC_8 for KIBs in which it exhibits the theoretical specific capacity of 279 mAh g^{-1}. The three-step intercalation mechanism of KIBs is represented as C → KC_{24} (Stage III) → KC_{16} (Stage II) → KC_8, which produces the capacity of 200–273 mAh g^{-1} with potential plateaus at 0.1–0.3 V for potassiation and 0.25–0.45 V for depotassiation. Safety is the major concern associated with low potential and high C-rates because of K-plating. Therefore, other than graphite, various carbonaceous electrode materials have been demonstrated as anode materials for KIBs. The various forms of carbonaceous materials include hard carbon, non-graphitic soft carbon, hetero-atom-doped carbons, composites of hard/soft carbons, carbon nanofibres (CNTs – Carbon nano tubes), reduced graphene oxides, and undoped graphenes.

Recently bio-waste-derived hard carbon anode material received much attention for KIBs, as reported by Chen et al. [10]. Hierarchically constructed porous nitrogen-doped hard carbon chitin (major seafood waste) (Figure 2.1) is investigated as an anode for KIBs which exhibit superior rate capability (154 mAh g^{-1} at 72C) and ultralong cycle life (4,000 cycles). This superior performance is attributed to the capacitive-driven surface-enhanced mechanism that enables storing a large amount of reversible K^+ storage. The carbonization of commercial melamine foams (MFCs) became oxygen-rich MFCs after oxidizing acid treatment, and they are investigated as anode material for KIBs [11]. These anodes deliver high reversible capacity (272 mAh g^{-1}), superior rate capability (160 mAh g^{-1} at 0.1 A g^{-1}), and long cycling life (150 mAh g^{-1} after 800 cycles) for KIBs. This enhanced performance is due to the presence of ether-type oxygen species and quinone in MFCs, boosting K^+ ion storage via surface-driven capacitive behaviour. Furthermore, 3D porous carbon materials provide a shorter reaction pathway for potassium ion transport, thereby enhancing the electrochemical

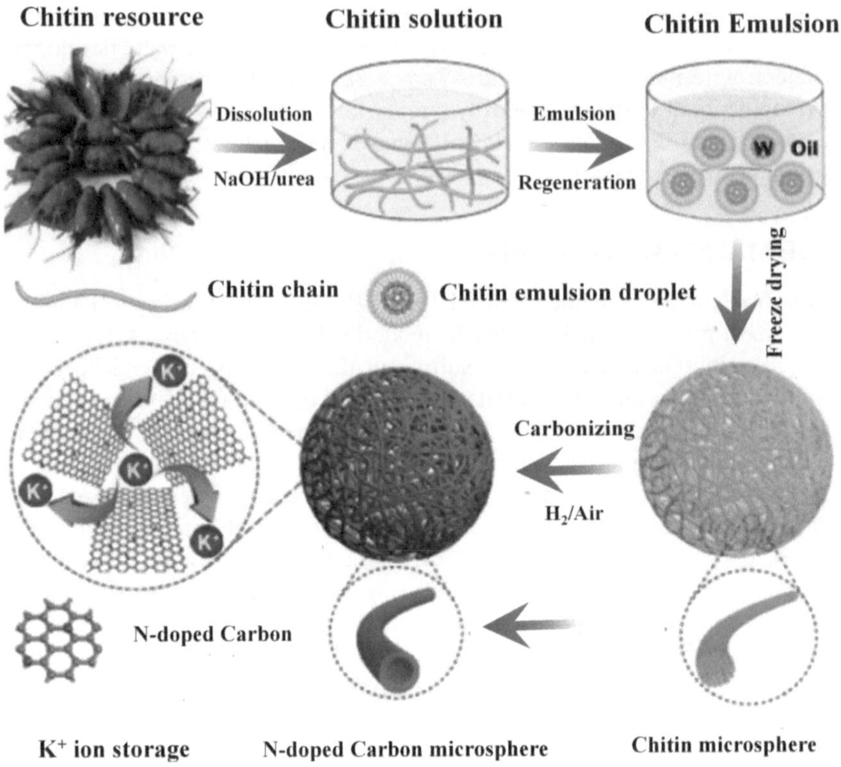

FIGURE 2.1 Schematic illustration for the large-scale fabrication of hierarchically porous nitrogen-doped carbon microspheres as anode material in KIBs. Adapted and reproduced from Ref. [10]. Copyright 2020 Elsevier.

performance of KIBs. This is attributed to the porous structure favour ion transport significantly inner-pore ion transport further reduces the resistance of ion transport. Further hetero atom doping (such as nitrogen, sulphur, and boron) into 3D porous carbonaceous materials has been successfully demonstrated to enhance the electrochemical performance (conductivity and reactivity) by adjusting the composition of surface functional groups. This is attributed to the strong interaction between the carbon atom and the doped hetero atom, which can adjust the local electronic structure of pores in carbon, thereby influencing the electrical conductivity. Among the heteroatom doping, nitrogen doping is particularly effective because the electronegativity of nitrogen is higher s compared to carbon, resulting in strong interaction with potassium ions. Further, different types of vacancies and defects are introduced by nitrogen doping, offering more electroactive sites in porous carbon. In this way Qi et al. [12] achieved high reversible capacity (296 mAh g^{-1} at 50 mA g^{-1}) by introducing nitrogen doping in hierarchically porous carbon materials. 3D nitrogen and sulphur co-doped porous carbon nanosheets (3D-NSCNs) also exhibit superior cycling stability (after 2,900 cycles, the reversible capacity is 171 mAh g^{-1}) [13]. Nitrogen doping in bamboo like carbon nanotubes [14] and carbon microtubes [15] is also shown remarkable electrochemical performance towards potassium-ion storage.

Nitrogen doping in graphene also exhibits remarkable electrochemical performance for potassium-ion storage. Ju et al. [16] investigated a few layers of N-doped graphene for KIBs in which it delivers the reversible K storage capacity of 320 mAh g^{-1} at 50 mA g^{-1} after 60 cycles and superior cycling stability (150 mAh g^{-1} at 500 mA g^{-1} after 500 cycles). This is attributed to the multiple synergistic effects of nitrogen doping, large specific surface area (479.21 m^2g^{-1}), and high nitrogen content (14.68 at%), as well as interconnected mesopores with enhanced pore volume not only favour the K^+ ion transport throughout the carbon matrix but also offer extensive active sites for K-ion storage. Graphene-like N-doped carbon on the 3D metal substrate [17] served as a binder-free anode material for KIBs. This binder-free anode delivers high reversible capacity and long cycle life for KIBs, which is attributed to the capacitive controlled K^+ ion storage. Nitrogen/oxygen dual-doped highly wrinkled carbon tubes (NO-WCTs) [18] delivered ultra-high reversible capacity, i.e., 536 mAh g^{-1} and superior cycling stability (after 2,000 cycles 92.8% capacity is retained). As a result, enhanced surface area offers more active sites for K^+ storage, thereby improving the electrochemical performance of KIBs.

2.2.2 Oxide Anode Materials

In the previous section, we discussed carbon-based anode materials for KIBs since they exhibit low redox potential, which is advantageous to achieve high energy density, but safety is a serious concern related to metal plating. In view of this concern, transition metal oxides are developed with moderately low redox potentials for KIBs. Various Ti-based intercalation anode materials such as $K_2Ti_4O_9$, $K_2Ti_8O_{17}$, and $K_2Ti_6O_{13}$ [19–21] are recently developed, which involve the reduction of Ti^{4+} to Ti^{3+}, accommodating $2K^+$ ions per formula unit. These anode materials exhibit acceptable reversible capacities in the range of 90–120 mAh g^{-1} and good cycling performance, but the working potential falls below ≤1 V. Low working voltage is beneficial to realize high energy density for KIBs. Materials with low working voltage require protection from the passivation layer; still safety is a challenge for high-power KIBs.

Various compositions of vanadium-based oxides are extensively studied as anode materials for KIBs. Electrode pulverization results in large volume variation and poor intrinsic electronic conductivity, the two main hurdles to realizing high-energy-density KIBs. To address these challenges, recently Chen et al. [22] synthesized hollow sphere carbon-coated V_2O_3 powder as a potential anode material for KIBs. It delivers promising cycling performance and acceptable rate capability in which it exhibits 330 mAh g^{-1} at 100 mA g^{-1} and realizes good cycling stability after 500 cycles. This is ascribed to the homogeneous carbon coating enhancing the conductivity and hollow sphere morphology buffering the volume variation during cycling. In addition to titanium and vanadium oxides, Nb- and Mo-based oxides are explored as negative electrode materials for KIBs. MoO_2 system [23] intercalated K composition with 0.25 only, and the average operating voltage is 0.75 V to realize safety and high energy density, but the delivered reversible capacity is much less compared to the theoretical capacity of 523 mAh g^{-1} for KIBs. Orthorhombic T-Nb_2O_5 contains unique structural properties in terms of layered framework structure and good chemical stability, which make them ideal candidates for KIBs. Li et al. [24] investigated hydrothermally synthesized T-Nb_2O_5 for KIBs, which shows the morphology of urchin-like interconnected nanowires with several microns length and diameter in the range of 20–40 nm. Based on reversible electrochemical redox couple of $Nb^{4+/5+}$, T-Nb_2O_5 delivered the reversible capacity of 152 mAh g^{-1} at the current density of <1 A g^{-1} and exhibited good cycling stability over 400 cycles. Many other transition metal oxides such as Fe_2O_3, Co_3O_4, and $Pb_3Nb_4O_{13}$ were also reported for KIBs based on conversion reaction mechanism and exhibit reversible capacities in the range of 200–400 mAh g^{-1} for K-ion cells. Recently, Luo et al. [25] successfully demonstrated ultra-small SnO_2 nanocrystals embedded in porous carbon for KIBs, which delivered the reversible capacity of 300 mAh g^{-1} over 100 cycles. The size of the nanocrystals in the range of 2–6 nm embedded in porous carbon facilitates the K^+ ion intercalation/deintercalation, and porous carbon accommodates volume expansion experienced by conversion reaction during cycling.

2.2.3 Transition/Non-transition Metal Chalcogenides

In the former section, we have briefly summarized the recent development of oxide anode materials based on intercalation and conversion mechanisms for KIBs. Like oxides, transition metal chalcogenides (TMCs) MX_2 (M=transition metal; X=S, Se, Te) also show the intercalation mechanism, but some transition metal dichalcogenides exhibit conversion mechanisms also. TMCs exhibit 2D layered structures and show larger interlayer distances than graphite and transition metal oxides. Recently K-ion intercalated 3D-transition metal dichalcogenides such as VS_2, TiS_2, MoS_2, WS_2, and $MoSe_2$ are explored and extensively studied for KIBs [26]. TiS_2 intercalates [27] potassium composition of 0.88 form $K_{0.88}TiS_2$ and exhibits the discharge capacity 145 mAh g^{-1}, which originated from the potential window between 2.3 and 2.5 V with a coulombic efficiency of 104.2%. Some of the K^+ ions trapped in the interlayers lower initial coulombic efficiency (69.2%), and also this is compensated by prepotassiation of TiS_2 with the potassium composition of 0.25. The working potential of TiS_2 is impressively high ~2 V vs K/K^+, which would be helpful to realize safe high-power KIBs. Vanadium-based chalcogenide VS_2 is also investigated as the negative electrode material for KIBs [28] in which VS_2 nanosheets delivered the reversible capacities 400 mAh g^{-1} for KIBs and also demonstrated good rate capability. The intercalation potential is ~1.5 V vs K/K^+, which is 0.5 V lower than TiS_2 and is also considered a promising anode material for KIBs.

4D and 5D transition metal dichalcogenides such as MoS_2, $MoSe_2$, and WS_2 are also extensively studied as the negative materials for KIBs in addition to 3D TMCs. Zheng et al. [29] investigated interlayer expanded, tubular MoS_2–N/O-doped carbon composite (E-MoS_2/NOC TC) for potassium-ion storage. E-MoS_2/NOC TC presented a reversible capacity of 220 mAh g^{-1} after 300 cycles with a voltage plateau of 1.3 V. Expanded interlayer spacing (0.92 nm) facilitates K-ion diffusion rate and N/O carbon matrix enhancing the electronic conductivity and buffering the volume expansion. $MoSe_2$/MXene embedded in carbon composite electrode material for KIBs is investigated by Huang et al. [30]. It exhibited a high charge capacity of 355 mAh g^{-1} at 200 mA g^{-1} and demonstrated excellent rate capability of 183 mAh g^{-1} at the high current density of 10 A g^{-1}. The strong chemical interaction between $MoSe_2$ nanosheets and MXene flakes is believed to be enhancing the charge transfer kinetics and carbon coating, enhancing the electronic conductivity. Shen et al. [31] investigated $MoSe_2$ incorporated into carbon fibres as a negative electrode material for KIBs. It delivers the reversible capacity of 81 mAh g^{-1} at 100 mA g^{-1} over 100 cycles in which carbon fibres improve the ionic and electronic conductivity. WS_2 has interlayer spacing of 6.18 Å, which is higher than that of MoS_2 and TiS_2, which is demonstrated as an anode material for KIBs [32]. It accommodates 0.62 K^+ per formula unit, resulting in the reversible capacity of 67 mAh g^{-1} and exhibiting the potential plateau at 1 V vs K/K^+, which is smaller than MoS_2 and TiS_2. WS_2 exhibits low gravimetric capacity, which is attributed to a large atomic weight of *W*, and is also considered a good anode material for KIBs. In contrast to classical TMCs, some other TMCs and chalcogenides based on non-transition metals exhibit conversion or conversion-based alloying mechanism. For instance, FeS_2, Fe_3S_4, SnS_2, Sb_2S_3, and $AgBiS_2$ based on conversion-based alloying mechanisms are investigated as negative materials for KIBs. Ren et al. [33] investigated $AgBiS_2$ nanocrystal anode for KIBs in which it exhibits the reversible capacity of 420 mAh g^{-1} at 0.5 A g^{-1} and also demonstrated as a superior rate capable anode material. Lakshmi et al. [34] first reported the electrochemical performance of nanocrystalline-coated SnS_2 on rGO for KIBs. This material delivers an impressive charge capacity of 350 mAh g^{-1}, which exceeds the capacity of graphite. Sb_2S_3 is also investigated by Lakshmi et al. [35] for KIBs, which is tested with two different electrolytes, namely, 1 M KPF_6 in ethylene carbonate (EC)–diethyl carbonate (DEC) and 1 M KPF_6 in 1,2-dimethoxyethane (DME) electrolytes. It exhibits a high charge capacity of 650 mAh g^{-1} for KIBs, which is ascribed to the conversion-alloying storage mechanism involving eight electrons per formula unit.

2.2.4 MXenes

In this section we discuss the new class of anode materials based on transition metal carbides/nitrides crystallizing into two-dimensional structures called MXenes. Generally these MXenes exhibit pseudocapacitive contribution to the overall capacity of batteries along with intercalation behaviour. It possesses high electronic conductivity and ionic conductivity resulting in excellent electrochemical performance and high rate capabilities towards metal ion batteries. Various classes of MXenes such as Ti_3C_2, Ti_3CNT, Nb2CT_z, and V_2CT_x [36,37] are successfully demonstrated as anode materials for KIBs and exhibit reversible capacities in the range of 130–400 mAh g^{-1} but do not show acceptable cycling performance for KIBs.

2.2.5 Metal-Based Anode Materials

Compared to intercalation-type anode materials, metal-based anode materials exhibit higher specific capacity than intercalation-type anodes through an alloying/de-alloying reaction with K anode. Nevertheless, these metal anodes suffered from large volume variations, resulting in electrode cracking occurring during cycling. Therefore, the incorporation of metals into carbonaceous materials offers high specific capacity, and the carbon matrix buffers the volume expansion during cycling. Recently, phosphorous-based anode materials have attracted KIB technology, which demonstrate capacities higher than those of graphite-based anode materials. Sultana et al. [38] investigated a black phosphorous anode for a potassium-ion battery in which it exhibits a reversible capacity of 617 mAh g^{-1}, more than twice that of graphite for K-ion cells. This high capacity is ascribed to the black phosphorous operates for K-ion cells via an alloying-dealloying mechanism. Phosphorous alloying with Sn forms Sn_4P_3 incorporated in N-doped carbon fibres, which are demonstrated as anode materials for KIBs [39]. This material demonstrates enhanced cycling stability and high rate capability (160 mAh g^{-1} after 1,000 cycles at 500 mA g^{-1}). The excellent electrochemical performance is attributed to the synergistic K-ion storage that occurred through sequential conversion (P to K_3P_{11} and K_3P) along with an alloying reaction (Sn to KSn). Sn- and Sb-based carbon composites, for example, Sb@CSN and Sn@C [40,41] also demonstrate high reversible capacity (500–700 mAh g^{-1}) and superior cycling stability, which are attributed to the alloying/dealloying mechanism.

2.2.6 Miscellaneous Anodes

At low potentials, various organic materials display their redox activity which is considered a suitable host to store K^+ ions. This is attributed to the organic molecules containing interstitial spaces and large voids that enable K-ion intercalation/deintercalation. For instance, Vitamin K [42] functions as ideal anode material for KIBs in which it exhibits the reversible capacity of 300 mAh g^{-1} and also demonstrates good cycling stability (200 mAh g^{-1} after 100 cycles with a coulombic efficiency of >90%). Recently 2D-based monolayer materials have possessed much attention in the field of semiconductors and electronics. For example, PC_6 monolayer [43] attracted KIB technology, owing to its small Young's modulus, and has shown potential application in flexible batteries. It is demonstrated as an anode material for KIBs in which it delivered the reversible potassium storage capacity of 781 mAh g^{-1}, which is superior to other anode materials for KIBs. More recently vanadium nitride-based (VN)/CNFs [44] synthesized by electro-spinning method are considered a very good anode material for KIBs. This composite delivered outstanding electrochemical performance with a reversible capacity of 266 mAh g^{-1} after 200 cycles at 0.1 Ag^{-1}. The reason for superior performance is attributed to the small VN nanocrystal confined on carbon nanofibres that offer pseudocapacitive type active sites that shorten K^+ diffusion and also carbon fibres enhancing the conductivity and buffering the volume expansion during cycling.

2.2.7 Challenges of Negative Electrode Materials

Several studies reported that graphite is an ideal intercalation host for KIBs, but inferior electrochemical performance is observed such as poor cyclability and low capacity at extremely low temperatures. Most of the soft/hard carbon-based anode materials exhibit slope-like plateaus, which enhance the average operation potential and decrease the energy density of full-cell KIBs. Even though heteroatom-doped carbon materials and graphene-like analogues offer promising capacity for KIBs, the cost of production of these kinds of hierarchical nanostructures is higher, and also they exhibit lower volumetric energy density and undesirable side reactions due to large surface area. Therefore, in the future, exploring low-cost production techniques and materials with significantly larger tap densities is needed. Some non-graphitic materials have shown superior electrochemical performance in terms of cycling stability, specific capacity, rate capability, and energy efficiency. Compared to graphite, these materials have experienced smaller volume expansion and shown high intercalation potential, which is beneficial for avoiding stripping/plating and dendrite formation during high current drains. TMCs, TMOs, metals, and MXenes-based composites also exhibit satisfactory electrochemical performance towards KIBs, and volume expansion is a serious concern to realize the practical applications of KIBs.

2.3 CATHODE MATERIALS

The cathode materials for KIBs are classified into potassiated transition metal oxides, polyanionic compounds (phosphates, fluorophosphates, oxyphosphates, and pyrophosphates), Prussian blue analogues (PBAs), organic cathodes, and miscellaneous materials. In the following section, we discuss the recent developments in transition metal oxides, polyanionic compounds, PBAs, and miscellaneous cathode materials. More significantly, the challenges associated with these cathode materials for potassium-ion batteries are also briefed.

2.3.1 Transition Metal Oxide-Based Cathodes

Layered transition metal oxides with P2 and P3 types are extensively studied as positive electrode materials for KIBs. P2-type K_xCoO_2 are generally produced from high-temperature calcination at 600°C than P3 type, which requires 400°C, and to the best of our knowledge O2 type K_xCoO_2 has not yet been reported to date, which may be because larger-size K^+ ions prefer to occupy P-site rather than O-site to maintain larger interlayer distances [45]. Recently Sada et al. [46] reported that K-ion intercalation in Na_xCoO_2 delivers 82 mAh g^{-1} with excellent rate capability. Apart from K_xCoO_2, many other transition metal oxides such as P2, P3-type $KScO_2$, $KMnO_2$, O3-$KCrO_2$, and O3-$KScO_2$ are reported for KIBs in which $KScO_2$ (all P2, P3, and O3 type) is electrochemically inactive [47]. Cr^{3+} preferred octahedral coordination sites in $KCrO_2$ crystal structure and delivered the reversible capacities of 90 mAh g^{-1} in the potential window between 1.5 and 4 V with stepwise voltage curves [48]. During cycling, multiple structural changes are observed for $KCrO_2$ in the sequence of O3–P3–P′3–P3–P′3–O′3; therefore, upper cut-off voltage is a significant parameter to overcome the irreversible structural changes. P2-type $KMnO_2$//hard carbon composite [49] exhibited the reversible capacity of 99 mAh g^{-1} in the potential plateau between 0.5 and 3.4 V with a coulombic efficiency greater than 99% without any structural degradation. Liu et al. [50] reported P2 and P3-type $K_{0.45}MnO_2$, which deliver 128.6 mAh g^{-1} at 20 mA g^{-1} with an average potential of 2.75 V. Compared to all other single transition metal oxide systems, K_xMnO_2 may be considered a suitable cathode for KIBs. The introduction of Ti in P2-type K_xMnO_2 layered framework largely prevents the sliding of TMO_2 slabs during cycling, hence thwarting the structural phase transition from P2 to other O2 phases. Recently Ti^{4+}-doped K_xMnO_2 cathode material with the formula $K_{5/9}Mn_{7/9}Ti_{2/9}O_2$ demonstrated an excellent high-capacity stable cathode material for K-ion batteries at high-voltage regions. This cathode material is charged up to 4.2 V; Ti^{4+} doping ultimately diminishes the P2- to O2-type transition [51].

P2- and P3-type binary and tertiary transition metal oxides such as $K_xFe_{0.5}Mn_{0.5}O_2$, P3-$K_xNi_{1/6}Co_{1/6}Mn_{2/3}O_2$, and P3-$K_xCo_{0.5}Mn_{0.5}O_2$ are synthesized by solid-state reaction method and investigated as a cathode material for KIBs and delivered the reversible capacities in the range of 100–180 mAh g^{-1}. Tai et al. [52] recently reported $KMn_{7.6}Co_{0.4}O_{16}$ nano-rod clusters for KIBs with a high discharge capacity of 246 mAh g^{-1} at 10 mAg^{-1} in which Co doping efficiently improved the structural stability and electrical conductivity of $KMnO_2$-based single transition metal systems. The above transition metal oxide systems deliver reversible capacities in the lower voltage region for KIBs; for practical application, these kinds of transition metal oxides in full cells reduce the overall voltage of the cell. Therefore designing high-voltage cathodes is a critical parameter to improve the average potential of K-ion cells. In this view, Masese et al. [53] synthesized a honeycomb layered framework P2-type $K_{2/3}Ni_{1/3}Co_{1/3}Te_{1/3}O_2$ cathode material for K-ion cells. This material delivers the reversible capacity of only 30 mAh g^{-1} but exhibits the highest voltage of 4 V; further improvement is still required to realize the high-energy-density KIBs.

2.3.2 Polyanionic Compound-Based Cathodes

Generally, polyanionic compounds have an open 3D framework structure like PBAs, which would be advantageous for fast ion diffusion. They consist of the general formula MO_x (M=transition metals) and $(XO_4)^{n-}$ (X=P, S, As, Si, Mo, and W) polyhedra. The inductive effect on the polyanions increases the reduction/oxidation potential of transition metal ions. The covalency of the M–O bond decreases due to the rich covalent nature of the X–O bond (X – electronegative element), which in turn increases the redox potential of transition metals. Polyanionic compounds have been successfully demonstrated as cathode materials for LIBs and NIBs since 1980. To date various polyanionic compounds such as $KVPO_4F$, $KVOPO_4$, KVP_2O_7, $K_3V_2(PO_4)_3$, $KMPO_4$ (M – Fe, Mn), $KFeSO_4F$, KMP_2O_7 (M=Cr, Fe, Mo, Mn), and $K_2M(SO_4)_3$ (M=Cu, Fe) are investigated as cathode materials for KIBs [54–58] and delivered the capacities in the range of 90–25 mAh g^{-1}. Chihara et al. [59] investigated structurally identical $KVPO_4F$ and $KVOPO_4$ polyanionic compounds for KIBs and delivered discharge capacities in the range of 92 and 84 mAh g^{-1}, respectively, with the average voltage reaching 4 V. The stable cycling performance is attributed to the lattice volume shrinkages happened while charging up to 5 V (5.8% for $KVPO_4F$ and 3.3% for $KVOPO_4$). Tailoring the particle morphology of new layered $KVOPO_4$ is investigated by Liao et al. [60]. This material contains two-dimensional ion diffusion active channels, which are beneficial for fast ion kinetics. It delivered the discharge capacity of 115 mAh g^{-1} at 0.2 C with 86.8% capacity retention over 100 cycles, and the observable volume changes are only 9.4%, as confirmed by ex-situ X-ray diffraction. $KVPO_4F$ is considered one of the high-voltage cathodes for KIBs; however, poor electronic conductivity and cycling performance are the key issues that dramatically hinder the practical application. To overcome this issue, Liu et al. [61] reported that plum-pudding model-inspired $KVPO_4F$ is covered with an amorphous carbon network (Figure 2.2) as a cathode for KIBs. The amorphous

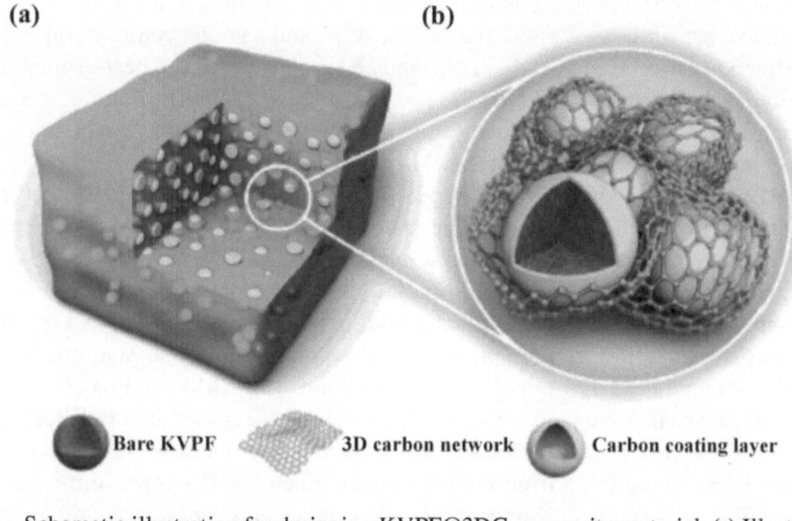

FIGURE 2.2 Schematic illustration for designing KVPF@3DC composite material. (a) Illustration of plum-pudding model-inspired design for a desired KVPF@3DC composite. (b) KVPF@3DC particles are in-situ armoured with a uniform amorphous carbon outer layer together with the large three-dimensional conductive network connecting each other. Adapted and reproduced from Ref. [61]. Copyright © 2020 Elsevier.

carbon layer homogeneously wrapped the $KVPO_4F$ particles, which are connected by cross-linked huge-area carbon sheets. This cathode exhibits a high operating voltage of 4.1 V and delivers a discharge capacity of 102.96 mAh g^{-1} at 20 mA g^{-1} and retains 85.4% of capacity after 550 cycles at 500 mAg^{-1}. The design of the plum-pudding structure, $KVPO_4F$@3DC, provides high electronic conductivity and stable cycling performance over prolonged cycling. Nanocubic $KTi_2(PO_4)_3$ is successfully fabricated using the hydrothermal method [62], and carbon coating is carried out with a normal cane sugar-assisted method. This $KTi_2(PO_4)_3$/C material delivers a discharge capacity of 75.6 mAh g^{-1}, and its capacity is continuously increased over 40 cycles, which may be due to continuous saturation of electrolytes, and fine capacity retention is observed due to carbon coating.

2.3.3 Prussian Blue Analogues

Prussian blue, $KFe[Fe(CN)_6]$, is first investigated as a cathode material for KIBs in 2004 by Eftikari et al. [63]. Various PBAs such as KFeFe-, KMnFe-, KCoFe-, KNiFe-, KCuFe-, and KTiFe-PBAs have displayed capacities in the range of 30–70 mAh g^{-1} at 3.75–4.0 V vs K/K^+ for KIBs. These charge–discharge curves constitute one electron redox $Fe^{2+/3+}$ couple, and Ni^{2+}, Co^{2+}, and Cu^{2+} are electrochemically inactive. On the contrary, KMnFe-PBA, KTiFePBA, and KFeFe-PBA correspond to two-electron redox couples, i.e., redox of $Mn^{2+/3+}$, $Ti^{2+/3+}$, and $Fe^{2+/3+}$ in the PBAs, respectively [64]. To date, KMnFe-PBA is considered an excellent cathode material for KIBs among the reported PBAs because its working potential is ~3.8 V vs K/K^+ and delivered the reversible capacity of 140 mAh g^{-1}. Compared to layered transition metal oxides and polyanionic compounds, some PBAs achieved high energy densities (for example, KFeFe-PBA and KMnFe-PBA).

2.3.4 Miscellaneous Cathodes

Many organic molecules containing redox-active centres, polymers, and organic phosphomolybdates have been explored as positive electrodes for KIBs, owing to their recyclability, eco-friendliness, low cost, and flexibility. An organic polymer hexaazatriphenylene-based molecule is investigated as a positive electrode material by Kapaev et al. [65]. This hexaazatriphenylene-based polymer cathode is synthesized from 3,3′ diaminobenzidine and triquinoyl. Using DME-based electrolytes in KIBs delivered the reversible capacity of 169 mAh g^{-1} at 10 A g^{-1} and retained the capacity after 4,600 cycles. Tang et al. [66] reported poly(pentacenetetrone sulphide) as a cathode for KIBs, which exhibited a reversible capacity of ~260 mAh g^{-1} at 0.1 A g^{-1} and a good cycling stability (retains 190 mAh g^{-1} even after 3,000 cycles at 5 A g^{-1}). This material is considered the best-ever cathode among the reported polymer-based cathode for KIBs. Polymeric aromatic amines are considered a good cathode host for LIBs, but surprisingly these polyamines also served as a good cathode candidate for KIBs. Poly(*N*-phenyl-5,10-dihydrophenazine) is investigated as a positive electrode by Obrezkov et al. [67] for KIBs. It delivers an impressive specific capacity of 162 mAh g^{-1} at 200 mAg^{-1} with 79% capacity retention after 1,000 cycles. The specific power exhibited by this cathode is $>3 \times 10^4$ WK g^{-1}, hence paving the way to design stable and ultrafast cathodes for KIBs. Shen et al. [68] investigated homogeneous tetra-*n*-butylammonium phosphomolybdate (TBAPM) as a cathode for KIBs. This novel TBAPM cathode delivers a discharge capacity of 232 mAh g^{-1} at 20 mA g^{-1} with little capacity decay in the subsequent cycles. The potassium-ion storage mechanism of this TBAPM cathode is found to be a reversible redox reaction of Mo^{4+}/Mo^{6+} ion pairs.

Mn-based oxide materials are extensively studied as positive electrode materials for KIBs because of their high theoretical capacity. Nonetheless, Mn-based fluoride materials have not received much attention towards KIBs owing to their poor electronic conductivity. To improve the electronic conductivity and cycling performance, Wang et al. [69] synthesized a new class of perovskite fluoride-type, carbon-coated, O-doped $KMnF_3$ cathodes (Figure 2.3) for KIBs. It delivers a discharge capacity of 113 mAh g^{-1} in a voltage range of 1.2–4.2 V and shows acceptable capacity retention over 40 cycles.

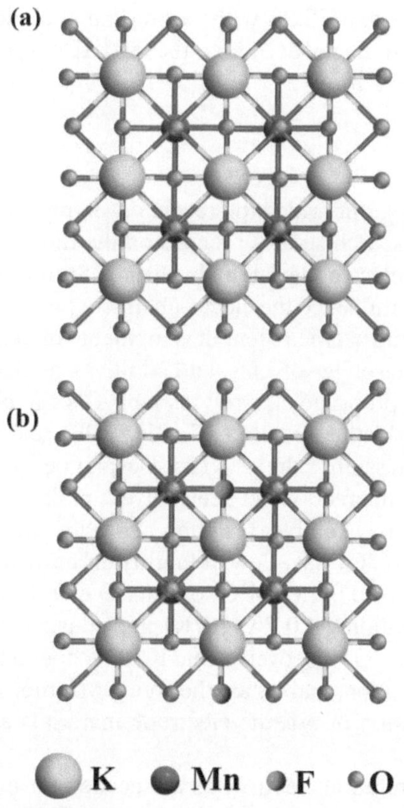

FIGURE 2.3 (a and b) Structural demonstration of $KMnF_3$ and O-doped $KMnF_3$, respectively. Adapted and reproduced from Ref. [69]. Copyright 2020 Elsevier.

2.3.5 Challenges of Cathode Materials

We have briefly summarized the recent progress of positive electrode materials for KIBs in the previous section. More significantly, reasonable progress has been achieved in layered transition metal oxides, polyanions, and PBAs; nevertheless, a lot of challenges are still associated with these cathode materials. For instance, layered transition metal oxides suffer from multiple phase transitions due to K^+ vacancy ordering, a decrease of K^+ ion stoichiometry after the first cycle, extremely small capacities, and low operating potentials. Therefore, to avoid multiple irreversible structural transitions, the limitation of upper cut-off voltage is required in layered transition metal oxides. Polyanionic compounds have large molecular weight, which decreases the volumetric energy density of a full K-ion cell when applied as a cathode in KIBs. Further, it is very difficult to obtain electrochemically active phases for intercalation/deintercaltion of K^+ ions. Polyanionic compounds also suffer from poor electronic conductivity and low working potential. Therefore, low-cost efficient synthesis methods and effective coating techniques are to be developed further to achieve good electrochemical performance. In spite of their excellent electrochemical performance, PBAs display two major drawbacks such as low bulk density and poor electronic conductivity, which are instigated by their distinctive crystal structure and nature of the chemical bond between transition metals and ligands. Low bulk density also causes low tap density, which is difficult to overcome; however, the addition of additives with electrode components and mixing conditions improves the bulk density of PBAs. To improve the electronic conductivity of PBAs, carbon coating was used, which in turn reduces the energy density of the full K-ion cell, and also limited thermal stability (<400°C) of PBAs hampered the carbon coating. Therefore, low-cost in-situ synthesis techniques are required to make the composite of PBAs with carbon/metal oxide. Although organic molecules are the most promising positive electrode materials for KIBs, typical organic molecules suffer from low capacity at high current drains.

2.4 ELECTROLYTES

In the previous section we have summarized the recent developments of anode and cathode materials for KIBs. Recently many research studies focused on the importance of electrolyte and critical side reactions originating from electrolyte salt and solvents. Similar to electrode materials, electrolyte materials also significantly influence the electrochemical performance of KIBs. In this section, we emphasize and briefly summarize the recent developments of electrolytes, the transport properties of ions, the influence of electrolyte solvents, and additives in SEI formation.

The choice or selection of appropriate electrolytes for KIBs can be gained from NIBs and LIBs. For a model, $KClO_4$, potassium bis(fluorosulfonyl)imide, KPF_6, and potassium bis(tri-fuoromethylsulfonyl)imide are common potassium salts of KIBs. Despite being different from the related salts in NIBs and LIBs, the unstable interaction between solvent molecules and K^+ ions makes a portion of the potassium salts inadequately soluble, bringing about low ionic conductivity and concentration. For example, KPF_6 and $KClO_4$ have low solubility in carbonate-ester-based solvents. Then again, KFSI can be highly dissolved in EC/DEC solvents, even though KFSI has a corrosion impact on Al foil during anodic polarization (>4.0 V vs K^+/K), and consequently KFSI is not appropriate for high-voltage KIBs. Furthermore, it is discovered that KFSI is bound to solvate with DME molecules than KPF_6 at a similar concentration, and hence the free DME molecules in KFSI electrolytes will be useful to reduce the dissolution of organic electrode materials and decrease the side reactions between K metal and solvent.

Now and again, high-concentration electrolytes are needed for improvement in electrochemical performance. For instance, some electrode materials are soluble in liquid electrolytes, while it has been discovered that increasing the concentration of the electrolytes can viably reduce the dissolution of the electrode materials because of the diminished free solvent molecules and in this manner improve the cyclability. Taking into account, the choice of proper electrolytes can altogether impact the performance. Then again, it ought to be noticed that there is no certain connection between the electrochemical performance and electrolyte concentration. As the concentration of electrolytes increased, the viscosity of the electrolytes would increase, while the wettability to the electrodes would reduce. The two of them held up the ionic transport, prompting more awful electrochemical performance [70,71]. For example, Hu et al. examined the potassium storage performance of a PI cathode in KFSI–DME electrolytes with various concentrations (the molar proportion among KFSI and DME was 1:1.5, 1:2.5, 1:5, and 1:10). They found that the PI cathode displayed best cycling performance in KFSI–DME with a molar proportion of 1:5. Subsequently, it is important to do viable trials to improve the concentration of electrolytes for best performance. Aside from the concentration, the wettability of the electrodes by the electrolytes is one of the key factors of the electrolytes. Generally, glyme-based electrolytes have good wettability of the electrodes over carbonate ester-based electrolytes and have ended up being ready to improve electrochemical properties. Other than the concentration and the wettability, the formation of a stable solid electrolyte interface is the third precondition for superior KIBs [72]. Recently Zhang et al. [73] demonstrated a highly concentrated electrolyte; i.e., 5M KTFSI (potassium bis(tri-fuoromethylsulphonyl)imide) in diethylene glycol dimethyl ether electrolyte for bismuth-carbon composite anode employed KIBs, which offer high capacity, excellent cycling stability, and high coulombic efficiency.

Lei et al. [74] discovered that the SEI films in carbonate ester-based electrolytes are thicker with breaks and harsh surfaces; conversely, the SEI in DME-based electrolytes demonstrated smooth surfaces. The distinction might be credited to the Lewis acid feature of K, prompting higher solubility and the formation of thicker and unstable SEI in carbonate ester-based electrolytes. On the one hand, the large charge transfer resistance that came from the thicker SEI in carbonate ester-based electrolytes prompted larger voltage polarization than that in DME-based electrolytes. The diverse voltage polarization ought to be answerable for the different discharge/charge voltage of a similar electrode material. Then again, the formed SEI in DME-based electrolytes was robust and had high K^+ ion conductivity, leading to the high capacity and rate capability. Further theoretical calculations also additionally affirmed that the ether-based electrolyte has advantages. The theoretical studies demonstrated that the lowest unoccupied molecular orbital (LUMO) energy level of bis(fluorosulfonyl)imide FSI was lower than that of DME molecules in the KFSI–DME electrolyte, showing that KFSI would decompose before DME solvent molecules. During the initial discharge, the decomposition products of KFSI formed a thick coating on the electrode surface, which has been affirmed to be highly impermeable to the electrolyte. Consequently, the ensuing decomposition of DME molecules can be successfully alleviated by the SEI layer, with the goal that the SEI can remain stable during the cycling process. In the KFSI–EC/DEC electrolyte, the LUMO energy levels of FSI, EC, and DEC were similar, which inferred that they could be decomposed simultaneously.

It has been accounted that a large amount of inorganic salts may prompt SEI failure, and hence the moderate organic–inorganic composite SEI in the KFSI–EC/DEC electrolyte was more stable than the KFSI–DME electrolyte. In addition, the calculated results likewise indicated that the solvation energy of KFSI–EC/DEC was lower than that of KFSI–DME, and thus the weaker interaction among KFSI and EC/DEC profited quicker K^+ ion transport. Subsequently, they suggested that a stable K^+ ion conductive SEI can form in the KFSI–EC/DEC electrolyte. At the point when the K metal was used as the counter electrode, the normal choice or selection of electrolytes likewise has the upsides of inhibiting the growth of K dendrites. It has been accounted that the addition of additives improved the coulombic efficiency (CE) and the cyclability of the batteries by controlling the solvation sheath in metal anodes. A limited quantity of proper electrolyte additives can form a stable SEI film, repressing further decomposition of the electrolytes and decreasing the side reactions. For example, the addition of fluoroethylene carbonate (FEC, just 2%) can successfully improve the CE of KIBs, and the charge–discharge profile of batteries with FEC indicated higher capacity over those without the addition of FEC. Other than FEC, presenting some low-concentration metal salts where the metal ions have more negative potential than working metal ions will likewise work for framing a non-dendrite metal anode by means of an electrostatic shield. LiTFSI (0.05 M) as an additive delivered few Li^+ ions with more negative potential than K^+ ions in 1 M KPF_6-DME as the electrolyte, and hence the electrostatic shielding of Li^+ ions empowered the formation of a smooth K metal surface and prompted less side reactions between K metal and electrolytes, which profited higher CE. Organic liquid electrolytes have the benefits of accomplishing higher capacity, a wider voltage window, higher energy density, and power density, compared to aqueous electrolytes and solid-state electrolytes. Notwithstanding, it ought to be noticed that organic liquid electrolytes likewise raised serious safety concerns because of combustible organic solvents [75].

In organic electrolytes, for example, EC:DEC solvents, K^+ ions additionally show a higher ionic conductivity than Li^+ and Na^+ ions. In addition, the lower desolvation energy of K^+ ions in liquid electrolytes can empower ion insertion into electrode materials, further improving the kinetics of potassiation and subsequently rate capability. Furthermore, the higher working voltage of KIBs will compensate for the decreased theoretical capacity and increase the energy density compared to NIBs in various organic electrolytes. Essentially, K^+ ions have higher ionic conductivity and mobility than Na^+ ions because of the small solvated cations and desolvation energy in liquid electrolytes, which gives these electrode materials improved kinetics and rate capacity in KIBs. Additionally, studies had demonstrated that K^+ could be stored in graphite and afterward different anode materials for KIBs were consistently reported. All the more noticeably, cathode materials are a vital factor in the full-cell system. New phyletic cathode materials of Prussian blue and its organic analogues give a suitable opportunity to K^+ insertion/extraction in KIBs, which are related well [76,77].

2.4.1 Solid Electrolyte Interphase

The SEI is taken into consideration because of its importance in LIBs, which is derived from the reaction between the anode and the electrolyte. The SEI layer permits the diffusion of metallic cations to preserve the reaction kinetics while insulating the electronic contact between the electrode and the electrolyte. Recent findings on the K-ion SEI were especially targeted on the characterization of the SEI composition, morphology, and structure, which the battery's overall performance is probably related to. Meanwhile, the figuring-out factors, which include the formation mechanism, stability, solubility, and ion transfer properties of the SEI, in answering the query of the way the SEI impacts the battery's overall performance had been hardly ever reported.

The Fermi energy-level difference between the LUMO of the electrolyte and the anode prevails the thermodynamic stability on the negative electrode to shape the SEI layer. An anode with μA > ELUMO can transfer electrons to the LUMO of the electrolyte, thereby triggering a spontaneous reaction. The SEI layer is understood to form below ~1 V vs Li^+/Li or Na^+/Na on metals, carbons, and oxides while they may be in contact with carbonate ester primarily based on total electrolytes. Similarly, anode materials with potassium storage potentials in the range of ~0–1 VK, such as potassium metal (0 VK) and graphite (~0.1 VK), in addition to alloy anodes (antimony (Sb) (~0.8 VK) and red phosphorus (P) (~0.7 VK)), have Fermi level above the LUMO level of non-aqueous electrolytes. Therefore, the electrochemical reduction of the electrolyte on the surface of those anodes is thermodynamically determined. Meanwhile, the character of the cation–solvent or cation–anion interaction has a sturdy effect on the reduction stabilities of the solvent–anion because it influences the solvation structure and the solvent coordination number across the cation centre. When solvents or anions coordinate with cations, their LUMO ranges lower because of electron pair donation to the cation. Consequently, solvation and ion-pair formation may even enhance the decomposition of electrolytes. The LUMO levels had been computed with the use of Density Functional Theory DFT calculations. All ion–solvent complexes, which include EC, DEC, FEC, propylene carbonate (PC), 1,3-dioxolane, and DME, exhibited a decrease in the LUMO level in comparison with those of the pure solvents. The magnitudes of the decrease in the LUMO level of the carbonate solvents had been within the order $Li^+ > Na^+ > K^+$, displaying a linear relationship with the binding energy. A larger LUMO level change was observed in the ether solvents, and the order was changed to $Na^+ > K^+ > Li^+$.

The one-of-a-kind guidelines for the ester and ether electrolytes name for careful research of every system, thinking about that the effect of the cations on the reduction stability of the solvents is complicated. It may be concluded that the reduction stability of the electrolyte and SEI formation are laid down with the cations in the electrolytes, stressing the need for expertise in the variations in the SEI layer among KIBs and broadly studied NIBs and LIBs. It is properly mounted that K-ions have a barely better redox potential ($E \sim 0.1$ V, K^+/K vs Li^+/Li) than Li-ions, and as a result the electrochemical driving force of electrolyte reduction is probably barely suppressed on the anode side from a thermodynamic view. The specific K chemistries may cause distinct chemical composition, morphology, mechanical stability, and structure of the K-ion SEI. Therefore, it will be tough to extrapolate the physical and chemical conduct of the K-ion SEI from its Li and Na counterparts. [75,78]. The effect of electrolytes on the formation of SEI is systematically studied by Wang et al. [78], which they have investigated using two different electrolytes, namely, KPF_6 and $KN(SO_2)F_2$ (KFSI) salts in EC/DEC solution. Significantly, SEI induced by KFSI is more stable (consists of homogeneously distributed inorganic species and polycarbonates) than the KPF_6-induced SEI layer (contains unstable alkyl carbonates) (Figure 2.4). This is attributed to the size of FSI^- being larger compared to PF_6^- and the decrease in the LUMO levels compared to solvents which efficiently overcome the damage of SEI caused by co-intercalation. This results in a highly stable SEI layer when using KFSI salt-based electrolytes compared to KPF_6-based electrolytes.

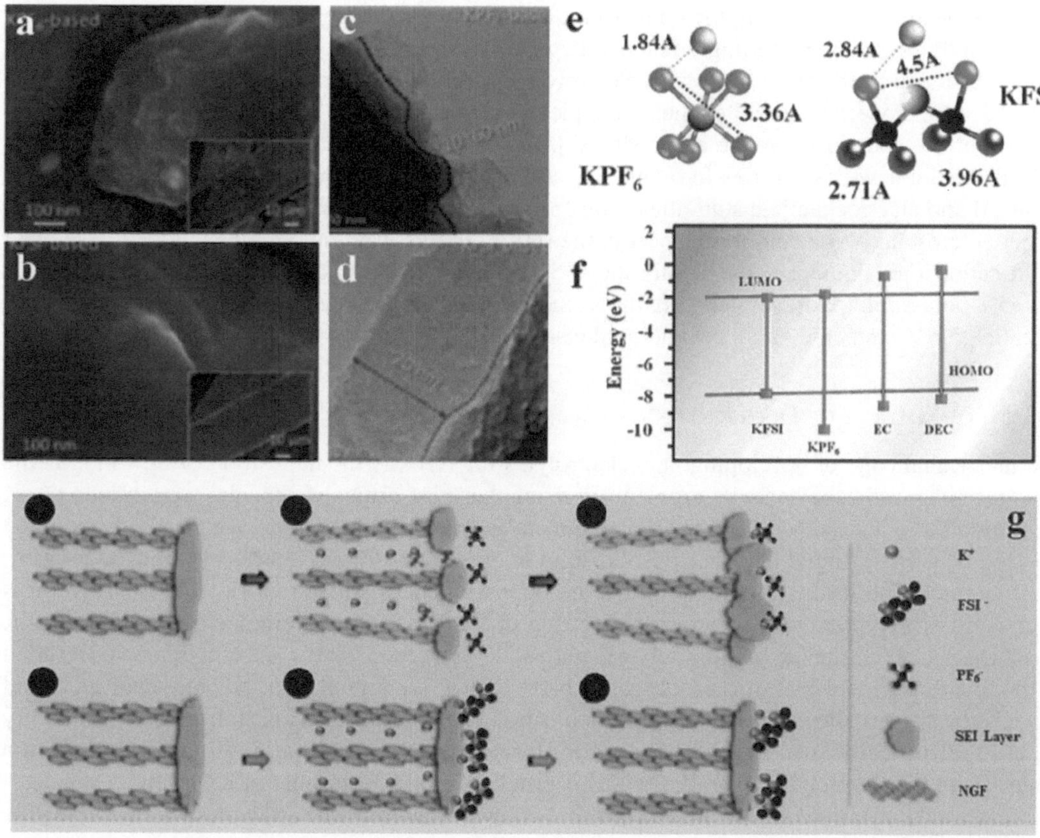

FIGURE 2.4 High-resolution SEM and TEM images of nitrogen–doped graphene fibres NGF electrodes in (a and c) KPF_6-based and (b and d) KFSI-based electrolytes at the discharge potential of 0.01 V (20th discharge). Insets in (a and b) are the low-magnification SEM/TEM images. (e) Molecular structure for KPF_6 and KFSI. (f) Calculated HOMO and LUMO energy levels for KPF_6, KFSI, and the solvents EC and DEC. (g) Schematic illustration of SEI layer formation in KPF_6- and KFSI-based electrolytes: (1) electrolyte reduction/decomposition in the potential range 0.8–0.2 V during the first discharge cycle; (2) K^+ intercalation below 0.2 V during the first cycle discharge; (3) electrolyte reduction/decomposition in the potential range 0.8–0.2 V during following discharge cycles. Adapted and reproduced from Ref. [78]. Copyright 2019 American Chemical Society.

Moshkovich et al. [79] in 2001 first identified that the solubility of the surface species has a sturdy impact on the stability of the SEI layer. It changed the surface films that formed on the noble metals in PC containing Li and K salts ($LiClO_4$ and $KClO_4$), which have been a lot more stable than the ones formed in the Na salt ($NaClO_4$) solution [79]. In general, the coulombic interaction of a stronger Lewis acid with a solvent or negatively charged species is extra notable. Consequently, its salts are extra soluble. Given that K^+ ions show far weaker Lewis acidity than Na^+ ions, it also has the potential that the solubility of K-ions within the SEI layer is lower than that of Na compounds. Yet, the query arises: is the K-ion SEI non-polymeric just like the case for the Na-ion SEI? And if now no longer, is the mechanical energy of the K-ion SEI similar to that of the Li-ion SEI? To date, little experimental or modelling outcomes are to be had in this subject matter.

Besides, the function of the SEI components in the SEI mechanical properties should be clarified. Electrolyte optimization may be a promising approach to deal with the troubles related to interfacial aspect reactions in KIBs. This fantastically clear-cut technique now no longer wants any current battery configuration and may be without difficulty implemented. The most critical challenge lies in understanding the competitive interfacial chemistry of potassium metal. One has to keep in thought that exceedingly reactive K metallic may also cause extreme electrolyte degradation. As a result, the effectiveness of electrolyte optimization cannot be fairly evaluated in half cells, because modern essential studies are typically carried out with half cells containing K metal because of the counter/reference electrode [80]. Recent successful attempts of using ionic liquids in KIBs create new possibilities for exploiting the overall performance of KIBs, concerning their chemical and electrochemical stabilities, film-forming, and Al anticorrosion properties. Therefore, unique interest has to be paid to this part in the future. In this way Yamamoto et al. [81] developed single cation ionic liquid, i.e., the molar ratio of 55% potassium bis(fluorosulphonyl)amide (K[FSA]) and 45% potassium (fluorosulphonyl)(trifluoromethylsulphonyl)amide (K[FTA]), operating at wide potential (5.6 V) and enhancing the energy density of KIBs.

2.4.2 Challenges of Electrolytes

The major challenge of developing new electrolytes for KIBs is the deposition of K, which is the basis for addressing the dendrite problem. Perhaps the most critical factor that determines the K deposition behaviour is the SEI formed on the metallic surface. Meanwhile, the transition from the conventional strong/liquid interface to solid/solid or even liquid/liquid interfaces might be a formidable attempt. The former method requires the improvement of solid-state electrolytes, regardless of ceramic or polymer, which lessens the wetting of the K anode and organic solvent. Generally, an alternative of flammable organic solvents with solid-state electrolytes is taken into consideration as the final answer for metallic anode-based batteries, while a number of recently evolved liquid Na–K alloy anodes show off dendrite-free deposition characteristics. Their inherent deformability ought to intrinsically keep away from dendrite formation while maintaining proper liquid/liquid contact with the electrolyte. Overall, the SEI is crucially a critical trouble in KIBs, and, each has a deep essential understanding of the interfacial phenomena and realistic strategies for enhancing the interfacial phenomena are in the initial stages. Intense research primarily on superior characterization and combined strategies is urgently required to facilitate in additional advancement of KIBs.

2.5 CONCLUSION AND PERSPECTIVE

We have discussed briefly the recent advancements and emerging challenges of electrode and electrolyte materials in this chapter. Although the working principles of LIBs, NIBs, and KIBs are similar except for the carrier ions, attention has to be paid to the energy density of KIBs. Because the atomic weight of potassium is more than five times heavier than lithium, which ultimately decreases the volumetric/gravimetric energy density of KIBs. Even though the polyanion compounds exhibit higher working voltage, the delivered capacity of these polyanionic compounds is still under debate. The high working voltage of polyanions is impressive, which maximizes the gravimetric/volumetric energy density of KIBs, which is comparable to that of LIBs. Therefore, future work should emphasis the design of new polyanionic compounds with diverse crystal structures to recognize the higher capacities and high working voltages. In the case of negative electrode materials, oxide electrode materials exhibit reasonable electrochemical performance, but a low working potential is not suitable to realize the safety of KIBs. Hence, future work should focus on the development of oxide-based electrode materials with working voltages between 1 and 1.5 V, which is more appropriate to realize safe KIBs. At the same time, the volume expansion of metal dichalcogenides is small compared to other negative electrode materials. Therefore, the synthesis of metal dichalcogenides with distinct crystal structures and diverse morphologies would pave the way to design high-power KIBs. The potential advantage of KIBs is faster K^+ ion kinetics in KPF_6-based electrolyte solutions, but the ionic conductivity of non-aqueous KPF_6 is lower compared to the ionic conductivity of non-aqueous $LiPF_6$ electrolytes. Therefore, the selection of appropriate electrolyte salt and solvent is much more significant to improve the ionic conductivity and prevent the ion association during cycling. Further, the development of non-aqueous electrolytes with superior ionic conductivity, solid electrolytes, and electrolyte additives provides new insights and opens the door to realizing high-power potassium-ion batteries.

ACKNOWLEDGEMENT

One of the authors Dr. C. Nithya wishes to thank the Science and Engineering Research Board, India, for the SERB-POWER Grant (SPG/2021/004543).

REFERENCES

1. Dunn B, Kamath H, Tarascon JM. (2011) Electrical energy storage for the grid: a battery of choices. *Science* 334, 928–935. https://doi.org/10.1126/science.1212741
2. Yabuuchi N, Kubota K, Dahbi M, Komaba S. (2014) Research development on sodium-ion batteries. *Chem. Rev.* 114, 11636–11682. https://doi.org/10.1021/cr500192f
3. Hwang JY, Myung ST, Sun YK. (2017) Sodium-ion batteries: present and future. *Chem. Soc. Rev.* 46, 3529–3614. https://doi.org/10.1039/C6CS00776G
4. Vaalma C, Buchholz D, Weil M, Passerini S. (2018) A cost and resource analysis of sodium-ion batteries. *Nat. Rev. Mater.* 3, 18013. https://doi.org/10.1038/natrevmats.2018.13
5. Hosaka T, Kubota K, Kojima H, Komaba S. (2018) Highly concentrated electrolyte solutions for 4 V class potassium-ion batteries. *Chem. Commun.* 54, 8387–8390. https://doi.org/10.1039/C8CC04433C
6. Jian Z, Xing Z, Bommier C, Li Z, Ji X. (2016) Hard carbon microspheres: potassium-ion anode vs sodium-ion anode. *Adv. Energy Mater.* 6, 1501874. https://doi.org/10.1002/aenm.201501874
7. Jian Z, Luo W, Ji X. (2015) Carbon electrodes for K-ion batteries. *J. Am. Chem. Soc.* 137, 11566–11569. https://doi.org/10.1021/jacs.5b06809
8. Zhang W, Liu Y, Guo Z. (2019) Approaching high-performance potassium-ion batteries via advanced design strategies and engineering. *Sci. Adv.* 5, 1–13. https://doi.org/10.1126/sciadv.aav7412
9. Adams RA, Varma A, Pol VG. (2018) Mechanistic elucidation of thermal runaway in potassium-ion batteries. *J. Power Sources* 375, 131–137. https://doi.org/10.1016/j.jpowsour.2017.11.065
10. Chen C, Wang Z, Zhang B, Mio L, Cai J, Peng L, Huang Y, Jiang J, Yunhui H, Zhang L, Xie J. (2017) Nitrogen rich hard carbon as a highly durable anode for high power potassium-ion batteries. *Energy Storage Mater.* 8, 161–168. https://doi.org/10.1016/j.ensm.2017.05.010
11. Zhang R, Li H, Li R, Wei D, Kang W, Ju Z, Xiong S. (2019) Boosting the potassium-ion storage performance of a carbon anode by chemically regulating oxygen-containing species. *Chem Commun.* 55, 14147–14150. https://doi.org/10.1039/C9CC07585B
12. Qi X, Huang K, Wu H, Zhao W, Wang H, Zhung Q, Ju Z. (2018) Novel fabrication of N-doped hierarchically porous carbon with exceptional potassium storage properties. *Carbon* 131, 79–85. https://doi.org/10.1016/j.carbon.2018.01.094
13. Zhang Y, Tian S, Yang C, Nan J. (2020) Three-dimensional nitrogen-sulfur codoped layered porous carbon nanosheets with sulfurregulated nitrogen content as a high-performance anode material for potassium-ion batteries. *Dalton Trans.* 49, 5108–5120. https://doi.org/10.1039/D0DT00697A
14. Liu Y, Yang C, Pan Q, Li Y, Wang G, Qu X, Zheng F, Xiaong X, Liu M, Zhang Q. (2018) Nitrogen-doped bamboo-like carbon nanotubes as anode material for high performance potassium ion batteries. *J. Mater. Chem. A* 6, 15162–15169. https://doi.org/10.1039/C8TA04694H
15. Wang B, Yuan F, Wang W, Zhang D, Sun H, Xi K, Wang D, Chu J, Wang Q, Li W. (2019) A carbon microtube array with a multihole cross profile: releasing the stress and boosting longcycling and high-rate potassium ion storage. *J. Mater. Chem. A* 7, 25845–25852. https://doi.org/10.1039/C9TA09598E
16. Ju Z, Li P, Ma G, Xing Z, Zhuang Q, Qian Y. (2018) Few layer nitrogen-doped graphene with highly reversible potassium storage. *Energy Storage Mater.* 11, 38–46. https://doi.org/10.1016/j.ensm.2017.09.009
17. An Y, Tian Y, Li Y, Xiong S, Zhao G, Feng J, Qian Y. (2019) Green and tunable fabrication of graphene-like N-doped carbon on a 3D metal substrate as a binderfree anode for high-performance potassium-ion batteries. *J. Mater. Chem. A* 7, 21966–21975. https://doi.org/10.1039/C9TA08721D
18. Li P, Hwang JY, Sun YK. (2019) Highly wrinkled carbon tubes as an advanced anode for K-ion full batteries. *J. Mater. Chem.* A 7, 20675–20682. https://doi.org/10.1039/C9TA08071F
19. Han J, Xu M, Niu Y, Li GN, Wang M, Zhang Y, Jia M, Li CM. (2016) Exploration of $K_2Ti_8O_{17}$ as an anode material for potassium-ion batteries. *Chem. Commun.* 52, 11274–11276. https://doi.org/10.1039/C6CC05102B

20. Dong S, Li Z, Xing Z, Wu X, Ji X, Zhang X. (2018) Novel potassium-ion hybrid capacitor based on an anode of $K_2Ti_6O_{13}$ microscaffolds. *ACS Appl. Mater. Interfaces* 10, 15542–15547. https://doi.org/10.1021/acsami.7b15314
21. Kishore BGV, Munichandraiah N. (2016) $K_2Ti_4O_9$: a promising anode material for potassium ion batteries. *J. Electrochem. Soc.* 163, A2551–A2554. https:/doi.org/10.1149/2.0421613jes
22. Chen F, Wang S, He XD, Liao JY, Hu Q, Dong JM, Chen CH. (2020) Hollow sphere structured V_2O_3@C as an anode material for high capacity potassium-ion batteries. *J. Mater. Chem. A* 8, 13261–13266. https://doi.org/10.1039/D0TA01057J
23. Rao YC, Yu S, Gu X, Duan XM. (2019) Prediction of MoO_2 as high capacity electrode material for (Na, K, Ca)-Ion Batteries. *Appl. Surf. Sci.* 479, 64–69. https://doi.org/10.48550/arXiv.1811.11029
24. Li N, Zhang F, Tang Y. (2018) Hierarchical T-Nb_2O_5 nanostructure with hybrid mechanisms of intercalation and pseudocapacitance for potassium storage and high-performance potassium dual-ion batteries. *J. Mater. Chem. A* 6, 17889–17895. https://doi.org/10.1039/C8TA07987K
25. Luo S, Wang T, Lu H, Xu X, Xue G, Xu N, Wang Y, Zhou D. (2020) Ultrasmall SnO_2 nanocrystals embedded in porous carbon as potassium ion battery anodes with long-term cycling performance. *New J. Chem.* 44, 11678–11683. https://doi.org/10.1039/D0NJ00323A
26. Mukherjee S, Singh G. (2019) Two-dimensional anode materials for non-lithium metal ion batteries. *ACS Appl. Energy Mater.* 2, 932–955. https://doi.org/10.1021/acsaem.8b00843
27. Tian B, Tang W, Leng K, Chen Z, Tan SJR, Peng C, Ning GH, Fu W, Su C, Zheng GW, et al. (2017) Phase transformations in TiS_2 during K intercalation. *ACS Energy Lett.* 2, 1835–1840. https://doi.org/10.1021/acsenergylett.7b00529
28. Zhou J, Wang L, Yang M, Wu J, Chen F, Huang W, Han N, Ye H, Zhao F, Li Y, et al. (2017) Hierarchical VS_2 nanosheet assemblies: a universal host material for the reversible storage of alkali metal ions. *Adv. Mater.* 29, 1702061. https://doi.org/10.1002/adma.201702061
29. Zheng N, Jiang G, Chen X, Mao J, Zhou Y, Li Y. (2019) Rational design of a tubular, interlayer expanded MoS_2-N/O doped carbon composite for excellent potassium-ion storage. *J. Mater. Chem. A* 7, 9305–9315. https://doi.org/10.1039/C9TA00423H
30. Huang H, Cui J, Liu G, Bi R, Zhang L. (2019) Carbon-coated $MoSe_2$/MXene hybrid nanosheets for superior potassium storage. *ACS Nano* 13, 3448–3456. https://doi.org/10.1021/acsnano.8b09548
31. Shen Q, Jiang P, He H, Chen C, Liu Y, Zhang M. (2019) Encapsulation of $MoSe_2$ in carbon fibers as anodes for potassium ion batteries and nonaqueous battery-supercapacitor hybrid devices. *Nanoscale* 11, 13511–13520. https://doi.org/10.1039/C9NR03480C
32. Zhang R, Bao J, Pan Y, Sun CF. (2019) Highly reversible potassium-ion intercalation in tungsten disulfide. *Chem. Sci.* 10, 2604–2612. https://doi.org/10.1039/C8SC04350G
33. Ren X, Yu D, Yuan L, Bai Y, Huang K, Liu J, Feng S. (2020) In situ exsolution of Ag from $AgBiS_2$ nanocrystal anode boosting high-performance potassium-ion batteries. *J. Mater. Chem. A* 8, 15058–15065. https://doi.org/10.1039/D0TA03964K
34. Lakshmi V, Mikhaylov AA, Medvedev AG, Zhang C, Ramireddy T, Rahman MM, Cizek P, Golberg D, Chen Y, Lev O, Prikhodchenko PV, Glushenkov AM. (2017) Nanocrystalline SnS_2 coated onto reduced graphene oxide: demonstrating the feasibility of a non-graphitic anode with sulfide chemistry for potassium-ion batteries. *Chem. Commun.* 53, 8272–8275. https://doi.org/10.1039/C7CC03998K
35. Lakshmi V, Mikhaylov AA, Medvedev AG, Zhang C, Ramireddy T, Rahman MM, Cizek P, Golberg D, Chen Y, Lev O, Prikhodchenko PV, Glushenkov AM. (2020) Probing electrochemical reactivity in an Sb_2S_3-containing potassium-ion battery anode: observation of an increased capacity. *J. Mater. Chem. A* 8, 11424–11434. https://doi.org/10.1039/D0TA03555F
36. Okubo M, Sugahar A, Kajiyama S, Yamada A. (2018) MXene as a charge storage host. *Acc. Chem. Res.* 51, 591–599. https://doi.org/10.1021/acs.accounts.7b00481
37. Wu X, Wang H, Zhao Z, Huang B. (2020) Interstratification-assembled 2D black phosphorene and V_2CT_x MXene as superior anodes for boosting potassium-ion storage. *J. Mater. Chem. A* 8, 12705–12715. https://doi.org/10.1039/D0TA04506C
38. Sultana I, Rahman, MM, Ramireddy T, Chen Y, Glushenkov AM. (2017) High capacity potassium-ion battery anodes based on black phosphorus. *J. Mater. Chem. A* 5, 23506–23512. https://doi.org/10.1039/C7TA02483E
39. Zhang W, Pang WK, Sencadas V, Guo Z. (2018) Understanding high energy density Sn_4P_3 anodes for potassium-ion batteries. *Joule* 2, 1534–1547. https://doi.org/10.1016/j.joule.2018.04.022
40. Zheng J, Yang Y, Fan X, Ji G, Ji X, Wang H, Hou S, Zachariah MR, Wang C. (2019) Extremely stable antimony-carbon composite anodes for potassium-ion batteries. *Energy Environ. Sci.* 12, 615–623. https://doi.org/10.1039/C8EE02836B

41. Zhao X, Yang Q, Quan Z. (2019) Tin-based nanomaterials: colloidal synthesis and battery applications. *Chem. Commun.* 55, 8683–8694. https://doi.org/10.1039/C9CC02811K
42. Xue Q, Li D, Huang Y, Zhang X, Ye Y, Fan E, Li L, Wu F, Chen R. (2018) Vitamin K as a high-performance organic anode material for rechargeable potassium ion batteries. *J. Mater. Chem. A* 6, 12559–12564. https://doi.org/10.1039/C8TA03921F
43. Dou K, Ma Y, Zhang T, Huang B, Dai Y. (2019) Prediction of two-dimensional PC_6 as a promising anode material for potassium-ion batteries. *Phys. Chem. Chem. Phys.* 21, 26212–26218. https://doi.org/10.1039/C9CP05251H
44. Xu L, Xiong P, Zeng L, Liu R, Liu J, Luo F, Li X, Chen Q, Wei M, Qian Q. (2020) Facile fabrication of a vanadium nitride/carbon fiber composite for half/full sodium-ion and potassium-ion batteries with long-term cycling performance. *Nanoscale* 12, 10693–10702. https://doi.org/10.1039/C9NR10211F
45. Hironaka Y, Kubota K, Komaba S. (2017) P2- and P3-K_xCoO_2 as an electrochemical potassiumIntercalation host. *Chem. Commun.* 53, 3693–3696. https://doi.org/10.1039/C7CC00806F
46. Sada K, Senthilkumar B, Barpanda P. (2017) Electrochemical potassium-ion intercalation in Na_xCoO_2: a novel cathode material for potassium-ion batteries. *Chem. Commun.* 53, 8588–8591. https://doi.org/10.1039/C7CC02791E
47. Kubota K, Kumakura S, Yoda Y, Kuroki K, Komaba S. (2018) Electrochemistry and solid-state chemistry of $NaMeO_2$ (Me=3d Transition Metals). *Adv. Energy Mater.* 8, 1703415. https://doi.org/10.1002/aenm.201703415
48. Kim H, Seo DH, Urban A, Lee J, Kwon DH, Bo SH, Shi T, Papp JK, Mccloskey BD, Ceder G. (2018) Stoichiometric layered potassium transition metal oxide for rechargeable potassium batteries. *Chem. Mater.* 30, 6532–6539. https://doi.org/10.1021/acs.chemmater.8b03228
49. Vaalma C, Giffin GA, Buchholz D, Passerini S. (2016) Non-aqueous K-ion battery based on layered $K_{0.3}MnO_2$ and hard carbon/carbon black. *J. Electrochem. Soc.* 163, A1295–A1299. https://doi.org/10.1149/2.0921607jes
50. Liu C, Luo SH, Huang H, Zhai Y, Wang ZW. (2019) Layered potassium-deficient P2- and P3-type cathode materials K_xMnO_2 for K-ion batteries. *Chem. Eng. J.* 356, 53–59. https://doi.org/10.1016/j.cej.2018.09.012
51. Xu YS, Zhang QH, Wang D, Gao JC, Tao XS, Liu Y, Sun YG, Gu L, Chang BB, Liu CT, Shi SQ, Cao AM. (2020) Enabling reversible phase transition on $K_{5/9}Mn_{7/9}Ti_{2/9}O_2$ for high-performance potassium-ion batteries cathodes. *Energy Storage Mater.* 31, 20–26. https://doi.org/10.1016/j.ensm.2020.06.002
52. Tai Z, Shi M, Chong S, Chen Y, Tan Q, Shu C, Liu Y. (2019) $KMn_{7.6}Co_{0.4}O_{16}$ nano-rod clusters with a high discharge specific capacity as cathode materials for potassium-ion batteries. *Sus. Energy Fuels* 3, 736–743. https://doi.org/10.1039/C8SE00550H
53. Masese T, Yoshii K, Kato M, Kubota K, Huang ZD, Senoha H, Shikano M. (2019) A high voltage honeycomb layered cathode framework for rechargeable potassium-ion battery: P2-type $K_{2/3}Ni_{1/3}Co_{1/3}Te_{1/3}O_2$. *Chem. Commun.* 55, 985–988. https://doi.org/10.1039/C8CC07239F
54. Manthiram A, Goodenough JB. (1989) Lithium Insertion into $Fe_2(SO_4)_3$ frameworks. *J. Power Sources* 26, 403–408. https://doi.org/10.1016/0378-7753(89)80153-3
55. Kim H, Seo DH, Bianchini M, Clément RJ, Kim H, Kim JC, Tian Y, Shi T, Yoon WS, Ceder G. (2018) A new strategy for high-voltage cathodes for K-ion batteries: stoichiometric $KVPO_4F$. *Adv. Energy Mater.* 8, 1801591. https://doi.org/10.1002/aenm.201801591
56. Hosaka T, Shimamura T, Kubota K, Komaba S. (2019) Polyanionic compounds for potassium-ion batteries. *Chem. Rec.* 19, 735–745. https://doi.org/10.1002/tcr.201800143
57. Lander L, Rousse G, Abakumov AM, Sougrati M, Van Tendeloo G, Tarascon JM. (2015) Structural, electrochemical and magnetic properties of a novel $KFeSO_4F$ Polymorph. *J. Mater. Chem.* 3, 19754–19764. https://doi.org/10.1039/C5TA05548B
58. Lander L, Rousse G, Batuk D, Colin CV, Dalla Corte DA, Tarascon JM. (2017) Synthesis, structure, and electrochemical properties of K-based sulfates $K_2M_2(SO_4)_3$ with M=Fe and Cu. *Inorg. Chem.* 56, 2013–2021. https://doi.org/10.1021/acs.inorgchem.6b02526
59. Chihara K, Katogi A, Kubota K, Komaba S. (2017) $KVPO_4F$ and $KVOPO_4$ toward 4 V-class potassium-ion batteries. *Chem. Commun.* 53, 5208–5211. https://doi.org/10.1039/C6CC10280H
60. Lia J, Hu Q, Che B, Ding X, Chen F, Chen C. (2019) Competing with other polyanionic cathode materials for potassium-ion batteries via fine structure design: new layered $KVOPO_4$ with a tailored particle morphology. *J. Mater. Chem. A* 7, 15244–15251. https://doi.org/10.1039/C9TA03192H
61. Liu Z, Wang J, Lu B. (2020) Plum pudding model inspired $KVPO_4F@_3DC$ as high-voltage and hyperstable cathode for potassium ion batteries. *Sci. Bull.* 65, 1242–1251. https://doi.org/10.1016/j.scib.2020.04.010

62. Han J, Niu Y, Bao SJ, Yu YN, Lu SY, Xu M. (2016) Nanocubic $KTi_2(PO_4)_3$ electrodes for potassium-ion batteries. *Chem. Commun.* 52, 11661–11664. https://doi.org/10.1039/C6CC06177J
63. Eftekhari, A. (2004) Potassium secondary cell based on Prussian blue cathode. *J. Power Sources* 126, 221–228. https://doi.org/10.1016/j.jpowsour.2003.08.007
64. Hosaka T, Kubota K, Hameed AS, Komaba S. (2020) Research development on K-ion batteries. *Chem. Rev.* 120, 6358–6466. https://doi.org/10.1021/acs.chemrev.9b00463
65. Kapaev RR, Zhidkov IS, Kurmaev EZ, Stevenson KJ, Troshin PA. (2019) Hexaazatriphenylene-based polymer cathode for fast and stable lithium-, sodium- and potassiumion batteries. *J. Mater. Chem. A* 7, 22596–22603. https://doi.org/10.1039/C9TA06430C
66. Tang M, Wu Y, Chen Y, Jiang C, Zhu S, Zhuo S, Wang C. (2019) An organic cathode with high capacities for fast charge potassium-ion batteries. *J. Mater. Chem. A* 7, 486–492. https://doi.org/10.1039/C8TA09960J
67. Obrezkov FA, Ramezankhani V, Zhidkov I, Traven VF, Kurmaev EZ, Stevenson KJ, Troshin PA. (2019) High-energy and high-power-density potassium ion batteries using dihydrophenazine-based polymer as active cathode material. *J. Phys. Chem. Lett.* 10, 5440–5445. https://doi.org/10.1021/acs.jpclett.9b02039
68. Shen D, Liu Z, Fan L, Lu B. (2020) Organic phosphomolybdate: a high capacity cathode for potassium ion batteries. *Chem. Commun.* 56, 12753–12756. https://doi.org/10.1039/D0CC05499B
69. Wang S, Chen Y, Cui B, Li B, Wang S, Cui Y, Ju Z, Zhuang Q. (2020) The combined first principles and experimental study of O-doped $KMnF_3$ as a cathode for potassium-ion batteries. *Appl. Surf. Sci.* 514, 145954–145962. https://doi.org/10.1016/j.apsusc.2020.145954
70. Xu S, Chen Y, Wang C. (2020) Emerging organic potassium-ion batteries: electrodes and electrolytes. *J. Mater. Chem. A* 8, 15547–15574. https://doi.org/10.1039/D0TA03310C
71. Hosaka T, Matsuyama T, Kubota K, Yasuno S, Komaba S. (2020) Development of KPF6/KFSA binary-salt solutions for long-life and high-voltage K-ion batteries. *ACS Appl. Mater. Interfaces* 12, 34873–34881. https://doi.org/10.1021/acsami.0c08002
72. Hu Y, Ding H, Bai Y, Liu Z, Chen S, Wu Y, Yu X, Fan L, Lu B. (2019) Rational design of a polyimide cathode for a stable and high-rate potassium-ion battery. *ACS Appl. Mater. Interfaces* 11, 42078–42085. https://doi.org/10.1021/acsami.9b13118
73. Zhang R, Bao J, Wang YH, Sun CF. (2018) Concentrated electrolytes stabilize bismuth-potassium batteries. *Chem. Sci.* 9, 6193–6198. https://doi.org/10.1039/C8SC01848K
74. Lei Y, Qin L, Liu R, Lau KC, Wu Y, Zhai D, Li B, Kang F. (2018) Exploring stability of nonaqueous electrolytes for potassium-ion batteries. *ACS Appl. Energy Mater.* 1, 1828–1833. https://doi.org/10.1021/acsaem.8b00214
75. Peljo P, Girault HH. (2018) Electrochemical potential window of battery electrolytes: the HOMO-LUMO misconception. *Energy Environ. Sci.* 11, 2306–2309. https://doi.org/10.1039/C8EE01286E
76. Qin L, Xiao N, Zheng J, Lei Y, Zhai D, Wu Y. (2019) Localized high-concentration electrolytes boost potassium storage in high-loading graphite. *Adv. Energy Mater.* 9, 1902618. https://doi.org/10.1002/aenm.201902618
77. Yu X, Manthiram A. (2018) Electrode-electrolyte interfaces in lithium-based batteries. *Energy Environ. Sci.* 11, 527–543. https://doi.org/10.1039/C7EE02555F
78. Wang H, Wang H, Chen S, Zhang B, Yang G, Gao P, Liu J, Fan X, Huang Y, Lin J. (2019) A depth-profiling study on the solid electrolyte interface: bis (fluorosulfuryl) imide anion toward improved K^+ storage. *ACS Appl. Energy Mater.* 2, 7942–7951. https://doi.org/10.1021/acsaem.9b01428
79. Moshkovich M, Gofer Y, Aurbach D. (2001) Investigation of the electrochemical windows of aprotic alkali metal (Li, Na, K) salt solutions. *J. Electrochem. Soc.* 148, E155–E167. https://doi.org/10.1149/1.1357316
80. Wang B, Yuan F, Li W, Wang Q, Ma X, Gu L, Sun H, Xi K, Zhang D, Wang W. (2020) Rational formation of solid electrolyte interface for high-rate potassium ion batteries. *Nano Energy* 75, 104979. https://doi.org/10.1016/j.nanoen.2020.104979
81. Yamamoto H, Chen CY, Kubota K, Matsumoto K, Hagiwara R. (2020) Potassium single cation ionic liquid electrolyte for potassium-ion batteries. *J. Phys. Chem. B* 124, 6341–6347. https://doi.org/10.1021/acs.jpcb.0c03272

3 Zinc-Ion Batteries
Materials to Mechanism of Energy Storage

Kothandaraman Ramanujam, Janraj Naik Ramavath, and L. K. Nivedha

3.1 INTRODUCTION

Among the various energy storage devices, batteries are one of the most efficient and reliable electrochemical devices to fulfil the energy demand at various levels starting from handheld electronics to grid-scale applications [1]. Different battery technologies such as lead-acid, sodium-metal halide, high-temperature sodium-sulphur, nickel-cadmium, and redox flow batteries are explored for various applications [2,3]. Among the batteries explored, the lead-acid battery ruled for nearly a century, as it was made of cheaper materials and provided reliable performance. Of late, lithium-ion batteries (LIBs) have been demonstrated as appealing candidates owing to their high energy and power densities [4]. Nevertheless, the high reactivity of Li, limited Li resources, thermal runaway, high-security risks, inadequate recycling infrastructure, and high fabrication cost severely limit LIB deployments on the multi-megawatt scale [5]. This propels scientists to explore novel battery chemistries which operate with high capacity besides being cost-effective and eco-friendly.

In this context, the unremitting efforts of the researchers have been devoted to investigating other metal ion batteries as an alternative to LIBs. Owing to the natural abundance of sodium and potassium, in addition to their akin chemistry as Li, the battery based on those two metals has become the first choice for the researchers. But, the lack of appropriate cathodes, the high reactivity of Na and K, and the usage of flammable organic electrolytes hinder their commercialization. To overcome some of these limitations, the new battery chemistry based on Mg^{2+}, Zn^{2+}, Ca^{2+}, and Al^{3+} multivalent cations was explored [6]. Among them, metal ion batteries of Mg, Ca, and Al are still facing challenges in finding an appropriate electrolyte. Therefore, in terms of development, they are still in their infancy. In contrast to other metal ion batteries, the exceptional qualities of rechargeable zinc-ion batteries (ZIBs) drew great attention in the scientific community [7], and they are as follows:

i. High abundance of Zn (approximately 300 times higher than Li)
ii. Reasonably high theoretical specific capacities of Zn (820 mAh g^{-1}/5, 855 mAh cm^{-3})
iii. Low cost of Zn (about US\$2.4 kg^{-1}) compared to reactive metals such as K (about US\$13.1 kg^{-1}), Li (about US\$19.2 kg^{-1}), and Na (about US\$3.1 kg^{-1})
iv. Compatibility of Zn with aqueous electrolytes, which are highly conducting (~1 S cm^{-1}) over non-aqueous counterparts (ca. 1 mS cm^{-1} to 0.01 S cm^{-1})
v. Usage of aqueous electrolytes in ZIBs, which overwhelms safety and cost issues that are associated with LIBs

Apart from the anode, the cost of typical cathode materials that are explored for ZIBs is much cheaper than that of LIB. For example, the price of MnO_2 and V_2O_5 is US\$1.7 kg^{-1} and US\$5.5 kg^{-1}, respectively, whereas the price of LIB cathodes such as $LiCoO_2$ and $Li(NiMnCo)O_2$ is US\$55 kg^{-1}

DOI: 10.1201/9781003310174-3

and US\$34 kg^{-1}, respectively. The estimated price of ZIBs (US\$65 kWh^{-1}) encompassing V_2O_5 cathode, Zn-anode, and aqueous $ZnSO_4$ electrolyte is much cheaper than that of LIBs (~US\$300 kWh^{-1}) [7]. A multiangle comparison of ZIBs over LIBs is depicted in Figure 3.1a. It is to be noted that the exploration of aqueous electrolytes for Zn-battery chemistry is not a new concept. For example, salty water (brine) was explored as an electrolyte in the 1800s by Italian physicist A. Volta in the Zn/Cu battery. Subsequently, an aqueous KOH electrolyte is widely used in commercialized

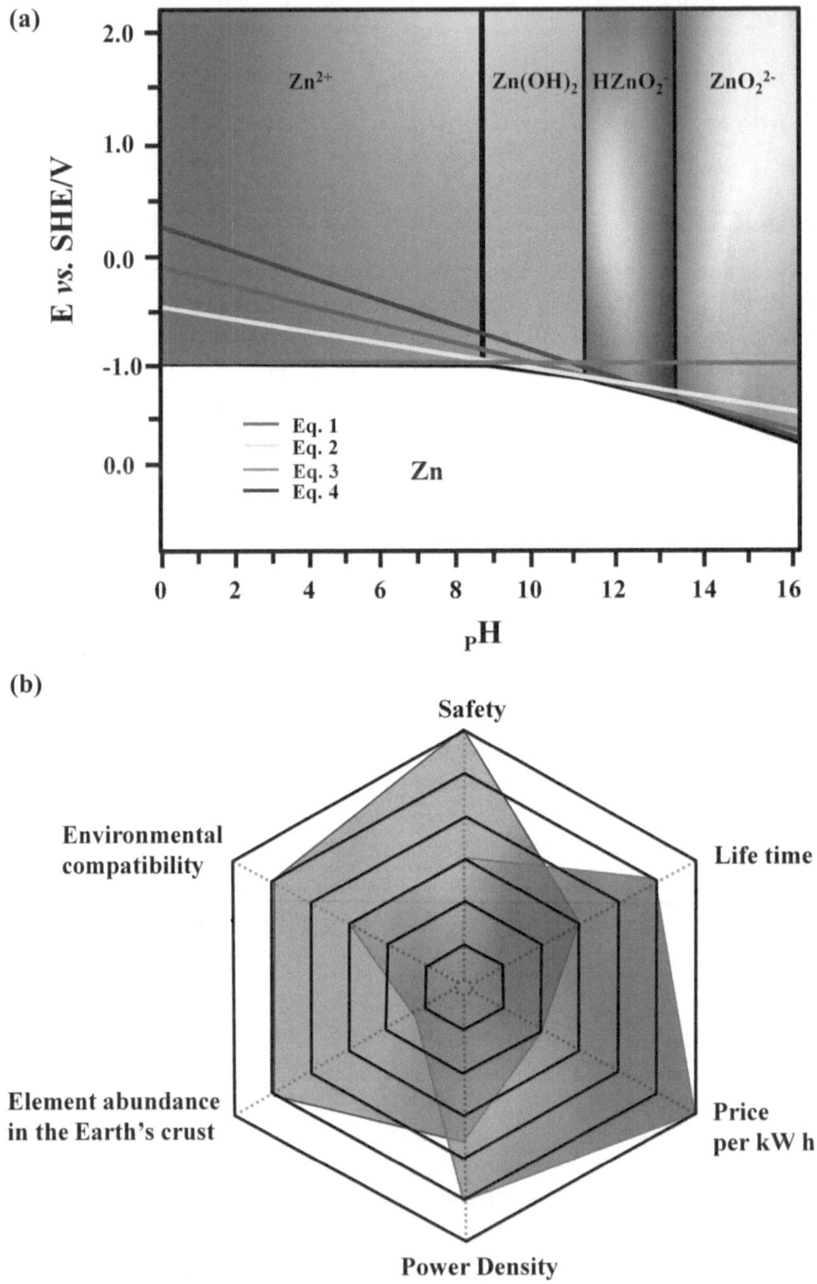

FIGURE 3.1 Multiangle comparison of (a) ZIBs with LIBs and the Pourbaix diagram of (b) Zn. (a, Adapted and reproduced from Ref. [7]. Copyright © 2021 Royal Society of Chemistry. b, Adapted and reproduced from Ref. [9]. Copyright © 2021 Elsevier.)

alkaline primary Zn-air and Zn-MnO_2 batteries. But the use of alkaline electrolytes in these batteries enriches hydroxide ions in the vicinity of the Zn-anode, leading to the formation of insulating zinc hydroxide/oxide by-products, which prevents them from recharging [7,8]. Interestingly, the Pourbaix diagram of the Zn system (Figure 3.1b) unveils the mechanism of avoiding such discharge products through the manipulation of the pH of the electrolyte [9].

The brief interfacial electrochemistry of Zn at various pH is elucidated in Section 3.3.1. It is realized that a near-neutral or mildly acidic pH of 4–6 can promote a direct oxidation pathway of Zn to Zn^{2+} (stripping) during the discharge at the anode. Hence, in the year 1986, the first report on the exploration of mildly acidic aqueous $ZnSO_4$ electrolytes for understanding the electrochemical behaviour of Zn (paired with MnO_2 cathode) is put forward by Yamamoto et al. [10]. In contrast to the alkaline medium, the absence of intermediate phases/discharge products (dendrites or passive film) on the Zn-anode in mild acidic media could facilitate the charging process. This revitalized the research enthusiasm for developing ZIBs as potential candidates for energy storage applications through facile Zn-ion chemistry. Following the work of Yamamoto et al., a multitude of works have been dedicated to the progress of mild acidic ZIBs. The first aqueous ZIB was built by Xu et al. [3] using a Zn-anode, a MnO_2 cathode, and a mild acidic $ZnSO_4$ solution. The brief evolution of ZIBs is summarized in Figure 3.2 [7]. In general, ZIB comprises Zn-metal foil as the anode, aqueous soluble zinc salts as an electrolyte component, and organic or layered inorganic materials as the cathodes. The Zn-anode and cathode are physically separated by a porous separator or ion exchange membranes, which are good ionic conductors but electronic insulators. During the discharge, metallic Zn is oxidized to Zn^{2+} at the anode (known as Zn stripping) and dissolves into the electrolyte. During this process, at the cathode Zn^{2+} ions from the electrolyte will be inserted into the cathode material (known as intercalation). During charging, Zn^{2+} ions are electrodeposited (also called plating) at the anode and simultaneously de-intercalated from the cathode. In short, Zn plating/stripping

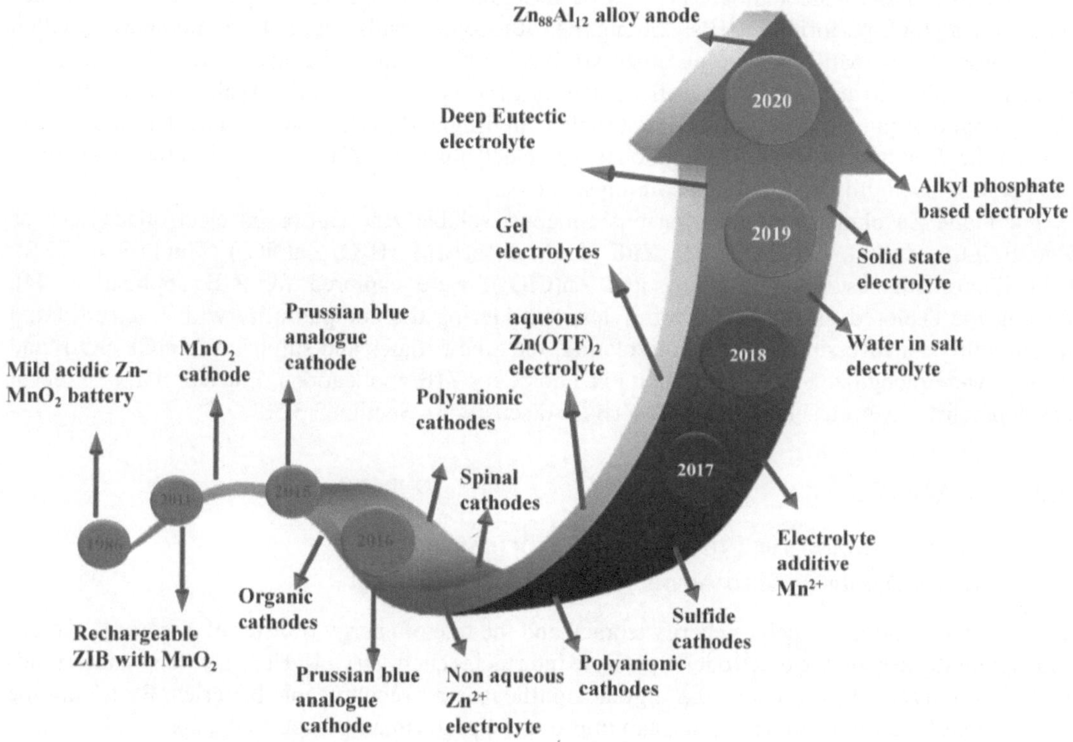

FIGURE 3.2 Brief historical development of ZIBs.

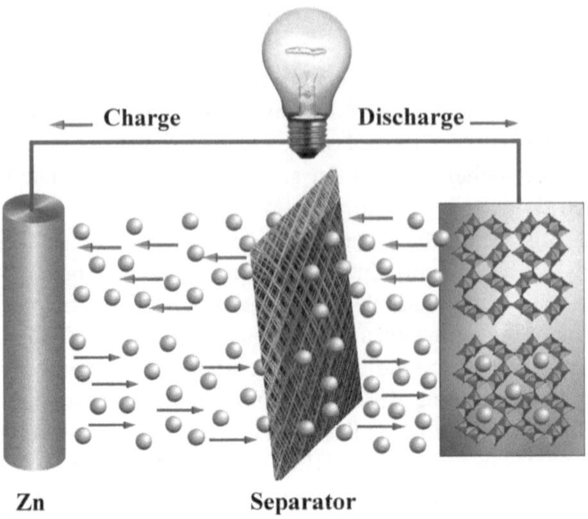

FIGURE 3.3 Working principles of ZIBs operated in mild acidic medium with α-MnO_2 cathode.

occurs at the anode, whereas reversible Zn^{2+} de-intercalation/intercalation occurs at the cathode during the charge/discharge. The working principle of ZIB is illustrated in Figure 3.3.

Other than Zn-anode and electrolytes, cathode materials play a crucial role in ZIBs to reversibly host the Zn^{2+} ions during the charge/discharge process. Various inorganic and organic cathode materials were explored to improve the intercalation/de-intercalation kinetics in ZIBs. The inorganic materials (such as oxides of Mn and oxides/sulphides of V, Chevrel phases, metal disulphides, and Prussian blue analogues (PBA)), and organic materials were explored as the cathodes for achieving high-performing ZIBs. Among the cathode materials, inorganic compounds based on manganese oxides (MnO_2, including other Mn-based oxides) and vanadium oxides (V_2O_5, defect-loaded vanadium oxides, etc.) were well studied for ZIB applications. A comprehensive overview of the rational design of various cathode materials is highlighted in Section 3.3. The detailed discussion on the development of anodes, cathodes, and electrolytes for ZIBs, along with the mechanism of operation, is highlighted in the upcoming sections.

For facile Zn plating/stripping, various aqueous, soluble zinc salt-based electrolytes such as $Zn(CF_3SO_3)_2$, $ZnSO_4$, $Zn(CH_3COO)_2$, ZnF_2, $ZnCl_2$, $Zn(BF_4)_2{\cdot}xH_2O$, $Zn(NO_3)_2$, $Zn(TFSI)_2$ [TFSI: bis(trifluoromethanesulphonyl)imide)], and $Zn(ClO_4)_2$ were explored for ZIB applications [4]. Among the explored mild acidic electrolytes, considering the compatibility with electrodes and favourability for reversible Zn stripping/plating, the zinc triflates and sulphates ($Zn(CF_3SO_3)_2$ and $ZnSO_4$) were recognized as the most suitable choice for ZIB applications. The summary of recent developments of various electrolytes for ZIBs is discussed in Section 3.3.5.

3.2 ZN-ANODE

3.2.1 Understanding the Interfacial Electrochemistry of Zn-Anode in Mild Acidic Media

Two critical phenomena such as energy storage and the rate of energy transfer of a working battery fundamentally rely on the electrode/electrolyte interfacial chemistry [4]. Hence, a deep understanding of the interfacial phenomena has a great significance on rechargeable batteries. By tuning the Zn electrode/electrolyte interface, one can improve the performance (such as life span and cyclability) of the rechargeable batteries [11]. Most importantly, the electrochemistry of the Zn-anode in terms of the type of product formed and its solubility in the given medium is highly dependent on the pH of the electrolyte (Pourbaix diagram Figure 3.1b) [8]; in other words, the mechanism and the product of the Zn-anode are not the same at all pHs. In the pH range of 0–8, zinc oxidation is independent of pH (Equation 3.1), which is shown as the horizontal line at –0.763 V vs Standard Hydrogen Electrode (SHE) in Figure 3.1b.

$$Zn^{2+} + 2e^{-} \leftrightarrow Zn \tag{3.1}$$

At pH > 8, Zn^{2+} hydrolysis produces various zinc species such as $HZnO_2^-$, ZnO_2^{2-}, and $Zn(OH)_2$, as in Equations 3.2–3.4 [9], and the equilibrium between them is shown as vertical lines in Figure 3.1b.

$$Zn^{2+} + H_2O \leftrightarrow Zn(OH)_2 + 2H^+ \text{ at pH} \sim 8.4-10 \tag{3.2}$$

$$Zn(OH)_2 \leftrightarrow HZnO_2^- + H^+ \text{ at pH} \sim 10.7-12 \tag{3.3}$$

$$HZnO_2^- \leftrightarrow ZnO_2^{2-} + H^+ \text{ at pH} \geq 13 \tag{3.4}$$

The Nernst equations for each of these reactants in equilibrium with metallic zinc can be expressed as follows. At pH >8, the electrodeopositon exhibits linear dependence on pH with negative gradients.

$$2H^+ + 2e^- + Zn(OH)_2 \leftrightarrow Zn + 2H_2O \; E = -0.339 \text{ V} - 0.0529 \text{ pH} \tag{3.5}$$

$$3H^+ + 2e^- + HZnO_2^- \leftrightarrow Zn + 2H_2O \; E = 0.54 \text{ V} - 0.0886 \text{ pH} + 0.0295 \log\left[HZnO_2^-\right] \tag{3.6}$$

$$4H^+ + 2e^- + ZnO_2^{2-} \leftrightarrow Zn + 2H_2O \; E = 0.414 \text{ V} - 0.1182 \text{ pH} + 0.0295 \log\left[ZnO_2^{2-}\right] \tag{3.7}$$

$$Zn^{2+} + 2e^- \leftrightarrow Zn \; E = -0.763 \text{ V} + 0.0295 \log\left[Zn^{2+}\right] \tag{3.8}$$

As we have discussed earlier, the choice of mild acidic electrolytes as an alternative to alkaline ones is to enhance the rechargeability of Zn-anode by eliminating critical issues such as dendritic growth and $Zn(OH)_2$/ZnO passivation at the electrode/electrolyte interface. It is clear from the Pourbaix diagram that, in comparison with alkaline medium, acidic or mild acidic electrolytes offer direct plating/stripping of Zn (Equation 3.1). But Zn^{2+} ions generally exist in the form of $[Zn(H_2O)_6]^{2+}$ complex in mild acidic media. During the charging process, due to water decomposition, Hydrogen Evolution Reaction (HER) could compete with the zinc electrodeposition process. Additionally, HER also induces the pH change at the surface of the Zn-anode due to H_2 gas and OH^- ion formation. This subsequently leads to the formation of undesirable and inactive by-products (such as oxides/hydroxides of Zn) on the Zn-anode surface. The aforementioned phenomenon inevitably disturbs the Zn^{2+} plating at the anode. Since a fraction of applied current is utilized for the HER, the coulombic efficiency (CE) of Zn-anode comes down. Moreover, HER blocks some of the regions (known as dead regions) from Zn plating, which eventually causes low CE and promotes dendritic growth on the surface of the Zn-anode [4,11]. Further, the continuous proton reduction in the charging process causes a rise in the local pH (due to the production of OH^- ions) near the surface of the Zn-anode. As a result of the accumulation of Zn^{2+}, SO_4^{2-}, and OH^- ions, zinc hydroxy sulphate (ZHS) readily forms over the metallic Zn-anode in mild acidic $ZnSO_4$ electrolyte [4]. The ZHS is an electrochemically inert by-product that passivates the metallic Zn-anode surface. Moreover, the build-up of the ZHS product promotes an uneven surface growth of Zn-anode, thereby inducing dendritic growth. Besides dendritic formation, the build-up of ZHS at the surface of the Zn-anode progressively chomps the electrolyte. This eventually lowers the concentration of active Zn^{2+} ions at the interface. This further diminishes the CE of the Zn-anode. So, CE is an important factor that decides the reversibility of the Zn-anode. Ideally, a high-performance Zn-anode should have 100% CE indicating its outstanding reversibility for the plating/stripping process. But the unwanted side reactions/formation of electro-inactive products (such as HER, surface passivation, and dendritic growth) on the surface of the Zn-anode drastically lower the Zn plating/stripping efficiency from the ideal value. Hence, to achieve smooth surfaces that are free from dendrites and to avoid HER, unremitting efforts are being made to improve the Zn-anode/electrolyte interfacial process.

3.2.2 Strategies to Improve the Zn-Anode/Electrolyte Interphase

To inhibit the formation of $ZnO/Zn(OH)_2$ side products and improve the rechargeability of Zn-anodes, systematic investigations are being carried out in various mild acidic aqueous electrolytes. For example, Gui et al. [12] studied the electrochemical behaviour of Zn stripping/plating in three electrolytes such as $Zn(NO_3)_2$, $Zn(CH_3COO)_2$, and $ZnSO_4$. The cyclic voltammetry (CV) studies reveal that, during the cathodic scan, the Zn plating occurs at –0.13 V, and, during the anodic scan, Zn stripping occurs at a potential of –0.034 V in $Zn(CH_3COO)_2$ aqueous electrolyte. Similarly, the Zn plating/stripping occurs at onset potentials of –0.11 V/–0.0055 V, respectively, in $ZnSO_4$. The suppression of redox peaks due to Zn dissolution/deposition in the zinc nitrate electrolyte signifies its reduced scope for ZIB applications. Additionally, owing to the strong oxidizing nature of NO_3^- ion, corrosion of Zn-anode/cathode is promoted, which adversely affects the cycle life of the ZIBs [12,13]. Despite its good reversibility, the lethargic kinetics of Zn plating/stripping in the $Zn(CH_3COO)_2$ electrolyte limits the applicability of acetate salts for ZIBs. On the contrary, $ZnSO_4$ shows superior kinetics for Zn plating/stripping compared to $Zn(CH_3COO)_2$ and is widely employed for ZIB applications. The dissolved oxygen in $ZnSO_4$ can detrimentally affect Zn-anode performance and its stability. Su et al. [14] investigated the effect of dissolved O_2 in mild acidic $ZnSO_4$ electrolytes on the cyclic stability and reversibility of Zn-anode. The symmetrical Zn/Zn cell exhibited outstanding cycling stability and reversibility for Zn plating/stripping in oxygen-free $ZnSO_4$ compared to the oxygen-rich electrolyte (Figure 3.4a). The thermodynamic instability of Zn towards oxygen leads to the corrosion of Zn-anode in the oxygen-rich electrolyte. The unusual behaviour such as an abrupt and irreversible increase in the voltage polarization signifies the instability of Zn-anode to the dissolved oxygen. The X-ray diffraction (XRD) studies revealed the appearance of more by-products (zincates, $Zn(OH)_2$, ZHS, and ZnO) at the Zn-anode surface that operated in oxygenated $ZnSO_4$ than oxygen-free one, indicating thermodynamic instability of Zn to oxygen (Figure 3.4b). Interestingly, the symmetrical Zn/Zn system demonstrated ~20-fold lifetime improvement in the oxygen-free electrolyte with reversible plating/stripping compared to the oxygen-enriched one. Apart from the dissolved oxygen, the presence of a large number of water molecules that are directly coordinating with zinc ions in the aqueous electrolytes also suppresses the plating efficiency of the Zn-anode and restricts the cycle life of ZIBs due to HER. In aqueous mild acidic electrolytes, Zn^{2+} cations usually exist in the form of $[Zn(H_2O)_6]^{2+}$ hydrated sheath. During the charging process, if the hydrated shell fails to get stripped from the Zn^{2+} ions, water decomposition takes place. This undesired HER eventually destabilizes the Zn-anode/electrolyte interface and hampers the plating efficiency. By carefully formulating the coordination environment of zinc cations, one can effectively retard the HER/undesired by-products at the Zn-anode [4,8,11]. Lately, highly concentrated electrolytes (water-in-salt electrolyte), zinc salts based on the bulkier anions ($Zn(CF_3SO_3)_2$ and $Zn(TFSI)_2$), etc., were explored to modulate the coordination environment of Zn^{2+} at the molecular level to eliminate the negative impact of water and achieve the smooth Zn-anode surfaces that are free from dendrites.

Wang et al. [15] demonstrated the tuning of the coordination environment of Zn^{2+} ions in an aqueous medium using highly concentrated Zn and Li salt-based electrolytes. The coordination environment of Zn^{2+} is investigated by varying the LiTFSI concentration (5, 10, and 20 m) in 1 m $Zn(TFSI)_2$ using molecular dynamics (MD). The MD simulation studies revealed that, in the dilute electrolyte (1 m $Zn(TFSI)_{2+5}$ m LiTFSI), Zn^{2+} readily coordinates with water and forms

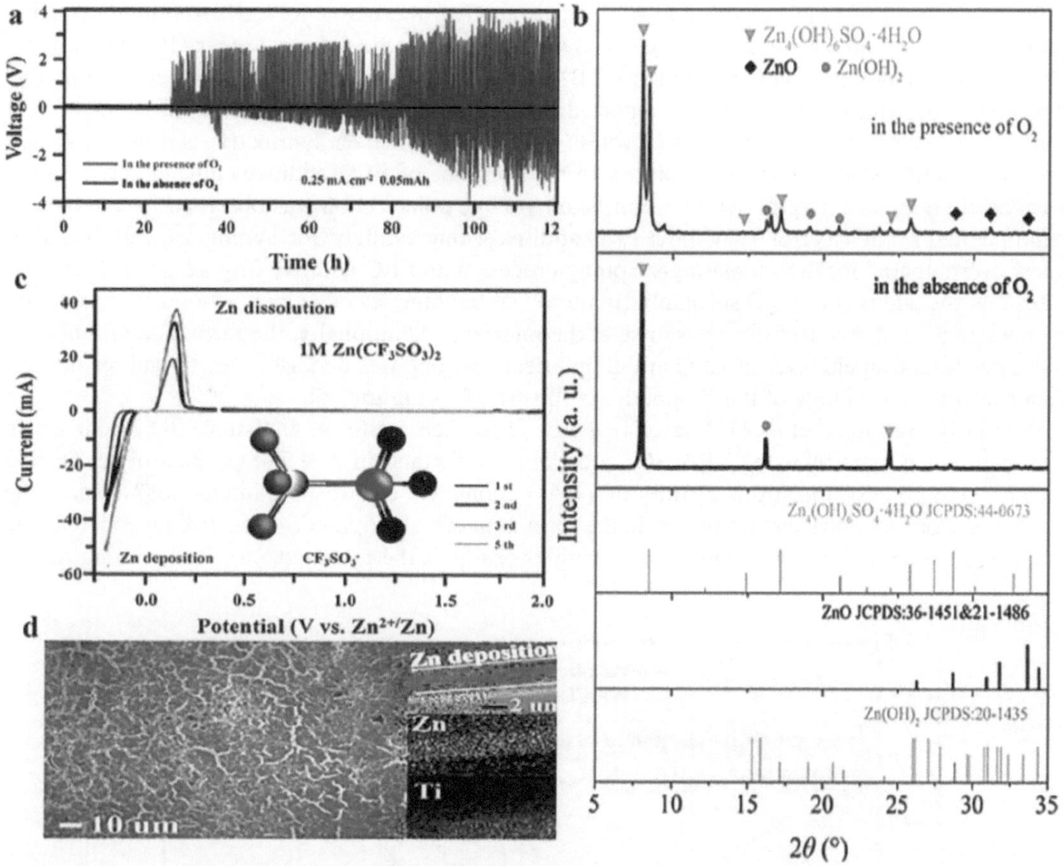

FIGURE 3.4 Electrochemical performance of (a) Zn/Zn symmetric cell and XRD patterns of Zn-anode (b) with and without oxygenated electrolyte. Adapted and reproduced from Ref. [14]. Copyright 2021 Elsevier. CV of (c) Zn electrode in 1 M $Zn(CF_3SO_3)_2$ and scanning electron micrographs of (d) Zn deposit on Ti foil in 3 M $Zn(CF_3SO_3)_2$. Adapted and reproduced from Ref. [13]. Copyright 2021 American Chemical Society.

$(Zn(H_2O)_6)^{2+}$ solvation sheath. As the concentration of supporting salt (LiTFSI) increases from 5 to 20 m in 1 m $Zn(TFSI)_2$ electrolyte, $TFSI^-$ starts to occupy the Zn^{2+} solvation sheath. Interestingly, in the most concentrated electrolyte (20 m LiTFSI in 1 m $Zn(TFSI)_2$), the Zn^{2+} solvation sheath is preferentially engaged by $TFSI^-$ and conquers the formation of $(Zn(H_2O)_6)^{2+}$. This special solvation sheath around the Zn^{2+} (with $TFSI^-$ anions), in the most concentrated electrolyte, significantly eliminates the undesired HER at the Zn-anode surface. Moreover, the absence of side products such as ZnO and dendritic growth in 1 m $Zn(TFSI)_2$ + 20 m LiTFSI electrolyte drastically enhanced Zn plating/stripping with ~100% CE, during cycling. Similarly, in another study, Zhang et al. [13] also proved the reduction of solvation sheath around Zn^{2+} by introducing larger anions that can effectively eliminate the formation of dendrites over the Zn-anode surface. Among the zinc triflate and sulphate electrolytes, the higher kinetics of Zn dissolution/deposition is noticed in $Zn(CF_3SO_3)_2$ than in $ZnSO_4$ (Figure 3.4c). The superior kinetics of Zn plating/stripping is due to the reduced solvation effect (with water) (owing to the large size of $CF_3SO_3^-$ anion) of Zn^{2+} in $Zn(CF_3SO_3)_2$ electrolyte that offers a low charge transfer resistance during cyclings. On increasing the concentration of $Zn(CF_3SO_3)_2$ to 3 M, smooth and dendrite-free Zn-anodes surfaces were realized (Figure 3.4d). These studies disclose that by carefully tuning the coordination environment of Zn^{2+} ions, one can improve the Zn plating/stripping efficiency without forming zinc oxides/hydroxide by-products.

The selection of suitable electrolyte additives is another alternative for ZIBs to inhibit dendritic growth on the Zn-anode surface. The electrolytic additives such as diethyl ether [16] and cationic surfactant(tetrabutylammonium sulphate (TBA_2SO_4)) [17] were intensively studied to improve Zn-anode performance. Xu et al. [16] reported the electrochemical behaviour of Zn-anode in 3 M $Zn(CF_3SO_3)_2$ with and without Et_2O (2 vol.%) additive. The Zn/Zn symmetric cell demonstrated excellent cycling stability and reversibility in the presence of Et_2O additives during entire galvanostatic charging/discharging while a significant voltage polarization was observed for the battery that operated in the electrolyte without Et_2O additives. Interestingly, the symmetric cell displayed a low overpotential for the Zn plating/stripping process with Et_2O additive (Figure 3.5a). This indicates that the addition of Et_2O substantially improved the kinetics of Zn deposition and dissolution by lowering the charge transfer resistance at the interface. Additionally, the formation of a protective electrostatic shield layer of Et_2O at the tip effectively opposes dendrite growth and results in a dendrite-free morphology of the Zn-anode, as illustrated in Figure 3.5b.

Similarly, Bayaguud et al. [17] developed a dendrite-free, uniform, and stable 3D Zn-anode on commercial Cu foam by using TBA_2SO_4, a cationic surfactant, in 2 M $ZnSO_4$ electrolyte. During the charging process, the strong affinity of TBA^+ cations (via electrostatic interaction) towards the Zn deposits or Zn-anode surface results in the formation of a zincophobic protective layer during the nucleation process. This zincophobic screening layer repels the hydrated Zn^{2+} ions coming in and

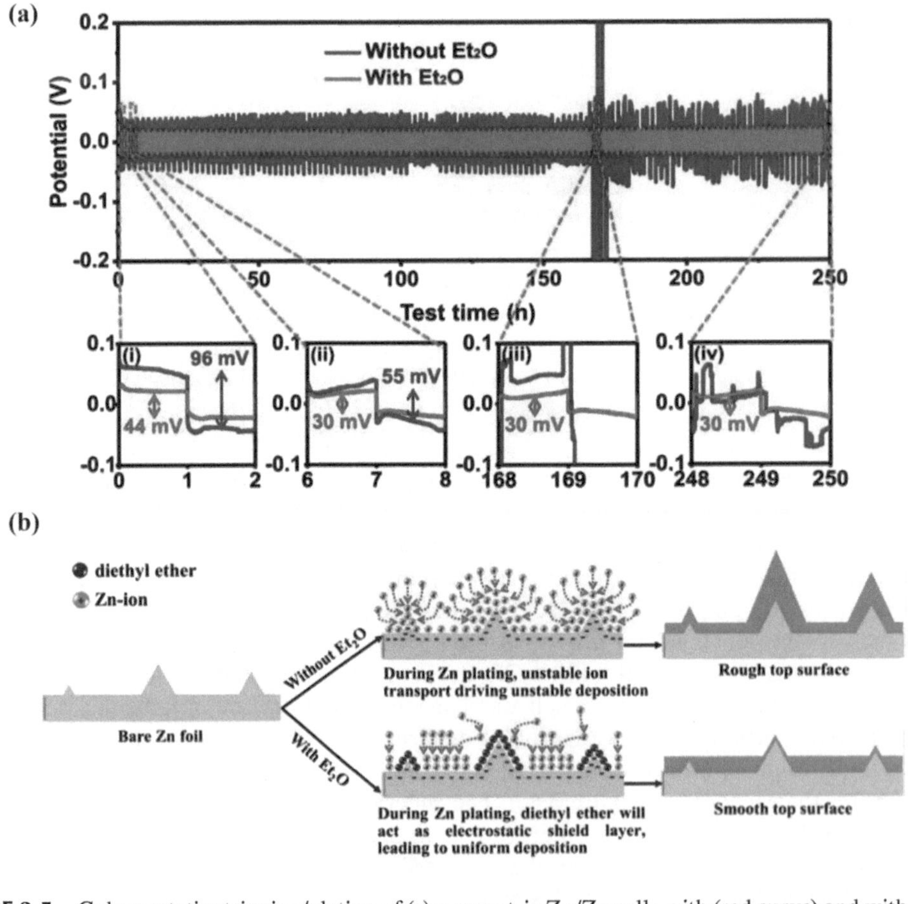

FIGURE 3.5 Galvanostatic stripping/plating of (a) symmetric Zn/Zn cells with (red curve) and without (blue curve) diethyl ether. The enlarged over-voltage profiles of the symmetric cells at different cycles (bottom), and the study of Zn-anode morphology (b) with and without Et_2O additive. Adapted and reproduced from Ref. [16]. Copyright 2021 Elsevier.

depositing further on the initial nuclei. This mechanism effectively protects these nuclei from the further formation of rough surfaces with massive dendrites. Other additives such as lignin, sodium dodecyl sulphate, sodium dodecyl benzenesulphonate, polyacrylamide, pyrazole, 1-ethyl-3-methylimidazolium dicyanamide (EMI-DCA) [18], fumed silica, polyethylene glycol, triethyl phosphate [19], and fatty methyl ester ethoxylate [17] were also explored for ZIB application for obtaining dendrite-free morphology.

Apart from electrolyte modification, surface modification is one of the most straightforward strategies to improve the Zn-anode/electrolyte interface [4,8,11]. Surface modification strategies such as coating the active Zn or mixing the active Zn with additives are widely adopted to achieve smooth and dendrite-free high-performance Zn-anodes. Cui et al. [20] developed an intelligent strategy for inhibiting the dendritic growth on the Zn-anode surface. In their work, uniform quasi-isolated gold nanoparticle (Au-NPs) deposits are decorated on the surface of the Zn-anodes initially by a sputtering method. During the charging process, these heterogeneous uniform Au-NP deposits direct the Zn-nuclei to form a uniform array of zinc flakes. In this process, decorated Au-NPs act as an effective heterogeneous seed to guide the morphology evolution at the surface of the Zn-anodes. This innovative strategy successfully produced a facile and uniform Zn-flake array on the Au-NP decorated Zn. Whereas an uneven and massively different morphological deposition is obtained on the bare Zn surface (without Au-NPs), as illustrated in Figure 3.6.

Similarly, the various surface coating strategies such as decorating/shielding the active Zn surface with carbon nanotubes, a layer of electrically insulating $CaCO_3$, polyamide/$Zn(CF_3SO_3)$ composite coating, and TiO_2 were proposed to promote the uniform Zn deposits without any undesirable gas evolution at the interphase [8]. Another interesting and novel strategy to address the non-planar irregular Zn electrodeposits at the surface of the anode is the "epitaxial growth" strategy, proposed by Zheng et al. [21]. In this mechanism, Zn electrodeposition preferentially occurs in the epitaxial

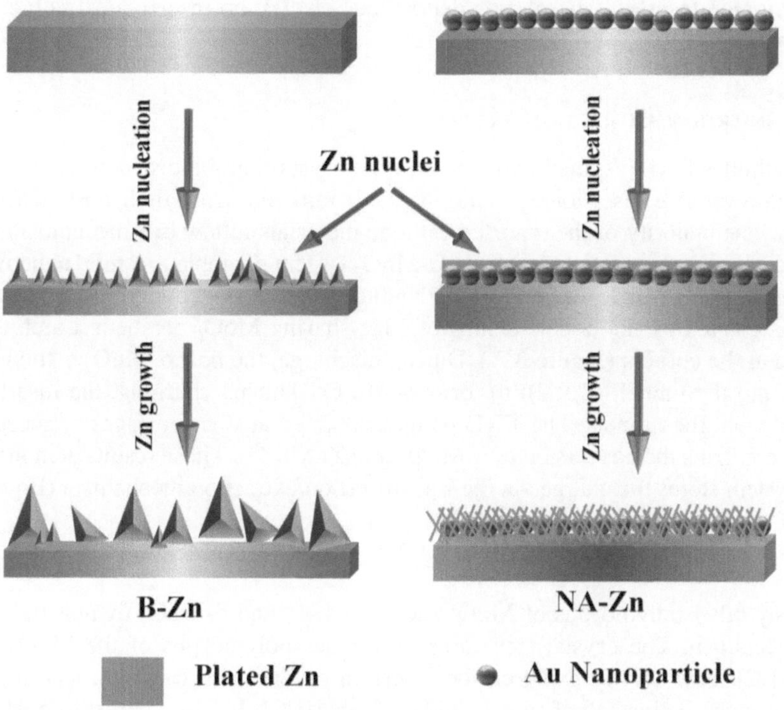

FIGURE 3.6 Schematic representation of morphological features of Zn plating/stripping on the bare (left) and Au-NPs decorated Zn (right). Adapted and reproduced from Ref. [20]. Copyright 2021 American Chemical Society.

fashion (in the direction of the (002) plane) as directed by graphene on the stainless steel (SS) substrate. Owing to the low lattice mismatch between Zn and graphene, Zn deposits formed on graphene are well directed. As a result, the highly ordered and preferential growth of thin plates of Zn deposits that are parallel to the substrate is realized. In contrast to graphene-coated SS, due to the absence of the epitaxial growth mechanism on the bare SS, the thin plates of Zn deposits are randomly oriented. The preferential growth of the (002) plane is identified using the grazing incidence X-ray diffraction pattern. This confirms that the graphene-coated SS directs Zn electrodeposition in the epitaxial fashion in contrast to bare SS (which promotes non-epitaxial and highly randomized thin plates of Zn electrodeposits). The epitaxial Zn-anode demonstrated an excellent Zn plating/stripping efficiency with CE>99% over 1,000 cycles.

Apart from the chemistry of Zn-anode, the biggest bottleneck for the aqueous ZIB for their commercialization is the lack of appropriate host cathode materials. The high polarization of the Zn^{2+} ion (0.74 Å) demonstrates a strong electrostatic/coulostatic interaction with the host structures that eventually leads to poor cyclability and sluggish intercalation kinetics. Hence, the search is on to find the appropriate cathodes for ZIBs to boost their performance in terms of cycling stability, rate capability, and energy density.

3.3 ENERGY STORAGE MECHANISMS OF ZIBS

Cathode materials play a critical role in boosting the performance of ZIB. The key parameters such as discharge capacity, operating voltage, energy density, and transport kinetics of Zn^{2+} ions are highly dependent on the cathode materials. Hence, developing cathode materials that reversibly accommodate Zn^{2+} ions with high transport kinetics is a great challenge in ZIBs. Moreover, the energy storage mechanisms of the proposed cathodes are much more complex, controversial, and still underdeveloped. According to the literature, ZIBs store energy via four basic routes, namely, the Zn^{2+} insertion/de-insertion, dissolution/deposition, Zn^{2+}/H^+ co-insertion/extraction, and chemical conversion.

3.3.1 Zn^{2+} Insertion/Extraction Mechanism

Most of the cathodes having tunnel and layered-type crystallographic structures demonstrate this mechanism via reversible insertion/extraction of Zn^{2+} ions into/from their host structures, during ZIB operation. The majority of the reported cathode materials follow this mechanism.

Xu et al. [3] constructed a rechargeable Zn/α-MnO_2 system operable in a mild aqueous electrolyte ($ZnSO_4$ or $Zn(NO_3)_2$). α-MnO_2 resembles a hollandite structure (space group I4/m) having $(2 \times 2)+(1 \times 1)$ tunnels created by the double chains of edge-sharing MoO_6 octahedral architectonic units that are linked at the corners (Figure 3.7a). During discharge, the host α-MnO_2 cathode intercalates the Zn^{2+} ions into the tunnel $((2 \times 2))$ to form $ZnMn_2O_4$. During charging, the inserted Zn^{2+} ions were extracted from the tunnels. The XRD pattern obtained at various stages of successive charge and discharge confirms the conversion of α-MnO_2 into $ZnMn_2O_4$. These results demonstrate that the Zn/α-MnO_2 system stores the charge via the Zn^{2+} insertion/extraction mechanism (Equation 3.9).

$$\text{Cathode}: Zn^{2+} + 2e^- + 2\alpha - MnO_2 \leftrightarrow ZnMn_2O_4 \tag{3.9}$$

Similarly, many other polymorphs of MnO_2 such as β-, γ-, and δ-MnO_2 exhibit the Zn insertion/extraction mechanism. The crystal structures of various polymorphs of the MnO_2 are depicted in Figure 3.7 [22]. Few studies explored the insertion of Zn^{2+} into α-MnO_2, leading to a layered Zn-birnessite or buserite instead of spinel $ZnMn_2O_4$ phase [23,24]. In contrast to α-MnO_2, γ-MnO_2 displays a multiphase transformation during the sequential Zn^{2+} ion insertion. Alfaruqi et al. [25] successfully investigated the structural conversion of the γ-MnO_2 cathode during the progressive insertion of Zn^{2+} ions. During the discharge process, Zn^{2+} progressively occupies (1×2) γ-MnO_2

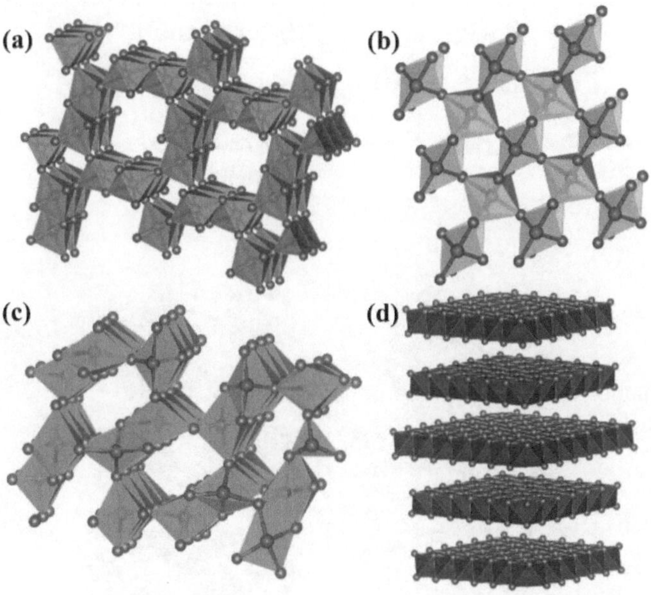

FIGURE 3.7 Crystal structures of (a) α-MnO_2, (b) β-MnO_2, (c) γ-MnO_2, and (d) δ-MnO_2. Note: red and blue atoms represent Mn and O, respectively. Adapted and reproduced from Ref. [22]. Copyright 2021 Elsevier.

tunnels. In the early stage of discharge, the tunnel-type γ-MnO_2 is partially converted into the spinel-type $ZnMn_2O_4$ (Figure 3.8). Further insertion of Zn^{2+} into γ-MnO_2 tunnels leads to the formation of a tunnel-type γ-Zn_xMnO_2 phase along with spinel. At the final stage of discharge, the tunnels tend to widen and open up the structural framework to form a layered-type (L-Zn_yMnO_2) phase along with spinel and tunnel-type structures. In other words, three phases (tunnel-type γ-Zn_xMnO_2, spinel-type $ZnMn_2O_4$ (ZMO), and layered-type L-Zn_xMnO_2) coexist at the end of discharge. Zn^{2+} insertion accompanies with change in the oxidation state of Mn from IV to III and finally to II. Upon complete charging, almost all the structural features of the parent orthorhombic lattice of γ-MnO_2 are restored. Some other Mn-based oxide compounds such as layer-type B-Zn_xMnO_2 (a first discharge product of β-MnO_2) [26], cation-deficient spinel $ZnMn_2O_4$ [13], and Mn_2O_3 [27] are also well known in the literature whose charge storage mainly attributed to the Zn^{2+} insertion/extraction mechanism.

In addition to Mn-oxide-based materials, a series of vanadium oxide-based materials such as $Zn_3V_2O_7(OH)_2{\cdot}2H_2O$, $V_2O_5{\cdot}nH_2O$, $Na_2V_6O_{16}{\cdot}3H_2O$, a layered-type NaV_3O_8, tunnel-type β-$Na_{0.33}V_2O_5$, and $Ca_{0.25}V_2O_5{\cdot}nH_2O$ exhibit reversible Zn^{2+} insertion/extraction mechanism. But the mechanism of energy storage is more complicated in the case of vanadium oxide-based cathodes. Very recently, Guo et al. [28] demonstrated the effect of the Zn^{2+} insertion/extraction mechanism on the structural stability of different sodium vanadates such as layered NaV_3O_8-type ($Na_5V_{12}O_{32}$ and $HNaV_6O_{16}{\cdot}4H_2O$) and tunnel-type β-$Na_{0.33}V_2O_5$ ($Na_{0.76}V_6O_{15}$). It is reported that during the reversible insertion/extraction of Zn^{2+}, a tunnelled β-$Na_{0.33}V_2O_5$ exhibits superior structural stability to that of layered NaV_3O_8 structures. The high structural stability of β-$Na_{0.33}V_2O_5$ is due to the bonding of inserted Na^+ ions into the tunnels (which are formed as a result of the association of VO_6 octahedra and VO_5 square pyramids) with single connected oxygen atoms (which are present in between the tunnels) and stabilizes the whole structure during repetitive charge/discharges (also known as "pillars" effect of sodium ion). In contrast, due to the poor interaction of Na^+ ions with tri-connected oxygen atoms in layered NaV_3O_8, some of the Na^+ ions could be easily displaced by the Zn^{2+}, leading to the formation of the new phase ($Zn_4V_2O_9$) (Figure 3.9). The formation of a new phase and the absence of the "pillars" effect of Na^+ result in the structural collapse of layered $Na_5V_{12}O_{32}$ and $HNaV_6O_{16}{\cdot}4H_2O$.

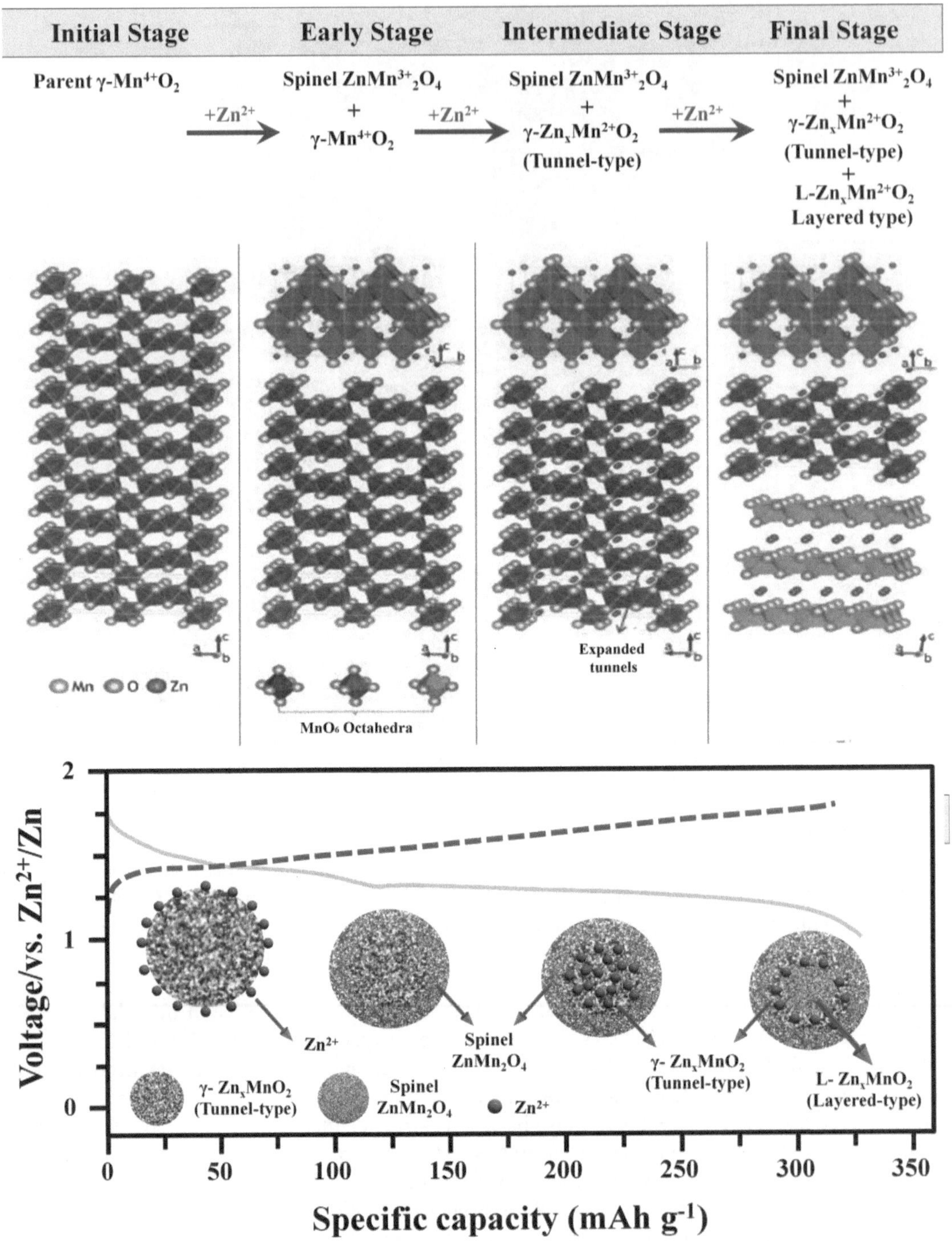

FIGURE 3.8 Schematic illustration demonstrates the structural transformation of the mesoporous γ-MnO_2 cathode during the progressive insertion of Zn^{2+} ions. Adapted and reproduced from Ref. [25]. Copyright 2021 American Chemical Society.

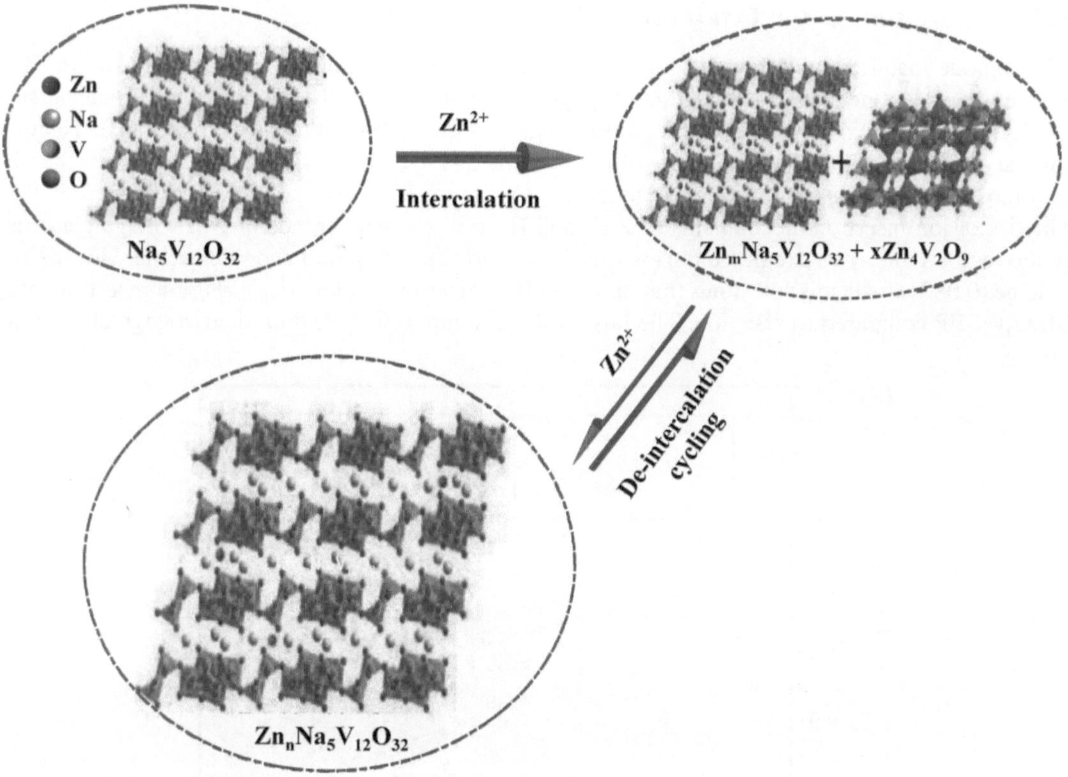

FIGURE 3.9 Representation of the mechanism of energy storage in $Na_5V_{12}O_{32}$ cathode during cycling.

The other inorganic compounds such as PBA (A_xM [M′ $(CN)_6$]y·$n$$H_2O$, (A=$Na^+$ or K^+), (M′=Ru, Co, Cr, Fe), (M=Mn, Mo, Ni, Fe, V, Cu, Co)) and polyanionic phosphates ($MM'PO_4$, (M=K, Na, Li) (M′=Co, Mn, Fe, Ni, V, etc.)) also stores the energy via the extraction/insertion of divalent Zn^{2+} from/into their large open-framework structures.

3.3.2 Chemical Conversion Mechanism

In addition to reversible Zn^{2+} insertion/extraction, the chemical conversion mechanism was realized for the Zn/α-MnO_2 system operated in mild acidic $ZnSO_4$ [29]. In a typical mechanism, α-MnO_2 interacts with the proton from the water and gets converted into the MnOOH phase during the discharge. The resulting hydroxyl ions react with an aqueous $ZnSO_4$ electrolyte and form ZHS to maintain charge neutrality in the system. Upon charging, the monoclinic MnOOH returns to its original α-MnO_2 phase, whereas ZHS subsequently undergoes decomposition according to Equations 3.10 and 3.11.

$$\text{Cathode}: 2\alpha - MnO_2 + H^+ + e^- \leftrightarrow MnOOH \tag{3.10}$$

$$6OH^- + SO_4^{2-} + 4Zn^{2+} + xH_2O \leftrightarrow ZnSO_4\left[Zn(OH)_2\right]_3.xH_2O \tag{3.11}$$

3.3.3 Co-Insertion and Extraction of Zn^{2+}/H^+

The cathode materials with layered- or tunnel-type structural framework probably enable the H^+/Zn^{2+} co-insertion mechanism. Sun et al. [30] systematically studied the energy storage mechanism of electrodeposited akhtenskite MnO_2 on the surface of the carbon fibre paper (MnO_2@CFP) during the discharge. The voltage response of the MnO_2@CFP electrode is examined by the discharge galvanostatic intermittent titration technique (GITT) (Figure 3.10a). Because of the difference in the kinetics of ion insertion mechanisms of Zn^{2+} and H^+ ions, the appearance of two voltage plateaus in the region I and II was realized during GITT measurement. Region I shows extremely low overvoltage (0.08 V), which is ~10 times that of region II, indicating fast kinetics of H^+ ion insertion into MnO_2@CFP compared to Zn^{2+} ion. The large voltage jump (0.6 V) and gradual voltage change in

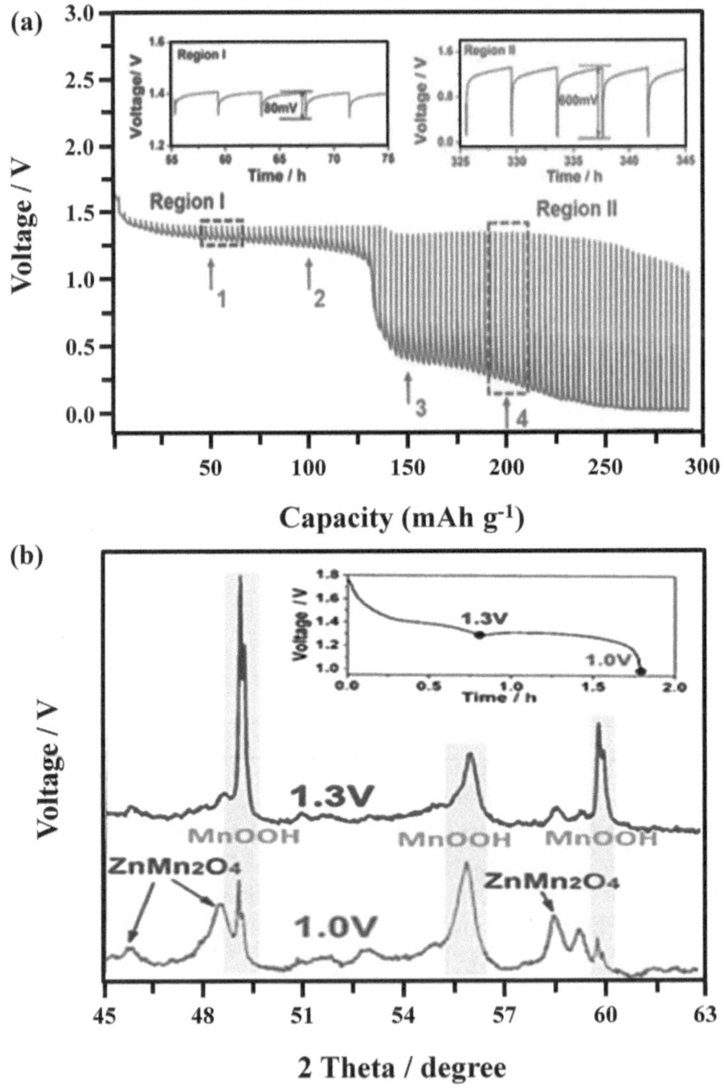

FIGURE 3.10 Discharge GITT profiles of (a) the Zn/MnO_2@CFP cell and ex-situ XRD patterns of (b) the MnO_2@CFP cathode at 1.0 and 1.3 V discharge voltages. Adapted and reproduced from Ref. [30]. Copyright 2021 American Chemical Society.

region II are attributed to high charge transfer and Zn^{2+} ion diffusion resistances, respectively. The XRD pattern of MnO_2@CFP cathode with or without $ZnSO_4$ in 0.2 M $MnSO_4$ electrolyte further confirms the Zn^{2+} and H^+ co-insertion mechanism (Figure 3.10b). The consecutive appearance of MnOOH and $ZnMn_2O_4$ phases in the XRD spectra after discharging the MnO_2@CFP electrodes to 1.3 and 1.0 V, respectively, reveals that the cathode experiences a sequential Zn^{2+} and H^+ ions co-insertion mechanism. In contrast to the MnO_2@CFP cathode, the simultaneous insertion/extraction of H^+ and Zn^{2+} ions is observed in the nanostructured $NaV_3O_8{\cdot}1.5H_2O$ cathode [31].

3.3.4 Dissolution/Deposition Mechanism

In this type of mechanism, the MnO_2 cathode upon dissolution produces Mn^{2+} ions during the discharge. During the subsequent charge process, the dissolved Mn^{2+} reverts to its original state per Equation 3.12.

$$\text{At positive electrode}: 2H_2O + Mn^{2+} \leftrightarrow MnO_2 + 4H^+ + 2e^- \tag{3.12}$$

Compared to the other three mechanisms that involve a one-electron transfer process with a low theoretical capacity of 308 mAh g^{-1}, the dissolution/deposition reaction mechanism with direct two electrons transfer imparts a ZIB battery with high energy density, high-voltage window (~2 V), and a high theoretical capacity of 616 mAh g^{-1} [32].

3.4 ELECTROCHEMICAL PERFORMANCE OF VARIOUS CATHODES

3.4.1 Manganese Oxide-Based Cathodes

Zn/α-MnO_2 system operated in mild aqueous electrolyte ($Zn(NO_3)_2$ or $ZnSO_4$) demonstrated a high capacity of 210 and 100 mAh g^{-1} at 0.5C and 6C, respectively, with ~100% of CE [3]. In another study, the Zn paired with layered buserite cathode delivered a good discharge capacity (225 mAh g^{-1}) with ~94% capacity retention over 2,000 charge/discharge cycles [26]. Similarly, λ-MnO_2 with limited 3D tunnels delivered a good specific capacity of 442.6 mAh g^{-1} at 13.8 mA g^{-1}, whereas it delivered a discharge capacity of 33.8 mAh g^{-1} at 408 mA g^{-1} in aqueous $ZnSO_4$ electrolyte [33]. Many other Mn-based oxides such as Mn_3O_4, Mn_2O_3, and MnO were also explored as cathode materials for high-performance ZIB applications. Low energy density, structural failure on repetitive cycling, and sluggish diffusion of Zn^{2+} ions in their host structure demand a new class of materials that can boost the ZIB performance.

3.4.2 Vanadium-Based Cathodes

Vanadium-based compounds have attractive features such as open-framework structures, the existence of multivalency states, and the high abundance of vanadium. Among the various vanadium-based cathodes, vanadium oxides (V_2O_5, V_6O_{13}, VO_2, $V_3O_7.H_2O$, etc.), vanadates ($M_xV_nO_m$), and vanadium sulphides (V_3S_4, and VS_2) were well explored as cathodes for ZIBs. Senguttuvan et al. [34] reported a multivalent Zn/V_2O_5 battery in 0.5 M acetonitrile-Zn(TFSI)$_2$, which demonstrated a reversible capacity of 130 mAh g^{-1} at a high rate of 20C. In another study, the V_2O_5 cathode was reported to exhibit a capacity of 238 mAh g^{-1} at 50 mA g^{-1} in the concentrated Li/zinc salt electrolyte [35]. The capacity was further improved using $V_2O_5{\cdot}nH_2O$/graphene (372 mA g^{-1}) and V_2O_5 (470 mA g^{-1}) cathodes [36]. Despite their high capacity and inherent structural stability compared to Mn-based compounds, the oxides and sulphides of V suffer from the narrow operating voltage of ~0.7 V. To overcome the narrow operating voltage, polyanionic materials were developed with superior features.

3.4.3 Polyanionic Materials

NASICON ($A_xMM'(XO_4)_3$ (A=alkali or alkaline earth metals; X=P, Si, S, etc.; M′, M=Fe, Cr, Ti, V, etc.) type cathodes belong to a class of polyanionic materials and have been recognized as a potential guest material for storing Zn^{2+} ions. Their attractive traits include (1) a rigid framework that ensures stability, (2) enough room for holding a large number of cations, and (3) channels that are favourable for smooth and rapid diffusion of cations. Zn/NASICON $Na_3V_2(PO_4)_3F_3$ system demonstrates a wide operating voltage window of ~1.62 V in 2 M zinc triflate electrolyte. The large electronegativity difference between F and V is the reason behind achieving high voltage. Besides, Zn/NASICON $Na_3V_2(PO_4)_3F_3$ exhibited a superior energy density of 97.5 Wh kg^{-1} with ~95% capacity retention over 4,000 cycles with excellent structural stability [37]. These investigations have now opened up an avenue for other NASICON-type materials.

3.4.4 PBA-based Cathodes

PBAs are transition-metal hexacyanoferrates (MHCFs), and the majority of them typically possess face-centred cubic structures. The various PBAs can be prepared by varying the M (Ni, Cu, Fe, Co, Mn). Among the explored MHCFs, the FeHCF, CuHCF, and NiHCF demonstrate the typical cubic structures, whereas ZnHCF exhibits the rhombohedral framework. Due to their open structure, PBAs facilitate facile intercalation/de-intercalation of Zn^{2+} ions into their framework and hence serve as potential cathodes for ZIBs. For example, the Zn/CuHCF system exhibited ~90% of theoretical capacity in 20 mM $ZnSO_4$. Besides its high capacity, Zn/CuHCF system showed good cyclic stability over the 100 cycles (~96.3% of capacity retention) with an attractive voltage of 1.73 V [38]. Similarly, a high capacity (64.4 mAh g^{-1}) with an attractive voltage of 1.7 V was also noticed with the Zn/ZnHCF system [39]. Despite high working voltages, Zn-anode paired with PBA cathodes suffers from low capacity and low energy density. The poor energy density of the PBA cathodes greatly restricts them from their commercialization.

3.4.5 Chevrel Phase

For the past three decades, Chevrel phase materials have been extensively studied as superconductors, catalysts, cathode of Li-ion and Mg-ion batteries, and thermoelectric materials. Chevrel phase with general formula $M_xMo_6T_8$ (M=Cd, Zn, Na, Cu, Mn, etc. and T=Te, Se, or S) is another class of host material explored for ZIB. These compounds are able to house (insertion) both monovalent (Li^+, Na^+) and divalent ions (Zn^{2+}, Cu^{2+}, Cd^{2+}, Co^{2+}, and Mg^{2+}) with outstanding electronic and ionic transport kinetics. The array of distorted MO_6 octahedron and cubic units (T_8) of chalcogenide atoms provides a unique three-dimensional framework for this cathode material. Chae et al. [40] studied the insertion/de-insertion properties of Zn^{2+} in $Zn_xMo_6S_8$ (X=1, 2) in 0.1 M $ZnSO_4$ aqueous electrolytes. The CV studies reveal two-step reductions of Mo_6S_8. The intercalation of Zn into Mo_6S_8 results in the formation of the $ZnMo_6S_8$ (~0.45–0.50 V) phase in the first step, and further the reduction of $ZnMo_6S_8$ at around ~0.35 V results in the formation of $Zn_2Mo_6S_8$. The Zn/Mo_6S_8 cell exhibited a moderate capacity of 57 mAh g^{-1} at 1C (the theoretical capacity concerning $Zn_2Mo_6S_8$ is 129 mAh g^{-1}). In the literature, Chevrel phase compounds were also treated as anodes for ZIBs. Cheng et al. [41] reported Mo_6S_8 nanocubes as anode material (an acid-leached product of $Cu_2Mo_6S_8$ nanocubes) with an average particle size of ~100 nm that host Zn^{2+} ions reversibly in both non-aqueous and aqueous media. The zinc poly iodide solution (aqueous 0.2 M I_2+1.5 M ZnI_2 mixture) was used as the catholyte. The high solubility and good redox features of I^-/I_3^- catholyte could enable the high capacity and energy density and hence possess great potential for practical application. The full cell is fabricated by integrating the Mo_6S_8 nanocubes (as the anode) with zinc poly iodide redox catholyte, which displayed a high capacity of 120 mAh g^{-1}, which is close to theoretical capacity (129 mAh g^{-1}) with good cycling stability over 150 galvano static charge-discharge (GCD) cycles (at 180 mA g^{-1}).

3.4.6 Metal Disulphides

These are layered materials and bonded by weak van der Waals forces. The weak forces between the layers enable the facile insertion/de-insertion and promote mass transport of Zn^{2+} ions during the cycling. Vanadium sulphides (VS_4 and VS_2) and molybdenum sulphide (MoS_2) have been used tremendously for ZIB applications owing to their unique layered structural framework. He et al. [42] fabricated a Zn/VS_2 full cell in a mild acidic $ZnSO_4$ electrolyte. Successful insertion of Zn^{2+} ions in between the layers of VS_2 leads to the formation of a new conductive $Zn_{0.09}VS_2$ phase during the discharge. Further addition of Zn^{2+} ions into the $Zn_{0.09}VS_2$ leads to the formation of the $Zn_{0.23}VS_2$ phase. The full cell demonstrates excellent capacity (190.3 mAh g^{-1} at 0.05 A g^{-1}) and retains ~98% of its original capacity over 200 GCD cycles. On the other side, Xu et al. [43] developed defect-rich MoS_{2-x} nanosheets pyrolysed at 250°C as promising cathodes for ZIB applications. The full cell with defect-rich MoS_{2-x} nanosheet cathode delivered an excellent capacity of ~135 mAh g^{-1} at 0.1 A g^{-1}. Interestingly, even after 1,000 GCD cycles, the full cell exhibits good capacity retention (~87.8%) and delivers an attractive discharge capacity of 88.6 mAh g^{-1} at 1 A g^{-1}. So, the defect engineering approach is a very promising strategy for boosting reversible Zn^{2+} ion storage capacity to enhance the energy density of the ZIB.

3.4.7 Organic Cathode Materials

Besides the inorganic cathode materials, the development of organic materials has gained massive attention to boost the performance of ZIB. The smart choice of organic cathodes enables achieving high voltages with attractive capacities. Moreover, organic cathodes store/deliver reversibly both mono (Li^+ and Na^+) and divalent (Mg^{2+}, Zn^{2+}, etc.) ions into/from their structural framework without structural collapse during repetitive cycling. Among the explored organic cathodes, quinones/quinone derivatives are employed as a sustainable and green electrode material for ZIB applications. Quinones mainly store/deliver the charge via the "ion-coordination" mechanism. Various quinone compounds such as 9,10-anthraquinone (9,10-AQ), 1,2-naphthoquinone (1,2-NQ), and 9,10-phenanthrenequinone (9,10-PQ) were exploited as promising cathodes for ZIB applications. For instance, Zhao et al. [44] reported an aqueous ZIB using sustainable quinone as the cathode. In a typical system, Zn serves as an anode and calix[4]quinone acts as the cathode. The fabricated battery had shown a high energy efficiency of 93%, with an extremely high capacity of 335 mAh g^{-1} at 0.02 A g^{-1}. The full cell also demonstrated superior capacity retention of 87% over the 1,000 GCD cycles at 0.5 A g^{-1} rate.

The attractive features such as good structural stability, ubiquitousness, eco-benignity, and lightweight made the electroactive carbonyl functional group–bearing quinone compounds the preferred choice for ZIBs. In general, the inorganic host cathodes suffer from poor structural stability (due to the dissolution of cathodes) in an aqueous medium and lead to rapid capacity fading during cycling. In contrast, the organic quinones are stable in aqueous electrolytes and exhibit high capacity. Furthermore, the working potential of the quinones can be tuned by incorporating suitable electron withdrawing or donating substituents in their structures. Very recently, Chinmaya et al. [45] reported an aqueous ZIB that operated at an extremely high rate (63 C) using an aromatic organic cathode material (pentacene-5,7,12,14-tetraone (PT)) that is encapsulated in mesoporous conductive carbon (CMK-3), which bestowed high rate charging to PT. The systematic investigation revealed that ZIB with PT@CMK-3 cathode offered low over-voltage compared to super P conductive additive with PT (Figure 3.11). The lower over-voltage for Zn/PT@CMK-3 system could be due to the highly ordered nature and high surface area of CMK-3 that offers the short diffusion path to Zn^{2+} ions to reach the redox-active quinone centre. Impressively, the superior structural robustness of PT@CMK-3 cathode with low over-voltage facilitates the fast charging/discharging behaviour of the Zn/PT@CMK-3 battery. Surprisingly, the full cell could even be operated at an extremely high current density of 20 A g^{-1} (63 C) without undergoing structural collapse during repetitive GCD

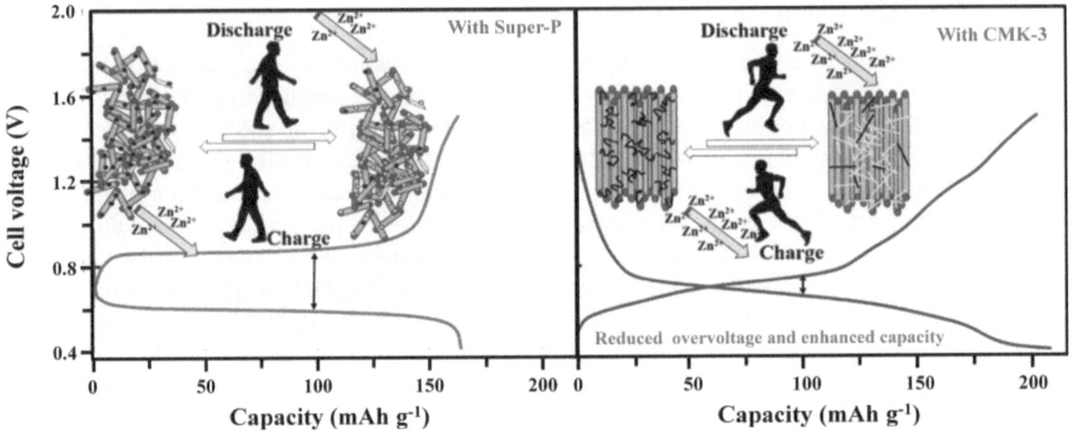

FIGURE 3.11 Comparative performance of aqueous ZIB with PT@Super-P (left) and PT@CMK-3 (right) cathodes. Adapted and reproduced from Ref. [45]. Copyright 2021 American Chemical Society.

cycles. Furthermore, the Zn/PT@CMK-3 system also demonstrated excellent cycling stability over the 5,000 GCD cycles with ≥95% capacity retention.

Despite the superior electrochemical performance, the serious dissolution of small organic quinones (which act as the cathode) during the discharge causes rapid capacity fading and eventually shortens the cycle life of ZIB in an aqueous medium. Strategies like the incorporation of redox-active species into the polymeric backbone, which is electron rich and able to conduct, are reported to combat the dissolution issue and improve the electronic conductivity of the electrode and cyclic stability of the battery. For example, Häupler et al. [46] developed a novel redox cathode for ZIBs based on 9,10-di (1,3-dithiol-2-ylidene)-9,10-dihydroanthracene (exTTF). The resulting Zn/exTTF-type battery demonstrated a high capacity of 109 mAh g^{-1}, which is close to the theoretical capacity (133 mAh g^{-1}) even at an extremely high rate (120C). Impressively, due to the superior structural stability of the cathode, the fabricated Zn-organic battery retained a good capacity of 81 mAh g^{-1} even after 10,000 GCD cycles. Due to superior benefits of polymeric cathodes such as ultra-high structural stability, high capacity, electrochemical reversibility, and good flexibility, many other organic/polymeric compounds such as poly(benzoquinonyl sulphide) [47], polyaniline [48], poly(3,4-ethylenedioxythiophene), pyrene-4,5,9,10-tetraone, and tetrachloro-1,4-benzoquinone (p-chloranil) [49] were explored for ZIB applications.

3.5 ELECTROLYTES FOR ZIBS

Electrolytes are one of the most influential components that play a crucial role in transporting the Zn^{2+} ions between Zn-anode and host cathodes during ZIB operation. The key parameters such as internal resistance, operating voltage, cycle life, rate performance, energy density, and CE of the ZIB are highly dependent on the electrolytes. Hence, designing the electrolytes that are most compatible with both anode and cathode chemistry, safe to handle, inexpensive, and eco-friendly are very significant to widen the scope of ZIB applications on a commercial scale.

Recently, many attempts were made to explore the various electrolytes (both aqueous and non-aqueous) to enhance the performance of ZIBs. As discussed earlier, to surmount the critical issues faced in the alkaline electrolytes such as poor CE of Zn-anode, the formation of inactive side products, rapid capacity decay, and poor rechargeability, the neutral or mild acidic electrolytes were preferred for ZIB applications. Due to their favourable interfacial chemistry and good compatibility with anode and cathode, the zinc salts based on sulphates and triflates gained good scope in ZIB applications. Despite their advantages, the narrow operating voltage window, irreversible side reactions on both anode and cathodes, cathodic dissolution, and rapid capacity fading greatly obstruct the wide utility of the aqueous mild acidic electrolytes for ZIB applications. As an alternative to aqueous mild acidic electrolytes, many organic electrolytes were developed for ZIB applications to enlarge the operating voltage window. Organic electrolytes not only offer a broad operating voltage window but also provide excellent compatibility with Zn-anodes, making them the most attractive choice for ZIBs. Han and co-workers studied the interfacial electrochemistry along with the transport and dynamical properties of Zn-anodes in various non-aqueous zinc electrolytes [50]. Among non-aqueous electrolytes used for this study, the three electrolytes (acetonitrile-$Zn(TFSI)_2$, acetonitrile-$Zn(CF_3SO_3)_2$, and propylene carbonate-$Zn(TFSI)_2$) demonstrated an excellent reversible Zn deposition/stripping behaviour (≥99% of CE) in the wide range of operating potential widow of ~3.4–3.8 V vs Zn/Zn^{2+}. Despite their (organic electrolytes) good compatibility with Zn-anodes, the toxicity, environment un-friendliness, high flammability, volatility, and poor transport kinetics of guest ions due to low ionic conductivity principally limit their applicability in ZIBs. Later, due to the striking features of ionic liquids (ILs) such as good thermal stability, non-flammability, wide operating potential/voltage window, and low volatile nature compared to organic electrolytes, ILs have been used as an alternative to organic electrolytes. In addition to that, electrolytes based on ILs are well compatible with Zn-anode and promote reversible zinc plating/stripping.

Ghazvini et al. [51] demonstrated the impact of the coordination around the Zn^{2+} on the Zn plating/stripping behaviour in 1-ethyl-3-methylimidazolium acetate ([EMIm]OAc) while varying the $Zn(OAc)_2$ concentration at elevated temperature (100°C). The lower coordination number of Zn^{2+} (1.8) in 4 M $Zn(OAc)_2$ dissolved in [EMIm]OAc facilitates highly reversible zinc deposition/stripping compared to 0.5 M $Zn(OAc)_2$ in [EMIm]OAc wherein Zn^{2+} coordination number is 4.79. Similarly, ZIB performance was also evaluated in many other ILs such as 1-butyl-1-methylpyrrolidiniumtrifluoromethylsulphonate, 1-ethyl-2,3-dimethylimidazoliumtrifluoromethylsulphonate, and 1-methylimidazoliumtrifluoromethylsulphonate. But their high cost, insufficient ionic conductivity, and high viscosity severely limit them from large-scale energy storage applications. The electrolyte leakage from the ZIB or any energy storage system ultimately reduces the shelf-life. To prevent the possible leakage of the liquid electrolyte and to improve the shelf-life of the ZIB, the gel/solid polymer electrolytes were explored. The attractive features of the solid polymer electrolytes, such as good mechanical strength, easy handling, non-flammability, low volatility, high physical flexibility, and less corrosive nature, enabled their practical use in the fabrication of flexible ZIB devices. The usage of gel/solid polymer electrolytes not only improves the performance of the battery but also prevents side reactions at the interface. There is much scope to study further to address the influence of various electrolytes on energy storage and the interfacial phenomenon in ZIBs. Hence, electrolyte development in the field of ZIBs has become a research hotspot in recent years.

3.6 CONCLUSIONS

In this chapter, we have discussed the fundamental strategies and chemistries behind the developments of anodes, cathodes, and electrolytes along with the energy storage mechanisms of low-cost and highly safe rechargeable ZIBs. The electroplating/stripping of Zn on the surface of the negative electrode is a highly complicated process. The unwanted side reactions such as HER, surface passivation, and dendritic growth on the surface of the Zn-anode dramatically lower the cycle life and CE of ZIB. The fundamental interfacial electrochemistry to inhibit the formation of side products and improve the rechargeability of Zn-anodes is systematically discussed in Section 3.2. The importance of host cathodes and their energy storage mechanisms were highlighted in Section 3.3. The various inorganic (such as oxides of Mn and oxides/sulphides of V, PBAs, Chevrel phases, and metal disulphides) and organic cathodes and the rationale behind their development are addressed in Section 3.3. The transport kinetics of Zn^{2+} ions and the electrochemically stable potential window of ZIB rely on electrolyte components. Hence, the electrolytes are considered to be one of the most influential components in ZIB application. The development of various electrolytes for ZIBs and the fundamental challenges associated with them are summarized in Section 3.3.5.

REFERENCES

1. Zhang Y, Wan F, Huang S, Wang S, Niu Z, Chen J (2020) A chemically self-charging aqueous zinc-ion battery. *Nat Commun* 11:1–10.
2. Kasiri G, Glenneberg J, Hashemi AB, Kun R, La Mantia F (2019) Mixed copper-zinc hexacyanoferrates as cathode materials for aqueous zinc-ion batteries. *Energy Storage Mater* 19:360–369.
3. Xu C, Li B, Du H, Kang F (2012) Energetic zinc ion chemistry: the rechargeable zinc ion battery. *Angew Chemie* 124:957–959.
4. Zhang T, Tang Y, Guo S, Cao X, Pan A, Fang G, Zhou J, Liang S (2020) Fundamentals and perspectives in developing zinc-ion battery electrolytes: a comprehensive review. *Energy Environ Sci* 13:4625–4665.
5. Cui J, Guo Z, Yi J, Liu X, Wu K, Liang P, Li Q, Liu Y, Wang Y, Xia Y (2020) Organic cathode materials for rechargeable zinc batteries: mechanisms, challenges and perspectives. *ChemSusChem* 13:2160–2185.
6. Ming J, Guo J, Xia C, Wang W, Alshareef HN (2019) Zinc-ion batteries: materials, mechanisms, and applications. *Mater Sci Eng R Reports* 135:58–84.
7. Zhang N, Chen X, Yu M, Niu Z, Cheng F, Chen J (2020) Materials chemistry for rechargeable zinc-ion batteries. *Chem Soc Rev* 49:4203–4219.
8. Shin J, Lee J, Park Y, Choi JW (2020) Aqueous zinc ion batteries: focus on zinc metal anodes. *Chem Sci* 11:2028–2044.
9. Khor A, Leung P, Mohamed MR, Flox C, Xu Q, An L, Wills RGA, Morante JR, Shah AA (2018) Review of zinc-based hybrid flow batteries: from fundamentals to applications. *Mater Today Energy* 8:80–108.
10. Yamamoto T, Shoji T (1986) Rechargeable Zn|$ZnSO_4$| MnO_2-type cells. *Inorganica Chim Acta* 117:L27–L28.
11. Cao Z, Zhuang P, Zhang X, Ye M, Shen J, Ajayan PM (2020) Strategies for dendrite-free anode in aqueous rechargeable zinc ion batteries. *Adv Energy Mater* 10:2001599.
12. Gui Y, Lei Y, Fan BA (2020) Investigation on electrode/electrolyte interface effect and the corresponding ZIBs performance among different aqueous mild acidic electrolyte. *Front Chem* 8:827.
13. Zhang N, Cheng F, Liu Y, Zhao Q, Lei K, Chen C, Liu X, Chen J (2016) Cation-deficient spinel $ZnMn_2O_4$ cathode in Zn $(CF_3SO_3)_2$ electrolyte for rechargeable aqueous Zn-ion battery. *J Am Chem Soc* 138:12894–12901.
14. Su L, Liu L, Liu B, Meng J, Yan X (2020) Revealing the impact of oxygen dissolved in electrolytes on aqueous zinc-ion batteries. *iscience* 23:100995.
15. Wang F, Borodin O, Gao T, Fan X, Sun W, Han F, Faraone A, Dura JA, Xu K, Wang C (2018) Highly reversible zinc metal anode for aqueous batteries. *Nat Mater* 17:543–549.
16. Xu W, Zhao K, Huo W, Wang Y, Yao G, Gu X, Cheng H, Mai L, Hu C, Wang X (2019) Diethyl ether as self-healing electrolyte additive enabled long-life rechargeable aqueous zinc ion batteries. *Nano Energy* 62:275–281.
17. Bayaguud A, Luo X, Fu Y, Zhu C (2020) Cationic surfactant-type electrolyte additive enables three-dimensional dendrite-free zinc anode for stable zinc-ion batteries. *ACS Energy Lett* 5:3012–3020.
18. Xu M, Ivey DG, Qu W, Xie Z (2015) Study of the mechanism for electrodeposition of dendrite-free zinc in an alkaline electrolyte modified with 1-ethyl-3-methylimidazolium dicyanamide. *J Power Sources* 274:1249–1253.
19. Naveed A, Yang H, Yang J, Nuli Y, Wang J (2019) Highly reversible and rechargeable safe Zn batteries based on a triethyl phosphate electrolyte. *Angew Chemie Int Ed* 58:2760–2764.
20. Cui M, Xiao Y, Kang L, Du W, Gao Y, Sun X, Zhou Y, Li X, Li H, Jiang F (2019) Quasi-isolated Au particles as heterogeneous seeds to guide uniform Zn deposition for aqueous zinc-ion batteries. *ACS Appl Energy Mater* 2:6490–6496.
21. Zheng J, Zhao Q, Tang T, Yin J, Quilty CD, Renderos GD, Liu X, Deng Y, Wang L, Bock DC (2019) Reversible epitaxial electrodeposition of metals in battery anodes. *Science* 366:645–648.
22. Ponnusamy R, Venkatesan R, Kandasamy M, Chakraborty B, Rout CS (2019) MnO_2 polymorph selection for non-enzymatic glucose detection: an integrated experimental and density functional theory investigation. *Appl Surf Sci* 487:1033–1042.
23. Lee B, Lee HR, Kim H, Chung KY, Cho BW, Oh SH (2015) Elucidating the intercalation mechanism of zinc ions into α-MnO_2 for rechargeable zinc batteries. *Chem Commun* 51:9265–9268.

24. Lee B, Yoon CS, Lee HR, Chung KY, Cho BW, Oh SH (2014) Electrochemically-induced reversible transition from the tunneled to layered polymorphs of manganese dioxide. *Sci Rep* 4:1–8.
25. Alfaruqi MH, Mathew V, Gim J, Kim S, Song J, Baboo JP, Choi SH, Kim J (2015) Electrochemically induced structural transformation in a γ-MnO_2 cathode of a high capacity zinc-ion battery system. *Chem Mater* 27:3609–3620.
26. Zhang N, Cheng F, Liu J, Wang L, Long X, Liu X, Li F, Chen J (2017) Rechargeable aqueous zinc-manganese dioxide batteries with high energy and power densities. *Nat Commun* 8:1–9.
27. Jiang B, Xu C, Wu C, Dong L, Li J, Kang F (2017) Manganese sesquioxide as cathode material for multivalent zinc ion battery with high capacity and long cycle life. *Electrochim Acta* 229:422–428.
28. Guo X, Fang G, Zhang W, Zhou J, Shan L, Wang L, Wang C, Lin T, Tang Y, Liang S (2018) Mechanistic insights of Zn^{2+} storage in sodium vanadates. *Adv Energy Mater* 8:1801819.
29. Pan H, Shao Y, Yan P, Cheng Y, Han KS, Nie Z, Wang C, Yang J, Li X, Bhattacharya P (2016) Reversible aqueous zinc/manganese oxide energy storage from conversion reactions. *Nat Energy* 1:1–7.
30. Sun W, Wang F, Hou S, Yang C, Fan X, Ma Z, Gao T, Han F, Hu R, Zhu M (2017) Zn/MnO_2 battery chemistry with H^+ and Zn^{2+} coinsertion. *J Am Chem Soc* 139:9775–9778.
31. Wan F, Zhang L, Dai X, Wang X, Niu Z, Chen J (2018) Aqueous rechargeable zinc/sodium vanadate batteries with enhanced performance from simultaneous insertion of dual carriers. *Nat Commun* 9:1–11.
32. Wang L, Zheng J (2020) Recent advances in cathode materials of rechargeable aqueous zinc-ion batteries. *Mater Today Adv* 7:100078.
33. Yuan C, Zhang Y, Pan Y, Liu X, Wang G, Cao D (2014) Investigation of the intercalation of polyvalent cations (Mg^{2+}, Zn^{2+}) into λ-MnO_2 for rechargeable aqueous battery. *Electrochim Acta* 116:404–412.
34. Senguttuvan P, Han S, Kim S, Lipson AL, Tepavcevic S, Fister TT, Bloom ID, Burrell AK, Johnson CS (2016) A high power rechargeable nonaqueous multivalent Zn/V_2O_5 battery. *Adv Energy Mater* 6:1600826.
35. Hu P, Yan M, Zhu T, Wang X, Wei X, Li J, Zhou L, Li Z, Chen L, Mai L (2017) Zn/V_2O_5 aqueous hybrid-ion battery with high voltage platform and long cycle life. *ACS Appl Mater Interfaces* 9:42717–42722.
36. Zhang N, Dong Y, Jia M, Bian X, Wang Y, Qiu M, Xu J, Liu Y, Jiao L, Cheng F (2018) Rechargeable aqueous Zn-V_2O_5 battery with high energy density and long cycle life. *ACS Energy Lett* 3:1366–1372.
37. Li W, Wang K, Cheng S, Jiang K (2018) A long-life aqueous Zn-ion battery based on Na_3V_2 $(PO_4)_2F_3$ cathode. *Energy Storage Mater* 15:14–21.
38. Trócoli R, La Mantia F (2015) An aqueous zinc-ion battery based on copper hexacyanoferrate. *ChemSusChem* 8:481–485.
39. Zhang L, Chen L, Zhou X, Liu Z (2015) Towards high-voltage aqueous metal-ion batteries beyond 1.5 V: the zinc/zinc hexacyanoferrate system. *Adv Energy Mater* 5:1400930.
40. Chae MS, Heo JW, Lim S-C, Hong S-T (2016) Electrochemical zinc-ion intercalation properties and crystal structures of $ZnMo_6S_8$ and $Zn_2Mo_6S_8$ chevrel phases in aqueous electrolytes. *Inorg Chem* 55:3294–3301.
41. Cheng Y, Luo L, Zhong L, Chen J, Li B, Wang W, Mao SX, Wang C, Sprenkle VL, Li G (2016) Highly reversible zinc-ion intercalation into chevrel phase Mo_6S_8 nanocubes and applications for advanced zinc-ion batteries. *ACS Appl Mater Interfaces* 8:13673–13677.
42. He P, Yan M, Zhang G, Sun R, Chen L, An Q, Mai L (2017) Layered VS_2 nanosheet-based aqueous Zn ion battery cathode. *Adv Energy Mater* 7:1601920.
43. Xu W, Sun C, Zhao K, Cheng X, Rawal S, Xu Y, Wang Y (2019) Defect engineering activating (Boosting) zinc storage capacity of MoS_2. *Energy Storage Mater* 16:527–534.
44. Zhao Q, Huang W, Luo Z, Liu L, Lu Y, Li Y, Li L, Hu J, Ma H, Chen J (2018) High-capacity aqueous zinc batteries using sustainable quinone electrodes. *Sci Adv* 4:eaao1761.
45. Mirle C, Medabalmi V, Ramanujam K (2021) Electrode and conductive additive compatibility yielding excellent rate capability and long cycle life for sustainable organic aqueous Zn-ion batteries. *ACS Appl Energy Mater* 4:1218–1227.
46. Häupler B, Rössel C, Schwenke AM, Winsberg J, Schmidt D, Wild A, Schubert US (2016) Aqueous zinc-organic polymer battery with a high rate performance and long lifetime. *NPG Asia Mater* 8:e283–e283.
47. Dawut G, Lu Y, Miao L, Chen J (2018) High-performance rechargeable aqueous Zn-ion batteries with a poly (benzoquinonyl sulfide) cathode. *Inorg Chem Front* 5:1391–1396.
48. Shi H, Ye Y, Liu K, Song Y, Sun X (2018) A long-cycle-life self-doped polyaniline cathode for rechargeable aqueous zinc batteries. *Angew Chemie Int Ed* 57:16359–16363.

49. Kundu D, Oberholzer P, Glaros C, Bouzid A, Tervoort E, Pasquarello A, Niederberger M (2018) Organic cathode for aqueous Zn-ion batteries: taming a unique phase evolution toward stable electrochemical cycling. *Chem Mater* 30:3874–3881.
50. Han S-D, Rajput NN, Qu X, Pan B, He M, Ferrandon MS, Liao C, Persson KA, Burrell AK (2016) Origin of electrochemical, structural, and transport properties in nonaqueous zinc electrolytes. *ACS Appl Mater Interfaces* 8:3021–3031.
51. Ghazvini MS, Pulletikurthi G, Lahiri A, Endres F (2016) Electrochemical and spectroscopic studies of zinc acetate in 1-ethyl-3-methylimidazolium acetate for zinc electrodeposition. *ChemElectroChem* 3:598–604.

4 Recent Advances and Trends in Al-Ion Batteries

Shakir Bin Mujib, Santanu Mukherjee, Zhongkan Ren, Davi Marcelo Soares, Carla Giselle Martins Real, Hudson Zanin, and Gurpreet Singh

4.1 INTRODUCTION

Developing an efficient energy storage system is a primary concern in the operation of modern energy systems. Ever-increasing energy demand and the growing environmental jeopardy brought about by diminishing fossil fuels (natural gas, oil, coal, etc.) are triggering the development of new types of renewable energy storage [1–3]. Electrochemical energy storage systems are cost-effective, reliable, and environmentally friendly and have a high energy density, which are critical in applications ranging from portable devices to electric vehicles and large grid storage [4,5]. Traditionally, lithium-ion batteries (LIBs) have been the most reliable and extensively utilized electrochemical batteries over the past decades due to their mature technology with $LiCoO_2$ and carbon as electrode materials [6,7]. Nonetheless, limitations regarding the stability and safety of LIBs, the high price and toxicity of cobalt oxide, and the shortage of lithium (Li) reserves have prompted the attention of scientists and researchers toward alternative battery systems [8,9]. Metal-ion batteries other than Li such as potassium (K), sodium (Na), magnesium (Mg), calcium (Ca), and aluminum (Al) have also received considerable attention as non-lithium energy storage systems [10–14].

Among these post-lithium technologies, aluminum-ion batteries (AIBs) are notably promising due to the low cost, practicality, and safety of the Al metal as shown in Figure 4.1. Al is one of the abundant metals in the earth's crust with a quantity of 82,000 ppm compared to 18 ppm for Li metal [15,16]. As a result, the price of Al (~1.40 USD kg^{-1}) is negligible in comparison with Li (~1,500 USD kg^{-1}) [17]. Additionally, Al is six times denser than Li and can contribute three electrons during the electrochemical process, yielding the highest theoretical volumetric capacity of 8,046 Ah L^{-1} and a high gravimetric capacity of 2,980 Ah kg^{-1} compared to other metal-ion batteries [16,18]. Al has a relatively higher redox potential (–1.676 V vs Normal Hydrogen Electrode (NHE)) than other metals [19]. Although Al has high oxidation potential, factors such as high volumetric and gravimetric capacities influence the performance of AIBs and contribute to high energy density. In addition, stability in the air, low flammability, high abundance, low cost, and relatively low atomic weight make aluminum one of the promising electrode materials for grid-scale applications.

For more than a century, scientists have been interested in Al as electrode material in batteries. The concept of Al as electrode material in a battery cell was first introduced in the 1850s by Hulot when he used Al as a cathode, Zn(Hg) as an anode, and H_2SO_4 as an electrolyte [18,20]. The first cell, where Al was used as an anode, was a Buff cell reported in 1857 [16,18,21]. In 1948, Heise et al. [23] reported amalgamated Al alloy with Zn as an anode in a chlorine-depolarized cell [22]. They used carbon as a cathode and reached a high open circuit voltage of 2.45 V. Sargent patented a Leclanche-type voltaic cell in 1951 with Al as an anode, carbon as a cathode, and aqueous $NaOH+ZnO_2$ solution as an electrolyte. Aluminum-air batteries have also been reported around the 1960s [24]. Despite being researched extensively, these Al batteries had not been commercialized due to the limitation of aqueous electrolytes. A passive oxide film forms on the surface of the Al metal during electrochemical reactions in an aqueous electrolyte, which limits the cell performance

DOI: 10.1201/9781003310174-4

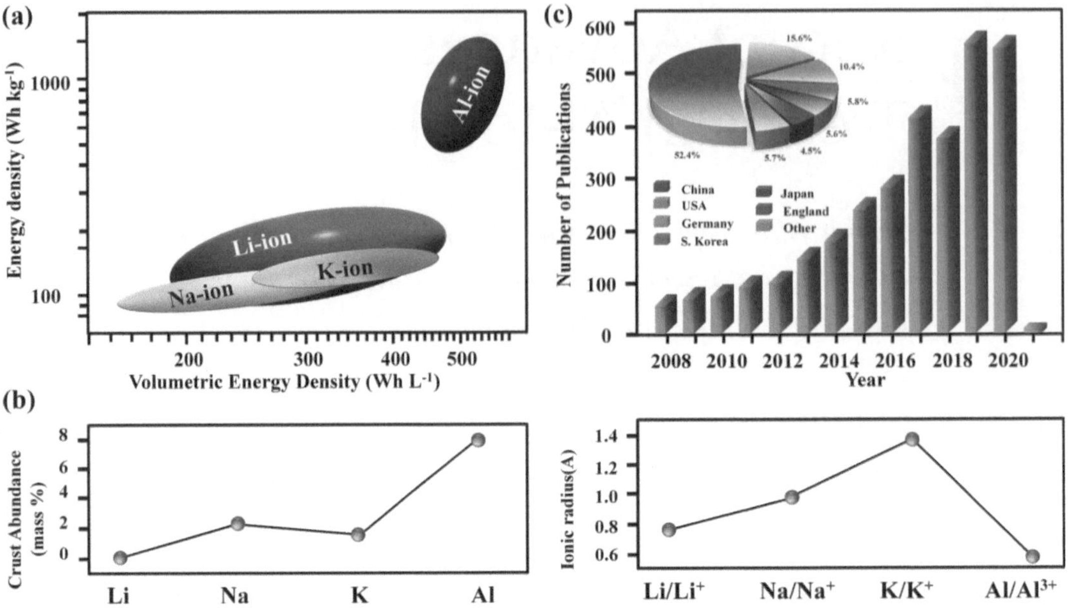

FIGURE 4.1 Opportunities of the aluminum-ion batteries (AIBs). (a) Comparisons of AIB with LIB, KIB, and SIB in terms of energy densities. (b) Abundance of the Al, Li, K, and Na metals in the earth's crust and comparison of their ionic radius. (c) Number of publications on AIBs according to Web of Science as of January 2021.

and voltage [25,26]. Also, hydrogen evolution due to water splitting and corrosion of the Al anode in the electrolyte causes the anodic efficiency to drop instantly [16,25]. In recent years, non-aqueous electrolytes have been developed to overcome these issues [27]. For example, non-aqueous electrolytes such as ionic liquids provide a wide electrochemical window, low vapor pressure, and electronic conductivity. The most frequently reported non-aqueous electrolytes in AIBs are $AlCl_3$ consisting of imidazolium-based ionic liquids [16,28–30]. Imidazolium plays a significant role in the separation of Al_2Cl_{4-} and Al_2Cl_{7-} ions, which are crucial for the deposition and stripping of Al anode [31].

Recent renewed interest in AIBs has highlighted the need for a comprehensive review of the strengths and weaknesses of the various research approaches in this field. This review provides an overview of recent AIB developments based on aqueous and non-aqueous electrolyte systems. We first summarize the recent developments of primary or aqueous AIBs and secondary AIBs. Then, a comprehensive summary of the aqueous and non-aqueous electrolytes with their correlation with Li^+ ion chemistry and charge storage mechanism is provided. Additionally, we have summarized the various cathode materials and their potential applications in AIBs. Finally, our perspective regarding the next-generation AIBs along with the major drawbacks and potential facing current AIB technologies are discussed.

4.2 AIB DEVELOPMENTS

In comparison to lithium, Al has several advantages, as well as disadvantages. Finding a suitable electrolyte that enables reversible plating/stripping of Al is difficult. Also, the high valence state of Al(3^+) contributes to slow ion intercalation kinetics. High reduction potential prevents Al from being used as the negative electrode, as a passivating oxide layer is formed in contact with the water-based electrolyte.

Nevertheless, the high abundance and low cost of Al, as well as its high volumetric capacity, make Al one of the promising negative electrodes for rechargeable batteries. In addition, the relatively mature smelting technology of Al with respect to other metals, easy-to-handle characteristics, and inherent safety compared to lithium make AIBs a reliable alternative for electrical energy storage systems.

4.2.1 Primary or Aqueous AIB

Batteries based on an aqueous electrolyte provide safe operation and limit the negative influence on the environment compared to toxic organic electrolyte systems. In addition, the higher ionic conductivity of the aqueous-based electrolytes than non-aqueous-based electrolytes enables increased power densities for aqueous batteries [26,32]. Another important aspect of the aqueous-based electrolyte is the low cost of the battery system. In an aqueous AIB, Al is an anode paired with a suitable cathode, and water is used as a solvent in the electrolyte, which facilitates necessary Al-ion insertions. Al has a redox potential of –1.676 V vs NHE, which is lower than the potential of electrochemical water splitting (1.23 V). As a result, an inherent hydrogen evolution reaction may occur even before Al stripping/plating, and the decomposition of aqueous electrolytes into H_2 and O_2 will limit the transportation of ions [33]. The small electrochemical potential window of the electrolyte restricts cell energy density and the range of electrode materials [31].

In addition, the formation of thin passivating films of alumina (Al_2O_3) hinders the solubility of Al and the transportation of Al^{3+} ions. To alleviate the formation of Al_2O_3, alkaline-based aqueous electrolytes are used in primary AIBs. The OH^- ions of the alkaline electrolyte can eliminate the passivation layer of Al_2O_3 from the Al anode [19]. Similar to a neutral system, the corrosion or self-discharge of the Al metal anode is inevitable in the alkaline system ($2Al+6H_2O \leftrightarrow 2Al(OH)_3+3H_2$). On the contrary, acidic electrolytes enhance H_2 evolution reactions and corrosion of the Al metal [34].

According to Yang et al. [19] primary AIBs can be divided into two groups based on either (a) combination reaction or (b) hybrid Al reaction. In the first group, a phase change occurs when Al participates in the cathodic reaction (e.g., Al–air and Al–H_2O_2), while Al does not participate in the cathodic reactions (e.g., Al–FeCN and Al–MnO_2) in the second group. In both cases, the reduction potential of Al/Al^{3+} in water-based electrolytes restrains aqueous AIBs from becoming rechargeable or secondary AIBs.

In 2014, Liu et al. [35] for the first time investigated the behavior of Al^{3+} ions in TiO_2 nanotube arrays in an aqueous AIB. The electrolyte was prepared by mixing 1 M $AlCl_3$ in various concentrations of NaCl solutions. Results showed that the presence of Cl^- ions significantly contributed to the successful insertion of Al^{3+} ions into TiO_2. Holland et al. [36] introduced an aqueous AIB in which TiO_2 was used as the anode, CuHCF was the cathode, and $AlCl_3$ solution was used as the electrolyte with KCl. The authors suggested that the storage capacity of TiO_2 was achieved due to the participation of Al^{3+} ions rather than H^+ ions. However, limited energy density, hydrogen evolution reactions, self-discharge of the electrodes, and a short shelf life limit the widespread use of aqueous AIBs. This chapter focuses primarily on the development of rechargeable or secondary AIBs. Further information on aqueous or primary AIBs is provided in Table 4.1 and other review articles [26,34].

4.2.2 Secondary AIBs

Secondary AIBs usually consist of Al metal as an anode, carbonaceous or metal compound as a cathode, and chloroaluminate-based ionic liquid as an electrolyte. Developing an appropriate electrolyte, which will be stable and support the reversible plating/stripping of Al, is essential for rechargeable AIBs. As mentioned, the electrolyte must be non-corrosive and highly resistant to oxidation.

TABLE 4.1
Electrochemical Performance of Different Cathode Materials for Primary and Secondary AIBs

Cathode Material	Electrolyte Chemistry	Voltage Range (V)	Performance[a]	Reference
3D graphitic foams	$AlCl_3$:[EMIm]Cl (1:3)	0.5–2.45	60/12/4000	[88]
Carbon nanoscrolls	$AlCl_3$:[EMIm]Cl (1.3:1)	0.4–2.3	101/50/55000	[89]
Carbon nanofibers	$AlCl_3$:[EMIm]Cl (1.3:1)	0.6–2.3	100/10/20000	[90]
Graphene film	$AlCl_3$:[EMIm]Cl (1.3:1)	0.5–2.5	120/400/250000	[47]
Defect-free graphene	$AlCl_3$:[EMIm]Cl (1.3:1)	0.7–2.51	100/5/25000	[78]
Mesoporous rGO	$AlCl_3$:[EMIm]Cl (1.3:1)	0.01–2.2	105/10/25000	[91]
Porous CuO microsphere	$AlCl_3$:[EMIm]Cl (1.3:1)	0.1–2.0	112/0.2/100	[92]
VO_2 nanorods	$AlCl_3$:[BMIm]Cl (1:1) and $C_{14}H_{14}OS$	0.01–0.9	116/0.05/100	[93]
V_2O_2	$AlCl_3$:[EMIm]Cl (1.1:1)	0.02–2.5	273/0.125/20	[25]
MoO_2	$AlCl_3$:[EMIm]Cl (1.3:1)	0.3–2.5	25/0.1/100	[94]
Co_3O_4	$AlCl_3$:[EMIm]Cl (1.3:1)	0.3–2.2	121.1/0.2/100	[95]
WO_{3-x}	$AlCl_3$:[EMIm]Cl (1.3:1)	0.01–2.2	64.7/0.1/100	[96]
SnS film	$AlCl_3$:[EMIm]Cl (1.3:1)	0.5–2.42	227/0.2/1000	[97]
SnS@G	$AlCl_3$:[EMIm]Cl (1.3:1)	0.4–2.1	70/0.2/100	[66]
TiS_2	$AlCl_3$:[BMIm]Cl (1.5:1)	0.1–1.35	70/0.005/50	[98]
NiS nanobelts	$AlCl_3$:[EMIm]Cl (1:1.3)	0.1–2	100/0.2/100	[99]
Mo_6S_8	$AlCl_3$:[BMIm]Cl (1.5:1)	0.3–1.0	70/0.012/50	[100]
Co_9S_8@CNT	$AlCl_3$:[EMIm]Cl (1.3:1)	0.45–1.62	87/1/6000	[85]
V_2CT_x	$AlCl_3$:[EMIm]Cl (1.3:1)	0.1–1.8	90/0.1/100	[101]
D-$Ti_3C_2T_x$@S@TiO_2	$AlCl_3$:[EMIm]Cl (1.3:1)	0.45–2.2	151.3/0.1/120	[102]
Mn_3O_4	5 M $Al(OTF)_3$	0.5–1.8	247/0.03/60	[103]
MnO_2	2 M $Al(OTF)_3$+ 0.5 M $MnSO_4$	0.5–1.7	320/0.1/65	[104]
MnO_2	2 M $Al(OTF)_3$	0.5–2.0	460/0.1/80	[105]
V_2O_5	1 M $AlCl_3$	−0.4 to 0.15	75/0.06/13	[106]
$FeVO_4$	1 M $AlCl_3$+NH_4OH	−0.7 to 0.9	50/0.06/20	[83]
TiO_2@G	0.5 M $AlCl_3$	−0.5 to −1.075	90/5.25/20	[107]
TiO_2@CNT	1 M $AlCl_3$	−0.5 to −1.2	170/0.335/100	[108]
MoO_3	1 M $AlCl_3$	−0.7 to 0.6	168/2.5/350	[54]
Bi_2O_3@G	1 M $AlCl_3$	0–1.2	103/1.5/50	[109]
Exfoliated graphite	1 M $AlCl_3$	0.0–2.0	88/0.5/50	[110]
$K_2CoFe(CN)_6$	1 M $Al(NO_3)_3$	0.0–1.2	50/0.1/1600	[111]
CuHCF	0.5 M $Al_2(SO_4)_3$	0.2–1.1	22.5/0.4/1000	[112]

[a] The performance data are in the following form: specific capacity (mA h g^{-1})/current density (A g^{-1})/number of cycles.

Because Al^{3+} has an ionic radius of 0.0535 nm and three positive charges, it has a surface charge density six times higher than a Li^+ ion [37]. Consequently, higher coulombic interaction between the anions and cations in the electrolyte occurs, meaning the solubility of Al salts in widely used organic solvents is significantly lower and the electronic conductivity of the electrolyte is reduced. Chloroaluminate-based ionic electrolytes advantageously facilitate the Al metal plating/stripping process with a stable window of electrochemical potential [27]. It is worth noting that, in ionic liquids, the ions exist in the forms of $AlCl_4^-$ and $AlCl_7^-$, not Al^{3+} [38]. The formation of $AlCl_7^-$ is crucial to the rechargeability of secondary AIBs.

The most successful ionic electrolyte for AIBs is $AlCl_3$ mixed with 1-ethyl-3-methylimidazolium chloride ([EMIm]Cl) or 1-butyl-3-methylimidazolium chloride ([BMIm]Cl) [14,39,40]. The mixture of $AlCl_3$ with imidazolium-based cations decreases the melting point of the electrolyte due to the reaction between $AlCl_3$ and Cl^-, which results in room-temperature ionic liquid (RTIL). In addition to efficient Al stripping/plating, a wide potential window of the imidazolium-based electrolyte made them the most predominantly used RTIL in AIBs since 1988 when Gifford et al. [41] introduced $AlCl_3$ into 1,2-dimethyl-3-propylimidazolium chloride to produce a room temperature–molten salt electrolyte for rechargeable AIBs. They used graphite as the cathode, resulting in a capacity of ~35 mAh g^{-1}. The low storage capacity and cycleability of the graphite were attributed to material degradation.

In 2015, Sun et al. [42] proposed carbon paper as a cathode and pure Al foil as an anode for a high-voltage rechargeable AIB. The electrolyte was a mixture of $AlCl_3$ and [(EMIm)Cl] with a molar ratio of 1.3:1. They reported a high-voltage plateau of 1.8 V with a reversible capacity of 69.92 mAh g^{-1} over 100 cycles. That same year, Lin et al. [28] introduced a three-dimensional graphitic foam as the cathode for AIB in an $AlCl_3$/[EMIm]Cl ionic electrolyte. They demonstrated an ultrafast AIB at a current density of 4,000 mA g^{-1} with cycling stability over 7,500 cycles. The superior cycleability of the graphite electrode was attributed to the strong covalent bonding between the graphene sheets, which restricted the volume expansion of the electrode during ion intercalation.

Finding a stable positive electrode is also essential for rechargeable AIBs. In addition to graphite or carbon-based materials, transition metal compounds have gained research attention due to their accommodation of Al^{3+} during intercalation. Wang et al. [43] proposed a binder-free V_2O_5 cathode on a Ni foam current collector for rechargeable AIBs. The V_2O_5 cathode was tested against an Al anode in $AlCl_3$/[BMIM]Cl ionic liquids as an electrolyte. The as-prepared Ni–V_2O_5 electrode showed a voltage plateau at 0.6 V, which was attributed to the properties of the binder-free cathode and the enhanced electrolyte diffusion.

4.3 ELECTROCHEMISTRY

4.3.1 Electrolyte

Low-cost AIBs also provide important advantages in terms of power and energy density. As such, these criteria open an important niche for their applications in large-scale energy storage, specifically grid-scale and distributed energy storage systems [44]. One of the most important components necessary for the proper functioning of an AIB is its electrolyte system [44]. The electrolyte helps determine the functioning voltage window and AIB system stability. Therefore, the proper choice of a suitable and stable electrolyte is imperative.

Ionic liquids (ILs), an essential class of electrolytes for non-aqueous AIB systems that consist almost entirely of ions, demonstrate advantageous low vapor pressure, low flammability, and safety over a large temperature range [45]. ILs are also specifically suited for AIBs due to the relatively low standard electrode potential of Al metal (–1.662 V vs SHE) [27]. $AlCl_3$ in imidazole chloride–based salts with alkyl side chains such as 1-butyl-3-methylimidazolium chlorides are commonly used ILs [27,46].

Previous research showed that a RTIL comprised $AlCl_3/Et_3NHCl$ in an AIB system, which had a working voltage window of approximately 2.6 V [47]. The authors demonstrated a specific capacity of 120 mAh g^{-1} at a current density as high as 400 A g^{-1} and a capacity retention of 91.7% even after 250,000 cycles of operation [47]. Similarly, Xia et al. [48] studied the performance of $AlCl_3/Et_3NHCl$ with added organic solvents in a secondary AIB by cycling the cell within a potential window of 0.6–2.4 V to observe a well-defined discharge plateau at 1.8 V. At a current density of 15 mA g^{-1}, the cell provided a first-cycle discharge capacity of 113.64 mAh g^{-1} with added solvents; the discharge capacity was 85.23 mAh g^{-1} without the solvents. The superior performance and stability of the cell were attributed to the IL, in general, and the performance enhancement was attributed to the addition of the organic solvent, in particular [48].

Other groups have investigated various aspects of IL application in AIBs. For example, Kamath et al. [49] studied IL performance from an atomistic standpoint in which kinetic and thermodynamic parameters were used to gain insight into IL behavior. The authors demonstrated that the superior performance of the ILs was due to considerably less free energy of solvation, lower coordination, and the formation of stable complexes that enhanced the transport of Al ions [49]. Zhu et al. [50] studied the effect of cations in ILs in AIB systems. They added different quantities of Al ions to the IL

1-methyl-1-propylpyrrolidinum chloride (Py13Cl) in molar ratios of 1.4–1.7 to produce a composite IL of $AlCl_3$/Py13Cl. A discharging energy and efficiency of 141 mWh g^{-1} and 89%, respectively, were obtained. The authors attributed this superior performance to the dimeric and monomeric anions in the IL and subsequent improved viscosity. Wang et al. [51] studied the performance of a non-corrosive IL AIB system, where the water-stable IL was a mixture of 1-butyl-3-methylmidazolium trifluoromethanesulfonate (BMIM|OTF) with a corresponding Al salt $Al(OTF)_3$ with an optimized concentration of 0.5 mol L^{-1}. The cell delivered a first-cycle capacity of 87 mAh g^{-1}, which dropped to 40 mAh g^{-1} with further cycling. During cycling, discharge voltage plateaus were 0.25 V, whereas the charge plateaus were observed at 1.2 and 2.7 V, respectively, indicating battery polarization. The IL electrolyte demonstrated a high oxidation voltage of 3.25 V vs Al^{3+}/Al and elevated electronic conductivity [51]. A summary of the results of Wang et al. [51] is provided in Figure 4.2.

ILs satisfy important performance criteria such as enhanced conductivity, stability, large voltage windows for cell cycling, and operation at room temperatures. Nevertheless, degradation of the

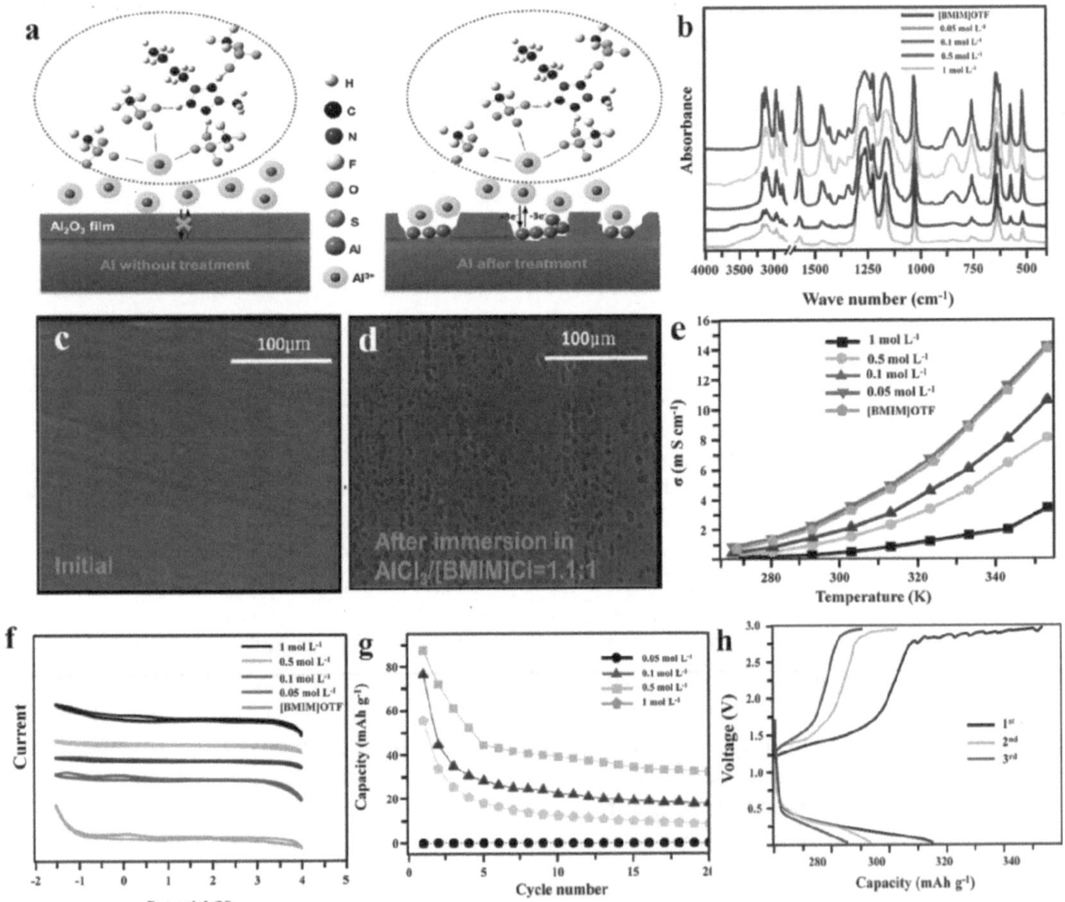

FIGURE 4.2 Structural characterization and electrochemical performance of IL in an AIB system. (a) Schematic representation of the interaction of the Al(OTF)3|[BMIM]OTF IL with Al cathode and its consequent passivation. (b) Full-range Fourier Transform Infrared Spectroscopy (FTIR) spectra of the IL (of varying concentrations) demonstrating the characteristic peaks. (c and d) Scanning electron micrographs of pristine Al anode after being subjected to IL for 24 h. (e) Plot of various concentrations of IL demonstrating conductivity variation with temperature. (f) Cyclic voltammetry (CV)curves demonstrating the potential operational window of the IL. (g) Cycling performance of the IL of varying concentrations in the AIB system. (h) Galvanostatic charge–discharge curves for the IL of 1 mol l^{-1} concentration exhibiting the characteristic charge–discharge plateaus. Reproduced with permission from Ref. [51]. Copyright 2016, American Chemical Society.

cathode material in the presence of the chloro-aluminate is a critically important concern, and, due to their corrosive nature, the chloroaluminates cause corrosion/degradation of the current collector species, especially stainless steels, which are rendered practically non-operational [27]. Varying the $AlCl_3$:IL ratio has been shown to provide some advantages, specifically optimal performance with a 1.5:1 ratio [27]. The application of novel organic electrolytes can help overcome this corrosion, but resolutions such as using Mo as the current collector are expensive (USD45 kg^{-1}) [27,52]. Therefore, a significant amount of optimization is necessary before AIB systems can be legitimate participants in large-scale and grid-scale energy storage. Detailed information on the application and performance of ILs in AIBs is presented in other research [40,53].

Aqueous electrolytes are emerging as a valuable remedy for the drawbacks of ILs. For example, aqueous electrolytes such as $AlCl_3$ have demonstrated an initial capacity of 680 mAh g^{-1} at a current density of 2.5 A g^{-1}, although a capacity fading of 25.4% was observed after 350 cycles of operation [54]. $Al(NO_3)_3$ fared even worse, with a capacity retention of only ~18.9% after 350 cycles [54,55]. However, the addition of diglyme additives has demonstrated a positive effect, especially when graphene has been added to the electrodes, and a specific capacity of 600 mAh g^{-1} has been obtained [54,55].

4.3.2 Correlation with Li-Ion Chemistry

LIBs have become the predominant electrochemical energy storage source, with bourgeoning applications in the global consumer electronics industry specifically. Nevertheless, pressing issues such as toxicity and flammability of the electrolyte system have necessitated the search for alternate electrochemistries, leading to increased utilization of AIBs. The monovalent Li^+ and the trivalent Al^{3+} ionic systems are the primary distinctions between LIB and AIB systems, respectively. The trivalent nature of the Al^{3+} ion results in more charge transported per cycle, leading to a higher theoretical capacity when connected to a suitable negative electrode. The theoretical gravimetric capacity of an Al system (2,980 mAh g^{-1}) is therefore comparable to a Li system (3,860 mAh g^{-1}), and the superior stability of Al increases its suitability for electrochemical systems [56].

One important distinction of an AIB system from the Li system is the –1.662 vs SHE (–1.676 V vs NHE) standard reduction potential of Al metal, which indicates that hydrogen evolution begins before Al plating starts to occur, resulting in the dissociation of aqueous electrolyte and making the AIB systems nearly unsuitable for practical use. Because ILs are primarily used as the electrolyte, the role of the anionic and/or cationic species is to facilitate the electrodeposition of Al or to allow alloying at the anode [14,57]. In addition, the electrolyte must be practically inert toward the negative electrode, as Al presents a stable passivation film. Complexing agents added to the electrolyte in AIB systems play an important role since the volume of the electrolyte determines the capacity that can be obtained. Unlike in LIB systems, large volumes of electrolytes tend to provide higher capacity in AIB systems [14,56,57].

4.3.3 Solid Electrolyte Interface Formation on Al Metal

One of the primary advantages of AIBs is their ability to withstand high-rate capabilities, especially for supercapacitors, and provide elevated coulombic efficiencies [58]. Consequently, AIBs are particularly suited for grid-scale and large-scale storage applications [44,58]. When graphite-based cathodes are used, the initial cycles of AIB cycling demonstrate a considerable drop in efficiency, resulting in considerably weakened performance [59]. Therefore, a fundamental understanding of the electrode/electrolyte interface in the AIB system must be improved.

In general, as passivation occurs, a robust oxide film develops on the Al surface, which prevents the smooth continuation of the electrochemical reaction since the metallic surface is hardly electrochemically available [58]. The first-principle results have indicated that decomposition of the ethylene carbonate on the Al surface proceeds very slowly, whereas data about a similar decomposition of chloroaluminates are unclear [58]. Therefore, no clearly established rigorous experimental observation is available for the existence of a classical solid electrolyte interface (SEI) layer on the Al metal surface on the anodic side. In this absence, however, the Al_2O_3 layer behaves as a functional SEI layer [60]. Although the Al_2O_3 layer on the Al surface becomes defective and consists of microscopic pores that permit limited ion transport and electron tunneling, research has shown that the removal of this Al_2O_3 layer enhances electrochemical performance and improves coulombic efficiency [60,61]. One of the ways this Al_2O_3 layer can be removed is by immersing or pretreating the Al metal in chloroaluminate [62]. Partial removal of this Al_2O_3 layer is essential, but complete removal does not result in a noticeable improvement in performance because the Al_2O_3 layer beneficially prevents dendrite growth [14,63].

The type of positive electrode material used in cathodes strongly influences the SEI layer [64]. If graphite is used as the cathode and $AlCl_4^-$ is the electrolyte, then no SEI forms, but few works in the literature highlight SEI formation and cathodes for AIB systems [64,65]. For example, Hu et al. [66] reported the formation of an SEI layer when a graphene oxide-supported SnS_2 was used as the cathode and a chloroaluminate species $Al_2Cl_7^-$ was the electrolyte. Likewise, Tu et al. [67] reported the formation of an SEI layer in an AIB system when $(Ni_{11}HPO_{-3})_8(OH)_6$/rGO was used as the cathode and $AlCl_4^-$ was the electrolyte at a potential of ~1.9 V.

4.3.4 Charge Storage Mechanisms

An increased understanding of charge storage mechanisms is essential to understanding AIB chemistry and developing efficient electrode–electrolyte systems that can reversibly employ aluminum ions. Recently, $AlCl_3$/[EMIM]Cl electrolyte has been widely employed in AIB studies [68]. This electrolyte is formed by mixing aluminum chloride ($AlCl_3$) and 1-ethyl-3-methylimidazolium chloride ([EMIM] Cl), an organic liquid species [61]. When unmixed, these salts (i.e., $AlCl_3$ and Cl^-) are solid at room temperature, but when mixed the Lewis acid–base interaction results in an RTIL reported as $AlCl_3$/ [EMIM]Cl [61]. The presented results are based on AIB studied in $AlCl_3$/[EMIM]Cl.

A starting point for the development of cathode materials for AIB lies in the progress made on LIB and sodium ion battery (SIB). The literature has shown that electrode materials such as metal sulfides that present conversion-type reactions for LIB and SIB also present these reactions for AIB systems [14]. However, the intercalation mechanism in AIB is more complex than in the mentioned monovalent ion batteries [68,69].

Lin et al. [28] showed that graphite provides stability in long-term cycling at high rates (4,000 mA g^{-1}). Presenting a different working mechanism, chroloroaluminate anions ($AlCl_4^-$) have been reversibly inserted and extracted from graphite layers during charge and discharge, respectively [28]. Ex-situ X-ray diffraction results confirmed the intercalation of $AlCl_4^-$ in pyrolytic graphite cathode. The characteristic peak at 2θ ~25.55° – plane (002) and *d* spacing of 3.35 Å – vanished upon charging. At the fully charged state, doubled peaks in the diffractogram at ~28.25° and ~23.66° appeared, corresponding to *d* spacings of approximately 3.15 Å and 3.77 Å, respectively. The doubled peak spacing corresponded to a layer separation of approximately 5.7 Å, indicating that a guest species was intercalated in graphite layers at a charged state [70]. A similar diffractogram was obtained for graphitic foam cathode [28]. During the discharge cycle, $AlCl_4^-$ was deintercalated from graphite electrodes and reacted with Al anode, generating $Al_2Cl_7^-$; concomitantly electrons flowed through the external circuit to the cathode. The generalized reaction equations for anode and cathode are presented in Equations 4.1 and 4.2, respectively.

$$4\ Al_2Cl_7^- + 3e^- \rightarrow Al + 7\ AlCl_4^- \tag{4.1}$$

$$C_n[AlCl_4] + e^- \leftrightarrow C_n + AlCl_4^- \tag{4.2}$$

To further elucidate the role played by $AlCl_4^-$ during intercalation in AIB, Bhauriyal et al. [71] investigated which geometry of $AlCl_4^-$ (tetrahedral or planar) yields the highest stability during intercalation in graphite-based electrodes. The authors used simulation techniques to determine that the intercalation of tetrahedral $AlCl_4^-$ allows a graphite structure to retain its stacking sequence. Furthermore, upon intercalation of $AlCl_4^-$, the gallery height was in the range of approximately 8.5 Å, in accordance with previous results [72]. According to Lin et al. [28], the large interlayer spacing has a fast charging process because of a diffusion barrier of approximately 0.01 eV [71].

V_2O_5, another example of intercalation-based material, delivered a specific charge capacity of 273 mAh g^{-1} after 20 cycles [25]. Reed and Menke initially proposed that V_2O_5 was electrochemically inactive toward Al^{3+} [73], followed by their conclusion that the observed charge storage was a result of reactions between Al^{3+} and the stainless steel current collector surface [73]; consequently, V_2O_5 played no role toward capacity. A few years later Gu et al. [74] reported that V_2O_5 can reversibly intercalate Al^{3+}; Al^{3+} storage in V_2O_5 is not a typical intercalation process because it also involves a phase-transition reaction. Based on the absence of Cl in elemental mapping analysis, the authors confirmed that Al^{3+} was the intercalant in V_2O_5, not the anion species as in graphite. From ex-situ characterizations, an amorphous layer (thickness of 10 nm) was identified after first discharge; yet at a fully charged state, a phase-transition interface ranged from 20 to 30 nm. The amorphous layer was still present after 10 cycles, which indicated that this amorphization is an irreversible process [74]. The overall proposed charge storage of Al^{3+} in V_2O_5 is firstly given by an intercalation and phase transition (Equation 4.3) with subsequent phase-transition reactions (Equation 4.4).

$$V_2O_5+4\times Al_2Cl_7^-+3\times e^- \rightarrow Al_xV_2O_{5\,(amorphous)}+7\times AlCl_4^- \quad (4.3)$$

$$4\ Al_zV_2O_5+3\ (x-z)\ e^-+4\ (x-z)\ Al_2Cl_7^- \leftrightarrow Al_xV_2O_{5\,(amorphous)}+7\ (x-z)\ AlCl_4^-\ (z\leq x) \quad (4.4)$$

To explore conversion-type reactions, previous research has investigated metal sulfides because of their high specific capacities for LIB and SIB. In the first study reporting FeS_2 as cathode material for AIB, Koura proposed a conversion reaction – which leads to the formation of a stable Al_2S_3 compound – at a high temperature (240°C) [75]. Mori et al. [76] later studied the charge/discharge reaction of FeS_2, as shown by Equation 4.5, to evaluate the charge storage mechanism of Al^{3+} in FeS_2 at a lower temperature (55°C).

$$FeS_2+2/3\ Al^{3+}+2\ e^- \leftrightarrow FeS+1/3\ Al_2S_3 \quad (4.5)$$

where Equation 4.4. corroborates the findings of Koura regarding the formation of Al_2S_3 [75]. A schematic illustration of the reversible reactions is presented in Figure 4.3.

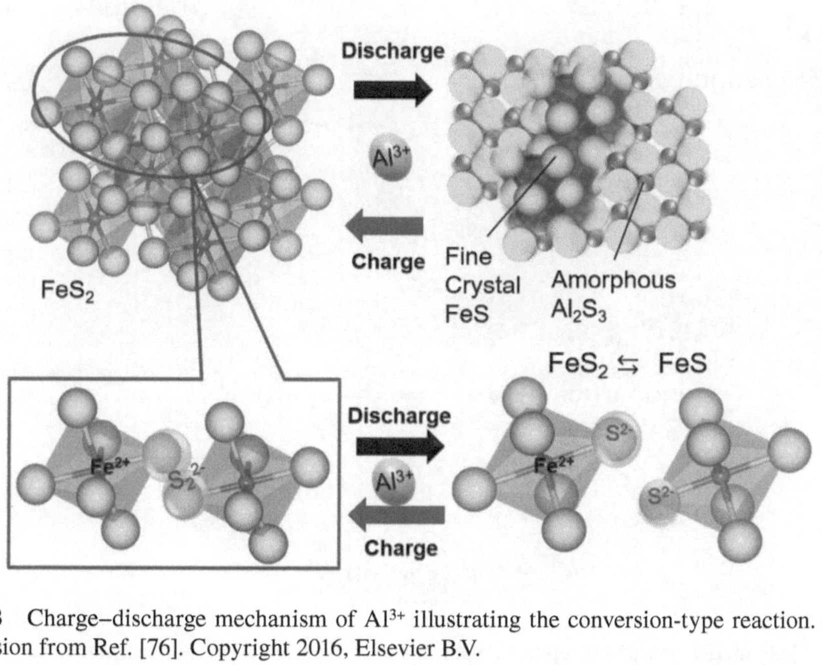

FIGURE 4.3 Charge–discharge mechanism of Al^{3+} illustrating the conversion-type reaction. Reproduced with permission from Ref. [76]. Copyright 2016, Elsevier B.V.

Although Mori et al. [76] highlighted the low cycling stability and dissolution of sulfides in FeS_2 AIB, a challenge that is also present in lithium–sulfur (Li–S) batteries, they also emphasized the safety, low cost, and potential of this sulfide-based material.

In summary, the recent interest in AIB has highlighted several challenges associated with this energy storage system that still exist. A better understanding of the complex intercalation mechanisms in AIB is essential. Another research trend has been to identify novel high-oxidation number transition metals that present conversion-type reactions in AIB [14]. Similar strategies applied in Li–S systems to mitigate polysulfide dissolution could also be employed in metal-sulfide electrodes for AIB.

4.4 CATHODE MATERIALS

Various cathode materials have been introduced to AIBs, especially secondary AIBs, based on the intercalation and reaction mechanisms. AIB performance is based on the electrochemical performances of the cathode materials, such as reversible capacity, rate capability, and long-term cycleability, and the limitations of these cathodes. Until recently, three types of cathode materials have shown satisfactory performance in AIBs: carbon-based materials, metal-based materials (e.g., transition metal oxide/sulfide), and composite materials, as shown in Figure 4.4. Challenges remain with finding a suitable cathode for AIBs, due to the low capacity of carbon-based materials, poor rate capability, and cycling performance of metal chalcogenides and oxides.

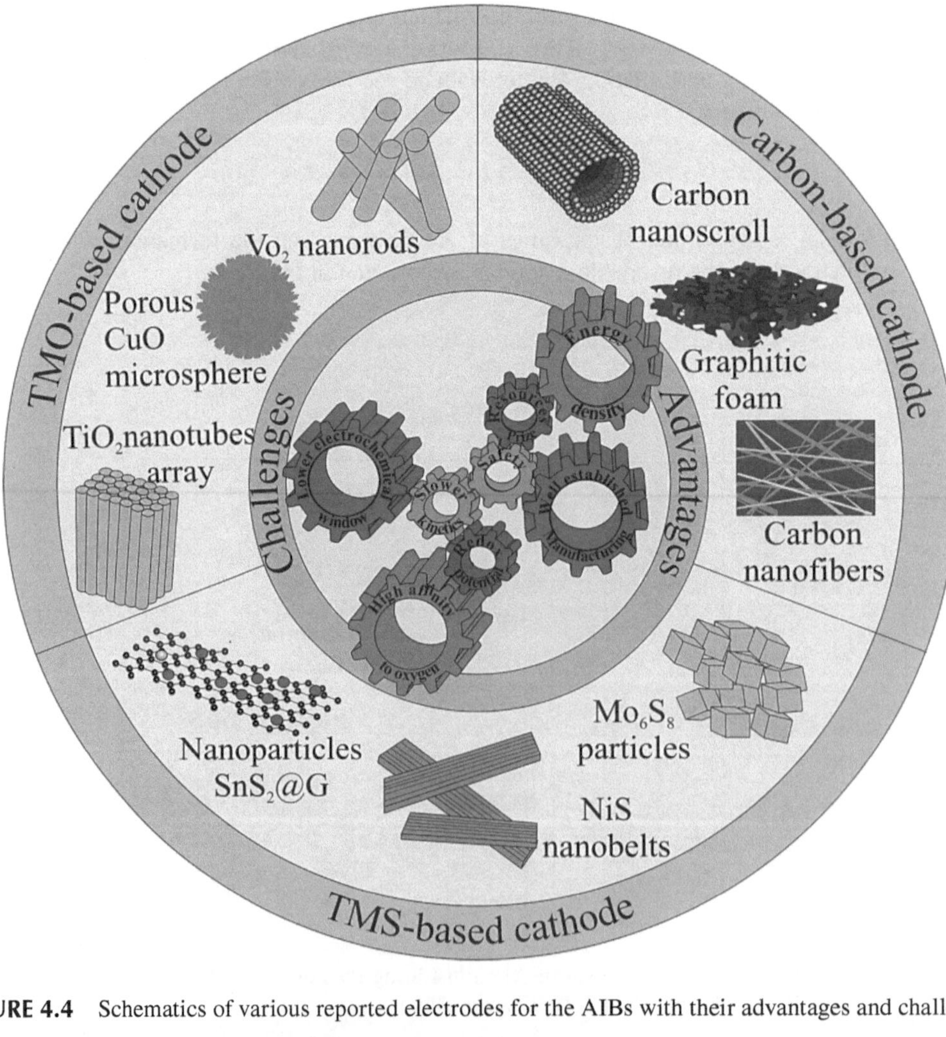

FIGURE 4.4 Schematics of various reported electrodes for the AIBs with their advantages and challenges.

4.4.1 Carbonaceous Materials

Developing effective and low-cost AIBs could theoretically benefit from their aluminum-based electrode and electrolyte materials for their natural abundance, inherent safety, and volumetric capacity (theoretical 8,040 mAh cm^{-3} or 2,980 mAh g^{-1}) over other types of commercialized batteries (lead-acid or LIBs) [77,78]. However, electrode material-related issues, such as disintegration, low discharge voltage (<1.2 V), capacitive behavior (lack of discharge plateaus), short cycle life (<100 cycles), and capacity decay, have suppressed the performance of rechargeable Al batteries over 30 years [28,41,78]. The current innovation of cathode materials for AIBs has been directed to carbonaceous [79] (or carbon-based) materials and metal-based materials. Carbonaceous cathodes have already exhibited several benefits from the open structures and high conductivity, such as outstanding operating voltage, rate capability, and cycling life. As shown in Equations 4.6 (carbonaceous cathode) and 4.7 (aluminum anode), because the Al storage in carbonaceous materials is $AlCl_4^-$ instead of trivalent aluminum (Al^{3+}), the storage of intercalating $AlCl_4^-$ anions has limited storage capabilities in these carbon-based materials [71,80].

$$C_n[AlCl_4] + e^- \rightleftharpoons C_n + AlCl_4^- \tag{4.6}$$

$$Al + 7AlCl_4^- \rightleftharpoons 4Al_2Cl_7^- \tag{4.7}$$

Bhauriyal et al. [71] used first-principle studies (Vienna Ab Initio with projected augmented wave method) to investigate the $AlCl_4$ intercalation mechanism and theoretical limits into bulk graphite cathode materials. The maximum $AlCl_4$ intercalation capacity was found to be ~70 mAh g^{-1} with an average voltage of 2.01–2.3 V, which closely matched the experimental results from other research groups. The following research works have shown the potential of other layered materials, specifically hexagonal BC_3 as the cathode of AIBs, with a specific capacity of ~74 mAh g^{-1} and an average voltage of about 2.4 V [81]. Jiao et al. [77] investigated AIB performance with natural graphite as the cathode material and molten $AlCl_3$/urea (optimized ratio to 1.5) as an electrolyte. The assembled batteries delivered a specific capacity of ~90 mAh g^{-1} with discharge plateaus of ~1.9 and 1.5 V and >99% coulombic efficiency at 100 mA g^{-1}. The battery was operated at 120°C to ensure the conductivity of the electrolyte, which suggested the high temperature (>100°C) capability of the AIB.

Lin et al. [28] revealed the superior rate capabilities of graphite foam cathodes in non-flammable liquid electrolytes of $AlCl_3$/1-ethyl-3-methylimidazolium chloride. The Al/graphite cell could continuously store an output of ~60 mAh g^{-1} at a high charge rate of 4 A g^{-1} without capacity decay for 7,500 cycles. In addition, the discharge voltage could reach ~2 V with a coulombic efficiency of ~98% under the fast charge conditions. Recently, Dong et al. [82] synthesized AIB using commercial expanded graphite as the cathode material to enhance the capacity and stability (Figure 4.5). The reported specific capacity could reach as high as ~80 mAh g^{-1} at 5 A g^{-1} with 30,000 cycling life. They investigated the mechanism of the excessive capacity beyond the theoretical capacity (~70 mAh g^{-1}) reported above and proposed the deposition behavior of $AlCl_4^-$ at the surface of the graphite cathode after interlayers are fully intercalated.

The performance of carbonaceous cathodes could be improved by synthesizing low-defected or defect-free carbons [78]. Chen et al. synthesized complete sp^2 few-layer graphene aerogels (GA) as cathode materials for AIBs via ultra-high temperature annealing of reduced graphene oxide aerogels [78]. Surface disordered defects (vacancies, polygons) were completely removed, and oxygen concentration was low at <2%. The refined GA cathodes have shown outstanding performance, for instance, 100 mAh g^{-1} at 5 A g^{-1} with an average discharge potential at 1.95 V and capacity retention of 97% after 25,000 cycles. The improvements resulted from the "defect-free" structure, which provided faster anion intercalations and enhanced conductivity.

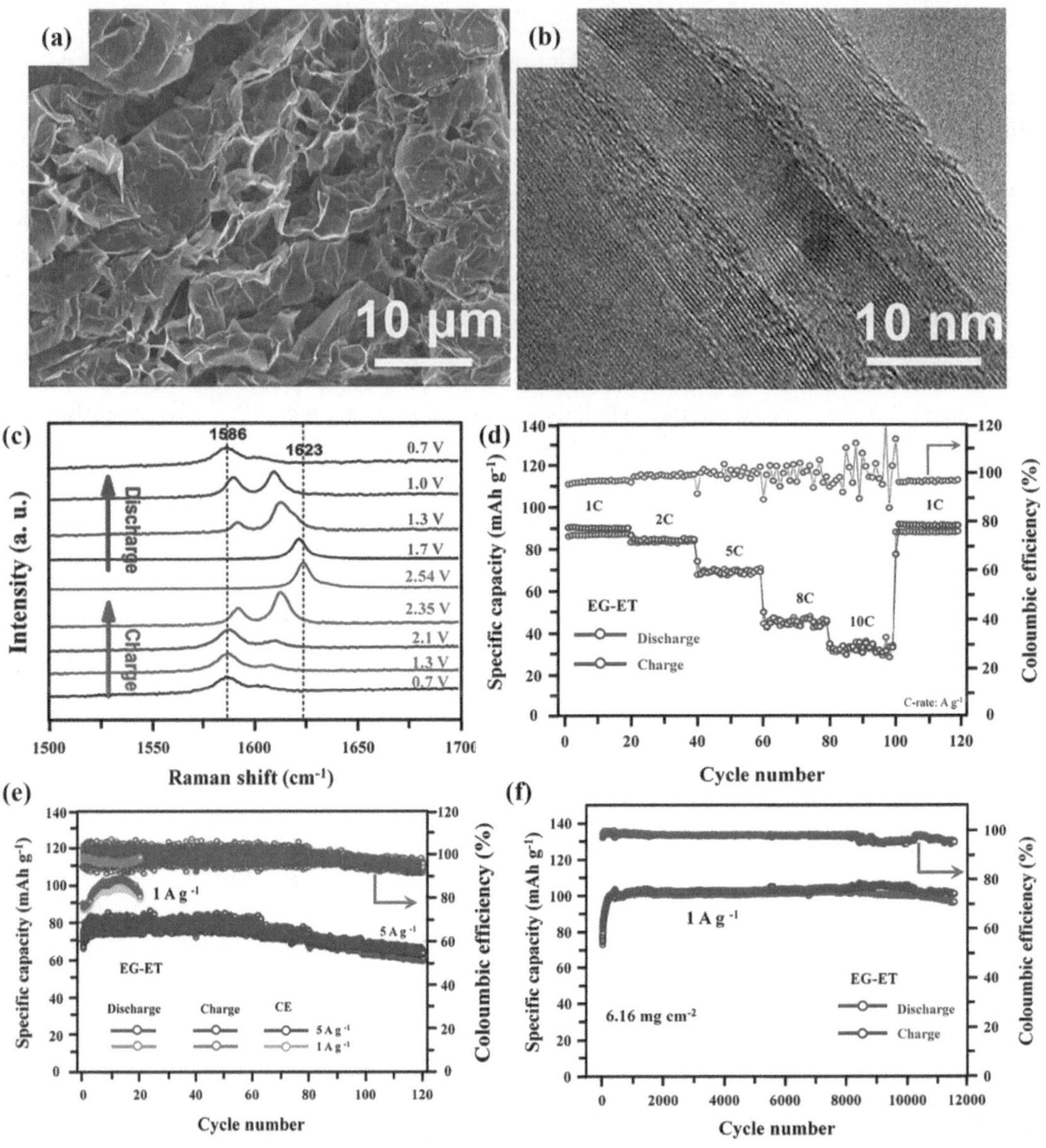

FIGURE 4.5 Electrochemical performance of a carbonaceous material in AIB. (a) Scanning electron microscopic image and (b) transmission electron microscopic image of the expanded graphite. (c) Raman spectra of expanded graphite electrode in $AlCl_3$/triethylamine hydrochloride electrolyte during charge and discharge. (d) Plot of rate capability of the AIB. (e) Plot of cycling performance of AIB. (f) Cycling performance of AIB under high mass loading. Reproduced with permission from Ref. [82]. Copyright 2019, Elsevier B.V.

4.4.2 Metal-based Materials

Metal-based materials provide high charge capacity due to their valence transition to high valence states, meaning a high Al^{3+} diffusion barrier and low conductivity decrease the cycling life and rate capability [79]. For example, Kumar et al. [83] demonstrated a high initial capacity of 350 mAh g^{-1} using $FeVO_4$ as a cathode material for aqueous AIBs (Figure 4.6). $FeVO_4$ could intake Al^{3+} with a two-step reaction and produce spinel $Al_xV_yO_4$ and amorphous Fe-O-Al; during this process, vanadium irreversibly dissolved into the electrolyte as V^{5+} and specific capacity rapidly

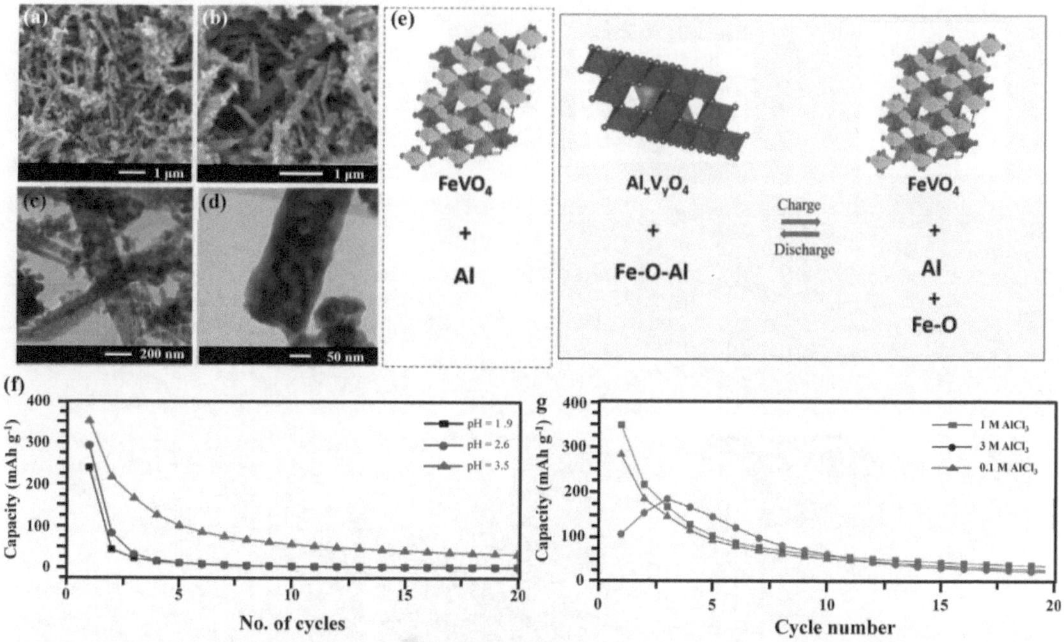

FIGURE 4.6 Electrochemical performance of a metal-oxide material in AIB. (a and b) SEM images and (c and d) TEM images of $FeVO_4$ nanorods. (e) Schematics of phase changes and chemical reactions during charge–discharge of AIBs with $FeVO_4$ cathodes. Electrochemical performance of $FeVO_4$ AIBs at (f) different pH values and (g) different $AlCl_3$ concentrations. Reproduced with permission from Ref. [83]. Copyright 2019, Elsevier B.V.

decayed to nearly ~50 mAh g^{-1}. The specific capacity and rate capability of a metal-based cathode also relate closely to the storing mechanisms. When studying δ-MnO_2 nanofiber cathodes, Almodovar et al. [84] found stable storage of ~37 mAh g^{-1} at 100 mAh g^{-1} after 100 cycles due to the Al storage mechanism as interlaminar $AlCl_4^-$, which is similar to the storage mechanism in carbonaceous materials.

Some 2D transition metal dichalcogenides have layered structures with weak interlayer van der Waals interactions similar to graphite [80]. Li et al. [80] investigated microsphere molybdenum disulfide (MoS_2) materials as cathodes for AIBs. They found that direct Al^{3+} storage could appear at the surface and bulk of the MoS_2 layers, leading to phase transformation that allows a higher initial specific capacity of ~250 mAh g^{-1} at 20mA g^{-1} compared to carbon-based cathodes (Figure 4.7). The capacity decayed rapidly to ~65 mAh g^{-1} after the charge density increased to 40mA g^{-1}.

4.4.3 Composites

Composite cathodes, which contain the benefits of both carbonaceous and metal-based materials, were proposed to break through the capacity cap of carbonaceous cathodes while maintaining outstanding rate capability and cycle stability. Metal-based particles may store Al in the form of Al^{3+}, and the porous structure of carbon materials allows rapid ion transportation. Hu et al. [85] reported a Co_9S_8/CNT-carbon nanofiber composite cathode for AIBs that showed a superior discharge capacity of 315 mAh g^{-1} at 100mA g^{-1} and a rate capability of ~150 mAh g^{-1} at 1 A g^{-1}. In addition, 87 mAh g^{-1} was retained after 6,000 cycles at the high charge speed.

Similarly, Wang et al. [86] proposed a composite of metal chalcogenide (CuS) and carbonaceous material as cathodes for AIBs (Figure 4.8). The research found that the cathode material could initially store a capacity of ~250 mAh g^{-1}; after a few cycles, it was able to reversibly store

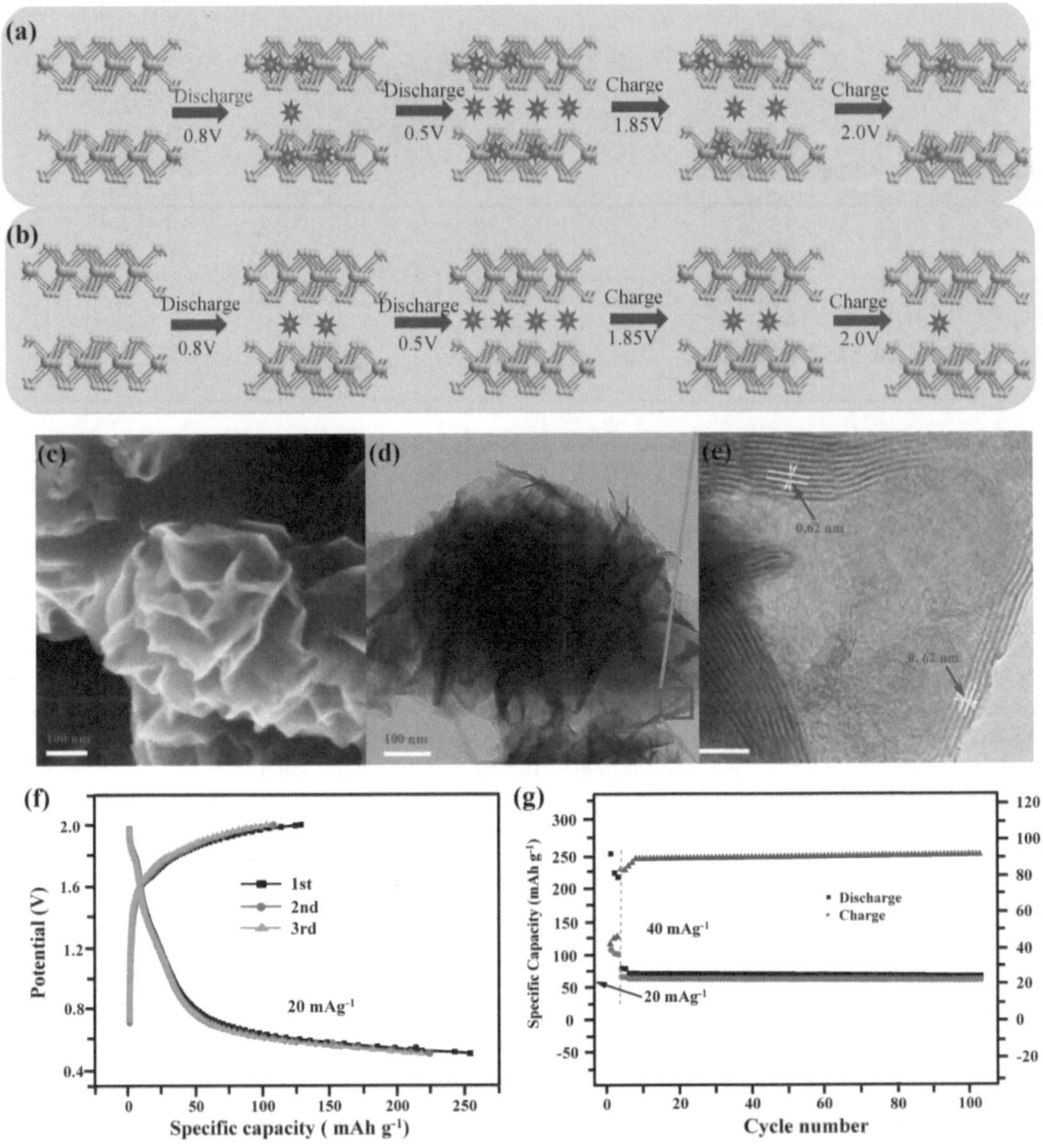

FIGURE 4.7 Electrochemical performance of MoS_2 in AIB. (a) Schematics of charge and discharge mechanisms of Al in MoS_2. (b–d) high-resolution transmission electron microscopy (HRTEM) images of spherical MoS_2 cathode material. (e) Initial three charge–discharge curves of the MoS_2 cathode. (f) Cyclability test of MoS_2 electrode at 40 mA g^{-1} up to 100 cycles. Reproduced with permission from Ref. [80]. Copyright 2018, American Chemical Society.

Al^{3+} with a specific capacity of ~90 mAh g^{-1} at 20 mA g^{-1}. The coulombic efficiency could maintain nearly 100% until 100 cycles.

A recent innovation by Wang et al. [79] reported a composite anode of carbonaceous material (single-wall carbon nanotube, or SWCNT) and organic polymer (polyaniline, or PANI) for AIB that showed excellent rate capability. The PANI nanorods were hyper self-protonated, which improved the $AlCl_2^+$ intercalation/deintercalation process and demonstrated a capacity of 200 mAh g^{-1} at 1 A g^{-1} and only 0.003% capacity decay after 8,000 cycles. The electrode was also flexible and self-standing due to the reinforced mechanical performance of the PANI/SWCNT composite.

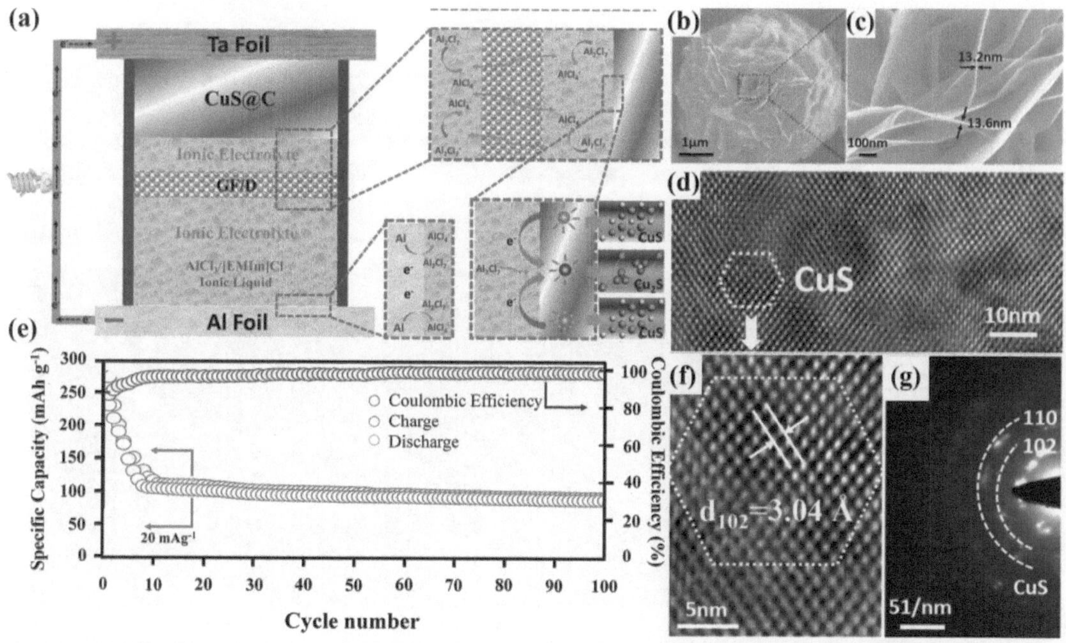

FIGURE 4.8 Electrochemical performance of CuS/C composite material in AIB. (a) Schematic of AIB discharging process with CuS/C composite cathode. (b and c) SEM images of CuS sphere at different magnifications. (d, f, and g) TEM and HRTEM images of CuS with corresponding selected area electron diffraction (SAED) pattern. (e) Electrochemical performance of AIB with CuS/C cathodes at 20 mA g^{-1}. Reproduced with permission from Ref. [86]. Copyright 2017, American Chemical Society.

More recently, Liao et al. [87] proposed a carbon-based composite cathode for AIBs where whisker-like PANI structures were grown in between the ordered mesoporous carbon rods (Figure 4.9). The battery performance investigations showed enhanced performance, ~140 mAh g^{-1} at 100 mA g^{-1} and ~120 mAh g^{-1} at 2 A g^{-1} with a capacity loss of 12% after 5,000 cycles, compared to neat carbonaceous anodes. The charging mechanism was revealed as the doping/de-doping of $AlCl_4^-$ into PANI structures.

The performances of various structured cathode materials for AIBs are presented in Table 4.1.

4.5 SUMMARY AND OUTLOOK

Considering the drawbacks and limitations of erstwhile rechargeable battery systems, such as the LIB and SIB systems, AIBs provide important advantages. In a post-lithium world, AIB demands increased research focus, especially due to its vast available resources, appreciable theoretical capacities, well-established Al metal processing and recycling ecosystem, and the relative safety of AIB systems.

The challenges of AIB include the weak mobility of the Al^{3+} ion in most electrolytes, which reduces the overall performance of the system. Although judicious use of novel ILs has demonstrated a solution to this challenge, further optimization is necessary to improve results. Another concern is the formation of a protective coating that forms on the Al metal surface during the initial cycling stage and the potential negative influence of this SEI layer on electron kinetics and overall ionic diffusivities. Also, when considering the metallic Al anode, non-corrosive organic electrolytes may be an interesting option, but such data are extremely limited.

Apart from using ILs, novel and creative manufacturing techniques for synthesizing electrodes could be an option for this case, especially thin film electrodes, in which considerable expertise can

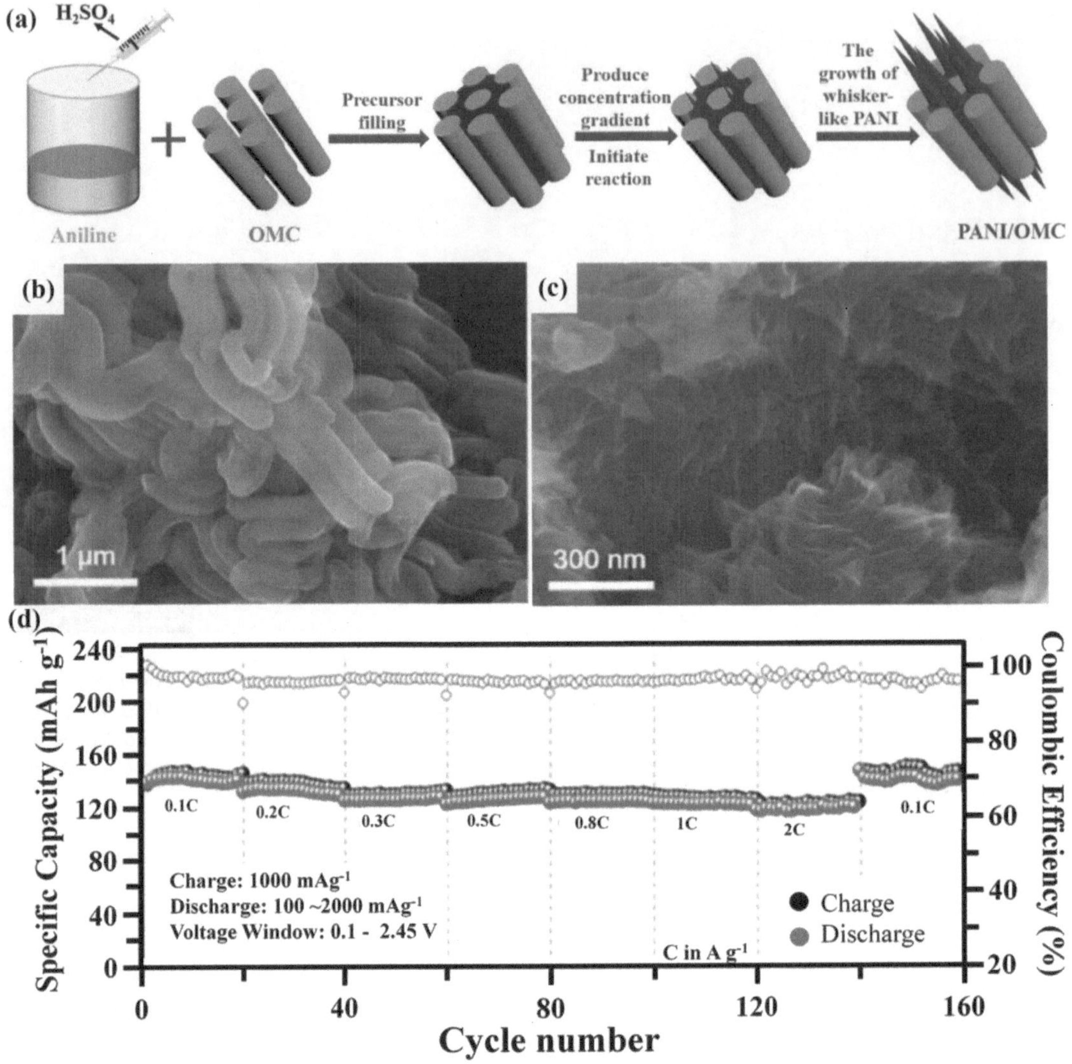

FIGURE 4.9 Electrochemical performance of a polymer-based composite material in AIB. (a) Schematic of the synthesis process of PANI/ordered mesoporous carbon cathode material. SEM images of (b) ordered mesoporous carbon and (c) PANI@ordered mesoporous carbon. (d) The rate capability of the cathode material. Reproduced with permission from Ref. [87]. Copyright 2020, Elsevier B.V.

be transferred from the semiconductor industry. The intercalation of Al^{3+} ions in cathode systems also needs further analysis, especially tailoring the crystallography and microstructure of such systems to enhance specific capacities as well as cycling stability and longer lifetimes.

Overall, although AIB systems are in their infancy, they provide important and interesting aspects for beyond-LIB systems, and they merit additional scrutiny. Improved and optimized AIB systems will develop as more research efforts materialize.

CONFLICTS OF INTEREST

The authors declare no conflict of interest.

ACKNOWLEDGMENTS

Financial support from the National Science Foundation (CMMI NSF CAREER Award) Grant No. 1454151 is gratefully acknowledged. Authors are also grateful to the financial support from the Brazilian funding agencies CNPq (PQ-2 grant 310544/2019-0), CAPES (374731/2019-00 & 572651/2020-00), and FAPESP (2014/02163-7 and 2017/11958-1) and to Shell and the strategic importance of the support given by ANP (Brazil's National Oil, Natural Gas, and Biofuels Agency) through the R&D levy regulation.

REFERENCES

1. Armand M, Tarascon JM (2008) Building better batteries. *Nature* 451 (7179):652–657. https://doi.org/10.1038/451652a
2. Grey CP, Tarascon JM (2017) Sustainability and in situ monitoring in battery development. *Nature Materials* 16 (1):45–56. https://doi.org/10.1038/nmat4777
3. Liang Y, Zhao C-Z, Yuan H, Chen Y, Zhang W, Huang J-Q, Yu D, Liu Y, Titirici M-M, Chueh Y-L, Yu H, Zhang Q (2019) A review of rechargeable batteries for portable electronic devices. *InfoMat* 1 (1):6–32. https://doi.org/10.1002/inf2.12000
4. Etacheri V, Marom R, Elazari R, Salitra G, Aurbach D (2011) Challenges in the development of advanced Li-ion batteries: a review. *Energy & Environmental Science* 4 (9):3243–3262. https://doi.org/10.1039/C1EE01598B
5. Dunn B, Kamath H, Tarascon J-M (2011) Electrical energy storage for the grid: a battery of choices. *Science* 334 (6058):928–935. https://doi.org/10.1126/science.1212741
6. Zeng X, Li M, Abd El-Hady D, Alshitari W, Al-Bogami AS, Lu J, Amine K (2019) Commercialization of lithium battery technologies for electric vehicles. *Advanced Energy Materials* 9 (27):1900161. https://doi.org/10.1002/aenm.201900161
7. Wu F, Maier J, Yu Y (2020) Guidelines and trends for next-generation rechargeable lithium and lithium-ion batteries. *Chemical Society Reviews* 49 (5):1569–1614. https://doi.org/10.1039/C7CS00863E
8. Wu X, Song K, Zhang X, Hu N, Li L, Li W, Zhang L, Zhang H (2019) Safety issues in lithium ion batteries: materials and cell design. *Frontiers in Energy Research* 7 (65):83–99. https://doi.org/10.3389/fenrg.2019.00065
9. Liu B, Jia Y, Yuan C, Wang L, Gao X, Yin S, Xu J (2020) Safety issues and mechanisms of lithium-ion battery cell upon mechanical abusive loading: a review. *Energy Storage Materials* 24:85–112. https://doi.org/10.1016/j.ensm.2019.06.036
10. Chayambuka K, Mulder G, Danilov DL, Notten PHL (2018) Sodium-ion battery materials and electrochemical properties reviewed. *Advanced Energy Materials* 8 (16):1800079. https://doi.org/10.1002/aenm.201800079
11. Zhang W, Liu Y, Guo Z (2019) Approaching high-performance potassium-ion batteries via advanced design strategies and engineering. *Science Advances* 5 (5):eaav7412. https://doi.org/10.1126/sciadv.aav7412
12. Kuang C, Zeng W, Li Y (2019) A review of electrode for rechargeable magnesium ion batteries. *Journal of Nanoscience and Nanotechnology* 19 (1):12–25. https://doi.org/10.1166/jnn.2019.16435
13. Gummow RJ, Vamvounis G, Kannan MB, He Y (2018) Calcium-ion batteries: current state-of-the-art and future perspectives. *Advanced Materials* 30 (39):1801702. https://doi.org/10.1002/adma.201801702
14. Leisegang T, Meutzner F, Zschornak M, Münchgesang W, Schmid R, Nestler T, Eremin RA, Kabanov AA, Blatov VA, Meyer DC (2019) The aluminum-ion battery: a sustainable and seminal concept? *Frontiers in Chemistry* 7:268. https://doi.org/10.3389/fchem.2019.00268
15. Zhang K, Han X, Hu Z, Zhang X, Tao Z, Chen J (2015) Nanostructured Mn-based oxides for electrochemical energy storage and conversion. *Chemical Society Reviews* 44 (3):699–728. https://doi.org/10.1039/C4CS00218K
16. Ambroz F, Macdonald TJ, Nann T (2017) Trends in aluminium-based intercalation batteries. *Advanced Energy Materials* 7 (15):1602093. https://doi.org/10.1002/aenm.201602093
17. Hu Y, Sun D, Luo B, Wang L (2019) Recent progress and future trends of aluminum batteries. *Energy Technology* 7 (1):86–106. https://doi.org/10.1002/ente.201800550

18. Li C, Zhang X, He W (2018) Design and modification of cathode materials for high energy density aluminum-ion batteries: a review. *Journal of Materials Science: Materials in Electronics* 29 (16):14353–14370. https://doi.org/10.1007/s10854-018-9478-1
19. Yang H, Li H, Li J, Sun Z, He K, Cheng H-M, Li F (2019) The rechargeable aluminum battery: opportunities and challenges. *Angewandte Chemie – International Edition* 58 (35):11978–11996. https://doi.org/10.1002/anie.201814031
20. Hulot M (1855) Comptes rendus hebdomadaires des séances de l'academie des sciences. *Comptes Rendus* 40:148
21. Tommasi D (1889) Traité des piles électriques. *Georges Carré* 131:1848–1907
22. Heise GW, Schumacher EA, Cahoon NC (1948) A heavy duty chlorine-depolarized cell. *Journal of the Electrochemical Society* 94 (3):99. https://doi.org/10.1149/1.2773830
23. Sargent DE (1951) Voltaic cell. Google Patents. US2554447A, United States.
24. Zaromb S (1962) The use and behavior of aluminum anodes in alkaline primary batteries. *Journal of the Electrochemical Society* 109 (12):1125. https://doi.org/10.1149/1.2425257
25. Jayaprakash N, Das SK, Archer LA (2011) The rechargeable aluminum-ion battery. *Chemical Communications* 47 (47):12610–12612. https://doi.org/10.1039/C1CC15779E
26. Verma V, Kumar S, Manalastas Jr. W, Satish R, Srinivasan M (2019) Progress in rechargeable aqueous zinc- and aluminum-ion battery electrodes: challenges and outlook. *Advanced Sustainable Systems* 3 (1):1800111. https://doi.org/10.1002/adsu.201800111
27. Zhang Y, Liu S, Ji Y, Ma J, Yu H (2018) Emerging nonaqueous aluminum-ion batteries: challenges, status, and perspectives. *Advanced Materials* 30 (38):1706310. https://doi.org/10.1002/adma.201706310
28. Lin M-C, Gong M, Lu B, Wu Y, Wang D-Y, Guan M, Angell M, Chen C, Yang J, Hwang B-J, Dai H (2015) An ultrafast rechargeable aluminium-ion battery. *Nature* 520 (7547):324–328. https://doi.org/10.1038/nature14340
29. Huang X, Liu Y, Liu C, Zhang J, Noonan O, Yu C (2018) Rechargeable aluminum-selenium batteries with high capacity. *Chemical Science* 9 (23):5178–5182. https://doi.org/10.1039/C8SC01054D
30. Jiang J, Li H, Fu T, Hwang B-J, Li X, Zhao J (2018) One-dimensional $Cu_{2-x}Se$ nanorods as the cathode material for high-performance aluminum-ion battery. *ACS Applied Materials & Interfaces* 10 (21):17942–17949. https://doi.org/10.1021/acsami.8b03259
31. Elia GA, Kravchyk KV, Kovalenko MV, Chacón J, Holland A, Wills RGA (2021) An overview and prospective on Al and Al-ion battery technologies. *Journal of Power Sources* 481:228870. https://doi.org/10.1016/j.jpowsour.2020.228870
32. Zhang H, Liu X, Li H, Hasa I, Passerini S (2021) Challenges and strategies for high-energy aqueous electrolyte rechargeable batteries. *Angewandte Chemie – International Edition* 60 (2):598–616. https://doi.org/10.1002/anie.202004433
33. Yuan D, Zhao J, Manalastas W, Kumar S, Srinivasan M (2020) Emerging rechargeable aqueous aluminum ion battery: status, challenges, and outlooks. *Nano Materials Science* 2 (3):248–263. https://doi.org/10.1016/j.nanoms.2019.11.001
34. Li Q, Bjerrum NJ (2002) Aluminum as anode for energy storage and conversion: a review. *Journal of Power Sources* 110 (1):1–10. https://doi.org/10.1016/S0378-7753(01)01014-X
35. Liu Y, Sang S, Wu Q, Lu Z, Liu K, Liu H (2014) The electrochemical behavior of Cl^- assisted Al^{3+} insertion into titanium dioxide nanotube arrays in aqueous solution for aluminum ion batteries. *Electrochimica Acta* 143:340–346. https://doi.org/10.1016/j.electacta.2014.08.016
36. Holland A, McKerracher RD, Cruden A, Wills RGA (2018) An aluminium battery operating with an aqueous electrolyte. *Journal of Applied Electrochemistry* 48 (3):243–250. https://doi.org/10.1007/s10800-018-1154-x
37. Shannon RD (1976) Revised effective ionic radii and systematic studies of interatomic distances in halides and chalcogenides. *Acta Crystallographica Section A: Crystal Physics, Diffraction, Theoretical and General Crystallography* 32 (5):751–767. https://doi.org/10.1107/S0567739476001551
38. Pan C, Shin M, Liu D, Kottwitz M, Zhang R, Nuzzo RG, Gewirth AA (2020) Energy storage mechanisms in high-capacity graphitic C_3N_4 cathodes for Al-ion batteries. *The Journal of Physical Chemistry C* 124 (19):10288–10297. https://doi.org/10.1021/acs.jpcc.0c00259
39. Pickup PG, Osteryoung RA (1985) Charging and discharging rate studies of polypyrrole films in $AlCl_3$: 1-methyl-(3-ethyl)-imidazolium chloride molten salts and in CH_3CN. *Journal of Electroanalytical Chemistry and Interfacial Electrochemistry* 195 (2):271–288. https://doi.org/10.1016/0022-0728(85)80048-6

40. Wang H, Gu S, Bai Y, Chen S, Zhu N, Wu C, Wu F (2015) Anion-effects on electrochemical properties of ionic liquid electrolytes for rechargeable aluminum batteries. *Journal of Materials Chemistry A* 3 (45):22677–22686. https://doi.org/10.1039/C5TA06187C
41. Gifford PR, Palmisano JB (1988) An aluminum/chlorine rechargeable cell employing a room temperature molten salt electrolyte. *Journal of the Electrochemical Society* 135 (3):650–654. https://doi.org/10.1149/1.2095685
42. Sun H, Wang W, Yu Z, Yuan Y, Wang S, Jiao S (2015) A new aluminium-ion battery with high voltage, high safety and low cost. *Chemical Communications* 51 (59):11892–11895. https://doi.org/10.1039/C5CC00542F
43. Wang H, Bai Y, Chen S, Luo X, Wu C, Wu F, Lu J, Amine K (2015) Binder-free V_2O_5 cathode for greener rechargeable aluminum battery. *ACS Applied Materials & Interfaces* 7 (1):80–84. https://doi.org/10.1021/am508001h
44. Yu Z, Jiao S, Tu J, Song W-L, Lei H, Jiao H, Chen H, Fang D (2019) Gel electrolytes with a wide potential window for high-rate Al-ion batteries. *Journal of Materials Chemistry A* 7 (35):20348–20356. https://doi.org/10.1039/C9TA06815E
45. Matsumoto K, Hwang J, Kaushik S, Chen C-Y, Hagiwara R (2019) Advances in sodium secondary batteries utilizing ionic liquid electrolytes. *Energy & Environmental Science* 12 (11):3247–3287. https://doi.org/10.1039/C9EE02041A
46. Kharlamov SM, Bobylev AV, Galkin PS, Dobroselsky KG, Vlasenko MG, Markovich DM (2015) The influence of aluminum anode alloys and alkaline electrolyte composition on performance of aluminum-air cells for portable applications. *Journal of Engineering Thermophysics* 24 (4):313–321. https://doi.org/10.1134/S1810232815040025
47. Chen H, Xu H, Wang S, Huang T, Xi J, Cai S, Guo F, Xu Z, Gao W, Gao C (2017) Ultrafast all-climate aluminum-graphene battery with quarter-million cycle life. *Science Advances* 3 (12):eaao7233. https://doi.org/10.1126/sciadv.aao7233
48. Xia S, Zhang X-M, Huang K, Chen Y-L, Wu Y-T (2015) Ionic liquid electrolytes for aluminium secondary battery: influence of organic solvents. *Journal of Electroanalytical Chemistry* 757:167–175. https://doi.org/10.1016/j.jelechem.2015.09.022
49. Kamath G, Narayanan B, Sankaranarayanan SKRS (2014) Atomistic origin of superior performance of ionic liquid electrolytes for Al-ion batteries. *Physical Chemistry Chemical Physics* 16 (38):20387–20391. https://doi.org/10.1039/C4CP02840F
50. Zhu G, Angell M, Pan C-J, Lin M-C, Chen H, Huang C-J, Lin J, Achazi AJ, Kaghazchi P, Hwang B-J, Dai H (2019) Rechargeable aluminum batteries: effects of cations in ionic liquid electrolytes. *RSC Advances* 9 (20):11322–11330. https://doi.org/10.1039/C9RA00765B
51. Wang H, Gu S, Bai Y, Chen S, Wu F, Wu C (2016) High-voltage and noncorrosive ionic liquid electrolyte used in rechargeable aluminum battery. *ACS Applied Materials & Interfaces* 8 (41):27444–27448. https://doi.org/10.1021/acsami.6b10579
52. Xie J, Zhang H, Liu Q, Liu X, Lu X (2020) Recent progress of molybdenum-based materials in aqueous rechargeable batteries. *Materials Today Advances* 8:100100. https://doi.org/10.1016/j.mtadv.2020.100100
53. Giffin GA (2016) Ionic liquid-based electrolytes for "beyond lithium" battery technologies. *Journal of Materials Chemistry A* 4 (35):13378–13389. https://doi.org/10.1039/C6TA05260F
54. Lahan H, Das SK (2019) Al^{3+} ion intercalation in MoO_3 for aqueous aluminum-ion battery. *Journal of Power Sources* 413:134–138. https://doi.org/10.1016/j.jpowsour.2018.12.032
55. Das SK, Mahapatra S, Lahan H (2017) Aluminium-ion batteries: developments and challenges. *Journal of Materials Chemistry A* 5 (14):6347–6367. https://doi.org/10.1039/C7TA00228A
56. Durmus YE, Zhang H, Baakes F, Desmaizieres G, Hayun H, Yang L, Kolek M, Küpers V, Janek J, Mandler D, Passerini S, Ein-Eli Y (2020) Side by side battery technologies with lithium-ion based batteries. *Advanced Energy Materials* 10 (24):2000089. https://doi.org/10.1002/aenm.202000089
57. Hu Z, Zhang H, Wang H, Zhang F, Li Q, Li H (2020) Nonaqueous aluminum ion batteries: recent progress and prospects. *ACS Materials Letters* 2 (8):887–904. https://doi.org/10.1021/acsmaterialslett.0c00208
58. Canever N, Hughson FR, Nann T (2020) Solid-electrolyte interphases (SEI) in nonaqueous aluminum-ion batteries. *ACS Applied Energy Materials* 3:3673–3683. https://doi.org/10.1021/acsaem.0c00132
59. Elia GA, Hasa I, Greco G, Diemant T, Marquardt K, Hoeppner K, Behm RJ, Hoell A, Passerini S, Hahn R (2017) Insights into the reversibility of aluminum graphite batteries. *Journal of Materials Chemistry A* 5 (20):9682–9690. https://doi.org/10.1039/C7TA01018D

60. Wu F, Zhu N, Bai Y, Gao Y, Wu C (2018) An interface-reconstruction effect for rechargeable aluminum battery in ionic liquid electrolyte to enhance cycling performances. *Green Energy & Environment* 3 (1):71–77. https://doi.org/10.1016/j.gee.2017.10.002
61. Kravchyk KV, Kovalenko MV (2020) Aluminum electrolytes for Al dual-ion batteries. *Communications Chemistry* 3 (1):120. https://doi.org/10.1038/s42004-020-00365-2
62. Qiu J, Zhao M, Zhao Q, Xu Y, Zhang L, Lu X, Xue H, Pang H (2017) Aluminum-based materials for advanced battery systems. *Science China Materials* 60 (7):577–607. https://doi.org/10.1007/s40843-017-9060-x
63. Chen H, Xu H, Zheng B, Wang S, Huang T, Guo F, Gao W, Gao C (2017) Oxide film efficiently suppresses dendrite growth in aluminum-ion battery. *ACS Applied Materials & Interfaces* 9 (27):22628–22634. https://doi.org/10.1021/acsami.7b07024
64. Yang H, Wu F, Bai Y, Wu C (2020) Toward better electrode/electrolyte interfaces in the ionic-liquid-based rechargeable aluminum batteries. *Journal of Energy Chemistry* 45:98–102. https://doi.org/10.1016/j.jechem.2019.10.003
65. Zhang X, Tang Y, Zhang F, Lee C-S (2016) A novel aluminum-graphite dual-ion battery. *Advanced Energy Materials* 6 (11):1502588. https://doi.org/10.1002/aenm.201502588
66. Hu Y, Luo B, Ye D, Zhu X, Lyu M, Wang L (2017) An innovative freeze-dried reduced graphene oxide supported $SnS_{(2)}$ cathode active material for aluminum-ion batteries. *Advanced Materials (Deerfield Beach, Fla)* 29:48. https://doi.org/10.1002/adma.201606132
67. Tu J, Lei H, Wang M, Yu Z, Jiao S (2018) Facile synthesis of $Ni11(HPO_3)_8(OH)_6$/rGO nanorods with enhanced electrochemical performance for aluminum-ion batteries. *Nanoscale* 10 (45):21284–21291. https://doi.org/10.1039/C8NR06380J
68. Faegh E, Ng B, Hayman D, Mustain WE (2021) Practical assessment of the performance of aluminium battery technologies. *Nature Energy* 6 (1):21–29. https://doi.org/10.1038/s41560-020-00728-y
69. Eftekhari A, Corrochano P (2017) Electrochemical energy storage by aluminum as a lightweight and cheap anode/charge carrier. *Sustainable Energy & Fuels* 1 (6):1246–1264. https://doi.org/10.1039/C7SE00050B
70. Acerce M, Voiry D, Chhowalla M (2015) Metallic 1T phase MoS_2 nanosheets as supercapacitor electrode materials. *Nature Nanotechnology* 10 (4):313–318. https://10.1038/nnano.2015.40
71. Bhauriyal P, Mahata A, Pathak B (2017) The staging mechanism of AlCl4 intercalation in a graphite electrode for an aluminium-ion battery. *Physical Chemistry Chemical Physics* 19 (11):7980–7989. https://doi.org/10.1039/C7CP00453B
72. Jung SC, Kang Y-J, Yoo D-J, Choi JW, Han Y-K (2016) Flexible few-layered graphene for the ultrafast rechargeable aluminum-ion battery. *The Journal of Physical Chemistry C* 120 (25):13384–13389. https://doi.org/10.1021/acs.jpcc.6b03657
73. Reed LD, Menke E (2013) The roles of V_2O_5 and stainless steel in rechargeable Al-ion batteries. *Journal of the Electrochemical Society* 160 (6):A915–A917. https://doi.org/10.1149/2.114306jes
74. Gu S, Wang H, Wu C, Bai Y, Li H, Wu F (2017) Confirming reversible Al^{3+} storage mechanism through intercalation of Al^{3+} into V_2O_5 nanowires in a rechargeable aluminum battery. *Energy Storage Materials* 6:9–17. https://doi.org/10.1016/j.ensm.2016.09.001
75. Koura N (1980) A preliminary investigation for an $Al/AlCl_3$-$NaCl/FeS_2$ secondary cell. *Journal of the Electrochemical Society* 127 (7):1529–1531. https://doi.org/10.1149/1.2129943
76. Mori T, Orikasa Y, Nakanishi K, Kezheng C, Hattori M, Ohta T, Uchimoto Y (2016) Discharge/charge reaction mechanisms of FeS_2 cathode material for aluminum rechargeable batteries at 55°C. *Journal of Power Sources* 313:9–14. https://doi.org/10.1016/j.jpowsour.2016.02.062
77. Jiao H, Wang C, Tu J, Tian D, Jiao S (2017) A rechargeable Al-ion battery: Al/molten $AlCl_3$-urea/graphite. *Chemical Communications* 53 (15):2331–2334. https://doi.org/10.1039/C6CC09825H
78. Chen H, Guo F, Liu Y, Huang T, Zheng B, Ananth N, Xu Z, Gao W, Gao C (2017) A defect-free principle for advanced graphene cathode of aluminum-ion battery. *Advanced Materials* 29 (12):1605958. https://doi.org/10.1002/adma.201605958
79. Wang S, Huang S, Yao M, Zhang Y, Niu Z (2020) Engineering active sites of polyaniline for $AlCl_2^+$ storage in an aluminum-ion battery. *Angewandte Chemie* 59 (29):11800–11807. https://doi.org/10.1002/ange.202002132
80. Li Z, Niu B, Liu J, Li J, Kang F (2018) Rechargeable aluminum-ion battery based on MoS_2 microsphere cathode. *ACS Applied Materials & Interfaces* 10 (11):9451–9459. https://doi.org/10.1021/acsami.8b00100
81. Bhauriyal P, Mahata A, Pathak B (2017) Hexagonal BC_3 electrode for a high-voltage Al-ion battery. *The Journal of Physical Chemistry C* 121 (18):9748–9756. https://doi.org/10.1021/acs.jpcc.7b02290
82. Dong X, Xu H, Chen H, Wang L, Wang J, Fang W, Chen C, Salman M, Xu Z, Gao C (2019) Commercial expanded graphite as high-performance cathode for low-cost aluminum-ion battery. *Carbon* 148:134–140. https://doi.org/10.1016/j.carbon.2019.03.080

83. Kumar S, Satish R, Verma V, Ren H, Kidkhunthod P, Manalastas Jr W, Srinivasan M (2019) Investigating $FeVO_4$ as a cathode material for aqueous aluminum-ion battery. *Journal of Power Sources* 426:151–161. https://doi.org/10.1016/j.jpowsour.2019.03.119
84. Almodóvar P, Giraldo DA, Chancón J, Álvarez-Serrano I, López ML (2020) δ-MnO_2 nanofibers: a promising cathode material for new aluminum-ion batteries. *ChemElectroChem* 7 (9):2102–2106. https://doi.org/10.1002/celc.202000425
85. Hu Y, Ye D, Luo B, Hu H, Zhu X, Wang S, Li L, Peng S, Wang L (2018) A binder-free and free-standing cobalt sulfide@ carbon nanotube cathode material for aluminum-ion batteries. *Advanced Materials* 30 (2):1703824. https://doi.org/10.1002/adma.201703824
86. Wang S, Jiao S, Wang J, Chen H-S, Tian D, Lei H, Fang D-N (2017) High-performance aluminum-ion battery with CuS@ C microsphere composite cathode. *ACS Nano* 11 (1):469–477. https://doi.org/10.1021/acsnano.6b06446
87. Liao Y, Wang D, Li X, Tian S, Hu H, Kong D, Cai T, Dai P, Ren H, Hu H (2020) High performance aluminum ion battery using polyaniline/ordered mesoporous carbon composite. *Journal of Power Sources* 477:228702. https://doi.org/10.1016/j.jpowsour.2020.228702
88. Wu Y, Gong M, Lin MC, Yuan C, Angell M, Huang L, Wang DY, Zhang X, Yang J, Hwang BJ (2016) 3D graphitic foams derived from chloroaluminate anion intercalation for ultrafast aluminum-ion battery. *Advanced Materials* 28 (41):9218–9222. https://doi.org/10.1002/adma.201602958
89. Liu Z, Wang J, Ding H, Chen S, Yu X, Lu B (2018) Carbon nanoscrolls for aluminum battery. *ACS Nano* 12 (8):8456–8466. https://doi.org/10.1021/acsnano.8b03961
90. Hu Y, Debnath S, Hu H, Luo B, Zhu X, Wang S, Hankel M, Searles DJ, Wang L (2019) Unlocking the potential of commercial carbon nanofibers as free-standing positive electrodes for flexible aluminum ion batteries. *Journal of Materials Chemistry A* 7 (25):15123–15130. https://doi.org/10.1039/C9TA04085D
91. Smajic J, Alazmi A, Batra N, Palanisamy T, Anjum DH, Costa PM (2018) Mesoporous reduced graphene oxide as a high capacity cathode for aluminum batteries. *Small* 14 (51):1803584. https://doi.org/10.1002/smll.201803584
92. Zhang X, Zhang G, Wang S, Li S, Jiao S (2018) Porous CuO microsphere architectures as high-performance cathode materials for aluminum-ion batteries. *Journal of Materials Chemistry A* 6 (7): 3084–3090. https://doi.org/10.1039/C7TA10632G
93. Wang W, Jiang B, Xiong W, Sun H, Lin Z, Hu L, Tu J, Hou J, Zhu H, Jiao S (2013) A new cathode material for super-valent battery based on aluminium ion intercalation and deintercalation. *Scientific Reports* 3 (1):1–6. https://doi.org/10.1038/srep03383
94. Wei J, Chen W, Chen D, Yang K (2017) Molybdenum oxide as cathode for high voltage rechargeable aluminum ion battery. *Journal of the Electrochemical Society* 164 (12):A2304. https://doi.org/10.1149/2.0411712jes
95. Liu J, Li Z, Huo X, Li J (2019) Nanosphere-rod-like Co_3O_4 as high performance cathode material for aluminium ion batteries. *Journal of Power Sources* 422:49–56. https://doi.org/10.1016/j.jpowsour.2019.03.029
96. Tu J, Lei H, Yu Z, Jiao S (2018) Ordered WO_{3-x} nanorods: facile synthesis and their electrochemical properties for aluminum-ion batteries. *Chemical Communications* 54 (11):1343–1346. https://doi.org/10.1039/C7CC09376D
97. Liang K, Ju L, Koul S, Kushima A, Yang Y (2019) Self-supported tin sulfide porous films for flexible aluminum-ion batteries. *Advanced Energy Materials* 9 (2):1802543. https://doi.org/10.1002/aenm.201802543
98. Geng L, Scheifers JP, Fu C, Zhang J, Fokwa BP, Guo J (2017) Titanium sulfides as intercalation-type cathode materials for rechargeable aluminum batteries. *ACS Applied Materials & Interfaces* 9 (25):21251–21257. https://doi.org/10.1021/acsami.7b04161
99. Yu Z, Kang Z, Hu Z, Lu J, Zhou Z, Jiao S (2016) Hexagonal NiS nanobelts as advanced cathode materials for rechargeable Al-ion batteries. *Chemical Communications* 52 (68):10427–10430. https://doi.org/10.1039/C6CC05974K
100. Geng L, Lv G, Xing X, Guo J (2015) Reversible electrochemical intercalation of aluminum in Mo_6S_8. *Chemistry of Materials* 27 (14):4926–4929. https://doi.org/10.1021/acs.chemmater.5b01918
101. VahidMohammadi A, Hadjikhani A, Shahbazmohamadi S, Beidaghi M (2017) Two-dimensional vanadium carbide (MXene) as a high-capacity cathode material for rechargeable aluminum batteries. *ACS Nano* 11 (11):11135–11144. https://doi.org/10.1021/acsnano.7b05350
102. Huo X, Wang X, Li Z, Liu J, Li J (2020) Two-dimensional composite of D-$Ti_3C_2T_x$@S@TiO_2 (MXene) as the cathode material for aluminum-ion batteries. *Nanoscale* 12 (5):3387–3399. https://doi.org/10.1039/C9NR09944A

103. Wu C, Gu S, Zhang Q, Bai Y, Li M, Yuan Y, Wang H, Liu X, Yuan Y, Zhu N (2019) Electrochemically activated spinel manganese oxide for rechargeable aqueous aluminum battery. *Nature Communications* 10 (1):1–10. https://doi.org/10.1038/s41467-018-07980-7
104. He S, Wang J, Zhang X, Chen J, Wang Z, Yang T, Liu Z, Liang Y, Wang B, Liu S (2019) A high-energy aqueous aluminum-manganese battery. *Advanced Functional Materials* 29 (45):1905228. https://doi.org/10.1002/adfm.201905228
105. Yan C, Lv C, Wang L, Cui W, Zhang L, Dinh KN, Tan H, Wu C, Wu T, Ren Y (2020) Architecting a stable high-energy aqueous al-ion battery. *Journal of the American Chemical Society* 142 (36):15295–15304. https://doi.org/10.1021/jacs.0c05054
106. González J, Nacimiento F, Cabello M, Alcántara R, Lavela P, Tirado J (2016) Reversible intercalation of aluminium into vanadium pentoxide xerogel for aqueous rechargeable batteries. *RSC Advances* 6 (67):62157–62164. https://doi.org/10.1039/C6RA11030D
107. Lahan H, Das SK (2018) An approach to improve the Al^{3+} ion intercalation in anatase TiO_2 nanoparticle for aqueous aluminum-ion battery. *Ionics* 24 (6):1855–1860. https://doi.org/10.1007/s11581-018-2530-6
108. Kazazi M, Zafar ZA, Delshad M, Cervenka J, Chen C (2018) TiO_2/CNT nanocomposite as an improved anode material for aqueous rechargeable aluminum batteries. *Solid State Ionics* 320:64–69. https://doi.org/10.1016/j.ssi.2018.02.034
109. Nandi S, Das SK (2020) An electrochemical study on bismuth oxide (Bi_2O_3) as an electrode material for rechargeable aqueous aluminum-ion battery. *Solid State Ionics* 347:115228. https://doi.org/10.1016/j.ssi.2020.115228
110. Nandi S, Das SK (2019) Realizing a low-cost and sustainable rechargeable aqueous aluminum-metal battery with exfoliated graphite cathode. *ACS Sustainable Chemistry & Engineering* 7 (24):19839–19847. https://doi.org/10.1021/acssuschemeng.9b05185
111. Ru Y, Zheng S, Xue H, Pang H (2020) Potassium cobalt hexacyanoferrate nanocubic assemblies for high-performance aqueous aluminum ion batteries. *Chemical Engineering Journal* 382:122853. https://doi.org/10.1016/j.cej.2019.122853
112. Liu S, Pan G, Li G, Gao X (2015) Copper hexacyanoferrate nanoparticles as cathode material for aqueous Al-ion batteries. *Journal of Materials Chemistry A* 3 (3):959–962. https://doi.org/10.1039/C4TA04644G

5 Characterization of Electrochemical Behavior of all Iron-Ion Batteries for Grid-Scale Applications

Anusree Thilak, Nikhil Medhavi, Anandu M. Nair, Devika S. Lal, Akhila Das, Jou-Hyeon Ahn, and Prasanth Raghavan

5.1 INTRODUCTION

Energy consumption matters to our environment and our economy. Every human being uses energy in a certain way, and it is at the core of contemporary civilization. Energy may be the "lifeblood" of the country's economic growth as it is the most basic and necessary element. The rise in global population and economic standards has led to greater energy use and, as a result, demand. However, the diminishing availability of fossil energy supplies necessitates the development of renewable energy sources and storage technologies to meet future demand. Renewable energy sources such as wind and solar are fostered to lessen dependency on non-renewable fossil fuels. But, still, switching to wind and solar energy for our daily requirements is not straightforward. The solution to this dilemma is energy storage. A power supply with prolonged life duration may be established without strain on earthly resources by storing solar and wind power energy in stable, robust batteries. Electrochemical storage technologies, particularly iron-ion batteries, have been offered as potential alternatives for grid-scale storage systems. Because of its low cost, ease of oxidation, various oxidation states, and ability to be electrodeposited from an aqueous electrolyte, iron is an appealing metal for a rechargeable battery. Redox processes use iron-ion batteries and are one of several types of electrochemical energy storage systems. The term "redox" refers to the chemical reduction and oxidation processes that aid in energy storage in electrolyte solutions that pass through an electrochemical battery during charge and discharge. During discharge, an electron is liberated from a high chemical potential state on the negative or anode side of the battery via an oxidation process. To perform beneficial work, the electron travels through an external circuit. Finally, on the positive or cathode side of the battery, the electron is accepted via a reduction process at a lower chemical potential state. During charging, the direction of the current and the chemical processes is reversed, although the system's response mechanism appears to be identical to and more or less connected to all other battery systems now in use. The iron-ion battery has several benefits compared to other battery systems.

5.1.1 Advantages of Iron-Ion Batteries Over Lithium-Ion Battery

The lithium-ion battery was commercialized widely and used in portable devices such as laptops and mobile phones. LiBs were first commercialized by Sony Energytec in 1990, with the battery constituting a graphitic anode and $LiCoO_2$ cathode. In 2012, the market share of LiBs was US$ 11.7, and it increased to US$ 33.1 by 2019 [1]. The first reported LiB with V_2O_5 anode and $LiMnO_2$

DOI: 10.1201/9781003310174-5

cathode in $LiNO_3$ electrolyte showed an energy density of 75 KW h^{-1}. The electrolyte stability is very poor during standard electrolysis conditions. In cathodes, the proton and water molecules block the insertion of Li ions, and the dissolution of electrode material in the electrolyte reduces long-term cyclability [2].

On moving to grid energy storage, LiBs are not satisfactory, and, hence, scientists are in search of suitable energy storage devices that fulfill future energy needs also are expected to be operated and work safely in restricted temperatures and voltages. An increase or decrease in the voltage or temperature can cause discrepancies in working conditions and can even be dangerous. The expected failures in LiBs are:

i. Due to over-voltage: The current flows excessively, which can increase the overheating of the cell. It can also lead to the accumulation of lithium ions on the surface of the electrode since excessive lithium ions can't be penetrated through the anode, which has an intercalated layer; thus reduction in free lithium ions and decomposition of anode occurs.
ii. Under-voltage: Electrode material decomposes; the level of electrolyte reduces, leading to the production of combustible gases. When the voltage drops beyond the optimum voltage (2 V), the anode material dissolves, and on a sudden voltage increase, it precipitates and forms dendrite, resulting in short circuits, whereas, in the cathode, the release of oxygen reduces the cell capacity, thereby reducing the cycle time.
iii. High temperature: Causes overall cell damage.
iv. Low temperature: The current capacity of the cell reduces [3,4].

Safety recyclability, cost, etc. force to find a new suitable material for energy storage device satisfying the requirement. The electrolyte reacts with the surface of lithium metal anodes, causing unstable surface reactions, which increase surface temperature and chances of inflammation. On repeated cycling, the capacity of the lithium battery fades, raising safety issues [5].

To accomplish future energy needs, we must look for suitable storage materials, and, hence, scientists leaped onto LiS_2 and LiO_2 from lithium-ion batteries. It is theoretically proved that cathode Li_2O_2 and LiOH showed excellent charge storage than graphitic anode of LiB.

Lithium–Oxygen Battery: O_2 being the cathode and lithium being the anode, the battery showed high charge density and reversibility and also delivered 1,200 mAh^{-1} capacity [6]. The electrodes must be porous, facilitating gas phase diffusion through electrolytes to electrodes. Electrolyte stability and the design of porous electrodes at low cost challenge the further application of Li–O_2 batteries [7]. Though ionic liquids, polymers, ionic liquids, and polymer compositions for electrolytes are proposed, the clarity regarding stability is still undefined [8,9].

Lithium–Sulfur Battery: Natural abundance, low cost of sulfur, and high theoretical charge density made it an attractive candidate. The battery operation takes place through the formation of LiS_2 by combining Li ions with sulfur reduced from the cathode. It exhibited significant degradation during the cycling and self-discharging. The main issues reported with this type are:

- Fast capacity fading
- Poor electrode rechargeability
- Poorly controlled Li/electrolyte interphase
- Finding a cathode with a porous structure, which is efficient for transporting S ions and capable of capturing polysulfide discharge, is a big challenge in designing Li–S batteries [7].

Moreover, the LiO_2 batteries require pure oxygen, and the synthesis of O_2 is an energy-expensive process. For Li-S batteries, sulfur is obtained as a side product in oil and gas industries by Clause process and is likely to phase out in the near future. Therefore, it is necessary to move on to reliable and abundant material, so new paths in finding suitable material from 3d transition metals like titanium, iron alkali, and alkali earth metals were investigated [10].

Among them, iron is found to be a better choice because it undergoes multiple oxidation states, is inexpensive, and can also be electrochemically deposited in the electrode through electrolyte [11]. Also, it is widely available and recyclable, thus causing no harm to the environment. A decrease in energy density due to high electrode potential can be overcome by opting divalent Fe-ion. The dendrite formation being very low is also found to be an attractive feature. It satisfies the requirement of the anode, i.e., the ability to electrodeposit reversibly and the small size of iron. It delivered a specific capacity of 207 $mAhg^{-1}$ when mild steel is used as an anode and vanadium pentoxide as a cathode in an ether electrolyte [12].

5.1.2 Monovalent and Multivalent Ion Battery

The lithium-ion battery has a low electrode potential, good specific charge density, and high specific energies (130 Wh kg^{-1}) and energy densities (300 Wh L^{-1}). It found its position in storage devices in the 1950s due to its storage capability, and this type of battery kept advancing over the years such that the rate of market share quadrupled from the 1990s to 2001, and found applications in many devices in 3C-cellular phones, portable computers, and camcorders [13]. It has been a better candidate for energy due to its long cyclability and low cost. After that sodium-ion batteries emerged to mitigate the challenges raised by LiBs. In sodium-ion batteries, sodium ions are renewable and abundant [14].

Sodium-Ion Batteries: It was a potent competitor of lithium-ion and lithium–sulfur batteries and showed similar intercalation chemistry. They are cheaper and have abundant availability of sodium. The anode materials usually used are germanium and red phosphorous, which provide high stability and discharge capacity and enable working at high current densities [15,16]. But, it forms a dendrite. Due to the large size of Na^+ ions, it is difficult to find anodes for the penetration of large-sized sodium ions through intercalated electrodes. It has a low volumetric density and poor cyclic stability. The difficulty in finding suitable anode material restricts its commercial application [17].

Similar to sodium metals, potassium is of low cost and abundant. The reported capacity of the cell is 591 mAh g^{-1} and has a potential similar to that of sodium-ion and lithium-ion. The cell consisting of potassium as an anode and Prussian blue as a cathode shows a high cycle life (500 cycles) [17].

5.1.2.1 Advantages and Disadvantages of Monovalent and Multivalent

The potential for maximum energy storage is a leading research field due to the requirement for suitable energy storage devices for the future. Metals yielding polyvalent ions are of great scope in the future [18]. Multivalent batteries such as Zn^{2+}, Ca^{2+}, and Mg^{2+} provide high charge storage capacities and charge density due to larger ions. It overcomes environment-related issues by utilizing electrodes with large, open structures, which paves a new dimension in energy storage devices. The low oxidation potential and multivalence are utilized to attain the required energy density [19]. Multivalent ion battery is an emerging class of batteries to mitigate the issues related to lithium-ion batteries and are potent enough to power up future energy requirements. Multivalent ion metals are abundant, recyclable, and ecofriendly. Moreover, it has a high charge density; hence, a smaller number of ions are required to achieve the required electrochemical capacity when compared with monovalent ions; i.e., greater capacity is observed [20]. Multivalent metals contribute two or more ions to the intercalated cathode, which improves the reversibility and volumetric charge capacity; the energy density increases as the capacity of the multivalent electrode increases [21]. Multivalent ions are safer as they exhibit minimum or no dendrite formation. Magnesium, aluminum, zinc, iron, etc. are examples of multivalent iron raw material. The strong columbic interaction between anions and cations hinders the intercalation/deintercalation of ions, leading to poor cycling, and this can be avoided in two ways: first by incorporating hydrated compounds, and thus the columbic attraction between anions and cations is shielded by hydroxyl ions [22,23]. Second, materials with smaller domain sizes improve intercalation of multivalent anions, or electrode materials are nano-sized and diffusion distances are shortened; otherwise, electrode structures are fabricated by large interlayer spacing, thus preventing the attraction of multivalent ions to polar molecules [24,25].

The reversibility of ions through the intercalated layer is the main disadvantage of the multivalent ion battery. Due to the large size of the ion, it gets trapped between the intercalated layer of the cathode, and as a result, reversing back is difficult. Also difficulty in choosing a suitable electrolyte should be mitigated such that cations must show high mobility and high desolvation energy [24,25]. The overcharging and overheating can cause the explosion of batteries due to the evaporation of the electrolyte [24]. It is found that the ion movement is sluggish in multivalent ion batteries. This is caused due to the columbic attraction and electrodeposition on the electrode and seems quite complicated due to the inability of sudden desolvation [18]. Though this class of batteries exhibited a large number of advancements from the lithium-ion battery, a complex behavior is found in electrode–electrolyte interaction. The high ionic radius of ions like Ca^{2+} prevents smooth accommodation in cathode layers, whereas ions like Mg^{2+} have high charge density and thus have reduced diffusion coefficient [19]. A stable or flat plateau is exhibited by the discharge profile of multivalent ions, which influence the energy density and voltage application, thereby inhibiting them from practical applications [26].

Though monovalent batteries exhibit a wide range of electrochemical properties, they seem unsatisfied when moving to grid-level storage applications such as electric cars. Thermal runaway and dendrite formation raise safety issues in monovalent ion batteries, since electrolyte dry-out takes place, leading to fire and explosion when the temperature of the cell increases, and capacity fading is also observed on continuous cycling [24].

5.1.2.2 Transition from Multivalent Ion to Iron-Ion Battery

Multivalent ion batteries are an expanding source of energy storage devices due to their improved performance when compared to monovalent types. Mg^{2+}, Ca^{2+}, Zn^{2+}, Al^{3+}, etc. are some examples. It provides high energy density, better safety, good cyclability, and so on. Intercalation cathode, metal anode, and electrolyte are the three components in the battery where charge is stored, and the multivalent ions are extracted at the intercalation of the cathode [5]. Since the valency is higher, electrochemical capacity can be achieved by less number of ions, and the chances of dendrite formation are less [20].

Magnesium ion batteries overcome most of the challenges posed by the previous findings; they are abundant; have low toxicity, low negative potential, and high charge storage capacity; and are suitable for anode material, thus making them safer [20]. They are also suitable for transport applications due to their high volumetric capacity. The inability to diffuse large magnesium ions through the intercalated cathode and the lack of a suitable cathode and electrolyte marks a limitation for the commercialization of magnesium ion batteries [20]. Impurities such as oxygen and water in small concentrations can lead to the formation of the passivating layer, making magnesium ions nonconductive. In addition as the impurity comes in contact with the Cl^- species in the electrolyte, oxides form and the electrode material gets blocked due to the polarization [20].

Calcium-Ion Battery: The next candidate was calcium, which is abundant, has uniform decomposition, and makes no dendrite growth. Although it promises various features, it has barely limited electrolyte choices due to the formation of a passive layer in an organic electrolyte.

Zinc-Ion Battery: High power/energy density, good safety, and high discharge capability make zinc-ion battery another class of suitable material. But it is disappointing that, even after certain pronounceable progression in findings, it still has drawbacks in terms of lower life, insufficient discharge capacity, and less choice of cathode materials.

Aluminum-Ion Battery: It demands to have better properties suitable for commercial applications. It is available in low-cost form and provides high charge capacity and high volumetric capacity due to its multi-valency. However, the electrode–electrolyte compatibility of an Al-ion battery must be well examined. Using graphite foam as an anode provides faster diffusion that can rectify problems due to difficulty in the penetration of large-size aluminum ions through the cathode.

The electrolyte used is anionic $[AlCl_4]$−complexes, from which Cl^- ions causes corrosion similar to that in magnesium ion batteries. Less specific capacity and densities during real-life applications restrict them from large-scale applications [16,20].

Iron-Ion Batteries: the scope of iron-ion batteries in large-scale energy storage is promising among the current energy storage battery systems. The structural benefits of a cathode in intercalation and speed movement of ions are add-ons to better deposition and dissolution of metal ions during charging and discharging. The capacity retention exhibited by the battery is 47% after 85 cycles, proving the suitability of the Fe-ion battery as the secondary battery.

5.2 CLASSIFICATION BASED ON TYPE OF ELECTRODE

Batteries, in general, are made up of three basic components: anode, cathode, and electrolyte. So they can therefore be classified based on the types of electrode and the electrolyte. The choice of the electrode is primarily dependent on the mass availability of the material. It can further be classified based on anode material and cathode material. There are also different types of electrolytes used according to which the batteries can also be differentiated.

5.2.1 Types of Anode Material

Iron is economical in terms of availability, cost, and recyclability. Multivalent metal as an anode provides high volumetric density (~3,833 mAH cm^{-3}). It also exhibits an excellent uniform deposition of ions in the cathode and can achieve high energy densities [27,28]. Moreover, the specific capacity and volumetric capacity are of much satisfying value around 960 mAh $^{-1}$ and 7,557 mAh cm^{-3}, respectively, thus proving theoretically that it has high electrochemical properties. It has a higher redox potential, providing better stability in the electrolyte. Therefore, easy intercalation of anodic ions onto the cathode occurs without reacting with the electrolyte, and also iron has less chance for dendrite formation.

It is reported that mild steel anode (99.33 wt.% Fe, 0.15 wt.% carbon, and 0.52 wt.% Mn) can be used in iron-ion batteries due to the following factors:

- Ease electrostripping of Fe^{2+} ions
- Less energy required for vacancy formation due to the presence of carbon [12,29,30].

5.2.2 Types of Cathode Materials

Because multivalent ions can donate two or more ions to the current carrier, it is supposed to have a higher current. The current collectors must have a suitable intercalation for the ions to fit in and are able to remove those ions from layers. The choice of cations is a challenging aspect of multivalent ion batteries as they should be capable of reversible storage [31]. Multivalent ions have a high polarizing nature, which causes strong attraction of anions and cations. Also, they find it difficult to diffuse through the close-packed cathode structure [32]. This impedes the reversible deintercalation eventually and results in poor electronic conductivity. Many classes of cathode suitable for metal ion batteries are reported, and they are divided into four classes. They are chalcogenide-based chevrel-phase cathodes, chalcogenide spinels, transition metal oxides, and polyanion cathodes [5,33]. The performance of cathode material is evaluated by the progress in energy density and electrochemical standard. The transition metal oxide increased the number of incorporated intercalation ions in the cathode [34]. The large interlayer distance in the cathode reduces the electrostatic interaction within the layer, thereby facilitating easy diffusion of ions, and thus kinetics is eventually increased [35]. The V_2O_5 is used as a cathode in the Fe-ion battery, having large interlayer spacing (437 pm), which eases intercalation and deintercalation of ions, resulting in increased conductivity [12]. V_2O_5 gained a lot of attention in energy storage devices, especially in multivalent energy storage devices [36]. They have layered framework appearing in different phases, which are capable of changing the layered structure when the cation is diffused into the framework. It is reported to have a large specific energy density [37].

5.3 CLASSIFICATION BASED ON TYPE OF ELECTROLYTE

The prime requirement of the electrolyte is to transport ions and make the cell conductive. Also, the salt must be highly dissociative and have a wide electrochemical stability [38]. It is equally important to consider the safety factor of the electrolyte so that it can function smoothly. External conditions such as heat, which may vary, increase, or decrease, can alter cell requirements. An increase in heat can lead to fire and other safety hazards [39]. Electrolyte must be chosen such that it does not catch fire rapidly or it must be non-flammable in its pristine state. The salvation structure and local dynamics have been studied, which helps in describing the battery performance and the nature of dendrite growth which passivates conductivity [40]. Also, the non-flammability was increased on new species formation during cyclization. Therefore designers must require clear-cut knowledge about the solvation, viscosity, and stability of the salvation species [41]. For multivalent ion batteries, the nature of the interaction of cation with corresponding anion species is an important factor in choosing a suitable electrolyte. Glyme-based electrolytes which exhibit high conductivity and cycling efficiency are a promising class of electrolyte [42]. The poor electrode passivating properties of such electrolytes can be rectified by the addition of film-forming additives like lithium nitrate [43]. The less reduction in capacity observed for a glyme-based electrolyte; that is, even after 300 cycles, the capacity is found to be reduced from 124 to 120 mAh^{-1}. TEGDME is the electrolyte used in Fe-ion batteries, which is non-flammable, is thermally stable having an ionic conductivity value at –49 to 200°C, and is 4×10^{-4} S cm^{-1} [44]. There are mainly three types of electrolytes: ionic, solid-state, and ionic liquids. The ionic electrolytes enhance conductivity by facilitating a conducting path between the anode and the cathode [45]. In solid electrolytes, simple cells can be designed, which are sealed naturally; have resistance to shock, vibrations, and temperature variations; and are electrochemically stable and safer [46]. The next category is ionic liquids where salts have melting points lower than room temperature. It shows a wide range of properties such as non-flammability, high thermal stability, good electrochemical stability, and low toxicity [47].

5.4 IRON-ION BATTERIES

An anode made of low-carbon steel and a cathode made of vanadium pentoxide is used in the newly created iron-ion battery. Vanadium pentoxide was chosen because of its layered structure and large interlayer spacing, which allows iron ions to quickly travel in and bond to cathode interlayers, as well as easily detach and return to the anode. Iron chlorate is used as an electrolyte here. An investigation on recent and relevant publications relating to this topic was carried out considering a period of 5–10 years. From the data available, it was understood that only one article on iron-ion battery is published so far. It was published by Sundara et al. in *Chem. Commun.*, Royal Society of Chemistry, 2019. (A room temperature multivalent rechargeable iron-ion battery with an ether based electrolyte: a new type of post-lithium-ion battery.)

5.4.1 Conventional Iron-Ion Batteries

Sustainable development is becoming a worldwide concern which has a global objective to overcome energy storage in a sustainable way. European Union has taken initiatives to move towards sustainability, for example; European Green cars were an initiative to promote green cars and trucks. The ongoing research is to convert batteries greener such that the conventional Lithium-ion batteries produced per year has to be recycled at the end of its life cycle. The problem with the recycling of

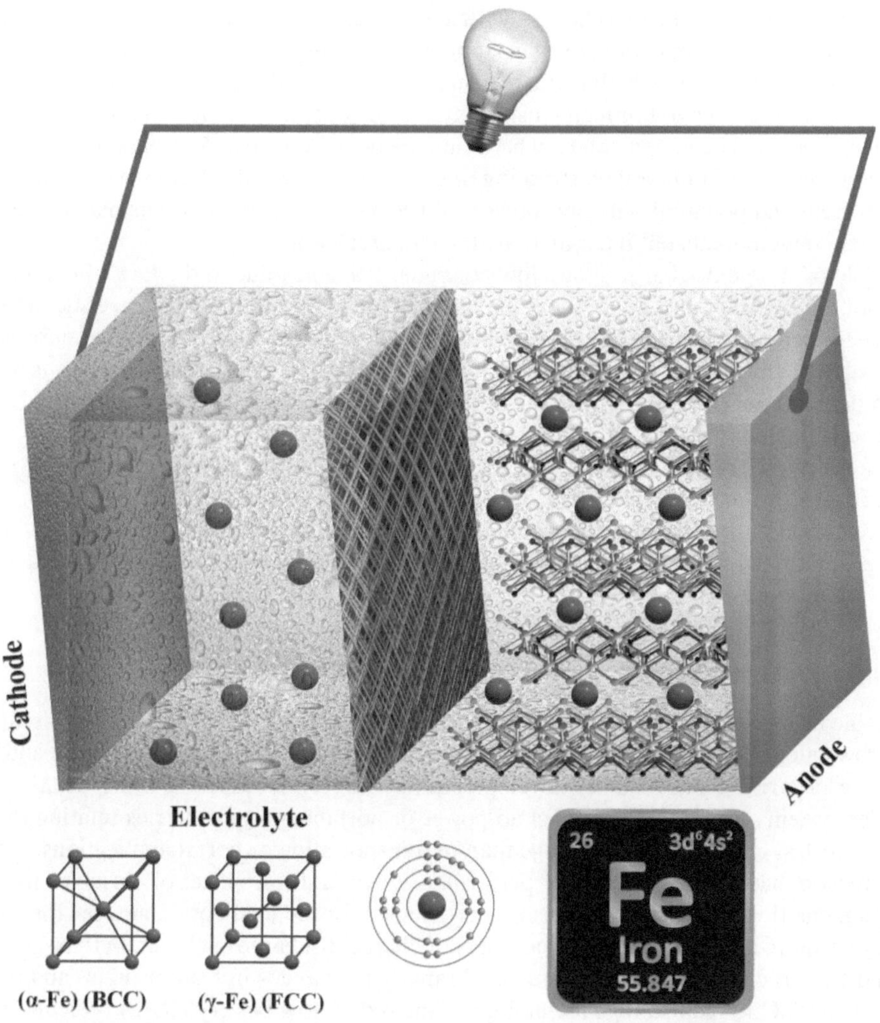

FIGURE 5.1 Schematic illustrations of iron-ion battery.

Lithium was addressed by companies, but at the same time, price and availability were not satisfied. Therefore magnesium and calcium gained more interest. But lately iron-ion battery has been identified with satisfactory property tackling environmental issues [48].

Iron-ion battery is recently identified multivalent ion battery (Figure 5.1) which mitigate the issues raised by conventional LIBs and other multivalent ion batteries. It is ecofriendly and cheaper; moreover the smaller size of Fe^{2+} (0.77 pm) allows intercalation through the cathode and facilitates reversible electrodeposition in the presence of electrolyte. The first iron-ion battery reported with mild steel as an anode and vanadium pentoxide as cathode in ether based electrolyte. Anode being mild steel anode contains 99.33 wt.% Fe, 0.15 wt.% C and 0.52 wt.% Mn. Carbon improves conductivity by facilitating ease stripping of Fe^{2+} ions from the electrode whereas it also decreases vacancy formation energy. V_2O_5 has a large interlayer spacing about 437 pm and non-flammable electrolyte solvent tetraethylene glycol dimethylether (TEGDME) is chosen but to enhance reversible electroplating and electrostripping, 1M $Fe(ClO_4)_2$ is dissolved in TEGDME solvent. The iron-ion battery has specific capacity of 208 mAh g^{-1} at specific current 30 mA g^{-1} and 47% of capacity retention even after 80 cycles.

The cathode have flake like morphology which indicates ease migration of Fe^{2+} ions. The layered VO_5 provide intercalation for cation, oxygen atom present at the apices and vanadium at the adjacent layer have a weak interaction which is the reason behind ease and fast intercalation.

With the help of cyclic voltammetry using three electrode, it is proved that reversible iron dissolution and deposition occurs in anode, where cathode peak indicates electrodeposition in bulk and anode peak gives the information on stripping of electrons. Charge discharge measurement done at different current and potential window shows 0–2.1 V. Voltage profile exhibits plateau and sloping where plateau region points out intercalation and deintercalation.

First cycle of charge/discharge shows low columbic efficiency due to the lack of deintercalation, and formation of a compound. Later on, a stable oxide layer formation takes place such that columbic efficiency increased from 36% to 58%. High specific capacity was observed at current density 30 mAg^{-1} such that the value of specific capacity is 208 mAh g^{-1}. The specific capacity is vulnerable to sudden decrease to 200 mAh g^{-1} and is observed when the formation of iron oxide layer takes place. It can be increased when the cell is cycled back to 30 mAg^{-1} current.

The durability of the cell is tested by rate capability test; it indicates that during first cycle it showed 250 $mAhg^{-1}$ and the specific capacity value after 50 cycles is noted to be 136 mAh g^{-1}. Capacity retention is thus found to be 54.4%. This cell cycling capacity is noted as 54.5% in 50 cycle and have normal functioning at ambient atmosphere. Fe-ion battery, hence is a promising candidate for the future purpose.

5.4.2 Introduction to All-Iron-Redox Flow Batteries

As renewable energy sources increasingly become a part of the grid, the need for large-scale energy storage grows alongside. Although conventional rechargeable batteries offer a simple and efficient way to store electricity, but to date, development has largely focused on transportation systems and smaller systems for intermittent backup power or portable power, metrics relating to size and volume are far less critical for grid storage than in transportation or portal applications. It therefore stands to reason that optimizing battery performance over a different set of variables might result in an implementation that delivers superior performance for reduced cost. Batteries for large-scale grid storage require durability for large numbers of charge/discharge cycles as well as calendar life, high round-trip efficiency, an ability to respond rapidly to changes in load or input, and reasonable capital costs [49]. One such storage technology is the redox flow battery (RFB) typically involving redox couples dissolved in an externally stored electrolyte. RFBs have seen a surge of interest over recent years for their ability to provide ancillary services to the electric grid which could lead to increased robustness, flexibility, and the ability to rely on intermittent renewable energy sources such as wind and solar [49,50]. Traditional RFBs, including the iron-chromium [51], iron-vanadium [52], bromide-polysulfide [53], and all-vanadium [54] systems that store energy chemically in electroactive species contained in electrolytes that are stored in external reservoirs. The amount of energy storage is solely based on the reservoir of electrolyte, and the power delivery is determined by the size of the battery pack. The architecture of RFBs shown in Figure 5.2 allows the energy storage capacity (dependent on the amount of active species) to be scaled independently of the power rating (dependent on the area of the RFB conversion device) leading to flexible operation and implementation. Despite the advantages, the cost, abundance, and toxicity of the active species have hindered the market penetration of these systems. An attractive alternative to the traditional RFB chemistries is the all-iron flow battery because iron, the only active element, is low cost, abundant, and environmentally benign.

Equations 5.1 and 5.2 represent the positive and negative reactions in an all-iron redox flow battery have a standard cell voltage of 1.21 V. Ferrous ions (Fe^{2+}) are oxidized to ferric ions (Fe^{3+}) on charge in the positive half of the cell while ferrous ions are reduced to iron metal in the negative half of the cell [54].

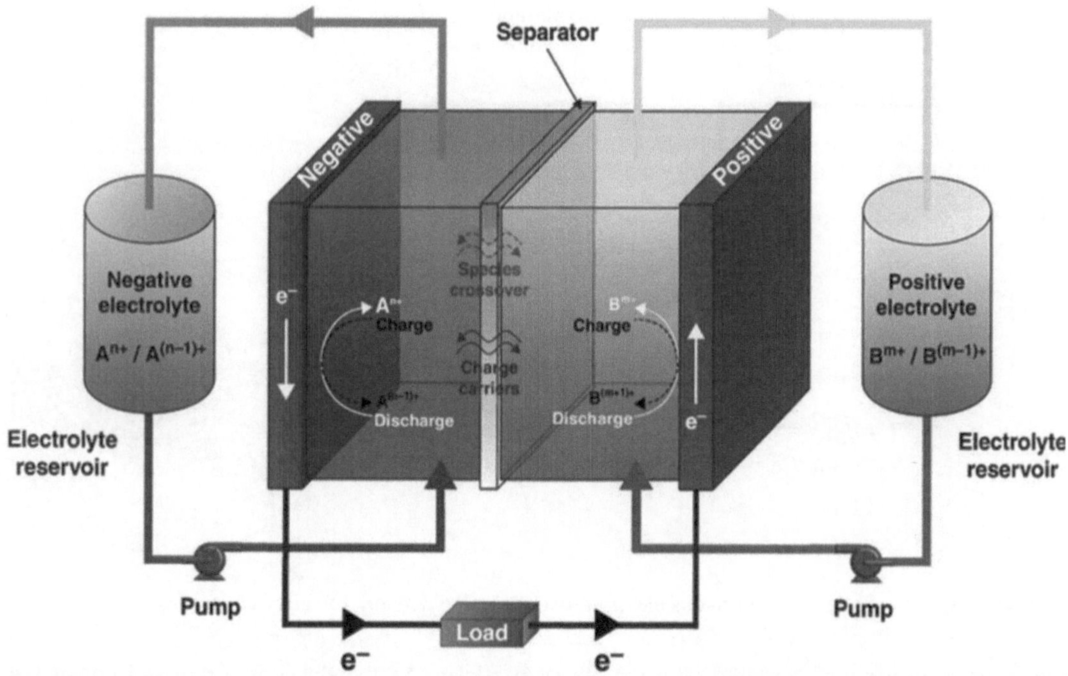

FIGURE 5.2 Schematic of all-iron redox flow battery with electron transport in the circuit, ion transport in the electrolyte and across the membrane, active species crossover, and mass transport in the electrolyte.

$$2Fe^{2+} \leftrightarrow 2Fe^{3+} + 2e^{-} \quad E^{0}=+0.77\text{V vs RHS} \tag{5.1}$$

$$Fe^{2+} + 2e^{-} \leftrightarrow Fe^{0} \quad E^{0}=-0.44\text{V vs RHE} \tag{5.2}$$

Other advantages of the iron-redox battery include low temperature aqueous electrolytes with inexpensive reactants that are readily available in large quantities, the use of inexpensive materials of construction, the use of electrodes which are not subjected to shape change or dendrites, the distinction of being the least hazardous of any advanced battery system [55].

5.4.2.1 Slurry Electrodes

RFBs typically employ electrodes made of porous carbon structures (e.g. felts, cloths, or papers) fixed in the electrochemical device. When using these stationary electrodes in the all-iron flow battery, iron metal is plated into their structure and subsequently stored in the electrochemical cell. Since iron metal is stored in the electrochemical cell, the conventional all-iron battery is limited to <4h (3–8h is required) [56] of energy storage at reasonably high current densities. In order to have a battery with the advantages inherent to traditional flow battery architectures and the advantages of the all-iron chemistry, a different electrode structure must be employed that enables higher energy storage to power ratios. One such option is the slurry electrode. Slurry electrodes are made by suspending solid, electronically conductive particles in the ionically conducting electrolyte and have been studied for a variety of electrochemical applications. During charge and discharge, the electrolyte is pumped through the electrochemical cell as in normal RFB operation. When using slurry electrodes, the solid particles are carried by the electrolyte into and out of the electrochemical cell. If the volume fraction of the solid particles is high enough, a continuous electrically conductive network forms [48,57–59] as shown in Figure 5.3. This allows the redox reaction to occur on the surface of the slurry particles. When slurries are employed as the negative electrode of the all-iron

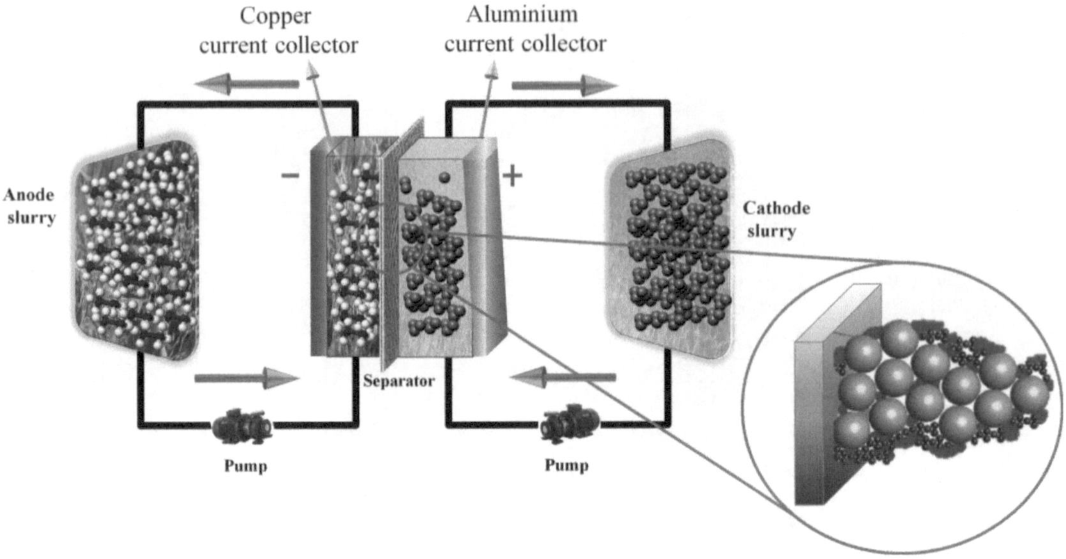

FIGURE 5.3 Schematic of a continuous electrically conducting network of slurry electrodes.

flow battery, iron metal is plated onto the slurry particles. As the slurry is circulated out of the electrochemical cell back into the external reservoir as shown in Figure 5.3, the iron metal is carried with the particles, enabling the energy storage capacity to be decoupled from the power rating.

Slurry electrodes are inherently dynamic. Particles are constantly moving and are continuously making and breaking contact with each other. Percolation theory describes a critical loading, *fc*, above which there are enough particles in the slurry that, at any point in time, a continuous network of particles is formed that spans the distance from one boundary to another. It is due to these networks that slurries are able to conduct electrons out into the electrode and away from the current collector to utilize the high surface area of the particles for electrochemical processes. There are many advantages to using slurry electrodes over conventional stationary electrodes. The use of slurry allows the electrode (1) surface area to be scaled independently of the separator area, (2) to be easily manufactured and replaced without disassembling the cell, (3) and to be easily recycled with filtration [60,61] (Figure 5.4).

5.4.2.1.1 Electrochemical and Rheological Studies

In the given study [62], slurry electrodes were made with MWCNTs s having an OD of 50–80 nm, length of 10–20 μm (aspect ratio of 125–400), a reported surface area of $40\,m^2g^{-1}$, and a carbon purity of 99.9%. Before tests were performed, the particles were mixed by hand into the electrolyte and then pumped through the system for at least 30 min. All-iron flow batteries were studied with slurries serving as both the positive and the negative electrodes. In all cases presented herein, the slurry electrodes were composed of 4.8 vol% MWCNTs in the electrolyte. The positive slurry was initially 0.5 M $FeCl_2$, 0.5 M $FeCl_3$, and 1.0 M NH_4Cl while the negative slurry was initially 0.5 M $FeCl_2$ and 1.0 M NH_4Cl. The ionic conductivity of these electrolytes is about 150 mS cm^{-1}. Adding ferric chloride to the positive electrolyte ensured that on discharge, the negative slurry would always be limiting. All solutions were made with deionized water. When using ferrous salts, solutions were made under a continuous nitrogen blanket and experiments were performed with the electrolyte under a continuous nitrogen purge to mitigate the air oxidation of ferrous to ferric iron [62]. Before the initial charging of the all-iron flow battery, EIS measurements were first conducted at the open circuit potential, confirming the conductivities of the slurry electrodes and separator. As the SOC of the negative slurry rises, the amount of metal deposited in the slurry increases as shown in Figure 5.5, and may therefore increase the slurry electronic conductivity.

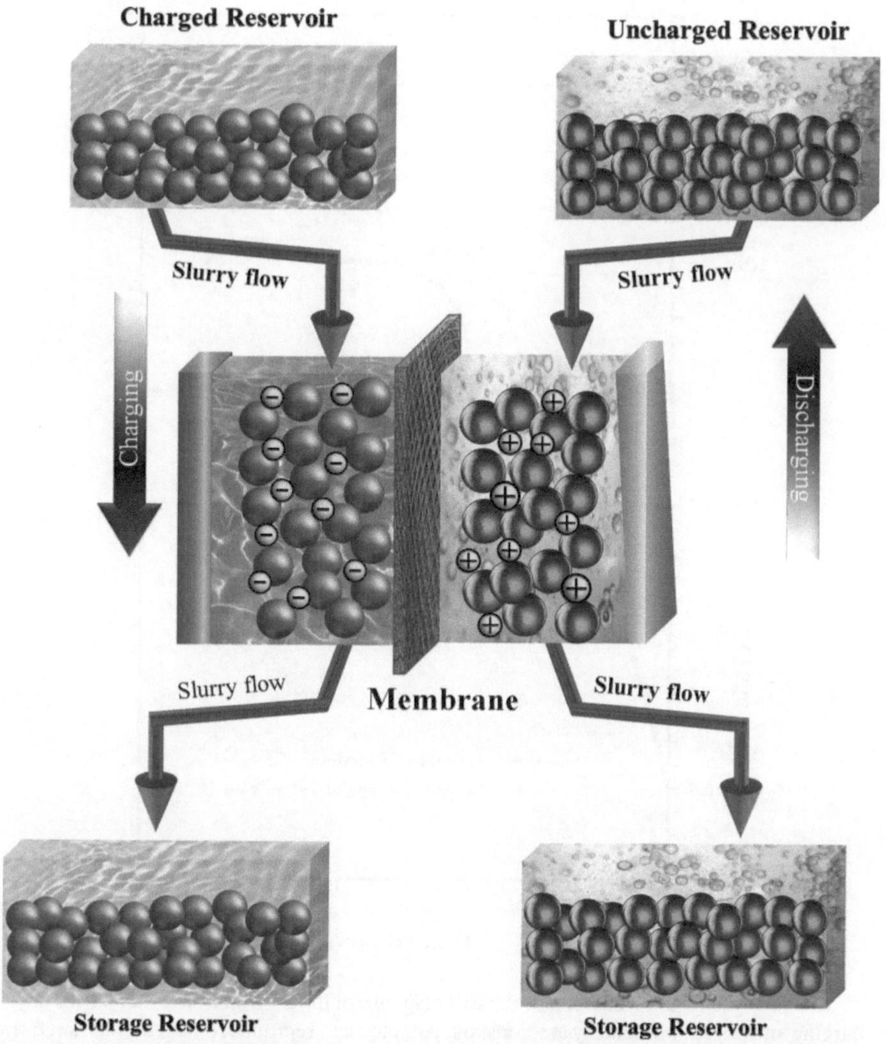

FIGURE 5.4 Operational schematic of an all-iron redox flow battery with electron transport in the circuit where uncharged slurry flows through polarized plates and the charged slurry is then pumped into external reservoirs for storage.

At 75 mA cm^{-2}, the battery voltage rose with the SOC before leveling off around 1.5 V as shown in Figure 5.5a. After the voltammogram was obtained as shown in Figure 5.5b, the charge was continued at 150 mA cm^{-2} for two additional hours. During this period, the cell charging voltage fell to 1.45 V after first going through a voltage maximum [63] (Figure 5.6).

Decrease in the overpotential on charge has an additional benefit. When charging at large potentials, the negative electrode can become negative enough for hydrogen evolution to occur. When this happens, the electrolyte pH will rise, potentially causing ferric or ferrous iron to precipitate out of solution as their respective hydroxides. The precipitated iron hydroxide can form a film on the slurry particles making discharge of the battery more difficult. Hence, two mechanisms thought to contribute to the improved performance are by increasing the slurry electronic conductivity (due to metal deposition on the particles) and an increase in plating kinetics [63].

Rheological studies were also conducted since the slurries in these studies are intended for use as flowing electrodes. The Ostwaldede Waele (power law) model was adopted to describe the non-Newtonian behavior of the slurry over the range of shear rates considered [64]. This model

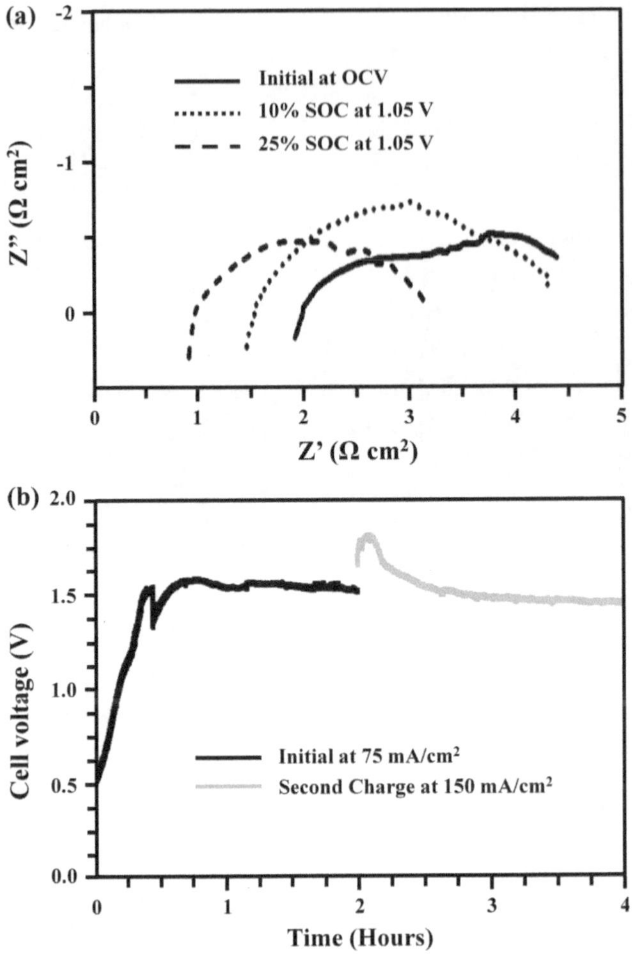

FIGURE 5.5 (a) The Nyquist representation of the EIS response of the all-iron flow battery. (b) The voltage profiles while charging an all-iron flow battery at a constant. Adapted and reproduced with permission from Ref. [62].

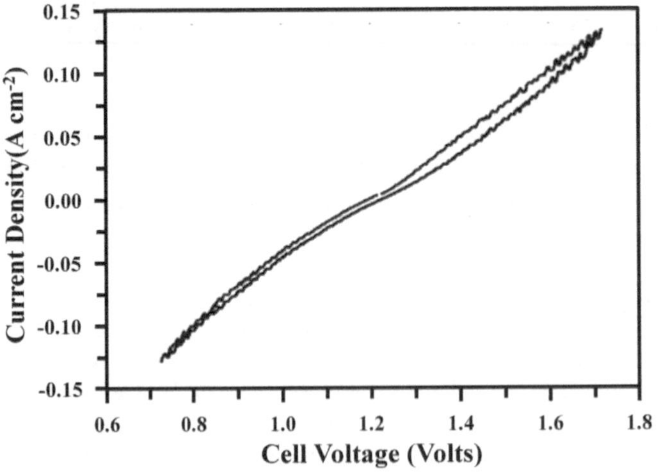

FIGURE 5.6 Cyclic voltammogram shows the open circuit potential of the all-iron flow battery, which was about 1.2 V. Adapted and reproduced with permission from Ref. [62].

TABLE 5.1
Flow Consistency Index, *K*, and Flow Behavior Index, *n*, for Various Weight Percentages of MWCNTs in 1 M Ferric Chloride

Vol %	n	K (cPs^{n-1})
0.5	0.99	1.92
1.9	0.64	59.6
2.8	0.49	351
4.3	0.40	1,470

describes the shear stress in terms of the shear rate, $\delta u/\delta y$, a flow consistency index, K, and a flow behavior index, n, as related through Equation 5.3 where K – a flow consistency index, n – flow behavior, $\delta u/\delta y$ – shear rate.

$$\tau = K\left[\frac{\partial u}{\partial y}\right]^n \tag{5.3}$$

When volume fraction of the solid particles was greater than some critical concentration, there was no longer enough liquid in the slurry to fully wet all of the particles. When volume fraction of solids was greater than critical volume fraction, slurry transitioned, quite quickly, from a shear thinning fluid to a semi-solid gel. At this point, the slurry is no longer able to flow [65]. For slurries made with the MWCNTs used in these studies, the viscosity critical volume fraction was found to be about 6 volume % (Table 5.1).

In another study [66], three different carbon particles were investigated. Asbury Carbons graciously provided 230U, a natural flake carbon with an average diameter of 20 μm, thickness of 100 nm, a reported surface area of 6.5 m^2g^{-1}, and a carbon purity of 99.4%, carbon platelets with an average diameter of 100 nm, thickness of 2.5–10 nm, a reported surface area of 250 m^2g^{-1}, and a carbon purity of >99%. MWCNTs had an OD of 50–80 nm, length of 10 μm, reported surface area of 40 m^2g^{-1}, and a carbon purity of >95%. For these studies, all carbon particles were used as received. Before tests were performed, particles were mixed by hand in electrolyte and then pumped through the system for at least 30 min. When using ferrous salts, solutions were made under a continuous nitrogen blanket and experiments were performed with the electrolyte under a continuous nitrogen purge to mitigate the air oxidation of ferrous to ferric iron. All tests were conducted at room temperature (20°C–25°C). The electrolyte used in these studies was 1.0 M $FeCl_2$, 1.0 M $FeCl_3$, and 1.0 M NH_4Cl. This electrolyte simulates the composition of iron electrolytes at 50% state-of-charge. The ferric/ferrous couple was investigated on slurry electrodes made with MWCNTs. Focus was put on these particles due to their higher electronic conductivity while suspended in a flowing slurry and due to their smaller primary particle size. As shown in Figure 5.7, at high frequencies (>10^3 Hz) - the impedance is dominated by the inductance and the parallel ohmic resistance. At intermediate frequencies (10–10^3 Hz) -the characteristics of the charge transfer reaction are present. The third region (<10^2 Hz) - the impedance plots are effects due to mass transfer and the advective current [67].

Of these three regions, only the effects due to mass transfer and advective current should depend on the flow rate of the slurry electrode. Impedance due to both mechanism decreases with increasing flow rate. Comparing the EIS results as a function of flow rate (curves b-d in Figure 5.7) to the analytical model shows that there are non-negligible low frequencies effects due to inductance, ohmic resistance or charge transfer reaction at all flow rates investigated. Mass transfer and the advective current are only significant at lower frequencies [60].

In another study [66], EIS has long been applied to a wide variety of fluidized bed and slurry electrodes. None of the papers, however, took into account the effect that the convection of the surface charge has on the resulting impedance [68] The work of Gabrielli et al. [68] for example,

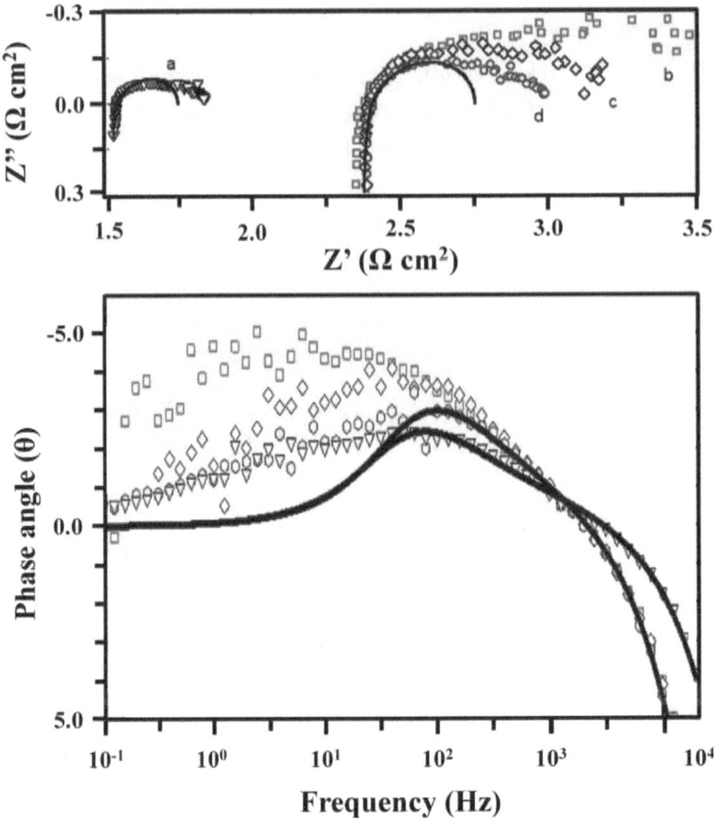

FIGURE 5.7 Nyquist (upper) and bode phase angle (lower) representations of EIS results on MWCNT slurry electrodes supporting the ferrous/ferric redox reaction. Adapted and reproduced with permission from Ref. [66].

looked exclusively at fluidized bed electrodes, where velocities are small and, hence, the convective effect is negligible. Similarly, Youssry et al. [69] examined slurry electrodes in a membrane-less cell with no inlets or outlets, thereby eliminating any effects of the convective current mechanism. Petek et al. [66] considered flowing electrosorptive slurries and pointed out that convective current does have an effect on the EIS spectra, but no attempt was made to quantitatively address the effect. Instead, the EIS spectra were analyzed at high frequencies only, where convection is irrelevant and the slurry acts as a quasi-steady porous electrode. To truly characterize a slurry for use in an IRFB or other slurry electrode application, the current induced by convection is of paramount importance as it controls the low-frequency behavior, and, hence, the steady-state performance. Any proper EIS model of a flowing slurry electrode system must therefore take this effect into account. This chapter develops the mathematical models that are necessary to fully characterize slurry electrode performance and introduces an equivalent circuit element that can be used to quantify slurry electrode properties. Local velocity gradient at the wall (G), as shown in Figure 5.8, is calculated from flow rate (Q), using the following formula (Equation 5.4) [70]:

$$G = \frac{2\,(1+2n)}{n}\frac{Q}{\omega\delta^2} \tag{5.4}$$

where w is the width of the channel, δ is the channel gap, and n is the flow index.

Figure 5.8 demonstrates how increased flow rates lead to lower impedance in the limit of low frequencies. At high frequencies, however, the flow rate does not affect the impedance (except perhaps

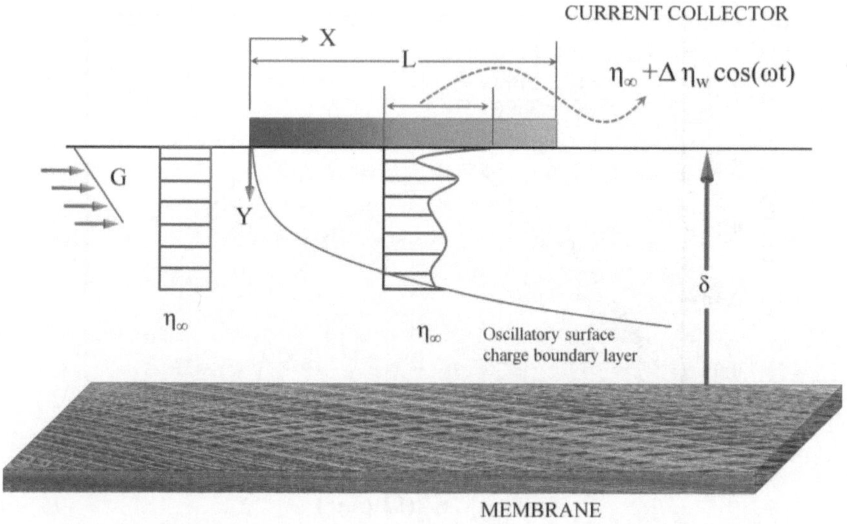

FIGURE 5.8 Schematic diagram showing oscillating EIS perturbation applied to flowing electrosorptive slurry. Adapted and reproduced with permission from Ref. [66].

through alteration to underlying intrinsic properties such as the electronic conductivity). This is the region where quasi-static behavior occurs such that the applied EIS perturbation frequency is so fast relative to the motion of the slurry particles that the conductive matrix acts identically to a stationary porous electrode [71]. The impedance element developed in these section only accounts for the thin surface charge boundary layer adjacent to the current collector surface. In Figure 5.9, presence of two loops indicates that the first loop is due to the contact resistance while other plots show the variation with flow rate.

Figure 5.10 shows how σ and α_e, both intrinsic properties of the slurry electrode, are altered by flow [72]. As can be seen, σ changes only very slightly with respect to flow rate (going from 1.30 S m^{-1} at 50 mL min^{-1} to 1.28 S m^{-1} at 250 mL min^{-1}). This suggests that the inter-particle percolation networks near the wall aren't altered much by changes in the flow rates in this range. There is a relatively large increase in the electronic diffusivity, α_e, with respect to flow rate (going from $2.13\times10^{-8} m^2 s^{-1}$ at 50 mL min^{-1} to $2.79\times10^{-8} m^2 s^{-1}$ at 250 mL min^{-1}). This suggests that a reduction in the specific capacitance of the slurry is occurring that is associated with underutilization of micropores in the activated carbon particles at higher flow rates.

5.4.2.2 Factors Affecting the Iron Plating Reaction

The slurry electrodes face challenges in terms of ohmic resistance which is proportional to slurry electronic conductivity. Even though efforts have been made to decrease ohmic resistance of the electrode, such attempts gave rise to advertent pressure drops leading to parasitic pumping losses. Due to the negative potential of the iron deposition/dissolution reaction, evolution of hydrogen can be a significant source of efficiency loss in the all-Iron redox flow battery [73]. The coulombic efficiency of iron plating will depend on the pH of the electrolyte, as well as the pH at the electrode surface [63]. Increasing the pH of the electrolyte will shift the equilibrium potential for hydrogen evolution more negative and decrease the diffusion limited current. However, Fe^{2+} will precipitate as a hydroxide species before the pH is sufficiently high to completely avoid hydrogen evolution in the potential range for iron plating. If the hydrogen evolution current is large enough, bubbles can cling to the electrode surface and cause the deposit to form around the bubble. Corrosion researchers have investigated the effect of anions on the kinetics of iron corrosion. Altering the anionic species and its concentration in the supporting electrolyte of an iron flow battery may lower the rate of hydrogen

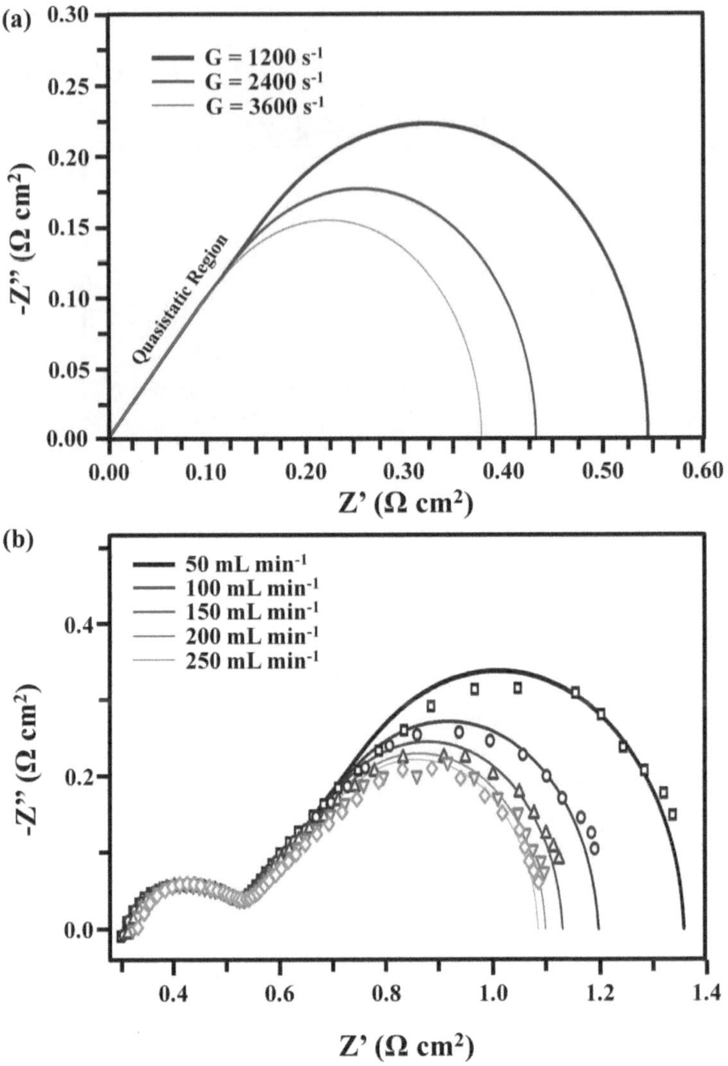

FIGURE 5.9 Nyquist plots for *Z* with (a) different indicated shear rates at the wall and (b) at different flow rates. Adapted and reproduced with permission from Ref. [71].

evolution occurring at the negative electrode and increase the plating efficiency. Raising the pH of the battery electrolyte will eventually result in precipitation of the iron species, especially for the ferric ions in the positive electrolyte. The addition of complexing ligands to an electrolyte has been shown to increase the ferric ion solubility at elevated pH values. Even if the ligand is only added to the positive electrolyte, crossover will occur throughout the life of the battery and the ligand will eventually appear in the negative electrolyte as well. It is likely that ligands will affect the iron plating reaction. The ferric-citrate complex is soluble at pH values greater than 5, though the $Fe^{2+/3+}$ redox reaction potential is shifted negative by about 600 mV. When the plating reaction with citrate was studied, it was found that the open circuit potential for the $Fe^{2+/0}$ reaction was shifted about 400 mV more negative [74]. Plating from an iron (II)-EDTA complex did not shift the plating potential negatively; the EDTA is electrochemically reduced at about 300 mV more positive than iron plating. Two ligands that showed promise in the positive half-cell for maintaining solubility of ions of $Fe^{2+/3+}$ redox couple are examined here to understand the effect of the ligand on the iron plating reaction [67]. The effect of the composition and concentration of the supporting electrolyte species on iron plating and

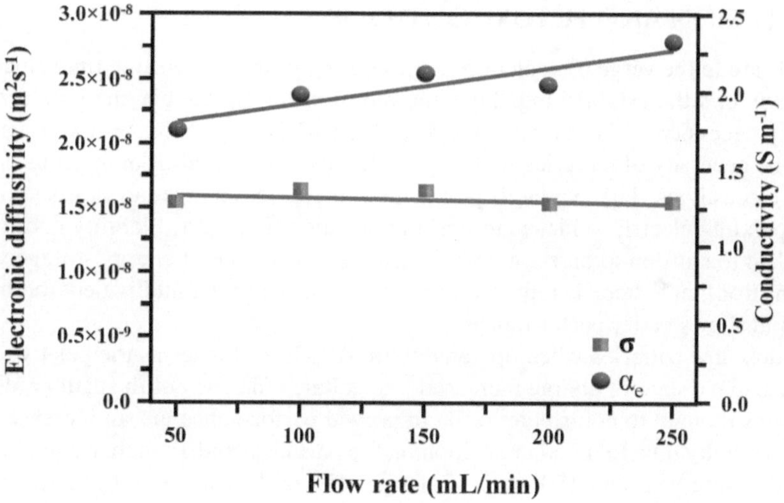

FIGURE 5.10 Flow rate dependence of σ and α_e. Adapted and reproduced with permission from Ref. [72].

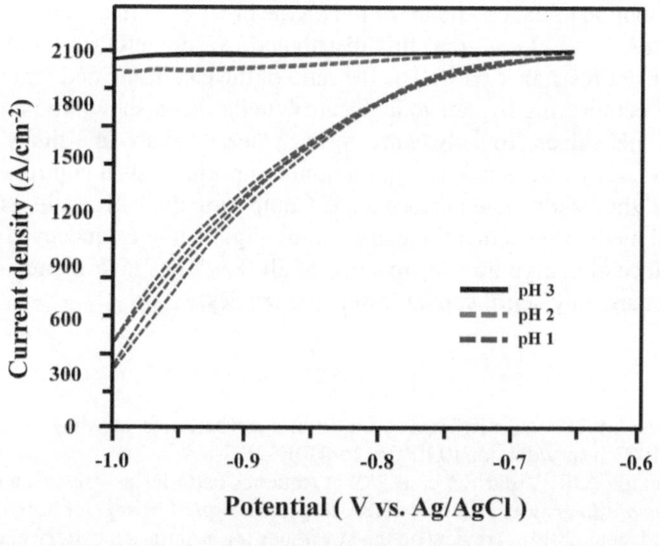

FIGURE 5.11 Cyclic voltammetry at 25 mV s^{-1} of hydrogen evolution on iron as a function of pH, with a constant 3 M Cl^- (with HCl and the remaining chloride from NH_4Cl). Adapted and reproduced from Ref. [72].

hydrogen evolution are also investigated with the goal of maximizing the plating efficiency in an iron flow battery. Due to the negative potential of the iron deposition/dissolution reaction, evolution of hydrogen can be a significant source of efficiency loss in the all-Iron flow battery. Hydrogen evolution unbalances the electrolyte, which raises the pH in both the negative and the positive electrolyte and causes precipitation of Fe^{+3} as $Fe(OH)_3$. Ultimately, the higher pH will cause precipitation of the reactants and cause failure in the battery. The hydrogen evolution reaction, in absence of iron plating, was measured directly through slow-scan cyclic voltammetry as a function of pH on an iron rod working electrode as shown in Figure 5.11. The current density for hydrogen evolution was 15 mA cm^{-2} at a pH of 1, 2.5 mA cm^{-2} at a pH of 2, while raising the pH to 3 further reduces the H_2 evolution to a current density of 0.7 mA cm^{-2} [75,76]. This suggest that even at a pH of 3, the hydrogen evolution will contribute to columbic efficiency losses, so additional approaches must be taken to suppress hydrogen evolution.

5.5 CONCLUSION AND FUTURE OUTLOOK

The fossil fuels are in the verge of depletion in the coming years, at the same time emerging renewable energy source is the establishing. Thus the requirement of suitable energy storage device is one of primary necessity in the modern world. Lithium-ion energy storage device gained a lot of acceptance. The necessity of batteries with higher density has shifted from aqueous electrolytes to non- aqueous electrolytes. The availability of lithium metal for storage devices is still a blank end in front of expanding electric vehicles and grid application. Lesser availability and higher cost of lithium provoked to shift on to an economical, also a green source of energy storage which can be recycled again. Iron-ion battery is reported to be the suitable among multivalent ion battery which is recyclable and shows better performance.

All-iron redox flow batteries when operated with MWCNT slurries as the positive and negative electrodes, the battery state-of charge increased, the voltaic efficiency of the battery also increased. Two mechanisms thought to contribute to the improved performance are an increase in the slurry electronic conductivity (due to metal deposition on the particles) and an increase in plating kinetics. After cycling the battery around 25% state-of-charge, <5% of the charge of the battery was observed as iron metal on the current collector. In addition to the traditional overpotentials associated with electrochemical electrodes at higher state-of-charge, there exists an overpotential due to the distributed nature of the current between the electronic and ionic phases of the slurry. This overpotential is a significant contribution to the cell polarization on the order of both the charge transfer and mass transfer overpotentials. It was found that this distributed overpotential is a function of the slurry electrode charge transfer resistance as well as the ratio of the electronic and ionic phase conductivities. The addition of complexing ligands to an electrolyte has been shown to increase the ferric ion solubility at elevated pH values. To truly characterize a slurry for use in a slurry electrode application, the current induced by convection is of paramount importance as it controls the low-frequency behavior, and, hence, the steady-state performance. Comparing the EIS results as a function of flow rate to the analytical model shows that there are non-negligible low-frequency effects due to inductance, ohmic resistance or charge transfer reaction at all flow rates investigated. Mass transfer and the advective current are only significant at lower frequencies.

REFERENCES

1. Mauger A, Julien CM (2017) Critical review on lithium-ion batteries: are they safe? Sustainable? *Ionics (Kiel)* 23:1933–1947. https://doi.org/10.1007/s11581-017-2177-8
2. Posada JOG, Rennie AJR, Villar SP, et al (2017) Aqueous batteries as grid scale energy storage solutions. *Renew Sustain Energy Rev* 68:1174–1182. https://doi.org/10.1016/j.rser.2016.02.024
3. Lu L, Han X, Li J, et al (2013) A review on the key issues for lithium-ion battery management in electric vehicles. *J Power Sources* 226:272–288. https://doi.org/10.1016/j.jpowsour.2012.10.060
4. Arora P, White RE, Doyle M, Carolina S (1998) Capacity fade mechanisms and side reactions in lithium - ion batteries. *J Electrochem Soc* 145:3647–3667
5. Canepa P, Sai Gautam G, Hannah DC, et al (2017) Odyssey of multivalent cathode materials: open questions and future challenges. *Chem Rev* 117:4287–4341. https://doi.org/10.1021/acs.chemrev.6b00614
6. Armand M, Tarascon J-M (2008) Building better batteries. *Nature* 451:652–657.
7. Bruce PG, Freunberger SA, Hardwick LJ, Tarascon JM (2012) LiO_2 and LigS batteries with high energy storage. *Nat Mater* 11:19–29. https://doi.org/10.1038/nmat3191
8. Mahne N, Fontaine O, Thotiyl O, et al (2017) Chemical Science batteries - a perspective. *Chem Sci* 8:6716–6729. https://doi.org/10.1039/c7sc02519j
9. Vegge T, Garcia-lastra JM, Siegel DJ (2017) Lithium - oxygen batteries : at a crossroads ? *Curr Opin Electrochem* 6:100–107. https://doi.org/10.1016/j.coelec.2017.10.014
10. Larcher D, Tarascon JM (2015) Towards greener and more sustainable batteries for electrical energy storage. *Nat Chem* 7:19–29. https://doi.org/10.1038/nchem.2085
11. McKerracher RD, Poncedeleon C, Wills RGA, et al (2015) A review of the iron-air secondary battery for energy storage. *Chemplusschem* 80:323–335. https://doi.org/10.1002/cplu.201402238

12. Vijaya Kumar Saroja AP, Samantaray SS, Sundara R (2019) A room temperature multivalent rechargeable iron ion battery with an ether based electrolyte: a new type of post-lithium ion battery. *Chem Commun* 55:10416–10419. https://doi.org/10.1039/c9cc04610k
13. Besenhard JO, Winter M (2002) Advances in battery technology: rechargeable magnesium batteries and novel negative-electrode materials for lithium ion batteries. *ChemPhysChem* 3:155–159. https://doi.org/10.1002/1439-7641(20020215)3:2<155::AID-CPHC155>3.0.CO;2–S
14. Palomares V, Serras P, Villaluenga I, et al (2012) Na-ion batteries, recent advances and present challenges to become low cost energy storage systems. *Energy Environ Sci* 5:5884–5901. https://doi.org/10.1039/c2ee02781j
15. Kulova TL, Fateev VN, Seregina EA, Grigoriev AS (2020) A brief review of post-lithium-ion batteries. *Int J Electrochem Sci* 15:7242–7259. https://doi.org/10.20964/2020.08.22
16. Walter M, Kovalenko MV, Kravchyk KV (2020) Challenges and benefits of post-lithium-ion batteries. *New J Chem* 44:1677–1683. https://doi.org/10.1039/c9nj05682c
17. Wang Y, Chen R, Chen T, et al (2016) Emerging non-lithium ion batteries. *Energy Storage Mater* 4:103–129. https://doi.org/10.1016/j.ensm.2016.04.001
18. Xu C., Chen Y., Shi S, et al (2015) Secondary batteries with multivalent ions for energy storage. *Sci Rep* 5: 14120. https://doi.org/10.1038/srep14120
19. Mukherjee S, Singh G (2019) Two-dimensional anode materials for non-lithium metal ion batteries. *ACS Appl Energy Mater* 2:932–955. https://doi.org/10.1021/acsaem.8b00843
20. Ponrouch A, Bitenc J, Dominko R, et al (2019) Multivalent rechargeable batteries. *Energy Storage Mater* 20:253–262. https://doi.org/10.1016/j.ensm.2019.04.012
21. Selvakumaran D, Pan A, Liang S, Cao G (2019) A review on recent developments and challenges of cathode materials for rechargeable aqueous Zn-ion batteries. *J Mater Chem A* 7:18209–18236. https://doi.org/10.1039/c9ta05053a
22. Novák P, Scheifele W, Joho F, Haas O (1995) Electrochemical insertion of magnesium into hydrated vanadium bronzes. *J Electrochem Soc* 142:2544–2550. https://doi.org/10.1149/1.2050051
23. Novák P, Desilvestro J (1993) Electrochemical insertion of magnesium in metal oxides and sulfides from aprotic electrolytes. *J Electrochem Soc* 140:140–144. https://doi.org/10.1149/1.2056075
24. Guduru RK, Icaza JC (2016) A brief review on multivalent intercalation batteries with aqueous electrolytes. *Nanomaterials* 6:41. https://doi.org/10.3390/nano6030041
25. Amatucci GG, Badway F, Singhal A, et al (2001) Investigation of yttrium and polyvalent ion intercalation into nanocrystalline vanadium oxide. *J Electrochem Soc* 148:A940. https://doi.org/10.1149/1.1383777
26. Yang, D, Zhou, Y, Geng, H, Liu, C, Lu, B, Rui, X, Yan, Q (2020) Pathways towards high energy aqueous rechargeable batteries. *Coord Chem Rev* 424:213521. https://doi.org/10.1016/j.ccr.2020.213
27. Aurbach D, Schechter A, Gofer Y (2000) Prototype systems for rechargeable. *Nature* 407:10–14. https://doi.org/10.1038/35037553
28. Aurbach D, Lu Z, Schechter A, et al (2010) ChemInform abstract: prototype systems for rechargeable magnesium batteries. *ChemInform* 32:407. https://doi.org/10.1002/chin.200103014
29. Schaefer JL, Merrill LC (2019) Multivalent metallic anodes for rechargeable batteries. *Encycl Inorg Bioinorg Chem* 5:1–14. https://doi.org/10.1002/9781119951438.eibc2678
30. Wu X, Xu Y, Markir A, et al (2019) A rechargeable battery with iron metal anode. *Adv Funct Mater* 29:1900911. https://doi.org/10.1002/adfm.201900911
31. Kravchyk KV, Kovalenko MV (2019) Rechargeable dual-ion batteries with graphite as a cathode: key challenges and opportunities. *Adv Energy Mater* 9:1901749. https://doi.org/10.1002/aenm.201901749
32. Liu M, Rong Z, Malik R, et al (2015) Environmental science spinel compounds as multivalent battery cathodes : a systematic evaluation based on ab initio. *Energy Environ Sci* 8:964–974. https://doi.org/10.1039/c4ee03389b
33. Tang H, Peng Z, Wu L, et al (2018) Vanadium - based cathode materials for rechargeable multivalent batteries : challenges and opportunities. *Energy Storage Mater* 50:21–46.
34. Parija A, Liang Y, Andrews J, et al (2016) Topochemically de-intercalated phases of V_2O_5 as cathode materials for multivalent intercalation batteries : a first-principles evaluation. *Chem Mater* 28:16.https://doi.org/10.1021/acs.chemmater.6b01006
35. Wang RY, Wessells CD, Huggins RA, Cui Y (2013) Highly reversible open framework nanoscale electrodes for divalent ion batteries. *Nano Lett* 13:5748–5752. https://doi.org/10.1021/nl403669a
36. Rong Z, Malik R, Canepa P, et al (2015) Materials design rules for multivalent ion mobility in intercalation structures. *Chem Mater* 27:6016–6021. https://doi.org/10.1021/acs.chemmater.5b02342

37. Wu F, Yang H, Bai Y, Wu C (2019) Paving the path toward reliable cathode materials for aluminum-ion batteries. *Adv Mater* 1806510:1–18. https://doi.org/10.1002/adma.201806510
38. Daniele DL (2016) Rechargeable lithium battery using non-flammable electrolyte basing on tetraethylene glycol dimethyl ether and olivine cathodes. *J Power Sources* 334:1–26
39. Deivanayagam R, Ingram BJ, Shahbazian-Yassar R (2019) Progress in development of electrolytes for magnesium batteries. *Energy Storage Mater* 21:1–44
40. Zhao C (2020) Designing solid-state electrolytes for safe, energy-dense batteries. *Nat Rev Mater* 5: 229–252. https://doi.org/10.1038/s41578-019-0165-5
41. Zhao Q, Nuli Y, Nasiman T, et al (2012) Reversible deposition and dissolution of magnesium from imidazolium-based ionic liquids. *Int J Electrochem* 2012:2090–3529. https://doi.org/10.1155/2012/701741
42. Kim HS, Arthur TS, Allred GD, et al (2011) Structure and compatibility of a magnesium electrolyte with a sulphur cathode. *Nat Commun* 2:427. https://doi.org/10.1038/ncomms1435
43. Park M, Zhang X, Chung M, et al (2010) A review of conduction phenomena in Li-ion batteries. *J Power Sources* 195:7904–7929. https://doi.org/10.1016/j.jpowsour.2010.06.060
44. Zhirong S (2019) Towards stable and efficient electrolytes for room-temperature rechargeable calcium batteries. *Energy Environ Sci* 12:3496–3501. https://doi.org/10.1039/c9ee01699f
45. Rajput NN, Seguin TJ, Wood BM, et al (2018) Elucidating solvation structures for rational design of multivalent electrolytes—a review. *Top Curr Chem* 376:19
46. Zhang J, Yao X, Misra RK, et al (2020) Progress in electrolytes for beyond-lithium-ion batteries. *J Mater Sci Technol* 44:237–257. https://doi.org/10.1016/j.jmst.2020.01.017
47. Wang P, NuLi Y, Yang J, Feng Z (2006) Mixed ionic liquids as electrolyte for reversible deposition and dissolution of magnesium. *Surf Coatings Technol* 201:3783–3787. https://doi.org/10.1016/j.surfcoat.2006.03.020
48. Baria DN, Hulburt HM (1973) Reduction of crotonic acid with hydrogen on a slurry electrode. *J Electrochem Soc* 120:1333. https://doi.org/10.1149/1.2403257
49. Ponce de León C, Frías-Ferrer A, González-García J, et al (2006) Redox flow cells for energy conversion. *J Power Sources* 160:716–732. https://doi.org/10.1016/j.jpowsour.2006.02.095
50. Skyllas-Kazacos M, Chakrabarti MH, Hajimolana SA, et al (2011) Progress in flow battery research and development. *J Electrochem Soc* 158:R55. https://doi.org/10.1149/1.3599565
51. Lopez-Atalaya M, Codina G, Perez JR, et al (1992) Optimization studies on a Fe/Cr redox flow battery. *J Power Sources* 39:147–154. https://doi.org/10.1016/0378-7753(92)80133-V
52. Wang W, Kim S, Chen B, et al (2011) A new redox flow battery using Fe/V redox couples in chloride supporting electrolyte. *Energy Environ Sci* 4:4068–4073. https://doi.org/10.1039/c0ee00765j
53. Wei X, Li L, Luo Q, et al (2012) Microporous separators for Fe/V redox flow batteries. *J Power Sources* 218:39–45. https://doi.org/10.1016/j.jpowsour.2012.06.073
54. Price A, Bartley S, Male S, Cooley G (1999) Novel approach to utility scale energy storage. *Power Eng J* 13:122–129. https://doi.org/10.1049/pe:19990304
55. Hawthorne KL, Wainright JS, Savinell RF (2014) Maximizing plating density and efficiency for a negative deposition reaction in a flow battery. *J Power Sources* 269:216–224. https://doi.org/10.1016/j.jpowsour.2014.06.125
56. Eyer J, Corey G (2011) Energy storage for the electricity grid: benefits and market potential assessment guide. *Light a Bottle Electr Energy Storage* 2011:1–191
57. Appleby AJ, Jacquier M (1976) The C.G.E. circulating zinc/air battery: a practical vehicle power source. *J Power Sources* 1:17–34. https://doi.org/10.1016/0378-7753(76)80003-1
58. Duduta M, Ho B, Wood VC, et al (2011) Semi-solid lithium rechargeable flow battery. *Adv Energy Mater* 1:511–516. https://doi.org/10.1002/aenm.201100152
59. Presser V, Dennison CR, Campos J, et al (2012) The electrochemical flow capacitor: a new concept for rapid energy storage and recovery. *Adv Energy Mater* 2:895–902. https://doi.org/10.1002/aenm.201100768
60. Granqvist CG, Hunderi O (1978) Conductivity of inhomogeneous materials: effective-medium theory with dipole-dipole interaction. *Phys Rev B* 18:1554–1561. https://doi.org/10.1103/PhysRevB.18.1554
61. Gao L, Zhou X, Ding Y (2007) Effective thermal and electrical conductivity of carbon nanotube composites. *Chem Phys Lett* 434:297–300. https://doi.org/10.1016/j.cplett.2006.12.036
62. Petek TJ, Hoyt NC, Savinell RF, Wainright JS (2015) Slurry electrodes for iron plating in an all-iron flow battery. *J Power Sources* 294:620–626. https://doi.org/10.1016/j.jpowsour.2015.06.050

63. Hawthorne KL, Petek TJ, Miller MA, et al (2015) An investigation into factors affecting the iron plating reaction for an all-iron flow battery. *J Electrochem Soc* 162:A108–A113. https://doi.org/10.1149/2.0591501jes
64. Campos JW, Beidaghi M, Hatzell KB, et al (2013) Investigation of carbon materials for use as a flowable electrode in electrochemical flow capacitors. *Electrochim Acta* 98:123–130. https://doi.org/10.1016/j.electacta.2013.03.037
65. Youssry M, Madec L, Soudan P, et al (2013) Non-aqueous carbon black suspensions for lithium-based redox flow batteries: rheology and simultaneous rheo-electrical behavior. *Phys Chem Chem Phys* 15:14476–14486. https://doi.org/10.1039/c3cp51371h
66. Petek TJ, Hoyt NC, Savinell RF, Wainright JS (2016) Characterizing slurry electrodes using electrochemical impedance spectroscopy. *J Electrochem Soc* 163:A5001–A5009. https://doi.org/10.1149/2.0011601jes
67. Hawthorne KL, Wainright JS, Savinell RF (2014) Studies of iron-ligand complexes for an all-iron flow battery application. *J Electrochem Soc* 161:A1662–A1671. https://doi.org/10.1149/2.0761410jes
68. Gabrielli C, Huet F, Sahar A, Valentin G (1994) Dynamic analysis of charge transport in fluidized bed electrodes: impedance techniques for electroactive beds. *J Appl Electrochem* 24:481–488. https://doi.org/10.1007/BF00249846
69. Dennison CR, Gogotsi Y, Kumbur EC (2014) In situ distributed diagnostics of flowable electrode systems: resolving spatial and temporal limitations. *Phys Chem Chem Phys* 16:18241–18252. https://doi.org/10.1039/c4cp02820a
70. Brunini VE, Chiang YM, Carter WC (2012) Modeling the hydrodynamic and electrochemical efficiency of semi-solid flow batteries. *Electrochim Acta* 69:301–307. https://doi.org/10.1016/j.electacta.2012.03.006
71. Hoyt NC, Wainright JS, Savinell RF (2015) Current density scaling in electrochemical flow capacitors. *J Electrochem Soc* 162:A1102–A1110. https://doi.org/10.1149/2.0031507jes
72. Bisquert J (2000) InÑuence of the boundaries in the impedance of porous Ðlm electrodes.pdf. *Phys Chem Chem Phys* 2:4185–4192.
73. Hoyt NC, Agar E, Nagelli EA, et al (2018) Electrochemical impedance spectroscopy of flowing electrosorptive slurry electrodes. *J Electrochem Soc* 165:E439–E444. https://doi.org/10.1149/2.0051810jes
74. Vukosav P, Mlakar M, Tomisi V (2012) Revision of iron(III)–citrate speciation in aqueous solution. Voltammetric and spectrophotometric studies. *Anal Chim Acta* 745: 85–91.
75. Horányi G (2004) Investigation of the specific adsorption of $HSO_4^-(SO_4^{2-})$ and Cl^- ions on Co and Fe by radiotracer technique in the course of corrosion of the metals in perchlorate media. *Corros Sci* 46:1741–1749. https://doi.org/10.1016/j.corsci.2003.11.002
76. Hackerman N and Stephens SJ (1954) The adsorption of sulfate ions from aqueous solutions by iron surfaces. *J Phys Chem* 58:904–908. https://doi.org/10.1021/j150520a025

6 Calcium Ion Batteries

Anjumole P. Thomas, Akhila Das, Anandu M. Nair, Devika S. Lal, Jou-Hyeon Ahn, M. V. Reddy, and Prasanth Raghavan

6.1 INTRODUCTION

An efficient and reliable energy storage system (ESS) is inevitable for driving the society from renewable resources such as hydro, wind, and solar energies [1–4]. High-energy-density rechargeable lithium-ion batteries (LIBs) at present dominate the portable electronics market and the electric vehicle (EV) market [5]. This innovation has been fruitful because of the generally high energy density of these batteries, their great cycle life, and dependable performance characteristics. In any case, development in the EV market has been tempered by the restricted driving range attainable with lithium-ion innovation, and there is a consistent strain to expand the range by expanding the energy density of the cells. High-energy-density LIBs are dominating the consumer electronics market and EVs [6]. In contrast, its application is limited by the low natural occurrence of lithium metal, high cost, and safety issues [7–10]. It is generally yielded that lithium-ion technology is arriving at the restrictions of its energy density capacities and that new "beyond lithium-ion" advances are needed to fill the gap. This thought, just as the drive to reduce costs, has incited a whirlwind of research activity in alternative chemistries, for example, Li-air [5], Li-S [11], Na-ion, and multivalent ion technologies [12].

Inspecting the practicality of calcium-ion batteries (CIBs), Ca is the fifth-most plentiful component in Earth's crust, more bountiful than both sodium and magnesium, and multiple times more plentiful than lithium. The ready accessibility of calcium would convert into low material cost for battery production. Ca is additionally nontoxic and along these lines would not represent an ecological peril whenever utilized in mass battery production. Furthermore, the deposition potential of calcium is just 0.17 V higher than that of lithium and 0.5 V lower than that of magnesium, conceivably prompting higher cell voltages (controlled by the distinction between the anode and the cathode potential) and, along these lines, higher energy densities contrasted with magnesium ion batteries (MIBs), for instance. Despite the fact that calcium is generally bigger in size and that we consider the mass on a for-each-electron basis, calcium is really lighter than sodium. The ionic size of Ca^{2+} particles is 1.00 Å, which is basically the same as that of Na^{+} (1.02 Å) yet essentially bigger than Li^{+} (0.76 Å) and Mg^{2+} (0.72 Å). Ca^{2+} is hence an alluring particle to consider for multivalent particle innovation. Investigation into CIBs is as yet in its earliest stages since the reversible deposition and stripping of calcium in common electrolytes were first shown in 2016, but matches with other multivalent particle frameworks, especially MIBs, which are generally well-informed in comparison, should prompt a fast increment in advancements in this area.

6.2 HISTORY OF CALCIUM ION BATTERIES

The first calcium-$SOCl_2$-based ESS was developed before 1990, but its technology was unknown to many. In 1964 Justus et al. developed a calcium-based thermal battery, which worked only in higher temperatures (450°C) [13]. Later several researches carried on to develop suitable electrolyte material for this thermal battery [14–16]. In 1980 Staniewicz et al. reported a calcium thionyl-chloride-based thermal battery where a passivation layer formed on the surface anode was first reported [17]. Later this Ca-$SOCl_2$ system was studied, and possible factors for the passivation layer were determined [18–20]. In 1988 Pujare et al. developed a Ca–O_2 secondary battery, which was

DOI: 10.1201/9781003310174-6

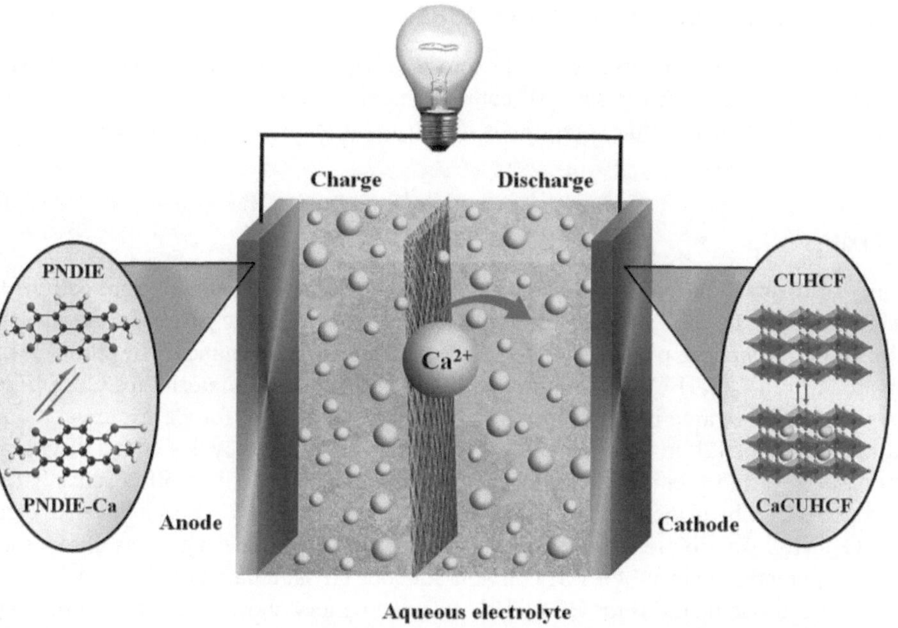

FIGURE 6.1 Schematic illustration of the polyimide–Ca_xCuHCF aqueous rechargeable Ca-ion battery [26].

operated for 52 h at 850°C [21]. In 1991 electrochemical studies of different electrolytes conducted in calcium batteries found that the passivation layer formed on the surface of the anode hampered the Ca^{2+} movement in the electrolytes system [22]. The reversible deposition/dissolution of Ca^{2+} in CIBs is impossible at that time, so it limited the application of calcium metal anode in CIBs [1]. See et al. [23] reported a Ca–S reversible battery with an initial discharge capacity of 600 mAh g^{-1}. In 2016, the first reversible CIB at moderate temperature was developed (100°C) [24]. Recently Bruce et al. reported plating/stripping of Ca^{2+} ion, which sheds light on the development of CIBs in the future [25] (Figure 6.1). The future of CIBs may rely on the development of suitable electrolytes and diffusion kinetics of Ca^{2+} ions in this system.

6.3 CHALLENGES FOR DEVELOPING CALCIUM ION BATTERIES

Among other multivalent metal ions, calcium ion technologies have attracted tremendous attention due to their low cost and natural abundance [27–29]. The standard reduction potential of Ca/Ca^{2+} is –2.87 V, and the volumetric capacity is 2,073 mAh cm^{-3}. The charge density and polarization of Ca^{2+} are relatively smaller than its counterparts (Mg^{2+}, Al^{3+}, and Zn^{2+}), and also Ca^{2+} ions show better diffusion and higher power density than other metal ions [27,30]. Despite the advantages of CIBs such as low cost and high charge density, there are several challenges to their development. Calcium metal possesses high capacity; however, the poor plating of Ca^{2+} ions makes them less feasible for the development of CIBs [1,31]. The columbic efficiency of current CIBs (96%) is not sufficient for rechargeable batteries. This is mainly attributed to the passivation layer formed on the surface of the anode, which constitutes $CaCO_3$, CaF_2, $Ca(OH)_2$, and so on, which block Ca^{2+} migration [22]. In addition to this, the high charge density of calcium makes it less dissociative in the electrolyte system and also less diffusive [31]. The strong association with the surroundings hence forms clusters, which act as a barrier to the diffusion mechanism and thus lead to poor power performance [32,33]. Although magnesium and calcium have equal charge densities, calcium has a larger effective ionic radius (Mg^{2+}; 0.66 Å, Ca^{2+}; 0.99 Å), and therefore, large channels are required for the Ca^{2+} diffusion. Similarly during intercalation, large volume expansion is found in Ca^{2+} hosts, which is always the reason for electrode failure [34].

6.4 CATHODE MATERIALS USED IN CIBS

In the case of CIB, several materials are experimented with for cathode, but only a few show good performance in electrochemical tests. CIB cathode materials are classified into four categories: (1) oxide materials, (2) chalcogenide compounds, (3) Prussian blue analogues (PBAs), and (4) others (fluorides, polyanions, etc.).

6.4.1 Oxides

As a family of diverse compositions and structures, oxides provide various applications for functional materials. The layered structure of some oxides provides large space for metal ion intercalation and offers promising possibilities for electrochemical performance. Xu et al. [35] reported that bilayered $Mg_{0.25}V_2O_5{\cdot}H_2O$ was identified as a potential cathode material for CIBs (Figure 6.2), which possesses a large area in between the layers, which is enough for Ca^{2+} intercalation. It has a high initial capacity of 120 mAh g^{-1}, stable cycling performance (50 cycles with a discharge capacity of 90 mAh g^{-1}), and relatively good capacity retention (86.9%) after 500 cycles. Tirado et al. [36] developed a CIB by using $CaCO_2O_4$ as a positive electrode and V_2O_5 as a negative electrode in 1 M $Ca(ClO_4)_2{\cdot}H_2O$ electrolyte. Within 30 cycles, a capacity of 80 mAh g^{-1} was observed, and this $CaCO_2O_4$ is a potential material for full-cell non-aqueous CIBs. Palacin et al. [37] reported α-V_2O_5 as a promising cathode material for CIBs. This full-cell battery showed electrochemical properties only above 100°C in dry alkyl-carbonate-based electrolyte (Figure 6.3). The author also proposed the effect of parasitic reactions in battery systems from the inference of adding moisture content (which enhances the capacity) in this CIB [1,38].

6.4.2 Prussian Blue Analogues

PBAs are a class of compounds having attractive abilities such as monovalent ion (Na^+, K^+, etc.) storage; researchers have also explored their divalent storage, particularly in energy storage devices

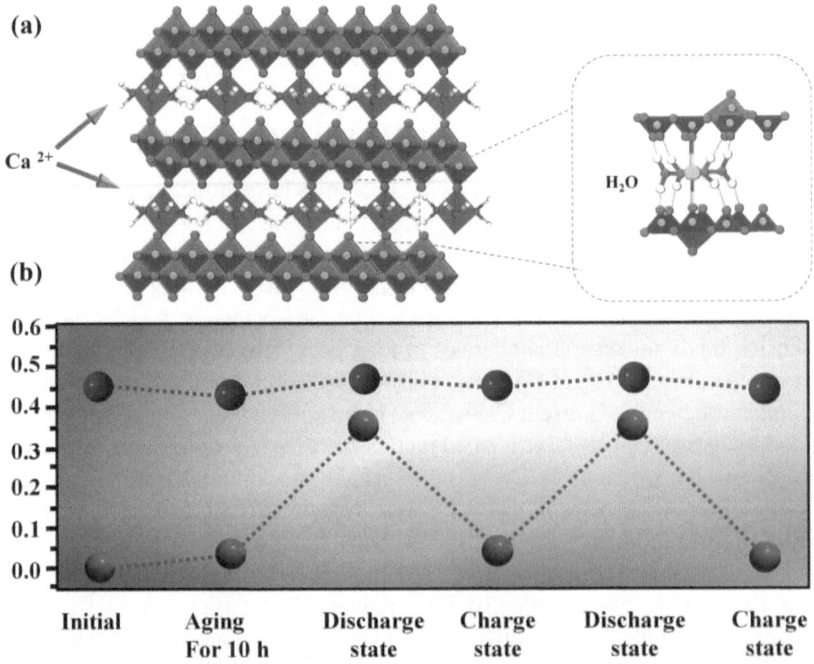

FIGURE 6.2 Bilayered $Mg_{0.25}V_2O_5{\cdot}H_2O$ as a stable cathode for rechargeable Ca-ion batteries [35].

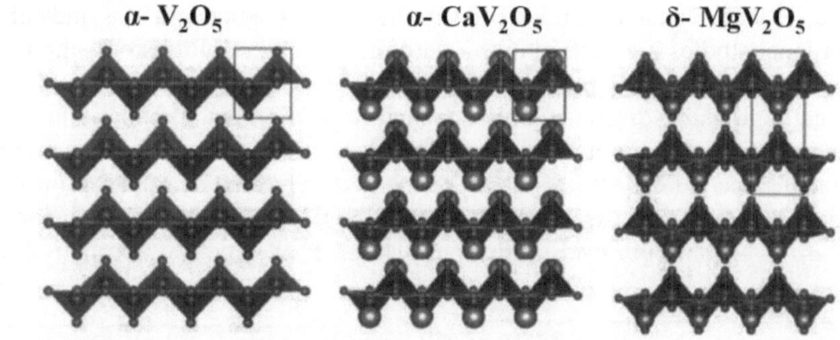

FIGURE 6.3 Crystal structure of the pristine α-V_2O_5, α-CaV_2O_5 polymorphs on the b–c plane, and δ-MgV_2O_5 polymorph on the a–b plane [37].

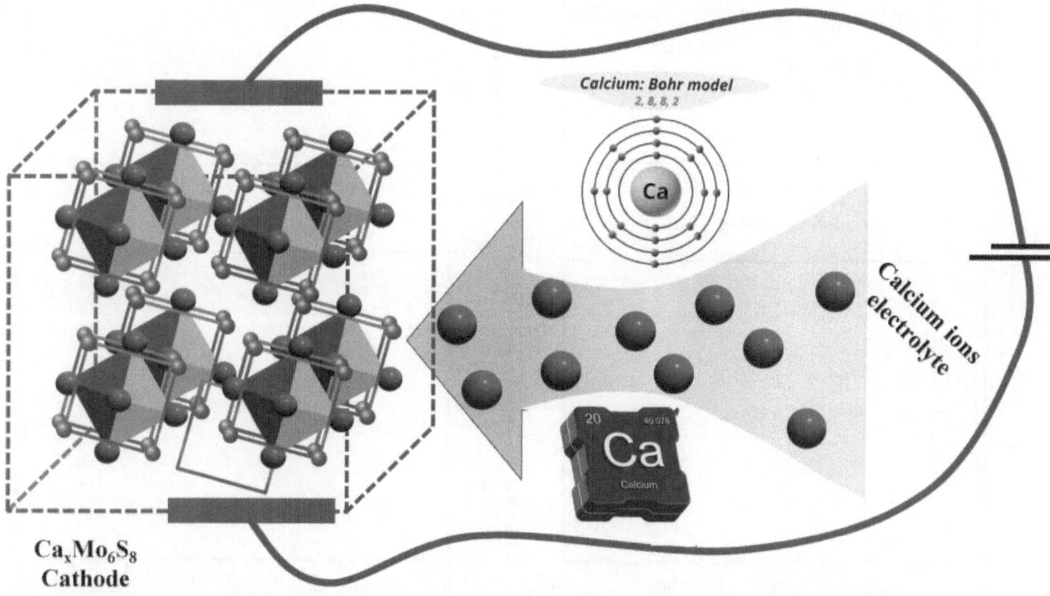

FIGURE 6.4 Theoretical investigation of cheveral phase materials for cathodes accommodating Ca^{2+} ions [44].

[26,39,40,41]. Yao et al. [26] studied the insertion of Ca^{2+} into $CuFe(CN)_6$ (CuHCF) in 2.5 M $Ca(NO_3)_2$, which showed an initial capacity of about 46 mAh and a retention of 88% after 2,000 cycles. Many researchers studied the electrochemical performance of PBAs in non-aqueous electrolytes. Ingram et al. [42] investigated the Ca^{2+} ion intercalation in de-sodiated NaMnFe $(CN)_6$ (MFCN) in 0.2 M $Ca(PF_6)$ in the ethylene carbonate : propylene carbonate (EC:PC) electrolyte system; it provided 75 mAh g^{-1} discharge capacity and 50% capacity retention after 30 cycles. Recently there have been several reports published on the PBA cathode material with mixed electrolytes. Lee et al. [43] developed a mixed solution of $Ca(CF_3SO_3)_2$ dissolved in water and PC, where CuHCF as the cathode material showed a capacity of 65 mAh g^{-1}. A columbic efficiency of 100% and a capacity retention of 94% (after 800 cycles) were observed. There are many reports on PBAs used in different electrolyte systems (aqueous, non-aqueous, and mixed electrolyte). Except for non-aqueous systems, others show good reversible deposition of Ca^{2+} ions and excellent capacity retention (Figure 6.4).

6.4.3 Chalcogenides

Chalcogenides are an important class of materials, widely used as cathode materials in multivalent ion batteries. For example, the cheveral phase Mo_6X_8 is an efficient cathode material used in magnesium batteries [45]. Due to the similarities in the size of magnesium ions and calcium ions, several computational studies are carried out to determine the possibilities of using the cheveral phase material as the cathode in CIBs. Zaghib et al. [46] investigated the Ca^{2+} interaction in the cheveral phase cathode using discrete Fourier transform (DFT) calculations. Smeu et al. [47] also studied the insertion of Ca^{2+} in Mo_6S_8 byDFT calculation and reported that the voltage of inserting the first and second Ca^{2+} was 2.1 and 1.8, respectively. Ren et al. [48] proposed layered CuS porous nanocages (CS-PNCs) as a cathode material in CIBs (Figure 6.5). It showed an initial capacity of 492 mAh g^{-1} but poor cycling reversibility. Li et al. reported that patronite VS_4 is a

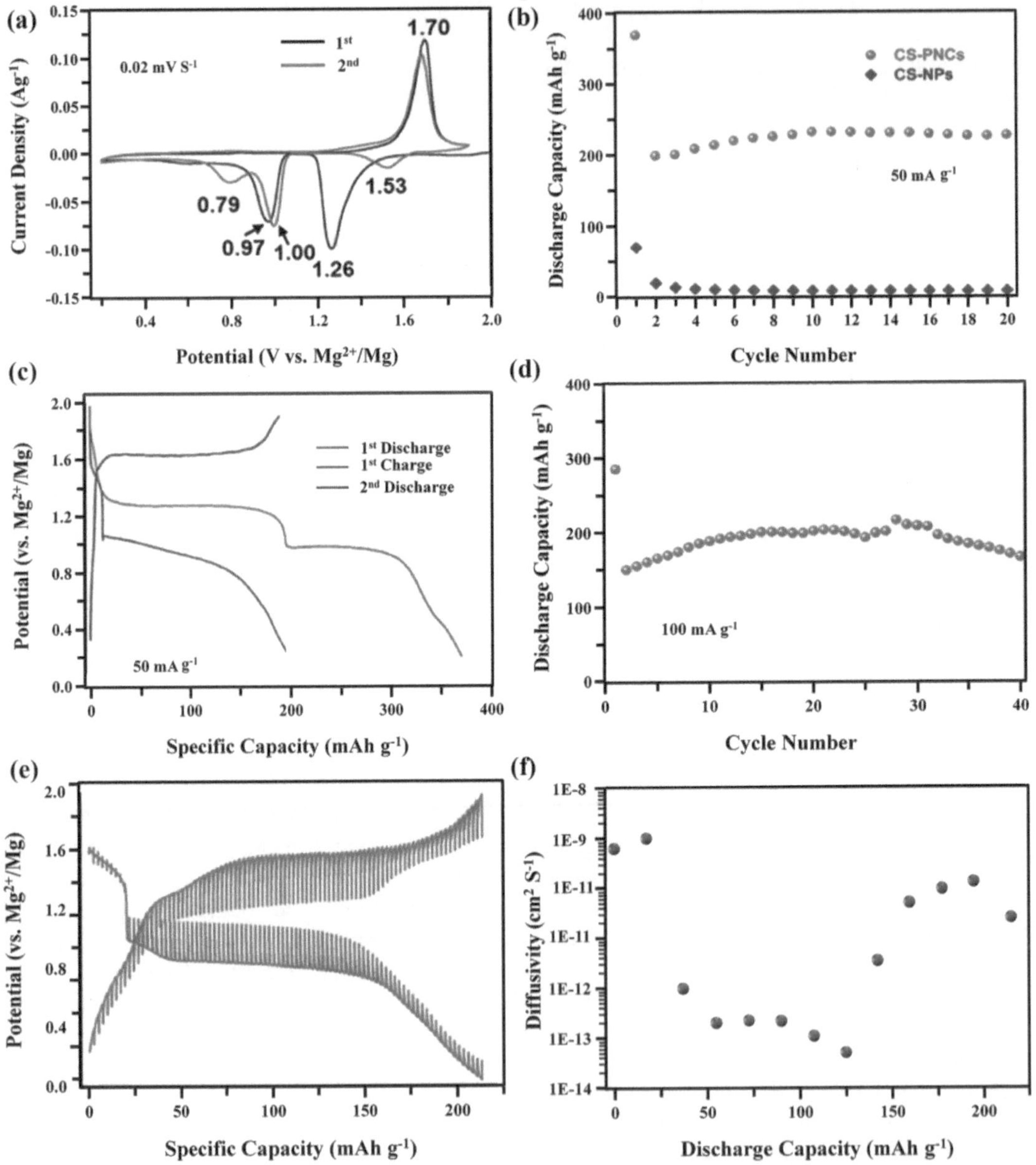

FIGURE 6.5 (a) CV curves of CS-PNCs at 0.02 mV s^{-1}; (b) cycling performance and (c) corresponding charge/discharge curves of CS-PNCs at 50 mA g^{-1}; (d) cycling performance of CS-PNCs at 100 mA g^{-1}. (e) galvanostatic Intermittent Titration Technique (GITT) curves (carried out after cycling at 20 mA g^{-1} for three cycles) and (f) Mg^{2+} diffusivity of CS-PNCs at various discharge states. Adapted and reproduced from Ref. [48].

promising cathode material for CIBs, delivering a capacity of 315 mAh g^{-1}. The cycling stability of VS_4 is lower, but further improvement of its composition definitely enhances the chances of future CIBs as an ESS.

6.4.4 Others

Sakurai et al. [49] reported $FeF_3{\cdot}0.33H_2O$-based carbon composite as an efficient cathode material for CIBs. The initial capacity of this composite was shown to be 120 mAh g^{-1}, which is thrice that of the simple $FeF_3.0.33H_2O$. Lipson et al. [50] reported polyanionic Na_2FePO_4 as the cathode material, which showed an initial capacity of 80 mAh g^{-1}. A recent study on $VOPO_4{\cdot}2H_2O$ shows that it has good cycling stability and relatively good capacity (100 mAh g^{-1}) [51]. There are several researches carried out on CIB cathode materials, but finding an ideal material is still a challenge. Computational simulations are a viable method to select material, without wasting resources and time.

6.5 ANODE MATERIALS

The use of calcium metal as an anode in CIBs with conventional electrolytes such as carbonates was not feasible at the practical level because of the SEI formation on the surface of the anode. In 2016 Palacin et al. [24] proposed the electrochemical behaviour of conventional electrolytes ($CaClO_4$, $CaBF_4$) in CIBs. No redox reaction was observed at room temperature, whereas at elevated temperature electrochemical process began in CIB. Later, Bruce et al. [25] observed room temperature–reversible deposition of Ca metal using three electrode cells in $Ca(BH_4)$ in THF electrolyte, where columbic efficiency of 94.8% was observed from cyclic voltammetry (CV) curves. Different classes of anode materials have been investigated for CIBs. In this section, we discuss the four categories of anodes used in CIBs. The latest developments and electrochemical characteristics of (1) intercalation anodes and (2) alloying anodes are explained here.

6.5.1 Intercalation Anodes

Emery et al. [52] developed the first-stage calcium-based graphite intercalation compound (GIC) called calcium carbide (CaC_6) (Figure 6.6). Later Lerner et al. [53] studied the calcium intercalation into graphite at high temperatures in $[Ca(EN)_2]C_{26}$-based GIC. In 2008 Ogumi et al. [54] reported the electrochemical characterization of graphite or highly oriented pyrolytic graphite (HOPG) in 2.0 M $Ca(CF_3SO_3)$ in dimethyl sulphoxide as an electrolyte. They claimed the intercalation of Ca^{2+} in graphite anode, but detailed information is not revealed. Electrochemical studies on graphite anode with $Ca(TFSI)_2$ in tetraglyme electrolyte showed a reversible capacity of 62 mAh g^{-1}. Kang et al. [55] proposed reversible calcium ion deposition in graphite as an anode and showed a reversible capacity of 85 mAh g^{-1} over 200 cycles.

Carbonaceous compounds such as multi-walled carbon nanotubes [56], 2D graphene [57–60], and mesocarbon microbeads (MCMBs) [61] are investigated for intercalation anodes in CIBs.

Spinel lithium titanate, $Li_4Ti_5O_{12}$, is proven as an effective anode for lithium batteries; Jeong et al. investigated the electrochemical behaviour of LTO electrodes for CIBs. LTO-based CIBs deliver a discharge capacity of ~85 mAh g^{-1}. By DFT calculations several research groups have proposed MXenes as high-capacity anodes for CIBs. Shenoy et al. [62] proposed a capacity of 318 mAh g^{-1} in Ti_3C_2 material; o-terminated and bare MXenes nanosheets also show good electrochemical performance in DFT calculations [63]. Several 2D materials like borophene [64] and GeP_3 [65] (Figure 6.7) are predicted by DFT calculations.

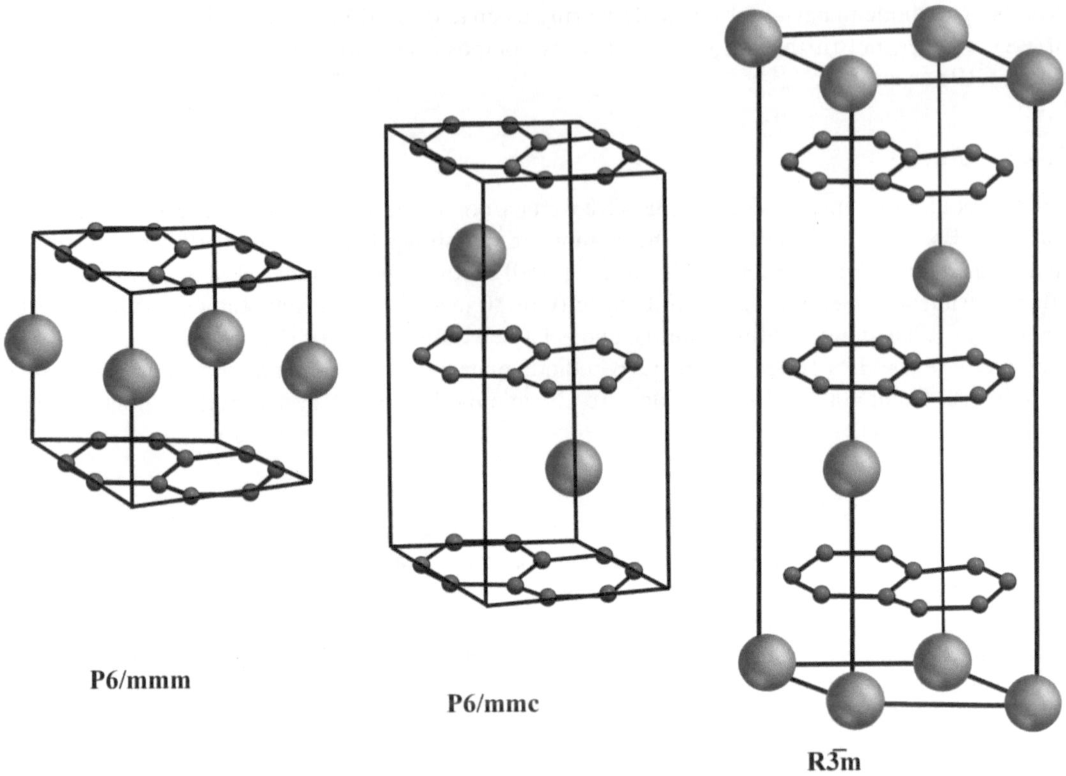

FIGURE 6.6 Possible structures for binary MC_6 graphite intercalation compounds [52].

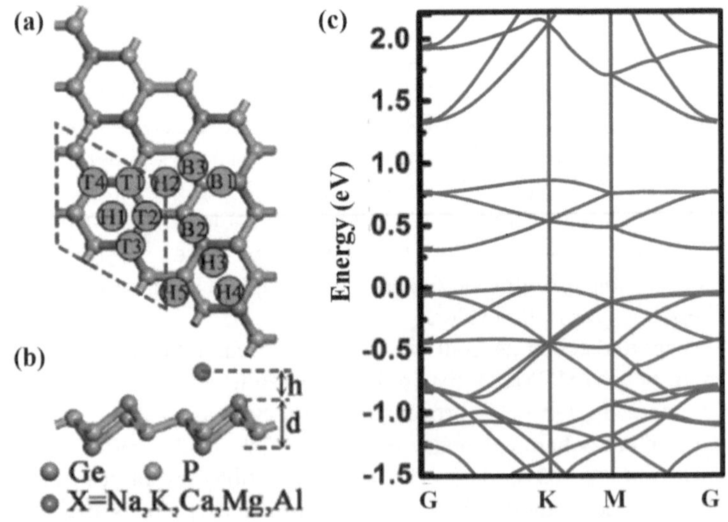

FIGURE 6.7 (a) Top and (b) side view of the crystal structure of a GeP_3 monolayer with possible adsorption sites for metal ions. The orchid and green balls represent the P and Ge atoms, respectively. The hexagonal unit cell is enclosed in the blue dashed line. (c) The calculated band structure of GeP_3 at the Perdew-Burke-Ernzerhof (PBE) level [65].

6.5.2 Alloy Anodes

Alloys are considered an efficient material for alkali metal ion batteries due to their high specific capacities and low reaction potentials [66]. Alloying systems show almost the same energy density as graphite-based anodes (in LIBs), suggesting their promising application in the CIB domain. Tran and Obrovac experimented with Ca alloying with Si, Al, Bi, Sn, and so on for forming Ca_xM phases [67]. Wolverton et al. [68] conducted DFT calculations for M–Ca systems considering calcination voltage and energy density. Ingram et al. [69] proposed CIBs that used tin alloy as an anode and PBAs as the cathode material. Palacin et al. [70] investigated the Si-based anode for CIBs using DFT simulations. Large volume expansion (306%) was observed in this experiment, resulting in structure collapse and large voltage hysteresis.

6.6 CONCLUSION

The construction of CIBs by adopting a normal pathway is difficult due to its high-temperature requirements. The main indispensable factors of CIBs are suitable electrode materials with reasonable differences in electrode potentials, electrolytes, and a better compatibility of these components. Computational investigation of materials before implementation in laboratories helps identify the potential of materials. The high-throughput computational design provides an effective programme for screening electrode materials in terms of their properties (adsorption energy, energy storage capacity, reaction potential, electronic structures, electronic conductivity, diffusion kinetics, etc.). There are only a limited number of works on CIBs; in this infancy stage, such computational design is beneficial for cost control and simplifying the subsequent experimental procedures, especially for cathode investigations. Nanotechnology can also offer a large number of materials with excellent properties such as an accessible surface area for ion exchange, short ion diffusion path length, and high electron transportation rate, which can be effectively used to modify CIBs. Exploring heterostructure materials can provide the unique properties of stable interface layer formation, synergetic effect, confined electrochemical reaction space, limited volume changes, good electrical conductivity, interfacial storage capacity, and so on, offering the additional opportunity for delivering better Ca-ion-storage ability.

Multivalent ion batteries like CIBs play an important role in the energy crisis. Despite the potential advantages of CIBs, it should be kept in mind that unless lithium price turns completely unaffordable, neither CIBs nor other multivalent metal ion batteries would be superior to LIBs regarding energy density, which is the dominant metric in portable electronics and EVs. The advantage of CIBs, as well as other multivalent metal ion batteries, is still their low cost and long-term developmental sustainability. Thus, potential application fields for CIBs lie in stationary energy storage. In addition, the development of CIBs based on rocking-chair mode is still in its infancy, and considerable efforts are required to solve fundamental scientific issues, further exploit new electrode materials, and optimize the electrolyte systems of CIBs. Meanwhile, multi-ion strategies have expanded the options of active electrode materials for a variety of CIB systems and provide new opportunities to achieve high-performance CIBs. Hence, both challenges and opportunities co-exist in the future practical applications of CIBs.

REFERENCES

1. Ji B, He H, Yao W, Tang Y (2020) Recent advances and perspectives on calcium-ion storage: key materials and devices. *Advanced Materials* 2005501:1–21. https://doi.org/10.1002/adma.202005501
2. Yan C, Lv C, Wang L, et al (2020) Architecting a stable high-energy aqueous Al-ion battery. *Journal of the American Chemical Society* 142:15295–15304. https://doi.org/10.1021/jacs.0c05054
3. Duan R, Liu Z, Wu Z, et al (2020) Quinoid conductive polymer for high-performance lithium- and sodium-ion batteries. *ACS Applied Energy Materials* 3:8179–8183. https://doi.org/10.1021/acsaem.0c01708

4. Jin Y, Zheng Z, Wei D, et al (2020) Detection of micro-scale Li dendrite via H_2 gas capture for early safety warning. *Joule* 4:1714–1729. https://doi.org/10.1016/j.joule.2020.05.016
5. Nitta N, Wu F, Lee JT, Yushin G (2015) Li-ion battery materials: present and future. *Materials Today* 18:252–264. https://doi.org/10.1016/j.mattod.2014.10.040
6. Yan R, Oschatz M, Wu F (2020) Towards stable lithium-sulfur battery cathodes by combining physical and chemical confinement of polysulfides in core-shell structured nitrogen-doped carbons. *Carbon* 161:162–168. https://doi.org/10.1016/j.carbon.2020.01.046
7. Goodenough JB, Kim Y (2010) Challenges for rechargeable Li batteries. *Chemistry of Materials* 22:587–603. https://doi.org/10.1021/cm901452z
8. Wu Y, Sun Y, Zheng J, et al (2021) MXenes: advanced materials in potassium ion batteries. *Chemical Engineering Journal* 404:126565. https://doi.org/10.1016/j.cej.2020.126565
9. Yu Y, Huang G, Wang JZ, et al (2020) In situ designing a gradient Li^+ capture and quasi-spontaneous diffusion anode protection layer toward long-life Li–O_2 batteries. *Advanced Materials* 32:1–9. https://doi.org/10.1002/adma.202004157
10. Armand M, Tarascon J-M (1997) Issues and challenges facing rechargeable lithium batteries. *Applied Radiology* 26:7–10
11. Canepa P, Sai Gautam G, Hannah DC, et al (2017) Odyssey of multivalent cathode materials: open questions and future challenges. *Chemical Reviews* 117:4287–4341. https://doi.org/10.1021/acs.chemrev.6b00614
12. Song J, Sahadeo E, Noked M, Lee SB (2016) Mapping the challenges of magnesium battery. *Journal of Physical Chemistry Letters* 7:1736–1749. https://doi.org/10.1021/acs.jpclett.6b00384
13. Selis SM, Wondowski JP, Justus RF (1964) A high-rate, high-energy thermal battery system. *Journal of the Electrochemical Society* 111:6. https://doi.org/10.1149/1.2426065
14. Clark RP, Grothaus KR (1971) An improved calcium anode for thermal batteries. *Journal of the Electrochemical Society* 118:10:1680–1683. https://doi.org/10.1149/1.2407811
15. Miles MH, Fine DA, Fletcher AN (1978) The calcium anode in molten nitrate electrolytes. *Journal of the Electrochemical Society* 125:1209–1214. https://doi.org/10.1149/1.2131651
16. Nissen DA (1979) A study of the anode-electrolyte interface in a thermal battery. *Journal of the Electrochemical Society* 126:176–180. https://doi.org/10.1149/1.2129001
17. Staniewicz RJ (1980) A study of the calcium-thionyl chloride electrochemical system. *Journal of the Electrochemical Society* 127:782–789. https://doi.org/10.1149/1.2129758
18. Peled E, Meitav A, Brand M (1981) Calcium thionyl chloride high-rate reserve cell. *Journal of the Electrochemical Society* 128:1936–1938. https://doi.org/10.1149/1.2127768
19. Meitav A, Peled E (1982) Calcium-Ca(AICI4)2-thionyl chloride cell: performance and safety. *Journal of the Electrochemical Society* 129:451.
20. Peled E, Cohen R, Melman A, Lavi Y (1992) Calcium – $SOCl_2$ cell: effect of can material and polarity on storage properties. *Journal of the Electrochemical Society* 139:1836–1839. https://doi.org/10.1149/1.2069507
21. Pujare NU, Semkow KW, Sammells AF (1988) A calcium oxygen secondary battery. *Journal of the Electrochemical Society* 135:260–261. https://doi.org/10.1149/1.2095574
22. Aurbach D, Skaletsky R, Gofer Y (1991) The electrochemical behavior of calcium electrodes in a few organic electrolytes. *Journal of the Electrochemical Society* 138:3536–3545. https://doi.org/10.1149/1.2085455
23. See KA, Gerbec JA, Jun YS, et al (2013) A high capacity calcium primary cell based on the Ca-S system. *Advanced Energy Materials* 3:1056–1061. https://doi.org/10.1002/aenm.201300160
24. Ponrouch A, Frontera C, Bardé F, Palacín MR (2016) Towards a calcium-based rechargeable battery. *Nature Materials* 15:169–172. https://doi.org/10.1038/nmat4462
25. Wang D, Gao X, Chen Y, et al (2018) Plating and stripping calcium in an organic electrolyte. *Nature Materials* 17:16–20. https://doi.org/10.1038/NMAT5036
26. Gheytani S, Liang Y, Wu F, et al (2017) An aqueous Ca-ion battery. *Advanced Science* 4:1–7. https://doi.org/10.1002/advs.201700465
27. Gummow RJ, Vamvounis G, Kannan MB, He Y (2018) Calcium-ion batteries : current state-of-the-art and future perspectives. *Advanced Materials* 1801702:1–14. https://doi.org/10.1002/adma.201801702
28. Konarov A, Voronina N, Jo JH, et al (2018) Present and future perspective on electrode materials for rechargeable zinc-ion batteries. *ACS Energy Letters* 3:2620–2640. https://doi.org/10.1021/acsenergylett.8b01552

29. Lin X, Liu J, Zhang H, et al (2020) General liquid-driven coaxial flow focusing preparation of novel microcapsules for rechargeable magnesium batteries. *Advanced Science* 2002298:1–7. https://doi.org/10.1002/advs.202002298
30. Huang H, Xu R, Feng Y, et al (2020) Sodium/potassium-ion batteries: boosting the rate capability and cycle life by combining morphology, defect and structure engineering. *Advanced Materials* 32:1–11. https://doi.org/10.1002/adma.201904320
31. Ponrouch A, Palacin MR (2018) On the road toward calcium-based batteries. *Current Opinion in Electrochemistry* 9:1–7. https://doi.org/10.1016/j.coelec.2018.02.001
32. Meng J, Liu X, Niu C, et al (2020) Advances in metal-organic framework coatings: versatile synthesis and broad applications. *Chemical Society Reviews* 49:3142–3186. https://doi.org/10.1039/c9cs00806c
33. Ponrouch A, Bitenc J, Dominko R, et al (2019) Multivalent rechargeable batteries. *Energy Storage Materials* 20:253–262. https://doi.org/10.1016/j.ensm.2019.04.012
34. Arroyo-De Dompablo ME, Ponrouch A, Johansson P, Palacín MR (2020) Achievements, challenges, and prospects of calcium batteries. *Chemical Reviews* 120:6331–6357. https://doi.org/10.1021/acs.chemrev.9b00339
35. Xu X, Duan M, Yue Y, et al (2019) Bilayered $Mg_{0.25}V_2O_5{\cdot}H_2O$ as a stable cathode for rechargeable Ca-ion batteries. *ACS Energy Letters* 4:1328–1335. https://doi.org/10.1021/acsenergylett.9b00830
36. Marta C, Francisco N, José RG, Gregorio O, Ricardo A, Pedro L, Carlos Pérez-Vicente JLT (2016) Advancing towards a veritable calcium-ion battery: $CaCo_2O_4$ positive electrode material. *Electrochemistry Communications* 67:59–64. https://doi.org/10.1016/j.elecom.2016.03.016
37. Verrelli R, Black AP, Pattanathummasid C, et al (2018) On the strange case of divalent ions intercalation in V_2O_5. *Journal of Power Sources* 407:162–172. https://doi.org/10.1016/j.jpowsour.2018.08.024
38. Chen Q, Jin J, Kou Z, et al (2020) Zn^{2+} pre-intercalation stabilizes the tunnel structure of MnO_2 nanowires and enables Zinc-ion hybrid supercapacitor of battery-level energy density. *Small* 16:1–9. https://doi.org/10.1002/smll.202000091
39. Lee C, Jeong SK (2018) Modulating the hydration number of calcium ions by varying the electrolyte concentration: electrochemical performance in a Prussian blue electrode/aqueous electrolyte system for calcium-ion batteries. *Electrochimica Acta* 265:430–436. https://doi.org/10.1016/j.electacta.2018.01.172
40. Kim H, Kim JC, Bianchini M, et al (2018) Recent progress and perspective in electrode materials for K-ion batteries. *Advanced Energy Materials* 8:1–19. https://doi.org/10.1002/aenm.201702384
41. Pan H, Hu YS, Chen L (2013) Room-temperature stationary sodium-ion batteries for large-scale electric energy storage. *Energy and Environmental Science* 6:2338–2360. https://doi.org/10.1039/c3ee40847g
42. Lipson AL, Pan B, Lapidus SH, et al (2015) Rechargeable Ca-ion batteries: a new energy storage system. *Chemistry of Materials* 27:8442–8447. https://doi.org/10.1021/acs.chemmater.5b04027
43. Lee C, Jeong S (2018) A novel strategy to improve the electrochemical performance of a prussian blue analogue electrode for calcium-ion batteries, *Electrochemistry* 86:10–13.
44. Smeu M, Hossain MS, Wang Z, et al (2016) Theoretical investigation of Chevrel phase materials for cathodes accommodating Ca^{2+} ions. *Journal of Power Sources* 306:431–436. https://doi.org/10.1016/j.jpowsour.2015.12.009
45. Yu X, Manthiram A (2018) Toward reversible room-temperature calcium-ion batteries. *Chem* 4:1200–1202. https://doi.org/10.1016/j.chempr.2018.05.009
46. Smeu M, Hossain MS, Wang Z, et al (2016) Theoretical investigation of Chevrel phase materials for cathodes accommodating Ca^{2+} ions. *Journal of Power Sources* 306:431–436. https://doi.org/10.1016/j.jpowsour.2015.12.009
47. Juran TR, Smeu M (2017) Hybrid density functional theory modeling of Ca, Zn, and Al ion batteries using the Chevrel phase Mo_6S_8 cathode. *Physical Chemistry Chemical Physics* 19:20684–20690. https://doi.org/10.1039/c7cp03378h
48. Ren W, Xiong F, Fan Y, et al (2020) Hierarchical copper sulfide porous nanocages for rechargeable multivalent-ion batteries. *ACS Applied Materials and Interfaces* 12:10471–10478. https://doi.org/10.1021/acsami.9b21999
49. Murata Y, Minami R, Takada S, Aoyanagi K (2017) A fundamental study on carbon composites of $FeF_3{\cdot}0.33H_2O$ as open-framework cathode materials for calcium-ion batteries. *AIP Conference Proceedings* 020005:8. https://doi.org/10.1063/1.4974787
50. Lipson AL, Kim S, Pan B, et al (2017) Calcium intercalation into layered fluorinated sodium iron phosphate. *Journal of Power Sources* 369:133–137. https://doi.org/10.1016/j.jpowsour.2017.09.081

51. Junjun W, Shuangshuang T, Fangyu X, Ruohan Y, Peijie W, Lianmeng C, Qinyou A (2020) $VOPO_4{\cdot}2H_2O$ as a new cathode material for rechargeable Ca-ion batteries. *Chemical Communications* 56:3805–3808. https://doi.org/10.1039/D0CC00772B
52. Lagrange P, Emery N, He C (2005) Structural study and crystal chemistry of the first stage calcium graphite intercalation compound. *Journal of Solid State Chemistry* 178:2947–2952. https://doi.org/10.1016/j.jssc.2005.05.031
53. Xu W, Lerner MM (2018) A new and facile route using electride solutions to intercalate alkaline earth ions into graphite. *Chemistry of Materials* 30:6930–6935. https://doi.org/10.1021/acs.chemmater.8b03421
54. Saya T, Yasutoshi I (2008) Electrochemical intercalation of calcium ion into graphite electrode. *ECS Meeting Abstracts* MA2008-02: 206. https://doi.org/10.1149/MA2008-02/1/206
55. Park J, Xu ZL, Yoon G, et al (2020) Stable and high-power calcium-ion batteries enabled by calcium intercalation into graphite. *Advanced Materials* 32:1–7. https://doi.org/10.1002/adma.201904411
56. Lee CH, Kim CS, Jeong YT, et al (2016) Electrochemical properties of multi-wall carbon nanotubes as a novel negative electrode for calcium secondary batteries. *Materials Science Forum* 859:75–78. https://doi.org/10.4028/www.scientific.net/MSF.859.75
57. Datta D, Li J, Shenoy VB (2014) Defective graphene as a high-capacity anode material for Na- and Ca-ion batteries. *ACS Applied Materials and Interfaces* 6:1788–1795. https://doi.org/10.1021/am404788e
58. Datta D, Li J, Shenoy VB (2013) Defective graphene as promising anode material for Na-ion battery and Ca-ion battery. *ACS Applied Materials and Interfaces* 6:1788–1795.
59. Farokh Niaei AH, Hussain T, Hankel M, Searles DJ (2018) Hydrogenated defective graphene as an anode material for sodium and calcium ion batteries: a density functional theory study. *Carbon* 136:73–84. https://doi.org/10.1016/j.carbon.2018.04.034
60. Salavati M, Rabczuk T (2019) First-principles investigation of N-triphenylene-graphdiyne nanosheets as an anode material for Na, K, Mg and Ca storage. *Computational Materials Science* 169:109093. https://doi.org/10.1016/j.commatsci.2019.109093
61. Dutta PK, Mitra S (2018) An aqueous Ca-ion full cell comprising BaHCF cathode and MCMB anode. *ChemistrySelect* 3:3687–3690. https://doi.org/10.1002/slct.201800419
62. Er D, Li J, Naguib M, et al (2014) Ti_3C_2 MXene as a high capacity electrode material for metal (Li, Na, K, Ca) ion batteries. *ACS Applied Materials and Interfaces* 6:11173–11179.
63. Xie Y, Dall'Agnese Y, Naguib M, et al (2014) Prediction and characterization of mxene nanosheet anodes for non-lithium-ion batteries. *ACS Nano* 8:9606–9615. https://doi.org/10.1021/nn503921j
64. Mortazavi B, Dianat A, Rahaman O, et al (2016) Borophene as an anode material for Ca, Mg, Na or Li ion storage: a first-principle study. *Journal of Power Sources* 329:456–461. https://doi.org/10.1016/j.jpowsour.2016.08.109
65. Deng X, Chen X, Huang Y, et al (2019) Two-dimensional GeP_3 as a high capacity anode material for non- lithium-ion batteries. *Journal of Physical Chemistry C* 8:4721–4728. https://doi.org/10.1021/acs.jpcc.8b11574
66. Liu BC, Li F, Ma L, Cheng H (2010) Advanced materials for energy storage. *Advanced Materials* 22:28–62. https://doi.org/10.1002/adma.200903328
67. Tran TT, Obrovac MN (2011) Alloy negative electrodes for high energy density metal ion cells. *Journal of the Electrochemical Society* 158:A1411. https://doi.org/10.1149/2.083112jes
68. He J, Yao Z, Hegde VI, et al (2020) Computational discovery of stable heteroanionic oxychalcogenides ABXO (A, B = Metals; X = S, Se, and Te) and their potential applications. *Chemistry of Materials* 32:8229–8242. https://doi.org/10.1021/acs.chemmater.0c01902
69. Lipson AL, Pan B, Lapidus SH, et al (2015) Rechargeable Ca-ion batteries: a new energy storage system. *Chemistry of Materials* 27:8442–8447. https://doi.org/10.1021/acs.chemmater.5b04027
70. Ponrouch A, Tchitchekova D, Frontera C, et al (2016) Assessing Si-based anodes for Ca-ion batteries: electrochemical decalciation of $CaSi_2$. *Electrochemistry Communications* 66:75–78. https://doi.org/10.1016/j.elecom.2016.03.004

7 Nuclear Batteries
An Overview

Neethu T. M. Balakrishnan, Anandu M. Nair, Akhila Das, Jabeen Fatima M. J., Jou-Hyeon Ahn, M. V. Reddy, and Prasanth Raghavan

7.1 INTRODUCTION

The evolution of energy storage devices enabled a wide array of technological approaches for balancing energy demand and consumption. In exploring the different energy storage devices, batteries are considered important owing to their flexible operation and convenient handling. Rechargeable batteries such as lithium-ion batteries (LIBs) overruled most of the consumer markets on account of their high energy storage capacity [1–3], even though it is notable that the requirement of batteries varies according to the area of their application. Considering LIBs, it is a fact that they cannot be used in every area of battery applications owing to their limitations especially due to safety problems [4,5]. When applying storage systems in an advanced area such as a space craft, the requirement is performance efficiency over a prolonged period of time without recharging. The attempt to power the space crafts with solar energy was examined initially but eventually failed due to the insufficiency of its availability. So exploring the outer planets was a risky task on account of the maintenance of a sustainable charged state [6–8]. Similar to the space crafts, another challenging area that required a prolonged operation without recharging is the implanted devices. The requirement of a powering system without consistent charging and maintenance is an important criterion for implanted medical device applications [9,10]. So, considering such area of application, the necessity of an alternative is very essential.

These essentialities exploited the outbreak of nuclear batteries, which are the energy storage devices that utilize radioactivity for the generation of power. Compared to other rechargeable batteries, nuclear batteries are operable for a long period of time, which is attributed to their higher volumetric energy density. Moreover, it has been observed that they have a stronger endurance in harsh conditions, which makes them distinctive from any other established batteries. Even though they are termed "batteries", unlike other batteries, nuclear batteries are not electrochemical and do not have the requirement of charging and recharging; hence, they are examined in devices that should be performed unguarded over a long period of time.

Nuclear batteries, also called atomic batteries, radioisotope batteries, or radioisotope generators, are devices that generate electricity by the decay of radioactive isotopes. They are similar to nuclear reactors on account of the generation of electricity by nuclear reaction, but, unlike nuclear reactors, nuclear batteries don't follow a chain reaction. The beta and alpha emission during the radioisotope decay is the basic energy generation mechanism observed in nuclear batteries (Figure 7.1a). Nuclear battery technology began to be used in 1913 when the English physicist, Henry Moseley, reported the production of electricity by the radioactive decay of the radioactive isotope radium, and he demonstrated a beta cell using this nuclear reaction [10]. Later in 1954 an American electronics company, Radio Corporation America, constructed a small nuclear battery for examining radio receivers and hearing aids. This battery was constructed using a minute quantity of radioactive isotope, along with a tiny transistor-like wafer. Even though the battery releases a small amount of energy, it might be a forerunner in power radios and hearing aids that could be performed for over 20 years without any further maintenance [11].

DOI: 10.1201/9781003310174-7

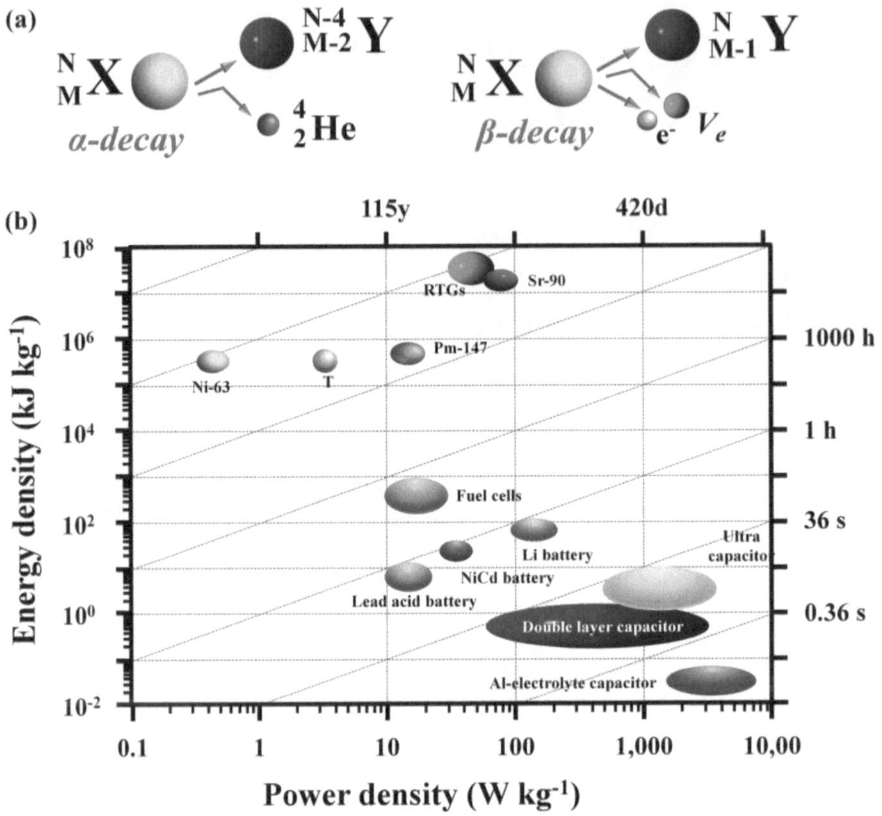

FIGURE 7.1 (a) Visualization of (left) alpha decay and (right) beta decay, (b) Ragone plot showing capacitors (grey), chemical batteries (blue), fuel cell (green), atomic batteries of various radioisotopes (red), and RTGs (purple). The sloped lines are constant timelines. Adapted and reproduced from Ref. [12]. Copyright © 2015 Scitechnol.

It is noteworthy that nuclear batteries can generate power in the range of 10–100 KW and have a life span of decades. They can perform 200 times more efficiently than any other currently used batteries and possess a potential to achieve a specific power of 1–50 mW g^{-1} and a high energy density compared to that of any other energy storage systems that can depicted from the Ragone plot (Figure 7.1b) [12]. Another versatile feature of nuclear batteries is that the radioisotopes employed in them burn at a rate of their half-life, thereby providing sufficient nuclear reaction for a long period [13].

The nuclear battery market is estimated to rise at a compound annual growth rate (CAGR) of 8.8% in the forecasted year of 2019–2026, and at this CAGR the global nuclear battery market is estimated to be $48.3–87.2 billion by 2019–2026. The global acceptance of nuclear batteries has arisen due to their smaller size, high energy generation, long durability, and no greenhouse effect attributed to the absence of radioactive waste production. The leading players in the current nuclear battery markets are Exide Technologies, Tesla Energy, GEV Attenfall AB, American Elements, Curtiss-Wright Corporation, Comsoll Inc., Thermo PV., HBL Power System, Luminous Power, and MarlowBut. Some of the issues like the requirement of high initial capital and strict regulation regarding the use of radioactive materials in some regions hamper the upright of the market value of this battery system, but, at the same time, the essentiality of long-life batteries lifts the innovations over the nuclear battery systems. Recently, Arkenlight, a company formed out of the University of Bristol, proposed the development of a self-charging nuclear battery called the diamond battery.

These batteries utilize graphite from nuclear plants that get harvested to carbon-14 and finally convert them into manmade diamonds. The interesting fact about the diamond battery is that it can be long-lasting for up to 12–5,000 years. California-based nano diamonds battery company claims that within three years there is a broad vision of nuclear battery technology that will power anything the current LIB could run.

7.2 BATTERY COMPONENTS AND ENERGY CONVERSION IN NUCLEAR BATTERIES

A nuclear battery is mainly composed of two parts: a radioactive source and a system that converts the produced energy to power. The energy conversion mechanisms employed in nuclear batteries depend on the radioactive mechanism of the radioactive isotopes. As mentioned in the previous section, the choice of radioactive isotopes is observed to be a vital part of determining the long-term performance of the nuclear battery employed. Table 7.1 gives the characteristics of frequently used radioactive isotopes in nuclear batteries [14]. The investigations emphasize that the radioactive isotopes are capable of exhibiting an energy density that is 100–1,000 times higher than that of chemical fuels or fossil fuels. Moreover, the high reactivity of the isotopes provides the provision to use a minute amount to deliver the desired best performance; hence, it aids in reducing the battery size.

Based on the conversion mechanisms, nuclear batteries are classified as thermal converters and non-thermal converters. The thermal converters follow the conversion of heat generated during the nuclear reaction into energy, while the non-thermal converters generate energy directly from the emitted radiation before it gets decomposed into heat. Further classification of thermal and non-thermal convertors is depicted in Figure 7.2. Non-thermal convertors such as direct charge nuclear batteries (DCNBs) utilize the mechanism of alpha and beta decay of radioactive isotopes, whereas the direct conversion batteries explore a capacitor to collect the charge emitted from the radioactive isotope and hence produce a high output voltage. The proper choice of the radioactive isotope is a significant factor in determining the performance of the battery. Even though there are naturally occurring radioisotopes, their difficulties in extraction and limitation in quantity make their choice challenging. The isotopes that are produced as nuclear reaction byproducts are considered to be more economical; at the same time, expensive products formed by irradiation are not feasible to use. So, a limited number of isotopes are considered to be the best that could satisfy the requirements such as half-life, radiation, energy density, cost, and availability.

TABLE 7.1
Some Radioisotopes Used in Nuclear Batteries

Isotope	Specific Activity ($g\ m^{-1}\ Ci^{-1}$)	Half-Life (Years)	Average Energy (keV)
^{63}Ni	1.8×10^{-5}	100.2	17.4
^{3}H	1.03×10^{-7}	12.3	5.7
^{147}Pm	1.06×10^{-6}	62	17.4
^{90}Sr	7.25×10^{-6}	28.8	195.8
^{85}Kr	2.56×10^{-6}	10.8	251.6
^{106}Ru	3.03×10^{-7}	1.06	93
^{45}Ca	5.06×10^{-8}	162	77
^{35}S	2.4×10^{-8}	87.2	49

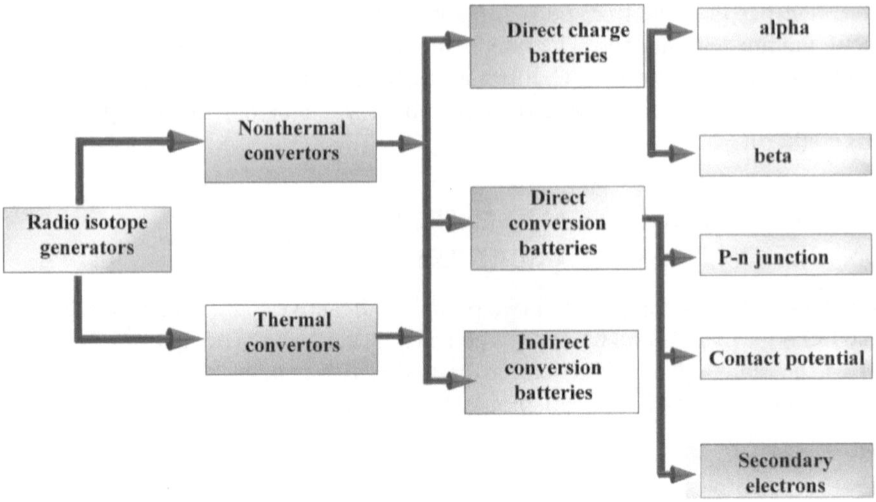

FIGURE 7.2 Types and classification of relevant radioisotope generators.

7.3 CLASSIFICATION OF NUCLEAR BATTERIES

The mechanism of energy conversion in nuclear batteries classifies them into different categories, as shown in Figure 7.3 [12]. As mentioned in Section 7.2, the non-thermal conversion mechanisms nuclear batteries undergo are betavoltaic, alphavoltaic, optoelectric, and direct charge, whereas, in thermal conversion batteries, the mechanism is thermoelectricity, thermionic, thermophotovoltaic (TPV), alkali metal thermo to electric conversion, Stirling generator, and piezoelectricity that will influence the temperature difference between the isotope and the constituents of energy conversion reaction. Some of the popular thermal and non-thermal conversion batteries are discussed in detail in further sections [15–17].

7.3.1 Radioisotope Thermal Electric Generator

Radioisotope thermal electric generator (RTGs) is a type of nuclear battery system that follows a thermal conversion mechanism of generating energy from heat produced by the decay of radioactive isotopes called the Seebeck effect. Seeback effect is the simple principle discovered by Thomas Seebeck in 1821, which states that the difference in temperatures between two ends can result in the formation of electric voltage and vice versa. The structural design of the radial RTG is depicted in Figure 7.4. It is considered the most reliable powering system for space craft, which is lightweight and compact. Similar to other nuclear batteries, RTGs constitute a heat source and energy conversion system [13]. A static or dynamic energy conversion system is utilized for converting the heat produced from the radioactive decay to electricity. The dynamic energy conversion system utilizes a rotating and reciprocating engine, whereas the static conversion system does not have any requirement for moving parts, and hence it possesses very little maintenance and operating cost compared to that of a dynamic energy conversion system [18]. The fuels chosen for RTG are those that have a long half-life and a smaller size. The presence of the radioisotope having a long half-life enables the battery to operate a long time without refuelling, which will result in sustainable energy production.

The small-sized isotopes are desired for developing a compact RTG. Every component including the fuel contents should be reasonably small. Based on these criteria, the most commonly used isotopes for RTGs are plutonium-238 (Pu-238), strontium-90 (Sr-90), and curium-244 (Cm-244) with Pu-238. The frequent use of Pu-238 is associated with its high radiation output including alpha

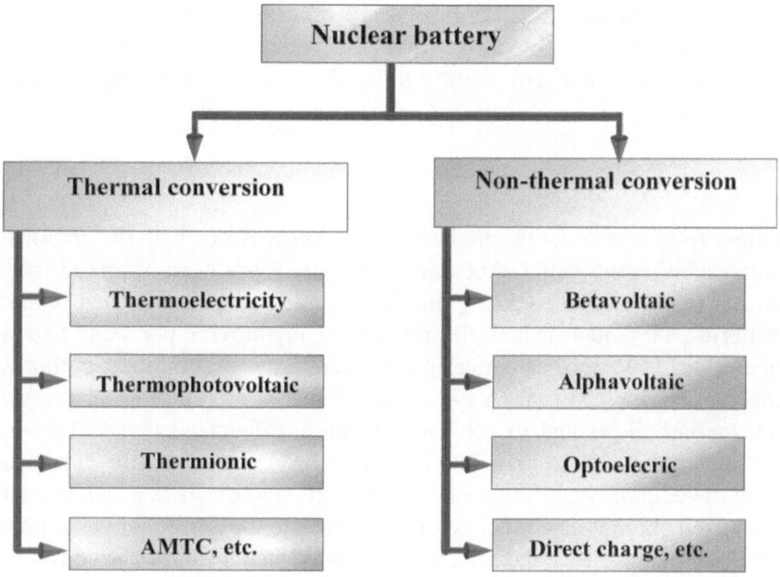

FIGURE 7.3 Classification of nuclear batteries.

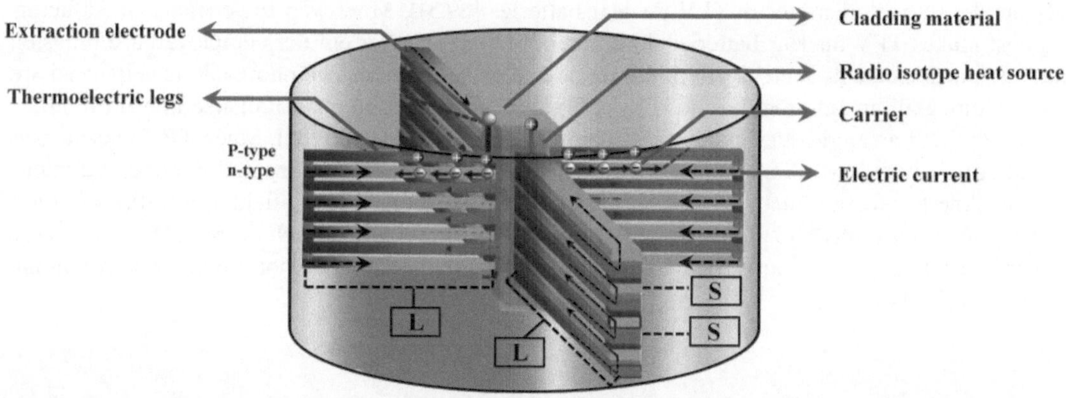

FIGURE 7.4 Structural design of the radial RTG. Adapted and reproduced from Ref. [27]. Copyright © 2017 Elsevier.

decay channels, and hence it needs high shielding; moreover, it possesses a long half-life of about 88 years [19,20]. On installing the isotope fuel pellets in RTG, the radioactive decay generates heat that gets collected by the heat distribution block. The heat distribution block sends the heat to a set of thermocouples that convert heat into electricity. The normally used thermocouple in RTG is bismuth telluride (BiTe), lead telluride (PbTe), tellurides containing antimony, germanium, and silver, and silicon germanium (SiGe) [19]. The simple design of RTGs enables their use in a variety of interstellar projects involving the space probes sent to the moon, Pioneer and Voyager which are the outer planets of the solar system, and most recently the robotic rover Curiosity sent to Mars [21–23]. Despite this healthier performance, RTG exhibits a defect of low conversion efficiency of heat to electrical energy. Even though the RTGs are utilized in environments with a sparse human population, the concern regarding safety is not resolved completely so far. Different safety measures are adopted for RTGs to overcome their safety aspects. The Cassini–Huygens probe, which is the NASA mission to Saturn, utilizes RTG isotope fuel stored in a high-strength block of graphite that is surrounded by a layer of indium metal to limit accidental explosions. The exploration of graphite block is observed to offer better protection, which is evident from the famous failed Apollo 13, which left its RTGs in the ocean after its return to Earth but didn't cause any contamination due to the presence of polonium [24–27].

7.3.2 TPV Nuclear Battery

TPV nuclear battery follows a direct thermal conversion process in which the photons generated by the thermal emitter convert the radiated heat to electricity. It has been observed that TPV nuclear batteries are similar to photovoltaic cells by means of their energy conversion process. Similar to other nuclear batteries, TPV nuclear batteries don't have any moving parts and hence require very low maintenance costs. TPV nuclear batteries constitute a thermal emitter, a photovoltaic cell, a filter, and a reflector (Figure 7.5) [28]. A filter and a reflector are employed to minimize the loss of non-converted radiated heat and to achieve maximum efficiency; about 20% of efficiency is reported for this nuclear battery system. The filter employed can reflect the non-ideal wavelength to the emitter, and the photons that didn't reflect efficiently to the collector can be taken for photon recycling. The reflected photons can either be directed back to the emitter or to the photovoltaic cell, and ultimately the photon recycling and selective emission can determine the conversion of photons to electricity [24,29].

The emitters employed in TPV nuclear batteries should possess high efficiency and temperature resistance. So, low-cost silicon carbide that can withstand thermal stability of up to 1,700°C is usually employed as the emitter in TPV nuclear batteries [29,31]. Moreover, the cheaper manufacturing cost makes TPV nuclear batteries commercially viable. Other emitters employed are tungsten and rare earth elements such as ytterbium oxide and erbium oxide. The photovoltaic cells used are germanium, gallium antimonide, and alloys of indium such as indium gallium arsenide antimonide, indium gallium arsenide, and indium phosphide arsenide antimonide [32]. Since TPV nuclear batteries require very low maintenance, they are efficiently used for military and commercial applications. Another feature that makes TVPs appropriate for tactic battery field applications is their smaller size and lightweight. The essential areas that utilize TVPs include remote areas where power line extensions are impractical and expensive and vehicles that do not use noisy conventional generators [33,34].

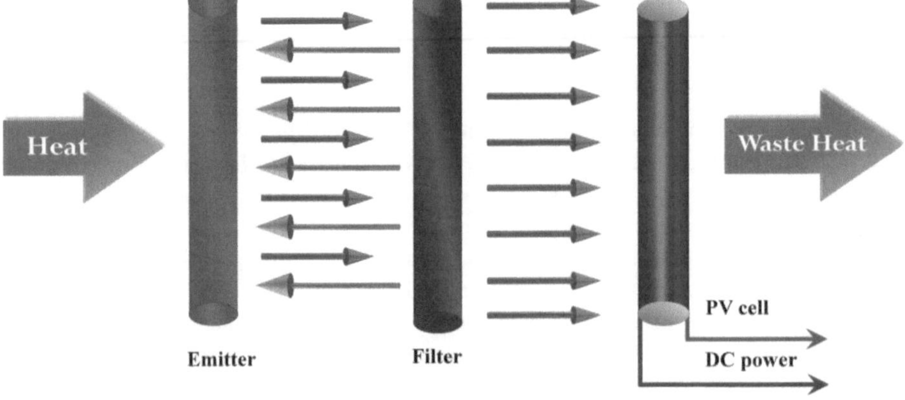

FIGURE 7.5 Schematic representation of thermophotovoltaic energy conversion. Adapted and reproduced from Ref. [30]. Copyright © 2019 Emarald Publishing Limited.

7.3.3 Betavoltaic Nuclear Battery

Betavoltaic batteries are the type of nuclear batteries that operate on the basis of the beta voltaic effect that involves the generation of electricity by means of beta particles generated from radioactive decay. The excess electron–hole pairs generated by the impinging of the beta particles onto the p–n junction semiconductor create electricity (Figure 7.6a). Various stages of energy conversion in the betavoltaic battery are depicted in (Figure 7.6b). The energy generation in betavoltaic cell can be similar to that of a solar cell that is composed of three different stages. In the first stage the beta particles emitted by the radioactive decay incident on the semiconductor to generate an electron–hole pair. The diffusion of electrons and holes to the Schottky junction results in their separation to form free holes and electrons in the second stage. Finally, both the holes and the electrons get collected at the anode and cathodes, respectively, to convert the radioactive energy into electrical energy [35]. The first explanation about the betavoltaic device was made by Paul Rapport in 1953 [36]. The device exhibited an efficiency of 0.2% and degraded rapidly by radioactive decay. Later in 1968 Dr. Larry Olsen at Donald W. Douglas Laboratories attempted to develop the most expensive viable betavoltaic battery [37–39]. The radioisotopes widely explored for betavoltaic nuclear batteries are radioactive isotopes of promethium and nickel (Pm-147 and Ni-63) [37,40]. The betavoltaic nuclear batteries constitute serially arranged unidirectional-source Si–n/p cells, and most of the volume

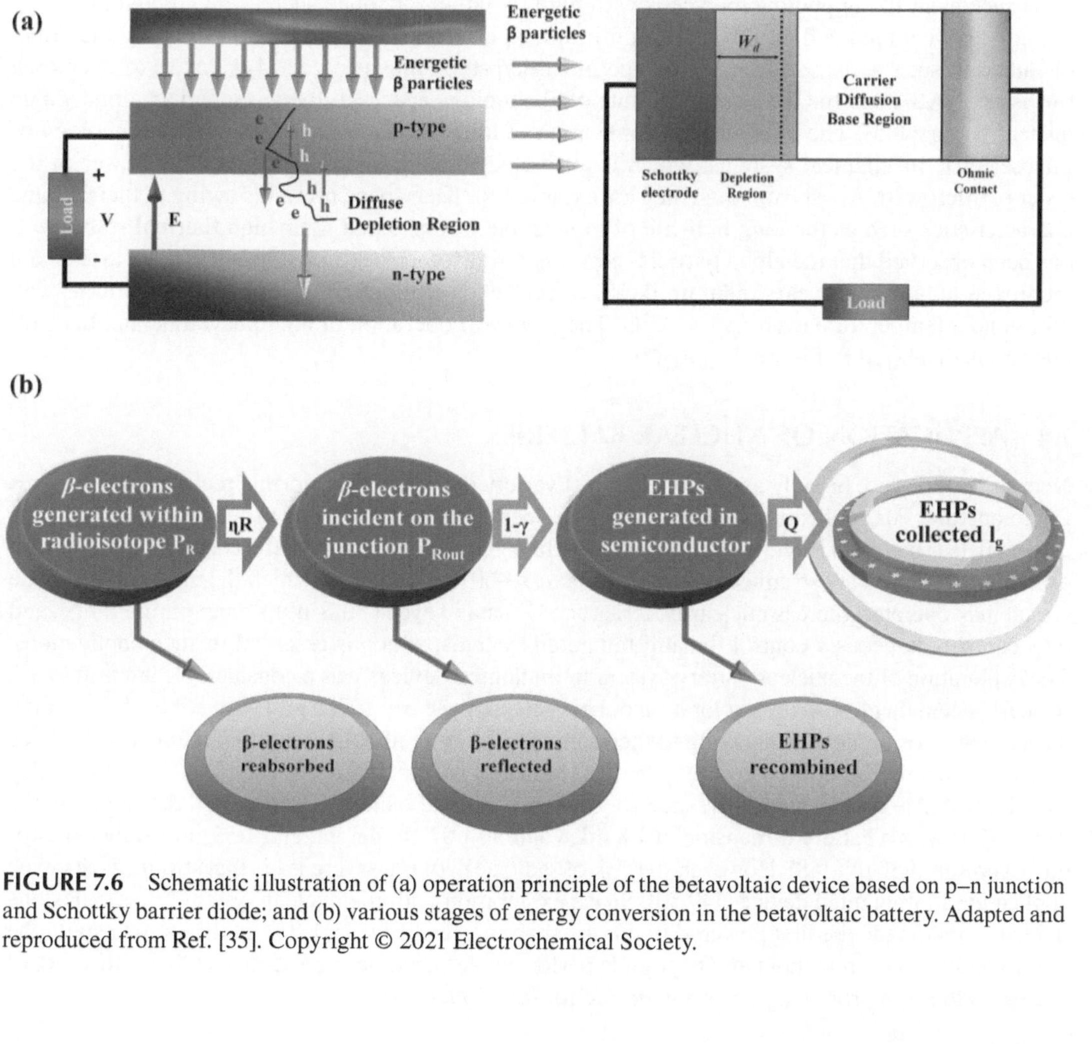

FIGURE 7.6 Schematic illustration of (a) operation principle of the betavoltaic device based on p–n junction and Schottky barrier diode; and (b) various stages of energy conversion in the betavoltaic battery. Adapted and reproduced from Ref. [35]. Copyright © 2021 Electrochemical Society.

inside the battery is utilized for the shielding of the radioactive gamma rays that are emitted from the Pm-146, which is the contaminant produced from the beta source of radioisotope Pm-147. The efficiency of these batteries based on a unidirectional beta source is about 4%, and if a bidirectional source is used, the efficiency is increased to 8%. In the early 1970s fabricated heart pacemakers were powered by betavoltaic batteries, constituting Pm-147 and Si–n/p cells but later replaced by LIBs, owing to the concern regarding the emission of gamma rays from the radioisotope. As an attempt to commercialize this nuclear battery, Dr. Olsen et al. [41] communicated a review of betavoltaic nuclear battery that suggests tritium as the alternate radioisotope that is reliable, safe, and abundant compared to other available isotopes [42]. The battery developed using this safer radioisotope was explored for applications such as exit signs for school buildings and theatres, illuminated gun or rifle sights, and wrist-watches [38].

7.3.4 Alphavoltaic Nuclear Battery

Alpha voltaic battery follows a similar operating mechanism as beta photovoltaic batteries, and the only difference is that it utilizes the alpha emitter isotope such as plutonium-238. This concept was introduced in early 1954 [43]. Initially alpha voltaic batteries are not successfully commercialized owing to the damage of the semiconductor material caused by the highly energetic alpha particles, leading to performance degradation. Different methods were introduced to overcome this issue such as utilizing photovoltaic cells with high radiation tolerance such as InGaP and introducing non-conventional cell design that can minimize performance degradation caused by the effect of radiation such as lateral junction n-type/intrinsic/p-type/intrinsic cell [44]. Later, to overcome this issue, NASA and the Rochester Institute of Technology successfully developed an alphavoltaic nuclear battery [45]. The most crucial application of alphavoltaic batteries is in the technology of microelectric mechanical systems, which is attributed to their capability to produce power in the order of microwatt. Apart from this, they are explored for deep space missions owing to their unique characteristics such as the long half-life of radioactive isotopes and their high thermal stability. It has been observed that the alpha particle–emitting radioisotopes exhibit a half-life that varies from months to hundreds of years, and, unlike the electrochemical batteries, alphavoltaic batteries can withstand a temperature even up to 135°C. The theory of operation of an alphavoltaic and betavoltaic cell is displayed in Figure 7.7 [46,47].

7.4 APPLICATION OF NUCLEAR BATTERIES

Nuclear batteries are broadly applicable in a vast variety of areas where normal rechargeable systems fail to operate. For applications such as deep-sea operations and front-line military applications, the use of nuclear batteries is really significant. In military-grade applications, the chemical batteries are difficult to maintain at extreme temperatures such as −60°C to 150°C at which the missiles and plane circuit may operate, whereas nuclear batteries can withstand even at this ultra-temperature range, and they can simultaneously control the anti-tampered systems, which is essential in such applications. The exploration of the nuclear battery system in implanted devices was a considerable attempt to use in heart pacemakers, with the exploration of beta cells in the early 1970s [47]. The applicability of the nuclear batteries in the area is owing to their long durability of about greater than 15 years of service life, which cannot be offered by any chemical battery systems without recharging. Moreover, they can offer a stable output energy irrespective of environmental factors. The beta cell developed in the early 1970s with a battery dimension of 1.8 mL volume, 1.02 cm height, and 1.52 cm height can produce a power of 50 μW [48]. However, the risk of radioactivity is a severe issue that still limits its wide application in implanted devices [49,50]. Space exploration with the nuclear system was a milestone in history that made the first powered flight of a space craft in 1961 [51]. The ability to withstand the abuse conditions even without any negligible performance variation is an attractive factor that makes nuclear batteries a promising powering device for future electronics.

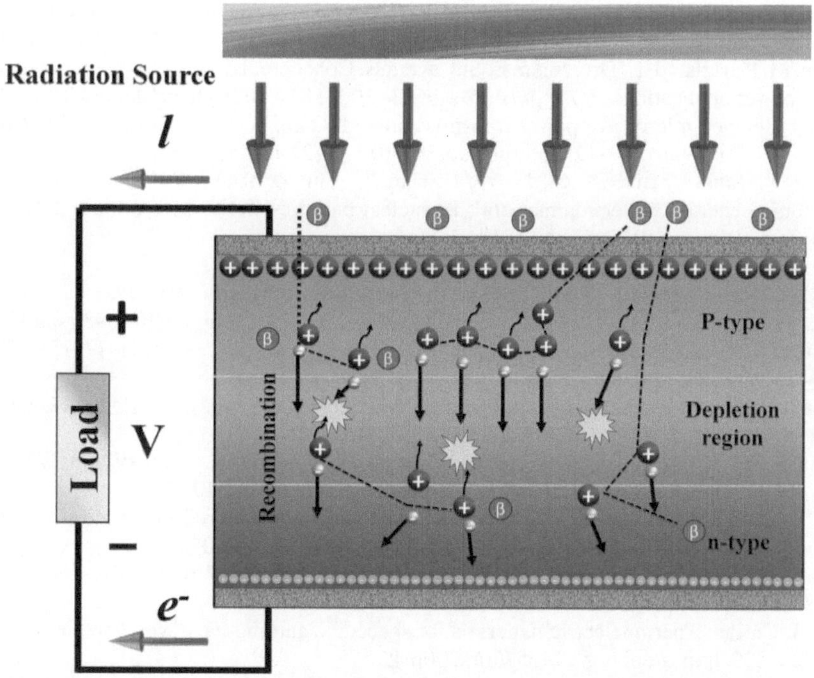

FIGURE 7.7 Theory of operation of an alphavoltaic and betavoltaic cell. Adapted and reproduced from Ref. [30]. Copyright © 2019 Emarald Publishing Limited.

7.5 CONCLUSION

Nuclear batteries seek considerable interest on account of their distinctive characteristics such as less maintenance, prolonged workability, and capability to operate in harsh environments. Nuclear batteries exploit the nuclear reaction taking place in the radioactive isotopes that can convert energy produced during the reaction into electricity. Based on the energy conversion, these batteries are classified as thermal and non-thermal, where the thermal converters can produce energy from the heat generated during the reaction, and the non-thermal converters cannot generate energy from the heat produced. The mechanism of radioactive reaction also categorizes nuclear batteries into different groups, namely, alpha voltaic, betavoltaic nuclear batteries, and so on. An advanced application such as space craft utilizes nuclear batteries, ensuring best performance even in an unsuitable environment. Even though there is concern regarding the safety issues of the radioactive isotopes, different safety precautions were implemented to ensure a better energy storage system with advanced safety features.

REFERENCES

[1] M.C. Dos Santos, O. Kesler, A.L.M. Reddy, Nanomaterials for energy conversion and storage, *J. Nanomater.* 2012 (2012) 2–4. https://doi.org/10.1155/2012/159249.

[2] S. Mogurampelly, V. Ganesan, Effect of nanoparticles on ion transport in polymer electrolytes, *Macromolecules.* 48 (2015) 2773–2786. https://doi.org/10.1021/ma502578s.

[3] P. Raghavan, J.W. Choi, J.H. Ahn, G. Cheruvally, G.S. Chauhan, H.J. Ahn, C. Nah, Novel electrospun poly(vinylidene fluoride-co-hexafluoropropylene)-in situ SiO_2 composite membrane-based polymer electrolyte for lithium batteries, *J. Power Sources.* 184 (2008) 437–443. https://doi.org/10.1016/j.jpowsour.2008.03.027.

[4] P.G. Balakrishnan, R. Ramesh, T.P. Kumar, Safety mechanisms in lithium-ion batteries, *J. Power Sources.* 155 (2006) 401–414. https://doi.org/10.1016/j.jpowsour.2005.12.002.

[5] J. Deng, C. Bae, J. Marcicki, A. Masias, T. Miller, Safety modelling and testing of lithium-ion batteries in electrified vehicles, *Nat. Energy.* 3 (2018) 261–266. https://doi.org/10.1038/s41560-018-0122-3.

[6] Y. Ronen, M. Kurtzhand, L. Droizman, E. Shwageraus, Conceptual design of Americium nuclear battery for space power applications, *J. Propuls. Power.* 23 (2007) 874–880. https://doi.org/10.2514/1.25837.

[7] E.E. Eves, Beamed microwave power transmission and its application to space, *IEEE Trans. Microw. Theory Tech.* 40 (1992) 1239–1250. https://doi.org/10.1109/22.141357.

[8] Z. Zhang, X. Tang, Y. Liu, Z. Xu, H. Ye, F. Tian, K. Liu, Z. Yuan, W. Chen, Application of liquid scintillators as energy conversion materials in nuclear batteries, *Sensors Actuators, A Phys.* 290 (2019) 162–171. https://doi.org/10.1016/j.sna.2019.03.024.

[9] L.E. Johnson, A. Kenneth, Nuclear battery, 1987.

[10] J.H. Fellow, The attainment of high potentials by the use of radium, *Proc. R. Soc. London. Ser. A, Contain. Pap. a Math. Phys. Character.* 88 (1913) 471–476. https://doi.org/10.1098/rspa.1913.0045.

[11] S. Kumar, Atomic batteries: energy from radioactivity, *Mater. Sci.* 1511 (2015) 87. https://doi.org/10.48550/arXiv.1511.07427.

[12] S. Kumar, Atomic batteries: energy from radioactivity, *J. Nucl. Energy Sci. Power Gener. Technol.* 05 (2015) 1–8. https://doi.org/10.4172/2325-9809.1000144.

[13] R.G. Lange, W.P. Carroll, Review of recent advances of radioisotope power systems, *Energy Convers. Manag.* 49 (2008) 393–401. https://doi.org/10.1016/j.enconman.2007.10.028.

[14] A. Lal, R. Duggirala, H. Li, Pervasive power: a radioisotppe-powered piezoelectric generator, *IEEE Pervasive Comput.* 4 (2005) 53–61. https://doi.org/10.1109/MPRV.2005.21.

[15] G.N. Yakubova, Nuclear batteries with tritium and promethium-147 radioactive sources, Diss. Univ. Illinois Urbana-Champaign. (2010) 1–162.

[16] J. Yang, T. Caillat, Thermoelectric materials for space and automotive power generation, *MRS Bull.* 31 (2006) 224–229. https://doi.org/10.1557/mrs2006.49.

[17] A. Sanchez-Torres, Radioisotope power systems for space applications, *Radioisot. - Appl. Phys. Sci.* 1 (2011) 457–471. https://doi.org/10.5772/20928.

[18] G.L. Bennett, Mission interplanetary: using radioisotope power to explore the solar system, *Energy Convers. Manag.* 49 (2008) 382–392. https://doi.org/10.1016/j.enconman.2007.06.051.

[19] D. Kramer, Shortage of plutonium-238 jeopardizes NASA's planetary science missions, *Phys. Today.* 64 (2011) 24–26. https://doi.org/10.1063/1.3541938.

[20] G.R. Schmidt, T.J. Sutliff, L.A. Dudzinski, Radioisotope power: a key technology for deep space exploration, *6th Int. Energy Convers. Eng. Conf. IECEC.* (2008) 28–30. https://doi.org/10.2514/6.2008-5640.

[21] M. Jiang, *An Overview of Radioisotope Thermoelectric Generators, PH241*, Stanford University, Stanford, CA (2013).

[22] G.L. Bennett, Space nuclear power: opening the final frontier, *Collect. Tech. Pap. -4th Int. Energy Convers. Eng. Conf.* 2 (2006) 1433–1449. https://doi.org/10.2514/6.2006-4191.

[23] D.M. Harland, *Apollo 12 - On the Ocean of Storms*, Praxis, New York (2011) 1–10. https://doi.org/10.1007/978-1-4419-7607-9.

[24] Y. Zhang, R. Vanbever, A. Ben-Jebria, J.D. Mintzes, R. Langer, D.A. Edwards, V. de M. Linck, A.L. da Silva, M. Figueiró, Â. Luis Piato, A. Paula Herrmann, F. Dupont Birck, E. Bastos Caramão, D. Sávio Nunes, P.R.H. Moreno, E. Elisabetsky, Z. Ma, B. Moulton, A. Romeu-Moreno, C. Aguilar, L. Arola, A. Mas, P. Perez, A.K. Renfrew, T. Simon-Yarza, M. Giménez-Marqués, R. Mrimi, A. Mielcarek, R. Gref, P. Horcajada, C. Serre, P. Couvreur, T. Loftsson, M. Sönmez, M. Çelebi, A. Levent, I. Berber, Z. Şentürk, Thermophotovoltaicelectrical generation systems, *Int. J. Pharm.* 56 (2009) 1986–2001. https://doi.org/10.1016/j.ijpharm.2017.08.087%0A. https://doi.org/10.1016/j.ccr.2011.01.031.

[25] A.K. Misra, Overview of NASA program on development of radioisotope power systems with high specific power, *Collect. Tech. Pap. 4th Int. Energy Convers. Eng. Conf.* 2 (2006) 1388–1394. https://doi.org/10.2514/6.2006-4187.

[26] F. Ritz, Craig E. Peterson, Multi-mission radioisotope thermoelectric generator (MMRTG) program overview. *2004 IEEE Aerospace Conference Proceedings (IEEE Cat. No. 04TH8720).* Vol. 5. IEEE, 2004.

[27] K. Liu, Y. Liu, Z. Xu, Z. Zhang, Z. Yuan, X. Guo, Z. Jin, X. Tang, Experimental prototype and simulation optimization of micro-radial milliwatt-power radioisotope thermoelectric generator, *Appl. Therm. Eng.* 125 (2017) 425–431. https://doi.org/10.1016/j.applthermaleng.2017.07.022.

[28] T.J. Coutts, M.C. Fitzgerald, Thermophotovoltaics, *Sci. Am.* 279 (1998) 90–95.

[29] R.E. Nelson, A brief history of thermophotovoltaic development, *Semicond. Sci. Technol.* 18 (2003) S141. https://doi.org/10.1088/0268-1242/18/5/301.

[30] O.L. Ayodele, K.O. Sanusi, M.T. Kahn, Nuclear battery: a source of environmentally friendly energy, *J. Eng. Des. Technol.* 17 (2019) 172–182. https://doi.org/10.1108/JEDT-02-2017-0011.

[31] A. Shakouri, Recent developments in semiconductor thermoelectric physics and materials, *Annu. Rev. Mater. Res.* 41 (2011) 399–431. https://doi.org/10.1146/annurev-matsci-062910-100445.

[32] B. Bitnar, Silicon, germanium and silicon/germanium photocells for thermophotovoltaics applications, *Semicond. Sci. Technol.* 18 (2003) 221–227. https://doi.org/10.1088/0268-1242/18/5/312.

[33] C. Wilt, D. Chhong, P. Wolford, D. Magari, P. Crowley, Termophotovoltaic generation of electricity, In: *AIP Proceedings 7th World Conference on Thermophotovoltaic Generation of Electricity*, AIP, Madrid, Spain (2007) 890, 335.

[34] S. Coutts, T. Wanlass, M. Ward, J. Johnson, A review of recent advances in thermophotovoltaics, In: *Proceeding of 25th IEEE Photovoltaic Specialists Conference*, IEEE, Washington, DC (1996) 25–30.

[35] C. Zhou, J. Zhang, X. Wang, Y. Yang, P. Xu, P. Li, L. Zhang, Z. Chen, H. Feng, W. Wu, Review-betavoltaic cell: the past, present, and future, *ECS J. Solid State Sci. Technol.* 10 (2021) 027005. https://doi.org/10.1149/2162-8777/abe423.

[36] P. Cabauy, L. Olsen, B. Elkind, T. Adams, Micropower betavoltaic hybrid sources, In: *44th Power Sources Conferences*, Las Vegas, NV (2010) 3–5.

[37] H. Flicker, J.J. Loferski, T.S. Elleman, Construction of a promethium-147 atomic battery, *IEEE Trans. Electron Devices*. 11 (1964) 2–8. https://doi.org/10.1109/T-ED.1964.15271.

[38] W.G. Pfann, W. Van Roosbroeck, Radioactive and photoelectric p-n junction power sources, *J. Appl. Phys.* 25 (1954) 1422–1434. https://doi.org/10.1063/1.1721579.

[39] A. Akimchenko, V. Chepurnov, M. Dolgopolov, A. Gurskaya, O. Kuznetsov, A. Mashnin, V. Radenko, A. Radenko, O. Surnin, G. Zanin, Betavoltaic device in por-SiC/Si C-nuclear energy converter, In: *EPJ Web Conferences* (2017) 1–9. https://doi.org/10.1051/epjconf/201715806004.

[40] G.H. Miley, B. Ulmen, P.D. Desai, S. Moghaddam, R.I. Masel, Long lived, low power betavoltaic nuclear battery for electronics, *7th Int. Energy Convers. Eng. Conf.* (2009) 1–6. https://doi.org/10.2514/6.2009-4601.

[41] L.C. Olsen, Review of betavoltaic energy conversion, *NASA Conf. Publ.* (1993) 256–267.

[42] P. Cabauy, Larry C. Olsen, Noren Pan, Tritium direct conversion semiconductor device for use with gallium arsenide or germanium substrates. U.S. Patent No. 9,799,419. 24 Oct. 2017.

[43] W.G. Pfann, W. Van Roosbroeck, Radioactive and photoelectric p-n junction power sources. *J. Appl. Phys.* 25.11 (1954) 1422–1434.

[44] S.G. Bailey, D.M. Wilt, S.L. Castro, C.D. Cress, R.P. Raffaelle, Photovoltaic development for alpha voltaic batteries, In: *Conference Record of the IEEE Photovoltaic Specialists Conference*, Lake Buena Vista, FL (2005) 106–109. https://doi.org/10.1109/pvsc.2005.1488080.

[45] H. Falls, P. Jenkins, D. Wilt, D. Scheiman, D. Chubb, O. Falls, T.U. States, Alphavoltac batteries and methods thereof, *Electric. Electron. Eng*. **2** (2011). https://ntrs.nasa.gov/citations/20110004872

[46] O.L. Ayodele, Nuclear battery : a source of environmentally friendly energy, *J. Eng. Des. Technol.* 17 (2019) 6. https://doi.org/10.1108/JEDT-02-2017-0011.

[47] T. Adams, S.T. Revankar, P. Cabauy, B. Elkind, D. Cheu, Betavoltaic per for mance under extreme tem per a tures, *Nucl. Technol. Radiat. Prot.* 31 (2016) 356–360. https://doi.org/10.2298/NTRP1604356A.

[48] X. Wei, J. Liu, Power sources and electrical recharging strategies for implantable medical devices, *Front. Energy Power Eng. China.* 2 (2008) 1–13. https://doi.org/10.1007/s11708-008-0016-3.

[49] J. Drews, G. Fehrmann, R. Staub, R. Wolf, Primary batteries for implantable pacemakers and defibrillators, *J. Power Sources*. 97–98 (2001) 747–749. https://doi.org/10.1016/S0378-7753(01)00649-8.

[50] V.S. Mallela, V. Ilankumaran, S.N. Rao, Trends in cardiac pacemaker batteries, *Indian Pacing Electrophysiol. J.* 4 (2004) 201–212.

[51] G.L. Bennett, E.W. Johnson, First flights: nuclear power to advance space exploration and exposition, *Int. Air Sp. Symp. Expo.* 18 (2003) 1–10.

8 Magnesium–Lithium Hybrid Batteries

Neethu T. M. Balakrishnan, Devika S. Lal, Akhila Das, N. S. Jishnu, Jou-Hyeon Ahn, Jabeen Fatima M. J., and Prasanth Raghavan

8.1 INTRODUCTION

The advancement of energy storage technology flourishes the economic and environmental benefits by stimulating the exploration of renewable resources and powering the energy grids [1]. The alternativeness of energy storage devices in regard to fossil fuel resources was a crucial attempt to decrease pollution. The commercially viable rechargeable systems such as lithium-ion batteries (LIBs) are vital and significant for advanced applications such as electric vehicles [2–5]. It is quite noticeable that, even now, the kingdom of LIBs is not occupied by any other battery technology owing to their best performance and wide consumer market acceptance [6]. The future success of LIBs in meeting the advanced requirements is quite complicated owing to safety issues, limitations in energy density, and most importantly the scarcity of lithium resources [7]. The extensive research on the different energy storage systems such as redox-flow batteries, sodium (Na)-metal halide batteries [8], Na-ion batteries [9], and high-energy-density metal-based batteries is the innovative attempt to replace the LIB technology.

The use of lithium metal as an anode in LIBs attracted major interest owing to its high energy density (3,861 mAh g^{-1}) along with the lowest redox potential (–3.04 V vs standard hydrogen electrode (SHE). The major concern that arises from its exploration of LIBs is their safety issues caused by the formation of dendrites during the cycling of the battery [10–12]. The dendrite formation can result in an internal short circuit that may finally lead to the explosion of the battery. Moreover, the limited coulombic efficiency resulting from the dendrite-occupied lithium anodes can affect the performance of LIBs and hence exhibit very poor cycling stability on cycling with high-capacity cathodes [13]. Different attempts such as modification of electrolytes and addition of different electrolyte additives were introduced to resolve this issue, but the dendrite formation remains a severe issue at high-rate charging conditions that can bring about a limited cycling performance in lithium metal batteries over 150 cycles [14–16].

Considering the beyond LIBs technology, magnesium (Mg)-based batteries excel in regard to their safety and availability of resources. The recent attentiveness of Mg comes from its considerably high theoretical gravimetric (2,205 mAh g^{-1}) and volumetric (3,832 mAh cm^3) energy density, low cost, and advanced safety features compared to that of lithium [17,18]. The dendrite resistive characteristics of Mg metal anode arise from its unique electrodeposition behaviour and the presence of a passivation-free electrode–electrolyte interface formed by the strong nucleophilic characteristics of Mg electrolyte. Even though Mg systems are facing major challenges due to the sluggish kinetics of Mg^{2+} ions that resulted from the high polarization and lack of proper cathode material to create better electrochemical performance [17,19]. The exploration of chevrel phase compounds like $Cu_2Mo_6S_8$ was reported effective for Mg-based batteries [20], but still their performance is limited at room temperature with a slow Mg^{2+} ions diffusion and poor rate capability. So, to resolve

DOI: 10.1201/9781003310174-8

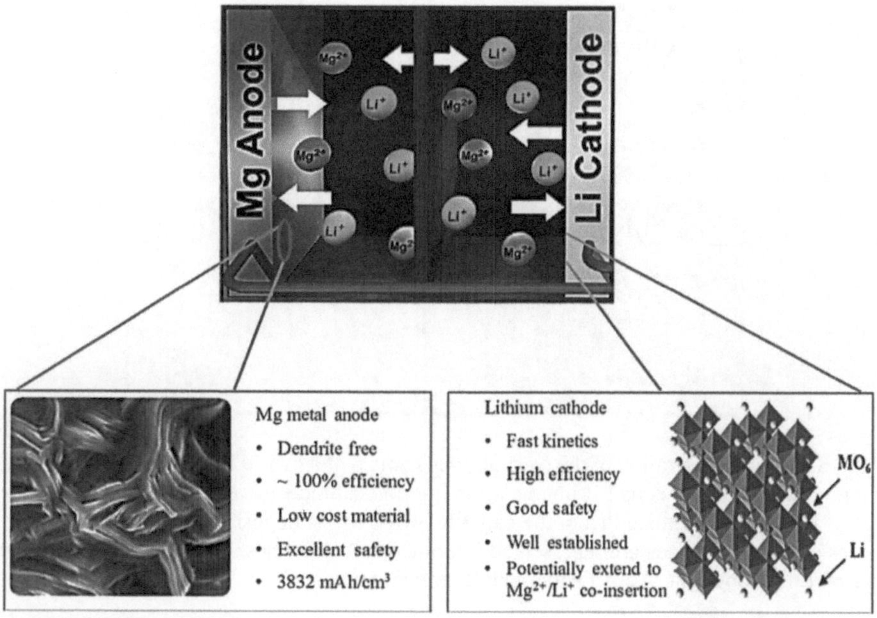

FIGURE 8.1 Schematic illustration of the hybrid Mg–Li battery design. The potential of combining the advantages of Mg metal anode and Li-ion cathode makes this design extremely attractive for high-rate and reliable batteries. Adapted and reproduced from Ref. [23]. Copyright © 2016 Springer.

the issue, Mg–Li hybrid batteries were explored that utilize the synergistic effect of both Mg metal anode and the well-established cathode materials from LIBs. The advantages of Mg anode and lithium-based cathodes are presented in Figure 8.1.

8.2 WORKING OF MG–LI HYBRID BATTERY

The working mechanism of the Mg–Li hybrid battery is explained based on the thermodynamic redox potential of lithium and magnesium (Mg^{2+}/Mg 0.67). The high reduction potential of Mg enables the Mg deposition at the anode before that of lithium. At the cathode side, lithium ions intercalate, rather than Mg ions, owing to their high diffusion compared to Mg^{2+} [21]. Owing to this asymmetric mechanism of ionic diffusion at both electrodes, the role of the electrolyte in a hybrid battery is very crucial to balance the ionic charge by providing enough lithium ions and to receive Mg ions during discharge and vice versa during the charging process. The working principle of the Mg–Li hybrid battery is demonstrated in Figure 8.2. The electrochemistry of the Mg–Li hybrid battery was firstly examined by Choi et al. [21] by exploring its theoretical and experimental aspects using the Mo_6S_8 cathode as the cathode. During the lowest lithium activity, the cathode was observed to be fully occupied by the Mg ions, and, when the lithium activity is enhanced, the co-insertion of both Mg and lithium ions is observed. However, a further rise in the lithium activity yields fully lithiated cathode material as $LiMo_6S_8$.

8.3 CATHODES FOR MG–LI HYBRID BATTERIES

The sluggish kinetics of Mg^{2+} ions can be solved by the introduction of lithium insertion-type cathode materials. Different types of cathode materials are employed for the improvement of Mg–Li hybrid batteries, which have been experimentally substantiated as a better candidate for rechargeable LIBs. The major cathode materials in Mg–Li hybrid batteries are $LiFePO_4$ (LFP), $LiMn_2O_4$,

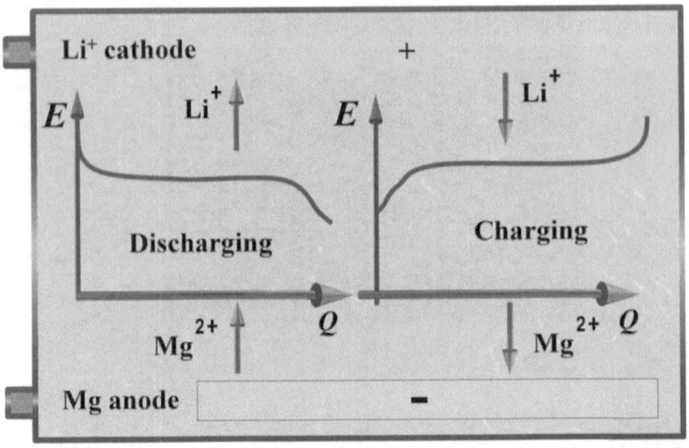

FIGURE 8.2 Schematic illustration of the working mechanism of a hybrid Mg-Li-ion batteries (MLIB). The cell is discharged with simultaneous Li^+ intercalation into the cathode and Mg^{2+} dissolution from the anode. Upon charging, Li^+ is de-intercalated from the cathode, while Mg is deposited on the anode. Corresponding voltage profiles of both electrodes are represented in orange and blue lines, respectively. Adapted and reproduced from Ref. [21]. Copyright © 2015 American Chemical Society.

LiV_3O_8 @GO, Mo_6S_8, etc. The performance of Li–Mg hybrid material depends on the energy density and charge–discharge capacity of cathode materials. The performance of Li–Mg hybrid material depends on the energy density and charge–discharge capacity of cathode materials [22,23]. The major types of cathode materials include conversion-type cathodes (FeS_2 and FeS), intercalation-type cathodes (chevrel phase), etc. Conversion-type cathodes offer high theoretical capacity in LIB compared to intercalated cathode materials. In contrast, the dendrite formation is not observed in Mg-Li battery since Li anode is displaced by Mg metal.

8.3.1 Chevrel-Type Cathodes for Mg–Li Hybrid Battery

The chevrel phase cathodes like Mo_6S_8 reveal only a low theoretical capacity of 128.8 mAh g^{-1} [24]. The investigation of Mo_6S_8 along with the dual salt electrolyte (0.4 M all phenyl complex (APC)) and (1.0 M LiCl) is reported to exhibit a specific capacity of 126 mAh g^{-1} at 0.1 C, which is closely similar to the theoretical capacity of the cathode. The dendrite-free mechanism of the anode Mg is confirmed by the high specific capacity achieved at a high C rate of 10 C that corresponds to only a 5% capacity fade. The dendrite-free morphology of Mg is confirmed by the scanning electron microscopic analysis of the electrode. By tuning the concentration of the salt in the electrolyte, the performance of the hybrid battery is observed to be varied. It has been reported that by controlling the LiCl concentration in APC electrolytes, it is possible to achieve a capacity of about 93.6% and 87.5% at 20 and 30 C, respectively [25].

Resembling Mo_6S_8 another cathode, explored for the hybrid battery system, is molybdenum sulphide (MoS_2). The horizontal sheet structure is formed by the covalently bonded molybdenum and sulphur atoms stacked by weak Van der Waals forces and allows for the accommodation of the guest ions [26,27]. This property of the layered structure provides high performance in a hybrid battery. The MoS_2 nano flower structure synthesized by Ju et al. [28] (Figure 8.3a) using the hydrothermal synthesis yielded the best electrochemical performance when treated with LiCl+[Mg_2Cl_3] + [$AlPh_2Cl_2$] @ (APC)/tetrahydrofuran (THF) hybrid electrolyte. The battery exhibits a high discharge capacity of about 243 mAh g^{-1} at 0.1 C and a rate capability of 108 mAh g^{-1} even at 5 C. It is notable that the battery can exhibit a high-capacity retention of about 87.2% even at 2,300 cycles (Figure 8.3b).

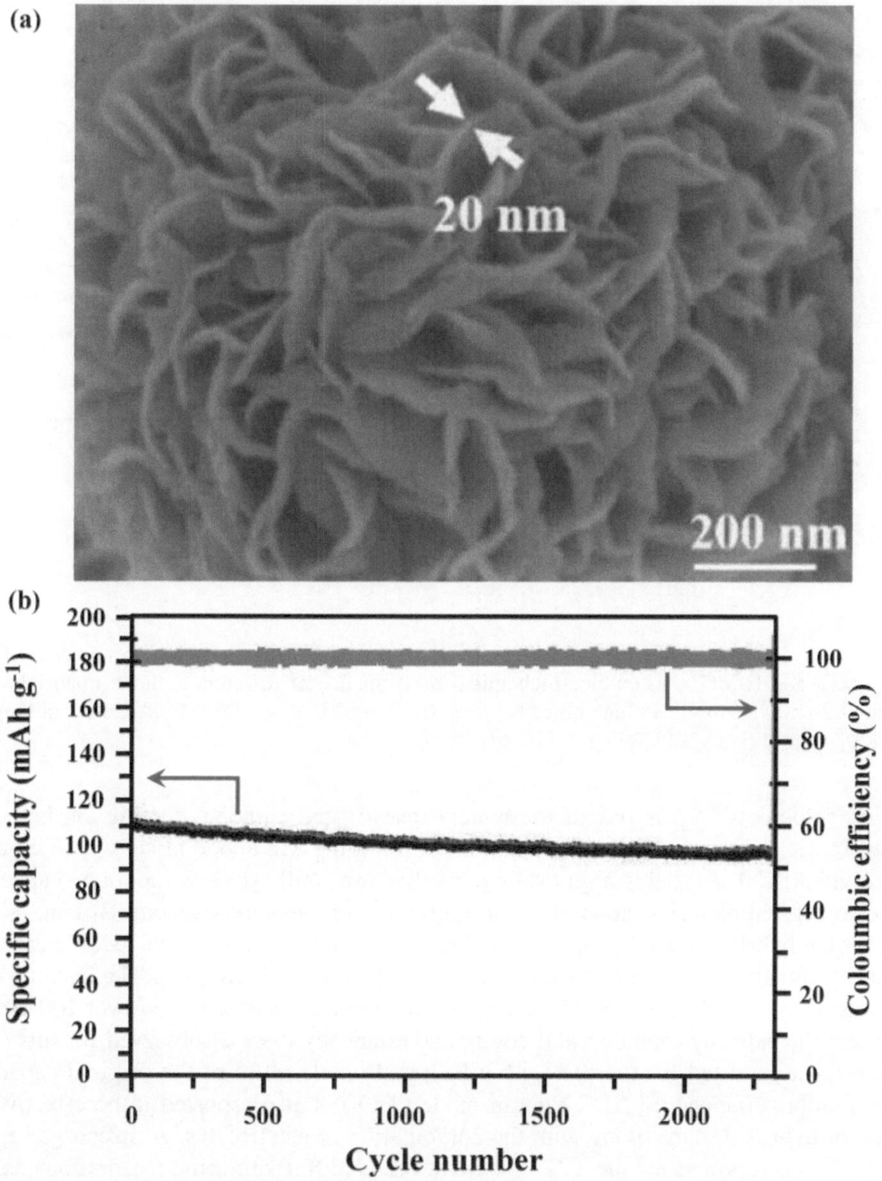

FIGURE 8.3 Structure and performance of MoS_2 nanoflowers. (a) Transmission electron microscopy images of MoS_2 nanoflowers and (b) cycle performance of the MoS_2//1.0 m LiCl + 0.4 M APC/THF//Mg cell at 1 Ag @1 current density. Adapted and reproduced from Ref. [28]. Copyright © 2016 Wiley.

For further modification of the structure, an interlayer expanded MoS_2–graphene composite was explored by Fan et al. [24] using an APC–LiCl electrolyte combination. The freestanding expanded MoS_2–graphene composite (E-MG) was directly grown over the graphene foam (GrF) using hydrothermal synthesis, followed by heat treatment (Figure 8.4), and revealed an upgraded interlayer spacing of 1.01 nm (which is 0.6 nm for bulk MoS_2) (Figure 8.4b) [29]. The fast intercalation mechanism of both Mg^{2+} and Li^+ enables the battery to exhibit a reversible capacity of 300 mAh g^{-1} at 20 mA g^{-1} that corresponds to a specific energy density of ≈316.9 Wh kg^{-1}. Even after 120 cycles, E-MG retained a capacity of 290 mAh g^{-1}. The value observed is reported to be better than that of non-expanded MoS_2/graphene (210 mAh g^{-1}) composite and bulk MoS_2 (175 mAh g^{-1}) (Figure 8.4c) [24].

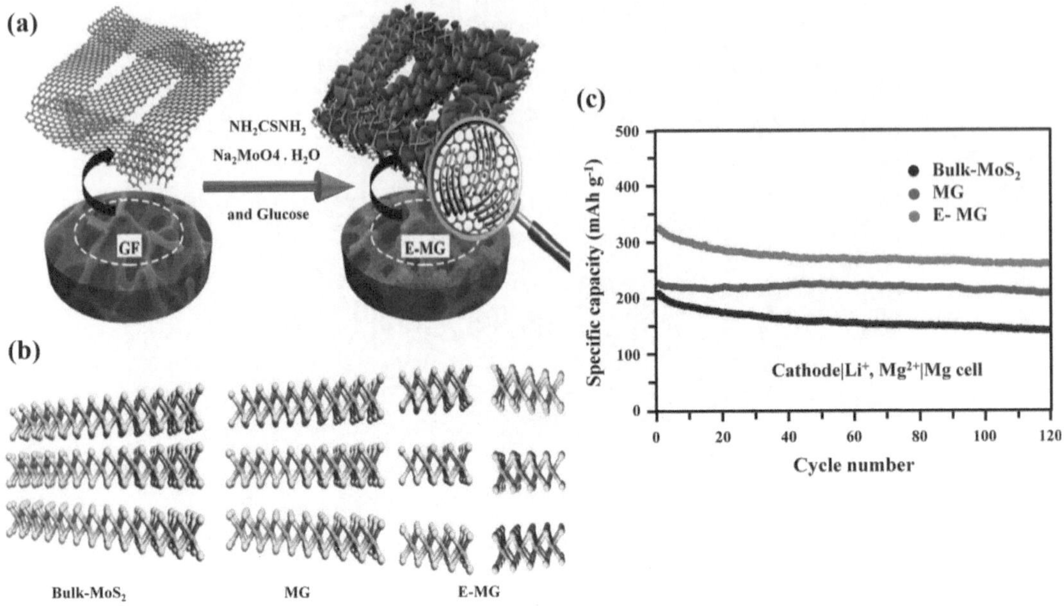

FIGURE 8.4 (a) Schematic illustration for the preparation of E-MG; (b) structure models of samples bulk-MoS_2, MG, and E-MG; and (c) electrochemical performance of different cathode materials in MLIB cells cycled at 20 mA g^{-1} in the voltage range between 0.01 and 2.0 V vs Mg^{2+}/Mg. Adapted and reproduced from Ref. [29]. Copyright © 2017 Wiley.

Titanium sulphide (TiS_2) is one of the widely investigated cathodes for Mg-ion batteries and is explored in Mg–Li hybrid battery systems. Examining the Mg anode along with TiS_2 and dual salt electrolyte APC–LiCl yields high cycling stability over 400 cycles without any capacity fade. Compared to Mg-ion batteries, about 32% of upgradation in specific capacity (161 mAh g^{-1}) was observed for the hybrid batteries [30]. Yoo et al reported the highest value of specific capacity and cyclic stability for the conventional Mg–Li hybrid [21] using TiS_2 cathode. The specific capacity reported was about 220 mAh g^{-1} at 0.1 C, and no capacity fade was observed over 100 cycles. It is attractive that the capacity retention and coulombic efficiency (99.9%) observed for this cell were higher than those reported for the cell with a Li metal anode (30% of the original capacity with 90% of coulombic efficiency) [21]. Titanium oxide (TiO_2) is also reported to be effective for the exploration of hybrid systems along with the combination of electrolytes. A discharge capacity of >140 mAh g^{-1} was reported for the TiO_2 system at 0.2 C [30]. Examining mesoporous nanoflakes of TiO_2 along with the dual salt electrolyte constituting $Mg(BH_4)$ and $LiBH_4$ in tetraglyme (TG) reveals a specific capacity of about 130 mAh g^{-1} at 1 C, which is better than that reported in the previous work. Apart from this, the cathode is able to retain an exceptionally high cyclic stability over 6,000 cycles even at a high C rate of 2 C, which is an attentive performance reported for the TiO_2 system [31]. Molybdenum dioxide (MoO_2) is also evaluated as the cathode along with the dual salt electrolyte owing to its high electrochemical and thermal stability. Hollow microspheres of MoO_2 nanoparticles are reported as an effective positive electrode with enhanced electrochemical performance in the Mg–Li hybrid system. Exploring along with the electrolyte system constituting 0.4 M APC/1 M LiCl/THF, the cell exhibits a discharge capacity of 217.2 mAh g^{-1} along with a coulombic efficiency of 88% at a current rate of 20 mA g^{-1} [32].

Considerable attention to lithium titanium oxide (LTO) as the cathode material arises from its unaltered lattice dimension during the lithiation and de-lithiation processes and, due to the same reason, it is known as a zero-strain material. LTO cathode is normally observed to be favourable for the Li^+ insertion rather than Mg^{2+} ions owing to the sluggish kinetics and the size [33]. A comparison of the electrochemical performances of both commercially available LTO and LTO/graphene (LTO/G) was made by examining the battery system along with the two different electrolytes such as $Mg(BH_4)_2$-based mixed electrolyte and APC-based mixed electrolyte. A better cycling performance was observed for the APC-based electrolyte attributed to its higher electronic conductivity of the electrolyte for the Mg deposition. Moreover, it has been observed that the electrode LTO/G is best in performance owing to the high specific surface area of the graphite that can promote high electron transport [34].

8.3.2 High-Voltage Cathodes for Mg–Li Hybrid Batteries

The insight into developing high-energy-density Mg–Li hybrid batteries leads to the examination of high-voltage cathodes for the system. The use of a high-voltage cathode was somewhat difficult for the hybrid system owing to the limited stability of electrolytes, poor control over the electrode–electrolyte interface, and the low coulombic efficiency attributed to the decomposition of electrolytes. So, the development of stable electrolyte composition is always considered to be the crucial factor for the high-voltage hybrid battery system. The well-established LIB cathodes such as lithium iron phosphate ($LiFePO_4$, simply LFP), lithium manganese oxide ($LiMn_2O_4$, simply LMN), lithium titanium oxide ($Li_4TI_5O_{12}$, LTO), etc. were widely studied for the performance enhancement of hybrid Mg–Li batteries.

LFP was investigated for the cathodes for the Mg–Li hybrid system using the dual salt electrolyte PhMgCl-$AlCl_3$. To facilitate the intercalation of Mg within the cathode, $LiBF_4$ was added as an inorganic additive, and the cell exhibited a specific capacity of 124 mAh g^{-1} along with a stable plateau at 2.7 V Mg/Mg^{2+} [35]. A freestanding film of LFP fabricated by using the rolling method exhibits an excellent discharge capacity of about 140 mAh g^{-1} at 0.1 C. Apart from this, the cell reveals a good rate performance and capacity retention of 120 mAh g^{-1} at 1 C [32]. About 90 mAh g^{-1} of capacity retention was observed even after 100 cycles. The energy density delivered by the Mg–Li/LFP cells (246 Wh kg^{-1}) is observed to be higher than that of the conventional Mg battery (134 Wh kg^{-1}) using Mo_6S_8 and LTO (143 Wh kg^{-1}) anode. Similarly, vanadium-based cathodes were reported as an effective candidate to enhance the electrochemical performance in hybrid Mg–Li batteries. Vanadium oxide cathode, on examination in the hybrid battery, was observed to exhibit a discharge capacity of 244.4 mAh g^{-1}, and at a current density of 100 Ah g^{-1}, they are able to exhibit a discharge capacity of about 206.8 mAh g^{-1} along with 75% of initial discharge capacity retention after 1,000 cycles. Moreover, the temperature adaptiveness of the cathode is not worthy, which can reveal a capacity of ~220 mA h g^{-1} at 55°C and ~35 mA h g^{-1} at 0°C [36] (Figure 8.5).

8.4 ELECTROLYTES IN MG–LI HYBRID BATTERIES

LIBs have a greater tendency to form dendrites with limited cycling stability, whereas Mg-ion batteries have poor ionic transport and structural stabilization. The Mg–Li hybrid batteries resolve these issues related to both Mg- and Li-ion batteries. Mg–Li hybrid batteries are constructed with dual salt electrolytes consisting of both Mg and Li ions as doping salt [23]. Electrolytes have a vital role in the electrode–electrolyte interface reaction, and hence electrochemically active species are required for their better performance. The first successful Mg-ion battery electrolytes belong to aprotic solvents, which form a surface passivation layer, and hence organo halo aluminates were later reported. But chloride-based electrolytes are corrosive in nature. Yagi et al. [35] introduced the concept of dual salt ions consisting of Li metal and a non-noble metal ion such as Ca, Mg, and Zn. The first prototype of the Mg–Li hybrid battery introduced $LiFePO_4$ as cathode material, Mg metal as anode material, and the mixture of two salts (PhMgCl and $LiBF_4$) dissolved in THF solvent (Figure 8.6). The reasons for choosing $LiFePO_4$ are as follows: (1) excellent cycling stability; (2) well-established electrolyte system; and (3) the dual salt ion exhibited following half-cell reaction:

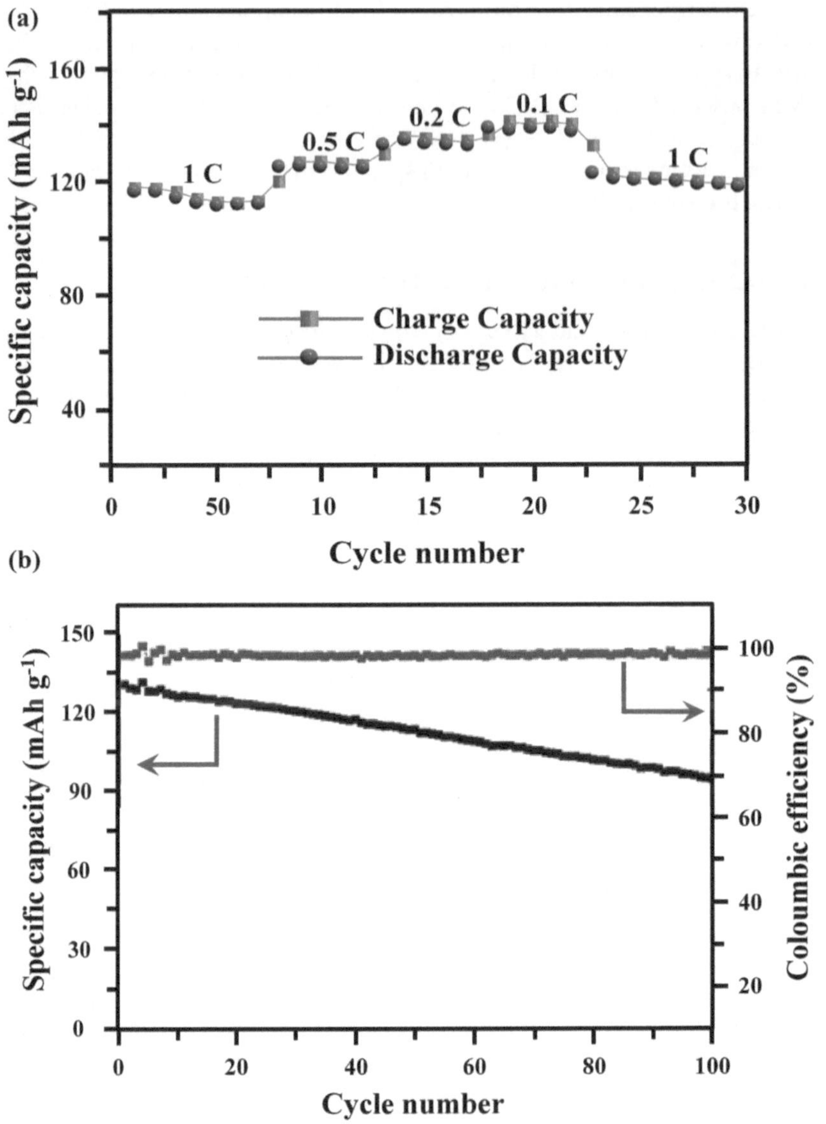

FIGURE 8.5 (a) Specific capacity (with respect to the weight of LFP) at different rates, and (b) cyclic stability at 0.5 C. Adapted and reproduced from Ref. [38]. Copyright © 2016 The Royal Society of Chemistry.

$$\text{PE: } FePO_4 + Li+ + e^- \rightarrow LiFePO_4 \tag{8.1}$$

$$\text{NE: } Mg \rightarrow Mg^{2+} + 2e^- \tag{8.2}$$

$$\text{Total: } FePO_4 + Li^+ + 1/2\ Mg \rightarrow LiFePO_4 + 1/2\ Mg^{2+} \tag{8.3}$$

Mg^{2+}/Li^+ mixed electrolyte shows high electrochemical stability, and specific capacity was reached up to 170 mAh g^{-1}. The major advantage of using dual salt ions in an Mg–Li hybrid battery is the enhancement in chemical kinetics by evading the strong interaction of cathode material and Mg^{2+} metal ions since intercalation of cathode materials preferentially chose monovalent ions, which have fast kinetics [37]. The ionic conductivity of the battery depends on the interaction of dual metal ions with the electrolytes, which determines ion pairs present in the solution and ionic

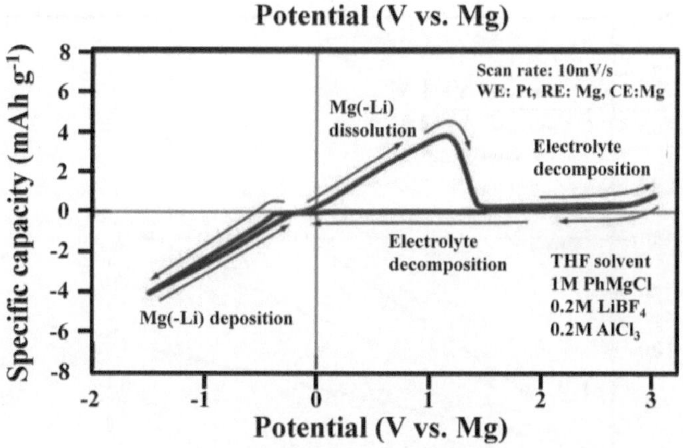

FIGURE 8.6 Cyclic voltammograms obtained in THF electrolyte containing PhMgCl, $LiBF_4$, and $AlCl_3$ using Pt as the working electrode. Adapted and reproduced from Ref. [35]. Copyright © 2014 The Royal Society of Chemistry.

species of the electrolyte, concentration of the salt, and nature of the solvent. Reports suggest that Mg and Li ions with the same anions decrease the rate of dissociation, whereas those with different anions increase the formation of large ionic species [37]. Cheng et al. [38] introduced APC with LiCl electrolytes dissolved in THF. This electrolyte took both the advantages of Li- and Mg-ion batteries. This system delivers cycle stability up to 3,000 cycles with the very least capacity fading (~5%) with 83% capacity retention. A full cell study was fabricated with Mo_6S_8 as the cathode material and Mg metal as the anode material, and the intercalation chemistry behind this reaction is vital [39]. The chapter mainly focuses on the intercalation mechanism of the Li host in Mo_6S_8 moiety. When LiCl and APC electrolytes were added to the system, Li insertion increased at a faster rate and depended on the concentration of Li ions in LiCl. Yoo et al. [21] also reported Mg–Li hybrid metal with the same APC electrolyte but different anodes and cathodes. The effect of TiS_2 as the cathode material was determined by two different Li salts, $LiBH_4$ and LiCl, and the highest energy density was found to be for $LiBH_4$ (200 Wh L^{-1}) compared to LiCl (80 Wh L^{-1}) (Figure 8.7a). The higher energy density of $LiBH_4$ and low electrochemical stability were attributed to the higher solubility of $LiBH_4$ toward diglyme solvent. Cyclic voltammograms of APC electrolytes with LiCl and pure $LiClO_4$ show that both electrolytes prefer lithium intercalation at the same rate (Figure 8.7b). Cho et al. [39–41] investigated dual salt electrolytes in Mg–Li hybrid batteries with LiTFSI as one of the dual salts. LiTFSI and $LiClO_4$ (LACC) mixture exhibit good anodic stability of 3 V and have a coloumbic efficiency of 89.7%. Recently the same group [40] modified this dual salt and obtained a coloumbic efficiency of 95%. The electrochemical process and productivity were improved by the conditioning-free process by the reaction of Mg metal with LACC along with $CrCl_3$. The initial discharge capacity was ~150 mAh g^{-1} at 2 C rate with a coloumbic efficiency of ~ 98%, which was stable up to 100 cycles. Later Bitenc et al. [42] investigated the effect of Cl- and bis(trifluoromethanesulfonyl)imide (TFSI) anions in dual salt electrolytes. Comparing the performance of LiTFSI and LiCl electrolytes in APC-based Mg doping salt reveals that TFSI anions have better performance than LiCl with an increased polarization on the Mg metal anode.

Grignard-based APC electrolytes offer almost 100% Mg deposition with high anodic stability (~3 V) [43]. Hence these types of electrolytes were further investigated in rechargeable Mg-ion batteries and Mg–Li hybrid batteries. Zhang et al. [44] fabricated an Mg–Li hybrid battery with LFP as the cathode material, Mg as the anode, pyrolytic graphitic film (GF) as the current collector, and APC electrolyte as the electrolyte. The overall reaction (Equations 8.4–8.6) in the anode and the cathode is given below:

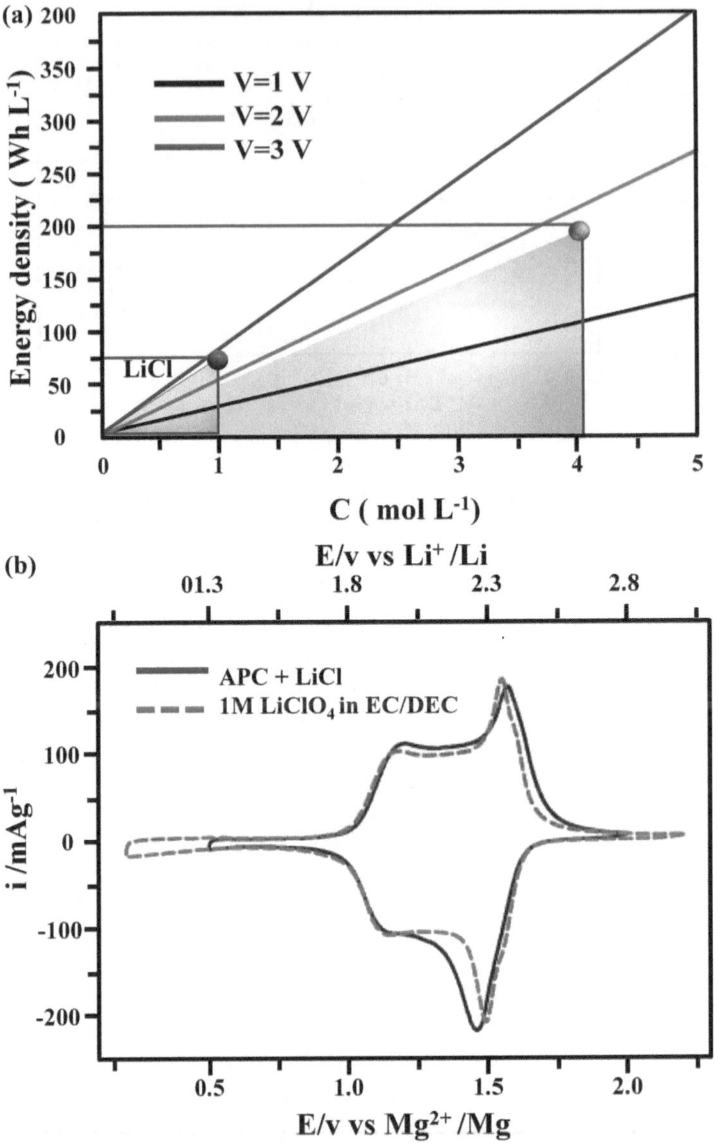

FIGURE 8.7 (a) The maximum energy density of a hybrid MLIB dependence on ion concentration and the electrochemical window of electrolyte (V). When LiCl is used as a Li-ion source, the highest energy density is 80 Wh L^{-1} (red dot), while, when $LiBH_4$ is used, the highest energy density is 200 Wh L^{-1} (blue dot) (b) Cyclic voltammograms obtained from three-electrode cells comprising TiS_2 as the working electrode and Mg or Li as a reference and counter electrode (scan rate = 0.1 mV s^{-1}). Adapted and reproduced from Ref. [21]. Copyright © 2015 American Chemical Society.

$$\text{Cathode: } FePO_4 + Li^+ + e^- \rightarrow LiFePO_4 \tag{8.4}$$

$$\text{Anode: } 2\,Mg + 3Cl^- + 6THF \rightarrow [(m\text{-}Cl)_3\,Mg_2(THF)_6]^+ + 4e^- \tag{8.5}$$

$$\text{Full cell: } 4FePO_4 + 4Li^+ + 2\,Mg + 3Cl^- + 6THF \rightarrow 4LiFePO_4 + [(m\text{-}Cl)_3Mg_2(THF)_6]^+ \tag{8.6}$$

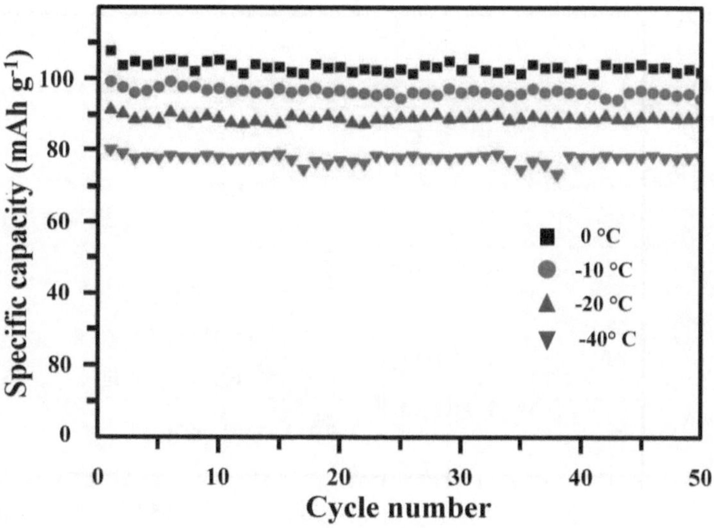

FIGURE 8.8 Schematic illustration of (a) LFP-hybrid electrolyte Mg battery during discharge and (b) cycling performance of the hybrid Mg^{2+}/Li^{+} battery. Adapted and reproduced from Ref. [44]. Copyright © The Royal Society of Chemistry 2016.

PhMgCl (APC electrolyte) reacted with $MgCl_2$ in THF solvent forms $[Mg_2Cl_3(THF)_6]^+[AlCl_2Ph_2]^-$. The complex compound shows that 3THF donors coordinated with each Mg ions, and three chloride ions are bridged to dimagnesium ions. The major advantages of fabricating with the proposed anode, cathode, electrolyte, and current collector (Figure 8.8a) are as follows: (1) the aforementioned current collector GF offers a wide electrochemical stability window (>2.8 V) and retards corrosion reactions in electrolytes; (2) this can operate with a wide temperature range mainly of low-temperature range (–20°C to 40°C) (Figure 8.8b), which will have wide application in polar regions; (3) Mg–Li hybrid batteries have superior cycling performance and coloumbic efficiency; and (4) Mg is more abundant and can be used cost-effectively.

8.5 CHALLENGES AND FUTURE PERSPECTIVES

Even though the Mg-ion battery suffers sluggish kinetics and the problem was rectified using a Mg–Li hybrid battery, the fabrication of Mg metal as the anode material and Li-based cathode material with dual salt electrolytes exhibits slow capacitance. Cheng et al. [38] fabricated an Mg–Li hybrid battery that has very a less specific capacitance (80 mAh g^{-1} at 0.1 C rate) compared to a pristine Mg- and Li-ion battery (Figure 8.4). TiS_2-based cathode material reported by Yoo et al. [21] offers comparatively high specific capacitance (220 mAh g^{-1}). The detailed mechanism of Mg metal with TiS_2 cathode in Mg–Li-anode and Li-ion battery suggests the favourable performance of the battery. Dendrite formation was absent in the Mg–Li hybrid battery whereas noticeable finger-like dendrites were observed in the Li-based battery. Reports on Mg–Li hybrid batteries reveal high coloumbic efficiency offered by the system. Among different reports, Cheng et al. [24] reported the high-coloumbic, efficient battery having very high cycling stability over 3,000 cycles.

Dual salt electrolytes have a vital role in the fabrication of Mg–Li hybrid batteries since dual salt offers both Mg and Li salt for conduction. In the case of cathode material, either lithiated cathode consisting of Li-based compounds (LFP, $LiCoO_3$, etc.) or non-lithiated cathodes which have no Li ions such as Mo_6S_8 and TiS_2. The most common anode materials which have been used in these types of batteries are Mg metal anode, which can inhibit the dendrite formation, the major challenge faced by Li-ion batteries. Considering the electrolyte as dual salt ions, the Mg^{2+} ions emerge from

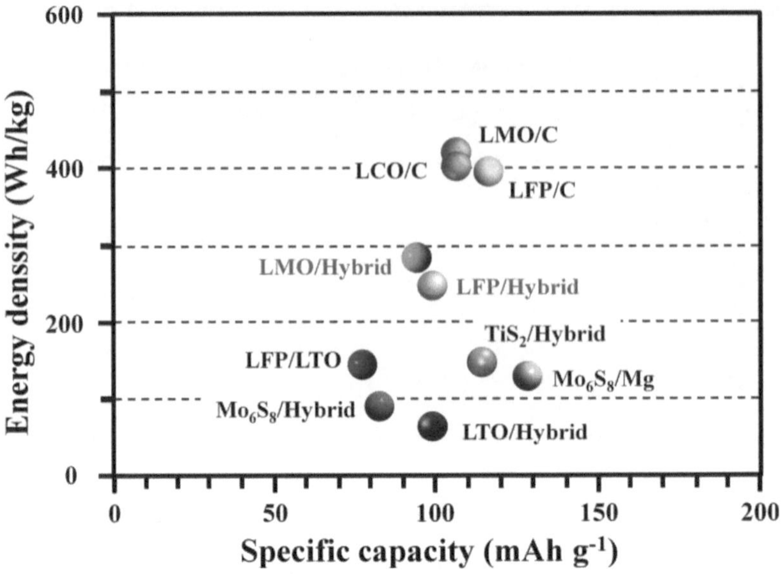

FIGURE 8.9 Specific energy density comparison for Mg/Mo_6S_8 battery, LIB, and Mg–Li hybrid batteries without considering the mass of solvents. Adapted and reproduced from Ref. [38]. Copyright © 2016 The Royal Society of Chemistry.

dual salt electrolytes, which should balance with Li-based cathode material. The amount of dual salt electrolyte needed was calculated by Equation 8.7:

$$V_h = 3{,}600{,}000.\, {C_{\mathrm{Li}}}\big/{zFC_{\mathrm{Mg}}} \tag{8.7}$$

where F is the Faraday constant, Z is the number of charges of Mg^{2+} ion (for Mg, its charge is 2), C_{Li} belongs to the capacity density of Li-based cathode material (mAh cm^{-2}), V_h is the volume of dual salt expressed in μL cm^{-2}, and C_{Mg} is the concentration of Mg^{2+} ion.

The concentration of Mg-ion and volume are inversely related in such a way that the lower concentration of Mg^{2+} requires more volume of dual salt electrolytes [38] in a Li/Mg hybrid battery evident from the above equation. Figure 8.9 shows that Mg–Li/LFP have an energy density of 246 Wh kg^{-1}, which is greater than conventional Mg-ion batteries with Mo_6S_8 as the cathode material (134 Wh kg^{-1}). As the energy density increases, there will be an increase in the output voltage greater than 2.5 V. At the same time, energy density can be increased by reducing the amount of electrolyte. This can be achieved by the process called the "precipitation-dissolution" mechanism. The mechanism depicts that during the discharge process, Mg salts get precipitated whereas during the charged process Li salts will be precipitated from the dual salt electrolyte due to the delithiation of Li salt in the cathode material. For this, pouch cell-type architecture is suitable rather than coin cell fabrication so that excess electrolytes can be added. Hence fabrication of an Mg–Li hybrid battery with a proper understanding of the "precipitation-dissolution" mechanism can effectively reduce the amount of electrolytes.

8.6 CONCLUSION

Mg–Li hybrid battery is in the infant stage, and research is investigating the suitable cathode, anode, and dual salt electrolytes. In the search for an Mg–Li hybrid battery, the most common cathode material that has been used is Mg metal anode, which can inhibit dendrite formation. The major class of cathodes belongs to lithiated and non-lithiated cathode materials, in which the most common lithiated cathode materials are $LiFePO_4$, $LCoO_3$, etc., and those of non-lithiated cathodes include TiS_2, MoS_2, Mo_6S_8, etc. Dual salt electrolytes are the major type of electrolytes used owing to the presence of both Mg and Li salts. The most common Mg-based electrolyte is Grignard-based APC electrolytes and Li doping salt belongs to LiCl, LiTFSI, etc. Even though its energy density is not considerably high compared to pristine LIBs, cycling stability and rate capability were higher in Mg–Li hybrid batteries.

REFERENCES

1. Smith SC, Sen PK, Kroposki B (2008) Advancement of energy storage devices and applications in electrical power system. *2008 IEEE Power and Energy Society General Meeting – Conversion and Delivery of Electrical Energy in the 21st Century*, Pittsburgh, PA, USA, pp. 1–8. https://doi.org/10.1109/PES.2008.4596436
2. Nishi Y (2014) *2 – Past, Present and Future of Lithium-Ion Batteries: Can New Technologies Open up New Horizons?* Elsevier, Amsterdam, The Netherlands.
3. Manthiram A (2017) An outlook on lithium ion battery technology. *ACS Cent Sci* 3:1063–1069. https://doi.org/10.1021/acscentsci.7b00288
4. Scrosati B, Hassoun J, Sun YK (2011) Lithium-ion batteries. A look into the future. *Energy Environ Sci* 4:3287–3295. https://doi.org/10.1039/c1ee01388b
5. Noorden RV (2014) A better battery: chemists are reinventing rechargeble cells to drive down costs and boost capacity. *Nature* 507:26–28.
6. Kurzweil P, Garche J (2017) *2 – Overview of Batteries for Future Automobiles.* Elsevier B.V, Amsterdam, The Netherlands.
7. Amon F, Andersson P, Karlson I, Sahlin E (2012) Fire risks associated with batteries. In: Francine A, Petra A, Ingvar Karlson ESS (eds) *SP Technic. Science Partner*, SP Technical Research Institute of Sweden, Science partner, p. 54.
8. Li G, Chang HJ, Lu X, et al (2017) Low cost and reliable sodium-metal halide(Na-MH) batteries for stationary energy storage application. *Adv Mater – TechConnect Briefs* 2:122–125
9. Eftekhari A, Kim DW (2018) Sodium-ion batteries: new opportunities beyond energy storage by lithium. *J Power Sources* 395:336–348. https://doi.org/10.1016/j.jpowsour.2018.05.089
10. Yan K, Lu Z, Lee HW, et al (2016) Selective deposition and stable encapsulation of lithium through heterogeneous seeded growth. *Nat Energy* 1:16010. https://doi.org/10.1038/NENERGY.2016.10
11. Xu W, Wang J, Ding F, et al (2014) Lithium metal anodes for rechargeable batteries. *Energy Environ Sci* 7:513–537. https://doi.org/10.1039/c3ee40795k
12. Gofer Y, Ben-Zion M, Aurbach D (1992) Solutions of $LiAsF_6$ in 1,3-dioxolane for secondary lithium batteries. *J Power Sources* 39:163–178. https://doi.org/10.1016/0378-7753(92)80135-X
13. Crowther O, West AC (2008) Effect of electrolyte composition on lithium dendrite growth. *J Electrochem Soc* 155:A806. https://doi.org/10.1149/1.2969424
14. Qian J, Henderson WA, Xu W, et al (2015) High rate and stable cycling of lithium metal anode. *Nat Commun* 6:6362. https://doi.org/10.1038/ncomms7362
15. Goodenough JB (2012) Rechargeable batteries: challenges old and new. *J Solid State Electrochem* 16:2019–2029. https://doi.org/10.1007/s10008-012-1751-2
16. Society J of the E (2000) Factors which limit the cycle life of rechargeable lithium (metal) batteries. *J Electrochem Soc* 147:1274.
17. Saha P, Datta MK, Velikokhatnyi OI, et al (2014) Rechargeable magnesium battery: current status and key challenges for the future. *Prog Mater Sci* 66:1–86. https://doi.org/10.1016/j.pmatsci.2014.04.001
18. Besenhard JO, Winter M (2002) Advances in battery technology: rechargeable magnesium batteries and novel negative-electrode materials for lithium ion batteries. *ChemPhysChem* 3:155–159. https://doi.org/10.1002/1439-7641(20020215)3:2<155::AID-CPHC155>3.0.CO;2–S
19. Muldoon J, Bucur CB, Oliver AG, et al (2013) Corrosion of magnesium electrolytes: chlorides-the culprit. *Energy Environ Sci* 6:482–487. https://doi.org/10.1039/c2ee23686a
20. Aurbach D, Lu Z, Schechter A, et al (2000) Prototype systems for rechargeable magnesium batteries. *Nature* 407:724–727. https://doi.org/10.1038/35037553
21. Yoo HD, Liang Y, Li Y, Yao Y (2015) High areal capacity hybrid magnesium-lithium-ion battery with 99.9% coulombic efficiency for large-scale energy storage. *ACS Appl Mater Interfaces* 7:7001–7007. https://doi.org/10.1021/acsami.5b01206

22. Li M, Pei C, Xiong F, et al (2019) A high energy density hybrid magnesium-lithium ion battery based on LiV_3O_8@GO cathode. *Electrochim Acta* 320:134556. https://doi.org/10.1016/j.electacta.2019.134556
23. Cheng Y, Chang HJ, Dong H, et al (2016) Rechargeable Mg-Li hybrid batteries: status and challenges. *J Mater Res* 31:3125–3141. https://doi.org/10.1557/jmr.2016.331
24. Cheng Y, Shao Y, Zhang JG, et al (2014) High performance batteries based on hybrid magnesium and lithium chemistry. *Chem Commun* 50:9644–9646. https://doi.org/10.1039/c4cc03620d
25. Nam KW, Kim S, Lee S, et al (2015) The high performance of crystal water containing manganese birnessite cathodes for magnesium batteries. *Nano Lett* 15:4071–4079. https://doi.org/10.1021/acs.nanolett.5b01109
26. Haycock DE, Urch DS, Wiech G (1979) Electronic structure of molybdenum disulphide. *J Chem Soc Faraday Trans 2 Mol Chem Phys* 75:1692–1702. https://doi.org/10.1039/F29797501692
27. Najmaei S, Liu Z, Zhou W, et al (2013) Vapour phase growth and grain boundary structure of molybdenum disulphide atomic layers. *Nat Mater* 12:754–759. https://doi.org/10.1038/nmat3673
28. Ju Y, Meng Y, Wei Y, et al (2016) Li^+/Mg^{2+} hybrid-ion batteries with long cycle life and high rate capability employing MoS^2 nano flowers as the cathode material. *Chem – A Eur J* 22:18073–18079. https://doi.org/10.1002/chem.201604175
29. Fan X, Gaddam RR, Kumar NA, Zhao XS (2017) A hybrid Mg^{2+}/Li^+ battery based on interlayer-expanded MoS2/graphene cathode. *Adv Energy Mater* 7:2–11. https://doi.org/10.1002/aenm.201700317
30. Gao T, Han F, Zhu Y, et al (2015) Hybrid Mg^{2+}/Li^+ battery with long cycle life and high rate capability. *Adv Energy Mater* 5:1–5. https://doi.org/10.1002/aenm.201401507
31. Su S, NuLi Y, Huang Z, et al (2016) A high-performance rechargeable Mg^{2+}/Li^+ hybrid battery using one-dimensional mesoporous TiO^2(B) nanoflakes as the cathode. *ACS Appl Mater Interfaces* 8:7111–7117. https://doi.org/10.1021/acsami.6b00106
32. Yingwen C, Daiwon C, Kee Sung H, Karl TM, Ji-Guang Z, Vincent LS, Jun L, Guosheng L (2013) Toward the design of high voltage magnesium-lithium hybrid batteries using dual-salt electrolytes. *R Soc Chem* 52:1–3. https://doi.org/10.1039/C6CC00986G
33. Miao Q, NuLi Y, Wang N, et al (2016) Effect of Mg^{2+}/Li^+ mixed electrolytes on a rechargeable hybrid battery with $Li_4Ti_5O_{12}$ cathode and Mg anode. *RSC Adv* 6:3231–3234. https://doi.org/10.1039/c5ra25827h
34. Wu N, Yang Z-Z, Yao H-R, et al (2015) Improving the electrochemical performance of the $Li_4Ti_5O_{12}$ electrode in a rechargeable magnesium battery by lithium-magnesium co-intercalation. *Angew Chemie* 127:5849–5853. https://doi.org/10.1002/ange.201501005
35. Yagi S, Ichitsubo T, Shirai Y, et al (2014) A concept of dual-salt polyvalent-metal storage battery. *J Mater Chem A* 2:1144–1149. https://doi.org/10.1039/c3ta13668j
36. Pei C, Xiong F, Sheng J, et al (2017) VO_2 nanoflakes as the cathode material of hybrid magnesium-lithium-ion batteries with high energy density. *ACS Appl Mater Interfaces* 9:17060–17066. https://doi.org/10.1021/acsami.7b02480
37. Yao HR, You Y, Yin YX, et al (2016) Rechargeable dual-metal ion batteries for advanced energy storage. *Phys Chem Chem Phys* 18:9326–9333. https://doi.org/10.1039/c6cp00586a
38. Liu SJ, Li G (2016) Toward the design of high voltage magnesium- lithium hybrid batteries using dual-salt electrolytes. *Chem Commun* 52:5379–5382. https://doi.org/10.1039/c6cc00986g
39. Cho J, Aykol M, Kim S, et al (2014) Controlling the intercalation chemistry to design high-performance dual-salt hybrid rechargeable batteries. *J Am Chem Soc* 136:16116–16119.
40. Cho J, Ha JH, Oh J, et al (2020) Facile modification of $LiAlCl_4$ electrolytes for Mg-Li hybrid batteries by the conditioning-free method. *J Phys Chem C* 124:25738–25747. https://doi.org/10.1021/acs.jpcc.0c07914
41. Cho J, Kim SJ, Oh J, et al (2018) Strategic design of highly concentrated electrolyte solutions for Mg^{2+}/Li^+ dual-salt hybrid batteries. *J Phys Chem C* 122:49. https://doi.org/10.1021/acs.jpcc.8b09080
42. Bitenc J, Firm M, Randon Vitanova A, Dominko R (2017) Effect of Cl^- and $TFSI^-$ anions on dual electrolyte systems in a hybrid $Mg/Li_4Ti_5O_{12}$ battery. *Electrochem Commun* 76:29–33. https://doi.org/10.1016/j.elecom.2017.01.013
43. Lv D, Xu T, Saha P, et al (2012) A scientific study of current collectors for Mg batteries in $Mg(AlCl_2EtBu)_2$/THF electrolyte. *J Electrochem Soc* 160:A351–A355. https://doi.org/10.1149/2.085302jes
44. Zhang Z, Xu H, Cui Z, et al (2016) High energy density hybrid Mg^{2+}/Li^+ battery with superior ultra-low temperature performance. *J Mater Chem A* 4:2277–2285. https://doi.org/10.1039/c5ta09591c

9 Metal-Air Batteries

Future of Hybrid or Electric Vehicles (HEV or EV)

S. K. Vineeth, Ajish Babu, Akhila Das, Abhilash Pullanchiyodan, Nutan Gupta, Vijayamohanan Pillai, and Prasanth Raghavan

9.1 INTRODUCTION

Eco-friendly, cost-effective, reliable energy power sources have always been the focus of research for electric vehicle (EV) technology categorized as follows: (1) Battery Electric Vehicle powered by electricity stored in a battery pack, (2) Hybrid Electric Vehicle (HEV), and (3) Plug-in Hybrid Electric Vehicle (PHEV) – combine a gasoline or diesel engine with an embedded electric motor and large packs of rechargeable batteries, (4) Fuel Cell Electric Vehicle (FCEV) – split electrons from hydrogen molecules to produce electricity to power the motor vehicles [1]. EVs rely on promising and sustainable safety features as well as minimizing costs and increasing the lifecycle of batteries, which is an important aspect of the current automotive industry. Advancements in the understanding of energy storage technology helped in moving a step further for research and development of batteries, supercapacitors, fuel cells, etc., and this has also led to a fast transition from FCEVs to HEVs, PHEVs, and full EVs. Moreover, with stringent environmental laws and government policies with the realization of the effects of global warming, there is a booming rate of interest in EVs.

Figure 9.1a shows battery capacity and other parametric comparisons in various batteries employed for current HEVs, PHEVs, and EVs. Figure 9.1b represents the Ragone plot of various battery technologies in terms of their electrochemical properties for comparing ranges in automotive applications. In general, the configuration of rechargeable LIBs consists of two electrodes: anode and cathode, immersed in an electrolyte and separated by a polymer membrane [2].

Briefly if we look back on the developments made on batteries, we see that in 1859 a lead–acid battery-based EV was developed by Gaston Plante, followed by the usage of nickel–cadmium battery, developed by Waldemar Jungner in 1899 [4]. These technologies faced limitations such as voltage suspension with the usage of these batteries. While considering various battery technologies such as nickel–cadmium (Ni–Cd), or nickel–metal halide (Ni–MH)-based batteries, LIBs overrule the other conventional batteries due to their high energy density of 250 Wh kg^{-1} and 650 Wh L^{-1}, longer cyclic stability, and superior electrochemical properties [5]. These batteries have undergone various transitions after their first commercialization by Sony Corporation in 1991, especially in the field of graphite as an anode and lithiated transition metal oxide as a cathode [6].

For improving the performance of EVs, several features such as energy density, battery weight ratio, cost, and safety play critical roles. To achieve these goals, high-capacity anodes and cathodes are certainly needed and recommended to be researched for a promising future [7]. Anode based on silicon and tin-based carbon composite electrodes showed promising results with good cyclic stability and high capacity. Cathode with nickel, manganese, and cobalt has better capacity and possesses

DOI: 10.1201/9781003310174-9

a

Mode of operation		Battery capacity needed	Energy density Wh Kg^{-1}	Battery weight Kg	Speed Km h^{-1}	Distance on one charge Km
Hybrid		< 3	40-50 Ni-MH	60 Ni-MH	100 +	15
Plug-in		5.6 - 18	90-100 Li-ion	60-200 Li-ion	100 +	10-60
Full EV		35 - 54	90-100 Li-ion	450 Li-ion	>100	150-200

b

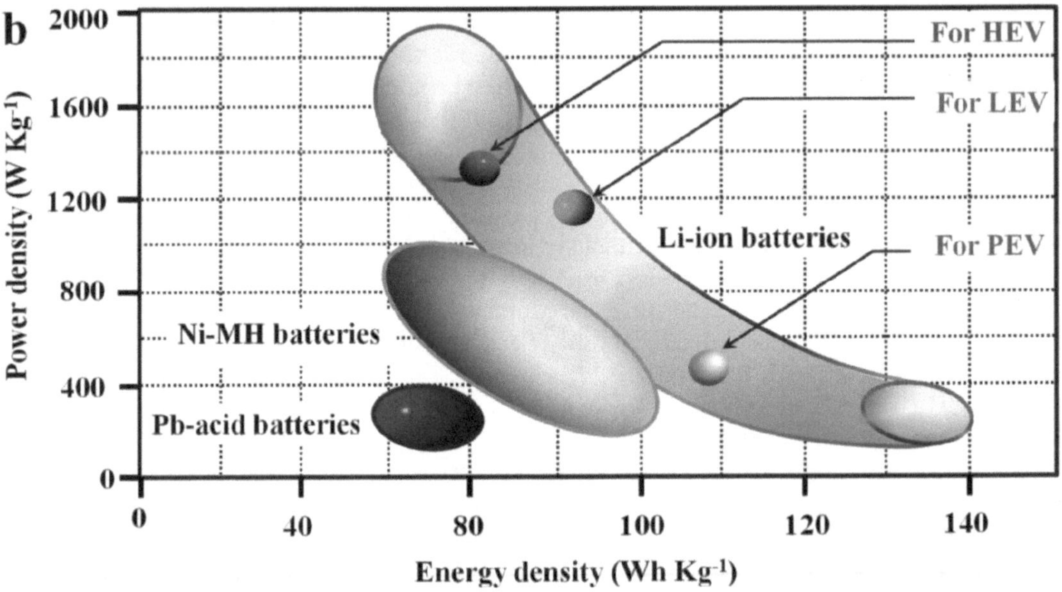

FIGURE 9.1 (a) Comparison of battery properties and performance of hybrid plug-in hybrid and full EVs. (b) Ragone plot for various battery technologies. Adapted and reproduced with permission from Ref. [3]. Copyright © 2011 The Royal Society of Chemistry.

the capability of further improvement [8,9]. Theoretical specific energy (Wh kg^{-1}) and energy density (Wh L^{-1}) for different battery electrochemical reactions have been shown in Table 9.1. The high energy density offered by metal-air batteries has also attracted researchers to explore various metal-air energy systems [10]. In recent years, the high theoretical specific energy of lithium-air batteries (LABs) opened a promising path to a transformational energy storage system for EVs, and Gallagher et al. [11] compared it with LIBs in terms of mass, volume, and cost. LABs have been identified as potential candidates for achieving long-term automotive objectives.

A graphical comparison of different types of LIBs used in EVs is demonstrated in Figure 9.2. The major factors for bigger coloured areas (outer hexagons) are mainly dependent on specific

TABLE 9.1
Electrochemical Reactions and Performance of Different Batteries

Battery	Cell Voltage (V)	Theoretical Specific Energy (Wh kg^{-1})	Theoretical Energy Density (Wh L^{-1})
Today's Li-ion	3.8	387	1,015
$\frac{1}{2}C_6Li+Li_{0.5}CoO_2 \leftrightarrow 3C+LiCoO_2$			
Zn-air	1.65	1,086	6,091 (ZnO)
$Zn+\frac{1}{2}O_2 \leftrightarrow ZnO$			
Li-S	2.2	2,567	2,199 ($Li+Li_2S$)
$2Li+S \leftrightarrow Li_2S$			
Li-O_2 (non-aqueous)	3.0	3,505	3,436 ($Li+Li_2O_2$)
$2Li+O_2 \leftrightarrow Li_2O_2$			
Li-O_2 (aqueous)	3.2	3,582	2,234 ($Li+H_2O+LiOH$)
$2Li+\frac{1}{2}O_2+H_2O \leftrightarrow 2LiOH$			

Source: Adapted and reproduced with permission from Ref. [10].

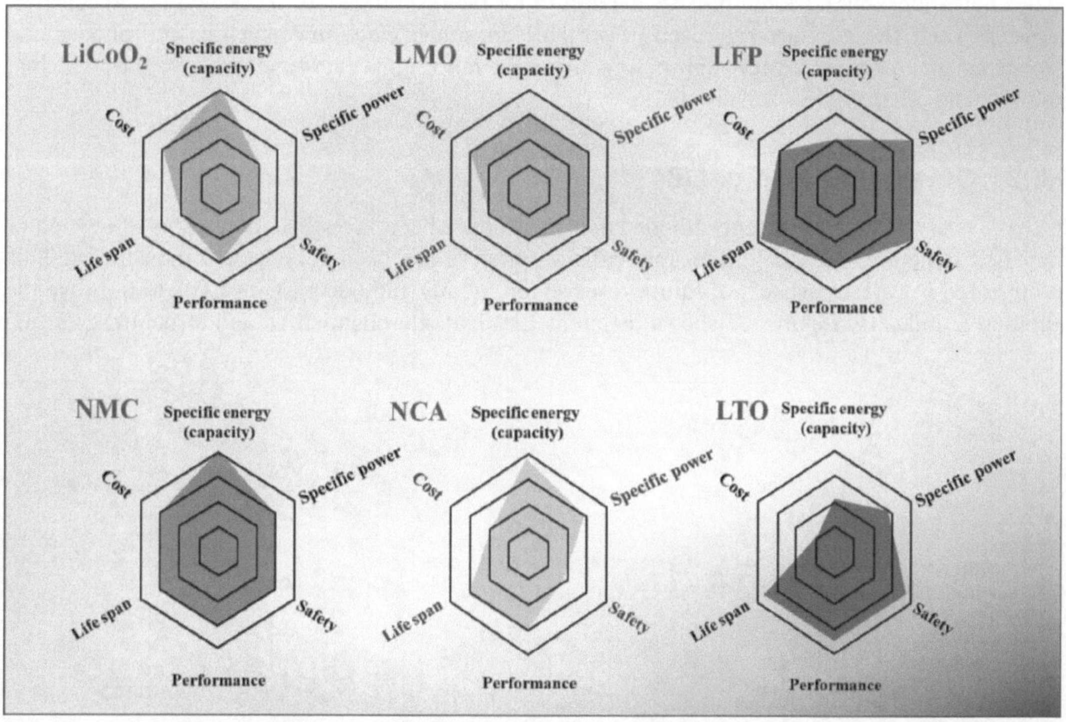

FIGURE 9.2 Comparisons of different LIBs used in EVs in terms of specific power, life span, cost, performance, and safety. Adapted and reproduced with permission from Ref. [12]. Copyright © 2019 MDPI.

energy, specific power, safety, performance, life span, and cost [12]. As clearly shown, safety parameters when driving the EV in extreme temperature conditions are one of the most important aspects when choosing a battery for EVs. Life span and lowering the cost of battery packs seem challenging for current EV technologies.

9.2 LITHIUM-ION BATTERIES

Green energy solutions providing no harmful gases, fumes or pollution motivates current research and developments in the area of lithium-ion batteries (LIBs) and LABs. After the introduction of LIBs in the 1970s, numerous researchers have been exploring the potential applicability of LIBs [13]. Replacement of metallic lithium anode by graphite and lithiated transition metal oxides as cathode led to its further exploration. Currently EVs use LIBs as the main energy source and are the main focus of research in this area to create batteries with more capacity and also with better power densities.

9.2.1 Working Principle of LIBs

A LIB works on the principle of a reversible Li-ion intercalation–deintercalation cycle. Figure 9.3 shows a brief overview of various materials used as anode and cathode in LIBs. It is through the electrolyte that the ionic conduction of Li ions takes place. The ionic conductivity of the electrolyte depends on the electrolyte material, electrodes and the interface formed between them also influences the ionic mobility. A LIB contains an organic solvent (various alkyl carbonates) and a mixture of lithium salts. An electrolyte should offer high ionic conduction. In addition to this a LIB contains a separator for ionic conduction while preventing internal short circuits. An ideal separator should also conduct ions, but it should possess the ability for rapid shutdown if in case a thermal runaway happens [12]. LIBs are currently used in portable consumer electronics such as cell phones and laptops as they possess a very high power-to-weight ratio, high energy efficiency, good thermal stability, and relatively low self-discharge.

9.2.2 Cathode Chemistry of LIBs

In LIBs, over the years different cathode chemistries have been investigated in terms of layered or tunnelled structure, but the lithium intercalation process has been very well demonstrated. This has resulted in various aspects of future research on mainly cathode parts of LIBs to improve the automotive industry. Figure 9.4 shows different LIB cathode chemistries and structures. On the

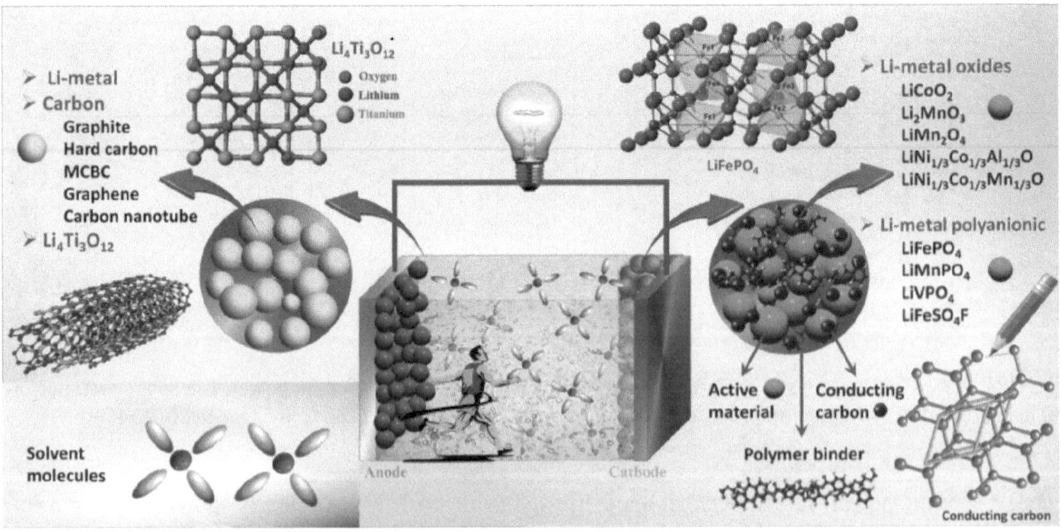

FIGURE 9.3 Lithium-ion battery: an overview of its structure and the materials used for various parts. Adapted and reproduced with permission from Ref. [14]. Copyright © 2016 Chinese Physical Society and IOP Publishing Ltd.

Cathode types	LCO	LFP	LMO	NCA	NMC
Chemical formula	$LiCoO_2$	$LiFeO_4$	$LiMn_2O_4$	$Li(Ni,Co,Al)O_2$	*[1]$LiNi_{0.33}Mn_{0.33}Co_{0.33}$ *[2]$LiNi_{0.5}Mn_{0.3}Co_{0.2}O_2$ *[3]$LiNi_{0.6}Mn_{0.2}Co_{0.2}O_2$ *[4]$LiNi_{0.8}Mn_{0.1}Co_{0.1}O_2$
Structure	Layered	Olivine	Spinel	Layered	Layered
Year introduced	**1991**	**1996**	**1996**	**1999**	**2008**
Safety	●●	●●●●●	●●●●	●●●	●●●
Energy density	●●●●	●●●	●●●	●●●●●	●●●●●
Power density	●●	●●●●	●●●●	●●●●	●●●
Calendar lifespan	●●●	●●●●	●●●●	●●●●	●●●●
Cycle lifespan	●●●	●●●●	●●●	●●●●	●●●
Performance	●●●●	●●●●	●●●	●●●●	●●●●
Cost	●	●●●●	●●●●	●●●	●●●
Market share	Obsolete	EVs	Small	Steady	Growing (from NMC 111 > 532 > 622 > 811 to no-cobalt chemistries

*[1] (NMC111), *[2] (NMC532), *[3] (NMC622), *[4] (NMC811)

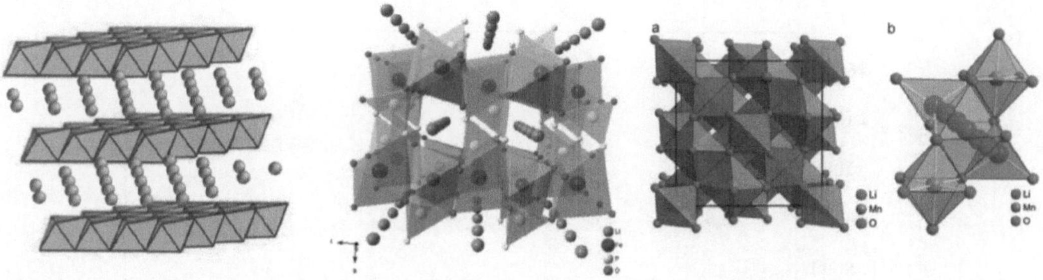

FIGURE 9.4 Development of LIBs cathode chemistry based on various parameters. Adapted and reproduced with permission from Ref. [16]. Copyright © 2019 Nature.

other side, high exothermic reactions inside LIBs have created a serious concern for EV industries and corporates. The main reasons for the high exothermic reactions are generalized as follows: (1) delithiation at cathode producing irreversible structural changes at the cathode with oxidation of organic solvents, (2) formation of dendrites in Li anode and reaction with electrolytes leading to an internal short circuit, (3) internal short circuit caused by the melting of the separator as temperature rises inside a battery, (4) decomposition of electrolyte due to the lower boiling point and flash point of organic solvents. As the amount of heat generated exceeds the heat dissipated, temperature build-up in LIBs shoots up, resulting in battery failure and explosion [15].

9.2.3 Recycling Challenges of LIBs

The cost of recycling components of LIBs remains a challenge for the battery industry. Although LIBs possess certain limitations, research on various materials for electrodes and electrolytes has been undertaken over the past decades and there has been a notable increase in energy density of approximately 15% per year [17]. Moreover, the requirement for efficient technologies for the recycling of LIBs is still in the developing stage. As the current commercial LIBs consist of harmful Li salts containing organic electrolytes, transition metals, its oxides, and other compounds, improper disposal of LIBs results in serious environmental consequences. Hence recycling of these needs to be taken into a serious context. To address this issue, both technological innovations and the participation of governments are required [18]. Compared to the successful lead–acid battery recycling industry, where there are standardized and simple procedures for disassembling and recycling with minimum cost, recycling LIBs should also take lead–acid battery recycling as a reference for the same development. The development of recycling techniques for batteries will be a key to the sustainability of the future automotive industry [16]. At present most of the designs for battery pack modules or cells within the automotive sector are mainly human-operated, leading to unsafe waste handling [19].

9.3 HIGH-ENERGY-DENSITY BATTERY: A FOCUS INTO THE METAL-AIR BATTERIES

Among the various energy storage devices, LIBs are the most popular in commercial applications and the best in performance. It is still a challenge to achieve a targeted energy density of 500 Wh kg^{-1} (or high capacity of >200 mAh g^{-1}) with a cost of <100 USD kWh^{-1} in LIBs and to meet safe energy storage and lightweight requirements for EVs [20,21]. Hence, very recently, among different potential energy storage devices, metal-air batteries (MABs) have revived interest as they can operate in an open-air atmosphere. The design aspects of MABs amalgamate with both fuel cells and conventional battery design and possess a high theoretical energy density, which is 3–30 folds that of LIB [22]. A comparison of theoretical energy densities for different types of MABs is shown in Figure 9.5.

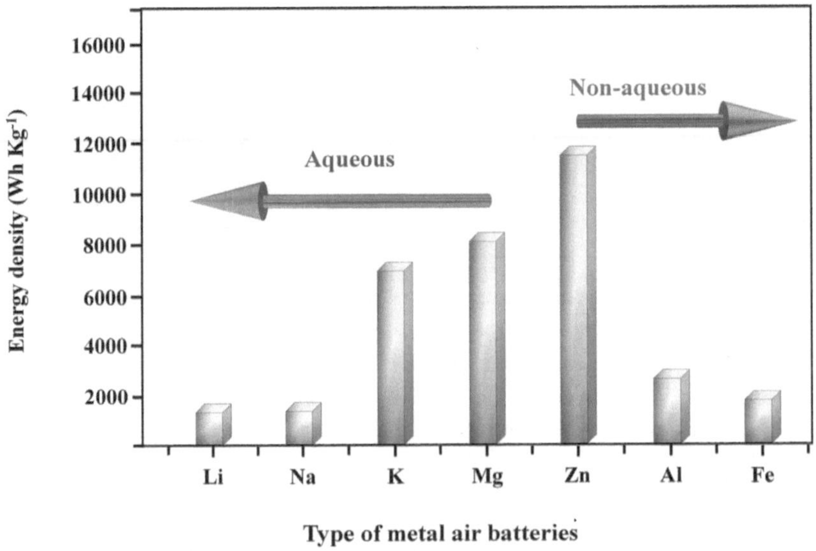

FIGURE 9.5 A comparison of the theoretical energy density for various metal-air battery systems. Adapted and reproduced with permission from Ref. [22]. Copyright © 2017 American Chemical Society.

TABLE 9.2
Overview of Different MABs with Respect to Their Theoretical Specific Energy and Open Circuit Voltage

Metal-Air Battery	Theoretical Specific Energy, Wh kg^{-1} (Including Oxygen)	Theoretical Specific Energy, Wh kg^{-1} (Excluding Oxygen)	Calculated Open Circuit Voltage (V)
Aluminium-air	4,300	8,140	1.2
Germanium-air	1,480	7,850	1
Calcium-air	2,990	4,180	3.12
Iron-air	1,431	2,044	1.3
Lithium-air	5,210	11,140	2.91
Magnesium-air	2,789	6,462	2.93
Potassium-air	935	1,700	2.48
Sodium-air	1,677	2,260	2.3
Silicon-air	4,217	9,036	1.6
Tin-air at 1,000 K	860	6,250	0.95
Zinc-air	1,090	1,350	1.65

Unlike LIB technology, the functioning of MABs is by the redox reaction between the metallic anode and oxygen in the cathode. Hence the values of theoretical energy density and specific capacities are very large [23]. One of the most explored MABs is the LABs. Like LIBs, in LAB, the lithium metal anode functions as a source of Li-ion, the cathode is basically carbon-based material which contains a catalyst, and oxygen is accumulated from the environment and stored at the cathode [24]. During discharge, Li ions dissolve from the anode and the electrolyte transports it to meet oxygen at the air cathode, where the catalyst supports the oxidation reaction [25]. While charging the battery, a reverse process takes place [26]. An overview of different MABs with respect to their theoretical specific energy and open circuit voltage (OCV) is represented in Table 9.2.

9.3.1 Major Components of MABs

Similar to batteries and supercapacitors, MABs usually contain four major components: an air-breathing electrode (cathode), metal electrode (anode), aqueous or non-aqueous electrolyte, and separator. The air cathode contains an electrocatalyst to rectify the sluggish electrochemical kinetics of oxygen reduction reaction (ORR), and the system is ionically connected to an electrolyte [27]. Overpotentials from ORR and oxygen evolution reaction (OER) suppress the round-trip efficiency and lower the power output of MABs [28]. The air cathodes in MAB use oxygen (O_2) from ambient air, which significantly reduces the weight of the battery pack, which has unprecedented advantages for many applications. Compared to other batteries, especially LIBs, which currently dominate the commercial battery market share, MABs are cheap, because the cathode source (O_2 from air) is abundant, and the anode can be made using low-cost first-row transition metals, alkaline metals, and alkaline earth metals. Table 9.3 summarizes the major advantages and disadvantages of MABs.

The metal anode in MABs can be alkali metals (e.g. Li, Na, and K), alkaline earth metals (e.g. Mg), or first-row transition metals (e.g. Al, Fe, and Zn) with good electrochemical equivalence; the electrolyte can be aqueous or non-aqueous – depending on the nature of the anode employed; the air cathode often has an open porous architecture that permits continuous O_2 supply to the cell from atmosphere air [29].

For most MABs, the air electrode reactant O_2 is continuously supplied from the ambient atmosphere air rather than encapsulated in the cell. Therefore, the air electrode (air-breathing cathode) usually consists of only an electrocatalyst layer and a GDL. The commonly investigated metals for

TABLE 9.3
The Major Advantages and Disadvantages of MABs

Advantages	Disadvantages
High energy density	Dependent on environmental conditions: Drying out limits shelf life once opened to air
Flat discharge voltage	
Long cycling stability	Electrolyte flooding limits power output
Higher rate capability	
Higher open circuit voltage	
Long shelf life (dry storage)	
Non-toxic (on a metal use basis)	Limited power density
Low-cost electrodes	Protection of metal anode from O_2 is difficult
Weightless compared to LIBs	

the metal electrode in MABs include zinc, lithium, aluminium, magnesium, potassium, and sodium [30]. Considering the stability and reactivity of metallic anode, both aqueous and organic electrolytes are employed in the MABs. For highly active metals that are unstable in aqueous systems, such as Li, Na, and K, a non-aqueous aprotic or room temperature ionic liquid (IL)-based electrolyte is usually required, while for relatively inactive metals such as Zn, Al, Mg, and Fe, alkaline aqueous electrolytes are widely employed [22]. Unlike in LIBs, the separator is an optional component in a metal-air battery, which is used to separate two different electrolytes, obstruct some mass transportation processes between the electrodes, and prevent the short-circuit caused by electronically conductive metal dendrites.

9.3.2 Achievements in the Development of MABs

Even before the investigations in LIB technology, the primary research on MABs commenced. The very first MAB was the primary zinc-air battery in 1878 by Maiche et al. [31]; the cell consisted of porous platinized carbon as the air cathode. Following this, in 1932, the National Carbon Company constructed alkaline-based zinc-air battery with the efforts of Heise and Schumacher [32]. As significant interest was received in zinc-air batteries, researchers expanded the field on MABs by introducing various other battery systems. These included aluminium-air batteries (AAB), iron-air batteries, magnesium-air batteries, LAB, sodium-air batteries (SAB), and potassium-air batteries (KAB) [33]. The first iron-air, AAB, and magnesium-air batteries developed in the 1960s are employed only for aqueous electrolytes [34–36]. Non-aqueous MABs first emerged about two decades ago, alkali-metal based batteries, initially for LAB, and more recently for SAB and KAB [37–39]. The major technical challenge that hampered the development or commercialization of MABs was associated with metal anodes, air catalysts, and electrolytes. The protection of metallic lithium as an anode especially for the LAB is still a great challenge [40]; hence, none of them are commercialized yet. Figure 9.6 is a general schematic illustration of MAB and timeline-based achievements in MABs.

9.3.3 Types of MABs

MABs are classified into different categories primarily based on the metal electrode employed. The major types of MABs are AAB, iron-air battery, magnesium-air battery, LAB, SAB, KAB, silicon-air battery, germanium-air battery, calcium-air battery, and tin-air battery [33]. Based on the rechargeability, MABs are again classified into primary and secondary MABs, and based on the physical state, they are classified as traditional static batteries, redox flow batteries, and novel flexible batteries. Figure 9.7 is the graphical representation of specific energy, specific energy density, and battery voltage of the major MABs.

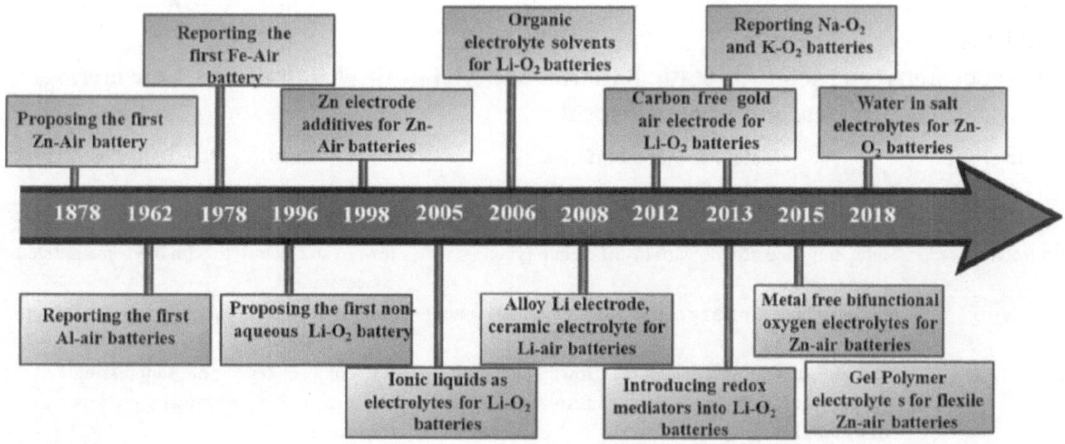

FIGURE 9.6 An overview of metal-air batteries. (a) The general architecture of a metal-air battery; (b) a timeline of the advances in metal-air battery technology, showing various milestones achieved. Adapted and reproduced with permission from Ref. [41]. Copyright © 2019 Elsevier.

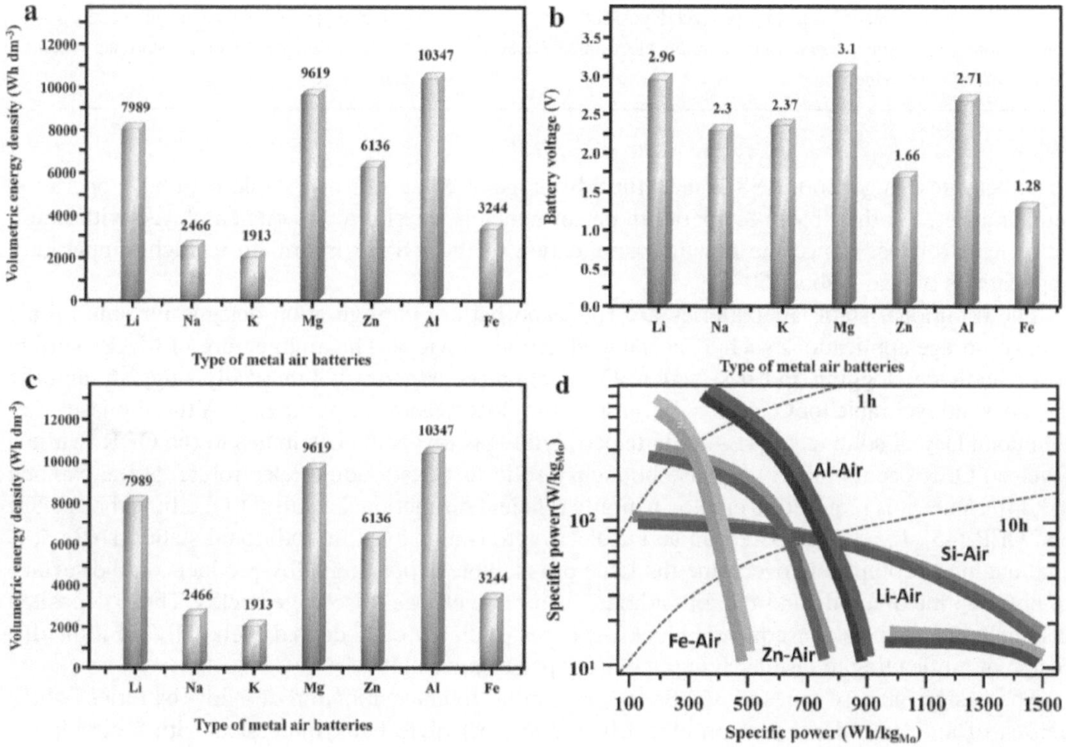

FIGURE 9.7 Graphical representation of specific energy, specific energy density, and battery voltage of the major MABs. Adapted and reproduced with permission from Ref. [33]. Copyright © 2019 MDPI.

Based on the physical configuration and packaging, MABs are classified as traditional static batteries, redox flow batteries, and novel flexible batteries [42]. These traditional static batteries are more popular and well-studied. Table 9.4 summarizes the difference between metal-air static batteries and metal-air flow batteries based on various aspects. Based on the type of electrolyte used in the MABs, these can either be an all-solid-state battery (ASSB) or traditional liquid/gel electrolyte type

TABLE 9.4
Difference between Metal-Air Static Batteries and Metal-Air Flow Batteries Considering Electrodes, Electrolytes, and Various Aspects

Characteristics	Metal-Air Static batteries	Metal-Air Flow Batteries
Cell structure	Challenge and complex to integrate with other energy technologies	Flexible design, facile to integrate with other energy technologies
Electrolyte	Static, finite electrolyte with insufficient oxygen concentration	Flow, fresh electrolyte with oxygen-saturated concentration
Cathode	Severe cathode clogging and passivation, unfavourable O_2 access	Eliminated cathode clogging and passivation, favourite O_2 access
Anode	Non-uniform deposition, dendrite growth, unexpected by-product formation, decrease in active material utilization	More compact deposition, suppressing dendrite growth, reducing by-product formation
Modelling parameters	Heat transfer, pore structure, pore size distribution of cathode, oxygen diffusion, metal ion diffusivity, etc.	Convection effect, mass transport, flow rate of electrolyte, flow field, heat transfer, etc.
Performance	Limited practical energy and power density, and energy efficiency	Decoupled energy or power density and larger energy efficiency
Stability	Relatively unstable, reduced cycle life, due to side reactions and by-product deposition	Long operation life, safety, due to flow electrolyte with activated side reactions
Application	Small-size devices, portable electronic devices	Large-scale energy storage, stationary plants
Development	Widely studied	Early stage

static batteries. In general, ASSBs have the advantage of being safe and stable with wide operation temperatures, but their room temperature performance is inferior to the one employed with liquid or gel electrolytes. Hence the practical application of the ASSBs is limited to high-temperature applications typically about 60°C.

The traditional static air batteries are rigid and battery configuration designs for small-scale energy storage applications such as portable electronic devices. The architecture of MABs suffers severe cathode clogging and passivation. O_2 reaction on cathodes is kinetically sluggish, and the design is unfavourable for O_2 access, because of the closed casing or packaging. A three-phase reaction boundary of solid (catalyst)–liquid (electrolyte)–gas (oxygen) contributes to the ORR, while a reversed OER occurs on a two-phase boundary solid (catalyst)–liquid (electrolyte) at the cathode [43,44]. Hence it is required to employ a highly efficient bifunctional catalyst to facilitate both ORR and OER [45]. In addition, since an active electrolyte is employed in traditional static MABs, it is challenging to completely overcome the issue of insoluble deposition of by-products on the surface of both the metal anode and the air cathode during the charge–discharge cycles. These deposited insoluble by-products consequently block the pores in the air cathode and eventually limit the diffusion of air, leading to fast decaying in battery performance [46].

The design of these types of MABs is very similar to conventional redox flow batteries, which consist of an electrode, separator, electrolyte, and electrolyte bank integrated with a circulation power pump for controlling the electrolyte into the cell installed as an additional part [47]. Metal-air flow batteries are preferable in large-scale energy storage applications and stationary plants where large energy efficiency is required [48]. Compared to the traditional static air batteries, a flowing-electrolyte configuration in the flow batteries addresses certain problems associated with the metal anode and the air cathode [48]. For example, the continuous flow of a large volume of electrolyte significantly increases the heat transfer, surpasses the electronically conducting dendrite formation and cathode clogging, and reduces the by-product formation and irregular shape changes of metal electrodes, thereby avoiding passivation by improving current distribution and reducing concentration gradients [49].

However, the complicated battery configuration has some shortcomings, including decreased volume density, and practically no application in portable electronic gadgets. The increased complications such as additional pumps and tubes are required to drive the flow of electrolytes during the charge–discharge process [48].

With the increasing demand, efforts are underway to make different lightweight and small form-factor flexible power sources including primary and rechargeable batteries for portable electronics in recent years [50]. The major components of a flexible battery include a cathode, anode, separator, and high-conductivity electrolyte, which are packed in flexible polymeric or sandwich substrates. These flexible packaging materials must have air, moisture, and electrolyte barrier capabilities; have resistance to hydrogen-fluoride attack; and be heat-sealable. Both primary and secondary MABs that are designed to be conformal and flexible. The flexible MABs should maintain their characteristic shape even against continual bending or twisting [51]. The increasing interest in portable and flexible electronics has led to the development of flexible batteries which can be implemented in products such as smart cards, wearable electronics, novelty packaging, flexible displays, and transdermal drug delivery patches. Conventionally flexible MABs are ASSBs, which employ solid-state electrolytes (SSE). A thin flexible metallic plate is used as a metal anode to impart flexibility and reduce the battery weight [51]. Various carbon materials including fibres, carbon nanotubes (CNTs), and graphene and their nanocomposites are being explored as a potential cathode material [52].

9.3.4 Principle of MABs

The working principle of MABs differs from that of traditional ionic batteries. Traditional ionic batteries involve the transformation of metallic ions from the anode to the cathode. In MABs, metals or alloys transform into metallic ions at the anode and oxygen transforms into hydroxide ions at the cathode [53]. In an aqueous electrolyte system, oxygen diffuses into batteries through the GDL and transforms into receiving electrons forming oxygen anions [54]. In a non-aqueous electrolyte system, oxygen receives electrons and transforms into oxygen anion [55]. Metals release electrons, transform into metallic ions, and dissolve them into electrolytes. These processes will be reversible during the charging procedure of a rechargeable MAB. Table 9.5 summarizes the standard potential of some metal anodes and air cathodes.

For MABs, oxygen and metals participate in electrochemical reactions. Specific reaction formulas are as Equations 9.1 and 9.2:

TABLE 9.5
Standard Potential of Some Metal Anodes and Air Cathode

Standard Potential of Some Metal Anodes		
Metal anodes	Half-reaction	Potential
Mg	$Mg^{2+}+2e^- \rightarrow Mg$	E^0=−2.363V
Al	$Al^{3+}+3e^- \rightarrow Al$	E^0=−1.662V
Zn	$Zn^{2+}+2e^- \rightarrow Zn$	E^0=−0.763V
Li	$Li^++e^- \rightarrow Li$	E^0=−3.040V
Oxygen Reduction Standard Electrode Potential		
Air cathodes		
Acidic condition	$O_2+4H^++4e^- \rightarrow 2H_2O$	E^0=1.230V
Neutral or alkaline conditions	$O_2+2H_2O+4e^- \rightarrow 4OH^-$	E^0=0.401V
Non-aqueous system oxygen reduction	$2O_2+2e^- \rightarrow O_2^-$	E^0=−0.130
Two-electron reduction under acidic condition	$O_2+2e^-+2H^+ \rightarrow H_2O_2$	E^0=0.680V
Two-electron reduction under alkaline condition	$O_2+H_2O+2e^- \rightarrow OH^-+HO_2^-$	E^0=0.070V

$$\text{Anode:}\quad M \rightarrow M^{n+} + ne^- \tag{9.1}$$

$$\text{Cathode:}\quad O_2 + 2H_2O + 4e^- \rightarrow 4OH^- \tag{9.2}$$

The reaction kinetics of iron-air batteries in the alkaline aqueous electrolyte are shown in Equations 9.1 and 9.6:

$$\text{Anode:}\quad Fe + 2OH^- \rightarrow Fe(OH)_2 + 2e^- \tag{9.3}$$

$$3Fe(OH)_2 + 2OH^- \rightarrow Fe_3O_4 + 4H_2O + 2e^- \tag{9.4}$$

$$\text{Cathode:}\quad O_2 + 2H_2O + 4e^- \rightarrow 4OH^- \tag{9.5}$$

$$\text{Overall reaction:}\quad 2O_2 + 3Fe \rightarrow Fe_3O_4 \tag{9.6}$$

The working principle of AABs in the alkaline aqueous electrolyte is shown in Equation 9.9.

$$\text{Anode:}\quad Al + 4OH^- \rightarrow Al(OH)_4^- + 3e^- \tag{9.7}$$

$$\text{Cathode:}\quad O_2 + 2H_2O + 4e^- \rightarrow 4OH^- \tag{9.8}$$

$$\text{Overall reaction:}\quad 3O_2 + 2Al \rightarrow Al_2O_3 \tag{9.9}$$

The working principle of zinc-air batteries in the alkaline aqueous electrolyte is shown in Equation 9.13.

$$\text{Anode:}\quad Zn + 4OH^- \rightarrow Zn(OH)_4^{2-} + 2e \tag{9.10}$$

$$Zn(OH)_4^{2-} \rightarrow ZnO + 2OH^- + H_2O \tag{9.11}$$

$$\text{Cathode:}\quad O_2 + 2H_2O + 4e^- \rightarrow 4OH^- \tag{9.12}$$

$$\text{Overall reaction:}\quad O_2 + Zn \rightarrow ZnO \tag{9.13}$$

The working principle of LABs in the non-aqueous electrolyte is shown in Equation 9.18 [1].

$$\text{Anode:}\quad Li \rightarrow Li^+ + e^- \tag{9.14}$$

$$\text{Cathode:}\quad O_2 + e^- \rightarrow O_2^- \tag{9.15}$$

$$O_2^- + Li^+ \rightarrow LiO_2 \tag{9.16}$$

$$LiO_2 + Li^+ + e^- \rightarrow Li_2O_2 \tag{9.17}$$

$$\text{Overall reaction:}\quad O_2 + Li \rightarrow Li_2O_2 \tag{9.18}$$

9.3.5 Components of Cathode

In MABs, the cathodic electrode or air-breathing electrode usually consists of only an electrocatalyst layer to reduce the electrode overpotential and a GDL to enhance the O_2 diffusion between the ambient atmosphere air and the catalyst surface. Structurally the practical air-breathing cathode or gas diffusion electrode has a highly porous metallic or non-metallic GDL, electronically conducting current collector and an active catalyst layer [46]. The current collector is placed on either side of the GDL, followed by the third layer of high surface area carbon and catalyst combined to form an active catalyst layer (as shown in Figure 9.8) [33]. The high surface area carbon has several advantages including the addition of a hydrophilic layer to the electrode. This highly amorphous carbon layer aids in catalyst wetting and can create air pockets needed for the ORR to take place. The current collector can be a metal or electronically conducting material. Metal current collectors are porous foam-like mesh, for example, Ni, Cu, or Al [56]. Non-metal current collectors are typically carbon-based materials, for example, conductive carbon paper, graphitic fibre, carbon cloth, conducting polymers, or their composites can be used. The electrochemical reactions occurring on the cathode are ORR and OER, where the oxidant is the ambient atmosphere O_2. A catalytic layer is required at the electrode surface to lower the overpotential of ORR and OER [57].

9.3.6 Gas Diffusion Layer

The most commonly used gas diffusion electrode or air-breathing cathode for MABs is very similar to that used in fuel cells. In MABs, the GDL has multi-folded functions: (1) supporting of the catalyst layer, (2) providing O_2 diffusion channels between air and catalyst layer, (3) preventing water from getting into the battery, and (4) electrolyte getting out of the battery. An ideal GDL should be thin, lightweight, highly porous, non-corrosive, and hydrophobic, which equips the GDL to serve better as a bridge between the air and the catalyst layer [58]. The GDL is a porous foil typically having a porosity of >50%. The most commonly used GDLs are carbon materials, porous silicon oxycarbide [58], or other conductive ceramic membranes or metal has been developed as a cathodic GDL. The commonly used and well-studied GDLs are carbon-based materials such as graphite, CNTs, grapheme, or carbon fibres [59]. Carbon fibre-based GDL, which is treated with polytetraflouroethylene to be hydrophobic, is reported for Zn-air batteries [60]. The hydrophobic carbon layer effectively prevents electrolyte leakage, while allowing easy diffusion of O_2 into the cell from the

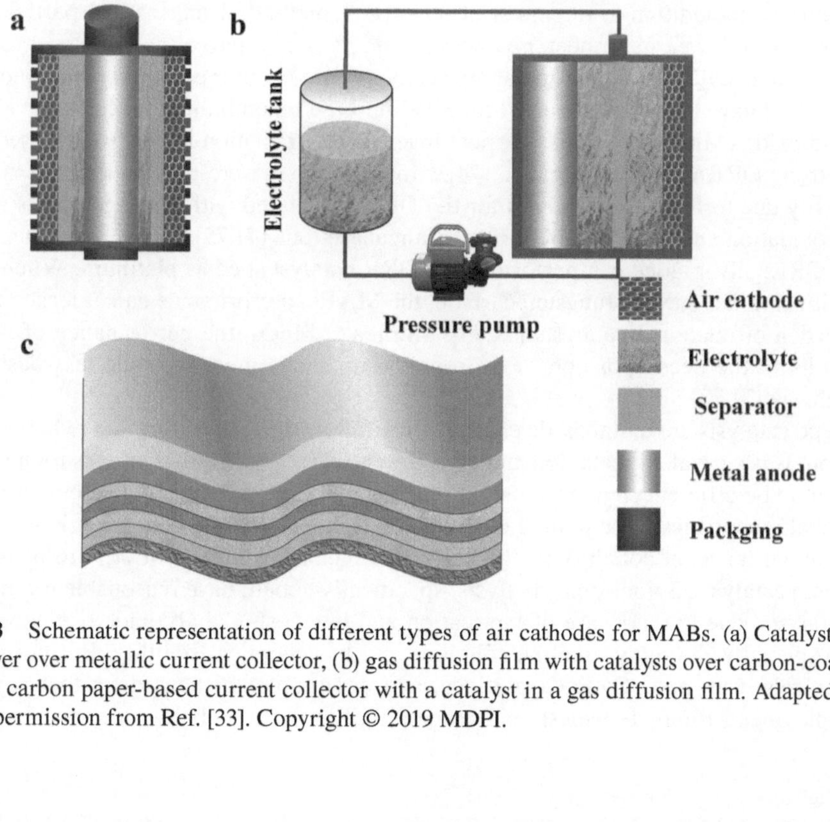

FIGURE 9.8 Schematic representation of different types of air cathodes for MABs. (a) Catalyst-coated gas diffusion layer over metallic current collector, (b) gas diffusion film with catalysts over carbon-coated current collector, (c) carbon paper-based current collector with a catalyst in a gas diffusion film. Adapted and reproduced with permission from Ref. [33]. Copyright © 2019 MDPI.

ambient air. The GDL can simultaneously provide hydrophilic micro-channels to the liquid electrolyte and hydrophobic layers to prevent electrolyte leakage and efficient gaseous O_2 diffusion [61].

One major technical challenge associated with gas diffusion electrodes in both fuel cells and with MABs is the corrosion of the carbon on which the catalyst is loaded [53]. During high potentials and in the presence of many harsh environments and operating conditions, such as OCV, be it caustic or acidic, the amorphous carbon corrodes and the bifunctional catalyst responsible for ORR and OER reactions are subsequently lost from the electrode, which leads to the quick degradation of full cell performance [62]. This has been extensively observed with platinum loaded on carbon and can also be seen through much testing of commercial primary catalysts responsible for ORR.

9.3.7 Catalyst Layer and Its Technical Challenges

One of the crucial parameters which control the performance of the cathode is the GDL and electrocatalyst employed. As the reaction kinetics for the O_2 is naturally slow, the electrocatalyst and its efficiency engineer the reaction kinetics for ORR and OER. Hence, bifunctional catalysts, which can accelerate ORR and OER, are required for the smooth functioning of MABs. Transition metal and metal oxides and their derivatives have been reported for their efficient ORR and OER performance in many MABs, including platinum [63,64], palladium [65], iridium oxides, and ruthenium oxide [66]; metal alloy (silver-copper) [67]; etc. Furthermore, nanostructured metal oxides such as manganese (II) oxide, cobalt (II) oxide, and nickel (II) oxide showed electrocatalytic activity [68]. Besides these, transition metal hydroxide and sulphides, spinel compounds, and carbon-based and nitrogen-doped carbon materials functioned as catalysts [28]. Considering the durability and performance among bifunctional catalysts, amorphous carbon, possessing high surface area, supported with precious metals has shown promising results [69]. High cost, coupled with the limited availability, hinders the usage of precious metals in MABs [70]. As research on electrocatalysts is further moving forward, currently many materials are under investigation, mainly cost-effective transition metal oxides and their derivatives, perovskites, spinel, etc., are under examination.

Per the reports, in addition to the material properties, method of fabrication, particle morphology (nanorods, nanoflower, nanosheet, nanowires), and its size (micro or nanorange) strongly influence the performance [71,72]. Noble-metal-free cathodes are another potential candidate which are cost-effective and have been well reported for alkaline-based membrane fuel cells and MABs [73]. Manganese dioxide (MnO_2) is one of the most investigated transition metal oxide-based catalysts, which has strong ORR/OER performance [74]. MnO_2 catalysts were used in the first zinc-air fuel cells, primarily due to their effectiveness for the ORR associated with hydrogen peroxide decomposition by oxidation and reduction of surface manganese ions [71,75]. Silver is another promising catalyst for ORR; silver alone can perform as an ORR catalyst such as platinum. When combined the silver with carbides such as tungsten carbide, the MABs' performance can at least equal that of platinum carbon bifunctional catalysts [71,76]. Similar to MnO_2, the performance of silver-based catalyst results in the decomposition of hydrogen peroxide formed from the less desirable two-electron pathway [71,77].

Spinel-type catalysts are metal oxide catalysts that follow the AB_2O_4 formula (where A is a divalent metal ion, B is a trivalent metal ion, and O is oxygen) [71]. Interestingly it was found that A and B do not need to be different elements, just follow the need for trivalent and divalent variations. For example, cobalt, manganese, titanium, etc. have both trivalent and divalent variations of ions [78]. Hence, the spinel forms of cobalt oxide (Co_3O_4) or manganese oxide (Mn_3O_4) are possible to use as bifunctional catalysts. Spinel-type catalysts, specifically cobalt, have reasonable promise as non-precious catalysts, due to their ease of fabrication and low cost. Cobalt oxide has good corrosion resistance, especially in highly alkaline solutions and has chemical stability. But poor electronic conductivity, relatively poor ORR capabilities, and environmental hazardous factors associated with its employment, limits its widespread usage [79]. Similar to cobalt oxide, manganese oxides are also non-precious semiconducting oxides and commercially affordable due to their low price and high abundance; however, they are environmentally friendly and non-toxic and have better catalytic activity towards ORR, making them more popular to be used often as electrocatalyst in MABs [80]. A catalyst with these properties would be suitable for use in zinc-air batteries, due to the reduced cell impedance associated with good conductivity, the increased durability associated with good corrosion resistance, and the reduced overpotentials in improved performance associated with the ability to effectively carry out ORR.

9.4 ELECTROLYTES FOR MABS

Electrolyte is a major component in any of the electrochemical energy storage devices and is generally known as the heart of the battery. Electrolytes serve as catalysts to make the battery conductive by promoting the movement of metal ions (which are electrically charged atoms that have lost or gained electrons) from the cathode to the anode on charge and in reverse on discharge. The electrolyte of a battery consists of soluble salts, acids, or other bases in liquid, gelled, and dry formats [81]. Molten salts and ILs are also popular as electrolytes in batteries. Typically based on the physical state, electrolytes are widely categorized as SSE, liquid electrolyte, or gel electrolytes. In MABs, both aqueous electrolytes and non-aqueous electrolytes are employed based on the nature of the metal electrode.

Among different electrolytes for MABs, neutral salt solutions, acidic solutions, and hybrid electrolytes are rarely used, while alkaline solutions and ILs are usually employed [28]. SSEs can work in all MABs, but their ionic conductivity and electrochemical performance are inferior to the liquid or gel electrolytes [82]. Compared to liquid and gel electrolytes, SSEs have a wide electrochemical window (>5 V) with the added advantage of safety [83]. Because of the high chemical reactivity and moisture sensitivity of lithium metal, aqueous electrolytes do not match in LABs [54]. It is well reported that metal electrodes in aqueous MABs suffer from serious challenges such as the formation of an ionically insulating passivation layer over electrodes, electronically conducting dendrite growth, and hydrogen evolution, which can be mitigated to some extent by the material design of the electrolyte [84].

9.4.1 Aqueous Electrolyte

Aqueous electrolytes are limited to be employed in MABs with water-stable metal electrodes, such as Zn-air batteries and Al-air batteries [85,86]. Aqueous alkaline electrolytes are commonly used aqueous electrolytes in MABs due to the fact that alkaline electrolytes have better oxygen electrocatalytic activity and metal electrodes show good corrosion resistance compared with acidic and neutral electrolytes [85]. For enhancing the electrochemical performance of anode and inhibiting metal corrosion, engineering by electrolyte additive showed improvement in cycling stability [87,88]. Commonly employed alkaline electrolytes are aqueous solutions of KOH, NaOH, and LiOH. Among the aqueous electrolytes, concentrated KOH solutions are most frequently employed in MABs due to their highest room temperature ionic conductivity, the lowest viscosity, and cost-effective [89]. The major bottleneck challenges of alkaline electrolytes are environmentally hostile and sensitive to CO_2, which results in the formation of carbonates and leads to a fast decrease in electrolyte concentration [90]. The adsorption of CO_2 and the electrolyte dilution seriously hampers the battery performance such as specific capacity, rate capability, and cycle life of MAB batteries [90]. Hence, neutral aqueous electrolytes for MABs have been studied [91]. NaCl solution or seawater is reported as a neutral electrolyte in MABs; the existence of complex components in seawater, in particular chloride ions, inevitably has a complex influence on air electrode processes, including ORR and OER [91]. Hence, the development of efficient chloride-resistant electrocatalysts with high ORR and OER activity in neutral conditions is highly desirable.

9.4.2 Organic Electrolytes

Aqueous electrolytes are not preferred for MABs using water-unstable metal electrodes such as Li and Na [5,29]. As a result, aprotic organic electrolytes have become prevalent in LAB and SAB. Organic electrolytes comprise dissolved solutes, such as Na or Li salts and additives, in a combination of two or even more organic solvents. These electrolytes have a wider electrochemical stability window, high ionic conductivity, and thermal stability and are thus considered for fabricating MABs. In addition, organic solvents (single or mixed), metal salt solutes (type and concentration), and electrolyte additives are used to tailor the electrochemical performance and electrolyte stability of organic electrolytes to improve the performance of an air/metal electrode [92].

Superoxide species generated on the air electrode decompose aprotic organic solvents, resulting in the formation of CO_2, H_2O, and metal carbonates, thus reducing the electrochemical performance of the MABs [93]. As a result, more attention was given to developing stable organic solvents for creating efficient electrolytes. In MABs, studies have been performed using carbonate-based [94], ether-based [95], sulphoxide-based, and amide-based organic solvents [96] as electrolytes. Carbonate-based aprotic solvents such as ethylene carbonate, propylene carbonate, and dimethyl carbonate were investigated in LABs [94]. They were found to be highly unstable in the presence of superoxide species, as well as the formation of lithium carbonate species has been revealed during the discharge process. In contrast to carbonate-based organic solvents, ether-based solvents were more stable against reduced O_2 species than carbonate-based solvents [95]. However, ether-based electrolytes were also decomposed on the Li electrode in Li metal batteries [97]. Although amide- and sulphoxide-based solvents possess high ambient-temperature electrochemical properties, their reliability on metal electrodes is poor, so a protective layer on the metal electrode is considered necessary for improved electrolyte stability, limiting their widespread use in MABs [96,98].

For enhancing the electrochemical behaviour of organic electrolytes, highly soluble metal salts with stable anions are widely used. Despite the fact that electrolytes are typically prepared by dissolving metal salt solutes in an aprotic organic solvent, the solubility and stability of metal salt solutes play an important role in their selection. For example, in LABs, lithium salts such as lithium hexafluorophosphate ($LiPF_6$), lithium perchlorate ($LiClO_4$), lithium bromide, lithium nitrate ($LiNO_3$), lithium bis(trifluoromethanesulphonyl)imide (LiTFSI), lithium bis(oxalate) borate, and lithium triflate have been prevalently employed [99]. These lithium salts also undergo decomposition caused by Li_2O_2 and other O_2 species, which is severe for metal slats with F-containing anions such as LiTFSI or $LiPF_6$ [100]. Although less hygroscopic susceptible metal salts such as $LiClO_4$ and $LiNO_3$ are reported to have superior electrochemical stability, their ionic conductivity and transference number are inferior to those of $LiPF_6$ [100]. Moreover, it is noteworthy that the stability of metal salts in MABs varies according to the type of electrolyte solvents, anion species, and the molar concentration [101].

9.4.3 IL Electrolytes

Room temperature ILs are a type of salts whose melting point is below 100°C and composed of large organic cations and organic/inorganic anions. ILs are considered an alternative to conventional electrolytes as they possess the advantage of a wide electrochemical window, better ionic conduction, lower volatility, and flammability [102]. ILs for MABs have been reported in the past decade [103]. Shiga et al. [104] explored the applicability of the IL in calcium-oxygen batteries and observed enhancement in the reversibility of plating and stripping. Kwon et al. [105] reported that ILs showed better stability than conventional organic electrolytes with no side reactions in LAB. Zhang et al. [106] incorporated ILs as an electrolyte additive in LAB. Functional groups of IL and redox mediators were catalysed for stable SEI formation, which enabled enhancement in performance.

Kang et al. [107], in 2016, reported a hybrid SAB based on bis(fluorosulphonyl) amide and Na[FSA–C_2C_1im][FSA] as an IL electrolyte. At a temperature of 50°C, the cell showed a higher value of capacity, 835 mAh g^{-1} (which is 99% of theoretical capacity), with 27.6 $mWcm^{-2}$ as power density. Zhang et al. [108] reported that with an increased concentration of NaTFSI salt in an IL electrolyte, *N*-butyl-*N*-methyl pyrrolidinium bis(trifluoromethylsulphonyl)imide [C4mpyr][TFSI], enhancement in discharge capacity was observed. Moreover, an increase in concentration led to a reduction in overpotential, contributing to enhanced cyclability in the SAB system. In the presence of a high content of Na^+ at higher concentrations, coordination bonds were observed between the O_2^- and Na^+, instead of with $[C4mpyr]^+$. Figure 9.9 shows the schematic representation of the deposition of byproducts' morphology on cathode.

9.4.4 Solid-State Electrolytes

Compared with liquid electrolytes, SSEs can avoid the problems of electrolyte evaporation, electrolyte leakage, and gas crossover [109]. Moreover, the application of SSEs is favourable for improving the energy density of MABs and possibilities for pouch cell fabrication and also enables the fabrication of flexible and wearable devices [110]. For the development of SSEs for MABs, suitable materials with high ionic conductivity, wide potential window, and chemical stability are required, and insufficient solid–solid interfacial contact with the electrolyte and electrodes should also receive much attention [41]. SSEs can be categorized as organic-based and inorganic or ceramic-based electrolytes based on the components. Polymer-based electrolytes constitute organic electrolytes. As polymeric materials are solution processable and can be tailor made, opening up options for further chemical modifications, with mechanical properties and flexibility, these are potential candidates for flexible devices and general applications [111–113]. Polymer electrolytes can be solid-polymer electrolytes (SPEs) and gel polymer electrolytes (GPEs)/quasi-SSEs, depending on the presence/absence of liquid electrolytes, respectively. GPEs consist of a 3D polymer skeleton with a liquid electrolyte expanded throughout its whole volume to afford fast ion transport similar to liquid electrolytes and high flexibility [2]. In recent years GPEs have been widely investigated in both aqueous

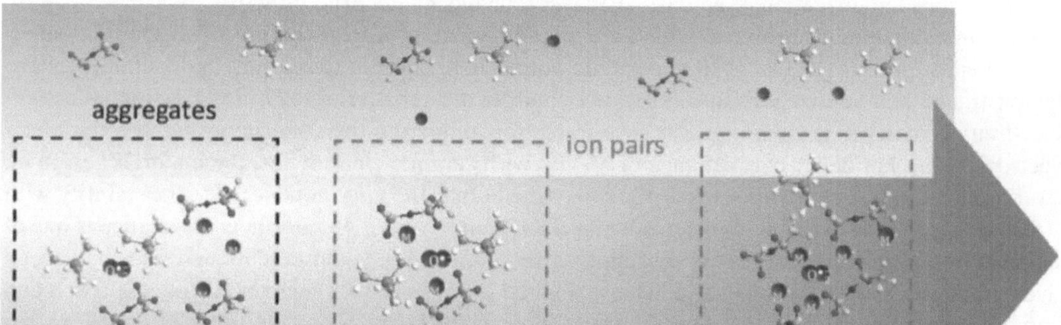

FIGURE 9.9 Schematic representation showing the relation between NaTFSI/[C4mpyr][TFSI] electrolyte concentration on ORR and the deposition. Adapted and reproduced with permission from Ref. [108]. Copyright © 2018 American Chemical Society.

and non-aqueous flexible MABs [114]. MABs with GPEs suffer from the limited stability of polymer frameworks and the quick loss of electrolyte solvents, resulting in short cycle life. Therefore, stable, solvent-retaining GPEs are highly desired for extending the battery lifetime.

Incorporating filler into GPE showed promising results with respect to ionic conduction and cation transference number [115]. For example, Chamaani et al. [116] observed enhanced electrochemical properties by incorporating 1 wt.% glass micro-filler to GPE. Polymer-blend GPE is another approach to improve ionic conduction. Yi et al. [117] used multi approach where nanofiller incorporation and blending approach were explored in LAB. Cell with fumed nano silica (SiO_2)-doped polypropylene-supported poly(methyl methacrylate) (PMMA)-poly(styrene) (PS) blend soaked in 1 M LiTFSI in tetraglyme showed enhanced electrochemical properties. In ambient air and at room temperature, the cell showed a stable cycling of 100 cycles.

Similar to GPE, an SPE is composed of an ion-conducting polymer matrix but not swollen with liquid solvents [118]. Due to the absence of liquid solvents, the ionic conductivity in SPE is limited [119]. Commonly investigated polymer materials for SPEs include polyacrylonitrile (PAN), poly(carbonate-ether), and poly(ethylene oxide) (PEO) [120]. Experimental and theoretical studies revealed that the ion-transport mechanism is accelerated when the polymer possesses a lower glass transition temperature and a more amorphous nature [121–123]. Ionic conduction hence can be tailored by designing the material by crosslinking, copolymerization, blending, filler incorporation, and plasticizer incorporation, which reduces crystalline nature and lowers glass transition temperature [124,125].

Huang et al. [126] designed a sodium polyacrylate hydrogel (PANa) for flexible Zn-air batteries. The synthesis involved solution polymerization of the acrylic acid monomer with ammonium persulphate initiator, followed by the incorporation of zinc acetate and potassium hydroxide to allow films to absorb water and anion, thereby ensuring lower viscosity and enhancing ionic conduction. The schematic representation showing the synthesis of PANa and architecture of Zn//NiCo battery and Zn-air battery with PANa electrolyte is shown in Figure 9.10. Zinc-air battery with PANa electrolyte exhibited stable cycling performance at 2 mA cm^{-2} for 160 h, significantly superior to traditional polyvinyl alcohol electrolyte, which showed a rapid drop of round-trip efficiency from 60.4% to 29.3% after only 16 h. Moreover, the electrostatic interactions between acrylate ions in PANa and Zn ions induce the formation of a quasi-SEI layer to avoid Zn dendrite growth.

Inorganic ceramic electrolytes, which are encompassed in SSE, including LISICON (Li^+ superionic conductor), NASICON (Na^+ superionic conductor), perovskite, garnet, and sulphides, have demonstrated their applications in solid-state Li and Na metal batteries [127,128]. The high mechanical strength of ceramic electrolytes is favourable for mitigating the dendrite growth on the metal electrode [129]. For better performance of MABs with ceramic electrolytes, the material design on ceramic electrolytes mainly focuses on the following points: the ionic conductivity, the stability with Li/Na metal, and the electrolyte–electrode interfacial contact [41]. As interfacial resistance is one of the main bottleneck challenges in SSEs, approaches to mitigate the aforementioned issue have been given great importance. Wang et al. [83] investigated the effect of the nano-polishing approach and employed ultra-fine surface $Li_{1.5}Al_{1.5}Ge_{1.5}P_3O_{12}$ as SSE for LAB. Nano-polishing removed micrograins in the pristine electrolyte and showed a reduction in surface roughness by 30 nm. A smooth electrode–electrolyte interface contributed to a longer cycling life (27 cycles, at room temperature) at 400 mA g^{-1} with a capacity limitation of 1,000 mAh g^{-1}.

Fillers can also reduce the interfacial resistance and contribute to smooth ionic conduction. Pan et al. [130] incorporated supramolecular filler crown ether Li salt complex into a membrane containing a $Li_{6.6}La_3Zr_{1.6}Ta_{0.4}O_{12}$ garnet and investigated the electrochemical properties of LAB. Enhancement in interfacial resistance was indicated by the improvement in ionic conduction (6.92×10^{-4} S cm^{-1} at 20°C and 2.85×10^{-3} S cm^{-1} at 80°C). Furthermore, the presence of redox mediators in the filler reduced the discharge-charge overpotential (~600 mV), which thereby enhanced the cycling stability of the solid LAB.

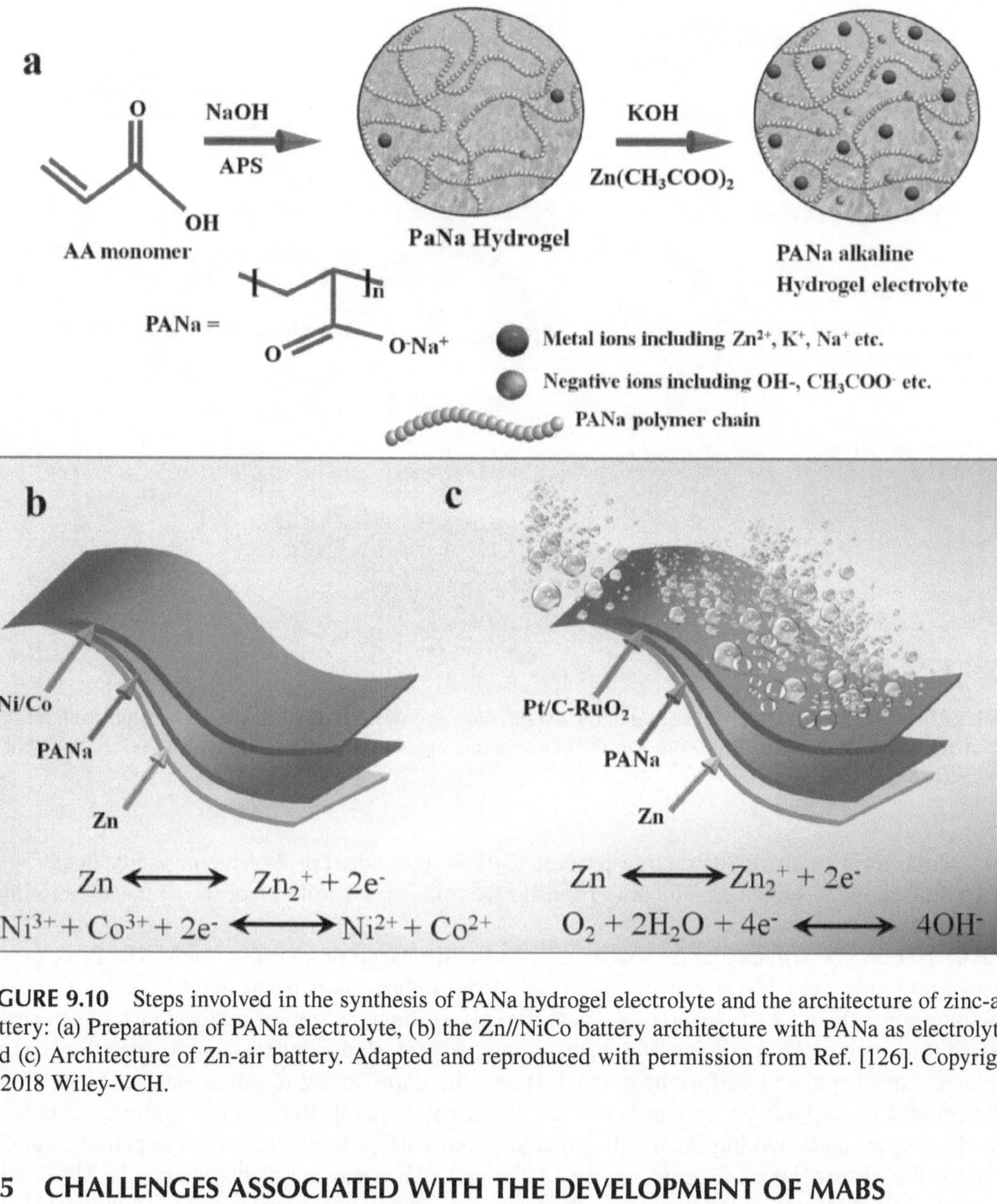

FIGURE 9.10 Steps involved in the synthesis of PANa hydrogel electrolyte and the architecture of zinc-air battery: (a) Preparation of PANa electrolyte, (b) the Zn//NiCo battery architecture with PANa as electrolyte, and (c) Architecture of Zn-air battery. Adapted and reproduced with permission from Ref. [126]. Copyright © 2018 Wiley-VCH.

9.5 CHALLENGES ASSOCIATED WITH THE DEVELOPMENT OF MABS

For evaluating the performance of MABs, there are mainly two aspects which need to be taken care of, namely, the rechargeability of the cell and its discharge properties. Therefore, the challenge which needs to be addressed is to extend the study on improving the aforementioned points. Rechargeability of MABs is defined by coulombic efficiency, round-trip efficiency, and cycle life. These parameters define the energy-utilizing efficiency of MABs and are dependent on the overpotential of charge and discharge reactions. It can be connected with the efficiency of the air cathode catalyst towards ORR and OER, referring to its bifunctional properties. Furthermore, mass transport during the cycling process impacts the reversibility. The coulombic efficiency of MAB indicated the cycle life. Irreversible consumption of metal anode dendritic growth, electrolyte decomposition, and parasitic reactions plague the efficiency and cycle life. In aqueous MABs, during the discharge process, the metal anode is coated with an insulating layer of metal oxide

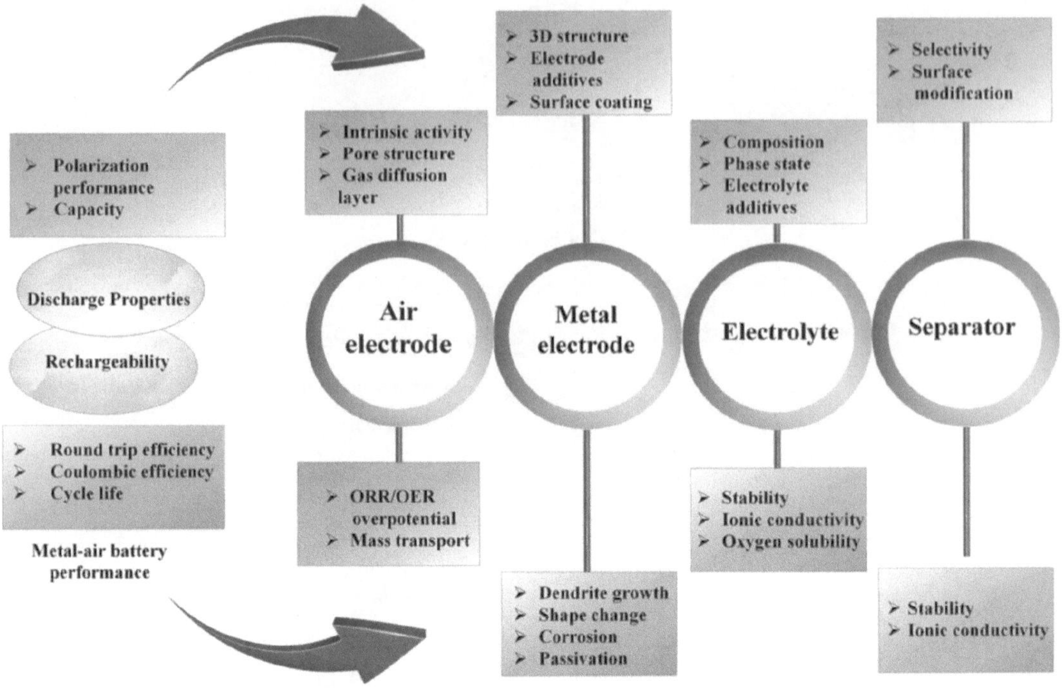

FIGURE 9.11 Schematic representation of the overall challenges in MAB with material design strategies for improving the performance of the cell. Adapted and reproduced with permission from Ref. [41]. Copyright © 2019 Elsevier.

film, which terminates the discharge process [30]. In non-aqueous MABs, capacity decay with the cycle is a main challenge. The major reason for this is the limited storage of metal peroxides or superoxide in the cathode [27].

Polarization performance and capacity constitute discharge properties of a MAB. Peak power density provided by a MAB can be implicated from polarization performance. Metal electrode reactions, air cathode reactions, and overpotential affect the polarization performance. As the rate-determining step is the reduction of oxygen in the cathode, sluggish kinetics impedes the electrochemical conversion and performance [131]. Hence there are strong requirements for bifunctional efficient electrocatalysts which can accelerate the kinetics [132]. Rationally designing electrodes and electrolytes and coupling them with suitable electrocatalysts and separators unequivocally can enhance the electrochemical performance of the cell. An overview of challenges in MAB with materials design strategies for improving the performance of cells is represented in Figure 9.11.

9.6 CONCLUSION

The past two decades have witnessed a proliferation in research for high-energy-density batteries, which fuelled development in MABs. As the ambient air supplies infinite oxygen, which is the air cathode reactant, the theoretical capacity of MAB is calculated from the quantity of metal encompassed in the cell. As the demand for storage technology for EVs and portable electronics is on the rise, MABs can be a potential battery technology. However, developments are plagued by bottleneck issues, which lower the capacity of the cell and depress efficiency. Designing MAB and tailoring the material property can be an effective solution for enhancing the electrochemical performance of a MAB. Extensive research has paved the way for exploring various kinds of MABs, which work on aqueous and non-aqueous media. Designing an air cathode and developing a bifunctional electrocatalyst can accelerate the sluggish inherent kinetics of MAB. Electrolyte engineering by additives and modifications on polymer-based and inorganic-based SSEs also improves the performance of the cell with a boost in safety aspects. Overall, by judicious selection of optimized cell design, materials engineering and strategies to rectify the bottleneck issues of a MAB can undoubtedly enhance the performance, thereby upgrading them as a candidate for future energy storage technology.

REFERENCES

1. Deng J, Bae C, Denlinger A, Miller T (2020) Electric vehicles batteries: requirements and challenges. *Joule* 4:511–515. https://doi.org/10.1016/j.joule.2020.01.013
2. Jishnu NS, Vineeth SK, Das A, et al (2021) Electrospun PVdF and PVdF-co-HFP-based blend polymer electrolytes for lithium ion batteries. In: Balakrishnan NTM, Prasanth R (eds) *Electrospinning for Advanced Energy Storage Applications* (1st ed., pp. 201–234) Springer, Singapore.
3. Etacheri V, Marom R, Elazari R, et al (2011) Challenges in the development of advanced Li-ion batteries: a review. *Energy Environ Sci* 4:3243. https://doi.org/10.1039/c1ee01598b
4. Morimoto M (2015) Which is the first electric vehicle? *Electr Eng Japan* 192:31–38. https://doi.org/10.1002/eej.22550
5. Placke T, Kloepsch R, Dühnen S, Winter M (2017) Lithium ion, lithium metal, and alternative rechargeable battery technologies: the odyssey for high energy density. *J Solid State Electrochem* 21:1939–1964. https://doi.org/10.1007/s10008-017-3610-7
6. Reddy MV, Mauger A, Julien CM, et al (2020) Brief history of early lithium-battery development. *Materials (Basel)* 13:1884. https://doi.org/10.3390/ma13081884
7. Vineeth SK, Sreeram P, Vlad, A, Joy, R, Raghavan, P, Pullanchiyodan A (2023) Polymer blend nanocomposite electrolytes for advanced energy storage applications. In: *Polymer Blend Nanocomposites for Energy Storage Applications* (pp. 203–238). Elsevier. https://doi.org/10.1016/B978-0-323-99549-8.00016-9
8. Xu M, Dalavi S, Lucht BL (2013) Electrolytes for lithium-ion batteries with high-voltage cathodes. In: *Lithium Batteries* (pp. 71–87). John Wiley & Sons, Inc., Hoboken, NJ.
9. Cui Z, Xie Q, Manthiram A (2021) A cobalt- and manganese-free high-nickel layered oxide cathode for long-life, safer lithium-ion batteries. *Adv Energy Mater* 11:2102421. https://doi.org/10.1002/aenm.202102421
10. Bruce PG, Freunberger SA, Hardwick LJ, Tarascon J-M (2012) Li-O_2 and Li-S batteries with high energy storage. *Nat Mater* 11:19–29. https://doi.org/10.1038/nmat3191
11. Gallagher KG, Goebel S, Greszler T, et al (2014) Quantifying the promise of lithium-air batteries for electric vehicles. *Energy Environ Sci* 7:1555–1563. https://doi.org/10.1039/c3ee43870h
12. Miao Y, Hynan P, von Jouanne A, Yokochi A (2019) Current Li-ion battery technologies in electric vehicles and opportunities for advancements. *Energies* 12:1074. https://doi.org/10.3390/en12061074
13. Li M, Lu J, Chen Z, Amine K (2018) 30 years of lithium-ion batteries. *Adv Mater* 30:1800561. https://doi.org/10.1002/adma.201800561
14. Gao J, Shi S-Q, Li H (2016) Brief overview of electrochemical potential in lithium ion batteries. *Chinese Phys B* 25:018210. https://doi.org/10.1088/1674-1056/25/1/018210
15. Feng X, Fang M, He X, et al (2014) Thermal runaway features of large format prismatic lithium ion battery using extended volume accelerating rate calorimetry. *J Power Sources* 255:294–301. https://doi.org/10.1016/j.jpowsour.2014.01.005
16. Harper G, Sommerville R, Kendrick E, et al (2019) Recycling lithium-ion batteries from electric vehicles. *Nature* 575:75–86. https://doi.org/10.1038/s41586-019-1682-5
17. Tarascon J-M (2010) Key challenges in future Li-battery research. *Philos Trans R Soc A Math Phys Eng Sci* 368:3227–3241. https://doi.org/10.1098/rsta.2010.0112
18. Huang B, Pan Z, Su X, An L (2018) Recycling of lithium-ion batteries: recent advances and perspectives. *J Power Sources* 399:274–286. https://doi.org/10.1016/j.jpowsour.2018.07.116
19. Shahjalal M, Roy PK, Shams T, et al (2022) A review on second-life of Li-ion batteries: prospects, challenges, and issues. *Energy* 241:122881. https://doi.org/10.1016/j.energy.2021.122881
20. Vineeth, S. K., Yamauchi, Y., & Kumar, V. (2023). Electrolytes for room-temperature sodium-sulfur batteries: A holistic approach to understand solvation. In *Room-temperature Sodium-Sulfur Batteries* (pp. 79–114). CRC Press. https://doi.org/10.1201/9781003388067-4
21. Pellow MA, Emmott CJM, Barnhart CJ, Benson SM (2015) Hydrogen or batteries for grid storage? A net energy analysis. *Energy Environ Sci* 8:1938–1952. https://doi.org/10.1039/C4EE04041D

22. Li Y, Lu J (2017) Metal-air batteries: will they be the future electrochemical energy storage device of choice? *ACS Energy Lett* 2:1370–1377. https://doi.org/10.1021/acsenergylett.7b00119
23. Durmus YE, Zhang H, Baakes F, et al (2020) Side by side battery technologies with lithium-ion based batteries. *Adv Energy Mater* 10:2000089. https://doi.org/10.1002/aenm.202000089
24. Kwak W-J, Rosy, Sharon D, et al (2020) Lithium-oxygen batteries and related systems: potential, status, and future. *Chem Rev* 120:6626–6683. https://doi.org/10.1021/acs.chemrev.9b00609
25. Imanishi N, Yamamoto O (2019) Perspectives and challenges of rechargeable lithium-air batteries. *Mater Today Adv* 4:100031. https://doi.org/10.1016/j.mtadv.2019.100031
26. Tan P, Jiang HR, Zhu XB, et al (2017) Advances and challenges in lithium-air batteries. *Appl Energy* 204:780–806. https://doi.org/10.1016/j.apenergy.2017.07.054
27. Lee J-S, Tai Kim S, Cao R, et al (2011) Metal-air batteries with high energy density: Li-air versus Zn-air. *Adv Energy Mater* 1:34–50. https://doi.org/10.1002/aenm.201000010
28. Wang Z-L, Xu D, Xu J-J, Zhang X-B (2014) Oxygen electrocatalysts in metal-air batteries: from aqueous to nonaqueous electrolytes. *Chem Soc Rev* 43:7746–7786. https://doi.org/10.1039/C3CS60248F
29. Vineeth, S. K., & Kumar, V. (2023). Introduction: Sodium-sulfur battery technology. In *Room-temperature Sodium-Sulfur Batteries* (pp. 1–16). CRC Press. https://doi.org/10.1201/9781003388067-1
30. Yi J, Liang P, Liu X, et al (2018) Challenges, mitigation strategies and perspectives in development of zinc-electrode materials and fabrication for rechargeable zinc-air batteries. *Energy Environ Sci* 11:3075–3095. https://doi.org/10.1039/c8ee01991f
31. Li Y, Dai H (2014) Recent advances in zinc-air batteries. *Chem Soc Rev* 43:5257–5275. https://doi.org/10.1039/C4CS00015C
32. Zhang J, Zhou Q, Tang Y, et al (2019) Zinc-air batteries: are they ready for prime time? *Chem Sci* 10:8924–8929. https://doi.org/10.1039/C9SC04221K
33. Wang C, Yu Y, Niu J, et al (2019) Recent progress of metal-air batteries-a mini review. *Appl Sci* 9:2787. https://doi.org/10.3390/app9142787
34. Öjefors L, Carlsson L (1978) An iron-air vehicle battery. *J Power Sources* 2:287–296. https://doi.org/10.1016/0378-7753(78)85019-8
35. Zaromb S (1962) The use and behavior of aluminum anodes in alkaline primary batteries. *J Electrochem Soc* 109:1125. https://doi.org/10.1149/1.2425257
36. Zhang T, Tao Z, Chen J (2014) Magnesium-air batteries: from principle to application. *Mater Horiz* 1:196–206. https://doi.org/10.1039/C3MH00059A
37. Abraham KM, Jiang Z (1996) A polymer electrolyte-based rechargeable lithium/oxygen battery. *J Electrochem Soc* 143:1–5. https://doi.org/10.1149/1.1836378
38. Hartmann P, Bender CL, Vračar M, et al (2013) A rechargeable room-temperature sodium superoxide (NaO_2) battery. *Nat Mater* 12:228–232. https://doi.org/10.1038/nmat3486
39. Ren X, Wu Y (2013) A low-overpotential potassium-oxygen battery based on potassium superoxide. *J Am Chem Soc* 135:2923–2926. https://doi.org/10.1021/ja312059q
40. Soni CB, Sungjemmenla J, Vineeth SK, Kumar V (2022) Challenges in regulating interfacial-chemistry of the sodium-metal anode for room-temperature sodium-sulfur batteries. *Energy Storage* 4:e264. https://doi.org/10.1002/est2.264
41. Wang H, Xu Q (2019) Materials design for rechargeable metal-air batteries. *Matter* 1:565–595. https://doi.org/10.1016/j.matt.2019.05.008
42. Wang Y, Kwok H, Pan W, et al (2019) Innovative paper-based Al-air batteries as a low-cost and green energy technology for the miniwatt market. *J Power Sources* 414:278–282. https://doi.org/10.1016/j.jpowsour.2019.01.018
43. Ma Z, Yuan X, Li L, et al (2015) A review of cathode materials and structures for rechargeable lithium-air batteries. *Energy Environ Sci* 8:2144–2198. https://doi.org/10.1039/C5EE00838G
44. Wu D, Guo Z, Yin X, et al (2014) Metal-organic frameworks as cathode materials for Li-O_2 batteries. *Adv Mater* 26:3258–3262. https://doi.org/10.1002/adma.201305492
45. Lu X, Zhang L, Sun X, et al (2016) Bifunctional Au-Pd decorated MnO_x nanomembranes as cathode materials for Li-O_2 batteries. *J Mater Chem A* 4:4155–4160. https://doi.org/10.1039/C5TA09781A
46. Cheng F, Chen J (2012) Metal-air batteries: from oxygen reduction electrochemistry to cathode catalysts. *Chem Soc Rev* 41:2172. https://doi.org/10.1039/c1cs15228a
47. Han X, Li X, White J, et al (2018) Metal-air batteries: from static to flow system. *Adv Energy Mater* 8:1801396. https://doi.org/10.1002/aenm.201801396
48. Yu W, Shang W, Tan P, et al (2019) Toward a new generation of low cost, efficient, and durable metal-air flow batteries. *J Mater Chem A* 7:26744–26768. https://doi.org/10.1039/C9TA10658H

49. Cheng J, Zhang L, Yang Y-S, et al (2007) Preliminary study of single flow zinc-nickel battery. *Electrochem Commun* 9:2639–2642. https://doi.org/10.1016/j.elecom.2007.08.016
50. Fu J, Zhang J, Song X, et al (2016) A flexible solid-state electrolyte for wide-scale integration of rechargeable zinc-air batteries. *Energy Environ Sci* 9:663–670. https://doi.org/10.1039/C5EE03404C
51. Tan P, Chen B, Xu H, et al (2017) Flexible Zn- and Li-air batteries: recent advances, challenges, and future perspectives. *Energy Environ Sci* 10:2056–2080. https://doi.org/10.1039/c7ee01913k
52. Liu Q, Chang Z, Li Z, Zhang X (2018) Flexible metal-air batteries: progress, challenges, and perspectives. *Small Methods* 2:1700231. https://doi.org/10.1002/smtd.201700231
53. Olabi AG, Sayed ET, Wilberforce T, et al (2021) Metal-air batteries-a review. *Energies* 14:7373. https://doi.org/10.3390/en14217373
54. Lai J, Xing Y, Chen N, et al (2020) Electrolytes for rechargeable lithium-air batteries. *Angew Chemie – Int Ed* 59:2974–2997. https://doi.org/10.1002/anie.201903459
55. Sun Q, Yang Y, Fu ZW (2012) Electrochemical properties of room temperature sodium-air batteries with non-aqueous electrolyte. *Electrochem Commun* 16:22–25. https://doi.org/10.1016/j.elecom.2011.12.019
56. Liu Y, Gao D, Xiang H, et al (2021) Research progress on copper-based current collector for lithium metal batteries. *Energy Fuels* 35:12921–12937. https://doi.org/10.1021/acs.energyfuels.1c02008
57. Suen N-T, Hung S-F, Quan Q, et al (2017) Electrocatalysis for the oxygen evolution reaction: recent development and future perspectives. *Chem Soc Rev* 46:337–365. https://doi.org/10.1039/C6CS00328A
58. Moni P, Deschamps A, Schumacher D, et al (2020) A new silicon oxycarbide based gas diffusion layer for zinc-air batteries. *J Colloid Interface Sci* 577:494–502. https://doi.org/10.1016/j.jcis.2020.05.041
59. Tran C, Yang X-Q, Qu D (2010) Investigation of the gas-diffusion-electrode used as lithium/air cathode in non-aqueous electrolyte and the importance of carbon material porosity. *J Power Sources* 195:2057–2063. https://doi.org/10.1016/j.jpowsour.2009.10.012
60. Bing Y, Liu H, Zhang L, et al (2010) Nanostructured Pt-alloy electrocatalysts for PEM fuel cell oxygen reduction reaction. *Chem Soc Rev* 39:2184. https://doi.org/10.1039/b912552c
61. Gu P, Zheng M, Zhao Q, et al (2017) Rechargeable zinc-air batteries: a promising way to green energy. *J Mater Chem A* 5:7651–7666. https://doi.org/10.1039/C7TA01693J
62. Kubannek F, Turek T, Krewer U (2019) Modeling oxygen gas diffusion electrodes for various technical applications. *Chemie Ing Tech* 91:720–733. https://doi.org/10.1002/cite.201800181
63. Guo S, Zhang S, Sun S (2013) Tuning nanoparticle catalysis for the oxygen reduction reaction. *Angew Chemie – Int Ed* 52:8526–8544. https://doi.org/10.1002/anie.201207186
64. Liu J, Jiao M, Lu L, et al (2017) High performance platinum single atom electrocatalyst for oxygen reduction reaction. *Nat Commun* 8:1–9. https://doi.org/10.1038/ncomms15938
65. Luo M, Sun Y, Qin Y, et al (2018) Palladium-based nanoelectrocatalysts for renewable energy generation and conversion. *Mater Today Nano* 1:29–40. https://doi.org/10.1016/j.mtnano.2018.04.008
66. Jörissen L (2006) Bifunctional oxygen/air electrodes. *J Power Sources* 155:23–32. https://doi.org/10.1016/j.jpowsour.2005.07.038
67. Jin Y, Chen F, Lei Y, Wu X (2015) A silver-copper alloy as an oxygen reduction electrocatalyst for an advanced zinc-air battery. *ChemCatChem* 7:2377–2383. https://doi.org/10.1002/cctc.201500228
68. Cao R, Lee J-S, Liu M, Cho J (2012) Recent progress in non-precious catalysts for metal-air batteries. *Adv Energy Mater* 2:816–829. https://doi.org/10.1002/aenm.201200013
69. Perini L, Durante C, Favaro M, et al (2015) Metal-support interaction in platinum and palladium nanoparticles loaded on nitrogen-doped mesoporous carbon for oxygen reduction reaction. *ACS Appl Mater Interfaces* 7:1170–1179. https://doi.org/10.1021/am506916y
70. Stamenkovic VR, Fowler B, Mun BS, et al (2007) Improved oxygen reduction activity on Pt 3 Ni(111) via increased surface site availability. *Science* 315:493–497. https://doi.org/10.1126/science.1135941
71. Neburchilov V, Wang H, Martin JJ, Qu W (2010) A review on air cathodes for zinc-air fuel cells. *J Power Sources* 195:1271–1291. https://doi.org/10.1016/j.jpowsour.2009.08.100
72. Yu J, Ran R, Zhong Y, et al (2020) Advances in porous perovskites: synthesis and electrocatalytic performance in fuel cells and metal-air batteries. *Energy Environ Mater* 3:121–145. https://doi.org/10.1002/eem2.12064
73. Zhang Y-L, Goh K, Zhao L, et al (2020) Advanced non-noble materials in bifunctional catalysts for ORR and OER toward aqueous metal-air batteries. *Nanoscale* 12:21534–21559. https://doi.org/10.1039/D0NR05511E
74. Yan S, Xue Y, Li S, et al (2019) Enhanced bifunctional catalytic activity of manganese oxide/perovskite hierarchical core-shell materials by adjusting the interface for metal-air batteries. *ACS Appl Mater Interfaces* 11:25870–25881. https://doi.org/10.1021/acsami.9b06141

75. Llompart S, Yu LT, Mas JC, et al (1990) Oxygen-regeneration of discharged manganese dioxide electrode: I. General phenomena observed on thick powder electrodes. *J Electrochem Soc* 137:371–377. https://doi.org/10.1149/1.2086448
76. Zaikovskii A V., Mal'tsev VA, Novopashin SA (2013) Synthesis of tungsten carbide nanoparticles in WO_3 pyrolysis in a plasma arc. *J Eng Thermophys* 22:77–85. https://doi.org/10.1134/S1810232813010098
77. Khaji Z, Klintberg L, Palmer K, Thornell G (2020) Catalytic effect of platinum and silver in a hydrogen peroxide monopropellant ceramic microthruster. *Propuls Power Res* 9:216–224. https://doi.org/10.1016/j.jppr.2020.08.003
78. Reith L, Triana CA, Pazoki F, et al (2021) Unraveling nanoscale cobalt oxide catalysts for the oxygen evolution reaction: maximum performance, Minimum Effort. *J Am Chem Soc* 143:15022–15038. https://doi.org/10.1021/jacs.1c03375
79. Moysiadou A, Lee S, Hsu C-S, et al (2020) Mechanism of oxygen evolution catalyzed by cobalt oxyhydroxide: cobalt superoxide species as a key intermediate and dioxygen release as a rate-determining step. *J Am Chem Soc* 142:11901–11914. https://doi.org/10.1021/jacs.0c04867
80. Osgood H, Devaguptapu S V., Xu H, et al (2016) Transition metal (Fe, Co, Ni, and Mn) oxides for oxygen reduction and evolution bifunctional catalysts in alkaline media. *Nano Today* 11:601–625. https://doi.org/10.1016/j.nantod.2016.09.001
81. Vineeth SK, Soni CB, Sun Y, et al (2022) Implications of Na-ion solvation on Na anode-electrolyte interphase. *Trends Chem* 4:48–59. https://doi.org/10.1016/j.trechm.2021.11.002
82. Yi J, Guo S, He P, Zhou H (2017) Status and prospects of polymer electrolytes for solid-state Li-O_2 (air) batteries. *Energy Environ Sci* 10:860–884. https://doi.org/10.1039/C6EE03499C
83. Wang S, Wang J, Liu J, et al (2018) Ultra-fine surface solid-state electrolytes for long cycle life all-solid-state lithium-air batteries. *J Mater Chem A* 6:21248–21254. https://doi.org/10.1039/C8TA08095J
84. Egan DR, Ponce De León C, Wood RJK, et al (2013) Developments in electrode materials and electrolytes for aluminium-air batteries. *J Power Sources* 236:293–310. https://doi.org/10.1016/j.jpowsour.2013.01.141
85. Liu Q, Pan Z, Wang E, et al (2020) Aqueous metal-air batteries: Fundamentals and applications. *Energy Storage Mater* 27:478–505. https://doi.org/10.1016/j.ensm.2019.12.011
86. Mainar AR, Leonet O, Bengoechea M, et al (2016) Alkaline aqueous electrolytes for secondary zinc-air batteries: an overview. *Int J Energy Res* 40:1032–1049. https://doi.org/10.1002/er.3499
87. Li Q, Chen J, Fan L, et al (2016) Progress in electrolytes for rechargeable Li-based batteries and beyond. *Green Energy Environ* 1:18–42. https://doi.org/10.1016/j.gee.2016.04.006
88. Yang H, Li X, Wang Y, et al (2020) Excellent performance of aluminium anode based on dithiothreitol additives for alkaline aluminium/air batteries. *J Power Sources* 452:227785. https://doi.org/10.1016/j.jpowsour.2020.227785
89. Wu S, Hu S, Zhang Q, et al (2020) Hybrid high-concentration electrolyte significantly strengthens the practicability of alkaline aluminum-air battery. *Energy Storage Mater* 31:310–317. https://doi.org/10.1016/j.ensm.2020.06.024
90. Liu D, Zhao S (2021) An efficient combination strategy for high-performance asymmetric-electrolyte metal-air batteries. *Matter* 4:1090–1092. https://doi.org/10.1016/j.matt.2021.03.015
91. Yu J, Li B-Q, Zhao C-X, Zhang Q (2020) Seawater electrolyte-based metal-air batteries: from strategies to applications. *Energy Environ Sci* 13:3253–3268. https://doi.org/10.1039/D0EE01617A
92. Lin Z, Xia Q, Wang W, et al (2019) Recent research progresses in ether- and ester-based electrolytes for sodium-ion batteries. *InfoMat* 1:376–389. https://doi.org/10.1002/inf2.12023
93. Feng N, He P, Zhou H (2016) Critical challenges in rechargeable aprotic Li-O_2 batteries. *Adv Energy Mater* 6:1502303. https://doi.org/10.1002/aenm.201502303
94. Ogasawara T, Débart A, Holzapfel M, et al (2006) Rechargeable Li_2O_2 electrode for lithium batteries. *J Am Chem Soc* 128:1390–1393. https://doi.org/10.1021/ja056811q
95. Bryantsev VS, Giordani V, Walker W, et al (2011) Predicting solvent stability in aprotic electrolyte li-air batteries: nucleophilic substitution by the superoxide anion radical (O_2 •–). *J Phys Chem A* 115:12399–12409. https://doi.org/10.1021/jp2073914
96. Song K, Agyeman DA, Park M, et al (2017) High-energy-density metal-oxygen batteries: lithium-oxygen batteries vs sodium-oxygen batteries. *Adv Mater* 29:1606572. https://doi.org/10.1002/adma.201606572
97. Assary RS, Lu J, Du P, et al (2013) The effect of oxygen crossover on the anode of a Li-O_2 battery using an ether-based solvent: insights from experimental and computational studies. *ChemSusChem* 6:51–55. https://doi.org/10.1002/cssc.201200810
98. Li Y, Wang X, Dong S, et al (2016) Recent advances in non-aqueous electrolyte for rechargeable Li-O_2 batteries. *Adv Energy Mater* 6:1600751. https://doi.org/10.1002/aenm.201600751

99. Younesi R, Veith GM, Johansson P, et al (2015) Lithium salts for advanced lithium batteries: Li-metal, Li-O_2, and Li-S. *Energy Environ Sci* 8:1905–1922. https://doi.org/10.1039/C5EE01215E
100. Zhang P, Zhao Y, Zhang X (2018) Functional and stability orientation synthesis of materials and structures in aprotic Li-O_2 batteries. *Chem Soc Rev* 47:2921–3004. https://doi.org/10.1039/C8CS00009C
101. Du P, Lu J, Lau KC, et al (2013) Compatibility of lithium salts with solvent of the non-aqueous electrolyte in Li-O_2 batteries. *Phys Chem Chem Phys* 15:5572. https://doi.org/10.1039/c3cp50500f
102. Sun H, Zhu G, Xu X, et al (2019) A safe and non-flammable sodium metal battery based on an ionic liquid electrolyte. *Nat Commun* 10:3302. https://doi.org/10.1038/s41467-019-11102-2
103. Ha TA, Li H, Wang X, et al (2021) Functional binders based on polymeric ionic liquids for sodium oxygen batteries using ionic liquid electrolytes. *ACS Appl Energy Mater* 4:434–444. https://doi.org/10.1021/acsaem.0c02352
104. Shiga T, Kato Y, Hase Y (2017) Coupling of nitroxyl radical as an electrochemical charging catalyst and ionic liquid for calcium plating/stripping toward a rechargeable calcium-oxygen battery. *J Mater Chem A* 5:13212–13219. https://doi.org/10.1039/C7TA03422A
105. Kwon H-M, Thomas ML, Tatara R, et al (2017) Stability of glyme solvate ionic liquid as an electrolyte for rechargeable Li–O_2 batteries. *ACS Appl Mater Interfaces* 9:6014–6021. https://doi.org/10.1021/acsami.6b14449
106. Zhang J, Sun B, Zhao Y, et al (2019) A versatile functionalized ionic liquid to boost the solution-mediated performances of lithium-oxygen batteries. *Nat Commun* 10:602. https://doi.org/10.1038/s41467-019-08422-8
107. Kang Y, Liang F, Hayashi K (2016) Hybrid sodium-air cell with Na[FSA-C_2C_1im][FSA] ionic liquid electrolyte. *Electrochim Acta* 218:119–124. https://doi.org/10.1016/j.electacta.2016.09.113
108. Zhang Y, Ortiz-Vitoriano N, Acebedo B, et al (2018) Elucidating the impact of sodium salt concentration on the cathode-electrolyte interrrolidinium-based ionic liquid electrolyte in a half-cell has been explored in this work. Increasing the concentration of sodium salt in an ionic liquid electrolyte produc. *J Phys Chem C* 122:15276–15286. https://doi.org/10.1021/acs.jpcc.8b02004
109. Vineeth SK, Tebyetekerwa M, Liu H, Soni CB, Zhao XS, Kumar V (2022) Progress in the development of solid-state electrolytes for reversible room-temperature sodium–sulfur batteries. *Mater Adv*, 3(16):6415–6440. https://doi.org/10.1039/D2MA00428C
110. Shinde SS, Jung JY, Wagh NK, et al (2021) Ampere-hour-scale zinc-air pouch cells. *Nat Energy* 6:592–604. https://doi.org/10.1038/s41560-021-00807-8
111. Dhawale PV, Vineeth SK, Gadhave RV, et al (2022) Tannin as a renewable raw material for adhesive applications: a review. *Mater Adv* 3:3365–3388. https://doi.org/10.1039/D1MA00841B
112. Gadhave RV, Vineeth SK, Dhawale PV, Gadekar PT (2020) Effect of boric acid on poly vinyl alcohol- tannin blend and its application as water-based wood adhesive. *Des Monomers Polym* 23:188–196. https://doi.org/10.1080/15685551.2020.1826124
113. Vineeth SK, Gadhave RV, Gadekar PT (2022) Investigation of crosslinking ability of sodium metabisulphite with polyvinyl alcohol-corn starch blend and its applicability as wood adhesive. *Indian Chem Eng* 64:197–207. https://doi.org/10.1080/00194506.2021.1887769
114. Tan MJ, Li B, Chee P, et al (2018) Acrylamide-derived freestanding polymer gel electrolyte for flexible metal-air batteries. *J Power Sources* 400:566–571. https://doi.org/10.1016/j.jpowsour.2018.08.066
115. Vineeth SK, Soni CB, Sanjaykumar C, Yamauchi Y, Han M, Kumar V (2023) A quasi-solid state polymer electrolyte for high-rate and long-life sodium-metal batteries. *Journal of Energy Storage*, 73, 108780. https://doi.org/10.1016/j.est.2023.108780
116. Chamaani A, Safa M, Chawla N, El-Zahab B (2017) Composite gel polymer electrolyte for improved cyclability in lithium-oxygen batteries. *ACS Appl Mater Interfaces* 9:33819–33826. https://doi.org/10.1021/acsami.7b08448
117. Yi J, Liu X, Guo S, et al (2015) Novel stable gel polymer electrolyte: toward a high safety and long life li-air battery. *ACS Appl Mater Interfaces* 7:23798–23804. https://doi.org/10.1021/acsami.5b08462
118. Xi G, Xiao M, Wang S, et al (2021) Polymer-based solid electrolytes: material selection, design, and application. *Adv Funct Mater* 31:2007598. https://doi.org/10.1002/adfm.202007598
119. Qiao L, Judez X, Rojo T, et al (2020) Review-polymer electrolytes for sodium batteries. *J Electrochem Soc* 167:070534. https://doi.org/10.1149/1945-7111/ab7aa0
120. Zhou D, Shanmukaraj D, Tkacheva A, et al (2019) Polymer electrolytes for lithium-based batteries: advances and prospects. *Chem* 5:2326–2352. https://doi.org/10.1016/j.chempr.2019.05.009
121. Wang Y, Fan F, Agapov AL, et al (2014) Examination of the fundamental relation between ionic transport and segmental relaxation in polymer electrolytes. *Polymer (Guildf)* 55:4067–4076. https://doi.org/10.1016/j.polymer.2014.06.085

122. Aziz SB, Woo TJ, Kadir MFZ, Ahmed HM (2018) A conceptual review on polymer electrolytes and ion transport models. *J Sci Adv Mater Devices* 3:1–17. https://doi.org/10.1016/j.jsamd.2018.01.002
123. Eng AYS, Soni CB, Lum Y, et al (2022) Theory-guided experimental design in battery materials research. *Sci Adv* 8:eabm2422. https://doi.org/10.1126/sciadv.abm2422
124. He R, Kyu T (2016) Effect of plasticization on ionic conductivity enhancement in relation to glass transition temperature of crosslinked polymer electrolyte membranes. *Macromolecules* 49:5637–5648. https://doi.org/10.1021/acs.macromol.6b00918
125. Arya A, Sharma AL (2017) Polymer electrolytes for lithium ion batteries : a critical study. *Ionics (Kiel)* 23:497–540. https://doi.org/10.1007/s11581-016-1908-6
126. Huang Y, Li Z, Pei Z, et al (2018) Solid-state rechargeable Zn//NiCo and Zn-air batteries with ultralong lifetime and high capacity: the role of a sodium polyacrylate hydrogel electrolyte. *Adv Energy Mater* 8:1802288. https://doi.org/10.1002/aenm.201802288
127. Chawla N, Safa M (2019) Sodium batteries: a review on sodium-sulfur and sodium-air batteries. *Electronics* 8:1201. https://doi.org/10.3390/electronics8101201
128. Wang Y, Zhou H (2010) A lithium-air battery with a potential to continuously reduce O_2 from air for delivering energy. *J Power Sources* 195:358–361. https://doi.org/10.1016/j.jpowsour.2009.06.109
129. Ruan Y, Guo F, Liu J, et al (2019) Optimization of $Na_3Zr_2Si_2PO_{12}$ ceramic electrolyte and interface for high performance solid-state sodium battery. *Ceram Int* 45:1770–1776. https://doi.org/10.1016/j.ceramint.2018.10.062
130. Pan X, Sun J, Jin C, et al (2020) A flexible composite electrolyte membrane with ultrahigh LLZTO garnet content for quasi solid state Li-air batteries. *Solid State Ionics* 351:115340. https://doi.org/10.1016/j.ssi.2020.115340
131. Pan J, Xu YY, Yang H, et al (2018) Advanced architectures and relatives of air electrodes in Zn-air batteries. *Adv Sci* 5:1700691. https://doi.org/10.1002/advs.201700691
132. Chang Z, Xu J, Zhang X (2017) Recent progress in electrocatalyst for Li-O_2 batteries. *Adv Energy Mater* 7:1700875. https://doi.org/10.1002/aenm.201700875

10 Lithium-Air Batteries

Batteries for Emerging Electric or Hybrid Electric Vehicles (EV or HEV), a Focus into General Aspects and Electrolyte Engineering

S. K. Vineeth, Akhila Das, Jabeen Fatima M. J., Vijayamohanan Pillai, Nutan Gupta, and Prasanth Raghavan

10.1 INTRODUCTION TO LITHIUM-ION BATTERIES: POTENTIAL CANDIDATE FOR POST-LITHIUM-ION BATTERY ERA

Advancements in understanding energy storage technology help move a step further for research and development of batteries, supercapacitors, fuel cells, etc. [1]. This has also led to invigorating a fast transition from fuel cell electric vehicles to hybrid electric vehicles (HEVs), plug-in hybrid electric vehicles, and full electric vehicles (EVs) [2]. Coupled with the concerns over global warming and pollution, and strict governmental policies to reduce the carbon footprint, there has been a great interest in the EV sector. André et al. [3] have reported that energy consumption and emissions of battery EVs are lowest for lithium-ion batteries (LIBs), 314–374 Wh km^{-1} and 76–90 gCO_2eq km^{-1} (assuming 593 gCO_2 kWh^{-1} for European electricity mix), compared to 450–760 Wh km^{-1} and 150–170 gCO_2eq km^{-1} for petrol and diesel cars. LIBs eliminate the possibility of using conventional lead-acid batteries in EVs at half of the weight but multiple times in terms of power, thus considered a promising solution for current EVs [4]. Even though LIBs are a commercialized energy storage technology, where great developments have been reported in the past few decades, proliferation in energy demand requires advanced energy storage solutions [5]. Current LIB research is mainly focused on better recycling solutions with optimum battery designs and technology.

To meet the demand for newer battery technologies for smarter energy storage needs, it is worth looking at the "post-LIB era", where the main idea is to look at systems that are much more efficient and have less impact on the environment [6]. Metal-air batteries/metal-oxygen batteries (MAB), considering the usage of pure or ambient oxygen (O_2), hence step into the limelight as they offer high energy density, heat capacity, and power density compared to the other competent battery technologies for future electric mobility [7]. The MAB consists of main four components, namely, metallic anode, electrolyte, separator, and air cathode with electrocatalyst. Depending on the type of electrolyte used, a MAB can be either an aqueous or non-aqueous battery. During discharge, at the anode, oxidation occurs with the dissolution of the metal in the electrolyte medium, and the oxygen reduction reaction (ORR) proceeds at the air cathode [8]. The reason the MAB possesses higher specific capacity is due to the usage of air as the reagent. In this context, lithium-based MAB, lithium-air battery (LAB), can provide a minimum of three times higher energy density and specific energy density than conventional LIBs and can be a sustainable solution that could possibly power EVs to drive 300 miles and beyond with single time charge, which is almost comparable to gasoline-powered vehicles [9]. Figure 10.1 shows a comparative analysis of various MABs based on the theoretical energy density.

DOI: 10.1201/9781003310174-10

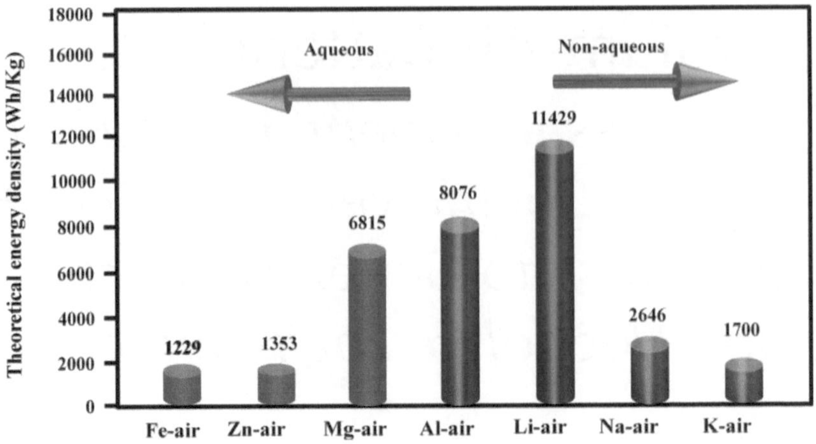

FIGURE 10.1 A comparison of the theoretical energy density of various metal-air batteries. Adapted and reproduced with permission from Ref. [10]. Copyright © 2020 American Chemical Society.

In the current scenario, the developments in LAB technology are in their infancy. Developments are plagued by various challenges associated with the materials, design, and efficient electrochemical conversion. As the demand for high-energy-density batteries spikes, the momentum of research in LAB should be accelerated, focusing on the booming advancements in portable electronics and the EV sector. Hence to regain the momentum, unveiling the complexities and rectification of the challenges associated with the LAB needs prime importance. Multifaced strategies that can provide better insights are required in unraveling the basic principles of the electrochemical mechanism. This chapter hence aims to unveil the complexities of the basic electrochemical mechanism, working principles, and the critical role of cell components, focusing mainly on electrolytes. Furthermore, the chapter provides a brief overview of the development of the LAB and its potential role in EVs. It is expected that by exploring multifaced strategies for designing the cell, selecting battery components, and monitoring the electrochemical properties, the LAB can be updated as a candidate for enabling EVs and advanced electronics.

10.2 WORKING PRINCIPLE OF A LAB

High values of specific energy and energy density (~3,500 Wh kg^{-1}) of LABs are the attractive features to revolutionize the next-generation battery technology [11,12]. Energy density denotes the amount of energy that can be extracted from a unit mass (1 kg) of a material, due to the presence of air cathode in LABs accounting for these high values. This active cathode can be "inhaled" from the environment and can be reduced using a catalytic reaction at the cathode [13]. Moreover, the lithium anode has a high specific capacity of 3,862 mAh g^{-1}, with the lowest electrochemical potential (–3.04 V vs standard hydrogen electrode) on reaction with O_2 [14,15]. The LAB consists of a metallic lithium-based anode, and the composite air cathode consists of a porous carbonaceous matrix, which allows O_2 to permeate through it (Figure 10.2) [16]. The porous carbon-cathode contains a catalyst that supports the oxidation/reduction reaction during the charging and discharging of the battery [17]. It ensures the contact between O_2 and Li-ion (Li^+) in the electrolyte. Lithium undergoes a redox reaction, and the mobile Li^+ is transported through the electrolyte medium toward the air cathode, where it reacts with O_2, forming mainly two products, lithium oxide (Li_2O) and lithium peroxide (Li_2O_2) [18]. Sluggish kinetics during cycling, in the generation and oxidation of Li_2O_2, and degradation of the electrolyte and the cathode by parasitic reactions are the bottleneck challenges while using the LAB. To rectify this, electrocatalysts including carbonaceous nanomaterials

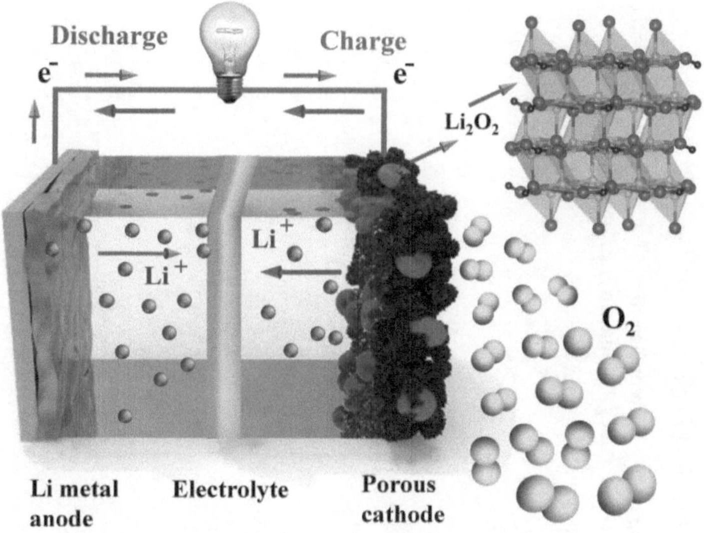

FIGURE 10.2 Schematic representation of a LAB. Adapted and reproduced with permission from Ref. [16]. Copyright © 2020 American Chemical Society.

and doped carbon materials using heteroatoms, metal oxides, and redox mediators are employed in the LAB [19]. The general mechanism (Equations 10.1–10.3) can be represented as follows:

$$\text{Anode reaction:} \quad Li(s) \leftrightarrow Li^{+}+e^{-} \tag{10.1}$$

$$\text{Cathode reaction:} \quad Li^{+}+e^{-}+0.5\ O_2 \leftrightarrow 0.5\ Li_2O_2 \tag{10.2}$$

$$\text{Cathode reaction:} \quad Li^{+}+e^{-} \leftrightarrow 0.5\ Li_2O \tag{10.3}$$

10.3 LITHIUM-AIR BATTERIES: A BRIEF HISTORY OF VARIOUS DEVELOPMENTS

The history of developments in LABs dates back to the middle of the 1970s. Littauer and Tsai [20] proposed LABs in 1976; in 1996, Abraham and Jiang [21] first serendipitously reported non-aqueous LABs, with lithium metal as an anode, carbon composite cathode containing cobalt phthalocyanine as a catalyst, and a gel polymer as an electrolyte. The work was successful and obtained a specific energy between 250 and 350 Wh kg^{-1}. In 2002, Read et al. [22] demonstrated the role of the electrolyte and air cathode in the rechargeability and performance of LAB. A major development occurred in 2006, which showed that the electrochemical reaction between Li^{+} and O_2 is reversible during charging and discharging, showing the feasibility of the O_2 cathode [23]. The work opened up the possibilities to enhance the performance of LAB by optimizing the design and materials for the O_2 cathode. These works attracted researchers and laid the foundation for the development of the LAB. Following these works, the past two decades have seen an increased number of research articles, which focused on enhancing the stability and performance of LABs [24]. A notable development in the field of LAB is the launching of the Battery 500 project by IBM for evaluating the automotive application of LABs [25]. IBM announced its partnership with Asahi Kasei, a Japan-based global supplier of membrane separators of LIB, and Central Glass, a leading electrolyte manufacturer for LIB, to design and enhance the efficiency of LABs. Table 10.1 depicts some of the developments associated with the development of LABs.

TABLE 10.1
Developments in LAB Regarding Cathode and Electrolyte Materials with Its Respective Electrochemical Performance

Electrolyte	Cathode Material	η_{charge} (V)	Discharge Depth (mAh g^{-1})	Cycling Interval (*V*)	Current Density (mA g^{-1} or mA cm^{-2})	Cycle Life (Number)	Reference
1 M $LiClO_4$ in DMSO with tetrathiafulvalene	Nanoporous Au	~0.6	300	2.3–4.0	1 mA cm^{-2}	100	[26]
1.3 M lithium bis(trifluoromethane) sulfonimide (LiTFSI) in tetraethylene glycol dimethyl ether (TEGDME)	Au nanoparticles coated Ni nanowires	~1.3	500	2.3–4.3	500	110	[27]
1.0 M LiTFSI in TEGDME	Porous $La_{0.75}Sr_{0.25}MnO_3$ Nanotubes	~1.24	1,000	2.2–4.4	0.15 mA cm^{-2}	124	[28]
0.1 M $LiClO_4$/DMSO	Ru@nanoporous graphene catalysts	0.64	1,000	2–4.4	200	200	[29]
0.5 M $LiClO_4$ in DMSO	TiC	~0.64	500	2–4	1 mA cm^{-2}	100	[30]
1.0 M $LiClO_4$ in TEGDME	Pd/FeO_x/ordered mesoporous carbon	0.48	500	2.0–4.3	100	68	[31]
LiTFSI/triglyme (1:5)	Ru nanoparticles on a multiwalled carbon nano tube paper	~0.78	5,000	2.3–4.6	2,000	50	[32]
1.0 M LiTFSI in TEGDME with 0.05 M LiI	Hierarchical CNT fibril	0.25	1,000	2–4.7	2,000	900	[33]
1.0 M LiTFSI in TEGDME	Carbon-free $MnCo_2O_4$@Ni	0.94–1.04	1,000	2–4.2	500	100	[34]
1.0 M LiTFSI in TEGDME	CoO mesoporous spheres + super P	0.79	1,000	2–4.3	0.04 mA cm^{-2}	300	[35]
1.0 M $LiCF_3SO_3$ in TEGDME	Hierarchical rutile TiO_2 nanowire arrays grown onto carbon textiles	~1.54	500	2.0–4.7	100	500	[36]

Source: Adapted and reproduced with permission from Ref. [12]. Copyright © 2016 Wiley-VCH.

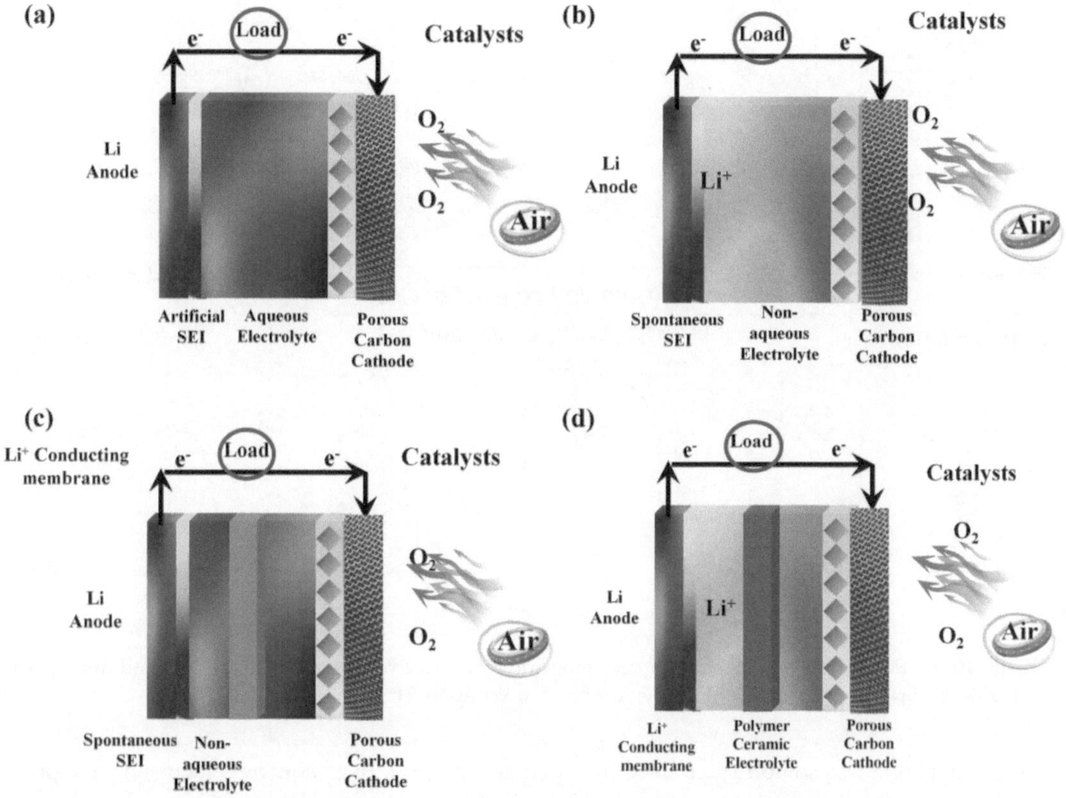

FIGURE 10.3 Types of lithium-air batteries: (a) Aqueous electrolytic type, (b) aprotic/non-aqueous electrolytic type, (c) mixed (aprotic-aqueous) type, and (d) solid-state type. Adapted and reproduced with permission from Ref. [39]. Copyright © 2014 Springer Nature.

10.4 ELECTROLYTES FOR LAB

LABs consist of Li metal as the anode, and porous carbonaceous air cathode with the catalyst can be classified into four types based on the electrolyte used, as shown in Figure 10.3: (1) aqueous electrolytic type, (2) aprotic/non-aqueous electrolytic type, (3) mixed/hybrid electrolytic type, and (4) solid-state electrolytic type [37]. The electrolyte in LABs functions by transporting the Li^+ and transferring dissolved O_2 permeated to the cathode before getting reduced. Electrochemical properties (mainly rate capability and discharge capacity) are determined by the rate of transport of O_2 through the electrolyte [38].

Apart from the parasitic reactions at the anode and cathode, electrolytes are prone to these unwanted chemical reactions resulting in poor stabilities and low charge/discharge round-trip efficiency [40]. Figure 10.4 depicts various possibilities for parasitic reaction by reactive O_2 species and the intermediates formed.

10.4.1 Aqueous and Non-Aqueous

Aqueous and non-aqueous LABs vary depending on the type of electrolyte used to separate the Li metal anode and air electrode. Non-aqueous electrolyte-based LAB employs Li^+ salt and organic solvents, namely, dimethyl sulfoxide (DMSO), tetra(ethylene) glycol dimethyl ether (tetraglyme), alkene-ester, etc. Ideally, a non-aqueous electrolyte should possess high ionic conduction for Li^+ transportation, stability with respect to chemicals and the electrochemical window, lower vapor

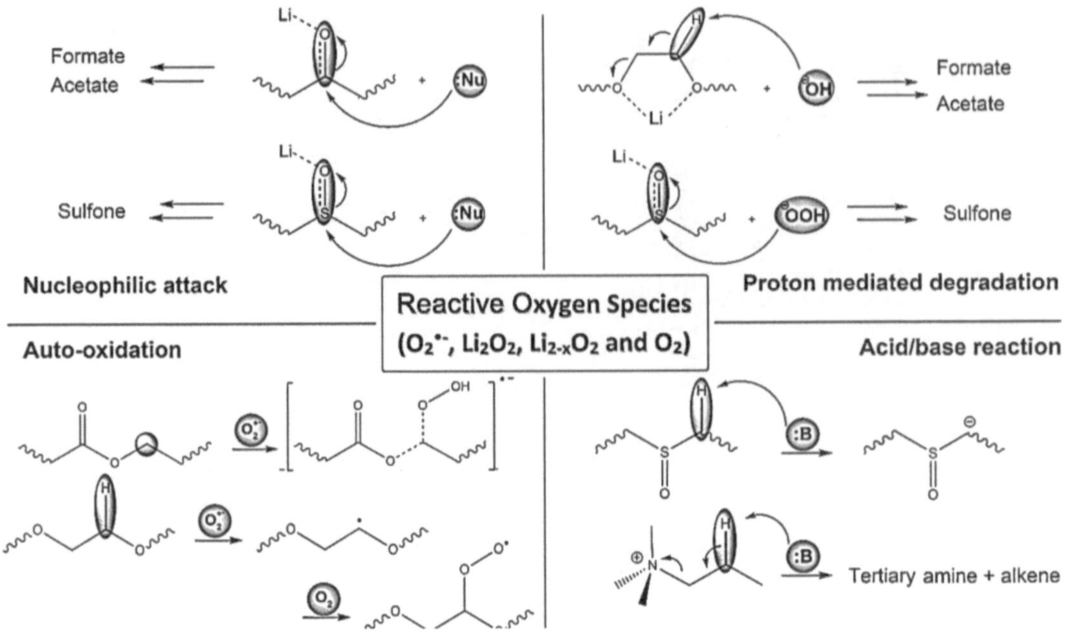

FIGURE 10.4 Possibilities of electrolyte decomposition by reactive oxygen species. Adapted and reproduced with permission from Ref. [40]. Copyright © 2016 Wiley-VCH.

pressure, and higher O_2 solubility, coupled by O_2 diffusivity [37]. In an aqueous electrolyte, the presence of water molecules acts as a reactant and joins in the electrochemical conversions. The electrochemical reaction at the negative electrode is the same for both systems, as shown in Equation 10.1, but the reaction at the positive electrode depends on the type of LABs. Equations 10.4 and 10.5 show the reaction at positive electrodes for alkaline (neutral) and acidic systems, respectively.

$$\text{Anode: Li} \leftrightarrow \text{Li}^+ + e^- \tag{10.4}$$

$$\text{Cathode: Alkaline} - O_2 + 2H_2O + 4e^- \leftrightarrow 4OH^- \ (E_0 = 3.43 \text{ V vs Li/Li}^+) \tag{10.5}$$

$$\text{Acid} - O_2 + 4e^- + 4H^+ \leftrightarrow 2H_2O \leftrightarrow (E_0 = 4.26 \text{ V vs Li/Li}^+) \tag{10.6}$$

$$\text{Non-aqueous} - 2\text{Li}^+ + 2e^- + O_2 \leftrightarrow \text{Li}_2O_2 \ (E_0 = 2.96 \text{ V vs Li/Li}^+) \tag{10.7}$$

The reaction that happens at the cathode is called an ORR, while the electrochemical decomposition of Li_2O_2 to Li and O_2 on charging is known as oxygen evolution reaction. Apart from these reactions, various products can be formed; for a successful rechargeable LAB, the formation and decomposition of Li_2O_2 is the most crucial step [41]. Any products other than this can cause inferior cyclability, the rate of charging becoming lower, and high overpotential, affecting the performance of LAB. Overpotential can degrade the cathode and electrolyte. In an aprotic electrolyte, during the ORR, an intermediate product, LiO_2, is formed [41]. The formation of the intermediate product depends on the catalyst, the electrolyte, and the solvent used. Therefore, to mitigate this issue and thereby favor the desired reaction, various catalysts are mostly used to avoid unwanted by-products, which also include lithium carbonates and carboxylates. O_2-selective membranes are installed to prevent the entrance of impurities into the cell. Since the anode consists of highly reactive Li, reaction with H_2O, even at the ppm level, causes huge risk. Moreover, leakage of air and chemical diffusing from the cathode and getting in contact with the anode needs to be prevented. Hence, in addition to electrodes and electrolytes, a membrane that protects the anode is another crucial component of LABs.

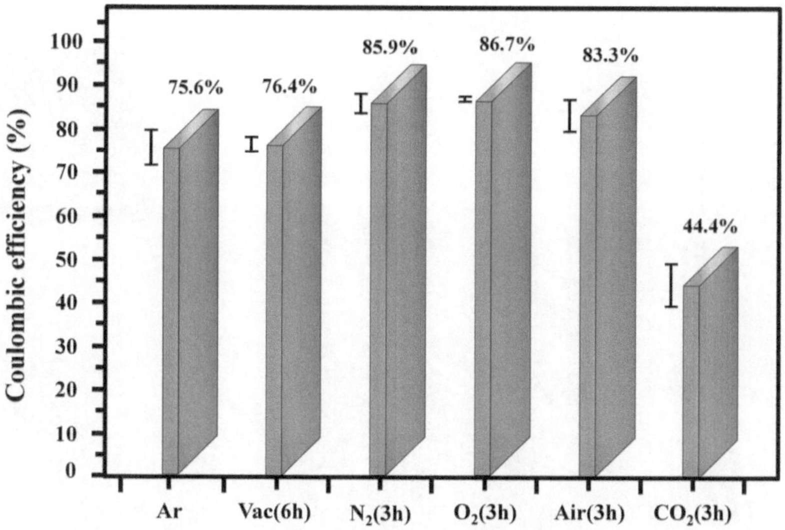

FIGURE 10.5 Effects of the presence of various gases and their corresponding coulombic efficiencies. It can be seen that the coulombic efficiencies are sensitive to the atmosphere in which the SEI formation occurs. Adapted and reproduced with permission from Ref. [44]. Copyright © 2013 American Chemical Society.

In non-aqueous LABs, capacity fading is one of the main issues that needs rectification. With this aim, Leskes et al. [42] employed nuclear magnetic resonance spectroscopy to monitor the evolution of discharge products. During the charging process of LABs, in 1 M LiTFSI in 1,2-dimethoxyethane electrolyte, the air cathode surface showed the presence of carbonates, which accumulated in the following cycles. Furthermore, the presence of lithium hydroxide and formate was detected. It was concluded that continuous cycling showed electrolyte decomposition and the subsequent accumulation of lithium salts on the cathode led to an increment in overpotential and capacity decay. Safanama et al. [43] fabricated aqueous and hybrid LAB with low overpotentials and high energy efficiency of up to 98% when NASICON-type $Li_{1+x}A_xGe_{2-x}(PO_4)_3$ (LAGP) ceramic was employed as the anode-protecting membrane.

Cycling efficiency is one of the parameters that indicates the electrochemical performance of a battery system. As metallic Li reacts with electrolyte counterparts, compatibility and the electrode–electrolyte interface determine the nature of solid–electrolyte interface (SEI) formed. Besides the stability of SEI, its thickness affects the electrochemical properties. Stark et al. [44] showed the variation of the thickness of SEI on coulombic efficiency for a secondary Li metal battery. The thickness and composition of SEI measured by a secondary ion mass spectroscopy showed that a thick SEI layer was formed when CO_2 was bubbled, and thin SEI was produced by N_2 and O_2. Furthermore, regardless of the adsorbed gas, the major composition of SEI consisted of LiF. It was observed that the cell with a thick SEI layer showed the least coulombic efficiency and the cell with a thin SEI showed high coulombic efficiency. As the thickness of SEI increases, reversible deposition of Li becomes uneasy, which reduces the performance of the cell. Figure 10.5 shows the representation of data, indicating the relation between the presence of various gases and their effect on coulombic efficiency.

10.4.2 Mixed/Hybrid Electrolytes

Zhou and Wang [45], in 2010, proposed the integration of O_2 catalytic reduction in an aqueous solution and the Li anode in an organic electrolyte using a super-ionic conductor glass film (LISICON). Figure 10.6 is a schematic representation of a hybrid electrolyte-based LAB with a LISICON film. The cathode used in the study was a Mn_3O_4-based air diffusion electrode and a Li metal anode. The cell showed stable cycling for 500 h, indicating the ability to deliver energy like a fuel cell. Moreover,

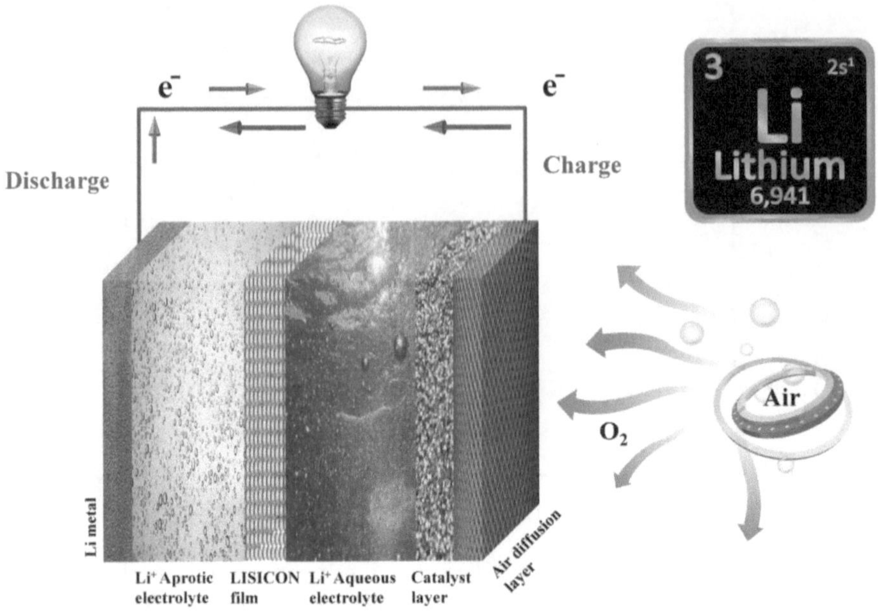

FIGURE 10.6 Schematic representation showing hybrid electrolyte LAB with a LISICON film. Adapted and reproduced with permission from Ref. [46]. Copyright © 2016 American Chemical Society.

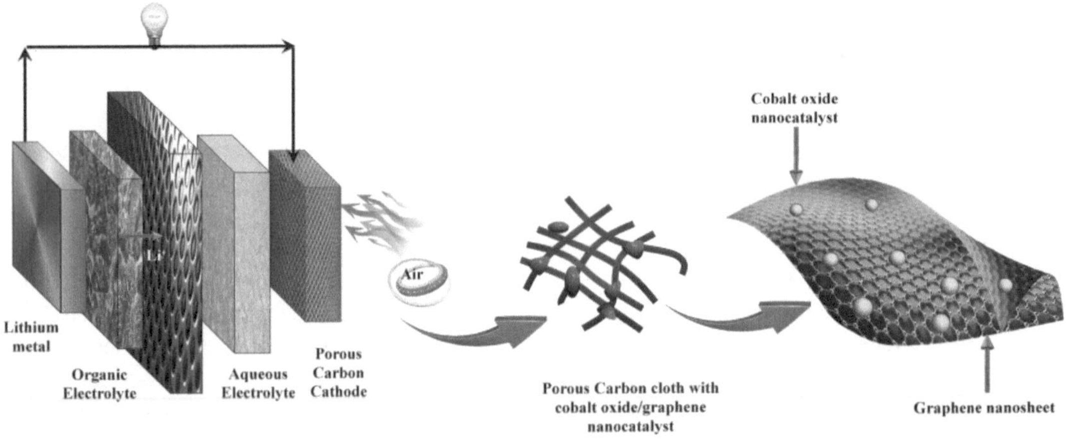

FIGURE 10.7 Schematic representation of the working principle of hybrid electrolyte lithium-air battery at discharge state. Adapted and reproduced with permission from Ref. [47]. Copyright © 2020 Elsevier.

the cathode delivered a capacity of 50,000 mAh g^{-1} based on the total mass of the catalytic electrode (carbon+binder+catalyst), reporting a higher value than what was previously reported.

Peng et al. [47] fabricated a hybrid electrolyte LAB with an aqueous/aprotic hybrid electrolyte assembled with a Li metal/organic electrolyte/ceramic Li^+ conductive membrane/aqueous electrolyte/air cathode, for a high relative humidity environment. Figure 10.7 is the schematic representation of the working principle of the hybrid system at the discharge state. The enhanced catalytic activity of cobalt oxide/graphene nanocomposite led to enhanced long-term stability over 120h when fed with a high (70%–100%) relative humidity ambient atmosphere.

To ensure larger operation and cycling stability, Li et al. [12] developed a deep eutectic electrolyte based on *N*-methylacetamide and LiTFSI. Polar groups of *N*-methylacetamide interact with both anions and cations of Li salt, forming deep eutectic electrolytes. The aforementioned interactions enhanced the electrochemical properties such as ionic conduction (3.75×10^{-3} S cm^{-1} at 60°C), with a boost in electrochemical performance, and with increased thermal and chemical stability. Full cell analysis showed a discharge capacity of 8,647 mAh g^{-1} with an enhanced cyclic life of 280 cycles.

10.4.3 IL Electrolytes

IL electrolytes have been employed in electrochemical storage applications, as it has a wide electrochemical window, non-flammability, non-volatility, and chemical and thermal stability compared to conventional liquid electrolytes [48]. Ionic conductivity ranges from 10^{-3} to 10^{-2} S cm^{-1} at room temperature, which accounts for its potential application for commercial-level products [49]. In addition to ionic conduction, the cation transference number in the electrolyte is another parameter that is the basis of the selection of electrolytes. It is preferably ideal to use solvating IL with the same anion as that of lithium salt, to ensure better electrochemical properties [50].

ILs have been used in LABs from the initial times [51–53]. Kuboki et al. [54] reported LABs based on IL, 1-ethyl-3-methylimidazolium tetrafluoroborate, which prevented hydrolysis at metallic Li anode. The cell with an air cathode containing carbon materials showed a high discharge capacity of 5,360 mAh g^{-1}. Nakamoto et al. [55] found that the incorporation of IL provided higher O_2 redox reversibility and radical stability in LABs. Elia et al. [56] fabricated a lithium-oxygen battery with novel IL, *N*-butyl-*N*-methylpyrrolidinium *bis*-(tri fluoromethanesulfonyl)imide-LiTFSI and showed that the formation of a stable electrode–electrolyte interface with reversible charge–discharge cycling behavior. Moreover, ORR showed lower overpotential, with an improvement in energy efficiency to 82%.

Bai et al. [57] fabricated an IL-based quasi-solid-state electrolyte for LABs with poly(vinylidene fluoride-co-hexafluoropropylene) (PVdF-HFP) as the polymer host in 1-ethyl-3-methyl-imidazolium *bis*(trifluoromethanesulfonyl)imide-LiTFSI. The polymer showed less crystalline nature upon the addition of IL, which contributed to the enhanced ionic conduction of 4.30×10^{-4} S cm^{-1} with a wider electrochemical stability window of 4.9 V. Coupling electrolytes with electrodes showed better compatibility, which resulted in enhancement in discharge capacity and reduction in overpotential and no capacity fading up to the 20th cycle.

Olivares-Marín et al. [58] reported that the formation of secondary products was negligible in IL electrolytes compared to ether-based electrolytes in LAB. The analysis was performed using energy-dependent soft x-ray transmission microscopy. Carbonate compounds were insignificantly less in IL, which indicated the stability of the system. Crystallites of LiOH were observed in IL, which was dependent on the concentration of Li salt (LiTFSI) and the storage time of electrodes that are discharged.

10.4.4 Solid-State Electrolytes

Research in solid-state electrolytes (SSEs) mitigates the bottleneck challenges of liquid electrolytes such as electrolyte leakage, flammability, unstable SEI formation, dendrite formation and its growth, and a narrower electrochemical window. Replacement of liquid electrolyte by SSE mitigates the safety concerns, as it is free from volatile organic solvents. Solid-state electrolytes are broadly divided into two categories depending on the nature of materials employed, inorganic/ceramic electrolytes, and organic/polymer electrolytes. Although inorganic electrolytes show better ionic conduction (~10^{-3} to 10^{-4} S cm^{-1}), their rigidness and relatively larger thickness contribute to interfacial resistance (>103 Ω) [59]. On the contrary, as polymers are flexible and are thermally and chemically stable, with the possibilities for modifications such as blending, filler incorporation, grafting, and copolymerization, the potential offered by these materials is huge [60–62]. Polymer electrolytes can be classified as solid polymer electrolytes and gel polymer electrolytes (GPE) or quasi-state electrolytes depending on the presence of liquid electrolytes [5].

The main drawback of solid electrolytes is the high interfacial instability between electrodes and electrolytes, which hampers the electrochemical performance. Poor interface catalyzes the growth of dendrites by the non-uniform deposition of Li and increases the ionic resistance, which negatively contributes to the cyclability and overall working of a battery [63]. GPE, on the contrary, functions in between liquid electrolyte and solid electrolyte, where the liquid components are encapsulated in a polymer host matrix. Furthermore, to enhance ionic conduction and Li transference number, the filler incorporation approach showed promising results in LAB [64]. Chamaani et al. [65] showed that incorporating 1 wt.% glass micro-filler into a polymer matrix showed enhancement in ionic conduction and Li^+ transference number, with improved electrochemical properties for LAB. The cell showed enhanced cycling performance with a capacity of 500 mAh g^{-1}.

Yi et al. [66] reported LAB with GPE based on blending and nano-filler incorporation approach. Polypropylene-supported poly (methyl methacrylate) (PMMA) blended with poly (styrene) (PS) and doped with fumed nano silica (SiO_2) was used as a polymer host. A porous polymer membrane was fabricated by immersing the commercially available Celgard separator in tetrahydrofuran solution containing PMMA, PS, polyethylene glycol (which acted as a pore generator), and SiO_2, followed by a phase inversion technique. GPE was obtained by immersing the polymer host in 1 M LiTFSI in tetraglyme. Coin cells fabricated with GPE showed stable cycling of 100 cycles in ambient air and at room temperature. Zou et al. [67] reported the applicability of nonwoven fabric-supported aerogel SiO_2-filled thermoplastic polyurethane matrix with 1 M LiTFSI tetraglyme as a potential candidate for LAB. The prepared GPE showed a room temperature ionic conduction of 1.02×10^{-3} S cm^{-1} with enhanced electrochemical properties. GPE-based Li-O_2 cells showed superior cycling stability to liquid electrolytes with 250 cycles (1,000h) (at room temperature). At 50°C, with commercial carbon nanotube (CNT) as cathode, the cell showed stability for 145 cycles (580h). Furthermore, the GPE showed high mechanical flexibility, flame retardancy, and suppressed dendrites, showing its durability. In another work, Liu et al. [68] fabricated LAB with PMMA–SiO_2 composite GPE over a separator using the phase inversion method and subsequently gelation. Interaction with LiTFSI in tetraglyme and GPE contributed to enhancement in electrolyte uptake by 120% with lower leakage capacity. The system showed improvement in the electrochemical window to 4.9 V and a cation transfer number of 0.54.

To increase cationic transference number and improve ionic conduction in GPE, Kufian et al. [69] employed lithium salt having a bigger anion size, LiTFSI (having an anionic radius of 3.3 Å) with PMMA as the polymer host, and tetraglyme as the electrolyte solvent. At 25 wt% of LiTFSI in PMMA, the sample showed a transference number of 0.47 and ionic conduction of 2.8×10^{-3} S cm^{-1}. LAB fabricated using GPE showed the formation of Li_2O_2 and showed good electrode–electrolyte interfacial stability. Zhao et al. [70] fabricated GPE consisting of PMMA for LAB with IL electrolyte, which showed ionic conduction of 0.92×10^{-3} S cm^{-1}, and a wide electrochemical window of 4.7 V. The usage of GPE protected the anode from O_2 attack; hence, the cell showed a longer cycle life of over 120 cycles with a lower charge overpotential of <2 V.

Wang et al. [71] rectified the issue of the poor interface by a nano-polishing approach, wherein they employed a $Li_{1.5}Al_{1.5}Ge_{1.5}P_3O_{12}$ (UFSLAGP) electrolyte with an ultra-fine surface. Nano-polishing the pristine electrolyte, abbreviated as LAGP, ensured a smooth electrolyte interface, which prevented dendritic growth and ensured Li^+ conduction by the increment in surface density of electrolyte, which was achieved by filling the voids in the electrolyte surface (Figure 10.8). Micro-sized grains present in the pristine LAGP electrolyte on polishing showed reduced surface roughness (30 nm), with a large decline in surface voids. Therefore, with a smooth interface, LABs fabricated with such electrolyte showed longer cycling life (27 cycles, at room temperature) at 400mA g^{-1} with a capacity limitation of 1,000 mAh g^{-1}, in comparison with LAB employing pristine electrolyte LAGP.

Chi et al. [72] prepared solid-state LABs with Li^+ exchanged zeolite membrane as the SSE. As zeolite possesses chemical stability, electrolyte degradation and the issues with the anode or its environment were prevented. Pan et al. [73] developed a flexible composite SSE membrane containing a $Li_{6.6}La_3Zr_{1.6}Ta_{0.4}O_{12}$ (LLZTO) garnet as a matrix with a supramolecular filler crown ether Li salt complex through a tap casting/phase inversion process and a simple solution infiltration method for

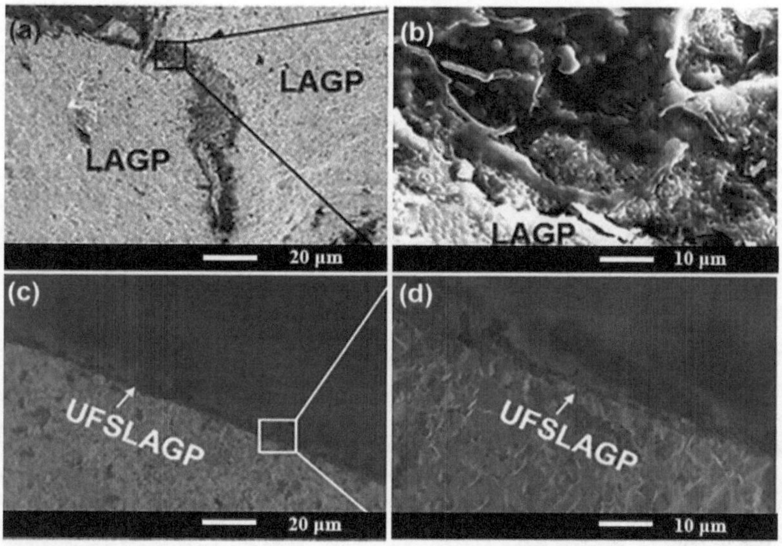

FIGURE 10.8 Cross-sectional scanning electron microscopic images of the cycled solid-state electrolyte. (a and b) Cycled LAGP with protrusions. (c and d) Cycled UFSLAGP. Adapted and reproduced with permission from Ref. [71]. Copyright © 2018 The Royal Society of Chemistry.

LAB. Introduction of filler-enhanced Li^+ conductivity (6.92×10^{-4} S cm^{-1} at 20°C and 2.85×10^{-3} S cm^{-1} at 80°C) and interfacial compatibility between the anode and the electrolyte. In addition, redox mediators in the filler contributed to reducing the discharge–charge overpotential (~600 mV), thereby enhancing the cycling stability of the solid LAB. In the view of creating high-performance solid-state LABs, Wang et al. [74] developed a porous plastic crystal electrolyte, which ensured high ionic conduction of 3.87×10^{-4} S cm^{-1}, with a smooth flow of O_2. The electrolyte was prepared using a thermally induced phase separation technique, using succinonitrile, and contained LiTFSI as Li salt. Enhanced ionic conduction with lower interfacial resistance was contributed by the lower Young's modulus value (than polyethylene oxide (PEO)), with better adhesion properties (Figure 10.9).

10.5 POTENTIAL OF LABS IN EVS: PAST, PRESENT, AND FUTURE

Researchers worldwide view the high theoretical specific energy of the LABs over conventional LIBs as a promising path to a transformational energy storage system for EVs in terms of cost, safety, lifespan, and performance. LABs seem to have tremendous potential to completely wipe out current LIB technology because of their ability to utilize O_2 as a free natural resource. Every research finding related to rechargeable LAB is of great significance because it advances the reduction of risks associated with large commercial rechargeable battery packs in HEVs. In the past decade, several efforts have been made to optimize the electrochemical properties of different components of batteries, and various researchers have observed that the performance of existing LIBs is heavily dependent on electrode chemistry and battery design. New chemical designs involving cathodes inhaling outside air during charge/discharge cycles, as well as its related chemical phenomena, are still ongoing. Non-aqueous LABs rule over solid-state and aqueous batteries because they minimize unwanted discharge products and parasitic electrochemical reactions on the anode–electrolyte and electrolyte–cathode interfaces. In this chapter, an attempt has been made to analyze and present a brief overview of the current EV technology. Finally, trends and challenges of future developments of LIBs and LABs have been summarized, providing a clear picture of EVs and the areas in need of future research. Pursuing fundamental research to overcome these barriers provides new directions for automotive battery technology.

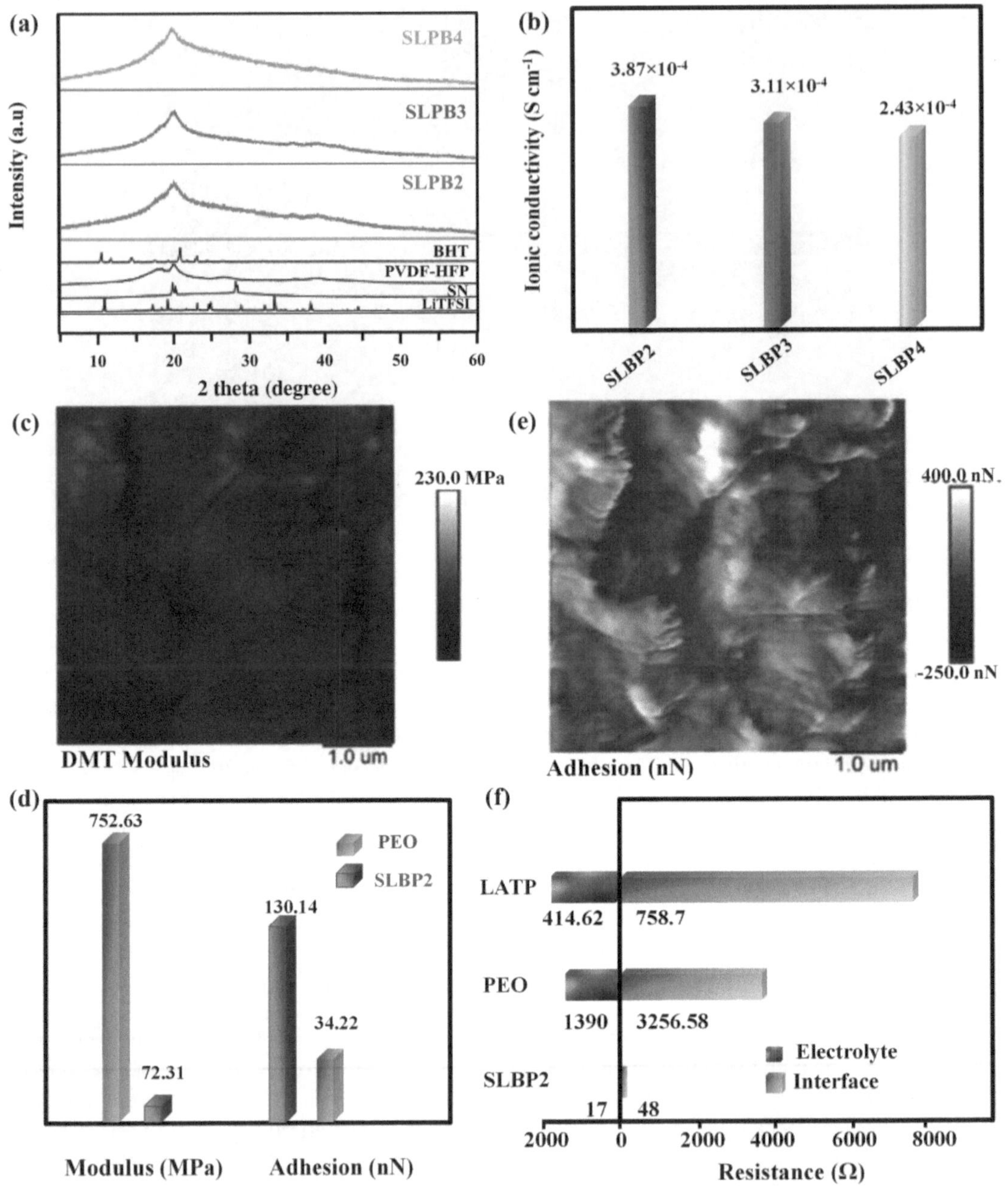

FIGURE 10.9 Characterization and properties of succinonitrile-based porous plastic crystal electrolyte. (a) Overlaid X-ray diffraction pattern of electrolyte, (b) ionic conductivity of the electrolyte, (c) Young's modulus of the electrolyte, (d) adhesion mapping, (e) comparison of Young's modulus and adhesion of succinonitrile-based crystal electrolyte and polyethylene oxide (PEO), and (f) a comparative resistance of electrolytes at the interface for various electrolytes. Adapted and reproduced with permission from Ref. [74]. Copyright © 2020 Wiley-VCH.

10.6 CONCLUSION

EVs significantly impact the environment, power systems, and the current economy, thus gaining popularity globally. As the present trends suggest, current EV technology uses LIBs as the main energy source, yet not capable enough to provide the amount of energy required in terms of cost, safety, lifespan, battery size, and performance. LABs are believed to have the capacity to hold up to five times more energy than the same LIBs powering today's phones, laptops, and EVs. Thus, LABs could be seen as the future direction where the EV manufacturers will head toward. Besides the attractive electrochemistry, LAB needs optimization to reduce charge–discharge overpotential, thereby increasing round-trip efficiency and enhancing cycling stability and rate capability. Sound momentum can be gained only by addressing the bottleneck challenges on stabilizing anode–electrolyte interface, rational design of air cathodes, and electrolyte engineering. This chapter mainly focused on reviewing all the useful data available on battery configurations for future EV technology developments. In addition, developments in electrolyte engineering have been discussed. There are so many open-ended questions related to flammability or explosion associated with lithium anode embedded in LABs that are still unsolved, which certainly motivates researchers to take future steps to scale up LABs at the level of small cells to bigger battery packs and test them in EVs to operate in a natural air environment.

REFERENCES

1. Sungjemmenla J, Soni CB, Vineeth SK, Kumar V (2022) Exploration of the unique structural chemistry of sulfur cathode for high-energy rechargeable beyond-Li batteries. *Adv Energy Sustain Res* 3:2100157. https://doi.org/10.1002/aesr.202100157
2. Yong JY, Ramachandaramurthy VK, Tan KM, Mithulananthan N (2015) A review on the state-of-the-art technologies of electric vehicle, its impacts and prospects. *Renew Sustain Energy Rev* 49:365–385. https://doi.org/10.1016/j.rser.2015.04.130
3. Gerssen-Gondelach SJ, Faaij APC (2012) Performance of batteries for electric vehicles on short and longer term. *J Power Sources* 212:111–129. https://doi.org/10.1016/j.jpowsour.2012.03.085
4. Chen W, Liang J, Yang Z, Li G (2019) A review of lithium-ion battery for electric vehicle applications and beyond. *Energy Procedia* 158:4363–4368. https://doi.org/10.1016/j.egypro.2019.01.783
5. Jishnu NS, Vineeth SK, Das A, et al (2021) Electrospun PVdF and PVdF-co-HFP-based blend polymer electrolytes for lithium ion batteries. In: Balakrishnan NTM, Prasanth R (eds) *Electrospinning for Advanced Energy Storage Applications*, 1st ed. Springer, Singapore, pp. 201–234
6. Sungjemmenla J, Soni CB, Vineeth SK, Kumar V (2021) Unveiling the physiochemical aspects of the matrix in improving sulfur-loading for room-temperature sodium-sulfur batteries. *Mater Adv* 2:4165–4189. https://doi.org/10.1039/D1MA00247C
7. Durmus YE, Zhang H, Baakes F, et al (2020) Side by side battery technologies with lithium-ion based batteries. *Adv Energy Mater* 10:2000089. https://doi.org/10.1002/aenm.202000089
8. Zhang X, Wang XG, Xie Z, Zhou Z (2016) Recent progress in rechargeable alkali metal-air batteries. *Green Energy Environ* 1:4–17. https://doi.org/10.1016/j.gee.2016.04.004
9. Gallagher KG, Goebel S, Greszler T, et al (2014) Quantifying the promise of lithium-air batteries for electric vehicles. *Energy Environ Sci* 7:1555–1563. https://doi.org/10.1039/c3ee43870h
10. Li Y, Lu J (2017) Metal-air batteries: will they be the future electrochemical energy storage device of choice? *ACS Energy Lett* 2:1370–1377. https://doi.org/10.1021/acsenergylett.7b00119
11. Bruce PG, Freunberger SA, Hardwick LJ, Tarascon J-M (2012) Li-O_2 and Li-S batteries with high energy storage. *Nat Mater* 11:19–29. https://doi.org/10.1038/nmat3191
12. Li C-L, Huang G, Yu Y, et al (2022) A low-volatile and durable deep eutectic electrolyte for high-performance lithium-oxygen battery. *J Am Chem Soc* 144:5827–5833. https://doi.org/10.1021/jacs.1c11711
13. Balaish M, Jung J, Kim I, Ein-Eli Y (2020) A critical review on functionalization of air-cathodes for nonaqueous Li-O_2 batteries. *Adv Funct Mater* 30:1808303. https://doi.org/10.1002/adfm.201808303
14. Xu K, von Cresce A (2011) Interfacing electrolytes with electrodes in Li ion batteries. *J Mater Chem* 21:9849. https://doi.org/10.1039/c0jm04309e
15. Placke T, Kloepsch R, Dühnen S, Winter M (2017) Lithium ion, lithium metal, and alternative rechargeable battery technologies: the odyssey for high energy density. *J Solid State Electrochem* 21:1939–1964. https://doi.org/10.1007/s10008-017-3610-7
16. Kwak W-J, Rosy R, Sharon D, et al (2020) Lithium-oxygen batteries and related systems: potential, status, and future. *Chem Rev* 120:6626–6683. https://doi.org/10.1021/acs.chemrev.9b00609
17. Wen Z, Shen C, Lu Y (2015) Air electrode for the lithium-air batteries: materials and structure designs. *Chempluschem* 80:270–287. https://doi.org/10.1002/cplu.201402104
18. Kraytsberg A, Ein-Eli Y (2011) Review on Li-air batteries-opportunities, limitations and perspective. *J Power Sources* 196:886–893. https://doi.org/10.1016/j.jpowsour.2010.09.031

19. Chang Z, Xu J, Zhang X (2017) Recent progress in electrocatalyst for Li-O_2 batteries. *Adv Energy Mater* 7:1700875. https://doi.org/10.1002/aenm.201700875
20. Littauer EL, Tsai KC (1976) Anodic behavior of lithium in aqueous electrolytes: I. Transient passivation. *J Electrochem Soc* 123:771–776. https://doi.org/10.1149/1.2132931
21. Abraham KM, Jiang Z (1996) A polymer electrolyte-based rechargeable lithium/oxygen battery. *J Electrochem Soc* 143:1–5. https://doi.org/10.1149/1.1836378
22. Read J (2002) Characterization of the lithium/oxygen organic electrolyte battery. *J Electrochem Soc* 149:A1190. https://doi.org/10.1149/1.1498256
23. Ogasawara T, Débart A, Holzapfel M, et al (2006) Rechargeable Li_2O_2 electrode for lithium batteries. *J Am Chem Soc* 128:1390–1393. https://doi.org/10.1021/ja056811q
24. Tan P, Jiang HR, Zhu XB, et al (2017) Advances and challenges in lithium-air batteries. *Appl Energy* 204:780–806. https://doi.org/10.1016/j.apenergy.2017.07.054
25. Girishkumar G, McCloskey B, Luntz AC, et al (2010) Lithium–air battery: promise and challenges. *J Phys Chem Lett* 1:2193–2203. https://doi.org/10.1021/jz1005384
26. Chen Y, Freunberger SA, Peng Z, et al (2013) Charging a Li-O_2 battery using a redox mediator. *Nat Chem* 5:489–494. https://doi.org/10.1038/nchem.1646
27. Kim ST, Choi N-S, Park S, Cho J (2015) Optimization of carbon- and binder-free Au nanoparticle-coated Ni nanowire electrodes for lithium-oxygen batteries. *Adv Energy Mater* 5:1401030. https://doi.org/10.1002/aenm.201401030
28. Xu J-J, Xu D, Wang Z-L, et al (2013) Synthesis of perovskite-based porous $La_{0.75}Sr_{0.25}$ MnO_3 nanotubes as a highly efficient electrocatalyst for rechargeable lithium-oxygen batteries. *Angew Chemie – Int Ed* 52:3887–3890. https://doi.org/10.1002/anie.201210057
29. Sun B, Huang X, Chen S, et al (2014) Porous graphene nanoarchitectures: an efficient catalyst for low charge-overpotential, long life, and high capacity lithium-oxygen batteries. *Nano Lett* 14:3145–3152. https://doi.org/10.1021/nl500397y
30. Ottakam Thotiyl MM, Freunberger SA, Peng Z, et al (2013) A stable cathode for the aprotic Li-O_2 battery. *Nat Mater* 12:1050–1056. https://doi.org/10.1038/nmat3737
31. Xie J, Yao X, Cheng Q, et al (2015) Three dimensionally ordered mesoporous carbon as a stable, high-performance Li-O_2 battery cathode. *Angew Chemie – Int Ed* 54:4299–4303. https://doi.org/10.1002/anie.201410786
32. Li F, Chen Y, Tang D-M, et al (2014) Performance-improved Li-O_2 battery with Ru nanoparticles supported on binder-free multi-walled carbon nanotube paper as cathode. *Energy Environ Sci* 7:1648–1652. https://doi.org/10.1039/C3EE44043E
33. Lim H-D, Song H, Kim J, et al (2014) Superior rechargeability and efficiency of lithium-oxygen batteries: hierarchical air electrode architecture combined with a soluble catalyst. *Angew Chemie* 126:4007–4012. https://doi.org/10.1002/ange.201400711
34. Kalubarme RS, Jadhav HS, Ngo DT, et al (2015) Simple synthesis of highly catalytic carbon-free $MnCo_2O_4$@Ni as an oxygen electrode for rechargeable Li-O_2 batteries with long-term stability. *Sci Rep* 5:13266. https://doi.org/10.1038/srep13266
35. Shang C, Dong S, Hu P, et al (2015) Compatible interface design of CoO-based Li-O_2 battery cathodes with long-cycling stability. *Sci Rep* 5:8335. https://doi.org/10.1038/srep08335
36. Liu Q-C, Xu J-J, Xu D, Zhang X-B (2015) Flexible lithium-oxygen battery based on a recoverable cathode. *Nat Commun* 6:7892. https://doi.org/10.1038/ncomms8892
37. Balaish M, Kraytsberg A, Ein-Eli Y (2014) A critical review on lithium-air battery electrolytes. *Phys Chem Chem Phys* 16:2801. https://doi.org/10.1039/c3cp54165g
38. Lai J, Xing Y, Chen N, et al (2020) Electrolytes for rechargeable lithium-air batteries. *Angew Chemie – Int Ed* 59:2974–2997. https://doi.org/10.1002/anie.201903459
39. Rahman MA, Wang X, Wen C (2014) A review of high energy density lithium-air battery technology. *J Appl Electrochem* 44:5–22. https://doi.org/10.1007/s10800-013-0620-8
40. Yao X, Dong Q, Cheng Q, Wang D (2016) Why do lithium-oxygen batteries fail: parasitic chemical reactions and their synergistic effect. *Angew Chemie – Int Ed* 55:11344–11353. https://doi.org/10.1002/anie.201601783
41. Peng Z, Freunberger SA, Hardwick LJ, et al (2011) Oxygen reactions in a non-aqueous Li+ electrolyte. *Angew Chemie – Int Ed* 50:6351–6355. https://doi.org/10.1002/anie.201100879
42. Leskes M, Moore AJ, Goward GR, Grey CP (2013) Monitoring the electrochemical processes in the lithium-air battery by solid state NMR spectroscopy. *J Phys Chem C* 117:26929–26939. https://doi.org/10.1021/jp410429k

43. Safanama D, Adams S (2017) High efficiency aqueous and hybrid lithium-air batteries enabled by $Li_{1.5}A_{l0.5}Ge_{1.5}(PO_4)_3$ ceramic anode-protecting membranes. *J Power Sources* 340:294–301. https://doi.org/10.1016/j.jpowsour.2016.11.076
44. Stark JK, Ding Y, Kohl PA (2013) Role of dissolved gas in ionic liquid electrolytes for secondary lithium metal batteries. *J Phys Chem C* 117:4980–4985. https://doi.org/10.1021/jp4001303
45. Wang Y, Zhou H (2010) A lithium-air battery with a potential to continuously reduce O_2 from air for delivering energy. *J Power Sources* 195:358–361. https://doi.org/10.1016/j.jpowsour.2009.06.109
46. He P, Zhang T, Jiang J, Zhou H (2016) Lithium-air batteries with hybrid electrolytes. *J Phys Chem Lett* 7:1267–1280. https://doi.org/10.1021/acs.jpclett.6b00080
47. Peng SH, Chen TH, Lee CH, et al (2020) Optimal cobalt oxide (Co_3O_4): graphene (GR) ratio in Co_3O_4/GR as air cathode catalyst for air-breathing hybrid electrolyte lithium-air battery. *J Power Sources* 471:228373. https://doi.org/10.1016/j.jpowsour.2020.228373
48. Vineeth SK, Sreeram P, Vlad A, Joy R, Raghavan P, Pullanchiyodan A (2023) Polymer blend nanocomposite electrolytes for advanced energy storage applications. In *Polymer blend nanocomposites for energy storage applications* (pp. 203–238). Elsevier. https://doi.org/10.1016/B978-0-323-99549-8.00016-9
49. Vineeth SK, Tebyetekerwa M, Liu H, Soni, CB, Zhao XS, Kumar V (2022) Progress in the development of solid-state electrolytes for reversible room-temperature sodium–sulfur batteries. *Mater Adv*, 3(16):6415–6440. https://doi.org/10.1039/D2MA00428C
50. Frömling T, Kunze M, Schönhoff M, et al (2008) Enhanced lithium transference numbers in ionic liquid electrolytes. *J Phys Chem B* 112:12985–12990. https://doi.org/10.1021/jp804097j
51. Allen CJ, Hwang J, Mukerjee S, Abraham KM (2011) Oxygen electrode processes in ionic liquids for the Li air battery. *ECS Meet Abstr* MA2011-02:1331–1331. https://doi.org/10.1149/ma2011-02/17/1331
52. McCloskey BD, Bethune DS, Shelby RM, et al (2012) Limitations in Rechargeability of Li-O_2 batteries and possible origins. *J Phys Chem Lett* 3:3043–3047. https://doi.org/10.1021/jz301359t
53. Grande L, Paillard E, Kim GT, et al (2014) Ionic liquid electrolytes for Li-air batteries: lithium metal cycling. *Int J Mol Sci* 15:8122–8137. https://doi.org/10.3390/ijms15058122
54. Kuboki T, Okuyama T, Ohsaki T, Takami N (2005) Lithium-air batteries using hydrophobic room temperature ionic liquid electrolyte. *J Power Sources* 146:766–769. https://doi.org/10.1016/j.jpowsour.2005.03.082
55. Nakamoto H, Suzuki Y, Shiotsuki T, et al (2013) Ether-functionalized ionic liquid electrolytes for lithium-air batteries. *J Power Sources* 243:19–23. https://doi.org/10.1016/j.jpowsour.2013.05.147
56. Elia GA, Hassoun J, Kwak WJ, et al (2014) An advanced lithium-air battery exploiting an ionic liquid-based electrolyte. *Nano Lett* 14:6572–6577. https://doi.org/10.1021/nl5031985
57. Bai J, Lu H, Cao Y, et al (2017) A novel ionic liquid polymer electrolyte for quasi-solid state lithium air batteries. *RSC Adv* 7:30603–30609. https://doi.org/10.1039/c7ra05035f
58. Olivares-Marín M, Sorrentino A, Pereiro E, Tonti D (2017) Discharge products of ionic liquid-based Li-O_2 batteries observed by energy dependent soft x-ray transmission microscopy. *J Power Sources* 359:234–241. https://doi.org/10.1016/j.jpowsour.2017.05.039
59. Vineeth SK, Soni CB, Sanjaykumar C, Yamauchi Y, Han M, Kumar V (2023) A quasi-solid state polymer electrolyte for high-rate and long-life sodium-metal batteries. *J Energy Storage*, 73:108780. https://doi.org/10.1016/j.est.2023.108780
60. Gadhave RV, Vineeth SK, Mahanwar PA, Gadekar PT (2021) Effect of addition of boric acid on thermo-mechanical properties of microcrystalline cellulose/polyvinyl alcohol blend and applicability as wood adhesive. *J Adhes Sci Technol* 35:1072–1086. https://doi.org/10.1080/01694243.2020.1832775
61. Gadhave RV, Vineeth SK, Mahanwar PA, Gadekar PT (2020) Effect of boric acid on poly vinyl alcohol-tannin blend and its application as water-based wood adhesive. *Des Monomers Polym* 23:188–196. https://doi.org/10.1080/15685551.2020.1826124
62. Dhawale PV, Vineeth SK, Gadhave RV, et al (2022) Tannin as a renewable raw material for adhesive applications: a review. *Mater Adv* 3:3365–3388. https://doi.org/10.1039/D1MA00841B
63. Vineeth SK, Soni CB, Sun Y, et al (2022) Implications of Na-ion solvation on Na anode-electrolyte interphase. *Trends Chem* 4:48–59. https://doi.org/10.1016/j.trechm.2021.11.002
64. Liu T, Chang Z, Yin Y, et al (2018) The PVDF-HFP gel polymer electrolyte for Li-O_2 battery. *Solid State Ionics* 318:88–94. https://doi.org/10.1016/j.ssi.2017.08.001
65. Chamaani A, Safa M, Chawla N, El-Zahab B (2017) Composite gel polymer electrolyte for improved cyclability in lithium-oxygen batteries. *ACS Appl Mater Interfaces* 9:33819–33826. https://doi.org/10.1021/acsami.7b08448
66. Yi J, Liu X, Guo S, et al (2015) Novel stable gel polymer electrolyte: toward a high safety and long life Li-air battery. *ACS Appl Mater Interfaces* 7:23798–23804. https://doi.org/10.1021/acsami.5b08462

67. Zou X, Lu Q, Zhong Y, et al (2018) Flexible, flame-resistant, and dendrite-impermeable gel-polymer electrolyte for Li-O_2/air batteries workable under hurdle conditions. *Small* 14:1801798. https://doi.org/10.1002/smll.201801798
68. Liu X, Xin X, Shen L, et al (2021) Poly(methyl methacrylate)-based gel polymer electrolyte for high-performance solid state Li-O_2 battery with enhanced cycling stability. *ACS Appl Energy Mater* 4:3975–3982. https://doi.org/10.1021/acsaem.1c00344
69. Kufian MZ, Ramesh S, Arof AK (2021) PMMA-LiTFSI based gel polymer electrolyte for lithium-oxygen cell application. *Opt Mater (Amst)* 120:111418. https://doi.org/10.1016/j.optmat.2021.111418
70. Zhao H, Liu X, Chi Z, et al (2021) Designing a new-type PMMA based gel polymer electrolyte incorporating ionic liquid for lithium oxygen batteries with Ru-based Binder-free cathode. *Appl Surf Sci* 565:150612. https://doi.org/10.1016/j.apsusc.2021.150612
71. Wang S, Wang J, Liu J, et al (2018) Ultra-fine surface solid-state electrolytes for long cycle life all-solid-state lithium-air batteries. *J Mater Chem A* 6:21248–21254. https://doi.org/10.1039/C8TA08095J
72. Chi X, Li M, Di J, et al (2021) A highly stable and flexible zeolite electrolyte solid-state Li-air battery. *Nature* 592:551–557. https://doi.org/10.1038/s41586-021-03410-9
73. Pan X, Sun J, Jin C, et al (2020) A flexible composite electrolyte membrane with ultrahigh LLZTO garnet content for quasi solid state Li-air batteries. *Solid State Ionics* 351:115340. https://doi.org/10.1016/j.ssi.2020.115340
74. Wang J, Huang G, Chen K, Zhang X (2020) An adjustable-porosity plastic crystal electrolyte enables high-performance all-solid-state lithium-oxygen batteries. *Angew Chemie – Int Ed* 59:9382–9387. https://doi.org/10.1002/anie.202002309

11 Lithium-Air Batteries

State-of-the-Art Developments in Anode, Cathode, and Electrocatalyst

S. K. Vineeth, Akhila Das, Abhilash Pullanchiyodan, Vijayamohanan Pillai, Nutan Gupta, and Prasanth Raghavan

11.1 INTRODUCTION

The development of rechargeable battery technology has revolutionized portable electronics, electric vehicles (EVs), and advanced smart devices [1]. Lithium-ion battery (LIB) technology is the most explored and commercialized system, but it cannot meet the demand of proliferating population energy demands and the requirement of EVs; hence, there are formidable challenges associated with LIB. Alternative energy storage systems which gained acceptance are the high-energy-density batteries, in which metal-air systems are one of the prime candidates. In metal-air batteries (MABs), lithium-air battery (LAB) relishes high values of theoretical specific energy and practical specific energy [2]. A general architecture of LAB consists of metallic lithium (Li) as the anode and a porous carbonaceous air cathode with an electrocatalyst catalyst and electrolyte that transport lithium ion (Li^+). For LAB, based on the product mass of Li_2O_2, the theoretical specific energy can be calculated as 3,500 Wh kg^{-1}, and for gasoline it can be only 700 Wh kg^{-1} and the value is ten times higher than LIB [3,4].

The developments are hindered by the inherent sluggish kinetics of oxygen reduction reaction (ORR) and oxygen evolution reaction (OER), the insulating behavior of discharge products at the cathode, and dendrite growth during cycling. Moreover, interfacial instability and anode passivation, which result in voltage hysteresis and reduction in the cycling life, negatively contribute to the wide acceptance of MABs. As materials used as electrodes, electrolytes, air cathodes, and separators play a determinant role in the efficiency of cells, judicious strategies for the selection of materials, engineering of the interface, and tailoring properties are utmost important [5,6]. Moreover, unveiling the structure–property relation elucidates the electrochemical properties of a cell. Combining theoretical predictions and simulations and relating them with experimental observations further add insights into the cell performance [7].

As the judicious selection of electrode, electrolyte, and catalyst decides the electrochemical properties of LAB, it is important to explore the material properties which can catalyze the performance of the system. Therefore, to unveil the electrochemical properties of LAB and to minimize the interfacial instabilities, the chapter provides insights into the materials used as electrodes and air cathode catalysts. A detailed discussion on various works reported in engineering the materials and interfaces has been highlighted. Strategies to stabilize anode passivation, mitigate corrosion of anodes, and prevent dendrite growth in anodes have been discussed in the initial part. Bifurcation of LABs based on the type of electrolytes used and the development in electrolyte modification has been overviewed. In the final section tailoring cathodes and developments of bifunctional cathode catalysts and their electrochemical performances have been evaluated. Overall, the chapter is a complete collection of recent works reported, focusing on the material engineering for LAB.

DOI: 10.1201/9781003310174-11

11.2 DEVELOPMENTS IN ANODES FOR LABS

Anodes in LABs have always been a popular topic of research for enhancing/maximizing the electrochemical performance [8]. Since anodes are basically made of Li metal, various problems have been observed associated with dendrite growth and unstable solid–electrolyte–interface (SEI) film formation, which hampers the electrochemical properties [9,10]. Quasi-reversible electrochemical reactions at the electrode-electrolyte interface catalyze the formation of the SEI layer [11]. The nature and thickness of SEI decide the ionic migration and reversibility of cells [12]. Furthermore, the with the completion of each cycles, intermediates are formed. Parasitic reaction between the anode and intermediates along with water and oxygen (O_2) catalyzes cracks, corrosion, and pulverization at the anode, thereby deteriorating the performance of a cell [13]. Approaches to rectify the challenges include the creation of alloy interface, electrolyte engineering, separator modification, and the creation of artificial film which protects anode [14–16].

Shui et al. [17] found that the charging and discharging cycle of a non-aqueous LAB was accompanied by the formation and accumulation of a lithium hydroxide (LiOH) layer over the anode. The chemical reaction of Li with H_2O catalyzed by the electrolyte decomposition was the main cause of the formation of LiOH. However, through the pores of the LiOH layer, Li^+ transport occurred until the Li was completely consumed, which is shown by its degradation. Younesi et al. [18] studied the chemistry and morphology of the SEI layer over the anode. It was seen that the formation of SEI was due to the decomposition of $LiPF_6$ and the propylene carbonate solvent. Moreover, the surface of Li contained a decomposed Kynar binder (degraded on the cathode) [18]. The anode modification approach has been found to have a profound effect on the electrochemical performance of a battery [19]. Kang et al. [20] found that the usage of anodized porous alumina as the model nanoporous separator improved the cycle efficiency of Li electrodes in lithium–oxygen cells. The electrolyte used was 1 M lithium bis(trifluoromethane)sulfonimide and $Li(CF_3SO_2)_2N$ (LiTFSI) in dimethoxyethane. The surface roughening of the metallic Li electrode decreased, and uniform nanopores of anodized porous alumina facilitated homogeneous plating/stripping of the ions. Lee et al. [21] reported that enhancement in the cycling stability of LAB when a composite protective layer (CPL) containing Al_2O_3 and polyvinylidene fluoride-hexafluoropropylene (PVdF-HFP) was used for Li metal anode. Protective layer coating mitigated the issue of anodic surface decomposition, which occurs by the decomposition of electrolyte molecules over the Li metal anode. The LAB with CPL-coated anode showed three times the discharge capacity at the 80th cycle in comparison to a cell without the coating. Morphological images with the performance properties of the LAB are shown in Figure 11.1.

Guo et al. [22] prepared a uniform SEI film over a lithiated aluminum carbon composite electrode and then coupled it with an O_2 catalytic electrode to form a rechargeable LAB. The work compared the electrochemical performance of the prepared anode with a metallic Li anode in a dry and humid atmosphere as the cathode. Lithiated aluminum carbon composite electrode outperformed metallic Li anode in a humid atmosphere with a relative humidity of ~40%, whereas the reverse was seen when dry air was the active cathode material.

As the CPL approach showed enhancements in electrochemical properties, Lee et al. [23] fabricated a LAB and used CPL consisting of Al_2O_3/PVdF-HFP over Li anode. The CPL prevented the dendritic growth of Li metal and the formation of a porous mossy Li layer. In the work, the LAB was fabricated with 2,2,6,6,-tetramethylpiperidinyl-1-oxyl (TEMPO) as a redox mediator. The CPL prevented the reaction of TEMPO at the Li metal anode, hence reducing self-discharge and thereby extending cycling. Figure 11.2 is the schematic representation of the work. Cai et al. [24] developed a double-gas-path structure LAB with a porous anode. The effect of porosity on the electrochemical properties showed that the cyclic stability increased with a battery with an enhancement in a specific capacity of 2,510 mAh g^{-1} and a specific energy density of 7,200 Wh kg^{-1}. The porous anode contributed to the easy achievement of a dynamic gas-phase equilibrium, which reduced the dissolution of the electrolyte and O_2, thereby preventing the contact of O_2 at the anode and reducing the dendrite formation.

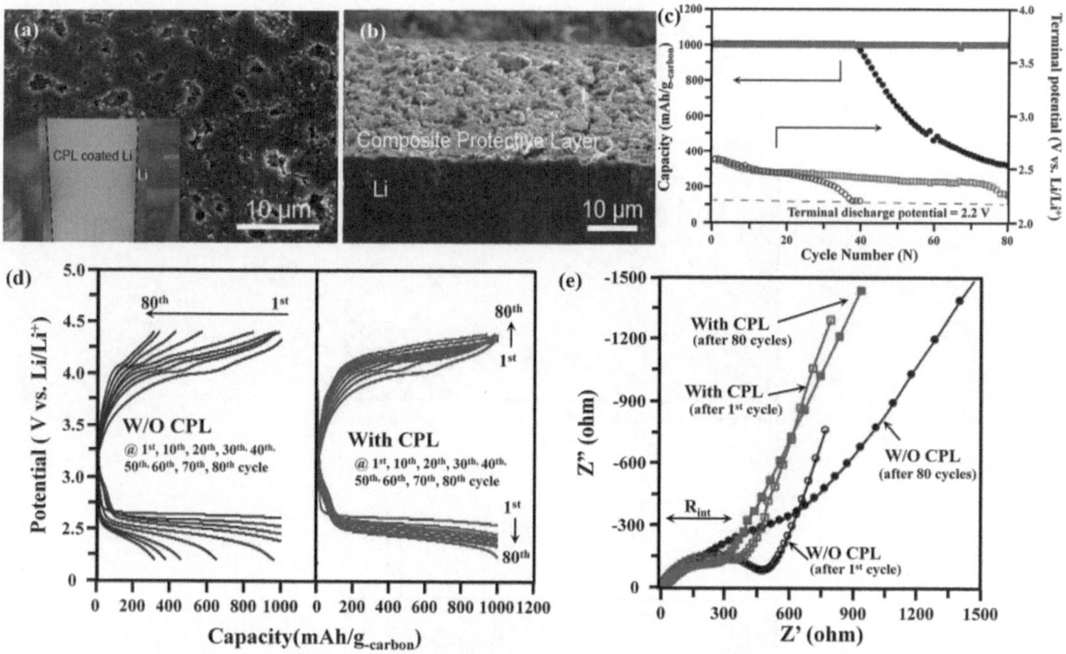

FIGURE 11.1 SEM images of (a) surface and (b) cross-sectional images of Li electrode after CPL coating (the inset shows a photograph of CPL-coated Li electrode). (c) The cycle performance and the potential profiles of Li–O_2 cells fabricated using Li electrode (d) without CPL and (e) with CPL at 0.1 mA cm^{-2}. (f) Impedance spectra of Li–O_2 cells after charging in the 1st cycle and 80th cycle. Adapted and reproduced with permission from Ref. [21]. Copyright © 2014 Elsevier.

In LABs as the Li metal anode comes in contact with a non-aqueous liquid, the parasitic reaction between atmospheric O_2 causes a reduction in cycling stability and coulombic efficiency. Therefore, developing LABs with high-rate capability became one of the primary topics in current battery research. The usage of solid electrolyte separators avoids the dendritic growth of Li by acting as a barrier. Hence researchers identified the importance of protecting the anode to improve the efficiency of LABs [25]. Bao et al. [26] reported LABs based on hybrid electrolytes with liquid lithium biphenyl as an anode and sputtered Ti as a protective layer, preventing side reactions and thereby protecting the anode. The cell showed enhanced cycling performance of up to 120 cycles with a discharge voltage of 2.81 V at the current density of 0.5 mA cm^{-2} and a specific power of 1,638 W kg^{-1} at 6 mA cm^{-2} current density. Li et al. [27] found that using superhydrophobic hydrogel-derived hierarchical porous carbon layer over a Li anode can prevent a reaction with water and dendrite formation during the Li plating/stripping process, ensuring the long running of the LABs. The conversion of O_2/Li_2O_2 affects the decomposition of the byproduct, and electrolyte was still a problem addressed by the work.

Guiding the uniform Li plating/stripping ensures the inhibition of Li dendrites. Hence Guo et al. [28] employed an iodine-containing alloy (IPA) of polymer as a layer multifunctional layer, which inhibited dendrite growth, prevented the anode from oxygen/water attack, and enhanced the electrochemical properties of the LAB. IPA/polymer was prepared by the reaction between zinc iodide and lithium and subsequent addition polymerization with ethyl α-cyanoacrylate. The detailed mechanism and schematic representation are shown in Figure 11.3. The thickness of the coating was measured to be ~10 μm, and the layer showed mechanical flexibility, which mitigated internal stresses on the anode (Figure 11.4). Morphological characterization and elemental mapping showed the presence of Zn, C, N, O, and I and observed the even distribution. The coating showed Young's modulus value of >25 GPa, which inhibited dendritic growth. The presence of an alkyl chain from

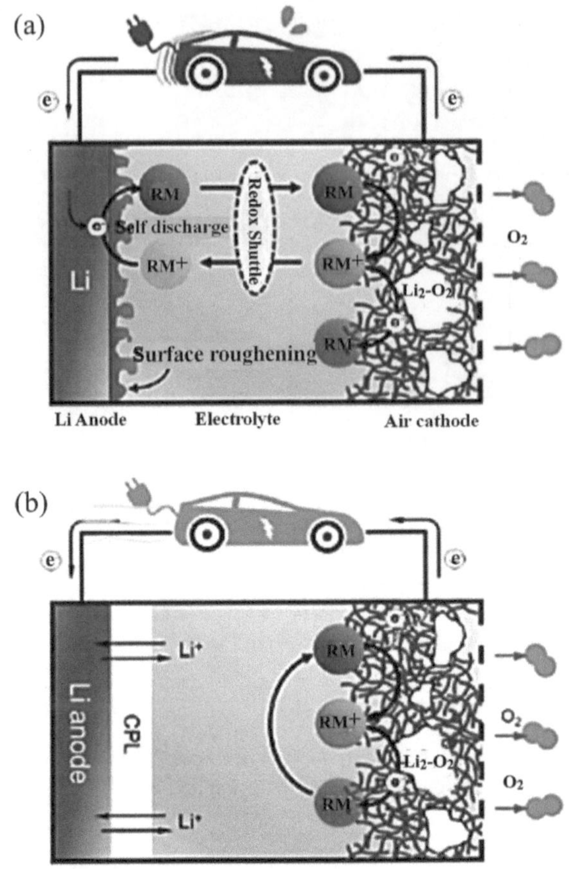

FIGURE 11.2 Schematic illustration of (a) the self-discharge of the redox mediator in the Li–O_2 battery and (b) the CPL-coated Li electrode, which is acting against the reaction between the redox mediator and the Li metal electrode. Adapted and reproduced with permission from Ref. [23]. Copyright © 2016 Wiley-VCH.

Fourier transform infrared spectroscopy (FTIR) suggested the addition polymerization, which was further confirmed by ^{1}H NMR spectra. The introduction of a layer contributed to the reversible decomposition of discharge products at a lower charge voltage. The measurement of electrochemical properties showed that LAB exhibited longer cycling, with lower overpotential and higher ionic mobility, with a suppression of dendritic growth.

11.3 EXPLORATION OF CATHODE ENGINEERING

The cathode in LABs consists of a porous material containing a catalyst, where a polymer binder holds the catalyst with the cathode. As polymer materials have inherent binding properties and have the advantage of solution processability and mechanical and thermal properties, polymer-based binders are preferred [29–31]. Material characteristics and the microstructure of the air cathode determine the round-trip efficiency, capacity cycling life, overpotential, and transport kinetics [32]. During discharge, at the air cathode, ORR occurs, which oxidizes metal. During charging, OER occurs at the air cathode and metal ions get reduced [33]. The challenge faced by the catalyst in an air cathode is the sluggish ORR/OER kinetics [34]. As ORR and OER occur at the air cathode, designing and optimizing the performance of the cathode has a finite role in the overall performance [35]. Discharge product formation and its deposition over an air cathode impede the electrochemical performance, and the nature of discharge products is dependent on the electrolyte used [36].

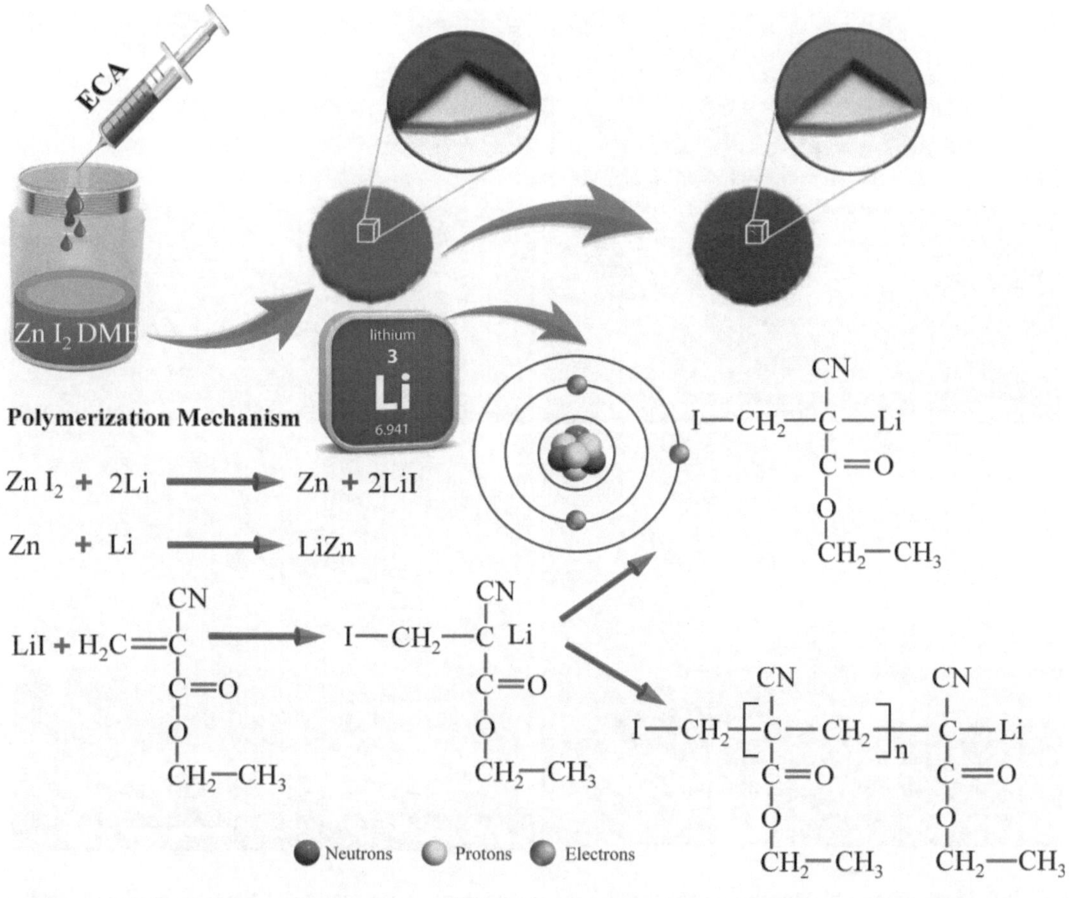

FIGURE 11.3 Preparation of iodine-containing polymer/alloy lithium composite; the inset shows the reaction mechanism. Adapted and reproduced with permission from Ref. [28]. Copyright © 2022 Wiley-VCH.

Read et al. [37] reported a cathode capacity of 2,120 mAh g^{-1} for lithium/O_2 organic electrolyte battery. Ether-based organic electrolytes facilitated the smooth transport of O_2, maintaining a stable electrolyte interface with Li anode, chemical stability toward the products (Li_2O_2 and LiO_2), and oxidative stability. Dobley et al. [38] showed that air cathodes with Mn-based catalysts had specific capacities approaching 4,000 mAh g^{-1}, which is more than 250% greater than the highest specific capacities reported in literature till 2005. Kuboki et al. [39] reported a LAB having a carbon electrode with a high discharge capacity of 5,360 mAh g^{-1} using a hydrophobic IL electrolyte, 1-ethyl-3-methylimidazoliumbis (trifluoromethylsulfonyl)amide. By the usage of IL electrolyte, vaporization of the electrolyte and hydrolysis of the anode were prevented. Beattie et al. [40] reported a LAB catalyst-free air cathode with an average voltage of 2.33 V and a gravimetric capacity of 5,813 mAh g^{-1} of carbon, achieved at a current density of 0.1 mA cm^{-2}. Děbart et al. [41] observed capacity retention of 6.67% per cycle, upon the usage of catalysts Fe_3O_4, CuO, and $CoFe_2O_4$ on the air cathode in a non-aqueous LiO_2 cell. Co_3O_4 showed the best compromise between initial capacity (2,000 mAh g^{-1}) with capacity retention (6.5% per cycle) and the lowest charging voltage of 4 V.

Eswaran et al. [42] observed discharge capacity of 3,100 mAh g^{-1} in a Li–O_2 cell with dual layer carbon electrodes consisting of an active layer and a diffusion layer using a high surface area carbon. Impedance spectra recorded at various stages of cell discharge suggested that the cell resistances increase due to the accumulation of reaction products Li_2O_2 and Li_2O inside the pores of the carbon electrode. Tran et al. [43] investigated the effect of the porosity of carbon in the cathode for a non-aqueous electrolyte-based LAB. The work pointed out that the formation of

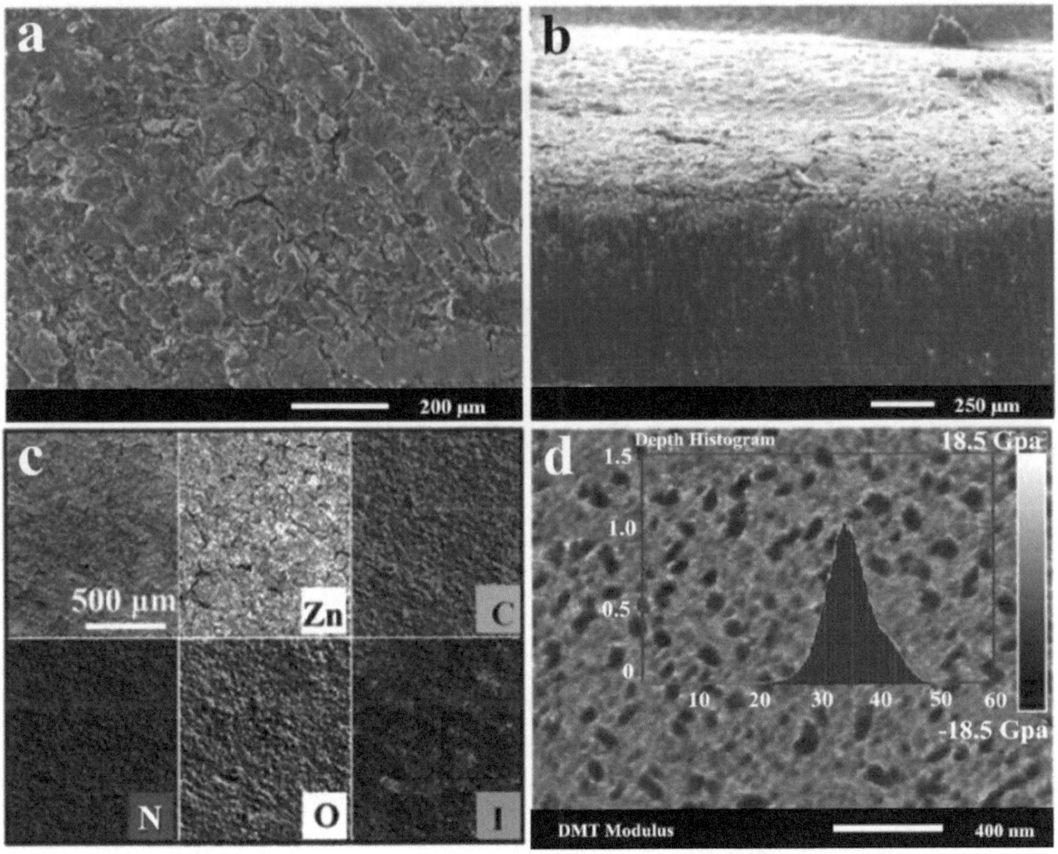

FIGURE 11.4 (a and b) Morphological analysis of surface and cross-section of the coating, (c) elemental mapping images of the film, and (d) Young's modulus mapping of the film, (e) overlaid FTIR spectra for lithium and film-coated anode, (f) ^1H NMR spectra, (h) XPS spectra, and (i) overlaid X-ray diffraction patterns of bare Li electrodes and coated electrodes. Adapted and reproduced with permission from Ref. [28]. Copyright © 2022 Wiley-VCH.

lithium oxides during ORR could block pores, thereby preventing them for electrochemical reactions. Moreover, these oxides accumulate, and density builds up and prevents mass transfer, hence pointing out the fact that porosity plays an important role in the performance of LAB. Xiao et al. [44] found that area-specific capacity is an important parameter to optimize the performance of the air electrode in LAB. Area-specific capacity is the product of specific capacity per unit weight of the carbon source mAh g^{-1} and the carbon locating per unit area g cm^{-2} and is expressed in mAh cm^{-2}. The work showed that the porosity of the cathode and its uniformity has profound effects on the electrochemical performances of the LAB. Optimization of the electrode porosity/thickness and electrolyte amount on LAB operated in a dry environment showed a high capacity of 1,756 mAh g^{-1} carbon corresponding to an area-specific capacity of 26.5 mAh cm^{-2}.

Zhang et al. [45] developed LAB which delivered a high energy density of 779 Wh kg^{-1}, which is twice that of the conventional graphite/$LiCoO_2$ cell. The anode used was a water-stable multilayer Li metal, acetic acid–water electrolyte, and a carbon sheet (320 mm thick) cathode containing platinum (1–4 nm) as the catalyst. The polymer electrolyte used was poly(ethylene oxide) with LiTFSI, and the cell retained a discharge–charge capacity of 250 mAh g^{-1}. Lu et al. [46] showed the bifunctional catalytic efficiency of platinum–gold nanoparticles, on the enhancement of the kinetics of the ORR and OER in rechargeable Li–O_2 cells. Compared to carbon cathode and gold/carbon cathode, platinum–gold nanoparticle/carbon electrode showed higher discharge voltage during ORR. The usage of hollandite $Ag_2Mn_8O_{16}$ and its electrochemical catalytic properties for the application in LAB was studied by Zhang et al. [47]. A composite air electrode made with $Ag_2Mn_8O_{16}$/single-wall carbon nanotube/carbon nanofiber (CNF) was studied using galvanostatic discharge method. The catalyst enhances the electrode reaction rate and decreases electrochemical polarization with a reduction in discharge overvoltage and a large discharge capacity, showing its applicability as a potential catalyst in LAB. Compared to Li–air cell with the α-MnO_2 catalyst on carbon nanotube/CNF composite as the air electrode with first cycle discharge capacity of 56%, Hollandite $Ag_2Mn_8O_{16}$ catalyst showed a ratio of charge and discharge capacities close to only 12% [48].

Kumar et al. [49] studied the electrochemical properties of solid-state, rechargeable Li–O_2 cells. The cell exhibited excellent thermal stability and rechargeability in the range of 30°C–105°C. The cell showed 40 charge–discharge cycles at current densities ranging from 0.05 to 0.25 mA cm^{-2} with good reversibility. The anode consisted of Li metal, an electrolyte consisting of a membrane laminate fabricated from glass ceramic and polymer ceramic materials, and a solid-state composite O_2/air cathode prepared from carbon and glass ceramic powder. The ion-conducting polymer ceramic electrolyte made from dried PEO, $LiNSO_2CF_2CF_{32}$ LiBETI salt, Li_2O, and boron nitride facilitated good charge–discharge reversibility, which enhanced the electrochemical performance of the cell.

Xiao et al. [50] reported an extremely high capacity of 15,000 mAh g^{-1} in primary Li–air batteries containing hierarchically porous functionalized graphene sheets without any catalyst. The porous structure facilitated rapid O_2 diffusion and reactive sites for reaction between Li and O_2. Moreover, the defects in the graphene sheets prevented the blockage of air in the electrode. For reducing the electrolyte decomposition, the work was carried out using ether-based electrolyte (LiTFSI in tri(ethylene glycol) dimethyl ether (Triglyme)).

To enhance the ORR activity of the cathode, Kichambare et al. [51] employed nitrogen-doped carbon cathode for lithium–oxygen cells. It was seen that there was a shift in the values of surface area, percentage of porosity, pore size, and pore volume upon doping with nitrogen. Electrochemical impedance spectroscopy results indicated that nitrogen doping of the carbon cathode suppressed the charge transfer resistance, thereby enhancing the ORR activity of the cathode. Li et al. [52] observed high electrocatalytic activities and reversibility when nitrogen-doped carbon nanotube (N-CNT) was used as cathode material for LAB. N-CNT was prepared by a floating catalyst chemical vapor deposition method; the scanning electron microscopy (SEM) and transmission electron microscopy (TEM) images of doped and undoped CNTs are shown in Figure 11.5, which shows uniformity in its diameter. Compared to undoped CNT, cells with N-CNT cathode showed a specific discharge capacity of 866 mAh g^{-1}, which accounted for 1.5 times the specific discharge capacity of undoped CNT cathode.

Li et al. [53] were the first to introduce graphene nanosheets (GNS) as superior cathode materials for LAB. The cell showed a very high discharge capacity of 8,705.9 mAh g^{-1} (at a current density of 75 mA g^{-1}) compared to carbon-based commercial cathodes such as BP-2000 and Vulcan XC-72 having a discharge capacity of 1,909.1 and 1,053.8 mAh g^{-1}, respectively (Figure 11.6). Morphological characterization showed that GNSs hold a thin and wrinkled structure with a curly structure. The unique morphology of GNSs provided porosity, which facilitated better electrolyte wetting and O_2 diffusion pathways. Furthermore, presence of unsaturated carbon was observed from infrared spectroscopy, which acted as a reaction site for O_2, thereby ensuring ORR activity. Overall, the uniqueness in the morphology of GNS, with a 3D three-phase electrochemical area, accounted for the high discharge capacity and efficient catalytic reaction.

Following the potential of GNS, Li et al. [54] further modified GNS by doping with nitrogen and used it as the cathode material. It was found that the cell delivered a discharge capacity of 11,660 mAh g^{-1} (40% higher than undoped GNS) for non-aqueous LAB. Moreover, doped GNS showed 2.5 times electrocatalytic activity for ORR. Doping created defects and functional groups, which acted as an active site for the nucleation of discharge products. Carbon atom neighboring to nitrogen-doped carbon showed increment in positive charge density, which enhances ORR by adsorbing O_2 and intermediates. Moreover, morphological analysis showed uniformity in the distribution of discharge products for the doped GNS, which enhanced the electrochemical performance

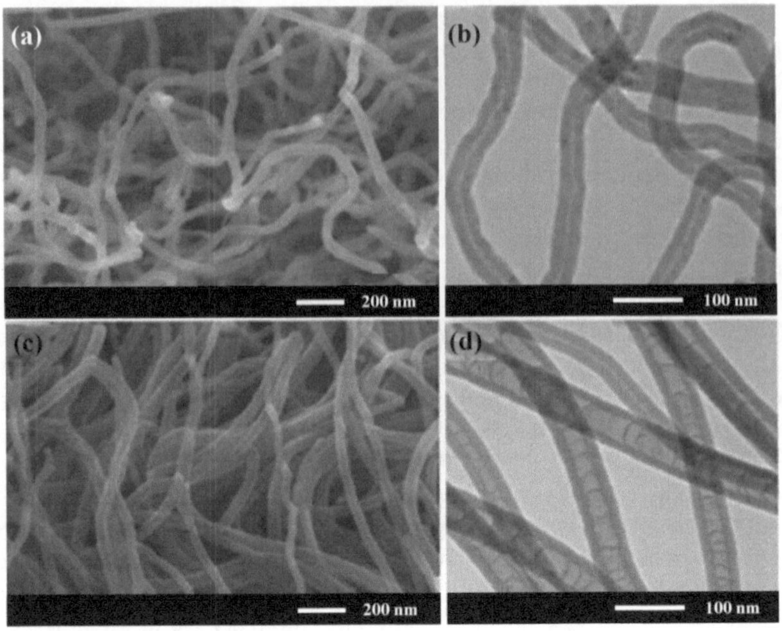

FIGURE 11.5 Morphological images showing (a) and (b) CNT (c) and (d) N-CNT. Adapted and reproduced with permission from Ref. [52]. Copyright © 2011 Elsevier.

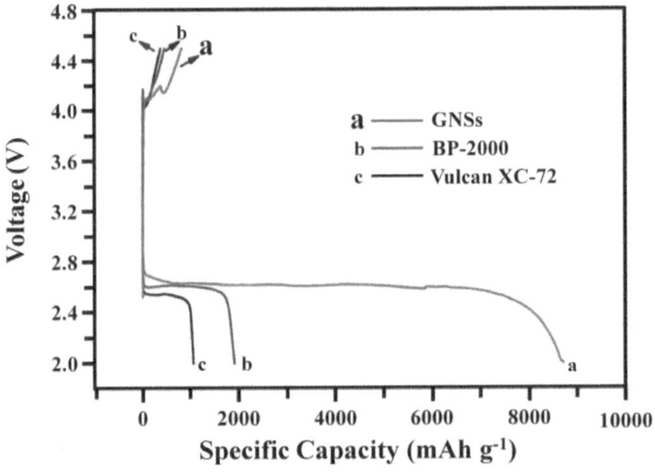

FIGURE 11.6 Discharge–charge curves for non-aqueous LABs at a current density of 75 mA g^{-1} with various carbon-based electrodes. Adapted and reproduced with permission from Ref. [53]. Copyright © 2011 The Royal Society of Chemistry.

of LAB. Figure 11.7 shows the results of various characterizations, namely, morphology (TEM), nitrogen adsorption–desorption isotherms taken at 77 K, X-ray photoelectron spectroscopy (XPS), and Raman shift for doped and undoped GNSs.

11.3.1 Cathode Catalysts

Enhancing the round-trip efficiency of a LAB needs a reduction in the overpotential with acceleration in reaction kinetics [55]. The catalytic activity of a catalyst, toward ORR and OER, decides the round-trip efficiency and overpotential [56]. Hence, great efforts have been dedicated to the

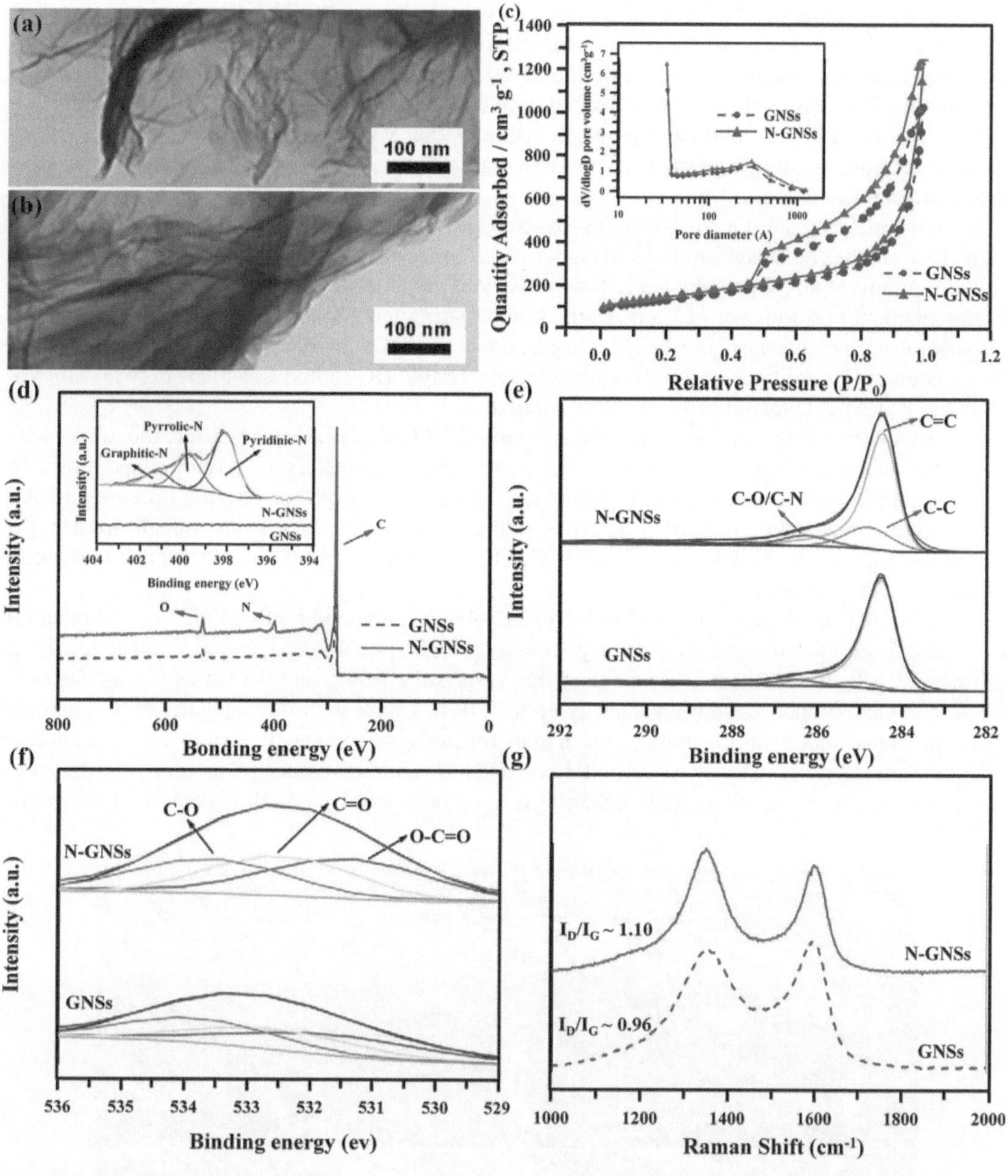

FIGURE 11.7 Results of various characterizations of GNSs and N-GNSs. (a and b) TEM images of GNSs and N-GNSs; (c) N_2 adsorption–desorption isotherms at 77 K; the inset shows pore size distribution; (d–f) XPS spectra; the inset of (d) shows the N1s spectra; (g) overlaid Raman spectra. Adapted and reproduced with permission from Ref. [54]. Copyright © 2011 Elsevier.

research on catalysts for LAB. Per the reported works, catalysts can be subdivided based on the material chemistry, namely, carbon-based materials, metallic oxides and their derivatives, precious metals, and their alloys [57]. Sun et al. [58] showed that GNSs prepared by the modified Hummers method functions as a cathode catalyst for LAB, which can reduce overpotential. For comparison with GNSs, a commercial Vulcan XC-72 carbon with a broad pore size distribution was also studied. The mesoporous structure of GNSs evidenced by Field emission scanning electron microscopy (FESEM), TEM, and Raman spectroscopy facilitated the O_2 diffusion and electrolyte impregnation. FESEM images of GNSs and Vulcan XC-72 before and after cycling are shown in Figure 11.8, and it can be observed that after cycling discharge, pores are filled by reduction products, leading to the reduction in O_2 diffusion and building up on polarization of the cathode. The lower value of overpotential with enhanced cyclability was achieved mainly due to the presence of defects and carbon vacancies on GNSs. Moreover, compared with the Vulcan XC-72 electrode, the discharge capacity of GNS was 2,332 mAh g^{-1}, which enhances the electrochemical performance of LAB.

Soavi et al. [59] developed a lithium–oxygen battery with *N*-butyl-*N*-methyl pyrrolidinium bis(trifluoromethanesulfonyl)imide as IL electrolyte and catalyst-free meso-macroporous carbon cathode. Adams et al. [60] conducted a study of various cathode structures prepared to enhance the electrochemical performance of LAB. Cathode-fabricating techniques such as (1) soaked electrolyte separate cast cathodes; (2) soaked electrolyte dual cast (DC) cathodes; (3) embedded electrolyte DC cathodes; (4) room temperature IL embedded electrolyte DC cathodes; and (5) poly tetrafluoroethylene (PTFE)-calendared cathodes were studied.

As it was found that metals act as a catalyst for the ORR and OER, metal-based cathodes not only reduce the overpotential of LABs but also function by changing the crystal structure and morphology of discharge products. Peng et al. [61] observed the kinetics of the charging process on LAB enhanced ten times more upon using a porous gold electrode than carbon electrode. Moreover, the capacity retention from cycles 1 to 100 was 95%. The cell used 0.1 M lithium perchlorate in dimethylsulfoxide as liquid electrolyte.

Lu et al. [62] developed Au- and Pd-decorated MnO_x nanomembranes (AMP) as cathodes for non-aqueous $Li–O_2$ batteries. The battery showed a high specific capacity of 3,200 mAh g^{-1} at 500 mA g^{-1}, which is 3.2 times higher than that of pristine MnO_x and 9.8 times that of Super P. Electrochemical properties showed stability up to 120 life cycles with lower polarization. The high cost of precious metals was the main concern in its further development. The same author in another work fabricated $Li–O_2$ batteries based on Pd/NiO (PNO) nanomembranes with ordered macropores and nanopores, by photolithography technology, as cathode catalysts [63]. Figure 11.8a shows the

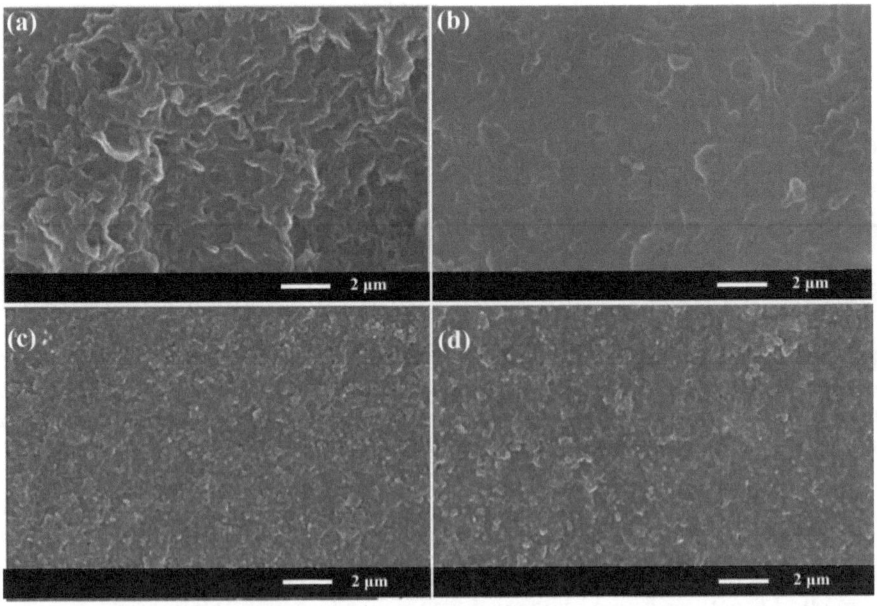

FIGURE 11.8 FESEM images various cathode catalysts before and after cycling. (a and b) GNS electrode (c and d) Vulcan XC-72 electrode. Adapted and reproduced with permission from Ref. [58]. Copyright © 2012 Elsevier.

schematic images of preparing PNO nanomembranes. The pores of micro- and nanometer sizes functioned as channels for transporting Li^+ and diffusion of O_2 and oxygen-containing intermediates, while lowering the overvoltage, which led to enhanced performance and rate capability with stability. Figure 11.9b–d and f represents morphological images of PNO nanomembranes.

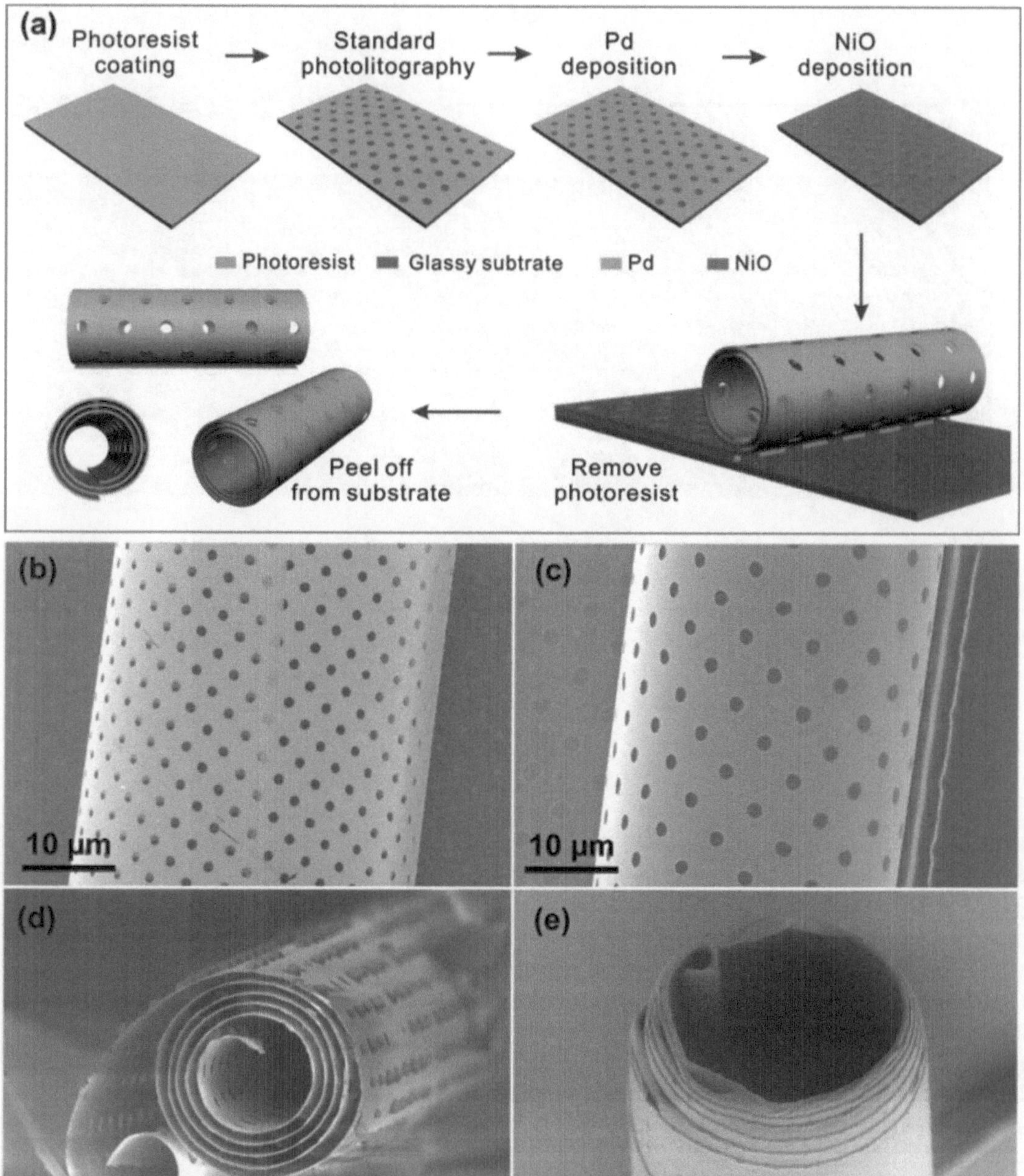

FIGURE 11.9 (a) Preparation of PNO nanomembranes. (b–d) Morphological images of PNO nanomembranes of varying macropore sizes. (e) Morphology of PNO nanomembrane without macropores. Adapted and reproduced with permission from Ref. [63]. Copyright © 2016 Elsevier.

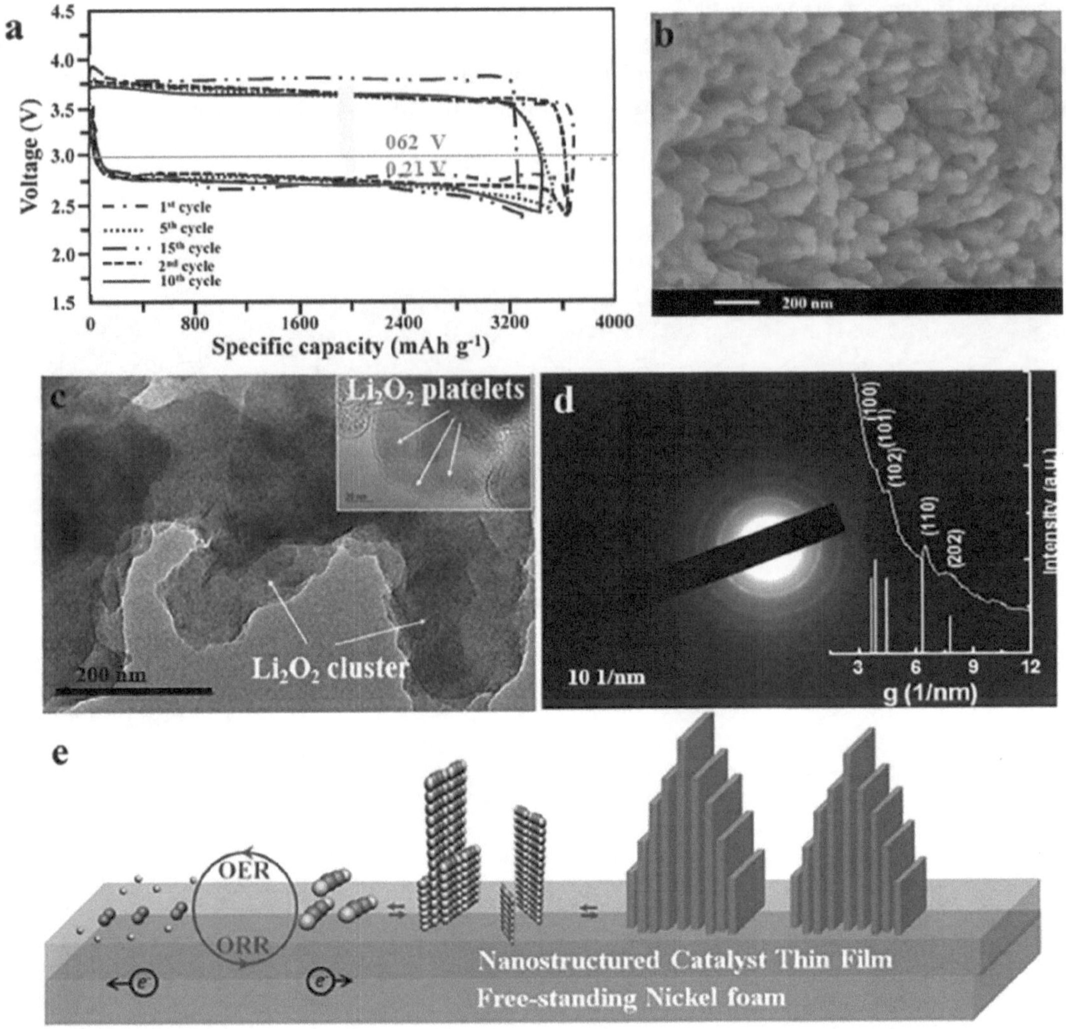

FIGURE 11.10 (a) Cycling discharge–charge curves of PG-PTF, (b–d) morphology of PG-PTF, and (e) schematic diagram on the mechanism of discharge–charge process associated with PG-PTF. Adapted and reproduced with permission from Ref. [64]. Copyright © 2016 Wiley-VCH.

Luo et al. [64] prepared 3D nickel foam containing Pt–Gd alloy polycrystalline thin film (PG-PTF), deposited by pulsed laser deposition method, as a carbon-free air electrode for LAB. The porous structure of Pt–Gd alloy facilitated not only electrolyte diffusion but also catalytic activity with good cycle performance in the full discharge–charge model. The work showed the importance of the deposition of electrocatalyst over the porous substrate and the pulsed laser deposition technique for further developing lithium–oxygen batteries. Figure 11.10 depicts cycling discharge–charge curves, morphological images, and a schematic diagram of the mechanism of the discharge–charge process associated with PG-PTF.

11.4 CONCLUSION

The search for post-LIB technology and the booming demand for high-energy-density batteries catalyzed the growth of MABs. LAB constitutes one of the most reported MABs, with a high energy density of 3,500 Wh kg^{-1}, which is tenfold larger than that of LIB. The growth in LAB is plagued by many issues such as dendrite growth and unstable SEI formation, intermediate formation, and reaction between the anode and intermediates along with water and O_2, which catalyzes cracks, corrosion, and pulverization at the anode. Therefore, it is important to protect the metallic Li anode. Many strategies have been postulated to alleviate the issue. Guiding the uniform Li plating/stripping, artificial SEI creation, employment of CPL, and so on are a few approaches. On the contrary, material characteristics and the microstructure of the air cathode also determine the performance of cells. The insulating nature of discharge products and their insolubility and deposition over the air cathode lead to lowering the cycle life. Properties such as round-trip efficiency, capacity cycling life, overpotential, and transport kinetics are dependent on the cathode and the electrocatalyst employed. A bifunctional catalyst mitigates the sluggish ORR/OER kinetics. Therefore, the development of such materials is crucial for the realization of high-performance LABs. The inclusion of multi-faced strategies, with engineering and tailoring materials for electrodes, electrolytes, and electrocatalysts can unequivocally enhance electrochemical performance, thereby updating the industrial viable stage for LAB.

REFERENCES

1. Sungjemmenla J, Soni CB, Vineeth SK, Kumar V (2021) Unveiling the physiochemical aspects of the matrix in improving sulfur-loading for room-temperature sodium-sulfur batteries. *Mater Adv* 2:4165–4189. https://doi.org/10.1039/D1MA00247C
2. Lee J-S, Tai Kim S, Cao R, et al (2011) Metal-air batteries with high energy density: Li-air versus Zn-air. *Adv Energy Mater* 1:34–50. https://doi.org/10.1002/aenm.201000010
3. Li F, Zhang T, Zhou H (2013) Challenges of non-aqueous Li-O_2 batteries: electrolytes, catalysts, and anodes. *Energy Environ Sci* 6:1125. https://doi.org/10.1039/c3ee00053b
4. Bruce PG, Freunberger SA, Hardwick LJ, Tarascon J-M (2012) Li-O_2 and Li-S batteries with high energy storage. *Nat Mater* 11:19–29. https://doi.org/10.1038/nmat3191
5. Vineeth SK, Tebyetekerwa M, Liu H, Soni CB, Zhao XS, Kumar V (2022) Progress in the development of solid-state electrolytes for reversible room-temperature sodium–sulfur batteries. *Adv Mater* 3(16):6415–6440. https://doi.org/10.1039/D2MA00428C
6. Sungjemmenla J, Soni CB, Vineeth SK, Kumar V (2022) Exploration of the unique structural chemistry of sulfur cathode for high-energy rechargeable beyond-Li batteries. *Adv Energy Sustain Res* 3:2100157. https://doi.org/10.1002/aesr.202100157
7. Eng AYS, Soni CB, Lum Y, et al (2022) Theory-guided experimental design in battery materials research. *Sci Adv* 8:eabm2422. https://doi.org/10.1126/sciadv.abm2422
8. Chen K, Huang G, Zhang X (2021) Efforts towards practical and sustainable Li/Na-air batteries. *Chinese J Chem* 39:32–42. https://doi.org/10.1002/cjoc.202000408
9. Ma J ling, Meng F lu, Yu Y, et al (2019) Prevention of dendrite growth and volume expansion to give high-performance aprotic bimetallic Li-Na alloy-O_2 batteries. *Nat Chem* 11:64–70. https://doi.org/10.1038/s41557-018-0166-9
10. Vineeth SK, Sreeram P, Vlad A, Joy R, Raghavan P, Pullanchiyodan A (2023) Polymer blend nanocomposite electrolytes for advanced energy storage applications. In *Polymer blend nanocomposites for energy storage applications* (pp. 203–238). Elsevier. https://doi.org/10.1016/B978-0-323-99549-8.00016-9
11. Vineeth SK, Soni CB, Sun Y, et al (2022) Implications of Na-ion solvation on Na anode-electrolyte interphase. *Trends Chem* 4:48–59. https://doi.org/10.1016/j.trechm.2021.11.002
12. Ghazi ZA, Sun Z, Sun C, et al (2019) Key aspects of lithium metal anodes for lithium metal batteries. *Small* 15:1900687. https://doi.org/10.1002/smll.201900687
13. Chen K, Yang D-Y, Huang G, Zhang X-B (2021) Lithium-air batteries: air-electrochemistry and anode stabilization. *Acc Chem Res* 54:632–641. https://doi.org/10.1021/acs.accounts.0c00772
14. Choudhury S, Wan CTC, Al Sadat WI, et al (2017) Designer interphases for the lithium-oxygen electrochemical cell. *Sci Adv* 3:1–12. https://doi.org/10.1126/sciadv.1602809
15. Vineeth SK, Soni CB, Sanjaykumar C, Yamauchi Y, Han M, Kumar V (2023) A quasi-solid state polymer electrolyte for high-rate and long-life sodium-metal batteries. *J. Energy Storage*, 73:108780. https://doi.org/10.1016/j.est.2023.108780
16. Zhang XP, Sun YY, Sun Z, et al (2019) Anode interfacial layer formation via reductive ethyl detaching of organic iodide in lithium-oxygen batteries. *Nat Commun* 10:1–12. https://doi.org/10.1038/s41467-019-11544-8

17. Shui J-L, Okasinski JS, Kenesei P, et al (2013) Reversibility of anodic lithium in rechargeable lithium-oxygen batteries. *Nat Commun* 4:2255. https://doi.org/10.1038/ncomms3255
18. Younesi R, Hahlin M, Roberts M, Edström K (2013) The SEI layer formed on lithium metal in the presence of oxygen: a seldom considered component in the development of the Li-O_2 battery. *J Power Sources* 225:40–45. https://doi.org/10.1016/j.jpowsour.2012.10.011
19. Soni CB, Sungjemmenla J, Vineeth SK, Kumar V (2022) Challenges in regulating interfacial-chemistry of the sodium-metal anode for room-temperature sodium-sulfur batteries. *Energy Storage* 4:e264. https://doi.org/10.1002/est2.264
20. Kang SJ, Mori T, Suk J, et al (2014) Improved cycle efficiency of lithium metal electrodes in Li-O_2 batteries by a two-dimensionally ordered nanoporous separator. *J Mater Chem A* 2:9970. https://doi.org/10.1039/c4ta01314j
21. Lee DJ, Lee H, Song J, et al (2014) Composite protective layer for Li metal anode in high-performance lithium-oxygen batteries. *Electrochem Commun* 40:45–48. https://doi.org/10.1016/j.elecom.2013.12.022
22. Guo Z, Dong X, Wang Y, Xia Y (2015) A lithium air battery with a lithiated Al-carbon anode. *Chem Commun* 51:676–678. https://doi.org/10.1039/C4CC07315K
23. Lee DJ, Lee H, Kim Y-J, et al (2016) Sustainable redox mediation for lithium-oxygen batteries by a composite protective layer on the lithium-metal anode. *Adv Mater* 28:857–863. https://doi.org/10.1002/adma.201503169
24. Cai K, Xiao Y, Lang X, et al (2017) Porous anode of lithium-oxygen battery based on double-gas-path structure. *Int J Hydrogen Energy* 42:29944–29948. https://doi.org/10.1016/j.ijhydene.2017.06.114
25. Krauskopf T, Richter FH, Zeier WG, Janek J (2020) Physicochemical concepts of the lithium metal anode in solid-state batteries. *Chem Rev* 120:7745–7794. https://doi.org/10.1021/acs.chemrev.0c00431
26. Bao J, Li C, Zhang F, et al (2020) A liquid anode of lithium biphenyl for highly safe lithium-air battery with hybrid electrolyte. *Batter Supercaps* 3:708–712. https://doi.org/10.1002/batt.202000092
27. Li C, Wei J, Qiu K, Wang Y (2020) Li-air battery with a superhydrophobic Li-protective layer. *ACS Appl Mater Interfaces* 12:23010–23016. https://doi.org/10.1021/acsami.0c05494
28. Guo Z, Zhang Q, Wang C, et al (2022) I-containing polymer/alloy layer-based Li anode mediating high-performance lithium-air batteries. *Adv Funct Mater* 32:2108993. https://doi.org/10.1002/adfm.202108993
29. Vineeth SK, Gadhave RV, Gadekar PT (2022) Investigation of crosslinking ability of sodium metabisulphite with polyvinyl alcohol-corn starch blend and its applicability as wood adhesive. *Indian Chem Eng* 64:197–207. https://doi.org/10.1080/00194506.2021.1887769
30. Gadhave RV, Vineeth SK, Dhawale PV, Gadekar PT (2020) Effect of boric acid on poly vinyl alcohol-tannin blend and its application as water-based wood adhesive. *Des Monomers Polym* 23:188–196. https://doi.org/10.1080/15685551.2020.1826124
31. Dhawale PV, Vineeth SK, Gadhave RV, et al (2022) Tannin as a renewable raw material for adhesive applications: a review. *Mater Adv* 3:3365–3388. https://doi.org/10.1039/D1MA00841B
32. Ma Z, Yuan X, Li L, et al (2015) A review of cathode materials and structures for rechargeable lithium-air batteries. *Energy Environ Sci* 8:2144–2198. https://doi.org/10.1039/C5EE00838G
33. Pan J, Tian XL, Zaman S, et al (2019) Recent progress on transition metal oxides as bifunctional catalysts for lithium-air and zinc-air batteries. *Batter Supercaps* 2:336–347. https://doi.org/10.1002/batt.201800082
34. Wang YJ, Fang B, Zhang D, et al (2018) *A Review of Carbon-Composited Materials as Air-Electrode Bifunctional Electrocatalysts for Metal-Air Batteries*. Springer, Singapore.
35. Wen Z, Shen C, Lu Y (2015) Air electrode for the lithium-air batteries: materials and structure designs. *Chempluschem* 80:270–287. https://doi.org/10.1002/cplu.201402104
36. Olivares-Marín M, Sorrentino A, Pereiro E, Tonti D (2017) Discharge products of ionic liquid-based Li-O_2 batteries observed by energy dependent soft x-ray transmission microscopy. *J Power Sources* 359:234–241. https://doi.org/10.1016/j.jpowsour.2017.05.039
37. Read J (2006) Ether-based electrolytes for the lithium/oxygen organic electrolyte battery. *J Electrochem Soc* 153:A96. https://doi.org/10.1149/1.2131827
38. Dobley A, Morein C, Abraham KM (2006) Cathode optimization for lithium-air batteries. *ECS Meet Abstr* MA2016-02:823. https://doi.org/10.1149/MA2005-02/22/823
39. Kuboki T, Okuyama T, Ohsaki T, Takami N (2005) Lithium-air batteries using hydrophobic room temperature ionic liquid electrolyte. *J Power Sources* 146:766–769. https://doi.org/10.1016/j.jpowsour.2005.03.082
40. Beattie SD, Manolescu DM, Blair SL (2009) High-capacity lithium-air cathodes. *J Electrochem Soc* 156:A44. https://doi.org/10.1149/1.3005989

41. Débart A, Bao J, Armstrong G, Bruce PG (2007) An O_2 cathode for rechargeable lithium batteries: the effect of a catalyst. *J Power Sources* 174:1177–1182. https://doi.org/10.1016/j.jpowsour.2007.06.180
42. Eswaran M, Munichandraiah N, Scanlon LG (2010) High capacity Li-O_2 cell and electrochemical impedance spectroscopy study. *Electrochem Solid-State Lett* 13:A121. https://doi.org/10.1149/1.3447867
43. Tran C, Yang X-Q, Qu D (2010) Investigation of the gas-diffusion-electrode used as lithium/air cathode in non-aqueous electrolyte and the importance of carbon material porosity. *J Power Sources* 195:2057–2063. https://doi.org/10.1016/j.jpowsour.2009.10.012
44. Xiao J, Wang D, Xu W, et al (2010) Optimization of air electrode for Li/air batteries. *J Electrochem Soc* 157:A487. https://doi.org/10.1149/1.3314375
45. Zhang T, Imanishi N, Shimonishi Y, et al (2010) A novel high energy density rechargeable lithium/air battery. *Chem Commun* 46:1661. https://doi.org/10.1039/b920012f
46. Lu Y-C, Xu Z, Gasteiger HA, et al (2010) Platinum–gold nanoparticles: a highly active bifunctional electrocatalyst for rechargeable lithium–air batteries. *J Am Chem Soc* 132:12170–12171. https://doi.org/10.1021/ja1036572
47. Zhang GQ, Hendrickson M, Plichta EJ, et al (2012) Preparation, characterization and electrochemical catalytic properties of hollandite $Ag_2Mn_8O_{16}$ for Li-air batteries. *J Electrochem Soc* 159:A310–A314. https://doi.org/10.1149/2.085203jes
48. Zhang GQ, Zheng JP, Liang R, et al (2011) α-MnO_2/carbon nanotube/carbon nanofiber composite catalytic air electrodes for rechargeable lithium-air batteries. *J Electrochem Soc* 158:A822. https://doi.org/10.1149/1.3590736
49. Kumar B, Kumar J, Leese R, et al (2010) A solid-state, rechargeable, long cycle life lithium-air battery. *J Electrochem Soc* 157:A50. https://doi.org/10.1149/1.3256129
50. Xiao J, Mei D, Li X, et al (2011) Hierarchically porous graphene as a lithium-air battery electrode. *Nano Lett* 11:5071–5078. https://doi.org/10.1021/nl203332e
51. Kichambare P, Kumar J, Rodrigues S, Kumar B (2011) Electrochemical performance of highly mesoporous nitrogen doped carbon cathode in lithium-oxygen batteries. *J Power Sources* 196:3310–3316. https://doi.org/10.1016/j.jpowsour.2010.11.112
52. Li Y, Wang J, Li X, et al (2011) Nitrogen-doped carbon nanotubes as cathode for lithium-air batteries. *Electrochem Commun* 13:668–672. https://doi.org/10.1016/j.elecom.2011.04.004
53. Li Y, Wang J, Li X, et al (2011) Superior energy capacity of graphene nanosheets for a nonaqueous lithium-oxygen battery. *Chem Commun* 47:9438. https://doi.org/10.1039/c1cc13464g
54. Li Y, Wang J, Li X, et al (2012) Nitrogen-doped graphene nanosheets as cathode materials with excellent electrocatalytic activity for high capacity lithium-oxygen batteries. *Electrochem Commun* 18:12–15. https://doi.org/10.1016/j.elecom.2012.01.023
55. Tan P, Jiang HR, Zhu XB, et al (2017) Advances and challenges in lithium-air batteries. *Appl Energy* 204:780–806. https://doi.org/10.1016/j.apenergy.2017.07.054
56. Geng D, Ding N, Hor TSA, et al (2016) From lithium-oxygen to lithium-air batteries: challenges and opportunities. *Adv Energy Mater* 6:1502164. https://doi.org/10.1002/aenm.201502164
57. Cheng F, Chen J (2012) Metal-air batteries: from oxygen reduction electrochemistry to cathode catalysts. *Chem Soc Rev* 41:2172. https://doi.org/10.1039/c1cs15228a
58. Sun B, Wang B, Su D, et al (2012) Graphene nanosheets as cathode catalysts for lithium-air batteries with an enhanced electrochemical performance. *Carbon* 50:727–733. https://doi.org/10.1016/j.carbon.2011.09.040
59. Soavi F, Monaco S, Mastragostino M (2013) Catalyst-free porous carbon cathode and ionic liquid for high efficiency, rechargeable Li/O_2 battery. *J Power Sources* 224:115–119. https://doi.org/10.1016/j.jpowsour.2012.09.095
60. Adams J, Karulkar M, Anandan V (2013) Evaluation and electrochemical analyses of cathodes for lithium-air batteries. *J Power Sources* 239:132–143. https://doi.org/10.1016/j.jpowsour.2013.03.140
61. Peng Z, Freunberger SA, Chen Y, Bruce PG (2012) A reversible and higher-rate Li-O_2 battery. *Science* 337:563–566. https://doi.org/10.1126/science.1223985
62. Lu X, Zhang L, Sun X, et al (2016) Bifunctional Au-Pd decorated MnO_x nanomembranes as cathode materials for Li-O2 batteries. *J Mater Chem A* 4:4155–4160. https://doi.org/10.1039/C5TA09781A
63. Lu X, Yin Y, Zhang L, et al (2016) Hierarchically porous Pd/NiO nanomembranes as cathode catalysts in Li-O2 batteries. *Nano Energy* 30:69–76. https://doi.org/10.1016/j.nanoen.2016.09.034
64. Luo W-B, Gao X-W, Shi D-Q, et al (2016) Binder-free and carbon-free 3D porous air electrode for Li-O_2 batteries with high efficiency, high capacity, and long life. *Small* 12:3031–3038. https://doi.org/10.1002/smll.201600699

12 Emerging Magnesium-Air Battery Technology

Electrolyte and Anodic Materials

Aswith R. Shenoy, Akhila Das, S. K. Vineeth, Roshny Joy, Vijay Kumar Thakur, Nutan Gupta, and Prasanth Raghavan

12.1 INTRODUCTION

Among various metal-air batteries such as lithium-air (Li-air), zinc-air (Zn-air), magnesium-air (Mg-air), sodium-air (Na-air), and aluminum air (Al-air), Li-air is capable of providing the highest specific energy density, followed by Mg-air due to its tremendous energy density. Growing energy demands and the depletion of fossil fuels, along with the existential environmental pollution crisis, have motivated a rapid increase in the research and development of relatively cleaner and more efficient energy systems. According to the International Energy Agency's Global Energy Review 2021, renewable electricity generation recently is set to expand by more than 8% to reach 8,300 TWh (the fastest year-on-year growth since the 1970s). The transition from fossil fuels to alternative fuel sources would harvest energy in an efficient and sustainable way to deliver increasing energy demands for the future. In this context electrochemical energy is considered as a major alternative energy source which is relevantly more sustainable and environmentally friendly than other known systems.

Electrochemical storage and conversion systems mainly include dielectric capacitors, supercapacitors, fuel cells, and batteries (Figure 12.1). Among them, batteries offer intermediate energy and power density along with ease of fabrication and portability, providing better commercialization

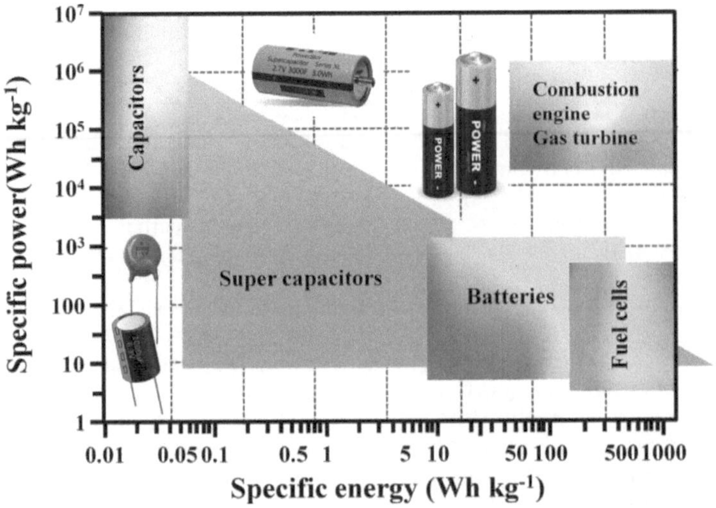

FIGURE 12.1 Various energy storage domains compared to an internal combustion engine, turbines, and conventional capacitors. Adapted and reproduced from Ref. [1]. Copyright © 2004 American Chemical Society.

DOI: 10.1201/9781003310174-12

and wider application prospects. If we compare batteries with other similar energy source systems, there is some overlap with supercapacitors in terms of energy and power as clearly evident by batteries utilizing thin-film electrodes. Other forms of battery technology have been identified in hybrid structures (metal-air batteries/cells) containing metal anode and air cathode.

The batteries constitute self-contained energy storage units that store chemical energy and convert it directly into electrical energy to power a variety of applications. Energy generation takes place by conversion of chemical energy using redox reactions at the electrodes. Anode reactions occur at a lower electrode potential compared with the cathode, making anode negative electrode and cathode positive ones. Winter et al. [1] highlighted that in batteries energy storage and conversion occur in the same compartment unlike it happens in the case of fuel cells where electrodes are just utilized as charge-transfer media, whereas active masses undergoing the redox reaction are delivered from outside the cell, either from the environment or from the storage tank. In supercapacitors, the orientation of electrolyte ions at the interface called electrical double layers is continuously formed and released, which results in a parallel movement of electrons in the external circuit enabling the energy-delivery process [1].

A metal-air flow battery is an emerging battery technology which possesses the unique structural design of conventional redox flow batteries and the high energy density offered by metal-air batteries, thus showing great potential. Different from the intercalation and conversion reactions of commercial lithium-ion batteries, metal-air batteries are based on the oxygen reduction reaction (ORR) and oxygen evolution reaction (OER) and usually operate in an open environment. Depending on the metal (Li, Zn, Al, Mg, K, Na) anode-utilized, metal-air batteries can be categorized into their subsequent metal-air batteries. Also compared to other existing batteries which currently dominate the market share, especially Li-ion batteries, metal-air batteries have better economic prospects as they utilize abundant atmospheric air as cathode and use the aforementioned metals as the anode, which are in larger abundance than Li. Linden et al. [2] (2001) mentioned that the major advantages driving the metal-air battery technology are their large theoretical energy density, which is about 3–30 times greater than that of lithium-ion batteries, and is seen as an alternative solution toward next-generation electrochemical energy storage for applications including electric vehicles (EVs) or grid energy storage [2].

Among various metal-air batteries, mechanically rechargeable zinc (Zn)-air, magnesium (Mg)-air, and aluminum (Al)-air batteries are receiving great interest as future energy storage technologies, especially from an EV development point of view. The several advantages such as environmental safety, low cost, and high abundance of materials utilized in these batteries can actually offer an efficient energy source for EVs compared to gasoline. Theoretically, energy density is an important factor in evaluating the performance of various battery configurations. In fact, metal-air batteries show considerably higher energy density because oxygen directly supplied from the surrounding environment acts as an oxidant during the discharge period. Figure 12.2 shows the theoretical energy density, specific energy, and nominal cell voltage of different metal-air batteries. The highly active metals like Li, Na, and K are unstable in aqueous systems and thus require a non-aqueous aprotic electrolyte, while for relatively inactive metals such as Zn, Al, Mg, and Fe, alkaline aqueous electrolytes are widely employed.

The lithium-air batteries offer the best combination of the highest theoretical energy density (5,928 Wh kg^{-1}) and high cell potential (nominally 2.96 V); iron-air batteries possess the smallest theoretical energy density and cell voltage (nominally 1.28 V). Aluminum-air batteries offer the highest volumetric energy density of 10,347 Wh L^{-1}, and magnesium-air batteries offer the highest nominal cell voltage of 3.09V. With comparable specific energy densities in (Wh kg^{-1}), the higher volumetric energy densities (Wh L^{-1}) for Mg and Al means that it can be smaller and compact while containing the same amount of energy. Among all the metal-air batteries, Li-air batteries are the most studied and advanced because they offer the best combination of the highest theoretical energy density (5,928 Wh kg^{-1}) and high cell potential (~2.96 V). But Li-air technology suffers from the intrinsic property of lithium to form needle-like lithium crystals, known as dendrites, which grow

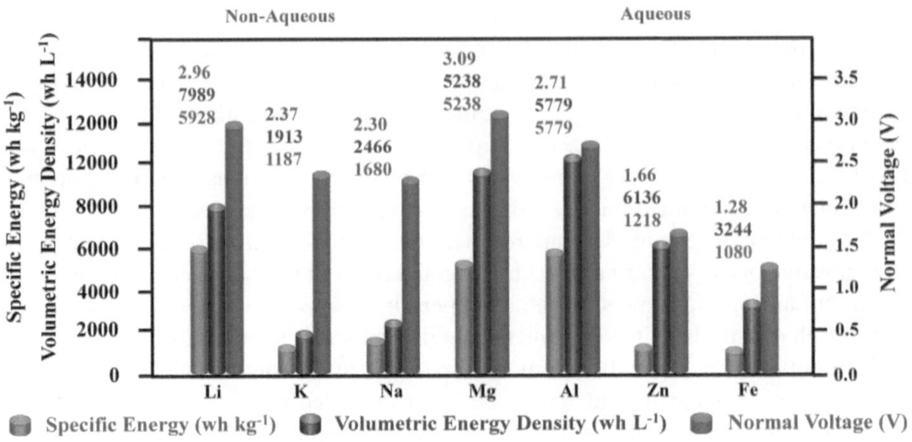

FIGURE 12.2 Theoretical specific/volumetric energy densities and nominal cell voltages for various metal-air batteries. Adapted and reproduced from Ref. [3]. Copyright 2014, American Chemical Society.

with subsequent plating/stripping cycles, resulting in an internal short circuit and fire hazards. Also, the limited availability and the high price of metallic lithium are preventing their large-scale development and production. Li-air batteries generally use non-aqueous electrolytes, which also raises the cost and causes safety concerns related to the flammable organic solvents.

In this list, Al-air batteries are also considered promising for several reasons. They offer the highest volumetric energy density (10,347 Wh L^{-1}) and theoretical energy density (5,779 Wh kg^{-1}) next to Li-air along with a comparable nominal cell voltage (~2.71 V). Also Al metal for the anode is extremely light and as the cathode is air, the overall package can be much lighter than most other battery types. Al is also cheap especially when compared to the rising costs of lithium. Goel et al. [4] reported that there are several problems being faced today by Al-air battery technology. These problems are (1) high rate of aluminum self-corrosion in alkaline solutions under both open-circuit and discharge conditions; (2) by-products such as Al_2O_3 and $Al(OH)_3$ accumulate at both the anode and the cathode, which also suppress electrochemical reactions; (3) complicated nature of trivalent Al^{3+} intercalation into the cathode; (4) irreversible nature of formation of hydrated alumina; (5) quick loss of charge on storage leading to meager shelf life [4]. Aqueous zinc-air battery is almost a century-old technology and was the first reported metal-air battery. They have moderate volumetric energy density (6,136 Wh L^{-1}) and theoretical energy density (1,218 Wh kg^{-1}) with a nominal cell voltage (~1.66 V). They typically have a very poor power output and are widely used in medical and telecommunication applications such as hearing aids and pagers. Li et al. [5] reported that, despite the early start, current Zn-air batteries are not commercialized for mass production or a range of wider applications. This is because (1) sluggish ORR and OER at the cathode involve multiple proton-coupled electron transfers, resulting in small current density and large electrode polarization. Also the corrosion and the dendrite growth of Zn metal upon recharge makes them poorly cyclable [5,6]. Mg-air battery possesses the highest nominal voltage (~3.09 V) among all the metal-air batteries with high volumetric density (9,619 Wh L^{-1}) only next to Al and excellent theoretical energy density (5,779 Wh kg^{-1}) at par with Al- and Li-air batteries.

The energy content of individual metals changes if the oxidized metal instead of the reduced material is considered for the evaluation. Owing to the uptake of oxygen atoms the specific energy of Li drops by more than 50%, other elements do not experience such a drastic change. Similarly the number of transferred electrons during the electrochemical reactions accounts for the reductions of specific energy of Al-air battery considerably compared to Mg-air battery because Al has three electrons, resulting in products containing three hydroxyl molecules ($Al(OH)_3$) compared to Mg having two electrons, resulting in two hydroxyl molecules ($Mg(OH)_2$). Accordingly, the application of other metals other than Li is gaining increased interest, if the specific energies in the reduced state are not the only parameter for the assessment of different systems, as shown in Figure 12.3.

Mg-air battery technology also delivers an economically viable alternative to Li-based batteries because of Mg metal's natural abundance (2.09%) in the earth's crust compared to Li (0.01%), as shown in Figure 12.4. However, the Mg-air battery offers the best combination of specific energy density and cost-effectiveness among all the metal-air batteries, as shown in Figure 12.5. Aurbach et al. [8] highlight that Mg metal doesn't pose the dendrite structure growth-related safety concerns like Li-air or Zn-air technology, making it a safer technology without fire hazards. Also magnesium metal is stable in atmospheric conditions, has low toxicity, and has relatively high safety. There are several factors that hamper the commercialization of Mg-air batteries, even after getting all favorable

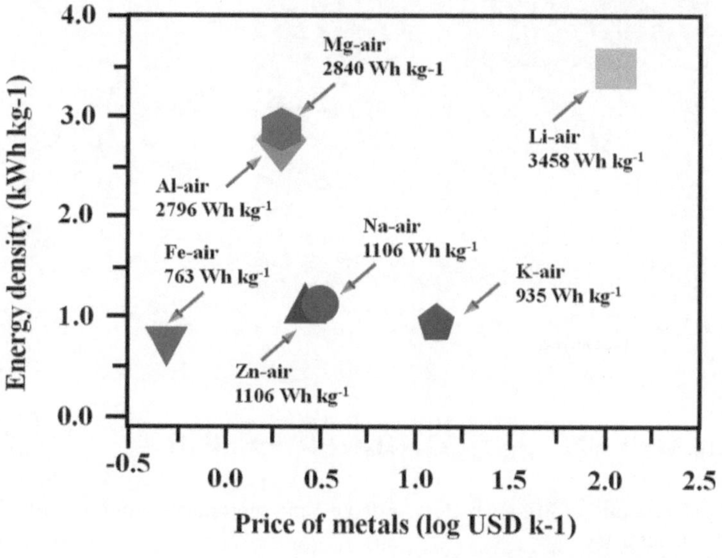

FIGURE 12.3 A comparison of theoretical specific energy densities of typical metal-air batteries as a function of prices of corresponding anode metals including oxygen uptake. Adapted and reproduced from Ref. [5]. Copyright © 2017 Wiley.

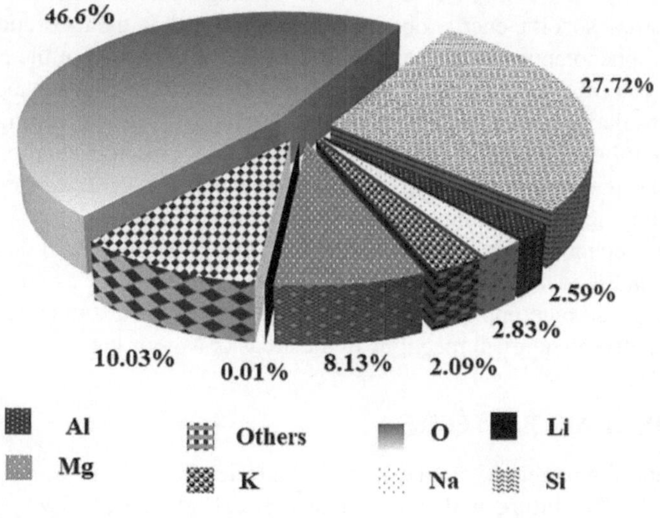

FIGURE 12.4 The elemental abundance of anode materials for metal-air batteries in the earth's crust. Adapted and reproduced from Ref. [7]. Copyright © 2017 Wiley.

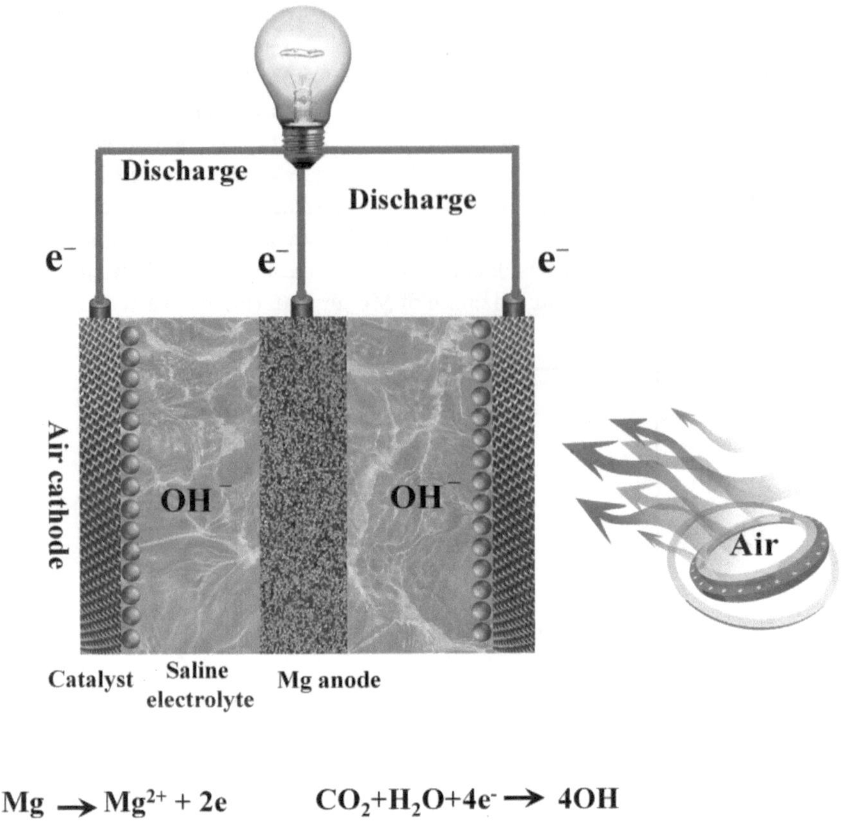

FIGURE 12.5 Basic structure of Mg-air battery. Adapted and reproduced from Ref. [9]. Copyright © 2013 The Royal Society of Chemistry.

advantages in electrochemical, economical, and safety hazards against the other existing battery technologies. The major cause of this is the high polarization and low coulombic efficiency associated with Mg-air batteries. This can be evident in the working voltage value, which is normally less than 1.2 V, and the actual specific energy obtained is 10% lower than the theoretical one. Li et al. [5] mentioned two major phenomena that cause these issues: (1) the corrosion of the Mg anode is caused by the reaction between Mg and the electrolyte; (2) the poor thermodynamics and kinetics owing to the MgO or MgO_2 insulating film as the initial discharge product on air-breathing, resulting in large polarization and a highly irreversible capacity. Keeping in line with high-quality academic reports, peer-reviewed articles, and patents. Starting with a concise and comprehensive review of systematic development and studying the working principles of metal-air batteries offer a viable solution for future energy-storing technology from renewables. The chapter addresses in detail the history and current status of the latest advances related to Mg-air battery technology as an alternative to future electrochemical energy storage applications. The review outlines the fundamental principles and understanding of the electrochemical reactions in the areas of Mg-air batteries.

12.2 MAGNESIUM-AIR BATTERIES

Among various metal-air batteries, magnesium-air batteries promise to be an alternative energy storage technology for the future with its high theoretical voltage (3.09 V), theoretical specific capacity (2,205 Ah kg^{-1}), energy density (2,840 Wh kg^{-1}), low cost, lightweight, environmental friendliness, and abundance in both the earth's crust and seawater [2]. These characteristics make magnesium-air battery technology a suitable power source for portable electronic devices and EV technology compared to other types of metal-air batteries.

Basic magnesium-air battery technology makes use of the chemical properties of magnesium metal, saline solution, and air to constitute a primary electrochemical cell. Magnesium metal or its alloys acts as the anode, air serves as the cathode source, and saline solution as the electrolyte, completing the electrochemical circuit. All of these components are highly stable in atmospheric conditions contributing a high level of safety.

12.3 REACTIONS/MECHANISMS/WORKING OF MAGNESIUM-AIR BATTERIES

A basic magnesium-air battery consists of a magnesium (or Mg alloy) anode, an air cathode, and a saline electrolyte. The basic structure and the working principle of the magnesium-air battery are shown in Figure 12.5.

The reaction mechanisms in the magnesium-air battery are as follows (Equations 12.1–12.3):

$$\text{Anode: } Mg \rightarrow Mg^{2+} + 2e- \tag{12.1}$$

$$\text{Cathode: } O_2 + 2H_2O + 4e^- \rightarrow 4OH^- \tag{12.2}$$

$$\text{Overall: } Mg + O_2 + 2H_2O \rightarrow 2Mg\,(OH)_2 \tag{12.3}$$

At the anode, the oxidation reaction occurs, and Mg transfers to Mg^{2+} with two electrons. At the cathode of the battery system, oxygen taken from the air combines with water to undergo reduction to produce hydroxyl (OH^-) ions. An electrolyte connects the anode and cathode to complete the electrochemical circuit. The electrons produced at the anode move through the external circuit. Both reactions at the anode and the cathode carry on until the magnesium electrode is entirely consumed.

As the reactions are exothermic, it causes the electrolyte to evaporate and consume during the reactions decreasing its volume. As the electrolyte volume decreases, the mass of the reaction products increases, which can cause supersaturation of electrolyte, leading to precipitation of the reaction products and eventually clogging of the cell. Thus, in general magnesium-air batteries constitute a primary cell. It can be made reusable by replacing the spent magnesium anode and electrolyte. New studies have shown promising results which aim to develop a rechargeable magnesium-air battery system with reversible oxygen reduction and evolution reactions. Magnesium-air batteries have the versatility to work with an aqueous or non-aqueous electrolyte. In aqueous electrolyte systems, saline electrolytes are used which are made up of water and salt solution. The anode and salt need to be replenished for continuous applications. Generally non-aqueous systems operate as a primary cell because magnesium oxide (MgO) forms a thermodynamically and electrochemically stable compound at the O_2 electrode in non-aqueous Mg–O_2 batteries. These magnesium oxide (MgO) compounds formed at the O_2 electrode cannot be decomposed by charging at ambient temperatures. A catalyst for accelerating the decomposition of Mg–O_2 would be required to realize secondary Mg–O_2 batteries. In recent times great strides have been made to create non-aqueous (organic) electrolytes by introducing polymer electrolyte membranes which serve the purpose of both the separator and the ion-transporting medium. The application of magnesium rechargeable batteries has had several problems regarding magnesium diffusion, formation of passivating surface film, and lack of efficient and stable electrolytes. There are only limited works available which address the rechargeable magnesium-air battery systems. Mg-air battery technology has been known since the 1960s, but its development from a technology into a commercial product has not been possible because of electrochemical and material property barriers arising due to hydrogen generation, precipitate control, electrode materials, and electrolyte composition.

12.4 RECHARGEABLE MG-AIR BATTERY WITH ORGANIC ELECTROLYTES

In comparison to the primary Mg-air battery and rechargeable Mg–hydride air battery detailed above, rechargeable $Mg–O_2$ batteries have been drawing a great deal of attention among researchers because of their high theoretical volumetric energy density and theoretical capacity. Among various rechargeable metal-air battery technologies, the rechargeable $Li–O_2$ battery has received significant attention in recent years because it offers favorable theoretical performances. But when it comes to the practical application of $Li–O_2$ batteries there are some persisting issues associated with lithium anode. $Li–O_2$ cells impart large quantities of hazardous Li_2O_2 deposition on the air cathode during discharging, which decomposes at high charging potentials. Also, the fact that Li_2O_2 is a highly explosive material raises serious safety hazards in practical commercial applications. The rechargeable Mg-air battery demonstrates an outstanding theoretical volumetric density of 14 kWh L^{-1} and a superior theoretical energy capacity of 3.9 kWh kg^{-1}, exceeding the corresponding theoretical data of $Li–O_2$ cells [5].

In the primary Mg-air batteries with aqueous electrolytes, $Mg(OH)_2$ is combined with Mg^{2+} ions and OH^- is reduced using O_2 in porous cathodes during discharge. Also the Mg anode is operated as an irreversible electrode in an aqueous solution. But in Mg-air batteries with non-aqueous electrolytes, when $Mg–O_2$ cells discharge, O_2 is catalytically reduced to O_2^- at the bifunctional air cathode to generate solid MgO_2 and MgO. As MgO_2 and MgO are both thermodynamically and electrochemically stable, this battery offers few concerns regarding safety hazards compared to $Li–O_2$ batteries. The MgO precipitates formed during discharging are not decomposed electrochemically at ambient temperatures this limits the non-aqueous Mg-air batteries only as primary cells. To develop non-aqueous Mg-air batteries as a secondary cell, a charging catalyst that accelerates the decomposition of MgO would be required [10].

To facilitate homogeneous precipitation of MgO, Shiga et al. [11] used iodine (I_2)–dimethylsulfoxide (DMSO) and 2,2,6,6 tetramethylpiperidine-oxyl (TEMPO) as electrocatalysts for air cathodes anion complex to catalyze the decomposition of MgO. Shiga et al. [11] used iodine (I_2)–dimethylsulfoxide (DMSO) as an electrocatalyst for the study of developing a rechargeable non-aqueous $Mg–O_2$ battery. The charging mechanism for the decomposition of MgO under the influence of the I_2–DMSO is illustrated in Figure 12.6. At first the magnesium iodide (MgI_2) formed during discharging at the cathode is electrochemically reduced to iodine. Iodine interacts with DMSO to form a molecular complex of I_2–DMSO. This molecular complex reacts with MgO formed by discharging at the cathode. Finally this MgO precipitate separates into Mg ions (as MgI_2) and oxygen.

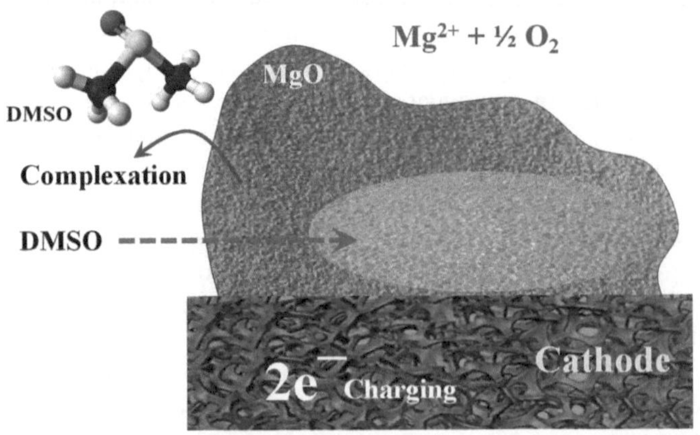

FIGURE 12.6 A proposed catalytic charging mechanism for the decomposition of MgO under the influence of the I_2–DMSO. Adapted and reproduced from Ref. [10]. Copyright © 2013 The Royal Society of Chemistry.

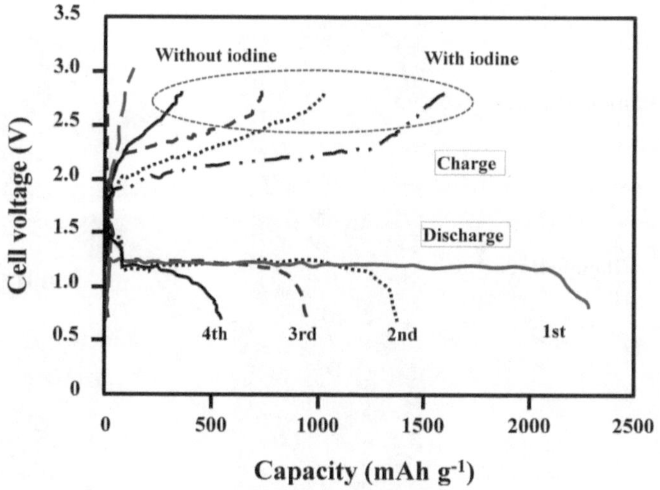

FIGURE 12.7 Discharge–charge curves of the non-aqueous Mg–O_2 battery with iodine at 60°C. Adapted and reproduced from Ref. [10]. Copyright © 2013 The Royal Society of Chemistry.

The catalyzing cycle continuously repeats until the MgO disappears from the cathode surface. The overall reaction (Equation 12.4) is as follows.

$$MgO + I_2 + 2e^- \rightarrow Mg^{2+} + 2I^- + \frac{1}{2} O_2 \tag{12.4}$$

The charging and discharging curves of the proposed Mg–O_2 battery at 60°C are shown in Figure 12.7. The discharge voltage of the Mg-iodine and Mg-air batteries was found to be lower than the theoretical values. The voltage drop was caused by the formation of a solid electrolyte interface on the Mg anode. When the polarity of the current was reversed during the charging process, the cell voltage increased to about 1.8 V, and a plateau was observed at about 2.2 V. The total charge capacity was 1,590 mAh g^{-1}, and the reversibility against discharging was 74.6%.

Shiga et al. [11] also demonstrated the ability of the TEMPO–anion complex ($PTMA^+ClO_4^-$) to electrochemically catalyze the oxidation of MgO to enable secondary Mg–O_2 batteries, shown in Figure 12.8. TEMPO reacts with an anion to generate a TEMPO–anion complex upon charging of the cell. Then the TEMPO–anion complex reacts with MgO immediately dissolving the solid MgO and regenerates the TEMPO units. TEMPO units once more form $TEMPO^+X^-$ by associating the regenerated TEMPO and another anion during cell charging. Thus TEMPO redox couple works as a mediator for MgO oxidation. Figure 12.9 shows the charging curves of cells with and without PTMA. It was found that the charging capacity for the cell with PTMA was 0.41 mAh, and without PTMA it was 0.074 mAh. Even though the results of these pioneering studies indicated that irreversible capacity underwent serious fading after the second cycle owing to the residual MgO and the degeneration of the PTMA electrocatalyst in the cathode, these studies showed new paths to the design and development of the rechargeable Mg–O_2 batteries.

12.4.1 Futuristic Rechargeable Mg-Air Battery with Hybrid Electrolytes

One of the main requirements for the successful development of rechargeable Mg-air batteries is the need for electrolytes that are stable with both the anodic and cathodic active materials remaining at their redox electrode potentials. This is the reason that makes aqueous electrolytes not suitable to be used with metallic active materials, such as Li, Na, Mg, or Al, because their electrode potentials are much less than that of H_2/H+. Also the use of organic electrolytes is difficult because of the wide reduction/oxidation potential range for these metals. These challenges have inspired the use of hybrid electrolytes for metal-air batteries.

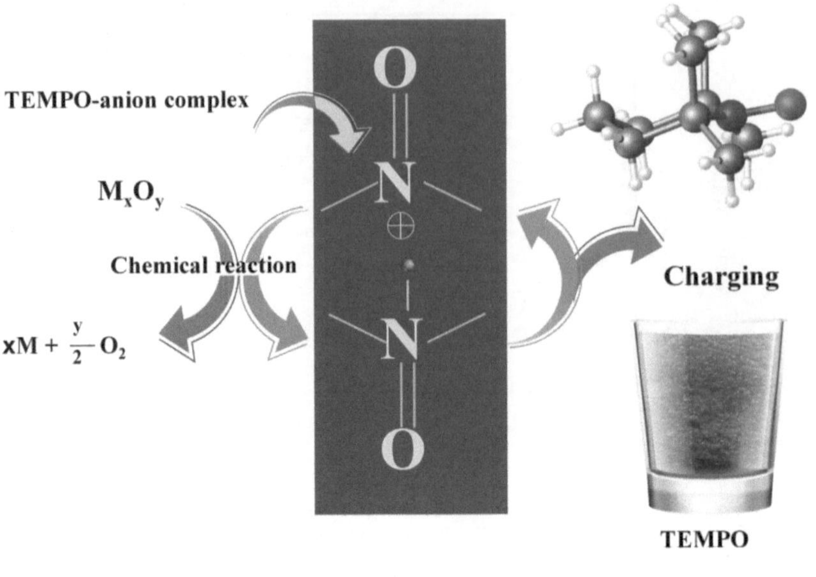

FIGURE 12.8 Catalytic cycle for the TEMPO–anion complex during the decomposition of metal oxide compounds. Adapted and Reproduced from Ref. [11]. Copyright © 2014 American Chemical Society.

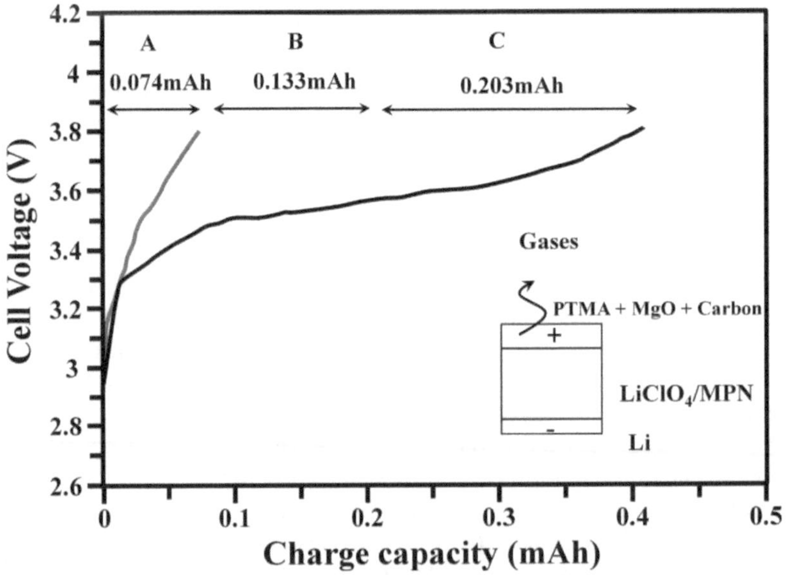

FIGURE 12.9 Charging curves for electrochemical cells incorporating cathodes with (black) and without (red) PTMA at 25°C. Adapted and Reproduced from Ref. [11]. Copyright © 2014 American Chemical Society.

The most common hybrid electrolyte design combines two or more electrolytes. Two electrolytes A and B form (A/B) type hybrid electrolyte, which may be solid/liquid or solid/solid. The other typical hybrid electrolyte design includes three electrolytes, A, B, and C, which form (A/B/C) type. These electrolytes may be of the following combinatorial forms: solid/liquid/solid or liquid/solid/liquid. The liquid electrolytes may be aqueous or organic. Possible solid electrolytes include lithium superionic conductor glass film (LISICON), sodium superionic conductor glass film (NASICON), and polymer. LISICON is used as a solid-state separator, which is different from the traditional polymer separators that have been used in rechargeable batteries to date. Generally, traditional polymer separators are used only to separate the active materials in the anode and cathode. In hybrid electrolyte batteries, LISICON is used to selectively separate the ions in the anode and cathode regions. This was successfully demonstrated by Wang et al. [12] with the Li–Cu rechargeable and recyclable battery with a hybrid electrolyte [13]. An illustration of the differences in the operating mechanisms of hybrid and traditional electrolyte-based batteries is shown in Figure 12.10. In hybrid electrolyte-based batteries, the electromigration is facilitated with two types of charge

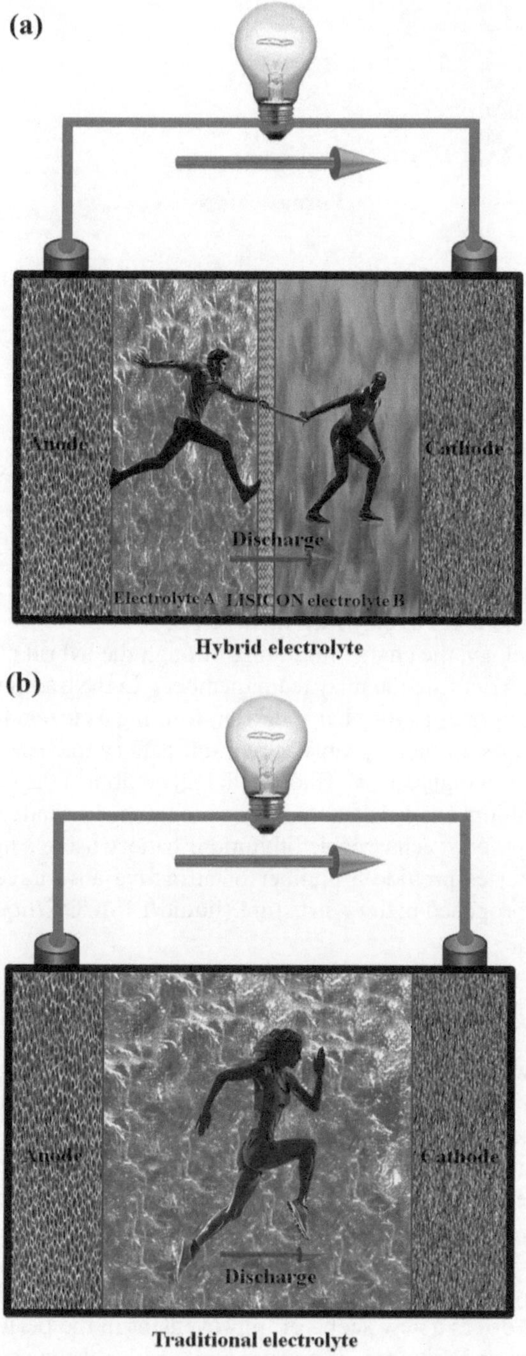

FIGURE 12.10 Schematic representation of the operating mechanism of batteries based on (a) a hybrid electrolyte and (b) a traditional electrolyte. Adapted and reproduced from Ref. [12]. Copyright © 2010 Wiley.

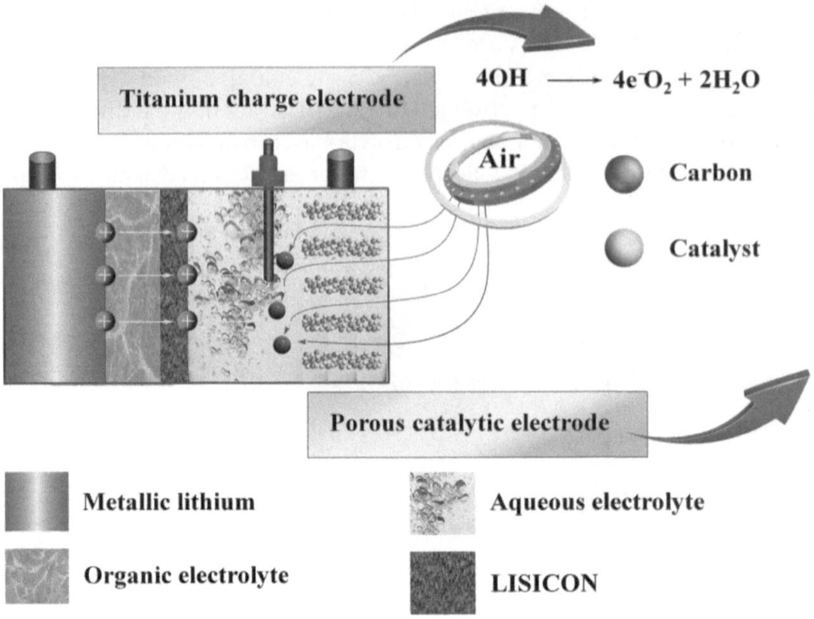

FIGURE 12.11 Structure and operating principle of the proposed rechargeable lithium-air battery, with an aqueous/LISICON/non-aqueous hybrid electrolyte design. Adapted and reproduced from Ref. [12].Copyright © 2010 Wiley.

carriers to generate a current. Taking the example of the Li–Cu battery mentioned above, Li ions and Cu ions play the role of charge carriers on either side of the LISICON during charge–discharge cycles. This mechanism resembles a relay race in athletics.

Using the relay race analogy, the charge migrating through the hybrid electrolyte-based battery is the baton, and the charge carriers are the relay team members. In the traditional electrolyte-based battery, only one type of ion migrates as the charge carrier, forming a current inside of the battery. Zhou et al. [12] proposed a new type of rechargeable lithium-air battery that uses an aqueous/non-aqueous hybrid electrolyte, as shown in Figure 12.11. Zhou et al. [12] combined the O_2 catalytic reduction in an aqueous solution and the lithium anode in a non-aqueous solution efficiently using a ceramic LISICON film to create a novel design for a rechargeable lithium-air battery using a hybrid electrolyte.

Hybrid electrolyte batteries provide a number of attractive advantages over traditional single electrolyte batteries. The proposed battery structure (liquid/LISICON/liquid) also has shown some challenges.

1. The ionic conductivity of the LISICON film was found to be much lower than that of liquid electrolyte.
2. LISICON film shows remarkable stability in neutral aqueous electrolytes for short-term use. In contrast, long-term immersion resulted in the slow corrosion of LISICON by an alkaline or acidic electrolyte.
3. During the charging and discharge process in the battery is accompanied by a large change in electrolyte concentration. This results in a large amount of electrolyte consumption which will lead to a lower available energy density.

Metal-air batteries with hybrid electrolytes are still at the conceptual level with very few published works. Hybrid electrolyte offers a new scope of improvement in the performance of metal-air batteries to extend the studies and research. The promising results of the use of hybrid electrolytes in Li-air batteries are opening a pathway for future prospects of developing a rechargeable Mg-air battery with hybrid electrolytes.

12.5 ANODES IN MAGNESIUM-AIR BATTERY

From the early stages of development of the Mg-air batteries, the researchers engaged in studies to overcome the challenges of developing novel magnesium anodes with high electrochemical activity. The anode, the negative electrode in the battery, hosts chemical activity between the magnesium metal and the electrolyte. This activity plays a vital role in determining the discharge capacity of the cell. Because of the intense chemical activity found in the anode, it also hosts some unwanted side reactions that will vary depending on the anode material and the nature of the electrolyte and cathode. Corrosion of anode material is the first huge challenge posed by this technology when used in aqueous electrolytes. The working principle of one of the most common corrosion reactions that take place in the anode is the hydrogen evolution reaction (Equations 12.5 and 12.6):

$$M + (2+x)_2\mathrm{O} \leftrightarrow 2M(OH) + H_2 \tag{12.5}$$

$$M + H_2\mathrm{O} \leftrightarrow MOX + H_2 \tag{12.6}$$

Hydrogen evolution reaction takes place in the metal anode during the charge and discharge processes of the cells. This hydrogen evolution reaction (HER) decreases the overall coulombic efficiency because it consumes useful electrons found in the metal anode. Also HER is the diffusion of hydrogen inside the electrolyte which can increase the internal pressure of the battery and cause some safety hazards for the cell setup. Hoey and Cohen studied the corrosion behavior of a polarized Mg anode in an aqueous NaCl solution. Their findings indicated that electrochemical corrosion occurs at the mechanically weakened lattice of the Mg anode. They observed to anode surface that when pressure caused by hydrogen gas generation exceeds the elastic strength of Mg, dislocations are formed resulting in embrittlement of Mg at the solution-metal interface. Their results indicated that the degree of embrittlement and, hence, the rate of corrosion, would be dependent on temperature and current density [14].

Another challenge is the passivation of the Mg anode surface. Passivated anodes cannot properly charge and discharge due to an insulating layer or film (called a passivation layer) that blocks the exchange of ions essential to the discharge process. Corrosion is also a problem as it accelerates anode dissolutions and decreases battery life. In some cases, corrosion leads to the formation of passivation and resistive layers. This depends on the anode metal composition and electrolyte nature.

Song et al. [15] studied the electrochemical behavior of magnesium in representative chloride and sulfate solutions including NaCl, Na_2SO_4, NaOH, and their mixed solutions, HCl, and H_2SO_4. His experiments showed that partially protective surface film plays an important role in the electrochemical dissolution processes for magnesium in NaCl, Na_2SO_4, and NaOH solutions. Also, results indicated the involvement of the intermediate species Mg^{2+} in the magnesium dissolution process [15]. Other factors also give rise to the corrosion of Mg anode. Like negative difference effect (NDE). NDE is the increasing rate of hydrogen evolving from magnesium (Mg) surfaces under anodic polarization. As the NDE increases the hydrogen evolution current, it accelerates the corrosion of Mg. Birbilis et al. [15] conducted a series of electrochemical tests to show that the corrosion rate increases in the Mg anode during oxidation, a phenomenon referred to as the NDE. Another factor that contributes to an increase in Mg anode corrosion is galvanic corrosion, which is caused by the impurities in Mg anode materials such as metallic iron, nickel, or copper. These elements construct micro-cells with Mg and expedite the cathodic reaction resulting in further corrosion of the Mg plate.

Song et al. [16] detailed the various mechanisms that cause corrosion in magnesium metals. He has indicated that the galvanic corrosion is directly associated with the impurity content. He reported a tolerance limit for impurity content. When the impurity content exceeds the "tolerance limit", the corrosion rate is greatly increased while with a lower impurity content, the corrosion rate remains low [16]. Makar and Kruger [17] published a detailed review of magnesium corrosion from the standpoint of corrosion science by discussing the thermodynamics, kinetics, and microstructural factors controlling corrosion behavior. They examined several types of corrosion that Mg and its alloys undergo: general, galvanic, pitting, crevice, intergranular, stress corrosion, corrosion fatigue, and high-temperature oxidation. They also discussed a number of approaches toward overcoming the corrosion of Mg alloys: (1) high purity or new alloys; (2) protective films and coatings; (3) rapid solidification processing; and (4) surface modification [17].

12.5.1 Magnesium Alloys

Alloying is an effective method for improving the discharge performance of Mg-based anodes as it can inhibit the HER. The commonly used alloying elements for Mg include Al, Zn, Mn, Li, Pb, Sn, Ca, In, Ga, Ce, Sm, and La. Ma et al. [18] reported that Mg–Li–Al–Ce alloys exhibited higher anode performance than commercial Mg alloys (pure Mg and AZ31) because discharge products on the anode surface became loose, which retained a large reaction surface during discharge. It has been observed that the Mg-air battery based on Mg–Li–Al–Ce offers higher operating voltage, anodic efficiency, and capacity than those with Mg and AZ31 [18]. Milusheva et al. [18] showed that Mg–Al alloys containing manganese are also an important option for Mg alloy anode option in Mg-air batteries. They showed that the existence of manganese reduces the particle size of the alloys. The performances of AM50, AM60, and MA8M06 as anode materials of Mg-air batteries have been investigated, among which MA8M06 is the best one with a higher voltage and more positive corrosion potential even than the AZ series alloys [19]. Zhao et al. [20] reported the superior discharge behavior of Mg–Ga–Hg alloy anodes compared with anodes of commercial Mg alloys because the addition of Ga accelerates the stripping of the corrosion film and the formation of Mg amalgams, thus promoting electrochemical activity [20].

Huang et al. [21] investigated the performance of an Mg-air battery based on AZ31 alloys with different densities. The microstructure and corrosion resistance of AZ31 alloys with different densities and surface morphologies were also investigated. The results clearly indicated that the AZ31 alloys decrease the density of corrosion products on the anode as shown in Figure 12.12. It was found that the AZ31 alloys improved the electrochemical activity of magnesium alloy and the electrochemical activity increased with the increase of density of the AZ31 alloys. Furthermore, the AZ31 alloys improved the discharge voltage of the Mg-air battery and reduced its voltage lag. The reduced voltage of the Mg-air battery was attributed to the porous corrosion products produced on the surface of the AZ31 alloy anode because this kind of corrosion product is easily broken down during discharging.

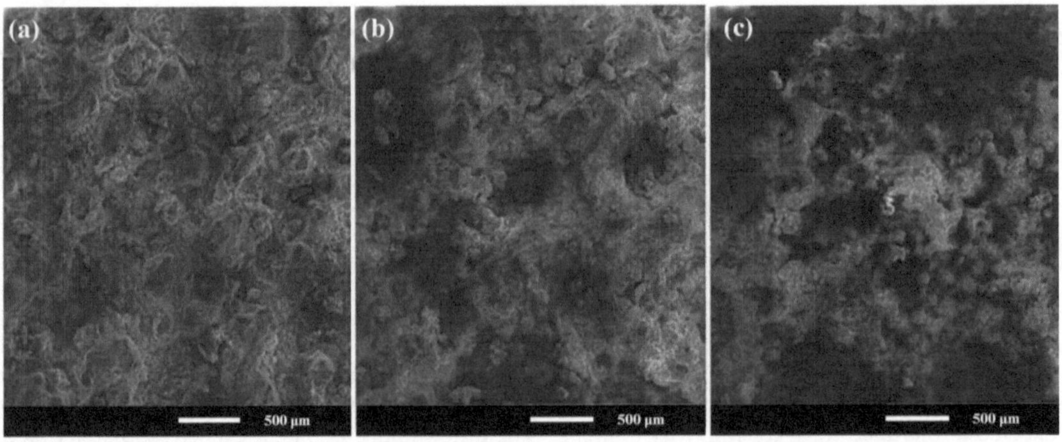

FIGURE 12.12 Scanning electron microscopy (SEM) micrographs of the AZ31 alloy anodes with different densities (0%, 3%, and 5%). Adapted and reproduced from Ref. [21]. Copyright © 2013 Elsevier.

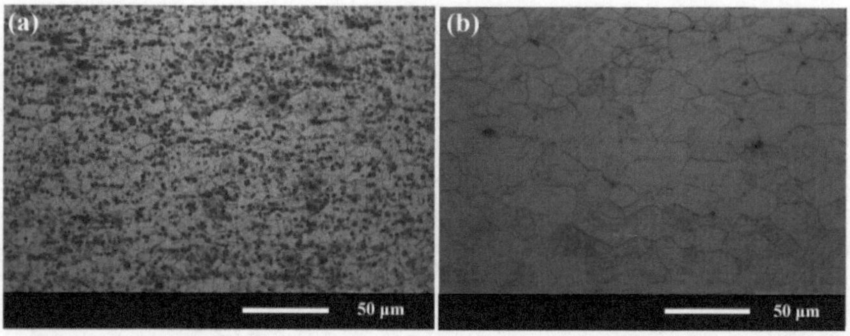

FIGURE 12.13 Optical micrographs of rolled (a) AMX602 and (b) AM60 alloys. Adapted and reproduced from Ref. [23]. Copyright © 2015 Elsevier.

Yuasa et al. [22] studied the effects of microstructure on the discharge behavior of AZ91 alloy as an anode for Mg-air batteries. Rolled Mg alloy sheets with 9 mass% Al and 1 mass% Zn (AZ91) were selected as the anode. It was confirmed that the finely and homogeneously distributed precipitant phase in the AZ91 specimens acted as a barrier to the dissolution of the Mg surface [22]. Yuasa et al. [22] also investigated discharge behaviors of rolled Mg–6 mass% Al–0.3 mass% Mn–2 mass% Ca (AMX602) and Mg–6 mass% Al– 0.3 mass% Mn (AM60) alloys used as anodes for magnesium-air batteries. The results showed that AMX602 alloy exhibited superior discharge properties compared to the AM60 alloy. The discharge products of the AMX602 alloy were dense and thin, and many cracks were observed at all current densities. Also, they were detached at some sites. The comparison of the discharge and corrosion indicated that the dense and thin discharge products of AMX602 were easily cracked and the cracks promoted the penetration of the electrolyte into the discharge products, retaining the discharge activity. In contrast, concerning the AM60 alloy, thick discharge products were formed on the surface during discharge, and cracking of the discharge products hardly occurred, degrading the discharge properties. Figure 12.13 shows the microstructures of the rolled AMX602 and AM60 alloys. Intermetallic particles were homogeneously distributed in the AMX602 alloy. In contrast, only a few intermetallic particles were observed in the AM60 alloy. The average grain sizes of the AMX602 and AM60 alloys were 13 and 29 µm, respectively, indicating that the distribution of intermetallic particles inhibited grain growth in the AMX602 alloy.

The addition of calcium (Ca) in Mg-based anodes can dramatically promote the anode efficiency as it leads to modification of the discharge products. Some recent literature which has reported promising results is detailed here. Deng et al. [24] investigated the effect of calcium content on self-corrosion and discharge performance of Mg anode primary Mg-air batteries. Mg–xCa (x=0.1, 0.5, 1, 2, 4 wt.%) binary alloys were used for the investigation. For comparison, pure Mg ingots were also prepared via the same casting procedures, named recast Mg. It was found that, for recast Mg and Mg–Ca alloys, the mass diffusion through the discharge products film can be easy so that the diffusion overpotential was to be low. This was proven by surface morphologies of recast Mg and Mg–Ca alloys after discharge at 2 mA cm^{-2} for 2 h, as presented in Figure 12.14a–f. The surface of Mg–0.1Ca (Figure 12.14b) and Mg–0.5Ca (Figure 12.14c) are covered by dense discharge products, which prevents the penetration of solution and reactants.

Liu et al. [25] reported the role of Al_2Ca and Al_2(Sm,Ca,La) particles in the microstructures and electrochemical discharge performance of the as-extruded Mg-3wt.% Al-1wt.% Zn AZ31 alloys for the anode design of Mg-air batteries. The results showed that Al_2Ca and Al_2(Sm,Ca,La) particles bring significant grain refinement to the AZ31 alloys. The grain refinement improved the corrosion resistance, and the fine Al_2Ca particles with dispersed distribution throughout the matrix are beneficial for the uniform dissolution of AZ31 alloys during discharge. Also there was an increase in output cell voltage and discharge capacity of the modified AZ31 alloy as anodes for Mg-air batteries [25].

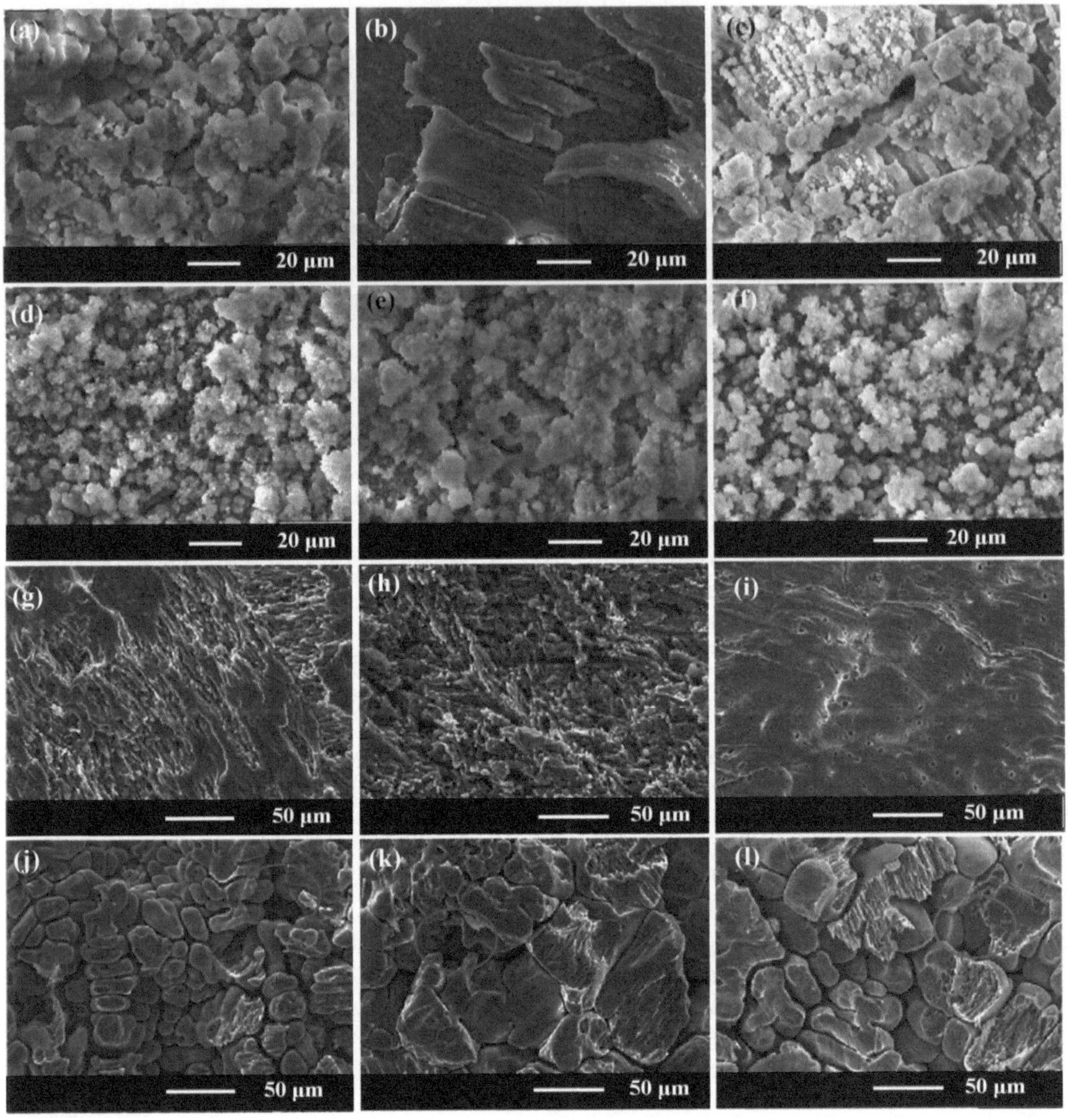

FIGURE 12.14 Surface morphologies of (a) recast Mg, (b) Mg-0.1Ca, (c) Mg-0.5Ca, (d) Mg-1Ca, (e) Mg-2Ca, (f) Mg-4Ca after discharge in 3.5 wt. % NaCl solution at 2 mA cm^{-2} for 2 h, and surface morphologies after discharge at 0.5 mA cm^{-2} for 16 h and removing all discharge products: (g) recast Mg, (h) Mg-0.1Ca, (i) Mg-0.5Ca, (j) Mg-1Ca, (k) Mg-2Ca, (l) Mg-4Ca. Adapted and reproduced from Ref. [24]. Copyright 2018 Elsevier.

Cheng et al. [26] in a recent study investigated the discharge behaviors of extruded low-alloyed Mg–Bi–Ca alloys in an aqueous electrolyte (3.5 wt% NaCl solution). The Mg–1Bi–0.5Ca alloy showed more negative discharge potential and higher anodic efficiency compared with the Mg–1Bi alloy during testing. In addition, the discharge morphologies of the Mg–1Bi–0.5Ca alloy showed loose and thin discharge products; along with numerous cracks, these features promoted diffusion of the NaCl solution into the discharge products, enhancing the discharge activity. The Mg-air battery with a Mg–1Bi–0.5Ca anode showed higher cell voltage and power density at all current densities. Mg–1Bi–0.5Ca alloy has great potential for further development as an anode material for Mg-air batteries [26]. Tong et al. [27] investigated four hypoeutectic Mg–*x*Zn (*x* ¼ 2, 4, 6, and 15 wt%) binary alloys are systematically investigated as anode materials for Mg-air batteries. Hypoeutectic

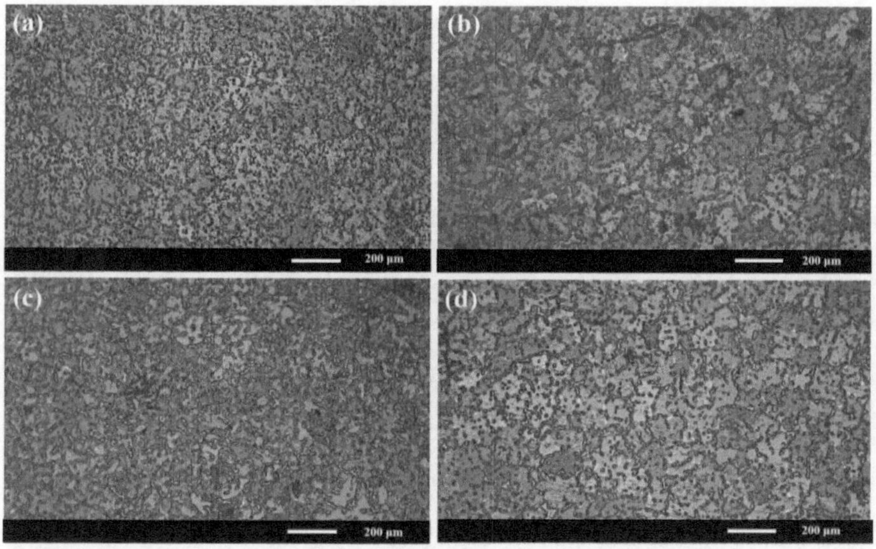

FIGURE 12.15 The microstructure of cast (a) AZ80, (b) AZ80–1Sm, (c) AZ80–2Sm, (d) AZ80–3Sm, and (e) grain size. Adapted and reproduced from Ref. [28]. Copyright © 2020 Elsevier.

Mg-*x*Zn binary alloys were prepared via casting. For comparison, pure Mg samples were also prepared via the same casting procedures and named recast Mg. It was seen that the addition of 2 wt% Zn significantly improves the anodic efficiency and specific capacity of Mg, while the addition of 6 wt% and 15 wt% Zn reduces its discharge properties. The results showed that increasing Zn content, anodic efficiency, and capacity deteriorated because of an increasing volume fraction of the intermetallic phase [27]. Chen et al. [28] investigated the influence of samarium (Sm) on the microstructure, electrochemical properties, and discharge performance of cast AZ80 as the anode for Mg-air batteries. Microstructure images of the investigated alloys are shown in Figure 12.15. AZ80 had large-sized grains, with developed dendrites. AZ80–1Sm had no dendrites. Increasing the Sm content to 2 wt% and 3 wt% further refined the Mg grains slightly shown in Figure 12.15e.

The surface morphology analysis revealed that there was an extremely uneven discharge surface for AZ80 with some large and deep holes (seen in Figure 12.16a), which were coral reef-like structures with many developed protrusions. During the discharge process, discharge products accumulate on the surface of the Mg anode, and it hinders the direct contact between fresh Mg and electrolyte. The complicated and rough surface structure of AZ80 increases the exfoliation difficulty of discharge products from the anodic surface, resulting in the continuous reduction of cell voltage. For AZ80–1 wt % of Sm, the number and the size of discharge holes were decreased. There were no large holes in the case of the AZ80–2Sm anode improving the discharge properties [27]. Various studies have been conducted on the reduction of the Mg particle size, and this has resulted in improving the capacity of Mg-air batteries accompanied by enhanced kinetics. The employment of Mg micro/nanostructures has been proven as a reasonable solution toward decreasing the thickness of the passivating surface films, which can increase ion diffusion in the anode and contribute toward improving anode utilization efficiency.

Chen et al. [28] identified Mg spheres, plates, nanorods, and sea-urchin-like nanostructures as anode material. These morphologies exhibited the superior properties of low corrosion rate and low anodic polarization for the Mg anode. Also, they displayed superior discharge performance such as long discharge plateaus and high specific discharge capacities because of the better electrochemical contact and lower electrode polarization. It was seen that uniform Mg microspheres were distributed homogeneously on both the surface and the holes of the screen mesh substrate. Increasing the deposition temperature indicated a high yield of microspheres as there was a thick layer of Mg

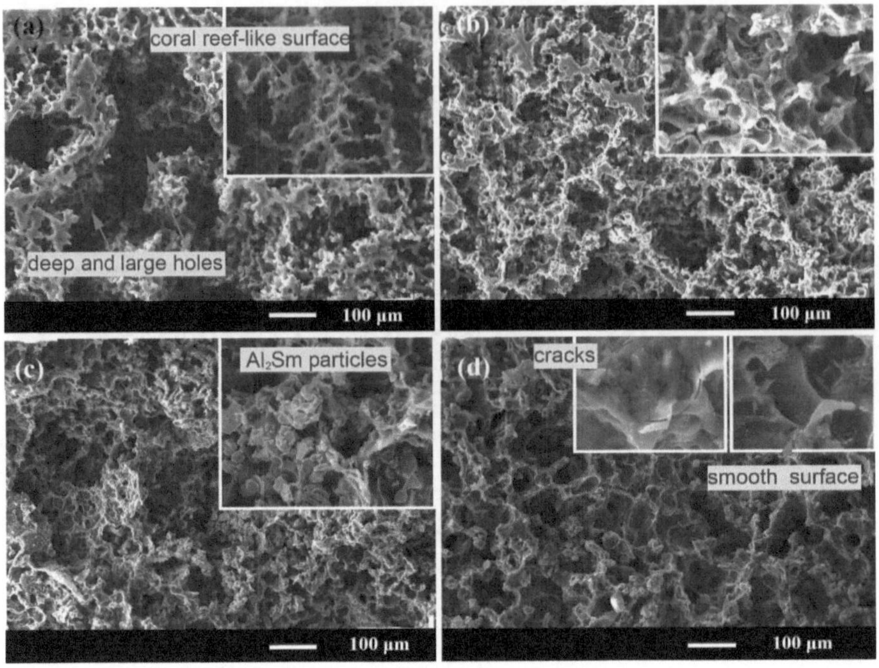

FIGURE 12.16 Surface morphologies of (a) AZ80, (b) AZ80–1Sm, (c) AZ80–2Sm, and (d) AZ80–3Sm anodes after the discharge in 3.5 wt % NaCl solution at 2.5 mA cm^2 for 10 h with the removal of discharge products. Adapted and reproduced from Ref. [28]. Copyright © 2020 Elsevier.

microspheres with different sizes covering the whole surface of the substrate. The results show that different deposition temperatures affected the size of Mg microspheres and that a mixture of Mg micro/nanospheres was formed by deposition at higher temperatures, as seen in Figure 12.16.

Li et al. [29] studied the electrochemical activity of various micro/nanostructured Mg anodes in Mg-air cells. He investigated anode samples with microspheres, microplates, nanorods, and sea-urchin-like nanostructures. His studies have shown that these nanostructures. Mg anodes have shown better performance in comparison with that of commercial Mg powders in the $Mg(NO_3)_2$+$NaNO_2$ electrolyte (Figure 12.17). The micro/nanostructures with a sea-urchin-like shape displayed the highest energy density and high-rate performance. The results demonstrated that porous and network nanostructures improve the performance of the Mg-air battery [11] (Figure 12.18).

Wang et al. [28] developed an alkaline primary Mg-air battery made from a porous Mg thin film. The battery was able to achieve a high open-circuit voltage of 1.41 V and a large discharge capacity of 821 mAh g^{-1}. The morphology of the thin film Mg anodes indicated that Mg films with porous structures showed excellent discharge properties, including stable discharge plateaus, high open-cell voltages, and large discharge capacities. The results indicated that the electrochemical performances of Mg-air batteries can be improved by controlling the Mg anode morphology. The study demonstrated porous Mg thin films as the anodes can improve the electrochemical performance of Mg-air batteries in the future [28].

12.5.2 Fabrication Methods in Mg Anodes

Numerous studies have shown that the electrochemical behavior of a component can be significantly changed by bringing about selective design modifications in the form of its structural changes such as altering the joining, processing, alloying, and selectively modifying geometric parameters of the

FIGURE 12.17 Scanning electron microscopy (SEM) images of the vapor-deposited Mg micro/nanoscale materials for different deposition temperatures. Adapted and reproduced from Ref. [29]. Copyright © 2009 Springer.

component. Jiang et al. [30] manufactured $Mg_{0.5}Zn_{0.2}Ca$ and $Mg_{0.5}Zn_{0.2}Ge$ alloys by casting and subsequently hot extrusion process to avoid the severe galvanic corrosion in Mg. Corrosion behavior, mechanical integrity, and fatigue behavior in two corrosive electrolytes of these two alloys were investigated. The results revealed a better short-term corrosion property for $Mg_{0.5}Zn_{0.2}Ge$ alloy but an improved long-term corrosion resistance for $Mg_{0.5}Zn_{0.2}Ca$ alloy, resulting from the different protective abilities of surface layers and microstructures of the alloys [30]. Luo et al. [31] found out that a minor addition of neodymium increased the corrosion resistance of as-extruded Mg–6Gd–2Y–(0–1.5) Nd–0.2Zr alloys.

The overall surface photographs of the four corroded specimens are seen in Figure 12.19. It can be seen that some loose and white corrosion products are distributed on the surface of the immersed specimens (as indicated in Figure 12.19f). The corrosion area of the four tested specimens decreases with the increase of Nd content from 0% to 0.5 wt%. But then it increases from 0.5 to 1.5 wt% (as shown in Figure 12.19b–e). As can be seen in Figure 12.19c, the corrosion on the surface of 0.5Nd is smaller than the other specimens. From the surface features of the four corroded specimens, the corrosion of the 1.5Nd specimen is the most serious. In addition, pitting corrosion occurred in

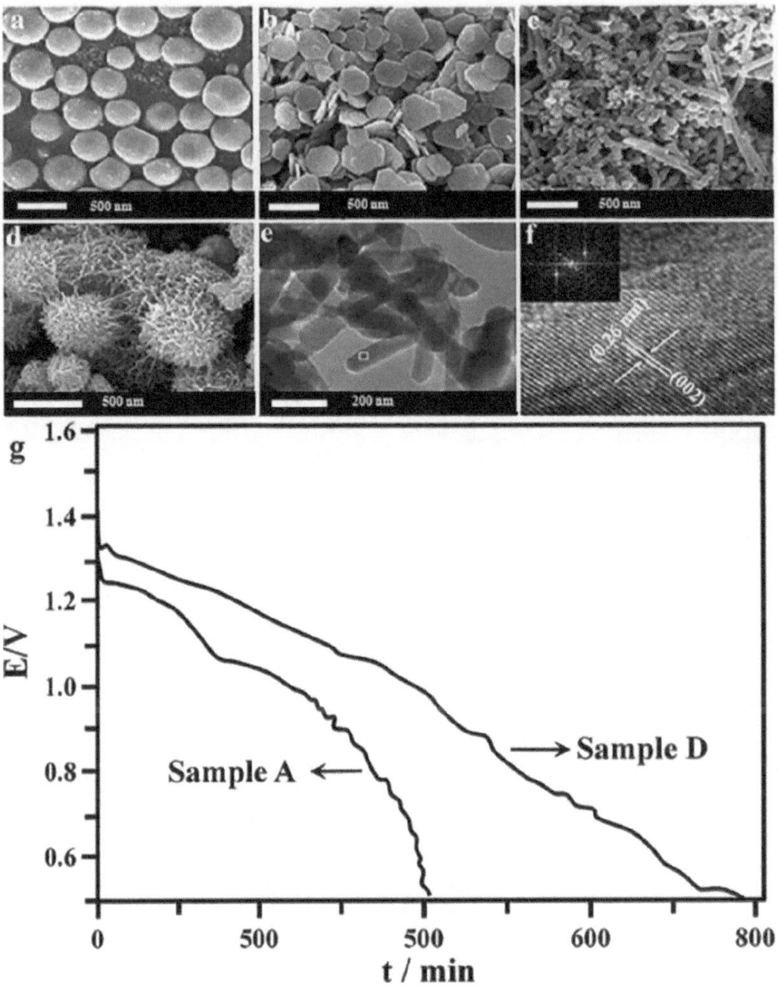

FIGURE 12.18 Typical scanning electron microscopy (SEM) images of the as-prepared Mg products: sample A (a), sample B (b), sample C (c), and sample D (d); representative transmission electron microscopy image of sample C (e) and high-resolution transmission electron microscopy image (f) of the white square marked in (e); the inset in (f) shows the corresponding fast Fourier transform pattern. (g) Representative discharge curves of the Mg-air batteries made from Mg microspheres and sea-urchin-like nanostructures at a constant current of 0.5 mA and a temperature of 25°C. Adapted and reproduced from Ref. [29]. Copyright © 2009 Springer.

Figure 12.19 [31]. Kaseem et al. [32] investigated the development of stable coral reef-like structures via self-assembly of functionalized polyvinyl alcohol for achieving better corrosion performance of AZ31 Mg alloy. The results showed that the synthesized coral reef-like structures have an excellent ability to block the infiltration of corrosive ions into the Mg alloy substrate and are expected to be useful in many other applications such as anode materials in batteries [32]. Studies have shown that some post-processing treatments during fabrication of metal anode components can also lead to refinement of the matrix grains, re-distribution of the second phase, and thus a homogeneous microstructure. These microstructural changes are favorable for better discharge performance of Mg-based anodes.

Xiong et al. [32] investigated the effects of the heat treatment process on the discharge behavior of Mg–6wt.%Al–1wt.%Sn alloy as anode for Mg-air batteries in 3.5 wt.% NaCl solution. It was seen that the Mg-air battery based on the heat-treated Mg–Al–Sn anode presents a higher operating voltage and more stable discharge process [32]. Wang et al. [33] used wrought Mg–Al–Pb–RE strips of

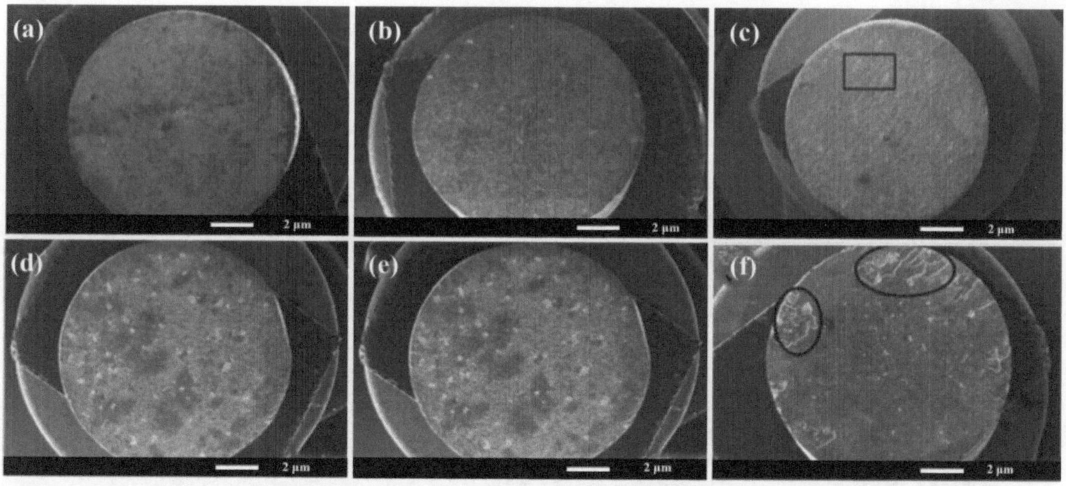

FIGURE 12.19 Overall surface photographs of the four corroded Mg-6Gd-2Y-χNd-0.2Zr alloys (χ=0, 0.5, 1, and 1.5 wt%) after immersion in 5% NaCl solution saturated with $Mg(OH)_2$ for 72 h. Before removal of the corrosion products of (a) 0Nd alloy and (f) 1.5Nd alloy; after removal of the corrosion products of (b) 0Nd alloy, (c) 0.5Nd alloy, (d) 1Nd alloy, and (e) 1.5Nd alloy. Adapted and reproduced from Ref. [31]. Copyright © 2019 Elsevier.

different thicknesses synthesized via extrusion as the anodes in Mg-air batteries. The anode performance was evaluated by observing the corroded surface at the outset of discharge. Figure 12.20 (left) showed the electrode morphologies discharged at 10 mA cm^2 for 10 min. The 5 mm-thick strip suffers active dissolution, and many small pores are distributed homogeneously on the corroded surface (Figure 12.20d). Hence, it would exhibit stronger activity and higher anodic efficiency during discharge. Based on the satisfactory performance, the wrought Mg–Al–Pb–RE strip of 5 mm thickness is an attractive anode for Mg-air batteries.

The study by Feng et al. [34] proposed the rolled Mg–Hg–Ga alloy as a promising anode material for application in low-power and long-term seawater batteries. His group investigated the effect of hot rolling on the microstructure and discharge properties of Mg–1.6 wt%Hg-2 wt.% Ga alloy anodes. The results showed that the hot rolling improves the corrosion resistance and anodic efficiency of Mg–Hg–Ga alloy [34]. Various post-processing treatments on the anode component fabrication such as rolling, extrusion, and heat treatment have been considered as a practical approach to enhance the discharge performance of Mg-based anodes, which can retain the lightweight advantage of the original Mg alloys. Studies have suggested that the extruded Mg–Al–Pb–La alloy is an appropriate anode for seawater-activated batteries. *Feng*'s group [35] investigated the microstructure and electrochemical corrosion behavior of extruded Mg–Al–Pb–La alloy as anode for seawater-activated battery and the results showed that the extrusion improves the discharge activity of Mg–Al–Pb–La alloy. Ma et al. [36] Studied the effect of post water annealing on the corrosion resistance and mechanical properties of a magnesium alloy. Their studies showed that the treated sample exhibits significantly improved corrosion resistance, presenting less corrosion current density and hydrogen evolution rate. Along with the corrosion resistance the ultimate strength and elongation and better-maintained mechanical integrity after corrosion was achieved simultaneously [36]. The discharge properties of Mg-based anodes can be reasonably enhanced by microstructure modification through severe plastic deformation. Studies by Song et al. [37] have also shown that die-casting, a widely used forming technique, may also influence the discharge performance of Mg-based anode. A die-casting usually experiences a rapid cooling process, having a very different microstructure from traditionally casted ingots. The surface layer of a die-cast Mg alloy forms finer grains and a more continuous secondary phase network, and hence it offers better corrosion-resistant properties as it dissolves more slowly than the interior of the casting. These results indicate that die-casted Mg-based anode may have superior discharge performance [37].

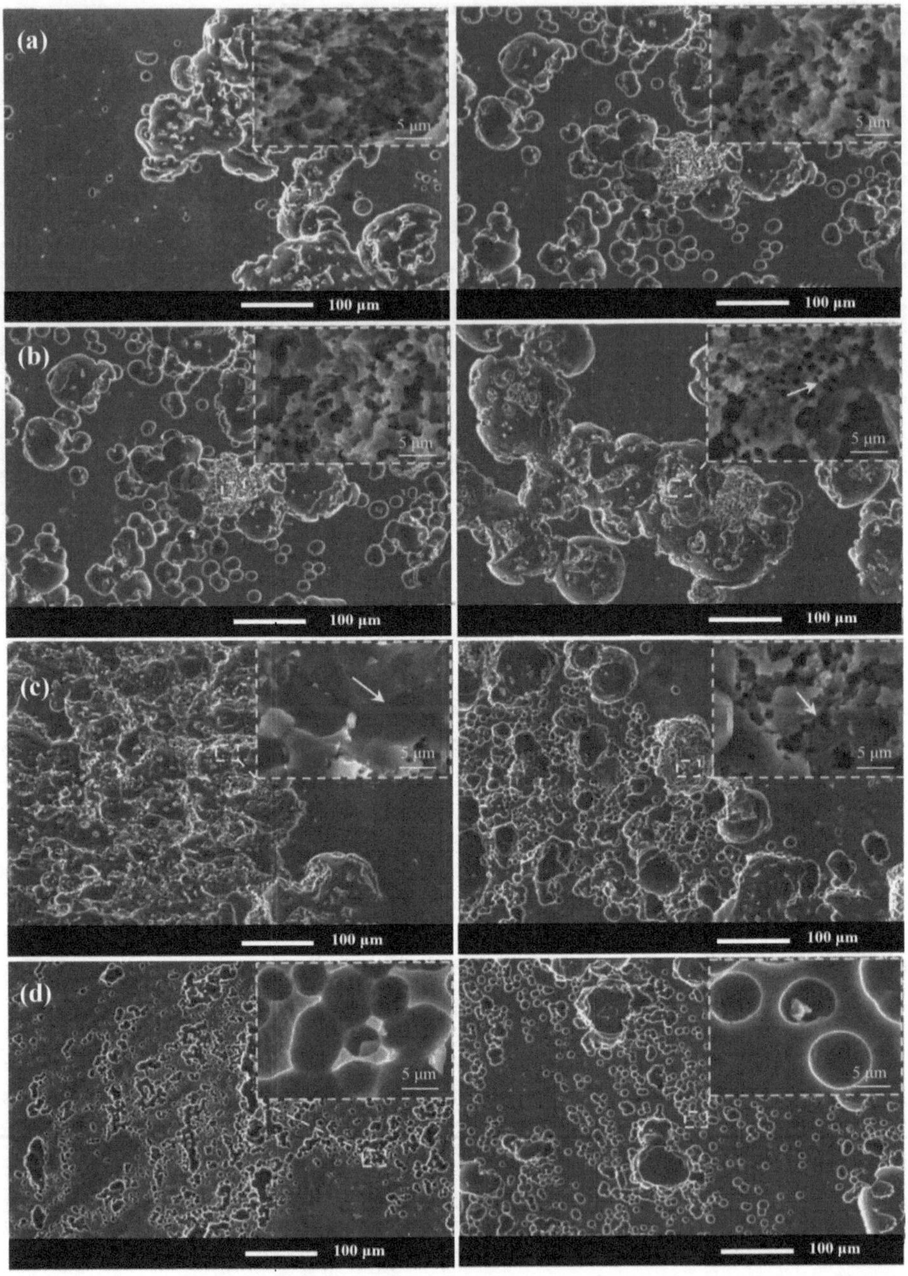

FIGURE 12.20 Surface morphologies of (a) homogenized sample, (b) 15 mm-thick strip,(c) 10 mm-thick strip, and (d) 5 mm-thick strip for Mg–Al–Pb–RE anodes discharged at 10 mA cm^{-2} (left) and 20 mA cm^{-2} (right) for 10 min after removing the oxidation products. The arrows indicate the grain boundaries. Adapted and reproduced from Ref. [33]. Copyright © 2019 Elsevier.

In a recent study by Xiao et al. [38] corrosion and anodic behavior between die-cast and hot-rolled Mg-9Al-1Zn alloy, the relationship between the microstructure and discharge behavior was compared and analyzed. The results indicated that the discharge performance of the Mg–9Al–1Zn anode in Mg-air batteries can be significantly improved by the die-casting process. It also implied that introducing some grain refiners in anode material design may further refine the microstructure of the alloy and will contribute to improved discharge performance in Mg-air batteries [38]. Doping magnesium with other elements effectively enhances its anode properties promotes the breakdown of the Mg $(OH)_2$ film and inhibits the corrosion on the electrode surface. Rare earth elements are proven as most suitable candidates for doping to improve corrosion resistance. Studies done by Feng et al. [39] have shown an improved corrosion resistance for Mg-air batteries with the Mg–Al–Pb–Ce–Y anode. Also an enhanced discharge performance is attributed to its modified microstructure. The finer dendrite crystals with a relatively homogeneous composition in the matrix promote uniform dissolution of the anode during discharge, hence suppressing the detachment of large metallic particles and improving the anodic efficiency [39].

Microalloying Mg anode with the combination of Ca/In is found to be a feasible strategy to improve the battery properties of primary Mg-air systems. In a recent study by Deng et al. [40], two micro-alloyed Mg anodes, with concentrations Mg–0.1%, Ca–0.2%In and Mg–0.2% Ca–0.4% in weight percent were fabricated and evaluated in terms of corrosion performance, wasteful discharge rate, and discharge activity. The conventional gravity casting method was utilized to prepare all the micro-alloyed Mg anodes reported in the experimental analysis. These micro-alloyed Mg–Ca–In anodes exhibited higher voltage than those based on commercially acceptable anode material for aqueous Mg primary batteries AZ31 and binary Mg–0.1Ca anode. Also battery life cycle is significantly enhanced by utilizing Mg–Ca–In anodes [40].

Recent studies by Chen [41] and his colleagues on anode material made up of quasicrystal of Mg–Zn–Y for its discharge and electrochemical behaviors as the anode for Mg-air battery. The anode was made by melting pure Mg (99.99%), pure Zn (99.99%), and Mg–30%Y master alloys in an iron crucible by a resistance furnace. The Mg–Zn–Y molten alloy was pulled into a copper mold to make the Mg ingots to be used as anode. The quasicrystal contained anodes has shown higher open-circuit potential and electrochemical activity but a low self-corrosion rate compared to ZK60 alloy (Zn-6.02% Zr–0.65% Mg–93.33% set as the control group given in wt. %). Mg–Zn–Y anodes display shallow discharge pits with smooth surface morphology and dense small holes, which are beneficial to the drop of discharge products [41].

12.6 ELECTROLYTES IN MAGNESIUM-AIR BATTERY

The basic electrolyte of an Mg-air battery is a neutral saline aqueous solution that contacts with the anode and cathode which highly influences the electrode reactions. There are several limitations for the use of aqueous media as electrolyte. Since the major deficiencies in Mg-air battery technology which are high polarization and low coulombic efficiency are not only related to the electrodes but also to the electrolyte, choosing the most suitable electrolyte lends itself to the better performance of Mg-air batteries. In terms of anode reaction electrolyte directly affects the corrosion of Mg anode. The extent of anode corrosion in magnesium anode is dependent upon the type of salt content and its concentration in the electrolyte. The higher the salt concentration greater the corrosion rate. As the magnesium corrosion progresses it increases the production of magnesium hydroxide and causes an increase in the pH of electrolyte. Additives are added to regulate the pH value by keeping magnesium hydroxide in solution.

Seawater is a natural electrolyte which has been used in magnesium-air battery studies for its abundance. Seawater has an average salt concentration of about 3.2% and also contains dissolved oxygen, which is the main cause of the corrosion of the magnesium anode and for reducing water to be the prevalent cathodic reaction. Aerobic bacteria are active in a high oxygen-content environment. But when the oxygen level is lowered aerobic bacteria are replaced by anaerobic bacteria. Some of these anaerobic bacteria have shown to be able to utilize hydrogen formed during the cathodic corrosion process which is beneficial [42]. Magnesium has a higher resistance in alkaline solutions than in acidic or neutral solutions. Some studies have shown that selecting or tuning the pH value of electrolytes above 10 promotes the corrosion resistance of Mg. In addition to the influence of pH, the anions in the aqueous solution also play an important role in this aspect. A large number of electrolytes containing different anions have been studied for the purpose of understanding the influence of the anions. It has been found that Cl^- and SO_4^{2-} are more likely to attack Mg and thus lead to high corrosion rates, while NO_3^- or NO_2^- ions, though they also attack magnesium, do not do so to the same extent as chlorides.

Recent Studies by Wang et al. [43], the addition of Mg^{2+} complexing agents into the electrolyte of Mg-air battery has shown an improved discharge performance of water-based Mg-air batteries. The aqueous Mg-air battery with an additive solution was tested and compared to a pure NaCl solution. The results show that the addition of Mg^{2+} complexing agents can efficiently increase the discharge voltage and specific energy of respective Mg-air batteries [43]. Vaghefinazari et al. [44] conducted studies toward enhanced aqueous Mg-air battery technology with electrolytes containing two strong complexing agents, nitrilotriacetic acid (NTA) and Tiron at different concentrations. The results showed that additives Tiron and NTA into water-based Mg-air battery electrolyte improves the battery discharge potential at concentrations in the ranges of 0.1–0.001 and 0.1–0.01 M, respectively. Also concentration of the additives was found to be an important factor in the trade-off between the improvement in the battery voltage and the utilization efficiency. A decrease in the concentration of the complexing agents in the battery electrolyte usually leads to a higher UE but a lower discharge potential [44].

There are several studies to characterize the magnesium electrode in a range of organic media. These investigations, including some polymeric complexes, have been conducted with a focus on understanding the interactions between Mg ions and the polymer solvent. Also the subsequent transport properties in these electrolytes. These studies have indicated promising physical and chemical properties of magnesium polymer electrolytes but any significant electrochemical characterization has not been achieved. The limited success with organic electrolytes has spurred interest in finding new media which can harness the full capability of magnesium electrodes. Room temperature ionic liquids have come out as promising alternative electrolytes in view of an increasing range of batteries with a unique combination of properties such as negligible vapor pressure, low flammability, high ionic conductivity, and high electrochemical stability. Khoo et al. [45] studied the potential for ionic liquid electrolytes to stabilize the magnesium interface for magnesium/air batteries depicted in Figure 12.21. The studies showed that when immersed in the ionic liquid, hydrogen evolution from the magnesium is not strongly observed compared to aqueous electrolyte systems.

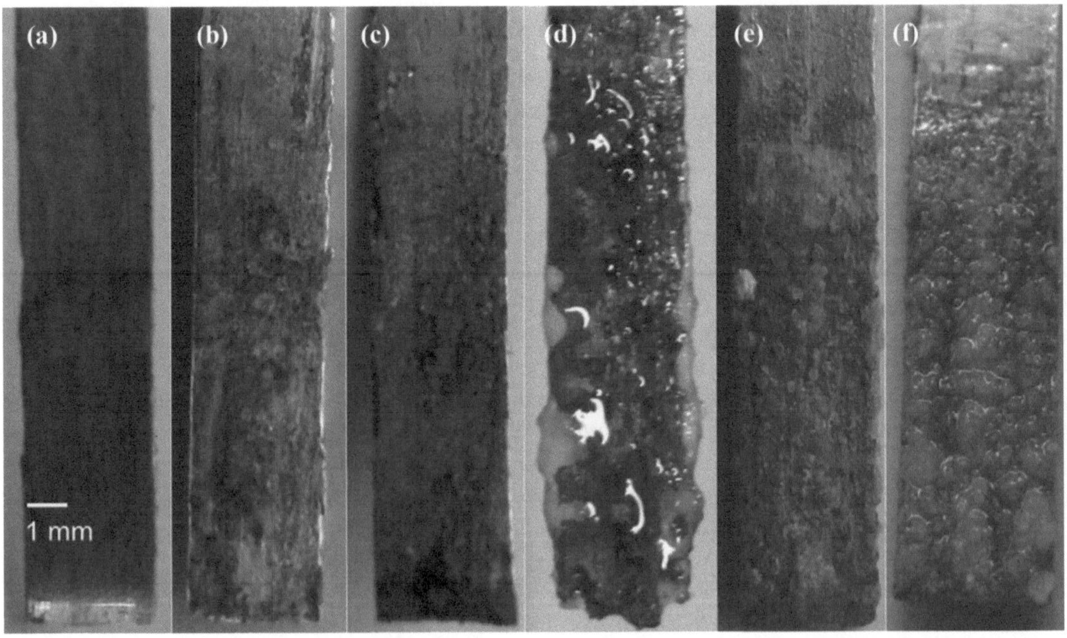

FIGURE 12.21 Images of the magnesium surface in (a) 0 and (b) 8 wt% H_2O [P6,6,6,14][Cl] for 48 h. (c) After 24 h discharge in 8 wt% H_2O and (d) after 96 h discharge in 8 wt% H_2O [P6,6,6,14][Cl]. (e) After 24 h discharge in 8 wt% H_2O [P4,4,4,10][Cl] and (f) after 10 h discharge in 3 wt% H_2O [P6,6,6,14]. Adapted and reproduced from Ref. [45].

An electrolyte based on a phosphonium chloride ionic liquid/water mixture has shown to be a promising candidate for magnesium/air batteries. Studies have shown that water-saturated trihexyl (tetradecyl) phosphonium chloride ([P6,6,6,14][Cl]) ionic liquid is a promising electrolyte for magnesium-air batteries. The electrolyte is composed of phosphonium chloride ionic liquid/water mixture. This electrolyte is able to stabilize the magnesium interface by forming a semi-passive interfacial layer in the form of an amorphous gel-like interface on Mg that leads to a level of passivation when the cell is at an open circuit. But, during discharge, this interfacial film becomes sufficiently conductive to allow stable discharge of Mg for long periods; this directly contributes toward the longevity of actual Mg/air batteries as the reactivity of the Mg is suppressed when the battery is in a dead state kept in backup applications [46].

In recent years, the replacement of aqueous electrolytes with gel polymer electrolytes appeared to be an encouraging approach to solve the aforementioned problems of leakage while maintaining similar ionic conductivity values to the liquid electrolytes. Using carbohydrate polymers such as starch, cellulose, chitosan, alginic acid, xanthan, agarose, and carrageenan has been broadly investigated for electrochemical devices. Jia et al. [47] reported the development of a biocompatible polymer electrolyte based on a biocompatible ionic liquid embedded in a host biopolymer, chitosan (CS). The integrated solid-state batteries are fabricated with the CS–[Ch][NO_3] film sandwiched between PPy–pTS cathode and Mg alloy anode, as shown schematically in Figure 12.22. In the proposed system the polymer electrolyte functions as an electrolyte, separator, and glue to hold all of the device components together.

With its small device dimensions and biocompatibility, this battery can be developed as a promising power source for miniaturized implantable medical devices [48]. Agar is a polysaccharide extracted from the cell wall of Rhodophyta (red algae), which is one of the most ancient lineages

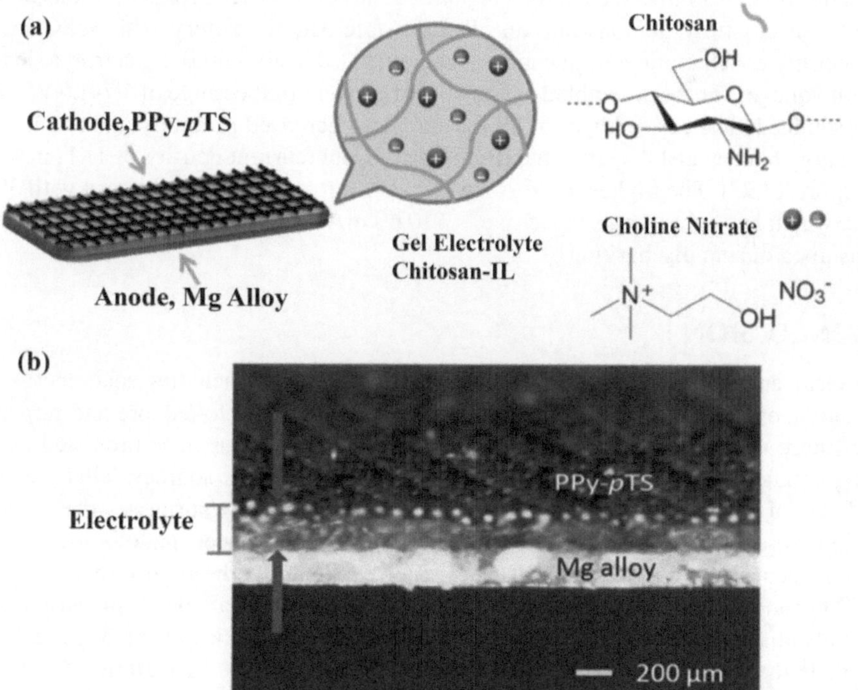

FIGURE 12.22 Schematic configuration (a) and optical image of cross-sectional view (20° tilted) (b) of the integrated solid-state Mg-air battery. Adapted and reproduced from Ref. [47]. Copyright © 2014 American Chemical Society.

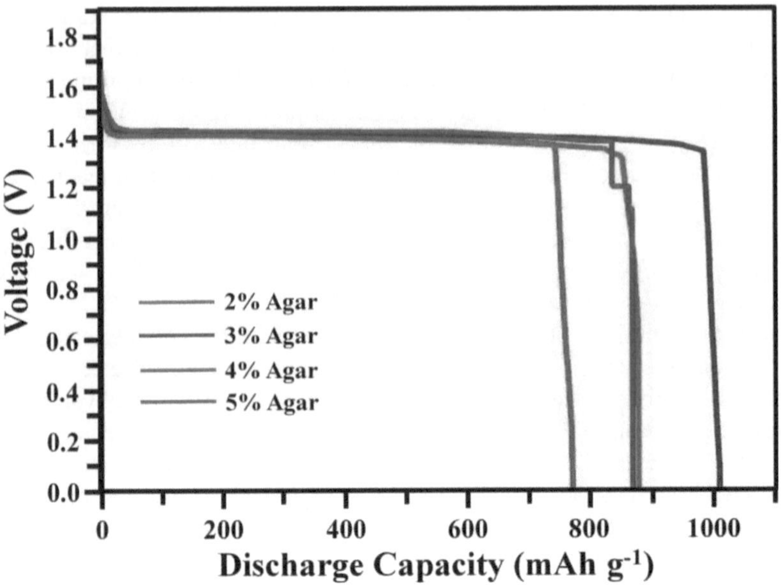

FIGURE 12.23 Discharge performance of Mg-air battery with different WSG incorporated agar gel electrolytes at the current density of 11.11 mA cm^{-2}. Adapted and reproduced from Ref. [49]. Copyright © Springer.

of photosynthetic eukaryotes. Industrially, agar is yielded from agarophyte genera *Gelidium* and *Gracilaria*. Liew et al. [49] investigated the effects on the electrochemical performance in magnesium-air batteries using environment-friendly water-soluble graphene (WSG)-incorporated gel electrolyte. WSG is synthesized and incorporated into the different concentrations of agar. They were able to successfully demonstrate an all-solid-state Mg-air battery with WSG-incorporated agar gel electrolyte, achieving a high capacity and energy density without electrolyte leakage. The open-circuit voltages for the assembled Mg-air batteries were in the range of 1.7–1.6 V just after the cell was assembled. The cell voltage for all the samples decreased promptly and reached a flat and stable discharge plateau at 1.4 V after the discharge current (current density of 11.11 mA cm^{-2}) was applied (Figure 12.23). The highest performance was given by the Mg-air battery with 3% agar gel electrolyte, which has a discharge capacity of 1,010.60 mAh g^{-1} with respect to the mass of the Mg anode consumed during discharging.

12.7 CONCLUSION

Search for clean energy alternatives and renewable energy sources, including solar energy, fuel cells, supercapacitors, and batteries, has become a current research trend to explore and improve performance for future applications. Magnesium-air battery technology has been promised as one of the outstanding candidates for the next generation of renewable power sources. This technology has been the focus of intensive research. In fact, its volumetric density surpasses the value of Li-air batteries. In this review, we have tried to highlight the recent advances made in the Mg-air battery technology in detail. For anode materials, the major challenge is the high level of corrosion. Mg alloys and Mg nanoparticles can improve the performance of the Mg anode. Also, various fabrication techniques also have shown improved corrosion resistance capabilities in the Mg anode. Different types of anode materials were scrutinized for the fabrication of Mg-air batteries. Commonly used alloys are Al, Zn, Ce, In, Ga, etc., exhibiting higher anodic performance compared to pure Mg metals. For the anodes, more studies should come up with the high hydrogen over potential elements for the alloy to better utilize Mg through structure design and various fabrication processes. There have been some results discussed with regard to the development of rechargeability of Mg-air batteries with organic electrolytes; the maintenance of rechargeability over the continuous cycle of operation has still not been able to be successfully demonstrated using organic electrolytes. The search is for the need for the development of electrolytes that are stable with both the anodic and cathodic active materials remaining at their redox electrode potentials. This has pointed toward the studies on hybrid electrolytes as a solution for the development of rechargeable Mg-air battery technology. Even though the research on the use of the hybrid electrolyte is still in its infancy, the success of this technology in Li-air batteries is an eye-opener for the future prospects of developing a rechargeable Mg-air battery with hybrid electrolytes. Although there are various technical hurdles that still remain to be identified and to be crossed in developing the rechargeable Mg-air battery, it is worth the effort as it may hold the key to developing a viable renewable energy storage system for future generations.

REFERENCES

1. Winter M, Brodd RJ (2004) What are batteries, fuel cells, and supercapacitors? *Chem Rev* 104:4245–4269. https://doi.org/10.1021/cr020730k
2. Linden D (1995) *Handbook of Batteries*. Elsevier, 36, 4, 265. ISSN :0140-6701.
3. Fu J, Cano ZP, Park MG, et al (2017) Electrically rechargeable zinc-air batteries: progress, challenges, and perspectives. *Adv Mater* 29:1604685. https://doi.org/10.1002/adma.201604685
4. Goel P, Dobhal D, Sharma RC (2020) Aluminum-air batteries: a viability review. *J Energy Storage* 28:101287. https://doi.org/10.1016/j.est.2020.101287
5. Li CS, Sun Y, Gebert F, Chou SL (2017) Current progress on rechargeable magnesium-air battery. *Adv Energy Mater* 7:1–11. https://doi.org/10.1002/aenm.201700869
6. Li Y, Dai H (2014) Recent advances in zinc-air batteries. *Chem Soc Rev* 43:5257–5275. https://doi.org/10.1039/c4cs00015c
7. Li F, Wei Z, Manthiram A, et al (2019) Sodium-based batteries: from critical materials to battery systems. *J Mater Chem A* 7:9406–9431. https://doi.org/10.1039/c8ta11999f
8. Aurbach D, Weissman I, Gofer Y, Levi E (2003) Nonaqueous magnesium electrochemistry and its application in secondary batteries. *Chem Rec* 3:61–73. https://doi.org/10.1002/tcr.10051
9. Zhang T, Tao Z, Chen J (2014) Magnesium-air batteries: from principle to application. *Mater Horizons* 1:196–206. https://doi.org/10.1039/c3mh00059a
10. Shiga T, Hase Y, Kato Y, et al (2013) A rechargeable non-aqueous Mg-O_2 battery. *Chem Commun* 49:9152–9154. https://doi.org/10.1039/c3cc43477j
11. Shiga T, Hase Y, Yagi Y, et al (2014) Catalytic cycle employing TEMPO-anion complex to obtain secondary Mg-O2 battery. *J Phys Chem Lett SI* 5:1648–1652
12. Zhou H, Wang Y, Li H, He P (2010) The development of a new type of rechargeable batteries based on hybrid electrolytes. *ChemSusChem* 3:1009–1019. https://doi.org/10.1002/cssc.201000123
13. Wang Y, Zhou H (2009) A new type rechargeable lithium battery based on a Cu-cathode. *Electrochem Commun* 11:1834–1837. https://doi.org/10.1016/j.elecom.2009.07.031
14. Hoey GR, Cohen M (1958) Corrosion of anodically and cathodically polarized magnesium in aqueous media. *J Electrochem Soc* 105:245. https://doi.org/10.1149/1.2428817
15. Song G, Atrens A, Stjohn D, Nairn J, Li Y (1997) The electrochemical corrosion of pure Magnesium 1 N NaCl. *Corros Sci* 39:855–875.
16. Song G, Atrens A (2003) Understanding magnesium corrosion. A framework for improved alloy performance. *Adv Eng Mater* 5:837–858. https://doi.org/10.1002/adem.200310405
17. Makar GL, Kruger J (1993) Corrosion of magnesium. *Int Mater Rev* 38:138–153.
18. Ma Y, Li N, Li D, et al (2011) Performance of Mg-14Li-1Al-0.1Ce as anode for Mg-air battery. *J Power Sources* 196:2346–2350. https://doi.org/10.1016/j.jpowsour.2010.07.097
19. Banov BI, Vasilchina HC (2000) Environmentally friendly cathode materials for Li-ion batteries. *Bulg Chem Commun* 43:7–16.
20. Zhao J, Yu K, Hu Y, et al (2011) Discharge behavior of Mg-4 wt%Ga-2 wt%Hg alloy as anode for seawater activated battery. *Electrochim Acta* 56:8224–8231. https://doi.org/10.1016/j.electacta.2011.06.065
21. Huang G, Zhao Y, Wang Y, et al (2013) Performance of Mg-air battery based on AZ31 alloy sheet with twins. *Mater Lett* 113:46–49. https://doi.org/10.1016/j.matlet.2013.09.041
22. Yuasa M, Huang X, Suzuki K, et al (2014) Effects of microstructure on discharge behavior of AZ91 alloy as anode for Mg-Air battery. *Mater Trans* 55:1202–1207. https://doi.org/10.2320/matertrans.MC201403

23. Yuasa M, Huang X, Suzuki K, et al (2015) Discharge properties of Mg-Al-Mn-Ca and Mg-Al-Mn alloys as anode materials for primary magnesium-air batteries. *J Power Sources* 297:449–456. https://doi.org/10.1016/j.jpowsour.2015.08.042
24. Deng M, Höche D, Lamaka SV, et al (2018) Mg-Ca binary alloys as anodes for primary Mg-air batteries. *J Power Sources* 396:109–118. https://doi.org/10.1016/j.jpowsour.2018.05.090
25. Tong F, Wei S, Chen X, Gao W (2021) Magnesium alloys as anodes for neutral aqueous magnesium-air batteries. *J Magnes Alloy* 9:1861–1883. https://doi.org/10.1016/j.jma.2021.04.011
26. Cheng S, Cheng W, Gu X, et al (2020) Discharge properties of low-alloyed Mg-Bi-Ca alloys as anode materials for Mg-air batteries : in fluence of Ca alloying. *J Alloys Compd* 823:153779. https://doi.org/10.1016/j.jallcom.2020.153779
27. Tong F, Chen X, Wang Q, et al (2021) Hypoeutectic Mg-Zn binary alloys as anode materials for magnesium-air batteries. *J Alloys Compd* 857:157579. https://doi.org/10.1016/j.jallcom.2020.157579
28. Chen X, Liao Q, Le Q, et al (2020) The influence of samarium (Sm) on the discharge and electrochemical behaviors of the magnesium alloy AZ80 as an anode for the Mg-air battery. *Electrochim Acta* 348:136315. https://doi.org/10.1016/j.electacta.2020.136315
29. Li C, Cheng F, Ji W, et al (2009) Magnesium microspheres and nanospheres: morphology-controlled synthesis and application in Mg/MnO_2 batteries. *Nano Res* 2:713–721. https://doi.org/10.1007/s12274-009-9075-y
30. Jiang P, Blawert C, Hou R, et al (2020) A comprehensive comparison of the corrosion performance, fatigue behavior and mechanical properties of micro-alloyed MgZnCa and MgZnGe alloys. *Mater Des* 185:108285. https://doi.org/10.1016/j.matdes.2019.108285
31. Luo Y, Deng Y, Guan L, et al (2020) The microstructure and corrosion resistance of as-extruded Mg-6Gd-2Y- (0–1.5) Nd-0.2Zr alloys. *Mater Des* 186:108289. https://doi.org/10.1016/j.matdes.2019.108289
32. Kaseem M, Hussain T, Baek SH, Ko YG (2020) Formation of stable coral reef-like structures via self-assembly of functionalized polyvinyl alcohol for superior corrosion performance of AZ31 Mg alloy. *Mater Des* 193:108823. https://doi.org/10.1016/j.matdes.2020.108823
33. Wang N, Li W, Huang Y, et al (2019) Wrought Mg-Al-Pb-RE alloy strips as the anodes for Mg-air batteries. *J Power Sources* 436:226855. https://doi.org/10.1016/j.jpowsour.2019.226855
34. Wu J, Wang R, Feng Y, Peng C (2018) Effect of hot rolling on the microstructure and discharge properties of Mg-1.6 wt%Hg-2 wt%Ga alloy anodes. *J Alloys Compd* 765:736–746. https://doi.org/10.1016/j.jallcom.2018.05.070
35. Shi Y, Peng C, Feng Y, et al (2017) Microstructure and electrochemical corrosion behavior of extruded Mg-Al-Pb-La alloy as anode for seawater-activated battery. *Mater Des* 124:24–33. https://doi.org/10.1016/j.matdes.2017.03.058
36. Song D, Li C, Liang N, et al (2019) Simultaneously improving corrosion resistance and mechanical properties of a magnesium alloy via equal-channel angular pressing and post water annealing. *Mater Des* 166:107621. https://doi.org/10.1016/j.matdes.2019.107621
37. Song G, Atrens A, Dargusch M (1998) Influence of microstructure on the corrosion of diecast AZ91D. *Corros Sci* 41:249–273. https://doi.org/10.1016/S0010-938X(98)00121-8
38. Xiao B, Song GL, Zheng D, Cao F (2020) A corrosion resistant die-cast Mg-9Al-1Zn anode with superior discharge performance for Mg-air battery. *Mater Des* 194:108931. https://doi.org/10.1016/j.matdes.2020.108931
39. Feng Y, Xiong W, Zhang J, et al (2016) Electrochemical discharge performance of the Mg-Al-Pb-Ce-Y alloy as the anode for Mg-air batteries. *J Mater Chem A* 4:8658–8668. https://doi.org/10.1039/c6ta02574a
40. Deng M, Wang L, Höche D, et al (2020) Ca/In micro alloying as a novel strategy to simultaneously enhance power and energy density of primary Mg-air batteries from anode aspect. *J Power Sources* 472:228528. https://doi.org/10.1016/j.jpowsour.2020.228528
41. Chen X, Zou Q, Le Q, et al (2020) The quasicrystal of Mg-Zn-Y on discharge and electrochemical behaviors as the anode for Mg-air battery. *J Power Sources* 451:227807. https://doi.org/10.1016/j.jpowsour.2020.227807
42. Jung SP, Yoon MH, Lee SM, et al (2014) Power generation and anode bacterial community compositions of sediment fuel cells differing in anode materials and carbon sources. *Int J Electrochem Sci* 9:315–326
43. Wang L, Snihirova D, Deng M, et al (2020) Tailoring electrolyte additives for controlled Mg-Ca anode activity in aqueous Mg-air batteries. *J Power Sources* 460:228106. https://doi.org/10.1016/j.jpowsour.2020.228106
44. Vaghefinazari B, Höche D, Lamaka SV., et al (2020) Tailoring the Mg-air primary battery performance using strong complexing agents as electrolyte additives. *J Power Sources* 453:227880. https://doi.org/10.1016/j.jpowsour.2020.227880

45. Khoo T, Howlett PC, Tsagouria M, et al (2011) The potential for ionic liquid electrolytes to stabilise the magnesium interface for magnesium/air batteries. *Electrochim Acta* 58:583–588. https://doi.org/10.1016/j.electacta.2011.10.006
46. Yan Y, Khoo T, Pozo-Gonzalo C, et al (2014) Roles of additives in the Trihexyl(tetradecyl) phosphonium chloride ionic liquid electrolyte for primary mg-air cells. *J Electrochem Soc* 161:A974–A980. https://doi.org/10.1149/2.031406jes
47. Jia X, Yang Y, Wang C, et al (2014) Biocompatible ionic liquid-biopolymer electrolyte-enabled thin and compact magnesium-air batteries. *ACS Appl Mater Interfaces* 6:21110–21117. https://doi.org/10.1021/am505985z
48. Ma J, Hu P, Jia X, et al (2021) Organic/inorganic double solutions for magnesium-air batteries. *RSC Adv* 11:7502–7510. https://doi.org/10.1039/d0ra10528g
49. Liew SY, Juan JC, Lai CW, et al (2019) An eco-friendly water-soluble graphene-incorporated agar gel electrolyte for magnesium-air batteries. *Ionics (Kiel)* 25:1291–1301. https://doi.org/10.1007/s11581-018-2710-4

13 Emerging Magnesium-Air Battery Technology

Cathodic Materials

Aswith R. Shenoy, Akhila Das, S. K. Vineeth, Roshny Joy, Vijay Kumar Thakur, Nutan Gupta, and Prasanth Raghavan

13.1 INTRODUCTION

Magnesium air batteries are promising energy storage devices in magnesium-based batteries. Though Li air batteries exhibit a leading role among other batteries, the major consequence remains due to the production of high explosive Li_2–O_2 deposited at the cathode material. Magnesium air battery is an alternative to the Li air battery because it has a higher theoretical capacity, lower cost, and lighter weight and is safer than Li air battery. Mg–O, which gets deposited in the cathode material, is an environmentally benign and safe compound. The key components of magnesium air batteries include magnesium alloys, saline electrolytes, and air cathodes, which exhibit a theoretical voltage of 3.1 V. Magnesium air battery is considered a "refuelable" cell since it can be reused by replacing Mg anode and a fresh electrolyte. The major scientific limitation faced by Mg air batteries is their low columbic efficiency and high polarization behavior. A keen inspection of the challenges faced by these magnesium air batteries is corrosion behavior of the anode materials and sluggish kinetics behavior. Table 13.1 compares different air batteries with energy density and voltage. This chapter scrutinizes the historical development of magnesium air batteries and the major cathode materials explored in magnesium air batteries.

TABLE 13.1
Energy Density and Voltage of Different Air Batteries

Battery Type	Cell Reaction	Cell Voltage (V vs SHE)	Specific Energy Density[a] (kW h kg^{-1})
Li ion	$0.5C_6Li+Li_{0.5}CoO_2 \leftrightarrow 3C+LiCoO_2$	3.8	0.387
Li air	$2Li+O_2 \leftrightarrow Li_2O$	2.91	13.0
Li sulfur	$2Li+S \leftrightarrow Li_2S$	2.2	2.6
Zn air	$2Zn+O_2 \leftrightarrow 2ZnO$	1.65	1.3
Mg air	$Mg+0.5O_{2+}H_2O \leftrightarrow Mg(OH)_2$	3.1	6.8

Source: Adapted and reproduced from Ref [1].

[a] The theoretical specific energy density $\varepsilon_M = {-nFE}/{\sum M}$, where F is the Faraday constant, E is the reaction potential, and M is the molar mass of reactants) is calculated excluding O_2 for metal air batteries.

DOI: 10.1201/9781003310174-13

13.2 SEQUENTIAL DEVELOPMENTAL HISTORY OF MAGNESIUM AIR BATTERIES

Magnesium air battery technology has been known since the 1960s, but its development from a technology into a commercial product has not been possible because of electrochemical and material property barriers arising due to hydrogen generation, precipitate control, electrode materials, and electrolyte composition.

In 1966 W. N. Carson Jr. and C. E. Kent [2] reported the theoretical and experimental studies of high-performance magnesium air primary cells. The "Magair" cell, as they reported it, was based on the use of air (oxygen) as cathode, a magnesium alloy as anode, and a neutral salt as electrolyte. The cell had an open-circuit voltage (OCV) of 1.60 V, with operating voltages of 1.30 V at 10 mA cm^{-2} of air electrode surface and 1.00 V at 60 mA cm^{-2}. Carson and Kent also reported the mechanical rechargeability of the Mg air battery technology by employing periodic replenishment of anode and salt for the electrolyte [2]. In 1966 Charles L. Opitz et al. [3] patented an invention of a galvanic cell, which uses a magnesium anode with seawater electrolyte. The cell used a porous annular basket containing steel wool or equivalent metal filament-type material serving as the cathode. A length of magnesium alloys in which magnesium is the dominant material rod supported generally concentrically in the annular cavity through the basket serves as the anode. The main aim of this invention was to provide an undersea source of electrical power, which is relatively independent of pressure (operating depth) and temperature and normally encounters variations in salinity [3].

In 1969 Hamlen et al. [4] conducted experiments with different anode electrolyte combinations for the operation of a magnesium air battery employing mechanical recharging with the replacement of the anode. Out of the different anode electrolyte combinations, AZ61 magnesium alloy and sodium chloride solutions, between 75% and 18%, were satisfactory electrolytes because of a lower resistive drop in the electrolyte, providing a stable voltage on changing current densities.

A novel type of magnesium air primary cell was developed by S. Sathyanarayana and N. Munichandraiah in 1981. The proposed cell is based on strips of magnesium alloy AZ-21 anodes. The magnesium anode utilization was found to be about 90% at a current density of 20 mA cm^{-2}. The anode showed a low open-circuit corrosion and a relatively uniform pattern of corrosion in the electrolyte. The cell was found to be capable of continuous discharge over several months at a constant cell voltage of about 1 V and a current density of 1 mA cm^{-2} at the cathode [5]. The development of magnesium air batteries is closely associated with the development of various seawater battery applications using seawater electrolytes for magnesium oxygen battery principles. The most important of them in the history of the development of magnesium air batteries are listed below. Hasvold et al. [6] listed the advantages of using magnesium/oxygen batteries with seawater electrolytes as having a high energy density, low price, and long storage potential. He described a power source for the autonomous control system of a subsea well (SWACS) in the Ionian Sea. The unit was deployed in January 1996 at a depth of 180 m. The 650-kWh sea water battery uses anodes made from commercial magnesium alloys, seawater as the electrolyte, and oxygen dissolved in seawater as the oxidant [6]. A joint research project between DGA/BEC (France) and FFI (Norway) focused on the design of a long-range, autonomous underwater vehicle using magnesium fuel and oxygen from the sea. The battery concept is based on the use of seawater as an electrolyte and oxygen dissolved in the ocean as an oxidant in a semi-fuel cell with magnesium anodes as the fuel. The cell delivered 133-W power for 504 h (or 67,032 Wh) and the charge delivered was 56,840 Ah [6].

Feng et al. [7] conducted experiments with cells that use alloys of the Mg–Hg–Ga type as the anode and seawater as the electrolyte. The best electrochemical activity occurs in Mg–5Hg–1Ga alloy, with an open circle potential of −2.124 V. The best corrosion resistance of the Mg alloys occurs in Mg–5Hg–5Ga alloy, with a corrosion current density of 0.42 mA cm^{-2} [7]. Shinohara et al. [8] reported on the development of a seawater battery system for practical application in deep water tanks. The cell consisted of a magnesium anode, which uses seawater as the electrolyte and oxygen dissolved in the seawater as the oxidant. The seawater battery system generated a power of 6 W, which was enough to power and continued to supply power for at least 5 years. The energy density is estimated to be 318 Wh kg^{-1} [8].

From the advent of the concept of magnesium air battery technology in the 1960s till the end of the twentieth century, there have been extensive studies conducted on various aspects of the development of magnesium air battery technology. However, most of them were mainly concerned with the development of primary backup or reserve-type batteries. These require activation by the addition of electrolytes before the battery needs to be switched on for use. Once the battery is activated by adding electrolyte, it produces a continuous discharge till the anode is fully consumed. The development of secondary magnesium air batteries has always been faced with several challenges. Major challenges in the development of secondary magnesium air batteries are:

1. The corrosion of the Mg anode arising from the reaction of the Mg anode with the electrolyte and the poor thermodynamics and kinetics properties of MgO or MgO_2 films, which lead to a large polarization and low coulombic efficiency.
2. High irreversible capability because of the nature of MgO and MgO_2 to act as insulating products, which are electrochemically inert and difficult to convert back into metallic Mg under moderate electrochemical conditions.
3. Finding the perfect combination of the efficient electrocatalysts for oxygen reduction reaction (ORR) in an air cathode [9]. Recent studies have shown some promising results in the development of both primary and secondary magnesium air battery technologies by overcoming various pre-existing challenges.

13.3 MAGNESIUM AIR BATTERY TYPES

The advancements in magnesium air battery technology can be broadly divided into four subcategories: (1) primary Mg air batteries, (2) rechargeable Mg_2Ni air batteries, (3) rechargeable Mg air batteries with organic electrolytes, and (4) futuristic rechargeable Mg air batteries with hybrid electrolytes.

13.3.1 Primary Mg Air Batteries

The magnesium air battery system was initially designed as a reserve-type battery where electrolytes are separated from the electrodes to only activate the cell when required. These were primarily developed as backup energy storage cells for emergency use, making them a potential power source for remote military and telecommunication or weather forecasting installations. As is already detailed in the previous section, a seawater electrolyte-based primary magnesium air battery with oxygen dissolved in seawater as a cathode reactant was also developed. Due to the limited oxygen solubility in seawater, this system was capable of delivering only low current density and low power, making it only to be used in low-power long-term underwater monitoring equipment, floats, observatories, and lighthouses.

The performance of the primary Mg air battery is closely associated with the electrolyte–anode and cathode combination, which dictates the nature and rate of ORR. The primary Mg air batteries with aqueous electrolytes possess many favorable characteristics for energy storage. But magnesium metal anode in an aqueous electrolyte also raises several challenges. Self-corrosion of magnesium anode in aqueous electrolytes results in the formation of insulating $Mg(OH)_2$ product, blocking further anode reactions. It also results in hydrogen gas generation, destroying the pressure balance in the sealed cell space limiting the efficiency of aqueous Mg air batteries.

Richey et al. [10] studied magnesium metal anode corrosion by employing aqueous electrolytes with different salts (NaCl, $NaNO_3$, $NaPO_4$, and a NaCl/$NaPO_4$ mixture) that provide various degrees of magnesium passivation. It was found out that hydrogen gas generation rates were highest in NaCl electrolytes and lowest in $NaNO_3$ electrolytes as shown in Figure 13.1.

But it was observed that there was a higher OCV of the Mg electrode in the $NaNO_3$ electrolyte compared to the NaCl electrolytes. The higher Mg potential in the $NaNO_3$ electrolyte is unfavorable for a battery, as it results in a lower discharge voltage as shown in Figure 13.2.

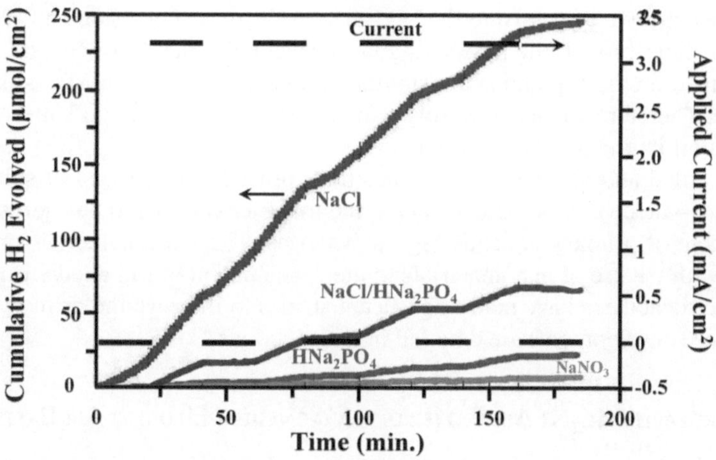

FIGURE 13.1 H_2 evolution from a Mg working electrode (1.2 cm^2 surface area) in 0.5 M NaCl (green), 0.5 M NaCl/0.05 M HNa_2PO_4 (blue), 0.5 M HNa_2PO_4 (red), and 0.5 M $NaNO_3$ (orange) electrolytes. Adapted and reproduced from Ref. [10]. Copyright 2016 Springer.

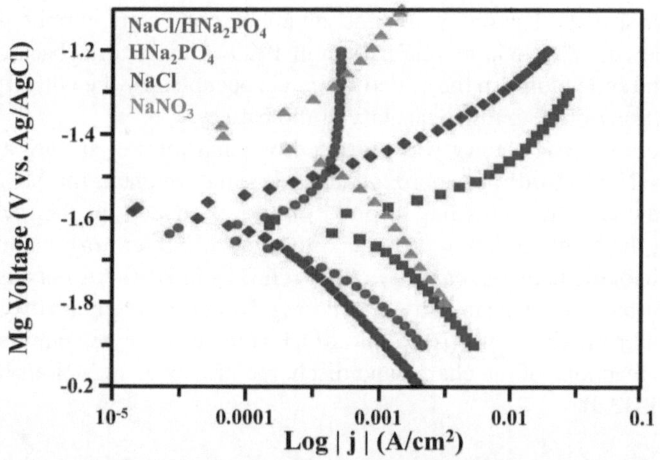

FIGURE 13.2 Apparent Tafel plots in the bulk electrolysis cell for 0.5 M NaCl (green data), 0.5 M NaCl/0.05 M HNa_2PO_4 (blue data), 0.5 M HNa_2PO_4 (red data), and 0.5 M $NaNO_3$ (orange data) electrolytes with a Mg foil working electrode, Pt wire counterelectrode, and Ag/AgCl reference electrode. Adapted and reproduced from Ref. [10]. Copyright 2016 Springer.

This showed that in NaCl electrolyte a weak passivating film existed on the Mg anode surface which facilitated Mg^{2+} dissolution at large current densities and simultaneously induced a large Mg corrosion rate. In contrast, the $NaNO_3$ electrolyte exhibited clear evidence of a significant mass transfer resistance inducing the lowest corrosion. So in essence $NaNO_3$ electrolyte seems to inhibit Mg corrosion, while simultaneously allowing Mg^{2+} dissolution during battery discharge. This phenomenon was attributed to the fact that $NaNO_3$ salts reduced the H_2-evolving corrosion rate at the expense of introducing a nitrate (NO_3^-) to nitrite (NO_2^-) corrosion reaction into the battery. But the total corrosion rate of both reactions (nitrate (NO_3^-) to nitrite (NO_2^-) and H_2-evolving corrosion) in $NaNO_3$ was still smaller compared to the overall corrosion rate in the NaCl and other electrolytes; hence, $NaNO_3$ was found to be a relatively favorable aqueous-based electrolyte for primary Mg air batteries. The studies including the one detailed above have proved that a trade-off exists between the ability of an electrolyte to passivate the Mg surface (decrease corrosion and H_2 evolution) and allow Mg^{2+} dissolution through the passivating layer during discharge. In general, the passivating surface films formed on the Mg anode are significant in corrosion and in Mg^{2+} dissolution during battery discharge. Therefore an ideal electrolyte for aqueous primary Mg air batteries would inhibit corrosion without inhibiting Mg^{2+} dissolution.

The results detailed and various other studies have proved that the lack of suitable and stable aqueous electrolyte–anode composition is one of the major challenges to be addressed for the successful development of primary aqueous Mg air batteries. There are numerous promising results in this direction with the use of nanomaterials in the development of Mg anode; various electrolyte additives and electrocatalysts have made significant strides to improve the primary Mg air battery performance. These developments are detailed in later sections.

13.3.2 Rechargeable Mg_2Ni Air Batteries – Magnesium Hydride Air Battery

Metal hydride (MH) air batteries have the merits of both Ni–MH batteries and metal air batteries. It's a new cell configuration, which is the combination of the rechargeable Ni–MH battery and the metal air battery in the strongly alkaline electrolyte. Ni–MH is a secondary battery consisting of a hydrogen storage alloy as the anode and $Ni(OH)_2$ as the cathode in KOH solution. They had been developed as a suitable substitute for nickel-cadmium cells with qualities like long charge and discharge cycle life, the lack of memory effect, and the low toxicity of the active materials. Ni–MH batteries have seen extensive application in cameras, notebook computers, mobile phones, and hybrid electric vehicles following their commercialization in 1990. A major drawback of the Ni–MH battery is that most of the cell volume in the sealed system is occupied by the $Ni(OH)_2$ electrode, which resulted in the reduction of the nominal capacity of the battery.

The novel concept of a MH battery was proposed by Chartouni et al. in 2002. In commercial Ni–MH batteries, the Ni electrode has approximately the same weight as the MH electrode but with only two-thirds of the capacity. This has inspired the novel concept of the rechargeable MH-air battery by replacing the heavy $Ni(OH)_2$ cathode with a much lighter air-breathing cathode to significantly raise the theoretical energy capacity of the cell [11]. In 2003 Arof et al. [12] developed the first magnesium MH battery. Mechanically alloyed Mg_2Ni and a single air (oxygen) electrode are used as the anode and cathode, respectively, and KOH as the electrolyte solution in a rechargeable MH air battery. The reactions of the charge and discharge processes in MH air batteries are as follows (Equations 13.1–13.3).

$$\text{At the positive electrode: } 4\ OH^- \leftrightarrow O_2 + 2H_2O + 4e^- \tag{13.1}$$

At the negative electrode:

$$Mg_2\,Ni + 4H_2O + 4e^- \xleftrightarrow[\text{Discharge}]{\text{Charge}} Mg_2\,NiH_4 + 4OH^- \tag{13.2}$$

Overall cell reaction:

$$Mg_2\,Ni + 2H_2O + \xleftrightarrow[\text{Discharge}]{\text{Charge}} Mg_2\,NiH_4 + O_2 \tag{13.3}$$

Mg_2Ni has been developed as one of the most important hydrogen storage alloys for the magnesium alloy anodes in MH air batteries. This is primarily because of the low cost of Mg due to its natural abundance, a high hydrogen storage capacity of 3.61 wt% calculated from Mg_2NiH_4, and a theoretical gravimetric energy density of 999 mAh g^{-1}. Arof et al. [12] reported polycrystalline Mg_2Ni alloy powders fabricated by a mechanical alloying method. He developed a MH air battery consisting of a Mg_2Ni anode and an air cathode using manganese as an electrocatalyst and a KOH electrolyte, which is shown in Figure 13.3.

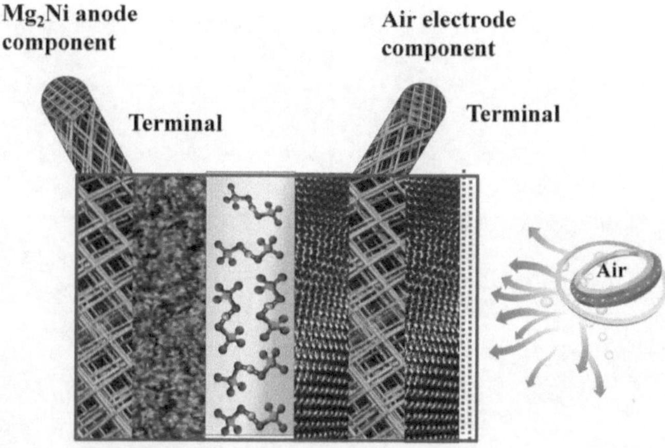

FIGURE 13.3 Cross-sectional view of the assembly of magnesium hydride air battery. Adapted and reproduced from Ref. [12].Copyright © 2003 Elsevier.

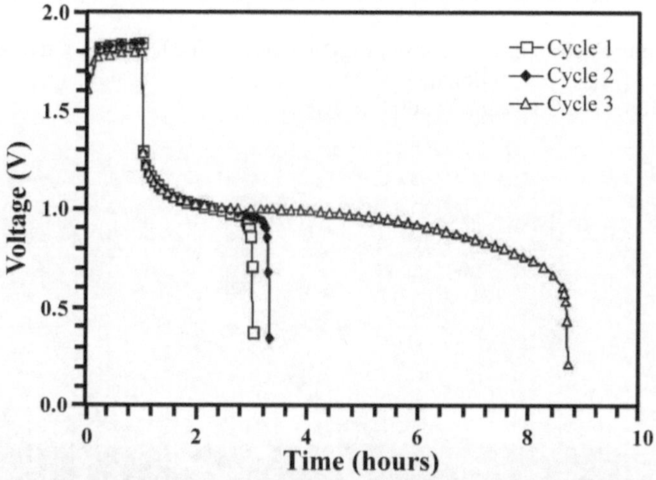

FIGURE 13.4 Charge–discharge curves of $Mg_2Ni|6$ M $KOH|O_2$ battery [12]. Cross-sectional view of assembly of magnesium hydride air battery. Adapted and reproduced from Ref. [12]. Copyright © 2003 Elsevier.

The first discharge–charge cycle of this MH air battery activates the Mg_2Ni alloy so that the discharge capacity gradually increases in the following two cycles illustrated in the charge–discharge characteristics reported by Arof et al. [12] in Figure 13.4.

The longest discharge was observed for cycle 3, which lasted for 8 h. The battery was subjected to 62 cycles. The average charging voltage was about 1.8 V, the average discharge voltage about 0.9 V, and the voltage drop about 0.9 V. This type of cycling phenomenon was strongly supported by Wang et al. [13] in 2009, who found that a part of the Mg_2NiH_4 formed spontaneously decomposes into Mg_2Ni and H_2 gas. It was observed that the anode was expanding probably due to the absorption of water from the electrolyte. The thickness of the anode was observed to expand from 0.4 to about 1.0 cm after cycling. Electron micrographs of the Mg_2Ni anode before and after cycling at different magnifications are presented in Figure 13.5a–d, respectively. The anode is seen as quite smooth and flat before cycling, but after cycling some portions of the surface it becomes thicker as seen in Figure 13.5d.

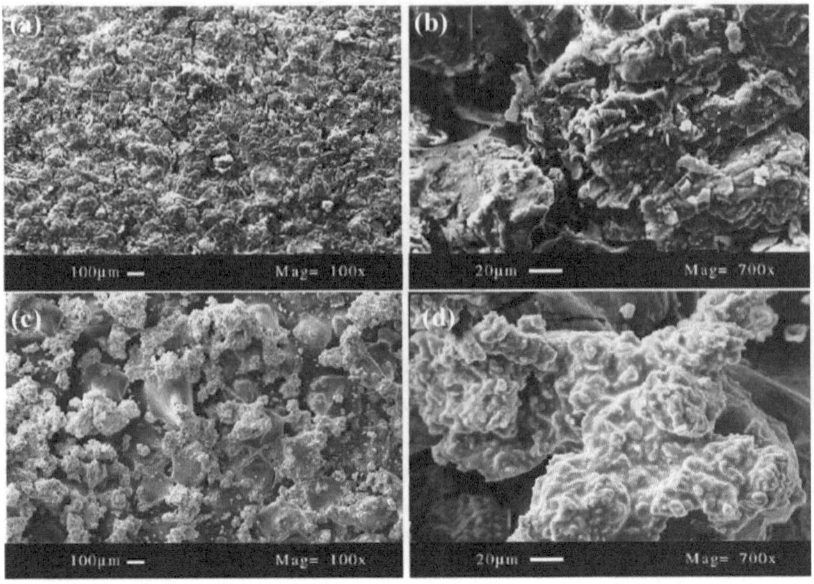

FIGURE 13.5 Scanning electron micrographs of Mg_2Ni anode: (a and b) before and (c and d) after cycling at different magnifications [12]. Cross-sectional view of the assembly of magnesium hydride air battery. Adapted and reproduced from Ref. [12]. Copyright © 2003 Elsevier.

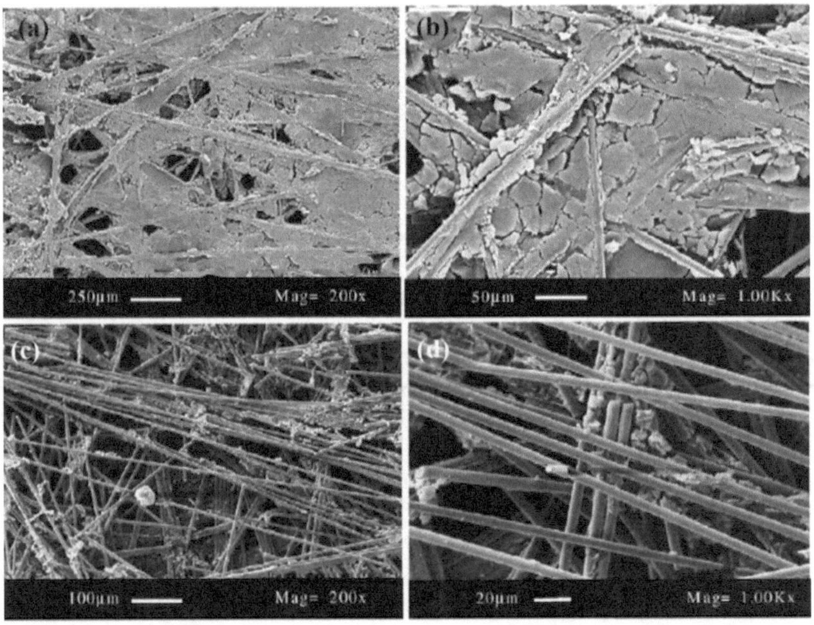

FIGURE 13.6 Scanning electron micrographs of the air cathode: (a and b) before and (c and d) after cycling at different magnifications. Cross-sectional view of the assembly of magnesium hydride air battery. Adapted and reproduced from Ref. [12]. Copyright © 2003 Elsevier.

Electron micrographs of the air electrode shown in Figure 13.6a and b indicate that the catalyst covers the surface of the electrolyte side. But after cycling it can be seen that the catalyst at the surface has peeled off after cycling, as shown in Figure 13.6c and d.

The above results indicate that major challenges regarding the development of mechanically alloyed Mg_2Ni can be the following: (1) the repeated charging and discharging in KOH solution causes continuous degradation of the alloy; (2) the anode degradation to $Mg(OH)_2$ and air electrode damage upon cycling also lead to the thickening of the anode, which might exert excess stress on the anode; and (3) the combined self-corrosion reactions of Mg_2NiH_4 produce $Mg(OH)_2$ and metallic Ni and H_2 gas, which exert stress on the bifunctional air cathode to be easily destroyed at the charging stage during cycling.

The development of practical rechargeable Mg_2Ni air batteries for implementation in portable devices should satisfy the mechanism to overcome the deficiencies reported above: (1) development of an anode with Mg_2Ni micro/nanostructures to control the hydrogen diffusion rate from the active surface to the interior of the MH electrode and to balance the anode expansion during the cycling process; (2) a suitable electrolyte which inhibits alloy corrosion; and (3) powerful electrocatalysts to control the ORR [9].

13.4 CATHODES IN MAGNESIUM AIR BATTERIES

The performance of the Mg air battery is closely related to the air cathode. In addition to the anode corrosion, the low coulombic efficiency of the Mg air battery is attributed to the overpotential of the air cathode caused by the poor kinetics of the ORR. This makes it most important to facilitate improvement in the performance of the air cathode.

The structure of a typical air cathode is composed of a layer, a catalyst layer, and a current collector layer with four layers: a waterproof breathable layer, a gas diffusion, and air diffusion (Figure 13.7).

The cathode is given a physical structure by providing an embedded electrically conductive current collector; it also provides the structure to this cathode but at the same time allows oxygen from the air to diffuse through. A typical collector is in the form of meshes or grids or foam made of metals such as aluminum, Inconel, stainless steel, copper, silver, titanium, or nickel. Graphite plates or sheets with channels are also used as the conductive current collector provided for facilitating air/oxygen flow. A waterproof layer is added to the collector, which is usually a water-repellent porous substance like Teflon or paraffin, which is used to separate the air cathode and electrolyte and, at the same time, is permeable only to O_2 and blocks CO_2 and H_2O.

The air diffusion cathode has its opposite faces exposed to two different environments, an atmosphere and an aqueous solution or an atmosphere and a solid, respectively. The cathode must form a three-phase (gas–solid–liquid) interface where gas, catalyst/carbon, and electrolyte are in contact

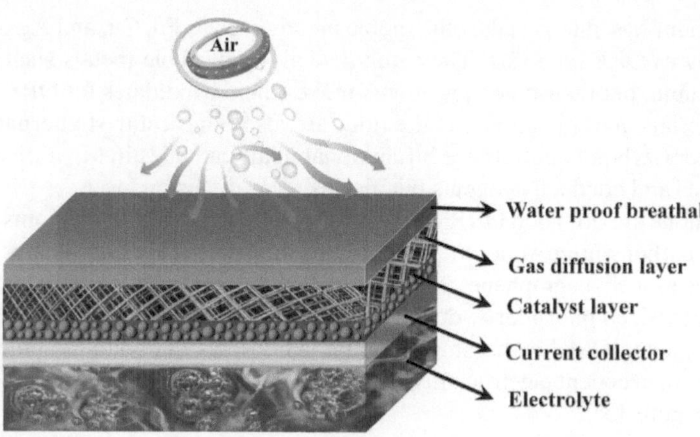

FIGURE 13.7 Structure of a typical air cathode with four layers. Adapted and reproduced from Ref. [1]. Copyright © 2014 The Royal Society of Chemistry.

to facilitate the reaction of gaseous oxygen. During operation, oxygen passes through the cathode by the difference in pressure between the outside and inside of the cell. The catalyst facilitates the reduction of oxygen to an anion via an electrochemical reaction with electrons generated from the oxidation of the magnesium anode. The hydroxyl ions migrate from the air cathode to the magnesium anode to complete the cell reaction. When the battery is in operation, air is drawn into and through the cathodes. The oxygen will be used in the oxidation–reduction reaction between the cathode and the anode. Both oxygen and nitrogen gases pass through the cathode gas diffusion layer together and react with the catalyst layer; nitrogen gas will diffuse into the electrolyte and thus form bubbles in the electrolyte and be released into the air. The electrolyte is composed of water that contains air by nature. When the water is agitated by a pumping mechanism, any air bubbles present will be dispersed from the electrolyte. Hydrogen is generated as a secondary reaction product on the anode during the operation of a magnesium air battery.

The catalyst layer consists of active catalysts for the ORR, which is dispersed on the surface of the diffusion layer near the electrolyte. The main aim of introducing a catalyst is to lower the overpotential and high polarization, which are usually observed for the ORR in neutral solution because it leads to bad performance of the Mg air battery. The choice of catalyst materials should also depend on their abundance and the cost to make it commercially feasible. The most common catalysts used in metal air batteries include noble metals, carbon-based materials such as carbon nanotubes, graphene, porous carbon, transitional metal oxides, and N-containing metallic complexes.

Platinum has shown an improved ORR activity when used as a catalyst consisting of carbon-supported 2–4 nm Pt particles with the surface. But the high price and the non-abundance of Pt can increase the overall cost of the metal air battery, which has bleak future commercial prospects [14]. To overcome the economic aspect of using pure platinum as a catalyst, alloying platinum with other cheap transition metals has gained more attention. In this regard Nørskov et al. [15] proposed a new set of ORR electrocatalysts consisting of Pd-palladium or Pt-platinum alloyed with early transition metals such as Sc-scandium or Y-yttrium. Platinum–yttrium alloy (Pt_3Y) has shown promising results with increased ORR activity. Lima et al. [16] studied the ORR on carbon-dispersed Pt3Co, PtCo, and Pt_3Ni alloys in KOH electrolyte PtCo alloy, indicating a lower variation of the platinum 5d orbital vacancy from lower to higher potentials due to the presence of cobalt. This increases the electronegativity of platinum, resulting in a higher catalytic activity.

Besides alloying, the modification in morphology of platinum has shown promising results in increasing ORR activity. A sponge-like structure with no ordered structure with numerous voids is obtained by implementing two dealloying methods (electrochemical and chemical) of intermetallic nanoparticles of Cu_3Pt, resulting in enhanced specific and mass activities and higher stability (Figure 13.8).

In addition to platinum-based materials, other noble metals such as Pd, Cu, and Ag or their alloys are studied extensively as ORR catalysts. The catalytic activity of noble metals such as Ag is not comparable with platinum, but their much lower cost makes them candidates for ORR catalysts.

Carbon-based materials and transition metal oxides are also other catalyst alternatives that are currently being studied. Carbon materials are bifunctional in air cathodes in Mg air batteries. They act not only as catalysts and conductive agents but also as a gas diffusion layer.

Carbon materials alone are not good ORR catalysts. Doping them with heteroatoms (P or N), the ORR activities of the carbons improve a lot, especially for high surface areas such as mesoporous carbons and nanotubes as well as graphene.

Yang et al. [17] synthesized phosphorus-doped ordered mesoporous carbons (POMCs) with different lengths. The resultant POMC with a small amount of phosphorus doping is demonstrated as a metal-free electrode with excellent electrocatalytic activity for ORR, coupled with much-enhanced stability, depicted in Figure 13.9.

The fabrication process involved a mixture of a phosphorus-containing carbon source, triphenylphosphine (TPP), and a primary carbon precursor (phenol) infiltrated into a SBA-15 template at room temperature and pyrolyzed at 900°C in an argon environment followed by the removal of the

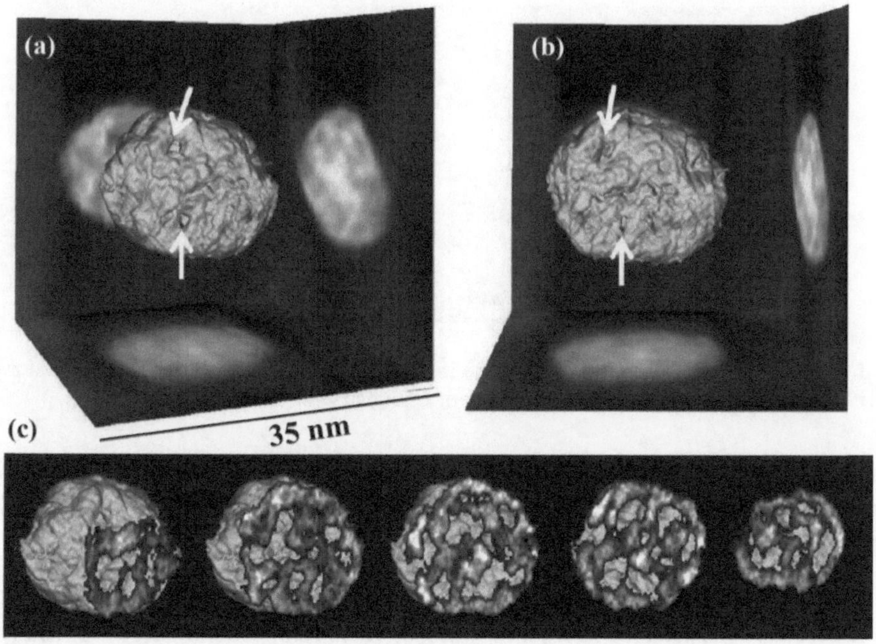

FIGURE 13.8 Three-dimensional tomographic reconstruction of (a and b) a chemically dealloyed nanoparticle (spongy) at two different viewing directions. The arrows indicate channels connecting to the exterior surfaces. (c) A series of consecutive slices through the isosurfaces of such nanoparticles, showing the porous networks. Adapted and reproduced from Ref. [16]. Copyright © 2012 American Chemical Society.

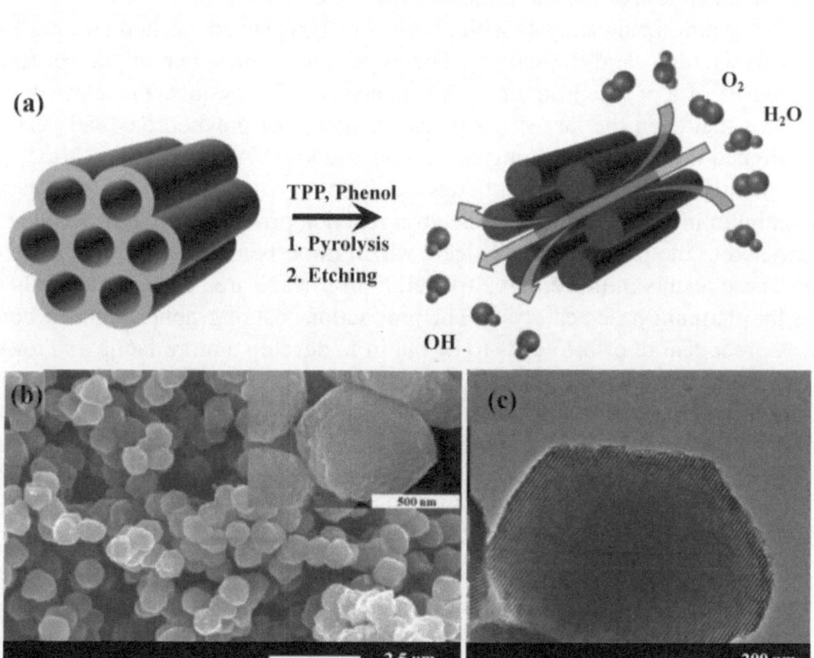

FIGURE 13.9 (a) Schematic illustration of POMC preparation, (b) FESEM image with an inset showing the magnified image of POMC-3, and (c) TEM image. Adapted and reproduced from Ref. [17].Copyright © 2012 American Chemical Society.

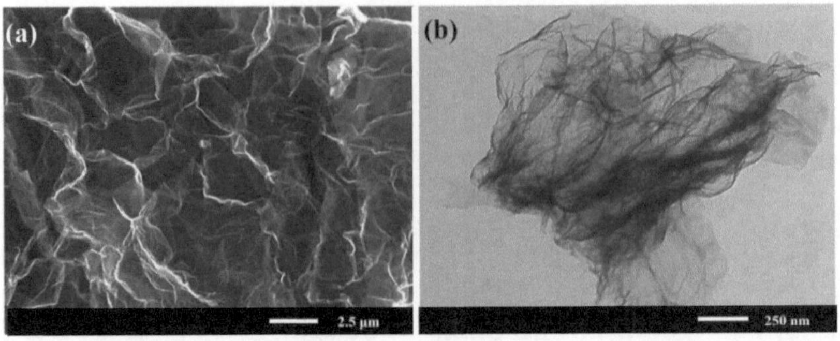

FIGURE 13.10 The typical SEM (a) and TEM (b) images for N-graphene (900) (Adapted and reproduced from Ref. [19]. Copyright © 2012 The Royal Society of Chemistry.

template by hydrofluoric acid (HF) etching to get silica-free POMC. The effect of the channel length of the prepared POMC on catalytic ORR performance was also investigated, showing an increase in activity with a decrease in the length of the POMC [17].

Yang et al. [18] made mesoporous nitrogen-doped carbon materials with high surface areas up to 1,500 m2 g^{-1} with narrow mesopore size distribution of ca. 12 nm diameter and local graphitic carbon structure. The results indicated that the resulting nitrogen-doped carbons show very high catalytic activity. Also, prepared materials exhibit a low onset voltage for ORR in an alkaline medium compared with those of commercial 20 wt% Pt/C catalysts. Geng et al. [19] synthesized nitrogen-doped graphene as a metal-free catalyst for oxygen reduction by heat treatment of graphene using ammonia. Morphological analysis of nitrogen-doped graphene is shown in Figure 13.10. It was found that the electrocatalytic activity and durability of this material are comparable to or better than the commercial Pt/C. The highest ORR activity in alkaline solution was obtained with the catalyst treated at 900°C.

Chen et al. [20] synthesized catalysts with porous Pt aggregates deposited on a carbon nanotube film, which has shown high catalytic activity. The Pt particles with spherical, flower-like, and cubic morphologies consist of a polyhedron and flake-shaped fine Pt crystals. The electrochemical testing results showed that even the largest particles consisting of polyhedrons had very good activity. Vertically aligned nitrogen-containing carbon nanotubes (VA-NCNTs) and N-doped graphene have shown promising results with much better catalytic activity and extended operation stability compared to platinum in alkali with a four-electron reaction process. The improvement in catalytic activity is attributed to the presence of nitrogen, which imparts a high positive charge on adjacent carbon atoms. These results indicate that N-doped high surface area carbon materials are promising substitutes for platinum-based catalysts. The production cost of graphene and carbon nanotubes isn't that low. A great deal of effort needs to be put in to develop a more facile and lower-cost synthesis of graphene and nanotubes (Figure 13.11).

Schematic representations of possible adsorption modes of an oxygen molecule at the coiled carbon nanotubes (CCNTs) (top) and NCNTs (bottom). The C atoms around the pyrrolic-like nitrogen could possess much higher positive charges than the C atoms around the pyridinic-like nitrogen [21]. Transition metal oxides are another important family of non-noble metal ORR catalysts. Studies by Morozan et al. [14] have investigated ORR activities of different types of transition metal oxides for their crystallographic structures, morphologies, and doping effects. Another group of candidates for the electrocatalyst that were studied earlier was the N-containing metal macrocyclic compounds. Cobalt and iron tetra methoxy phenyl porphyrin (CoTMPP and FeTMPP) catalysts loaded on carbons or with preheating show high ORR activity in alkaline solutions [22]. Another alternative is trinuclear metal N_4 complexes $[MN_4]_n$ on carbon black. They have shown much improved electrochemical activity and extended stability compared with commercially available Pt/C catalysts in an alkaline electrolyte (Figure 13.12) [23].

FIGURE 13.11 (a) SEM image of the as-synthesized VA-NCNTs on a quartz substrate. (b) TEM image of the electrochemically purified VA-NCNTs. (c) Digital photograph of the VANCNT array. (d) Calculated charge density distribution for the NCNTs. (e) Schematic representations of possible adsorption modes of an oxygen molecule at the CCNTs (top) and NCNTs (bottom). The C atoms around the pyrrolic-like nitrogen could possess much higher positive charges than the C atoms around the pyridinic-like nitrogen. Adapted and reproduced from Ref. [21]. Copyright © 2009 Science.

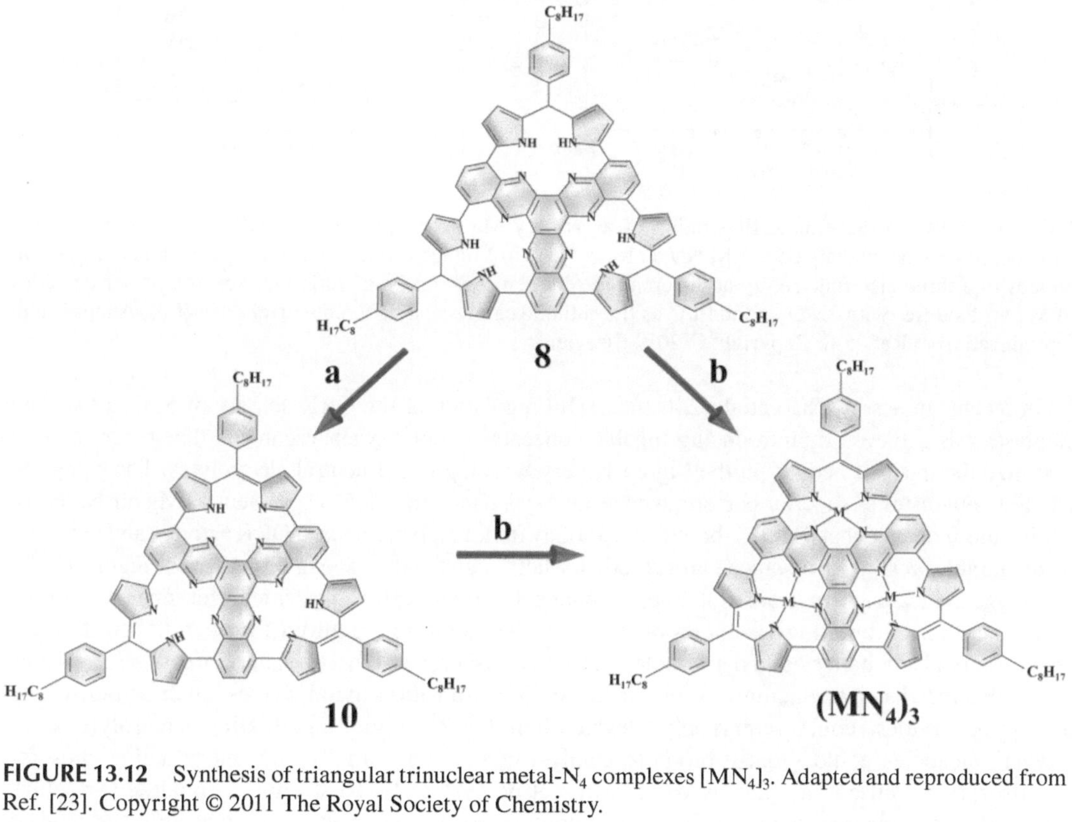

FIGURE 13.12 Synthesis of triangular trinuclear metal-N_4 complexes $[MN_4]_3$. Adapted and reproduced from Ref. [23]. Copyright © 2011 The Royal Society of Chemistry.

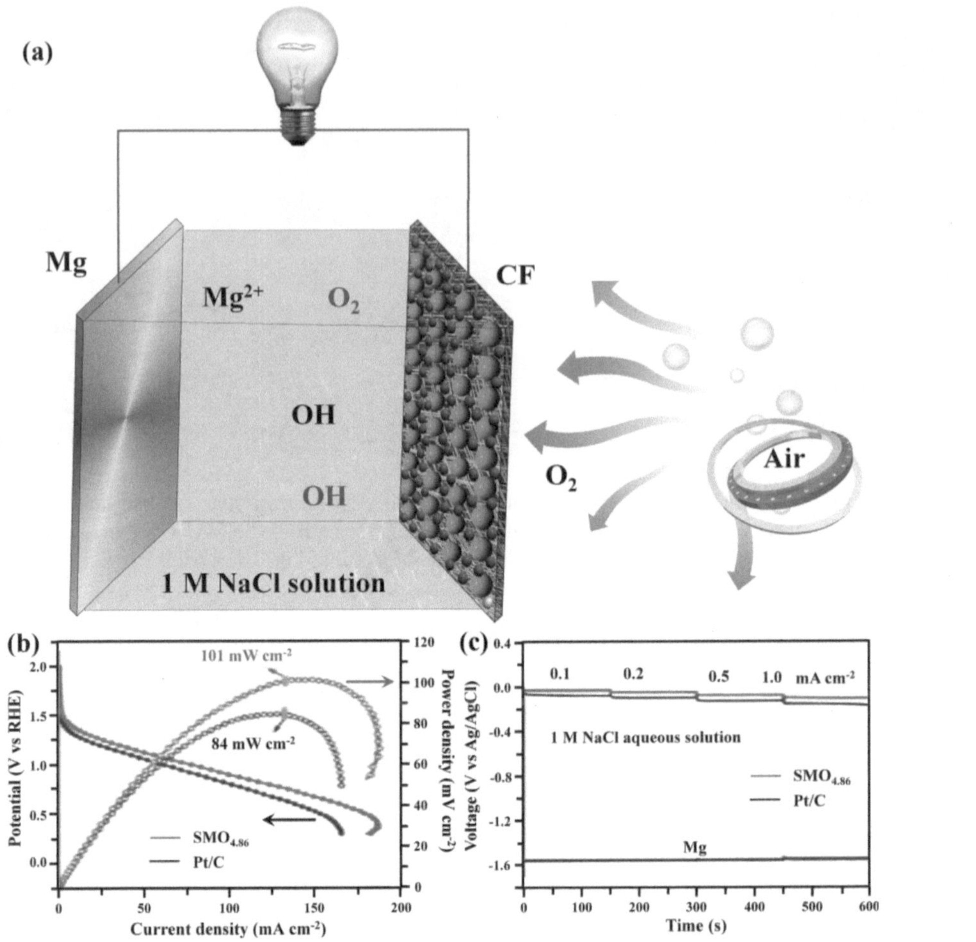

FIGURE 13.13 (a) Schematic illustration of a primary Mg air battery using $SMO_{4.86}$ catalyst as oxygen cathode, Mg anode, and aqueous 1 M NaCl electrolyte. (b) Voltage profiles of both oxygen cathodes and Mg anodes in a three-electrode set-up at different current densities. (c) Polarization curves and power densities of Mg air batteries with $SMO_{4.86}$ and Pt/C as the cathode catalysts at different current densities. Adapted and reproduced from Ref. [24]. Copyright © 2019 Elsevier.

In recent studies by Zhao et al. [24], successful regulation of the ORR activity of $SmMn_2O_5$ nano catalysts was achieved by introducing tunable concentration of oxygen vacancies. They were able to optimize the intrinsic activity in the Mg air battery working in pH-neutral electrolytes. The proposed catalyst demonstrates superior performance as an oxygen electrode for the pH-neutral Mg air batteries. The results indicated that $SMO_{4.86}$-based air cathode offers a high pH-neutral ORR activity, and therefore it can durably work as an advanced air cathode for both liquid and solid Mg air batteries (Figure 13.13).

Oxygen vacancies in transition metal oxides facilitate catalysis critical for energy storage and generation. Promoting vacancies at the lower temperatures required for operation in devices such as metal air batteries and portable fuel cells has proven elusive. There are several studies that indicated that introducing oxygen vacancies into transition metal oxides, such as perovskite and spinel oxides, could remarkably elevate their ORR activity in alkaline electrolyte since oxygen vacancies could greatly promote catalyst conductivity and supply more active sites for O_2 adsorption. Petrie et al. [25] used thin films of perovskite-based strontium cobaltite ($SrCoO_x$), a powerful tool for manipulating the oxygen content under conditions consistent with the oxygen evolution reaction, creating increased oxygen-deficient states. Results showed that the additional oxygen vacancies created through tensile strain enhance the cobaltite's catalytic activity and allow

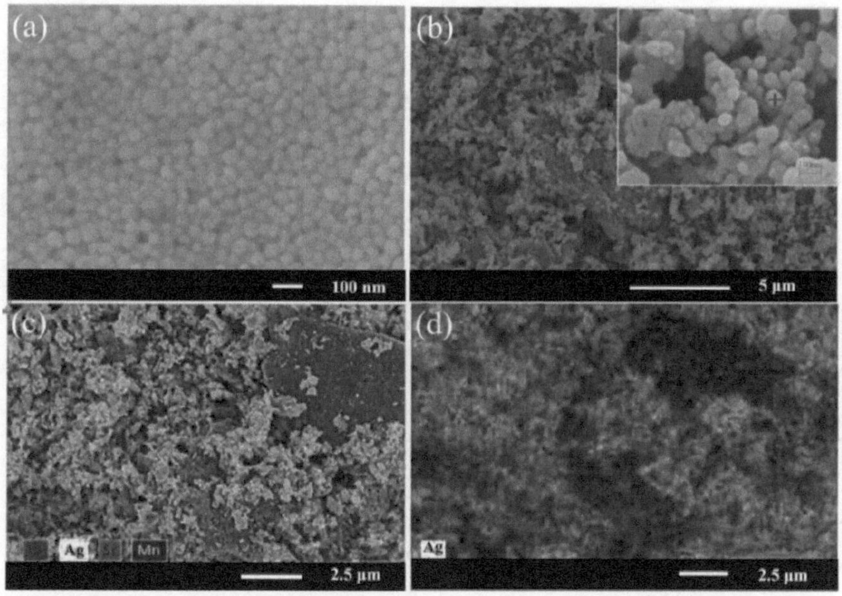

FIGURE 13.14 SEM images of (a) Ag nanoparticles and (b) LSM-14%Ag powder. Energy dispersive X-ray analysis images of element distribution in LSM-14% Ag powder: (c) all elements; (d) Ag. Adapted and reproduced from Ref. [27].

significant control over oxygen vacancies essential for catalysis near room temperature. However, very little is known about the effects of oxygen vacancies on the ORR process and its effects in pH-neutral electrolytes. There are studies which reported that manganese (Mn)-based mullites (i.e., $SmMn_2O_5$) are a promising cathode catalyst for Mg air batteries [26].

In a recent study Wu et al. [27] synthesized an Ag-$La_{0.8}Sr_{0.2}MnO_3$ (LSM-Ag) composite powder as cathodes in application in magnesium air battery. The results on the electrocatalytic performance of LSM-Ag cathodes with different amounts of silver loading revealed that the number of electrons transferred during the ORR of the cathode with an Ag content of 14% by weight reached 3.9, which is very close to that of commercial Pt/C catalysts. The surface morphological analysis and energy dispersive analysis graph of Ag nanoparticles are shown in Figure 13.14. The maximum power density of the Mg air battery made of LSM-14%Ag is 73 mW cm^{-2}, which is greater than that of 63 mW cm^{-2}, found for the pure LSM battery. Also increasing the amount of silver loading allowed improvement in the electrochemical performance of LSM catalysts. The best effect is achieved when the Ag loading exceeds 14%. The overall results indicated that the Ag nanoparticles created additional effective surface active sites for oxygen adsorption and dissociation, making the battery more stable, exhibiting a longer discharge time along with improved specific capacity density and energy density of the cathode (Figure 13.15).

13.5 APPLICATION OF MAGNESIUM AIR BATTERIES

The Mg air cell can be kept in a dead state by withholding the electrolyte into the cell for a long time. Upon call for use, it can be activated quickly by introducing electrolytes into the system [28], which makes it the most suitable backup system that can be employed in hospitals and schools for emergencies and also for expeditionary operations and emergency power for lifeboats. The history of applications with Mg air batteries dates back to the 1960s. American company General Electric was the first to introduce Mg air cells with a Neutral Na–Cl solution as an electrolyte in 1960.

A Canadian company MagPower Systems Inc. is positioned to enter consumer markets with its magnesium air fuel cell (MAFC) as a commercially viable, alternative energy source in the near

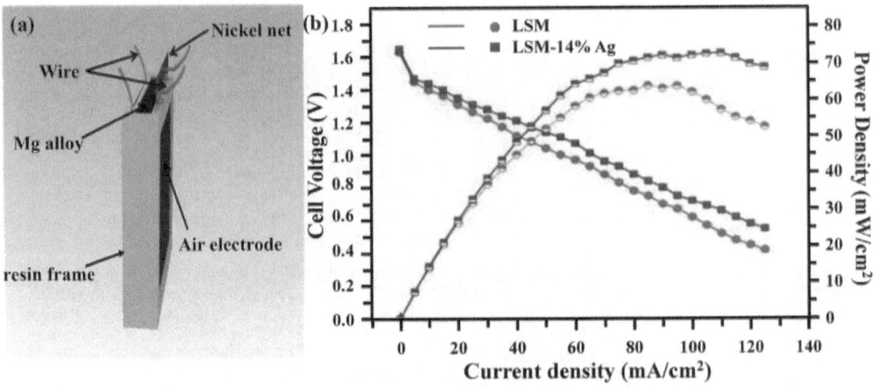

FIGURE 13.15 (a) Schematic diagram of the discharged Mg air battery; (b) cell voltage and power density vs current density. Adapted and reproduced from Ref. [27].

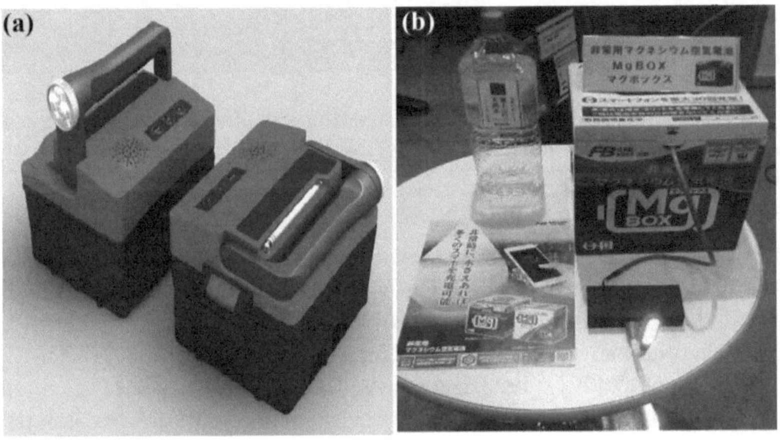

FIGURE 13.16 Pictorial representation of magnesium air fuel cell and (b) magnesium box type magnesium air battery.

future. In the MagPower MAFC system, hydrogen generation is prevented by adding hydrogen inhibitors, which increases their efficiency to 90%. The reported duration of the Mg anode is around 60 h, and fuel replacement is completed in 5 min. Another similar Mg air primary battery MgBOX is developed by Furukawa Battery Co., Ltd. in Japan (Figure 13.16). It is a portable emergency Mg air battery. It was originally developed as an emergency battery for use on land but is now available for use on ships. Once the battery is activated, it begins generating electricity after freshwater or saltwater is added. An MgBOX will provide a source of electricity for up to 5 days via a USB adaptor box or hub. Devices that can be connected to the USB adaptor box include small lights, mobile phones, and smartphones. The USB adaptor box is included inside the MgBOX and can also be used to recharge small portable devices multiple times.

Dalian Institute of Chemistry and Physics has developed a new MAFC system that is intended to serve as a power supply in emergency situations, disaster response, emergency communications, and expeditionary operations. It is designed as a portable backup power source with two Mg air single cells used in a series. The battery provides a power of 5 V in two standard USB charging ports designed for charging most of the portable electronic equipment on the go.

Mg air battery lends itself as a power source in undersea environments, as it can work with seawater as an electrolyte and dissolved oxygen as the oxidant. Mg air cells were used for automatic control systems in offshore oilfield exploration by Norway and Italy in 1996. It was designed to work with a total energy of 650 kWh for 15 years [29]. The US Navy also explored the potential of a novel hybridized MAFC and nickel-zinc battery as the ideal power source for USV sensor payloads. It was created to provide 25 kW of pulse power for more than 14 days [1].

13.6 FUTURE PROSPECTS OF MAGNESIUM AIR BATTERY TECHNOLOGY

Magnesium air battery technology is positioned as one of the promising candidates for the next generation of renewable power sources through the intensive research focus it has received over the years. It has the potential to be one of the power sources in emergency backup storage, defense equipment, electric vehicles, portable charging stations for electronic devices in emergency situations, and dispersed power stations. As mentioned earlier, the aqueous magnesium air battery already has some commercial products but is limited to novelty applications.

Only very few works on fundamental research findings have come out in recent years about secondary Mg air cells compared with primary Mg air batteries. Rechargeable Mg air batteries undergo reversible ORRs and oxygen evolution reactions. One of the major hurdles in the development of rechargeable Mg air batteries is water passivation of the anode surface due to the formation of $Mg(OH)_2$ during the discharge process as the metal gets exposed to water. This creates a surface layer which inhibits any electrochemical reaction accompanied by the generation of hydrogen.

Using magnesium-based alloys as anode instead of magnesium metal is to be explored to achieve a higher specific capacity, wider negative electrode potential, and reduced corrosion due to hydrogen evolution. Proper selection of alloy elements and specific post-treatment for the selected alloys remain the main objectives to improve the performance of the metal anode. Development of an air cathode with lower polarization and longer life has to be done by choosing a proper catalyst by understanding the relationship between the cathode and ORR mechanism in specific catalyst materials. Some studies have reported promising results which demonstrate complex catalysts for decomposing MgO during the discharging process. The poor kinetic efficiency needs to be improved, as it causes large polarization leading to a poor cycling property. The development of cathodes should go hand-in-hand with the development of noncorrosive and electrochemically stable electrolytes.

13.7 CONCLUSION

Search for clean energy alternatives and renewable energy sources, including solar energy, fuel cells, supercapacitors, and batteries, has become a current research trend to explore and improve performance for future applications. Magnesium air battery technology has been promised as one of the outstanding candidates for the next generation of renewable power sources. This technology has been the focus of intensive research. In fact, its volumetric density surpasses the value of Li air batteries. In this review we have tried to highlight the recent advances made in the Mg air battery technology in detail. For anode materials the major challenge is the high level of corrosion. Mg alloys and Mg nanoparticles can improve the performance of the Mg anode. Also, various fabrication techniques have shown improved corrosion resistance capabilities in Mg anode. The air cathode has sluggish kinetics of the ORR, which is a major drawback in its functionality. The catalysts and multilayer structures were shown to play a key role in enhancing the air cathode. Noble metals such as Pt and Ag are the most common catalysts used in the air cathode. But new research has shown that different types of catalysts such as N-doped carbonaceous metal oxides and metal oxide carbonaceous mixtures are used as low-cost substitutes for noble metals. In the neutral electrolyte $Mg(NO_3)_2 6H_2O$ and $NaNO_2$ shows a high performance. High polarization and low coulombic efficiency are the major disadvantages currently holding back the further development of Mg air batteries. To overcome this technical hurdle the studies are now concentrating on finding electrode materials with high activities and low corrosion rates. Further studies should come up with the high hydrogen over potential elements for alloys to better utilization of Mg. For the cathode, the main objective is to find cost-effective alternative catalysts by finding novel materials and improving the activity and durability of the catalysts through hybrid technology and the development of a multilayer structure to improve the contact area between oxygen, electrolyte, and catalyst. Also exploring and developing highly stable electrolytes is also important.

Most of the results are available on primary Mg air battery technology; only the development of rechargeable Mg air batteries is still in its very early stages due to major challenges in comparison with primary Mg air batteries. Rechargeable Mg air batteries offer unique challenges mainly due to the occurrence of reversible ORRs and oxygen evolution reactions. There have been some results discussed with regard to the development of rechargeability of Mg air batteries with organic electrolytes; the maintenance of rechargeability over a continuous cycle of operation has still not been able to successfully demonstrate using organic electrolytes. The search is for the development of electrolytes that are stable with both the anodic and cathodic active materials remaining at their redox electrode potentials. This has pointed toward the studies on hybrid electrolytes as a solution for the development of rechargeable Mg air battery technology. Even though the research on the use of the hybrid electrolyte is still in its infancy, the success of this technology in Li air batteries is an eyeopener for the future prospects of developing a rechargeable Mg air battery with hybrid electrolytes. Although various technical hurdles still remain to be identified and to be crossed over in developing the rechargeable Mg air battery, it is worth the effort as it may hold the key in developing viable renewable energy storage systems for future generations.

REFERENCES

1. Zhang T, Tao Z, Chen J (2014) Magnesium-air batteries: from principle to application. *Mater Horizons* 1:196–206. https://doi.org/10.1039/c3mh00059a
2. Zhang L, Shao Q, Zhang J (2021) An overview of non-noble metal electrocatalysts and their associated air cathodes for Mg-air batteries. *Mater Reports Energy* 1:100002. https://doi.org/https://doi.org/10.1016/j.matre.2020.11.001
3. Keith F. Blurton, Anthony F. Sammells (1979) Metal/air batteries: Their status and potential — a review, *J. Power Sources* 4:263–279. https://doi.org/10.1016/0378-7753(79)80001-4
4. Hamlen RP, Jerabek EC, Ruzzo JC, Siwek EG (1969) Anodes for refuelable magnesium-air batteries. *J Electrochem Soc* 116:1588. https://doi.org/10.1149/1.2411622
5. Sathyanarayana S, Munichandraiah N (1981) A new magnesium - air cell for long-life applications. *J Appl Electrochem* 11:33–39. https://doi.org/10.1007/BF00615319
6. Hasvold Ø, Lian T, Haakaas E, et al. (2004) CLIPPER: a long-range, autonomous underwater vehicle using magnesium fuel and oxygen from the sea. *J Power Sources* 136:232–239. https://doi.org/10.1016/j.jpowsour.2004.03.023
7. Feng Y, Wang R, Yu K, et al (2009) Activation of Mg-Hg anodes by Ga in NaCl solution. *J Alloys Compd* 473:215–219. https://doi.org/https://doi.org/10.1016/j.jallcom.2008.05.054
8. Shinohara M, Araki E, Mochizuki M, et al (2009) Practical application of a sea-water battery in deep-sea basin and its performance. *J Power Sources* 187:253–260. https://doi.org/10.1016/j.jpowsour.2008.10.105
9. Li CS, Sun Y, Gebert F, Chou SL (2017) Current progress on rechargeable magnesium-air battery. *Adv Energy Mater* 7:1–11. https://doi.org/10.1002/aenm.201700869
10. Richey FW, McCloskey BD, Luntz AC (2016) Mg anode corrosion in aqueous electrolytes and implications for Mg-air batteries. *J Electrochem Soc* 163:A958–A963. https://doi.org/10.1149/2.0781606jes
11. Chartouni D, Kuriyama N, Kiyobayashi T, Chen J (2002) Air-metal hydride secondary battery with long cycle life. *J Alloys Compd* 330–332:766–770. https://doi.org/10.1016/S0925-8388(01)01466-9
12. Mohamad AA, Mohamed NS, Alias Y, Arof AK (2003) Mechanically alloyed Mg_2Ni for metal-hydride-air secondary battery. *J Power Sources* 115:161–166. https://doi.org/10.1016/S0378-7753(02)00619-5
13. Wang H, Han L, Hu H, Northwood DO (2009) The hydrolysis behaviour of Mg_2Ni and Mg_2NiH_4 in water or a 6 M KOH solution and its application to Ni nanoparticles synthesis. *J Alloys Compd* 470:539–543. https://doi.org/10.1016/j.jallcom.2008.03.016
14. Morozan A, Jousselme B, Palacin S (2011) Low-platinum and platinum-free catalysts for the oxygen reduction reaction at fuel cell cathodes. *Energy Environ Sci* 4:1238–1254. https://doi.org/10.1039/c0ee00601g

15. Greeley J, Stephens IEL, Bondarenko AS, et al (2009) Alloys of platinum and early transition metals as oxygen reduction electrocatalysts. *Nat Chem* 1:552–556. https://doi.org/10.1038/nchem.367
16. Wang D, Yu Y, Xin HL, et al (2012) Tuning oxygen reduction reaction activity via controllable dealloying: a model study of ordered Cu 3Pt/C intermetallic nanocatalysts. *Nano Lett* 12:5230–5238. https://doi.org/10.1021/nl302404g
17. Yang D, Bhattacharjya D, Inamdar S, et al (2012) Phosphorus-doped ordered mesoporous carbons with different lengths as efficient metal-free electrocatalysts for oxygen reduction reaction in alkaline media. *J Am Chem Soc* 134:16127–16130.
18. Yang W, Fellinger TP, Antonietti M (2011) Efficient metal-free oxygen reduction in alkaline medium on high-surface-area mesoporous nitrogen-doped carbons made from ionic liquids and nucleobases. *J Am Chem Soc* 133:206–209. https://doi.org/10.1021/ja108039j
19. Wang H, Yang Y, Liang Y, et al (2012) Rechargeable Li-O_2 batteries with a covalently coupled MnCo2O4-graphene hybrid as an oxygen cathode catalyst. *Energy Environ Sci* 5:7931–7935. https://doi.org/10.1039/c2ee21746e
20. Chen Y, Wang XX, Li B, et al (2016) Porous Pt aggregates deposited on a carbon nanotube film with high catalytic activity and durability. *ChemElectroChem* 3:1908–1914. https://doi.org/10.1002/celc.201600349
21. Gong K, Du F, Xia Z, et al (2009) Nitrogen-doped carbon nanotube arrays with high electrocatalytic activity for oxygen reduction. *Science* 323:760–764. https://doi.org/10.1126/science.1168049
22. Gojković SL, Gupta S, Savinell RF (1998) Heat-treated iron(III) tetramethoxyphenyl porphyrin supported on high-area carbon as an electrocatalyst for oxygen reduction: I. Characterization of the electrocatalyst. *J Electrochem Soc* 145:3493–3499. https://doi.org/10.1149/1.1838833
23. Liu R, Von Malotki C, Arnold L, et al (2011) Triangular trinuclear metal-N_4 complexes with high electrocatalytic activity for oxygen reduction. *J Am Chem Soc* 133:10372–10375. https://doi.org/10.1021/ja203776f
24. Zhao X, Wang L, Chen X, et al (2020) Ultrafine $SmMn_2O_5$-δ electrocatalysts with modest oxygen deficiency for highly-efficient pH-neutral magnesium-air batteries. *J Power Sources* 449:227482. https://doi.org/10.1016/j.jpowsour.2019.227482
25. Petrie JR, Jeen H, Barron SC, et al (2016) Enhancing perovskite electrocatalysis through strain tuning of the oxygen deficiency. *J Am Chem Soc* 138:7252–7255. https://doi.org/10.1021/jacs.6b03520
26. Lv Z, Ling M, Yue M, et al (2021) Vanadium-based polyanionic compounds as cathode materials for sodium-ion batteries : toward high-energy and high-power applications. *J Energy Chem* 55:361–390. https://doi.org/10.1016/j.jechem.2020.07.008
27. Wu X (1967) Synthesis of Ag-$La_{0.8}Sr_{0.2}MnO_3$ (LSM-Ag) composite powder and its application in magnesium air battery. *Gastron Ecuatorianay Turlocal* 1:5–24
28. Poonam F, Bhatnagar JK, Sharma RC (2021) A review of magnesium air battery systems: From design aspects to performance characteristics. *ISSRG IJMSE*, Turkey, 7 Issue: 18-28
29. Henriksen H, Melvax E, Citi G, et al (2007) Sources 39. *Power* 65:253–261.

14 Advancements in Sodium-Air Batteries

General Electrochemical Mechanisms and Advances

Ajish Babu, S. K. Vineeth, Roshny Joy, Jabeen Fatima M. J., M. J. Nutan Gupta, and Prasanth Raghavan

14.1 INTRODUCTION

Renewable energy sources and growing demand for environmentally safe high-energy storage systems at the lowest cost with adequate material supply are critical for future research and innovation. Lithium-ion batteries (LIBs) with high theoretical capacity (~3,800 mAh g^{-1}), electrochemical potential (–3.04 V vs standard hydrogen electrode), and high energy density (350–400 Wh kg^{-1}) have a stronger foothold in versatile electronics applications [1,2]. In addition to the demand for high-energy-density systems, critical parameters such as recyclability, cost, and safety have been realized by various organizations and electric vehicle (EV) industries to meet future demands of high-performance EVs. To solve these issues, lithium-air batteries (LABs) were first reported by Abraham et al. [3] in 1996, and they have continuously investigated LABs over the past few decades. However, despite the high theoretical specific energy of these systems of 3,502 Wh kg^{-1}, complexity related to cyclability, large overpotentials, lithium deposition on solid electrolyte interphase, and unwanted discharge products such as lithium oxide and peroxide impeded the long-term applicability and realistic spectrum for the realization of EVs.

Coupled with the requirement for high-performance rechargeable batteries, the availability of raw materials and their cost-effective nature have a vital role with regard to the progress toward commercialization [4,5]. In this context, sodium-metal batteries (SMBs) stand over lithium-based batteries due to their huge natural availability (~2.6% by weight on the earth's crust) [6,7]. Sodium (Na) has an ionic radius of 1.02 Å, with a redox potential of –2.7 V vs standard hydrogen electrode, and is the second-lightest metal [8,9]. The availability of Na in the upper continental crust counts approximately as 25,670 ppm, while for lithium it is nearly 22 ppm [10]. Comparing specific discharge capacity, lithium has a higher value (3,800 mAh g^{-1}) than Na (1,165 mAh g^{-1}), but, considering the cost, Na holds a share of about 4% of Li, which makes SMBs an attractive candidate [11]. Per the 2019 U.S. Geological Survey report, the cost of sodium carbonate accounted for nearly 149 USD mt^{-1}, while lithium carbonate as raw material counted for 170,000 USD mt^{-1} [12].

Currently, LIBs are considered a valuable technology for energy storage applications and the major system for building sustainable electricity platforms [13]. However, considering the demand and the uncertainty regarding the availability and cost of lithium, the future of LIBs is ambiguous [14,15]. These drawbacks motivated researchers to find an alternative such as the selection of Na metal for energy storage applications, and the class of batteries which uses metallic Na as an anode are categorized as SMBs [16–18]. The metal-air battery is a major class of high-energy-density batteries (approximately 3–30 times more energy density than LIBs) which are fuelled by the metal oxidation and oxygen reduction processes [19,20]. In addition to the high energy density of metal-air systems, the added advantage is the availability and accessibility of atmospheric oxygen for the devices using

 DOI: 10.1201/9781003310174-14

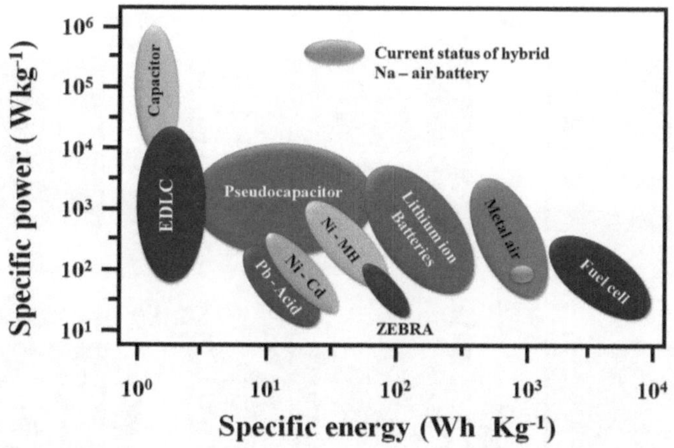

FIGURE 14.1 Ragone plot, depicting the current status of hybrid SAB with different energy storage systems. Adapted and reproduced with permission from Ref. [22]. Copyright © 2020 Wiley-VCH.

metal as an anode and oxygen as an active cathode material [21]. While comparing the specific energy with other metal-air batteries, sodium-air battery (SAB) lies in the considerable range, shown in Figure 14.1, which underlines the performance of SABs and their potential for future battery systems.

Although various approaches have been implemented for enhancing the electrochemical performance of SABs throughout the past few decades, the developments in SAB are still in their infancy. Accelerating the research momentum for the development of cost-effective and high-energy-density SAB involves a complete understanding of electrochemical processes in SAB. Unveiling the complexities of SAB involves understanding the basics of electrochemical mechanisms; recent research interests; exploration of anode, cathode, electrolyte engineering aspects; and identifying the research gaps which can catalyze the developments in SAB. Hence, with this aim, the chapter highlights the basic mechanisms, cutting-edge advancements in SAB technology, and ongoing research endeavors in the field.

14.1.1 Importance of Sodium-Air Batteries

Metal-air battery technology is anticipated to be a prominent next-generation technology; however, the development in this field is still in its infancy. In the past two decades, SMBs in various configurations have been investigated [23]. A solid β-alumina electrolyte and sulfur-embedded graphite cathode are used in sodium–sulfur rechargeable cells [24]. The reaction of Na^+ with molten sulfur produces sodium polysulfides on the cathode surface. Although sodium–sulfur batteries have the capacity for large-scale applications, Na metal halide and SAB are favored as alternatives due to safety concerns [25]. Coupled with the larger abundance and reliability of SMBs, developments in SABs gained momentum in the past two decades. Compared with LAB having a semi-irreversible electrochemical redox pair, oxygen-superoxide system (O_2^- O_2^{2-}), SAB possesses low charging overpotential mainly due to the reversibility of electrochemical redox pair (O_2^- O_2^{2-}) [26,27]. Moreover, SAB possess a high theoretical energy density of 1,108 Wh kg^{-1}, considering the formation of sodium superoxide (NaO_2) as a discharge product [28]. Although SAB relishes these advantages, one of the main challenges is the formation of sodium peroxide (Na_2O_2) as a discharge product, which is thermodynamically favored (standard potential (E^0) for Na_2O_2=2.33 V and for NaO_2=2.27 V) [29]. Moreover, the corrosive effect of electrolyte on the Na anode, dendritic growth, superoxide radical formation, accumulation of discharge products at the cathode, which impede the performance, designing of air cathodes, etc., are the major critical areas which needs to be rectified [28]. Different types of Na-based batteries and their possible cell chemical reactions are depicted in Figure 14.2.

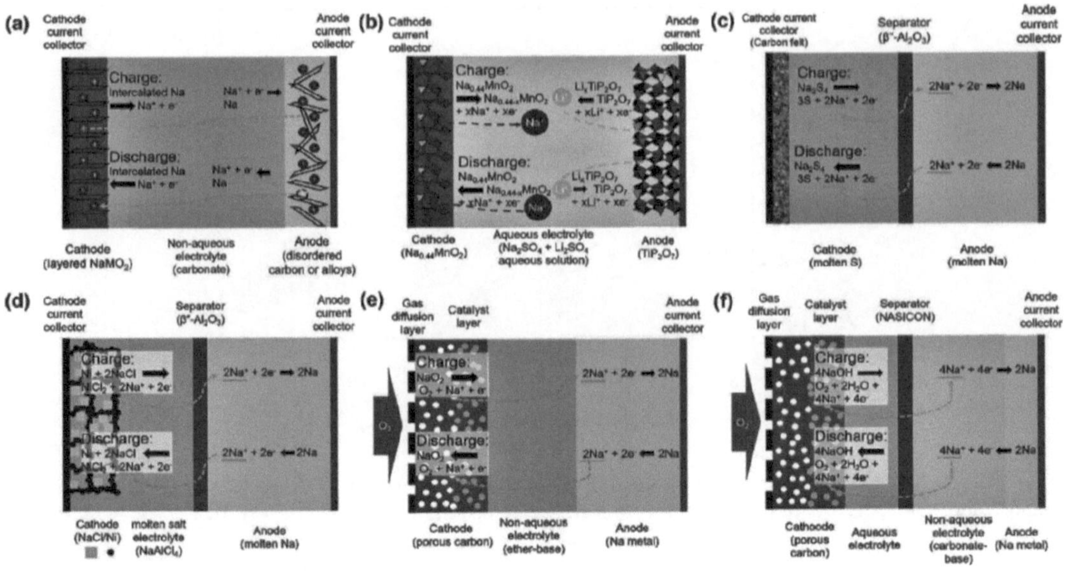

FIGURE 14.2 Schematic illustration showing the architecture and mechanism of various sodium-based battery systems (a) sodium ion battery (SIB), (b) sodium–aqueous electrolyte cell, (c) sodium–sulfur (Na–S) cell, (d) sodium-nickel (II) chloride (Na–$NiCl_2$) cell, (e) non-aqueous sodium-air battery (SAB), and (f) aqueous SAB. Adapted and reproduced with permission from Ref. [25]. Copyright © 2014 Wiley-VCH.

TABLE 14.1
A Comparative Analysis between SAB and LAB Technologies Considering the Cell Voltage, Overpotential, Theoretical Capacities, and Energy Density

	SAB	LAB
Cell chemistry	$Na^+ + O_2 + e^- \rightarrow NaO_2$ $2Na^+ + O_2 + 2e^- \rightarrow Na_2O_2$	$2Li^+ + O_2 + 2e^- \rightarrow Li_2O_2$
Cell voltage	E^0 ($2NaO_2$)=2.27 V (Δ^0=−437.5 kJ mol^{-1}) E^0 (Na_2O_2)=2.33 V (ΔG^0=−449.7 kJ mol^{-1})	
Overpotential (Discharge/charge)		η_{dis} ~ 300 mV η_{ch} ~ 1,300 mV
Theoretical capacity	1,165 $mAhg^{-1}$ (Na) 488 $mAhg^{-1}$ (NaO_2) 689 $mAhg^{-1}$ (Na_2O_2)	3,861 $mAhg^{-1}$ (Li) 1,168 $mAhg^{-1}$ (Li_2O_2)
Energy density	1,108 Wh kg^{-1} (NaO_2) 1,605 Wh kg^{-1} (Na_2O_2)	kg^{-1} (Li_2O_2)

Unlike LABs, the advantages of SABs include the high open-circuit voltage reported with low overpotential due to the formation of electronically conductive NaO_2. Table 14.1 shows a comparative study of various parameters between LAB and SAB.

14.2 AN OVERVIEW ON GENERAL MECHANISM AND THE STATE OF THE ART IN DEVELOPMENT OF SODIUM-AIR BATTERIES

Generally, in any metal-air system, rather than electrochemical reactions, dissolution–precipitation processes are more pronounced. SABs are designed with highly reactive metallic Na as an anode, carbon-based air electrode as a cathode, and an electrolyte which facilitates ionic conduction inside the cell. The electrolyte also functions as a separator between these two electrodes [30]. During the discharge cycle in the cell, Na^+ and electron (e^-), formed by the oxidation of Na anode, will migrate toward the air cathode, and the reduced oxygen molecules at the air cathode combine with these migrated Na^+ and form insoluble sodium oxide as a discharge product [31]. Also, depending on the number of e^- transferred, the discharge product can be NaO_2 and/or Na_2O_2. Once after the discharge cycle, the products formed will be deposited at the air cathode. Hence, the air cathode must be able to accommodate the discharge products so that the cell can deliver better efficiency.

The reaction involved is denoted as follows:

$$Na \rightarrow Na^+ + e^- \quad (14.1)$$

$$O_2 + 2e^- \rightarrow O_2^{2-} \quad (14.2)$$

$$\text{Overall, } Na + O_2 \rightarrow NaO_2 \quad (14.3)$$

$E° = 2.27$ V or

$$2Na + O_2 \rightarrow Na_2O_2 \quad (14.4)$$

$E° = 2.33$ V

The classification of SABs can be according to the electrolytes which the system uses [32]. SAB which utilizes sodium hydroxide (NaOH) solutions as electrolytes with a separator, usually Natrium Superionic Conductors, are referred to as aqueous SABs. In the case of non-aqueous-based SABs, carbonate or ether-based electrolyte solvents are employed. Based on solvents and salts, the composition of discharge products is decided [33,34]. Furthermore, SABs operated using ionic liquids and solid-state electrolytes have also been reported in the past decade.

14.2.1 An Outline on the Development in Sodium-Air Batteries

The general architecture of a SAB consists of a Na metal anode, an electrolyte, and an air cathode (generally carbon based), with a catalyst which facilitates and mitigates the sluggish oxygen reduction reaction (ORR) and oxygen evolution reaction (OER) [35]. The category of the catalyst consists of noble metals transition metals and metal oxides, which facilitate the OER and ORR activity [36]. Each component in a battery has a specific role, which can influence the electrochemical properties; hence, the selection should be judicious. Electrolyte plays a profound role in the stability and composition of the passivation layer formed at the electrode–electrolyte interface referred to as the solid electrolyte interface (SEI) layer [37]. SEI layer is ionically conductive but electronically non-conductive [38]. Ideal SEI permits ionic mobility, while restricting the electronic flow [39]. Unstable SEI catalyzes the growth of dendrites, which are formed by the non-uniform deposition of Na and adversely affect the coulombic efficiency and capacity of the cell [40]. In view of the higher reactivity of Na metal and its realistic handling difficulties, Peled et al. [41] reported the first SAB cell in 2011, which was based on a liquid Na anode, with polymer electrolyte and commercial E-TEK air electrode (Figure 14.3). This system was maintained at a temperature of 105°C–110°C, which is more than the melting temperature of Na metal (97.8°C), to avoid the accumulation on the air electrode [41]. The molten anode was to address challenges such as the eradication of dendrite formation and acceleration of slower cathode reactions which were already seen in Li-based

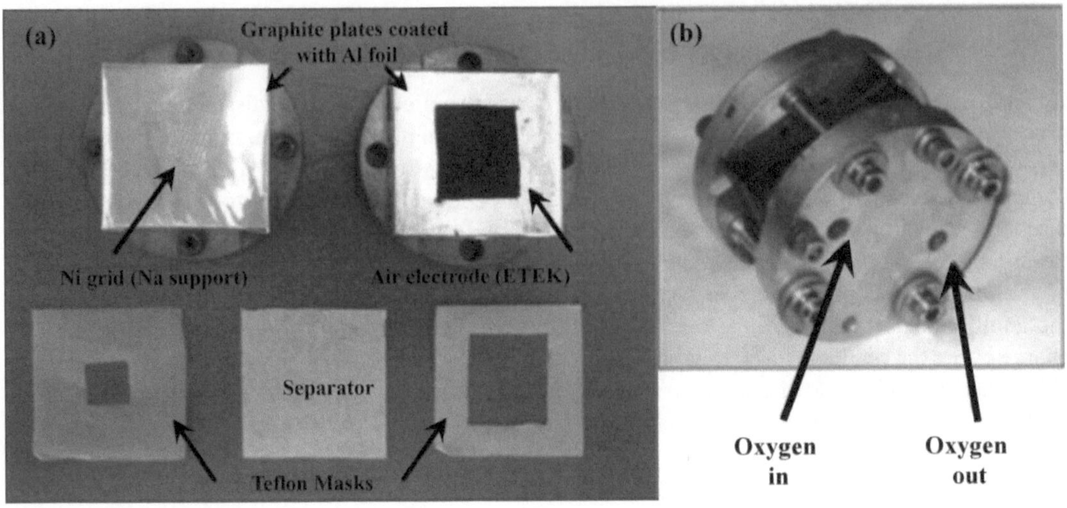

FIGURE 14.3 Images of the first SAB developed by Peled and colleagues. (a) The basic components of SAB and (b) assembled-cell and inlet and outlet passages for oxygen. Adapted and reproduced with permission from Ref. [41]. Copyright © 2010 Elsevier.

batteries. Charge/discharge analysis has shown the completion of about 140 cycles without any sort of dendrite appearance with 85% faradaic efficiency.

Sun et al. [42] crafted a room-temperature working SAB (in an H-shaped glass tube) considering the corrosive nature of liquid Na and overcoming the inconvenience of operating the battery at elevated temperatures as reported by Peled et al. [41]. The cell consisted of electrolyte-containing sodium hexafluorophosphate ($NaPF_6$) and diamond-like carbon film cathode. The authors also found a decrease in discharge capacity with a large current density. The cell obtained an average voltage of about 3.9 V and a maximum discharge capacity of 2,070 mAh g^{-1} with crystalline Na_2O_2 as a discharged product. In addition, the reported vanishing of Na_2O_2 in charged electrodes reveals better reversibility of the prepared SAB. However, the investigators also pointed out the poor cyclability and dendrite formation of the fabricated cell, which is possibly due to the disadvantage of using solid Na anode instead of liquid Na. Hartmann et al. [29] demonstrated, in 2013, that the substitution of lithium by Na offered an unexpected route toward rechargeable metal-air batteries. Scanning electron microscopy (SEM) images revealed that the crystalline and cubic NaO_2 (dimension ~ 10 μm) was formed as the discharge product on carbon-fiber gas diffusion layer cathode even without the addition of any catalyst (Figure 14.4). The prepared cell showed overpotential lower than 200 mV, which is less when compared with LABs.

To evaluate the effect of the SEI layer on electrochemical properties (including faradaic efficiency, which is the ratio of charge and discharge capacity) of a SAB, Peled et al. [43] compared the performance of SAB fabricated using ionic liquid electrolyte and polymer-based electrolyte. The anode consisted of molten Na (operated at 105°C). The ionic liquid electrolyte consisted of 0.5 M sodium trifluoromethanesulfonate ($NaCF_3SO_3$) and 0.5 M sodium sulfate as Na salt in 1-butyl-1-methylpyrrolidinium bis(trifluoromethylsulfonyl)imide solvent. The polymer electrolyte was consisted of $NaCF_3SO_3$ dissolved in polyethylene oxide (PEO), with additional inclusion of alumina nanoparticles and methyl methanesulfonate serving as a precursor for the formation of The SEI. X-ray photoelectron spectroscopy (XPS) showed that SEI consisted of oxosulfur compounds, which prevented the anode from corrosion and ensured larger cycling. Das et al. [44] reported that a mixture of oxygen and carbon dioxide (CO_2) can be utilized for non-aqueous SAB, which showed a high specific energy of 6,500–7,000 Wh kg^{-1}. In 2016, He et al. [45] communicated a work on the effect of concentration of electrolyte salt (sodium trifluoromethanesulfonimide–dimethyl sulfoxide)

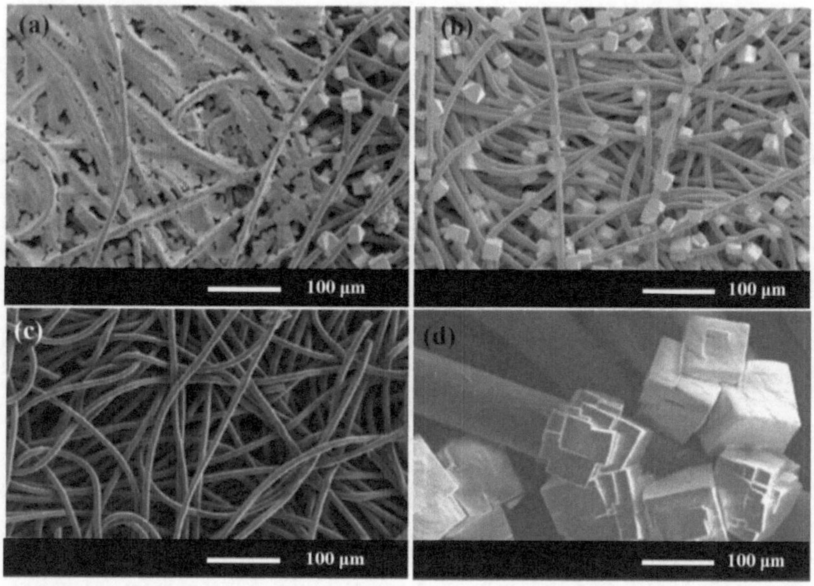

FIGURE 14.4 Morphological analysis (SEM) of air cathode. (a and b) post-mortem analysis of a cathode after discharge to 2 mAh at 80 μA cm^{-2} at the interface of oxygen/carbon-fiber gas diffusion layer cathode and carbon-fiber gas diffusion layer cathode/separator side, respectively, (c) pristine cathode, and (d) image of cubic sodium superoxide (NaO_2) as a discharge product formed on the surface of carbon cathode. Adapted and reproduced with permission from Ref. [29]. Copyright © 2012 Nature.

on the stability of Na metal anode. From Raman spectroscopy and theoretical calculations, it was revealed that at high concentrations of Na salt, lower solvent molecules are available. The reduction of free solvent molecules led to the higher stability of the Na anode in concentrated electrolytes. Moreover, dimethyl sulfoxide acted as a new solvent for SAB. A similar study by Park et al. [46] correlated the effect of the concentration of sodium perchlorate and diglyme system and the stability of NaO_2 in SAB. The work reported that, in concentrated electrolytes, the rate of decomposition reaction decelerated. This was due to the reduction of free solvent molecules in the electrolyte. Wei et al. [47] developed a film formed through an electro-initiated polymerization reaction of Na metal with imidazolium-type ionic liquid monomers, having unsaturated functional groups, in a liquid electrolyte. These formed membranes acted as a protective layer and reduced the electric field at the anode and also created a pathway for stable deposition of Na. Sun et al. [48] made experiments on chloroaluminate ionic liquid electrolyte involving aluminum chloride/1-methyl-3-ethylimidazolium chloride/sodium chloride ionic liquid spiked with additives such as ethyl aluminum dichloride and 1-ethyl-3-methylimidazolium bis(fluorosulfonyl)imide as factors for stabilizing the SEI. The activity of the interphase composition, which included sodium chloride (NaCl), aluminum oxide (Al_2O_3), and sodium fluoride (NaF), derived from the reaction of Na metal and ionic electrolyte, paved the way to obtain coulombic efficiency up to 99.9%. Following the aforementioned notable works, in the second half of the past decade, there has been a great effort in the development of efficient and high-performance SAB. An overall overview of developments in SAB is shown in Figure 14.5.

14.2.2 Sodium-Air Batteries: Limitations

Similar to the reactivity of lithium metal anode in Li-based batteries, Na metal also faces several challenges in operating conditions while acting as an anode [50]. Interfacial instability between the electrode and the electrolyte also contributes to lower performance properties of the battery [51]. Some of the identified limitations of SAB are listed as follows:

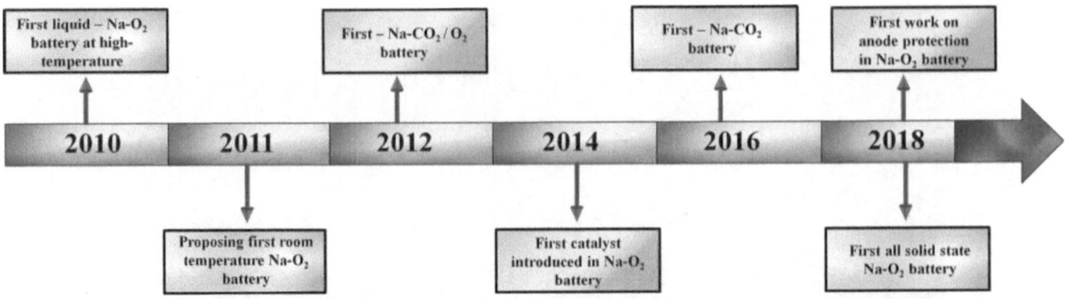

FIGURE 14.5 An overview of developments in SAB. (a) Schematic representation of a SAB and (b) timeline showing the progress made in the case of SAB. Adapted and reproduced with permission from Ref. [49].

i. The high reactivity of Na metal causes side reactions typically in the presence of oxygen, which paves the way for dendrite formation [34]. These uncontrollable dendrite formations could eventually connect with the cathode and lead to an internal battery short circuit, which severely affects the safety of batteries.
ii. Crack formation in the SEI layer (formed by the reaction of metallic Na with organic solvents in the liquid electrolyte) due to certain volume changes of Na metal that occur during the charging and discharging process which could lead to a collapse of coulombic efficiency and charge capacity [52]. Moreover, the instability of SEI layers brings down the chemical activity of electrolytes, which further affects the life of batteries [53].
iii. During the chemical reaction, the collective ionic movement of O_2^- ions from the cathode forms an oxide layer on the anode metal, which shows an insulating behavior, contributing negatively to the performance of the cell [54].
iv. Accumulation of discharge products (NaO_2/Na_2O_2) at the cathode, which are inherently electrically non-conductive and result in the termination of OER [55].

14.2.3 Applications/Innovations in SAB

From the early developments in SABs, anodic instability, electrode–electrolyte interfacial resistance, the efficiency of catalysts, and air cathode engineering have been focused on by the researchers for enhancing the performance of the cell [56,57]. Various methods like modification of electrolytes, tailoring of the interface, inducing bifunctionalities in the catalyst, and structural modification of electrodes have been adopted by the research community [58]. As mentioned above SEI layers are formed by the reaction of the anode and liquid electrolyte, on the anode surface [59]. These layers are expected to create physical isolation to prevent the passage of electrolytes to the metal anode surface and to avoid corrosion due to electrolyte reactions. In lithium-based batteries, the SEI has constituents as by-products of electrolyte decomposition and other products formed while the lithium and electrolyte reactions are stable and dense [60]. Computational and theoretical studies have gained acceptance as they can unfold the reasons for the low performance of SABs. It was found that by the contamination of CO_2 the performance of SAB could be affected negatively. A computational study by Benti et al. [61] using density functional theory showed the formation of carbonates as discharge products, which leads to overpotentials, influencing reaction pathways and blocking the nucleating sites, hence contributing negatively toward the performance of SABs.

Understanding of reaction mechanism, the nature of reaction products, and their correlation with the performance will open up new potential and tremendous possibilities in the area of SABs. Kwak et al. [62] studied a SAB cell with a carbon nanotube (CNT) as an air cathode and Na anode with a solid electrolyte of Na_2O and compared it with a SAB with a liquid electrolyte. It was evidenced from SEM analysis and X-ray diffraction analysis that both SABs produced similar discharge

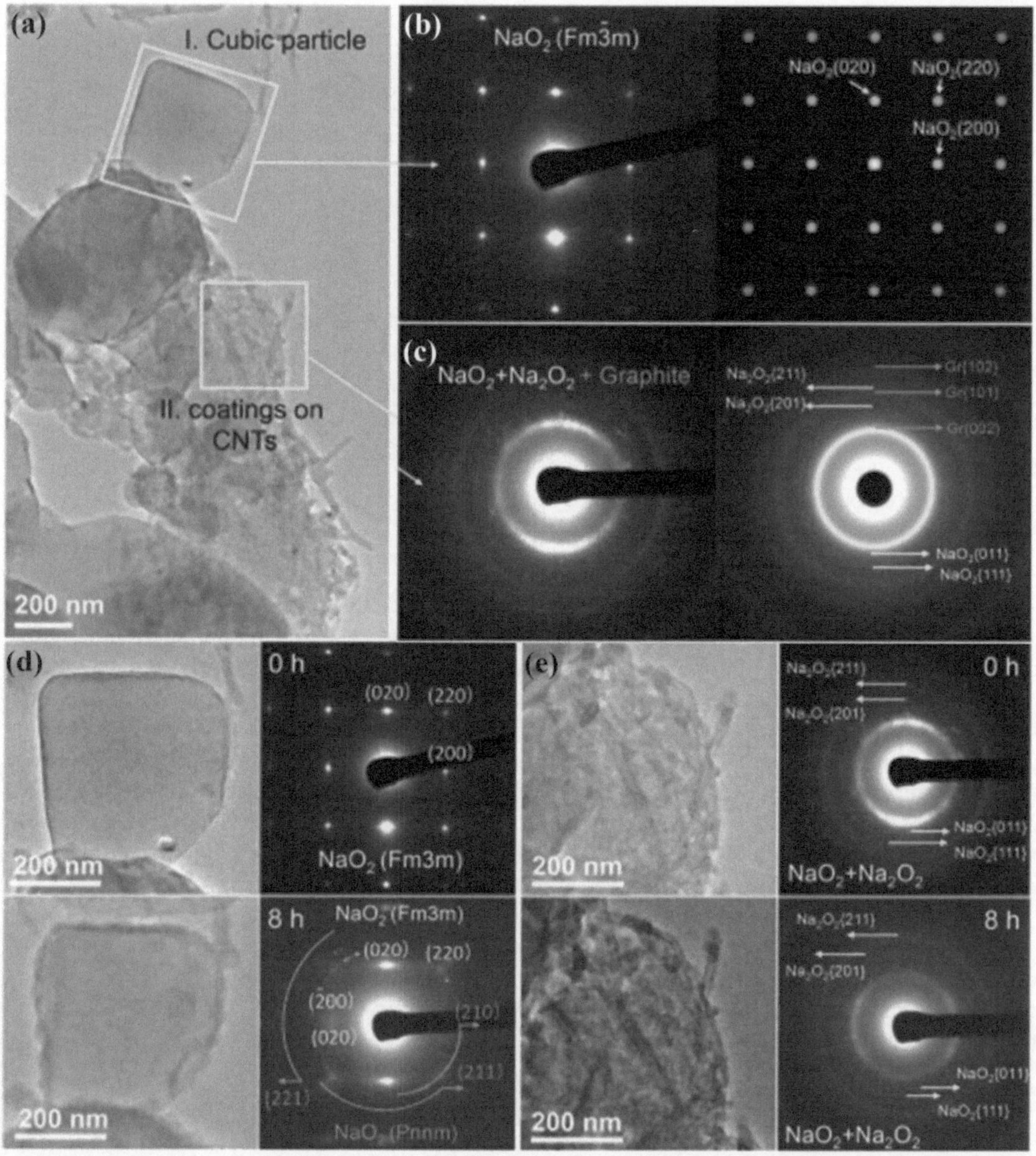

FIGURE 14.6 Time-based observation from the in-situ TEM analysis. (a) As formed cubic discharge product and CNTs; (b) and (c) NaO_2 (Fm3m) and $NaO_2+Na_2O_2$; (d) NaO_2 after 8 h; (e) conformal coating on CNTs. Adapted and reproduced with permission from Ref. [62]. Copyright © 2018 American Chemical Society.

products. While charging the cell, discharge products were observed to be decomposed and shrinkage occurred on discharging the growth of NaO_2. Na_2O_2 was observed over the cathode, which formed a thin coating. Moreover, after the first cycle, the discharge product showed an irreversible nature, which consisted of both NaO_2 and Na_2O_2. From the in-situ transmission electron microscopy (TEM) analysis, it is evident that the discharge product consisted of two types, one of which is large cubic-shaped particles and the other being conformal coatings. The cubic particles showed a NaO_2 (Fm3m) phase, and after 8 h, the NaO_2 (Pnnm) phase was also formed over the rough surface of cubic particles, as shown in Figure 14.6.

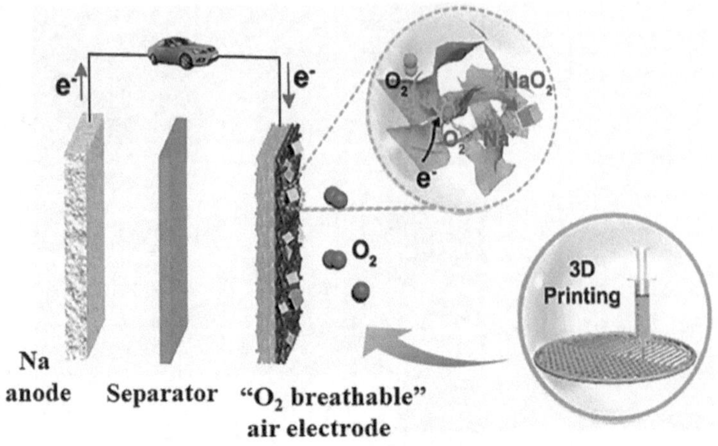

FIGURE 14.7 Schematic representation of 3D printed r-GO cathode and the architecture of SAB focusing on the reaction happening at the cathode. Adapted and reproduced with permission from Ref. [64]. Copyright © 2020 American Chemical Society.

To mitigate the issue of dendrites in the anode of a SAB, Lin et al. [63] fabricated a hybrid solid-state electrolyte (SSE). SSE ($Na_{3.25}Zr_2Si_{2.25}P_{0.75}O_{12}$) was prepared by sol–gel method with tetraethyl orthosilicate and zirconium (IV) propoxide solution as reactants, followed by annealing and sintering. Carbon paper was used as an air cathode and was also employed for covering the Na anode. The work was successful in preventing oxygen (O_2/O_2^-) crossover while preventing parasitic reactions when a glass fiber separator was used. From XPS analysis, a SEI layer with NaF-rich inorganic phases was evidenced. Hence, overall, the modification brought enhancement in the coulombic efficiency with a stable cycling performance. The same group designed a well-patterned, fine filamentous, stacked mesh fabrication of reduced graphene oxide (r-GO) air cathodes. Homogeneous graphene oxide ink was used to print the stacked arrangement of the r-GO sheets that underwent a freeze-drying and thermal annealing process to deliver the desired 3-D printed structure (Figure 14.7). This framework with macropores stimulates oxygen diffusion and broadens sufficient storage space for the discharge product. The high initial discharge capacity achieved was measured as 13,484.6 mAh g^{-1} with a cyclability of 120 cycles and 500 mAh g^{-1} as the desired cut-off capacity [64].

14.3 SODIUM-AIR BATTERIES: CHALLENGES AND FUTURE ADVANCEMENTS

Unlike Li-based batteries, the research based on SABs is still in its early stages; some challenges need to be addressed which help to understand and solve the problems for developing innovative technologies. One of the factors that inhibit the advancement of SAB is the reactivity and instability of the Na anode [49]. The migration of oxygen species dissolved in electrolytes and superoxide reactants formed in the cathode toward the negative electrode results in interactions that corrode and impede the reactivity of the Na anode [65]. Furthermore, these side reactions increase the overpotential and constrain cell performance, leading to battery failure. In terms of safety, dendrite formation on the Na anode and SEI instability should be blocked to alleviate safety concerns, particularly when it comes to EV technology. The ionic mobility and cycling performance of a cell is dependent on the stability of SEI [66]. Engineering the interface of the electrode and the electrolyte, through additive incorporation, solvents, and metal salts, mitigates the unstable SEI formation, which enhances the cycling of the battery [67]. For the stability and reversibility of Na metal anodes, the solvation–desolvation kinetics of electrolytes have a profound role [67]. Furthermore, the Na ion (Na^+) mobility and the viscosity of electrolytes are governed by solvation dynamics [68,69]. Na anode can be

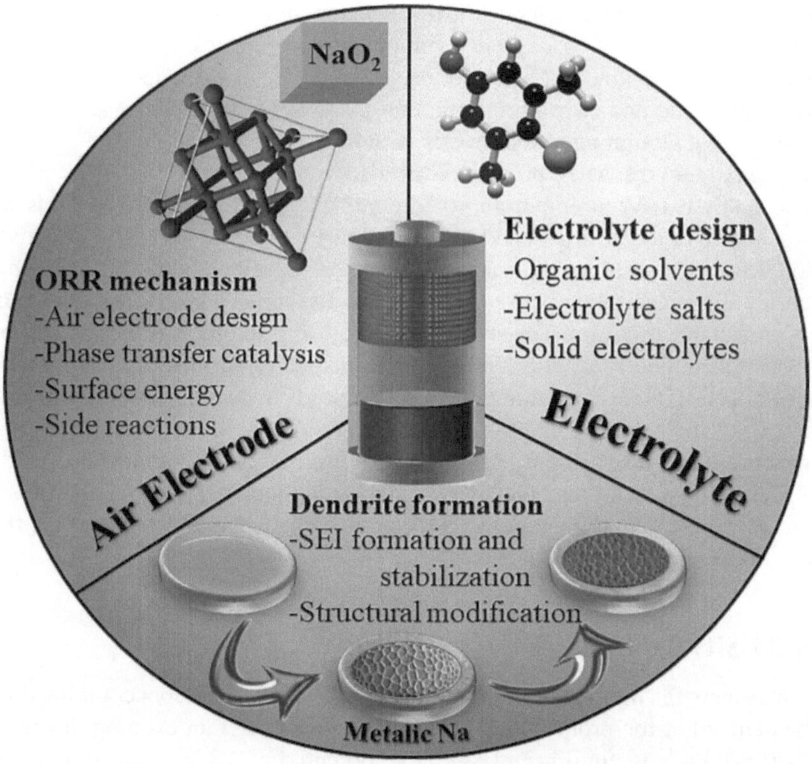

FIGURE 14.8 Schematic representation of the challenges in the development of SABs. Adapted and reproduced with permission from Ref. [75]. Copyright © 2019 Elsevier.

stabilized by many approaches including electrolyte additive, which can create tailorable SEI and suppress dendrite growth and its propagation, alloy anodes, guide the deposition of Na, create an anode host, etc. [70–74]. Protective layer coating over an anode also faces challenges in that it can affect the ionic conductivity; Na^+ transport, moisture, and oxygen crossover from the air cathode; non-uniformity in the nucleation; etc. [75]. To develop a successful, safer Na anode, methods to guide uniform Na^+ deposition, limiting oxygen crossover, stable SEI, and an advanced sodiophobic architecture to prevent dendrite growth on the Na anode should be effectively implemented [75]. A schematic representation of the challenges in the development of SABs is given in Figure 14.8.

Electrolytes, such as carbonate and ether-based electrolytes, are well accepted for Na-based cell operation and stability [33,76]. However, doing further research to improve factors such as compatibility with electrodes, increased stability against oxygen species, lower volatility, and high ionic conductivity may contribute to enhanced cell reactions. Furthermore, a liquid electrolyte subjected to high pressure could undergo leakage during a high-impact situation, particularly in an EV, raising safety concerns. This can be mitigated by developing appropriate electrolytes, such as solid, gel, or hybrid electrolytes, with high reliability and the ability to withstand impacts without any deformations. However, in the solid and gel electrolytes, poor interfacial stability between electrode–electrolyte interphase and low ionic conductivity compared to liquid electrolytes are the main concerns. A thorough understanding of the reactions that occur in SAB on the evolution of different discharge species such as oxides, peroxides, and carbonates is required. Although in the past decades, in-situ and operando characterizations have been attempted to study the mechanism, the basic facts are still unknown. Theoretical calculations and simulations predicted that the parasitic reactions increase the charge potential and catalyze the electrolyte decomposition, hence degrading the performance of the cell [77]. It should be noted that the combination of theoretical predictions and experimental validations can create an optimized framework, which can accelerate the development of battery technology [78]. Control over these discharge products, lack of space to accommodate these species, formation of singlet oxygen during cycling, slow reaction kinetics, and poor reversibility negatively affect the cell's capacity and cyclic stability. Rapid ionic and e^- transport and considerable electrocatalytic activity for ORR and OER are important [79]. Designing air electrodes with high catalytic activity and adequate space for accommodating maximum discharge products is needed to bridge these challenges. Moreover, cathode and cathode catalysts have an important role in the battery performance [35]. It is worth noting that the rechargeability of a cell is dependent on the ability of an air cathode to effectively catalyze both ORR and OER; hence, bifunctional catalysts have been developed [80–82]. The air cathode has a large difference in the polarization between charge and discharge, which affects the round-trip efficiency. Moreover, suitable modifications are needed in cycling stability. Although carbon-based air cathodes have been identified as a better option, the formation of various discharge products affects the performance of cells [56,83]. Designing of SAB also needs to be addressed, to mitigate the issues related to safety, cost, and recycling and to enhance the performance. For the cell design, it is critical to avoid the penetration of water impurities; to limit the formation of sodium superoxide dihydrate ($Na_2O_2.2H_2O$), designing a membrane that allows only air without any sort of impurities is challenging [84]. Moreover, severe contamination by CO_2 can be a challenge which can affect the reversible reactions, hence lowering the performance of SAB.

14.4 CONCLUSION

Emerging requirements for high-energy storage technology has been conceptualized by researchers to meet the demand of the proliferating population. SABs offer an exciting theoretical specific energy of ~1,600 Wh kg^{-1}, an equilibrium discharge potential of 2.3 V, coupled with cost-effective raw materials. Moreover, compared to LAB, SABs possess lower charging overpotential with higher values of energy efficiency, mainly due to the high reversibility of the electrochemical redox pair (O_2^- O_2^{2-}), which governs both ORR and OER mechanism in cathode. Effective approaches are to be considered to mitigate the migration of oxygen and superoxide intermediate by electrode and electrolyte engineering. In addition to the superoxide anion intermediates at the cathode, owing to the higher chemical reaction of metallic Na, dissolved oxygen molecules in the electrolytes migrate to anode, creating overpotential, and passivate the Na anode, thereby lowering the discharge capacity. The development of electrolytes which have higher oxidative decomposition stability with a wide electrochemical window and ionic conduction is also crucial for advanced SAB. Unveiling the complexities of electrochemical mechanisms in SAB needs advanced analytical techniques and theoretical predictions. In-situ and operando analysis hence have a vital role in revealing the intricated electrochemical processes and help in comprehending the real-time accurate information. Theory-guided experimental methods can accelerate the momentum in developing high-performance and efficient SAB. Hence, with muti-faced and holistic approaches including the designing, engineering, and tailoring of electrodes and electrolyte and developing bifunctional catalysts, an acceleration in the development of SAB is possible. Doubtlessly, realizing the potentials of SAB, with the ongoing research involving theorists and experimentalists, the aforementioned challenges can be rectified, thereby positively contributing to the requirement of high-energy density and cost-effective battery storage solutions.

REFERENCES

1. Xie J, Lu Y-C (2020) A retrospective on lithium-ion batteries. *Nat Commun* 11:2499. https://doi.org/10.1038/s41467-020-16259-9
2. Xu W, Wang J, Ding F, et al (2014) Lithium metal anodes for rechargeable batteries. *Energy Environ Sci* 7:513–537. https://doi.org/10.1039/C3EE40795K
3. Abraham KM, Jiang Z (1996) A polymer electrolyte-based rechargeable lithium/oxygen battery. *J Electrochem Soc* 143:1–5. https://doi.org/10.1149/1.1836378

4. Kim I, Kim CH, Choi SH, et al (2016) A singular flexible cathode for room temperature sodium/sulfur battery. *J Power Sources* 307:31–37. https://doi.org/10.1016/j.jpowsour.2015.12.035
5. Adelhelm P, Hartmann P, Bender CL, et al (2015) From lithium to sodium: cell chemistry of room temperature sodium-air and sodium-sulfur batteries. *Beilstein J Nanotechnol* 6:1016–1055. https://doi.org/10.3762/bjnano.6.105
6. Choudhury S, Wei S, Ozhabes Y, et al (2017) Designing solid-liquid interphases for sodium batteries. *Nat Commun* 8:898. https://doi.org/10.1038/s41467-017-00742-x
7. Hwang J-Y, Myung S-T, Sun Y-K (2017) Sodium-ion batteries: present and future. *Chem Soc Rev* 46:3529–3614. https://doi.org/10.1039/C6CS00776G
8. Li L, Zheng Y, Zhang S, et al (2018) Recent progress on sodium ion batteries: potential high-performance anodes. *Energy Environ Sci* 11:2310–2340. https://doi.org/10.1039/C8EE01023D
9. Yang J, Gao H, Kang X (2020) Nanomaterials application in Li-Se and Na-Se batteries. In: Ran F, Chen S (eds) *Advanced Nanomaterials for Electrochemical-Based Energy Conversion and Storage.* Elsevier, Amsterdam, The Netherlands, pp. 69–114.
10. Hans Wedepohl K (1995) The composition of the continental crust. *Geochim Cosmochim Acta* 59:1217–1232. https://doi.org/10.1016/0016-7037(95)00038-2
11. Abraham KM (2020) How comparable are sodium-ion batteries to lithium-ion counterparts? *ACS Energy Lett* 5:3544–3547. https://doi.org/10.1021/acsenergylett.0c02181
12. USG Survey (2019) *Mineral Commodity Summaries.* U.S. Geological Survey, Reston, VA.
13. Jishnu NS, Vineeth SK, Das A, et al (2021) Electrospun PVdF and PVdF-co-HFP-based blend polymer electrolytes for lithium ion batteries. In: Balakrishnan NTM, Prasanth R (eds) *Electrospinning for Advanced Energy Storage Applications*, 1st ed. Springer, Singapore, pp. 201–234.
14. Duffner F, Kronemeyer N, Tübke J, et al (2021) Post-lithium-ion battery cell production and its compatibility with lithium-ion cell production infrastructure. *Nat Energy* 6:123–134. https://doi.org/10.1038/s41560-020-00748-8
15. Goodenough JB, Kim Y (2010) Challenges for rechargeable Li batteries. *Chem Mater* 22:587–603. https://doi.org/10.1021/cm901452z
16. Soni CB, Sungjemmenla J, Vineeth SK, Kumar V (2022) Challenges in regulating interfacial-chemistry of the sodium-metal anode for room-temperature sodium-sulfur batteries. *Energy Storage* 4:e264. https://doi.org/10.1002/est2.264
17. Mandl M, Becherer J, Kramer D, et al (2020) Sodium metal anodes: deposition and dissolution behaviour and SEI formation. *Electrochim Acta* 354:136698. https://doi.org/10.1016/j.electacta.2020.136698
18. Sungjemmenla J, Soni CB, Vineeth SK, Kumar V (2021) Unveiling the physiochemical aspects of the matrix in improving sulfur-loading for room-temperature sodium-sulfur batteries. *Mater Adv* 2:4165–4189. https://doi.org/10.1039/D1MA00247C
19. Li Y, Lu J (2017) Metal-air batteries: will they be the future electrochemical energy storage device of choice? *ACS Energy Lett* 2:1370–1377. https://doi.org/10.1021/acsenergylett.7b00119
20. Cheng F, Chen J (2012) Metal-air batteries: from oxygen reduction electrochemistry to cathode catalysts. *Chem Soc Rev* 41:2172. https://doi.org/10.1039/c1cs15228a
21. Ahuja D, Kalpna V, Varshney PK (2021) Metal air battery: a sustainable and low cost material for energy storage. *J Phys Conf Ser* 1913:012065. https://doi.org/10.1088/1742-6596/1913/1/012065
22. Khan Z, Vagin M, Crispin X (2020) Can hybrid Na-air batteries outperform nonaqueous Na-O_2 batteries? *Adv Sci* 7:1902866. https://doi.org/10.1002/advs.201902866
23. Sungjemmenla J, Soni CB, Vineeth SK, Kumar V (2022) Exploration of the unique structural chemistry of sulfur cathode for high-energy rechargeable beyond-Li batteries. *Adv Energy Sustain Res* 3:2100157. https://doi.org/10.1002/aesr.202100157
24. Wang Y, Zhou D, Palomares V, et al (2020) Revitalising sodium-sulfur batteries for non-high-temperature operation: a crucial review. *Energy Environ* Sci 13:3848–3879. https://doi.org/10.1039/D0EE02203A
25. Ha S, Kim J-K, Choi A, et al (2014) Sodium-metal halide and sodium-air batteries. *ChemPhysChem* 15:1971–1982. https://doi.org/10.1002/cphc.201402215
26. Yadegari H, Sun X (2018) Recent advances on sodium-oxygen batteries: a chemical perspective. *Acc Chem Res* 51:1532–1540. https://doi.org/10.1021/acs.accounts.8b00139
27. Aurbach D, McCloskey BD, Nazar LF, Bruce PG (2016) Advances in understanding mechanisms underpinning lithium-air batteries. *Nat Energy* 1:16128. https://doi.org/10.1038/nenergy.2016.128
28. Sun B, Pompe C, Dongmo S, et al (2018) Challenges for developing rechargeable room-temperature sodium oxygen batteries. *Adv Mater Technol* 3:1800110. https://doi.org/10.1002/admt.201800110
29. Hartmann P, Bender CL, Vračar M, et al (2013) A rechargeable room-temperature sodium superoxide (NaO_2) battery. *Nat Mater* 12:228–232. https://doi.org/10.1038/nmat3486

30. Landa-Medrano I, Li C, Ortiz-Vitoriano N, et al (2016) Sodium-oxygen battery: steps toward reality. *J Phys Chem Lett* 7:1161–1166. https://doi.org/10.1021/acs.jpclett.5b02845
31. Liu S, Liu S, Luo J (2016) Carbon-based cathodes for sodium-air batteries. *New Carbon Mater* 31:264–270. https://doi.org/10.1016/S1872-5805(16)60012-4
32. Hashimoto T, Hayashi K (2015) Aqueous and nonaqueous sodium-air cells with nanoporous gold cathode. *Electrochim Acta* 182:809–814. https://doi.org/10.1016/j.electacta.2015.09.153
33. Sun Q, Yang Y, Fu ZW (2012) Electrochemical properties of room temperature sodium-air batteries with non-aqueous electrolyte. *Electrochem Commun* 16:22–25. https://doi.org/10.1016/j.elecom.2011.12.019
34. Das SK, Lau S, Archer LA (2014) Sodium-oxygen batteries: a new class of metal-air batteries. *J Mater Chem A* 2:12623–12629. https://doi.org/10.1039/c4ta02176b
35. Yin W-W, Fu Z-W (2017) A highly efficient bifunctional heterogeneous catalyst for morphological control of discharged products in Na-air batteries. *Chem Commun* 53:1522–1525. https://doi.org/10.1039/C6CC09348E
36. Reier T, Oezaslan M, Strasser P (2012) Electrocatalytic oxygen evolution reaction (OER) on Ru, Ir, and Pt catalysts: a comparative study of nanoparticles and bulk materials. *ACS Catal* 2:1765–1772. https://doi.org/10.1021/cs3003098
37. Lee J, Kim J, Kim S, et al (2020) A review on recent approaches for designing the SEI layer on sodium metal anodes. *Mater Adv* 1:3143–3166. https://doi.org/10.1039/D0MA00695E
38. Haber S, Leskes M (2018) What can we learn from solid state NMR on the electrode-electrolyte interface? *Adv Mater* 30:1706496. https://doi.org/10.1002/adma.201706496
39. Matios E, Wang H, Wang C, Li W (2019) Enabling safe sodium metal batteries by solid electrolyte interphase engineering: a review. *Ind Eng Chem Res* 58:9758–9780. https://doi.org/10.1021/acs.iecr.9b02029
40. Lee B, Paek E, Mitlin D, Lee SW (2019) Sodium metal anodes: emerging solutions to dendrite growth. *Chem Rev* 119:5416–5460. https://doi.org/10.1021/acs.chemrev.8b00642
41. Peled E, Golodnitsky D, Mazor H, et al (2011) Parameter analysis of a practical lithium- and sodium-air electric vehicle battery. *J Power Sources* 196:6835–6840. https://doi.org/10.1016/j.jpowsour.2010.09.104
42. Sun Q, Yang Y, Fu Z-W (2012) Electrochemical properties of room temperature sodium-air batteries with non-aqueous electrolyte. *Electrochem Commun* 16:22–25. https://doi.org/10.1016/j.elecom.2011.12.019
43. Peled E, Golodnitsky D, Hadar R, et al (2013) Challenges and obstacles in the development of sodium-air batteries. *J Power Sources* 244:771–776. https://doi.org/10.1016/j.jpowsour.2013.01.177
44. Das SK, Xu S, Archer LA (2013) Carbon dioxide assist for non-aqueous sodium-oxygen batteries. *Electrochem Commun* 27:59–62. https://doi.org/10.1016/j.elecom.2012.10.036
45. He M, Lau KC, Ren X, et al (2016) concentrated electrolyte for the sodium-oxygen battery: solvation structure and improved cycle life. *Angew Chemie – Int Ed* 55:15310–15314. https://doi.org/10.1002/anie.201608607
46. Park H, Kim J, Lee MH, et al (2018) Highly durable and stable sodium superoxide in concentrated electrolytes for sodium-oxygen batteries. *Adv Energy Mater* 8:1801760. https://doi.org/10.1002/aenm.201801760
47. Wei S, Choudhury S, Xu J, et al (2017) Highly stable sodium batteries enabled by functional ionic polymer membranes. *Adv Mater* 29:1605512. https://doi.org/10.1002/adma.201605512
48. Sun H, Zhu G, Xu X, et al (2019) A safe and non-flammable sodium metal battery based on an ionic liquid electrolyte. *Nat Commun* 10:3302. https://doi.org/10.1038/s41467-019-11102-2
49. Zheng Z, Wu C, Gu Q, et al (2021) Research progress and future perspectives on rechargeable Na-O_2 and Na-CO2 batteries. *Energy Environ Mater* 4:158–177. https://doi.org/10.1002/eem2.12139
50. Wang H, Wang C, Matios E, Li W (2018) Facile stabilization of the sodium metal anode with additives: unexpected key role of sodium polysulfide and adverse effect of sodium nitrate. *Angew Chemie* 130:7860–7863. https://doi.org/10.1002/ange.201801818
51. Kumar V, Wang Y, Eng AYS, et al (2020) A biphasic interphase design enabling high performance in room temperature sodium- sulfur batteries. *Cell Reports Phys Sci* 1:100044. https://doi.org/10.1016/j.xcrp.2020.100044
52. Liu T, Yang X, Nai J, et al (2021) Recent development of Na metal anodes: interphase engineering chemistries determine the electrochemical performance. *Chem Eng J* 409:127943. https://doi.org/10.1016/j.cej.2020.127943
53. Wang H, Matios E, Luo J, Li W (2020) Combining theories and experiments to understand the sodium nucleation behavior towards safe sodium metal batteries. *Chem Soc Rev* 49:3783–3805. https://doi.org/10.1039/D0CS00033G
54. Yadegari H, Sun Q, Sun X (2016) Sodium-oxygen batteries: a comparative review from chemical and electrochemical fundamentals to future perspective. *Adv Mater* 28:7065–7093. https://doi.org/10.1002/adma.201504373

55. Enterría M, Reynaud M, Paredes JI, et al (2022) Driving the sodium-oxygen battery chemistry towards the efficient formation of discharge products: the importance of sodium superoxide quantification. *J Energy Chem* 68:709–720. https://doi.org/10.1016/j.jechem.2021.12.014
56. Sun Q, Yadegari H, Banis MN, et al (2015) Self-stacked nitrogen-doped carbon nanotubes as long-life air electrode for sodium-air batteries: elucidating the evolution of discharge product morphology. *Nano Energy* 12:698–708. https://doi.org/10.1016/j.nanoen.2015.01.018
57. Ha TA, Pozo-Gonzalo C, Nairn K, et al (2020) An investigation of commercial carbon air cathode structure in ionic liquid based sodium oxygen batteries. *Sci Rep* 10:1–10. https://doi.org/10.1038/s41598-020-63473-y
58. Wu S, Qiao Y, Jiang K, et al (2018) Tailoring sodium anodes for stable sodium-oxygen batteries. *Adv Funct Mater* 28:1706374. https://doi.org/10.1002/adfm.201706374
59. Chawla N, Safa M (2019) Sodium batteries: a review on sodium-sulfur and sodium-air batteries. *Electronics* 8:1201. https://doi.org/10.3390/electronics8101201
60. Yuan H, Nai J, Tian H, et al (2020) An ultrastable lithium metal anode enabled by designed metal fluoride spansules. *Sci Adv* 6:eaaz3112. https://doi.org/10.1126/sciadv.aaz3112
61. Wang Z-L, Xu D, Xu J-J, Zhang X-B (2014) Oxygen electrocatalysts in metal-air batteries: from aqueous to nonaqueous electrolytes. *Chem Soc Rev* 43:7746–7786. https://doi.org/10.1039/C3CS60248F
62. Kwak W-J, Luo L, Jung H-G, et al (2018) Revealing the reaction mechanism of Na-O_2 batteries using environmental transmission electron microscopy. *ACS Energy Lett* 3:393–399. https://doi.org/10.1021/acsenergylett.7b01273
63. Lin X, Sun F, Sun Q, et al (2019) O_2/O_2- crossover- and dendrite-free hybrid solid-state Na-O_2 batteries. *Chem Mater* 31:9024–9031. https://doi.org/10.1021/acs.chemmater.9b03266
64. Lin X, Wang J, Gao X, et al (2020) 3D printing of free-standing "O_2 breathable" air electrodes for high-capacity and long-life Na-O_2 batteries. *Chem Mater* 32:3018–3027. https://doi.org/10.1021/acs.chemmater.9b05360
65. Chawla N (2019) Recent advances in air-battery chemistries. *Mater Today Chem* 12:324–331. https://doi.org/10.1016/j.mtchem.2019.03.006
66. Song J, Xiao B, Lin Y, et al (2018) Interphases in sodium-ion batteries. *Adv Energy Mater* 8:1703082. https://doi.org/10.1002/aenm.201703082
67. Vineeth SK, Soni CB, Sun Y, et al (2022) Implications of Na-ion solvation on Na anode-electrolyte interphase. *Trends Chem* 4:48–59. https://doi.org/10.1016/j.trechm.2021.11.002
68. Andreev M, de Pablo JJ, Chremos A, Douglas JF (2018) Influence of ion solvation on the properties of electrolyte solutions. *J Phys Chem B* 122:4029–4034. https://doi.org/10.1021/acs.jpcb.8b00518
69. Siegel DJ, Nazar L, Chiang YM, et al (2021) Establishing a unified framework for ion solvation and transport in liquid and solid electrolytes. *Trends Chem* 3:807–818. https://doi.org/10.1016/j.trechm.2021.06.004
70. Kumar V, Eng AYS, Wang Y, et al (2020) An artificial metal-alloy interphase for high-rate and long-life sodium-sulfur batteries. *Energy Storage Mater* 29:1–8. https://doi.org/10.1016/j.ensm.2020.03.027
71. Eng AYS, Kumar V, Zhang Y, et al (2021) Room-temperature sodium-sulfur batteries and beyond: realizing practical high energy systems through anode, cathode, and electrolyte engineering. *Adv Energy Mater* 11:2003493. https://doi.org/10.1002/aenm.202003493
72. Soni CB, Kumar V, Seh ZW (2022) Guiding uniform sodium deposition through host modification for sodium metal batteries. *Batter Supercaps* 5:e202100207. https://doi.org/10.1002/batt.202100207
73. Liu C, Carboni M, Brant WR, et al (2018) On the stability of NaO_2 in Na-O_2 batteries. *ACS Appl Mater Interfaces* 10:13534–13541. https://doi.org/10.1021/acsami.8b01516
74. Zhao Y, Yang X, Kuo L-Y, et al (2018) High capacity, dendrite-free growth, and minimum volume change Na metal anode. *Small* 14:1703717. https://doi.org/10.1002/smll.201703717
75. Yadegari H, Sun X (2020) Sodium-oxygen batteries: recent developments and remaining challenges. *Trends Chem* 2:241–253. https://doi.org/10.1016/j.trechm.2019.12.003
76. Wahlers J, Fulfer KD, Harding DP, et al (2016) Solvation structure and concentration in glyme-based sodium electrolytes: a combined spectroscopic and computational study. *J Phys Chem C* 120:17949–17959. https://doi.org/10.1021/acs.jpcc.6b06160
77. Raz K, Tereshchuk P, Golodnitsky D, Natan A (2018) Adsorption of Li_2O_2, Na_2 O_2, and NaO_2 on TiC(111) surface for metal-air rechargeable batteries: a theoretical study. *J Phys Chem C* 122:16473–16480. https://doi.org/10.1021/acs.jpcc.8b01983
78. Eng AYS, Soni CB, Lum Y, et al (2022) Theory-guided experimental design in battery materials research. *Sci Adv* 8:eabm2422. https://doi.org/10.1126/sciadv.abm2422

79. Baskar S, Murugesan C, Barpanda P (2017) Highly porous N-doped graphitic carbon from bio-waste as bifunctional electrocatalyst for hybrid Na-air battery. *ECS Trans* 80:425–430. https://doi.org/10.1149/08010.0425ecst
80. Kim M, Ju H, Kim J (2019) Single crystalline $Bi_2Ru_2O_7$ pyrochlore oxide nanoparticles as efficient bifunctional oxygen electrocatalyst for hybrid Na-air batteries. *Chem Eng J* 358:11–19. https://doi.org/10.1016/j.cej.2018.09.204
81. Zhang Y-L, Goh K, Zhao L, et al (2020) Advanced non-noble materials in bifunctional catalysts for ORR and OER toward aqueous metal-air batteries. *Nanoscale* 12:21534–21559. https://doi.org/10.1039/D0NR05511E
82. Zhao S, Wang C, Du D, et al (2021) Bifunctional effects of cation additive on Na-O_2 batteries. *Angew Chemie – Int Ed* 60:3205–3211. https://doi.org/10.1002/anie.202012787
83. Liu Q, Geng L, Yang T, et al (2019) In-situ imaging electrocatalysis in a Na-O_2 battery with Au-coated MnO_2 nanowires air cathode. *Energy Storage Mater* 19:48–55. https://doi.org/10.1016/j.ensm.2018.08.026
84. Xu X, Hui KS, Dinh DA, et al (2019) Recent advances in hybrid sodium-air batteries. *Mater Horizons* 6:1306–1335. https://doi.org/10.1039/C8MH01375F

15 Emerging Sodium-Air Batteries
Developments in Anode Stabilization

Ajish Babu, S. K. Vineeth, Roshny Joy, Abhilash Pullanchiyodan, Nutan Gupta, and Prasanth Raghavan

15.1 INTRODUCTION

Stringent governmental policies against non-renewable energy sources with environmental considerations catalysed the exploration of extracting energy from renewable sources. As renewable energy sources are intermittent and climate dependent, the requirement for energy storage technology is unequivocally important [1]. Hence, rechargeable batteries gained momentum with the proliferation in population with the growth of portable electronic devices and electrical vehicles [2]. Although lithium-ion battery (LIB) technology is incumbent and mature to meet the demand of the thriving population, it is still ambiguous [3]. LIB consists of the insertion of compound-based anodes and cathodes, which have limited capacity and energy density. Therefore, to enhance electrochemical properties in a LIB, alternate cathodes and anodes which can offer higher capacity need to be further explored. Another approach is to unveil a newer chemistry where, instead of insertion electrodes, materials that can undertake conversion reactions with the ability to accommodate more ions need to be explored. In this aspect, considering extraordinary energy density, metal-air batteries have gained wide acceptance [4,5].

Considering metal-air batteries, lithium-air batteries (LABs) and sodium-air batteries (SABs) show the highest theoretical specific energy values (considering the weight of negative electrode materials) of 11,430 and 3,451 Wh kg^{-1}, respectively [6]. Sodium (Na) is widely distributed on the earth, with elemental abundance as fourth, and accounts for 25,670 ppm in the upper continental crust [7,8]. The content of Na in water bodies counts up to 0.18 mg L^{-1}, and 20 mg kg^{-1} on the earth's crust [9,10]. Furthermore, after lithium, Na holds the second position with respect to the classification of lightest metal, with an ionic radius of 1.02 Å and a redox potential accounting for −2.7 V vs standard hydrogen electrode [11,12]. Regarding the cost of raw materials, Na overrules lithium. U.S. Geological Survey report of 2019 indicates that the cost of lithium carbonate is 170,000 U.S. dollars per metric ton, whereas for sodium carbonate it is approximately 149 U.S. dollars [13]. Hence, metallic Na, being employed as an anode for SABs, has the overall advantages of cost-effectiveness, high round-trip efficiency, and low overpotential with abundance in the availability of precursor, compared to LABs [6].

Sodium metal batteries constitute a class of alkali metal-based batteries, where metallic Na is used as an anode [14]. Examples of Na metal batteries include SAB, Na–sulphur batteries and Na–selenium batteries [15]. Despite the abundance and cost efficiency of Na metal, the progress of SABs is substantially hindered by the higher interfacial instability between the metallic sodium anode and electrolyte. [16]. As the electrolyte reacts with Na, the solid–electrolyte interphase (SEI) is formed [17]. Although the SEI layer further prevents Na from corrosive reaction, the stability of the SEI layer and its thickness regulate the electrochemical performance of the cell [18]. Hence, it is unequivocally important to implement electrode–electrolyte engineering to stabilize the anode part. This chapter highlights the state of the art in the development of SAB with a focus on anode.

DOI: 10.1201/9781003310174-15

A detailed discussion on protecting metallic Na anode using various approaches reported in recent years has been focussed. The mechanism of SEI formation with remedies to stabilize the SEI layer, along with a short discussion on monitoring interface by operando characterizations, has also been emphasized in the chapter.

15.2 PROTECTION OF NA METAL ANODES AND ITS STRUCTURAL CHANGES

The high theoretical specific energy density of SAB is calculated based on the Na anode. With a high theoretical capacity of 1,166 mAh g^{-1}, Na metal attracts the next-generation rechargeable battery technology. Compared to LAB technology, SAB overrules, as it has no side reactions; however, metallic Na is plagued by various intractable problems [10]. Uneven deposition occurs over the Na anode, during the electrochemical charging, as sodium ion (Na^+) combines with electrons [19]. The continual growth of the uneven deposition leads to a dendritic structure, which protrudes the separators and connects the anode and the cathode internally, leading to short circuits and thermal runaway. Furthermore, the Na anode experiences large volume changes due to the repeated charging and discharging cycle, leading to the crack formation of the SEI layer. The direct effect is experienced through capacity decay and a substantial drop in the coulombic efficiency [20,21]. Discharge products also influence the electrochemical processes in SABs. The discharge process involves the reduction of oxygen at the cathode, and Na^+ coming from the anode forms sodium peroxide (Na_2O_2) and sodium superoxide (NaO_2) and vice versa during charging [22]. Depending on the electrolyte type used, the type of carbon used in the air cathode, and the mass loaded, variation in capacity occurs [23]. Hence, to maximize the capacity, mitigate parasitic reactions, and attain high round-trip efficiency, various strategies are being recommended [24]. One of the most explored methods of protecting alkali metal anode is the creation of artificial, stable SEI over Na metal. It enhances the interface of Na, which protects it and enhances the coulombic efficiency and cycle life of a cell [25]. Wei et al. [26] created an ionic membrane coating over Na metal through in-situ electro-initiated polymerization of imidazolium-type ionic liquid (IL). The coating (of ~ 50 mm thickness) protected Na metal from the parasitic reaction, which was evidenced when the cell was cycled with carbonate electrolyte. Ionic conduction and transport properties were unaffected by the creation of the film, and the cell showed longer cycle life (even at high current densities) with enhanced coulombic efficiency than the pristine Na anode cell. Film formed using IL reduced the electric field at the Na metal anode and stabilized Na deposition. Similar to lithium metal batteries, another approach to suppress the dendrite growth in Na metal-based batteries is to incorporate a scaffold current collector [27,28]. Zhang et al. [29] demonstrated a 3D conductive carbon fibre paper scaffold as a current collector in Na anode showed a high affinity of graphite carbon for Na, thereby effectively inhibiting the dendrite formation in Na anode. The scaffold structure showed an electrically conductive nature, which resulted in a reduced local current and uniform Na deposition. Uniformity in deposition enhanced electrochemical properties, and the cell delivered a coulombic efficiency of 99.5% with a stable cycling of 100 cycles. Controlling the alkali metal nucleation to obtain uniform deposition of anode metal ions has been developed from LIB [30]. Inspired by that, Liu et al. [31] communicated the role of porous aluminium (Al) foil as a current collector, which effectively inhibited dendrite growth in the Na anode. The interconnected porous structure of Al provided abundant space for Na nucleation, with a decrease in the distribution of Na^+ flux. Compared to planar Al, porous Al also guided the uniform deposition of Na, which was evidenced by the post-mortem analysis (morphology) of the cycled anode. Figures 15.1 and 15.2 depict the morphological analysis performed using scanning electron microscopy (SEM), where the planar Al foil showed the formation and propagation of mossy Na dendrites, which were absent in the case of porous Al foil. The cell with porous Al foil displayed a coulombic efficiency of 99.9% for over 1,000 cycles, without any dendrite formation. The cell gives a promising utilization of low-cost and abundantly available Al metal.

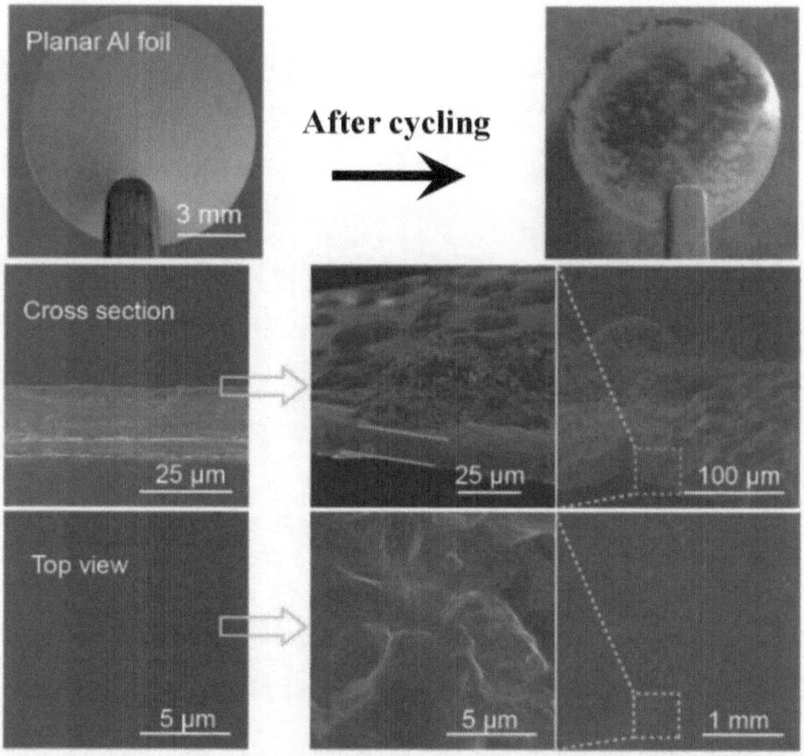

FIGURE 15.1 A comparative analysis of morphology showing the Na deposition on planar and porous aluminium (Al) foil current collector before and after the cycling process. (a) Planar Al foil, (c) cross-sectional scanning electron microscopy (SEM), and (e) top-view SEM for planar (left) and porous (right) Al foils taken before and after the cycling. Adapted and reproduced with permission from Ref. [31]. Copyright © 2017 American Chemical Society.

Sun et al. [32] employed a doped interlayer film over the Na anode, which protected it from dendritic growth and guided the Na nucleation, thereby enhancing the cycle life of SAB. Carbon nanotube (CNT) co-doped with nitrogen (N) and sulphur (S) (abbreviated as NSCNT) was dip-coated to cellulose-based paper, which acted as sodiophilic surface (Figure 15.3). The effect of doping was characterized by morphological analysis (high-resolution transmission electron microscopy (HRTEM)), elemental mapping, and X-ray photoelectron spectroscopy (XPS) (Figure 15.2). Doped CNT with a sodiophilic nature guided the nucleation of Na and uniform distribution of Na during the cell operation. The cell achieved a coulombic efficiency of 99.8% with excellent cyclic stability of about 500 cycles.

To inhibit the dendrite growth and volumetric expansion associated with conventional Na anode, Chi et al. [33] modified the pure Na anode to a composite anode containing Na and carbon, via melt infusion technique. The reported carbon embedded with Na consisted of high porosity, flexibility, and surface area enough to accommodate the Na deposits and to effectively reduce the polarization for achieving cyclic stability. The cell achieved high stability without any dendrite formation up to 120 cycles under a high current density of 5 mA cm^{-2}. Wang et al. [34] found that depositing a thin layer of silver nano paper (having a thickness of 2 μm) over the Na metal anode prevents dendritic growth and can be a tool for enhancing the cyclic life of the cell. The deposition was made possible by vacuum filtration of silver nanowire and nano cellulose suspension as a host for Na anode. A high conductivity, sodiophilic behaviour, high reversibility, and uniform Na deposition during plating

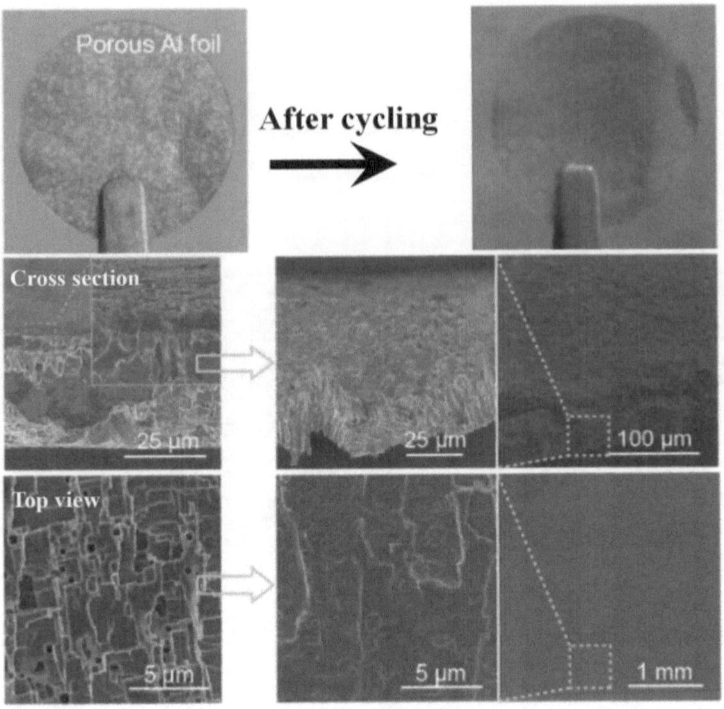

FIGURE 15.2 A comparative analysis of morphology showing the Na deposition on planar and porous aluminium (Al) foil current collector before and after the cycling process. (a) Porous Al foil, (b) cross-sectional scanning electron microscopy (SEM), and (c) top-view SEM for planar (left) and porous (right) Al foils taken before and after the cycling. Adapted and reproduced with permission from Ref. [31]. Copyright © 2017 American Chemical Society.

were observed, which was believed to be due to the electrochemical activity of silver nano paper. The cell achieved stability up to 400 cycles with a discharge capacity of 995 mAh g^{-1} at a current density of 1 mA cm^{-2}.

To protect the metallic Na anode from reacting with electrolyte solvents and to suppress dendrite formation, Ma et al. [35] coated the anode with a layer of bismuth. The formation of stable bismuth coating over Na anode was achieved by an ion exchange process in a diglyme electrolyte. The in-situ synthesis method for the bismuth layer over the anode reduced the instability of the anode. The thickness of the coating was measured to be 10 μm thick. Post-mortem analysis showed that the chemical composition of the bismuth layer was composed of fluorinated and carbonate species, and these remained unaffected during the stripping–plating process, indicating the stability of the film. The coated anode facilitated a smooth plating process by reducing charge transfer resistance, hindering O_2 crossover, and preventing electrolyte decomposition and dendrite formation, with porous morphology indicating the easiness of electrolytes to penetrate and transfer ions. Operando optical microscopy results (Figure 15.3) showed the suppression of dendrites. Results from a comparative electrochemical analysis employing symmetric cells with coated and pristine Na metal showed the stability of the coated electrode. Bismuth-coated symmetric cells showed a stable plating–stripping study for 1,000 h at a current density of 0.5 mA cm^{-2} (Figure 15.4). Furthermore, electrochemical properties were studied for full cell configuration with 1.0 M $NaSO_3CF_3$ in diglyme with super-P as an air cathode. SAB cell with pristine Na anode could sustain for ten cycles, whereas cell with coated anode sustained for 50 cycles with a high coulombic efficiency of 96.3%, delivering 1,000 mAhg^{-1} as discharge capability.

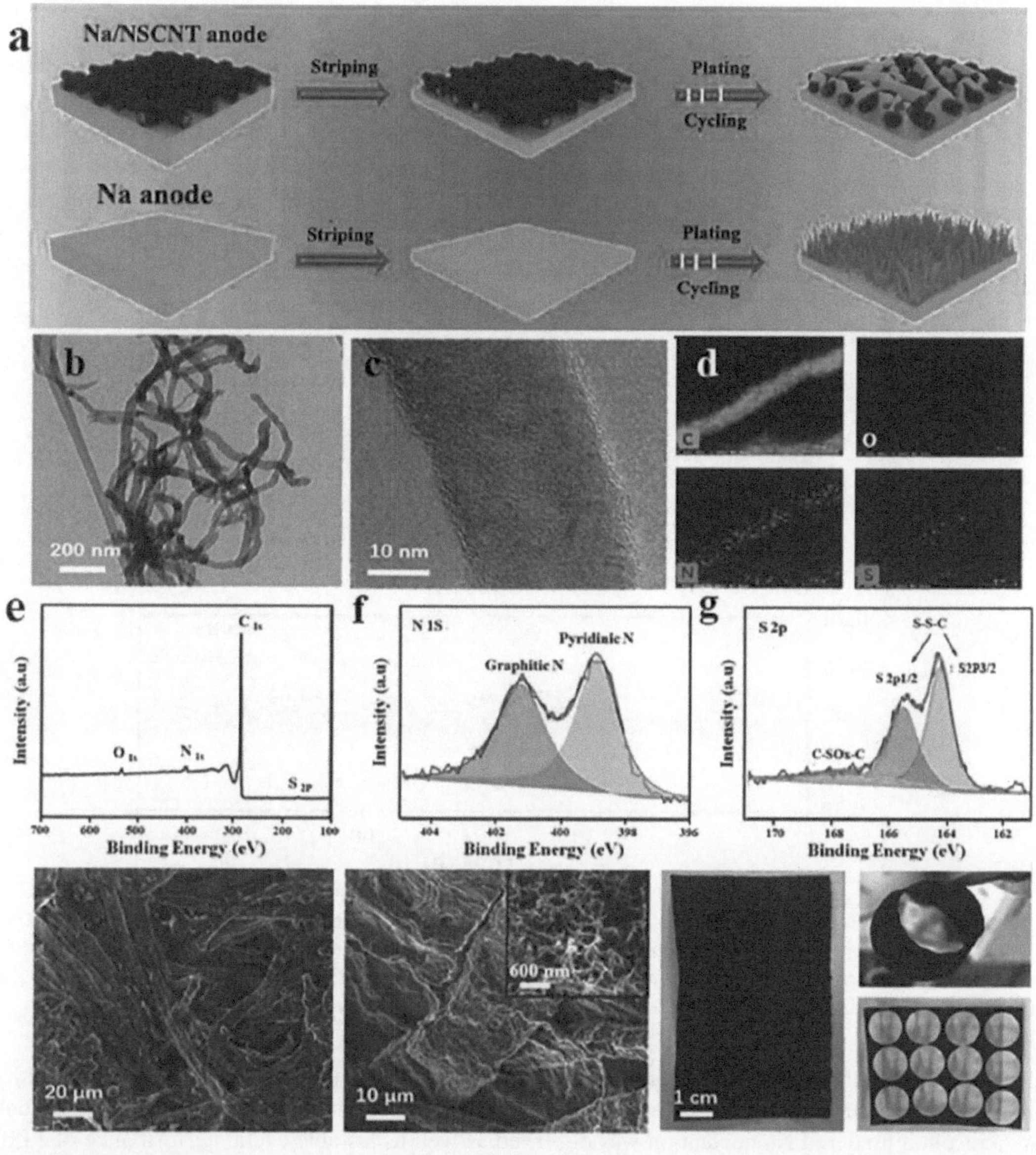

FIGURE 15.3 (a) Schematic illustration of mechanism showing the prevention of dendrites on Na metal anode with NSCNT anode and a bare Na anode, (b and c) morphology of NSCNT observed with a high-resolution TEM (HRTEM), (d) elemental mapping of co-doped CNT, (e–g) X-ray photoelectron spectroscopy (XPS) showing the doped samples, (h and i) SEM images of NSCNT paper, and (j) photographs of NSCNT paper demonstrating mechanical properties of the sample. Adapted and reproduced with permission from Ref. [32]. Copyright © 2018 Wiley-VCH.

Enhancing the cyclability of Na metal batteries through the employment of a host, which is sodiophilic, has evolved in Na metal batteries as well [36]. Wu et al. [37] adapted the application of 2D materials as a stable host for fabricating Na-based composite anode in carbonate electrolytes. The Na-embedded reduced graphene oxide aerogel (Na@rGa) framework was synthesized through oriented freeze drying of hydrogel and molten infusion technique of reduced graphene oxide aerogel; a schematic representation of the same is shown in Figure 15.5. 3D scaffolds with microcavities were developed when the aerogel was exposed to hydrothermal treatment. The large porosity, uniform

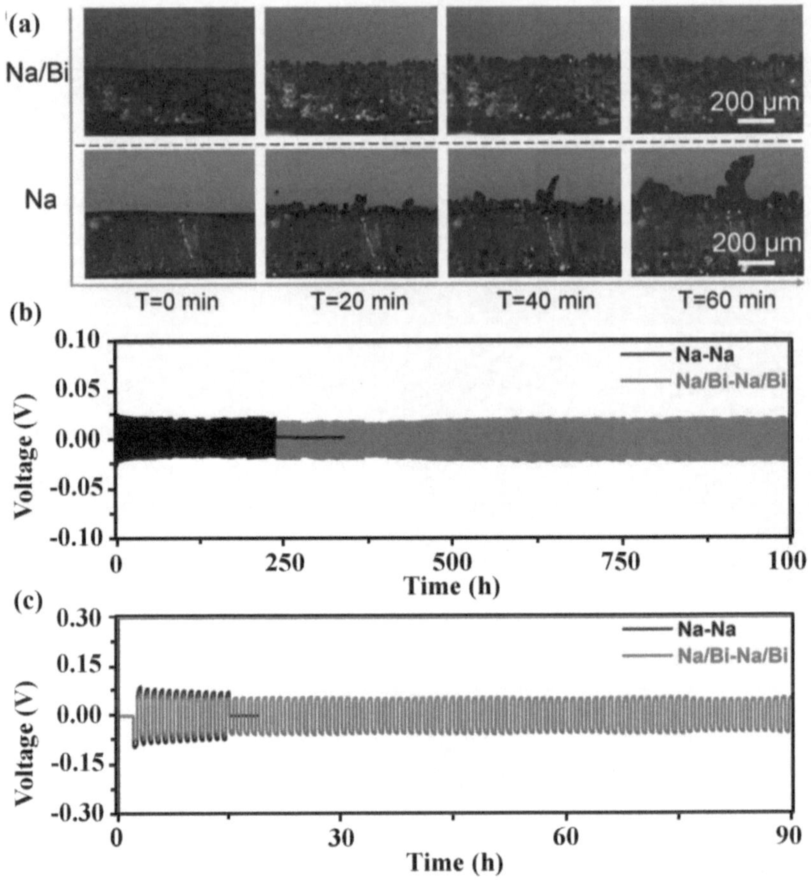

FIGURE 15.4 Comparative analysis of pristine Na anode and bismuth layer-coated Na anode. (a) Operando optical microscopy images showing the dendrite formation with respect to time (performed at 2 mA cm^{-2} for 1 h), (b) plating/stripping study of symmetric cells (for 1 mAh cm^{-2}) at 0.5 mA cm^{-2}, (c) 1 mA cm^{-2}. Adapted and reproduced with permission from Ref. [35]. Copyright © 2019 Wiley-VCH.

pore structure, enhanced good wettability with electrolyte, and high surface area of rGa provided a reduced local current density, homogeneous electrolyte distribution, and rapid transport of ions. As a result, a hindered Na nucleation was observed as well as a stable cyclic performance of 1,000 cycles at a current density of 5 mA cm^{-2} with a low overpotential of 50 mV.

Yu et al. [38] created a lightweight 3D-printed graphene lattice that used graphene oxide hydrogel and a direct inkjet approach. The framework with microstructure was engineered with a pore size of 500 μm and served as a host for moderating the Na deposit to obtain a stable Na anode. During the electroplating process, it was observed that the Na had a proclivity to deposit inside the pores available in the structure with a compact morphology, forming a uniform deposition of Na^+, and suppressing the formation of dendrites. This demonstrates that the highly ordered oriented microporous architecture could moderate the deposition behaviour of Na, thereby improving the performance. The cell showed a coulombic efficiency of 99.84% with stable cyclability for more than 500 cycles for approximately 1,000 h.

Guo et al. [39] designed a 3D superionic conductor as a host to solve the nucleation issues of Na metal batteries. Natrium superionic conductor $Na_3V_2(PO_4)_3$ (NVP) with microflower morphologies provides rapid ion transportation and abundant sites for Na nucleation. A $Na_{3+x}V_2(PO_4)_3$ phase is formed due to the increased affinity of Na and NVP during plating. This helps for a homogenous Na^+ deposition and hinders the dendrite growth as well as limits the volume fluctuation of the Na

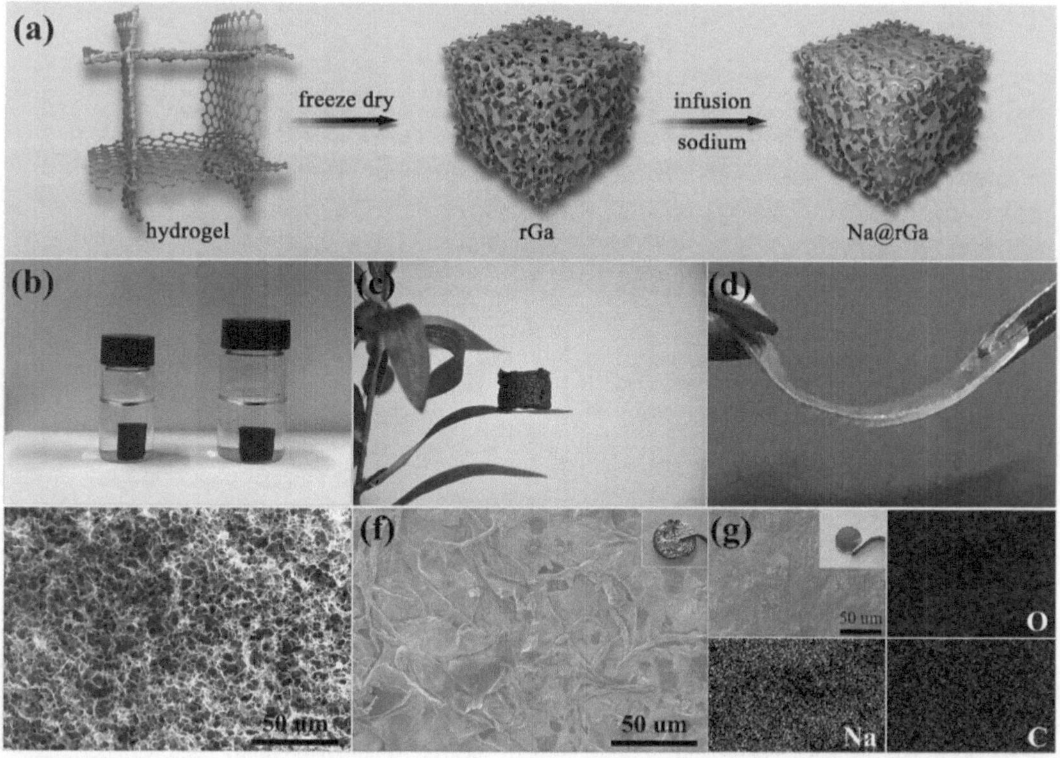

FIGURE 15.5 (a) Schematic representation of the synthesis of Na-embedded reduced graphene oxide aerogel (Na@rGa) anode; (b) graphene oxide after 4 h of the hydrothermal process; (c) picture depicting the ultralight weight of the prepared aerogel; (d) flexibility of Na@rGa strip; SEM images of (e) rGa-4h and (f) aerogel composite anode Na@rGa; and (g) images of SEM and elemental mapping after Na@rGa anode pressing. Adapted and reproduced with permission from Ref. [37]. Copyright © 2019 Elsevier.

anode to provide a stable nucleation and enhance the life of the cell. Liu et al. [40] developed a 3D porous carbon fibre structure homogeneously doped with nitrogen to protect the Na anode. Electrospinning was used to create the structure, which has a high specific surface area of 1,098 $m^2\ g^{-1}$. The researchers discovered that nitrogen doping increased the affinity for Na, resulting in a lower nucleation overpotential, and minimized the volume change of anode, thus limiting the dendrite growth. Furthermore, the greater specific surface area of nitrogen-doped carbon fibre contributed to uniform and mass loading of Na as well as a significant decrease in local current density, which is attributed to the improved cell performance with a coulombic efficiency of 99.8% at a current density of 4 mA cm^{-2} for up to 950 h.

He et al. [41] prepared a flexible fibrous hydroxylated MXene/CNT (h-Ti_3C_2/CNTs) composite film for anode protection. The one-dimensional fibrous h-Ti_3C_2 was prepared by the alkalization of 2D h-Ti_3C_2, whereas the 3D interconnected h-Ti_3C_2/CNTs film was fabricated using the vacuum filtration method. The investigators found that the film contained abundant "sodiophilic" sites for rapid electron-ion transfer, which reduced nucleation overpotential and provided uniform nucleation of Na^+ ions to regulate the dendrite growth. The cell performed with an average coulombic efficiency E of 99.2% up to 1,000 cycles. To reduce the corrosive effect on the Na anode, Liu et al. [42] developed a SAB with Li-modified Na composite anode instead of conventional pure metallic Na, which effectively prevented subsequent cell failure. The cell consisted of a gel electrolyte, in situ formed over the Na surface. Gelation of electrolyte was achieved when Li at the Na surface reacted with ethylenediamine forming a sacrificial lithium ethylenediamine film which further crosslinked with tetraglyme. A gas permeation study showed that the Na-gel electrolyte showed the ability

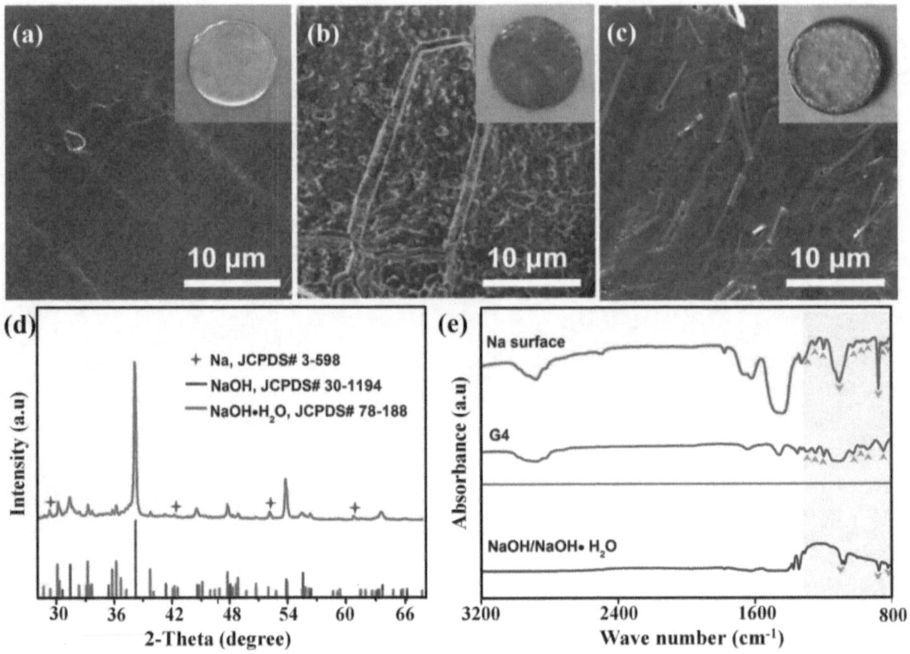

FIGURE 15.6 Characterization of Na metal anode after cycling. (a) Morphological analysis of Li-modified Na composite anode, (b) analysis of Na anode after cycling, (c) SAB anode after cycling at ambient conditions with gel electrolyte, (d) XRD peaks, and (e) overlaid FTIR spectra. Adapted and reproduced with permission from Ref. [42]. Copyright © 2021 American Chemical Society.

to suppress the O_2 and H_2O crossover in ambient conditions. This contributed to the delay in the anode corrosion and electrolyte decomposition. Furthermore, the Li ions (Li^+), showed an electro-shielding effect that hampered the dendrite growth to obtain a better electrochemical performance. The smooth morphology of the cycled Na anode in the gel electrolyte indicated the absence of dendrite growth and corrosion (Figure 15.6). Even after cycling for 500 cycles, the cell showed metallic Na with the presence of hydrated NaOH, this was confirmed by XRD and FTIR analysis (Figure 15.6). Although the presence of Na_2CO_3 was shown in FTIR spectra, it contributed to stabilizing the Na anode [43]. SAB with liquid electrolyte showed increasing electrode polarization; however, there was no increment in polarization in the case of gel electrolyte for 300 h. Moreover, SAB with gel electrolyte showed enhancement in cycling at ambient conditions for more than 2,000 h.

15.3 APPROACHES TO STABILIZE SEI LAYER

As metallic Na is used as an anode for post-lithium battery technology, the main attraction is due to its high earth abundance, lower cost, and high theoretical capacity [44]. The limitation is the reactivity at the electrode–electrolyte interface, which leads to lowering the performance of the system [19]. SAB, as Na is used as the anode, has an indispensable role in monitoring the electrochemical performance. The unstable SEI formation over the anode accelerates the parasitic side reaction between the electrolyte and the Na metal, thereby restricting ionic conduction [45]. Hence, various researchers have investigated rectifying this problem by creating a stable layer of SEI, to limit the chances of premature failures that could possibly occur due to undesired chemical reactions [46]. Formation of Na dendrite is another issue that causes short circuits and also thermal runways in the cell [44]. Moreover, dendrite formation creates cracks during the stripping/plating and allows metallic Na to react with the electrolyte, lowering the cyclic stability and causing safety issues [47,48]. The formation of dense films on the anode surface is difficult when it comes to the wide perspective of commercialization of

SAB. Hence, unveiling the mechanism of SEI formation needs to be well understood. Lutz et al. [49] reported chemical structure of salt anion and solvation dynamics of Na^+ influence the stability of SEI in SAB. The mechanism of SEI formation was characterized using ^{23}Na-NMR spectroscopy, showing that the Lewis basicity of the salt anion influences Na^+ solvation for a weakly solvating solvent like dimethoxyethane (DME). The decomposition of salt anions and solvent molecule polymerization are initiated by the attack of free radicals formed by the reduction of solvent molecules, which also contribute to the formation of SEI [49]. With the aim of creating a passivation layer, which is permeable to Na^+, Zhao et al. [50] reported an ultrathin Al_2O_3 layer of the coating through the atomic layer deposition technique, which suppressed dendrite formation and stabilized SEI. A thin layer of Al_2O_3 was formed using trimethylaluminium (TMA) and water (H_2O) as precursors. Morphological analysis and electrochemical characterizations showed the suppression of dendrites with a reduction in the side reaction between the anode and the electrolyte and the formation of a stable and homogenous SEI layer. Optimizing the thickness of deposited Al_2O_3 by 3.5 nm, the cell showed stable plating/stripping behaviour even after 500 cycles at a current density of 3 mA cm^{-2}. The same group [51] explored the molecular layer deposition of alucone for protecting the Na metal anode cycled in ethylene carbonate/propylene carbonate (EC/PC) electrolyte containing $NaPF_6$ salt. The sequential deposition of TMA and ethylene glycol for the inorganic–organic coating, to form a highly stable SEI layer and dendrite suppression over the Na anode. The stability was observed due to the high stability against and passivation behaviour of alucone coating against the electrolyte. Moreover, by passivizing anode to form the SEI layer facilitated a smooth stripping/plating process, resulting in a significant enhancement in the life cycle of the Na anode. Luo et al. [52] framed an Al_2O_3 coating on the Na metal anode through a low temperature-based plasma-enhanced atomic layer deposition (PEALD) process. TMA and O_2 plasma precursors were used for layer-by-layer growth of film over anode in ethylene carbonate/diethyl carbonate (EC/DEC)-based electrolyte containing sodium perchlorate ($NaClO_4$). This thin layer acted as an artificial SEI on the anode, which contributed towards a stable plating/stripping process. A comparison of Al_2O_3-coated Na metal anode with bare Na anode during the cyclic process of plating/stripping is shown in Figure 15.7.

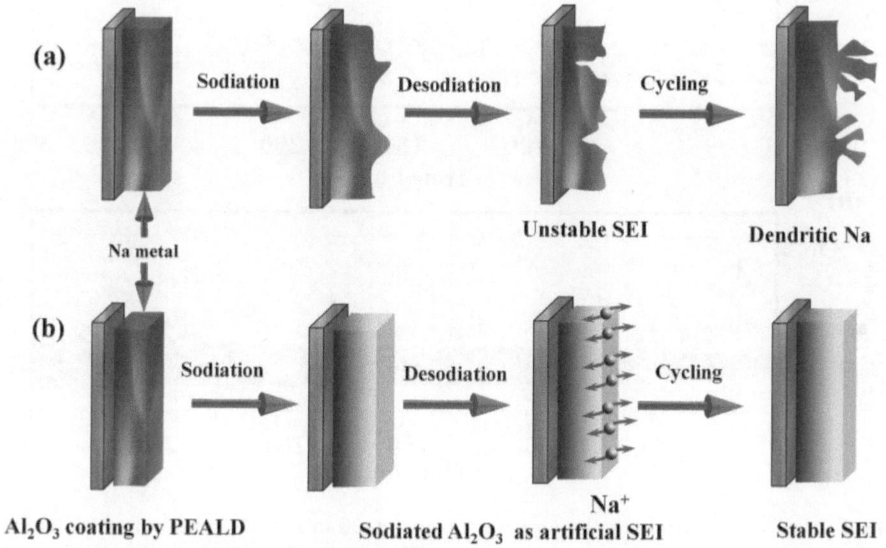

FIGURE 15.7 Schematic illustration of Na anode stabilization on platting/stripping process with carbonate-based electrolyte system. (a) Dendrite formation over Na metal anode, and (b) stable SEI formation over Al_2O_3 coated Na metal by PEALD method. Adapted and reproduced with permission from Ref. [52]. Copyright © 2017 Wiley-VCH.

To increase the cyclic stability and reduce the dendrite formation, Wang et al. [53] reported the usage of a protective layer of multilayer graphene with a thickness of ~5 nm, stabilized anode from dendritic growth, and ensured prolonged cycling (100 cycles without additives in carbonate electrolyte) at a high current density of 2 mA cm^{-2}. From various analyses, it was confirmed that the rate capability of the anode depends on the thickness of the protecting layer graphene. For cells operated at lower current densities (≤1 mA cm^{-2}), the best performance is shown when the thickness of the graphene film was between 2 and 3 nm. However, for a cell operated at high current density, a thick layer of approximately 5 nm showed the best performance.

As layered materials showed promising results in terms of protecting alkali metal anodes, Tian et al. [54], performed a theoretical investigation on the applicability of 2D materials for protecting the anode. The effect of protection showed dependency on various factors including the structure, defect pattern, metal proximity effect, bond angle, and bond length, which influenced not only the electrical but also the mechanical properties of the system. The introduction of defects and bond angle improved the ionic conductivity but showed an adverse effect on the stiffness of the material with respect to the dendrite growth suppression. Wu et al. [55] created a sodium fluoride (NaF)-rich protection passivation layer on the Na anode surface to restrict dendrite growth. The electrolyte for the study consisted of sodium trifluoromethanesulphonate ($NaCF_3SO_3$), tetraglyme, and fluoroethylene carbonate electrolyte (FEC). The researchers compared the cell performance of tailored Na anode with that of pristine Na anode on oxygen crossover at a current density of 0.1 mA cm^{-2} and noticed a steady stripping/plating process with no dendrite emergence for tailored Na anode-based cells, the result of which is shown in Figure 15.8. The oxygen crossover and parasitic side reactions were found to be suppressed due to the protective coating. The NaF-rich film that forms on the Na metal surface contributed to the stable performance and inhibited dendrite growth of tailored Na anode-based cells.

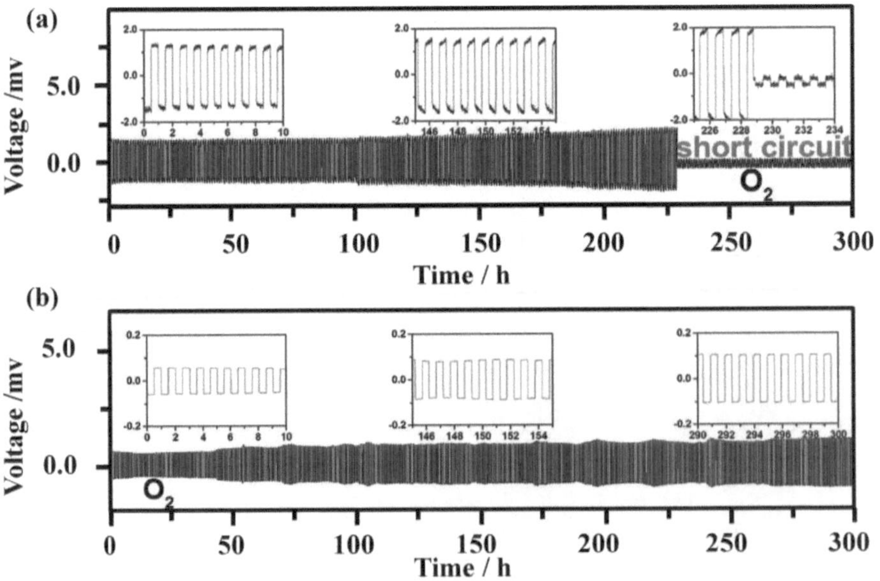

FIGURE 15.8 Analysis of stripping/plating performance on 30 min of O_2 crossover for tailored Na anode and pristine Na anode-based cells with a current density of 0.1 mA cm^{-2}. (a) Performance of pristine Na anode-based cell, showing a short circuit after 230 h. (b) Performance of a tailored Na anode-based cell showing stability up to 300 h. Adapted and reproduced with permission from Ref. [55]. Copyright © 2018 Wiley-VCH.

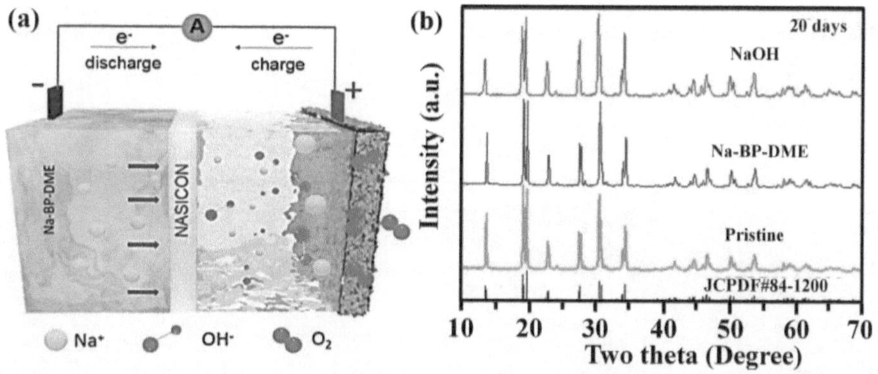

FIGURE 15.9 (a) Schematic illustration of liquid anode-based SAB and (b) overlaid XRD plots of pristine NASICON separator and separator immersed in Na–BP–DME liquid anode and catholyte consisting of 1 M NaOH solution (for 20 days). Adapted and reproduced with permission from Ref. [56]. Copyright © 2018 Elsevier.

Considering the safety issues involving exothermic reaction, reactivity with water, and volume expansion regarding the Na metal anode during the charging/discharging process, Liang et al. [56] reported an organic liquid anode-based hybrid rechargeable SAB for the first time. Dissolving the Na metal into a solution containing aromatic hydrocarbons as biphenyl (BP) and dimethoxyethane solvent was adapted for the preparation of a liquid anode. Despite the liquid anode, the battery consisted of a ceramic material with a molecular formula $Na_3Zr_2Si_2PO_{12}$ (NASICON) solid–electrolyte, NaOH catholyte, and commercial platinum-loaded carbon cathode (Figure 15.9). To confirm the compatibility of the NASICON separator with liquid anode and catholyte, the separator was immersed in both systems and analysed XRD patterns (Figure 15.8b). XRD pattern showed no significant changes which indicated the compatibility and benignity with both the systems. On analysing the safety parameters, the prepared anode Na–BP in DME solution reacted with water, and no hydrogen evolution was found. In addition, the investigators observed a negligible change in temperature which indicated a limited exothermic reaction, when comparing with the traditional Na metal used in SAB. Furthermore, the cell demonstrated a low voltage gap of 0.14 V with 95.3% round-trip efficiency and a high-power density reported as 39.0 mW cm^{-2}.

The incorporation of electrolyte additives is an effective strategy in enabling a stable SEI layer in alkali metal batteries. Lee et al. [57] investigated the role of FEC on anode metal stabilization and found that FEC along with 1 M sodium bis(fluorosulphonyl)imide (NaFSI) salt in electrolyte functioned as an effective additive. The work reported the formation of a uniform ionic interlayer constituting sodium carbonate (Na_2CO_3), NaF, anhydrides, and alkyl carbonates on the metal anode surface due to the reductive decomposition of FEC at the anode surface. From in-situ analysis such as differential electrochemical mass spectrometry and optical microscopy, the mechanism was revealed, and possible reactions are represented in Figure 15.10. The layer formed and acted as a shielding film on the Na metal surface, which further inhibited the electrolyte reactions and hence suppressed the dendrite growth and effectively stabilized the SEI layer.

Zhu et al. [58] found that tetramethylthiuram disulphide, an organosulphur-based additive functioned as an effective electrolyte for Na metal batteries. During electrochemical cycling, the additive showed the formation of an in-situ protective layer over Na metal. XPS analysis, coupled with Density Functional Theory (DFT) calculations revealed the SEI formation mechanism, where SEI consisted of organic sulphide salt from the tetramethylthiuram disulphide decomposition. The presence of SEI with sodium sulphide layer facilitated Na^+ transport and was found to suppress dendrite

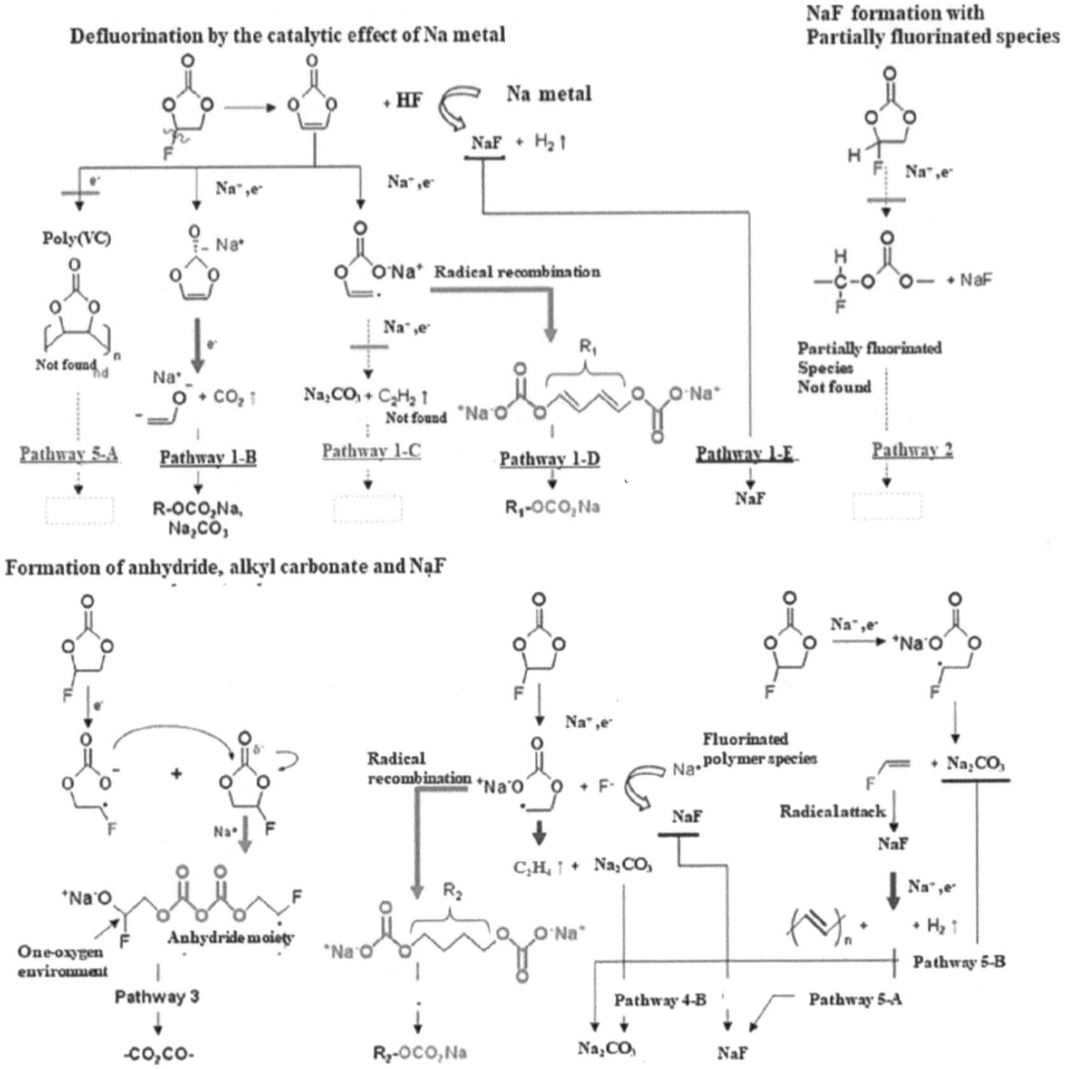

FIGURE 15.10 Chemical reaction mechanism showing the interlayer formation through reductive decomposition of FEC. The formation of carbonyl-functionalized moieties ionic compounds is highlighted in green. Adapted and reproduced with permission from Ref. [57]. Copyright © 2018 American Chemical Society.

formation, thereby contributing to a capacity retention of 80%, after 600 cycles at a C rate of 4 C. Furthermore, a symmetric cell with additive showed stability in cycling at 0.25 mAh cm^{-2} for 1,600 h with a coulombic efficiency of 94.25%.

Zhao et al. [59] examined the bifunctional impact of a tetrabutylammonium cation (TBA^+) additive for anode protection and superoxide stabilization in sodium trifluoromethanesulphonate (NaOTf) in diglyme electrolyte-based SAB. During Na plating, these TBA^+ become adsorbed on the electrode surface, resulting in an electrostatic shielding effect that directs a homogeneous aggregation of Na^+ rather than the irregular surface of the Na anode. As a result, dendrite formation is suppressed in the Na anode. On the contrary, at the cathode, the TBA^+ pairs up with O_2^- and generates $TBA_2(O_2)_2$ dimer and suppresses the disproportionation reaction of discharge products. On the cyclic voltammetry curves of SAB with TBA^+ additive, the reduction and oxidation peaks visibly increased,

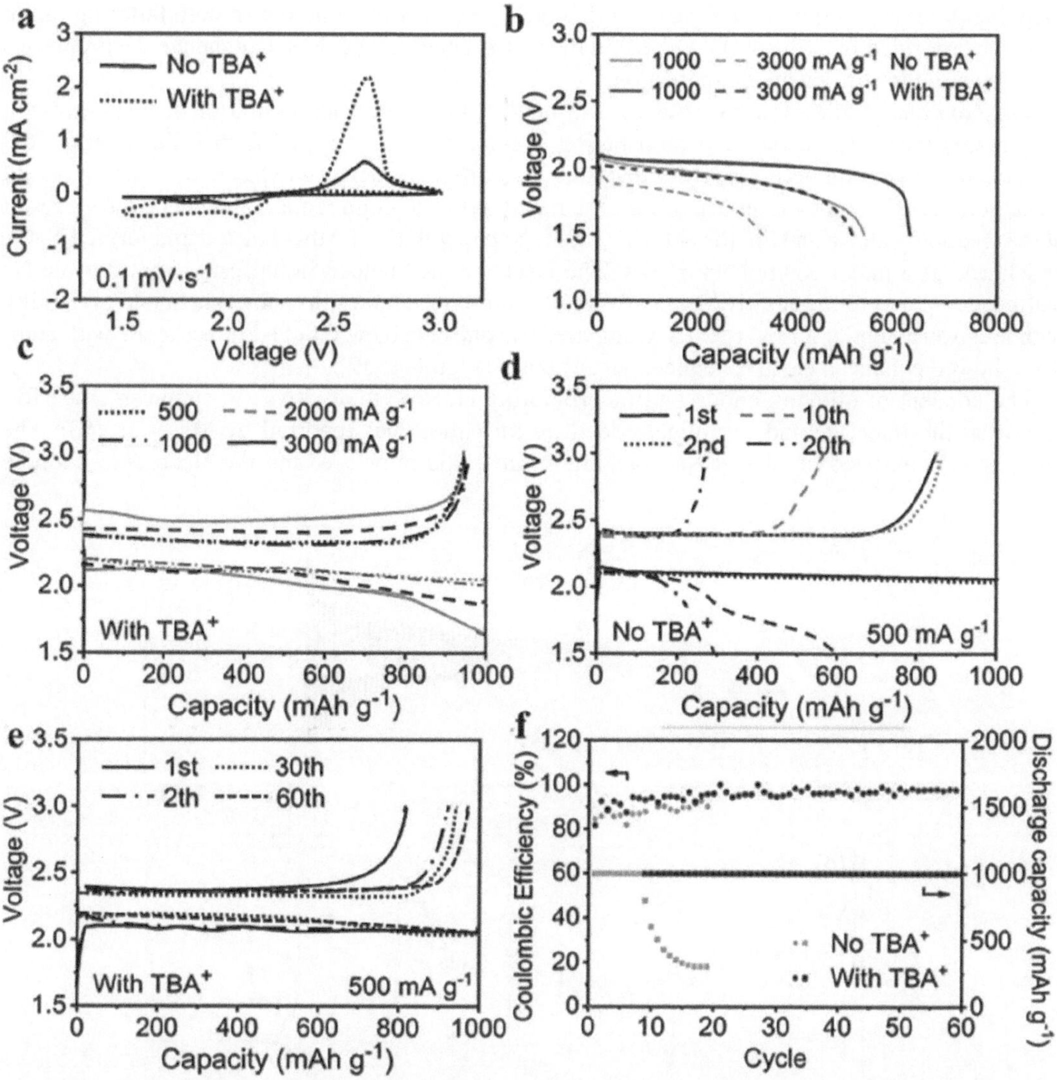

FIGURE 15.11 Electrochemical performance of SAB with and without tetrabutylammonium cation (TBA^+) additive. (a) Cyclic voltammetry curves of SAB, (b) first discharge profile, (c) overlaid discharge/charge profiles at various current densities for the fifth cycle, (d and e) discharge/charge profiles of various cycles with and without additive, and (f) cycling stability. Adapted and reproduced with permission from Ref. [59]. Copyright © 2021 Wiley-VCH.

indicating the enhancement in the kinetics for oxygen redox reaction (Figure 15.11). Incorporation of TBA^+ additive enhanced discharge–charge capacity of SAB with enhanced coulombic efficiency. The cell showed a coulombic efficiency of 95.5%, with high cyclic stability of more than 100 h in an oxygen atmosphere, which is attributed to the bifunctional effect of TBA^+.

The formation of a thin amorphous protective layer of Na_3PS_4 (NaPS) through an in-situ solution-based technique was developed by Zhao et al. [60] for counterattacking the challenges of Na metal anode protection. This thin layer was found to be ionically conductive and could make significant changes in the improvement of homogeneous SEI layers, limiting the side reaction of metal anode with electrolytes, and a considerable increase in Na^+ density during the plating and stripping process. Optimizing the reaction time and precursor amount and smooth tailoring of the NaPS layer formed in terms of thickness and the constituents can be done to enhance the properties to be favoured for the Na anode protection.

The creation of alloy interfaces at the Na anode allows the suppression of dendrite growth, minimizing the overpotential, and contributing to enhancement in cycle life [61]. Zheng et al. [62] found that the addition of tin (II) chloride ($SnCl_2$) in carbonate electrolyte paved the way for in-situ formation of Na–Sn alloy interlayer on the Na metal due to its high reduction tendency. As a result of the reaction with Na and Sn, the chlorine ions together with Na^+ further reacted and formed NaCl, which acts as a major component of SEI. The layers formed helped in uniform, dendrite-free Na plating, compared to the additive-less carbonate electrolyte. The results showed a stable cyclability with less overpotential for over 500 h. Compared to control cells without additives, cells with additives showed enhanced electrochemical performance (Figure 15.12).

The concept of alloying anode and incorporating electrolyte additives to suppress crack formation at the interface and to mitigate dendrite formation was reported by Ma et al. [63]. The alloy anode consisted of Li and Na, forming a bimetallic interface, and the electrolyte additive

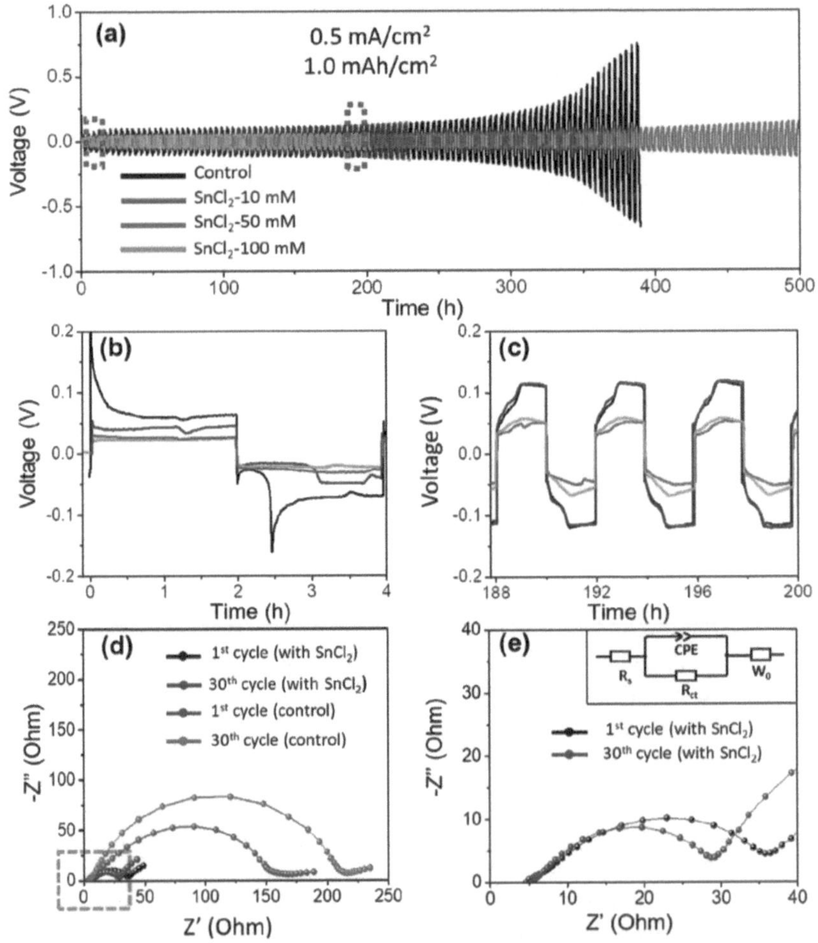

FIGURE 15.12 Various electrochemical performances of cells with $SnCl_2$ additive. (a) Cyclic stability of Na–Na symmetric cell with varying concentrations of $SnCl_2$ additive, (b and c) enlarged voltage profile, and (d and e) plot of electrochemical impedance spectroscopy for the control specimen and cell with $SnCl_2$ additive after various cycles. Adapted and reproduced with permission from Ref. [62]. Copyright © 2019 American Chemical Society.

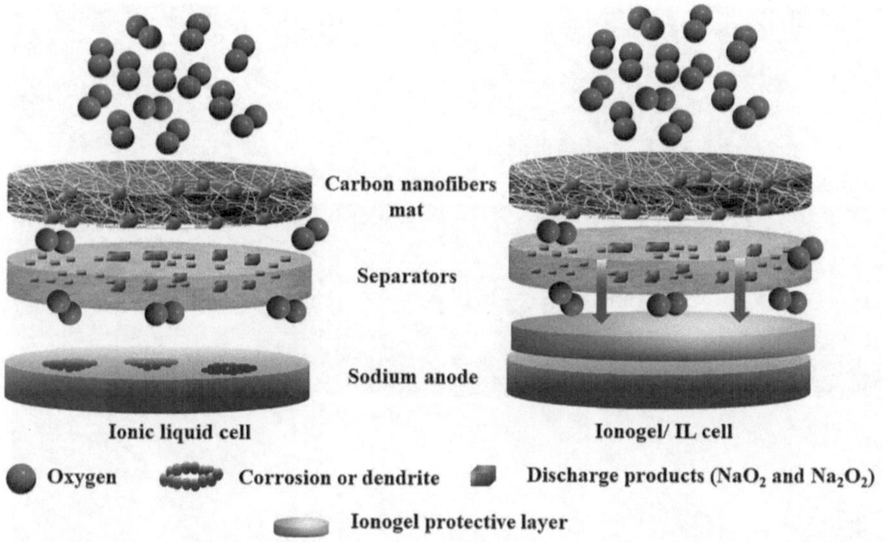

FIGURE 15.13 Schematic illustration showing a comparative mechanism inside SAB using liquid electrolyte and ionogel/IL during the discharge process. Adapted and reproduced with permission from Ref. [65]. Copyright © 2019 American Chemical Society.

consisted of 1,3-dioxolane for 0.5 M $NaCF_3SO_3$ in tetraglyme. Morphological analysis evidenced the absence of cracks and dendrites at the anode. Electrochemical characterization revealed that an aprotic bimetallic alloy–O_2 battery showed longer cycling stability. Compared to SAB with 31 cycles, the bimetallic alloy–O_2 battery showed a stable cycling of 137 cycles.

Ionogel (IL in a polymer matrix) was reported to have a vital role in suppressing the side reactions of metallic anode for stabilizing SEI layers in LAB [64]. With this as a background, Ha et al. [65] investigated a bilayer electrolyte consisting of ionogel and an IL mixture for SAB. This bilayer was formed by liquid electrolyte (sodium bis(trifluoromethanesulphonate) imide (NaTFSI) in [C4mpr][TFSI]) entrapped within a polymer matrix (poly(ethylene glycol) diacrylate). The discharge products that were produced in the process were able to pass through the separator/electrolyte to form dendrites on metal anode surfaces. Introducing the bilayer electrolyte limits the oxygen crossover and the discharge products to the Na metal, preventing them from corrosion and dendrite formation. The schematic representation of this mechanism is represented in Figure 15.13.

Wang et al. [66] reported the fabrication of SAB using an ether-based electrolyte along with $NaBF_4$ for enabling a dendrite-free Na anode. The cell achieved a high coulombic efficiency of about 99.93% over 400 cycles. The formation of an ultrathin and strong SEI layer, consisting mainly of NaF and B–O species, showed a low electrochemical impedance and high exchange current density. Moreover, the SEI formed provided high ionic conductivity with significant mechanical properties that accounted for the improved performance.

15.4 REAL-TIME MONITORING OF NA METAL ANODE INTERFACE

As metallic Na, being reactive, unveils the complexities of SEI formation and mitigates the dendritic growth at the anode, it is necessary to monitor the interface in real time [67]. Conventional characterizations fail to reveal vital information during the electrochemical process including platting and stripping of Na, transportation of Na^+, and the formation of discharge products. Hence, operando techniques, which are non-destructive, can deliver key information about the electrode–electrolyte interface [68]. Operando microscopic techniques constitute one of the widely used techniques to find the formation and growth of dendrites [69]. Ma et al. [70] incorporated in-situ optical microscopy

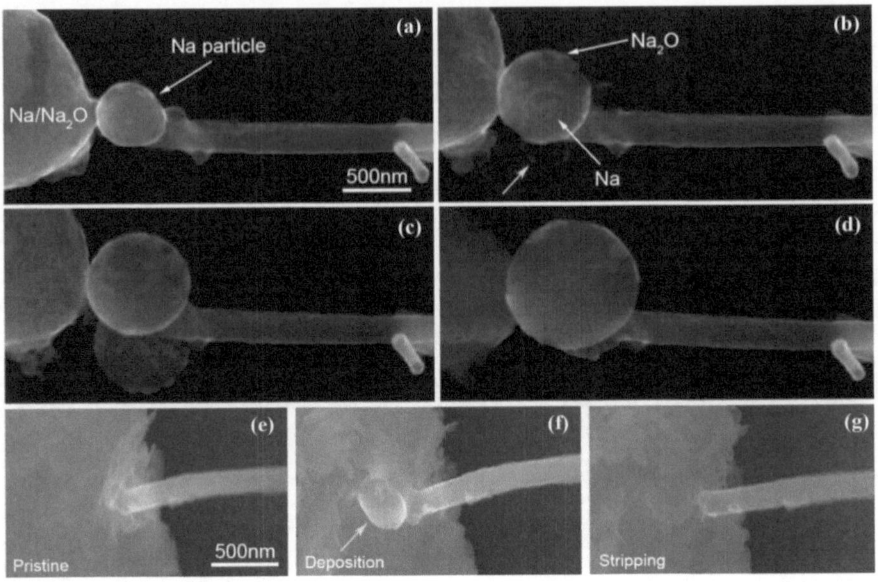

FIGURE 15.14 Evaluation of interface between CNF and Na_2O electrolyte during the electrochemical process, observed using in-situ SEM analysis. Adapted and reproduced with permission from Ref. [73]. Copyright © 2017 Elsevier.

experiments for Na metal anode battery, where the growth of Na dendrites was observed at higher moisture levels along with bubble formation. Similar observations were reported by Yui et al. [71], where the Na dendrites showed a needle-like morphology after deposition, and it extended to 36 μm.

Operando electron microscopic techniques have also been employed in monitoring the interface of Na metal batteries. One such technique is the operando TEM. Zeng et al. [72] observed the morphology of Na deposition, followed by a two-step growing process. The first step includes nucleation, followed by growth on the flat surface. As the thickness of SEI increases, the rate of Na deposition became slower, and when the thickness was at 130 nm, deposition completely halted. It directly indicated that thin SEI allowed the ionic transportation of Na^+, but as SEI thickness increased Na^+ was impermeable. Li et al. [73] incorporated in-situ TEM and in-situ SEM analysis to find the deposition of Na on carbon nanofibre (CNF) as a 3D current collector in Na_2O electrolyte. During the electrochemical plating and stripping, the mesoporous structure of CNF acted as a site for Na nucleation. In-situ SEM analysis showed the formation of spherical particles of Na, shown in Figure 15.14.

To unveil the discharge–charge mechanism and to gain insights into the discharge product formation, Banis et al. [74] employed in-situ X-ray absorption spectroscopy (XAS) in SAB. A heterogeneous and thick layer of discharge products was generated, showing the progressive degradation at the interface of products and electrolytes. From the XAS spectra, nucleation and growth of NaO_2 discharge products were confirmed by the peaks corresponding to 532 eV. As charging time increased, the intensity of XAS peaks amplified, and, during discharge, the intensity of peaks declined. Broad peaks from 532 to 534 eV signified the presence of NaO_2 discharge product. Han et al. [75] investigated an in-situ atomic force microscopy to study real-time Na electrodeposition during the electrochemical process. The addition of FEC as an electrolyte additive in EC/PC stabilized the SEI layers, which was indicated by a uniform morphology. In addition to the dendrite suppression, the cell showed a high coulombic efficiency of 88% over 50 cycles.

Cryogenic transmission electron microscope (cryo-TEM), being an advanced technique, has been reported to show the morphological and elemental analysis for the SEI [76,77]. Kumar et al. [61] exposed the formation and growth of dendrite over Na metal and found that the uniformity of the SEI layer as the bare Na anode was replaced with an alloy of Na–Sn. In another work, Kumar et al. [78], highlighted the usage of cryo-TEM as an advanced tool for observing the deposition of Na on the anode. Both works indicated the potentiality of cryo-TEM as a tool for monitoring the electrode–electrolyte interface during electrochemical processes.

15.5 CONCLUSION

As the demand for sustainable energy storage solutions is the need of the hour, SABs have received immense research attention considering sustainability and cost-effective technology. The abundance of Na sources, environmental benignity, and high capacity of SAB make it a suitable candidate for industrial mass production technology. As metallic Na, being a reactive soft metal is used as an anode for SABs, it has inherent limitations. Reaction with electrolytes, formation of dendrites, unstable SEI formation, and non-uniform Na deposition are the major drawbacks. Dendritic growth and unstable SEI also lower the cycling life and serve as a serious safety concern. Hence, various strategies have been postulated, which preferentially protect Na from corrosion and guide uniform deposition. The factors affecting the performance of a battery relating to the metallic anode can be summarized by the following points: judicious and meticulous selection of electrolyte salts and solvents, concentration of salt in the solvent, optimum content of electrolyte additive, the affinity of the anodic host towards Na (sodiophilic nature), smoothness of electrode surfaces, electrode–electrolyte interfacial affinity, etc. Unveiling the complex SEI formation mechanism needs monitoring of the interface in real time. Operando or in-situ techniques such as operando optical microscopy, NMR spectroscopy, in-situ TEM, operando XRD, and XPS are among a few. The composition of the discharged product depends on the electrolyte solvent and salt used, which determines the reversibility and cycle life of the cell. Hence, the engineering of the electrode–electrolyte interface also catalyses the performance. To unveil the complexities of SABs, a complete understanding of each battery component is prime important. It is expected that by a multifaced approach to stabilizing anode, engineering electrolyte, and tailoring cathode, there can be a proliferating advancement in the SAB technology.

REFERENCES

1. Yang Z, Zhang J, Kintner-Meyer MCW, et al (2011) Electrochemical energy storage for green grid. *Chem Rev* 111:3577–3613. https://doi.org/10.1021/cr100290v
2. Jishnu NS, Vineeth SK, Das A, et al (2021) Electrospun PVdF and PVdF-co-HFP-based blend polymer electrolytes for lithium ion batteries. In: Balakrishnan NTM, Prasanth R (eds) *Electrospinning for Advanced Energy Storage Applications*, 1st ed. Springer, Singapore, pp. 201–234.
3. Sungjemmenla J, Soni CB, Vineeth SK, Kumar V (2021) Exploration of the unique structural chemistry of sulfur cathode for high-energy rechargeable beyond-Li batteries. *Adv Energy Sustain Res* 2100157:2100157. https://doi.org/10.1002/aesr.202100157
4. Zhang X, Wang XG, Xie Z, Zhou Z (2016) Recent progress in rechargeable alkali metal-air batteries. *Green Energy Environ* 1:4–17. https://doi.org/10.1016/j.gee.2016.04.004
5. Ahuja D, Kalpna V, Varshney PK (2021) Metal air battery: a sustainable and low cost material for energy storage. *J Phys Conf Ser* 1913:012065. https://doi.org/10.1088/1742-6596/1913/1/012065
6. Yadegari H, Sun Q, Sun X (2016) Sodium-oxygen batteries: a comparative review from chemical and electrochemical fundamentals to future perspective. *Adv Mater* 28:7065–7093. https://doi.org/10.1002/adma.201504373
7. Hwang J-Y, Myung S-T, Sun Y-K (2017) Sodium-ion batteries: present and future. *Chem Soc Rev* 46:3529–3614. https://doi.org/10.1039/C6CS00776G
8. Hans Wedepohl K (1995) The composition of the continental crust. *Geochim Cosmochim Acta* 59:1217–1232. https://doi.org/10.1016/0016-7037(95)00038-2
9. Kim I, Kim CH, Choi SH, et al (2016) A singular flexible cathode for room temperature sodium/sulfur battery. *J Power Sources* 307:31–37. https://doi.org/10.1016/j.jpowsour.2015.12.035
10. Adelhelm P, Hartmann P, Bender CL, et al (2015) From lithium to sodium: cell chemistry of room temperature sodium-air and sodium-sulfur batteries. *Beilstein J Nanotechnol* 6:1016–1055. https://doi.org/10.3762/bjnano.6.105

11. Li L, Zheng Y, Zhang S, et al (2018) Recent progress on sodium ion batteries: potential high-performance anodes. *Energy Environ Sci* 11:2310–2340. https://doi.org/10.1039/C8EE01023D
12. Yang J, Gao H, Kang X (2020) Nanomaterials application in Li-Se and Na-Se batteries. In: Ran F, Chen S (eds) *Advanced Nanomaterials for Electrochemical-Based Energy Conversion and Storage*. Elsevier, Amsterdam, The Netherlands, pp. 69–114.
13. USG Survey (2019) *Mineral Commodity Summaries*. U.S. Geological Survey, Reston, VA.
14. Lee K, Lee YJ, Lee MJ, et al (2022) A 3D hierarchical host with enhanced sodiophilicity enabling anode-free sodium-metal batteries. *Adv Mater* 34:2109767. https://doi.org/10.1002/adma.202109767
15. Wang T, Hua Y, Xu Z, Yu JS (2022) recent advanced development of artificial interphase engineering for stable sodium metal anodes. *Small* 18:2102250. https://doi.org/10.1002/smll.202102250
16. Matios E, Wang H, Wang C, Li W (2019) Enabling safe sodium metal batteries by solid electrolyte interphase engineering: a review. *Ind Eng Chem Res* 58:9758–9780. https://doi.org/10.1021/acs.iecr.9b02029
17. Wang H, Wang C, Matios E, Li W (2018) Facile stabilization of the sodium metal anode with additives: unexpected key role of sodium polysulfide and adverse effect of sodium nitrate. *Angew Chemie* 130:7860–7863. https://doi.org/10.1002/ange.201801818
18. Vineeth SK, Soni CB, Sun Y, et al (2022) Implications of Na-ion solvation on Na anode-electrolyte interphase. *Trends Chem* 4:48–59. https://doi.org/10.1016/j.trechm.2021.11.002
19. Zheng Z, Wu C, Gu Q, et al (2021) Research progress and future perspectives on rechargeable Na-O_2 and Na-CO2 batteries. *Energy Environ Mater* 4:158–177. https://doi.org/10.1002/eem2.12139
20. Soni CB, Kumar V, Seh ZW (2022) Guiding uniform sodium deposition through host modification for sodium metal batteries. *Batter Supercaps* 5:e202100207. https://doi.org/10.1002/batt.202100207
21. Wang H, Matios E, Luo J, Li W (2020) Combining theories and experiments to understand the sodium nucleation behavior towards safe sodium metal batteries. *Chem Soc Rev* 49:3783–3805. https://doi.org/10.1039/D0CS00033G
22. Han S, Cai C, Yang F, et al (2020) Interrogation of the reaction mechanism in a Na-O_2 battery using in situ transmission electron microscopy. *ACS Nano* 14:3669–3677. https://doi.org/10.1021/acsnano.0c00283
23. Hartmann P, Bender CL, Vračar M, et al (2013) A rechargeable room-temperature sodium superoxide (NaO_2) battery. *Nat Mater* 12:228–232. https://doi.org/10.1038/nmat3486
24. Lee J, Kim J, Kim S, et al (2020) A review on recent approaches for designing the SEI layer on sodium metal anodes. *Mater Adv* 1:3143–3166. https://doi.org/10.1039/D0MA00695E
25. Peled E, Menkin S (2017) Review-SEI: past, present and future. *J Electrochem Soc* 164:A1703–A1719. https://doi.org/10.1149/2.1441707jes
26. Wei S, Choudhury S, Xu J, et al (2017) Highly stable sodium batteries enabled by functional ionic polymer membranes. *Adv Mater* 29:1605512. https://doi.org/10.1002/adma.201605512
27. Zhuang Y, Deng D, Lin L, et al (2022) Ion-conductive gradient sodiophilic 3D scaffold induced homogeneous sodium deposition for highly stable sodium metal batteries. *Nano Energy* 97:107202. https://doi.org/10.1016/j.nanoen.2022.107202
28. Yang C-P, Yin Y-X, Zhang S-F, et al (2015) Accommodating lithium into 3D current collectors with a submicron skeleton towards long-life lithium metal anodes. *Nat Commun* 6:8058. https://doi.org/10.1038/ncomms9058
29. Zhang Q, Lu Y, Zhou M, et al (2018) Achieving a stable Na metal anode with a 3D carbon fibre scaffold. *Inorg Chem Front* 5:864–869. https://doi.org/10.1039/C7QI00802C
30. Liu T, Yang X, Nai J, et al (2021) Recent development of Na metal anodes: interphase engineering chemistries determine the electrochemical performance. *Chem Eng J* 409:127943. https://doi.org/10.1016/j.cej.2020.127943
31. Liu S, Tang S, Zhang X, et al (2017) Porous Al current collector for dendrite-free Na metal anodes. *Nano Lett* 17:5862–5868. https://doi.org/10.1021/acs.nanolett.7b03185
32. Sun B, Li P, Zhang J, et al (2018) Dendrite-free sodium-metal anodes for high-energy sodium-metal batteries. *Adv Mater* 30:1801334. https://doi.org/10.1002/adma.201801334
33. Chi S-S, Qi X-G, Hu Y-S, Fan L-Z (2018) 3D flexible carbon felt host for highly stable sodium metal anodes. *Adv Energy Mater* 8:1702764. https://doi.org/10.1002/aenm.201702764
34. Wang Z, Zhang X, Zhou S, et al (2018) Lightweight, thin, and flexible silver nanopaper electrodes for high-capacity dendrite-free sodium metal anodes. *Adv Funct Mater* 28:1804038. https://doi.org/10.1002/adfm.201804038
35. Ma M, Lu Y, Yan Z, Chen J (2019) In situ synthesis of a bismuth layer on a sodium metal anode for fast interfacial transport in sodium-oxygen batteries. *Batter Supercaps* 2:663–667. https://doi.org/10.1002/batt.201900020

36. Tao L, Hu A, Mu L, et al (2021) A self-sodiophilic carbon host promotes the cyclability of sodium anode. *Adv Funct Mater* 31:2007556. https://doi.org/10.1002/adfm.202007556
37. Wu F, Zhou J, Luo R, et al (2019) Reduced graphene oxide aerogel as stable host for dendrite-free sodium metal anode. *Energy Storage Mater* 22:376–383. https://doi.org/10.1016/j.ensm.2019.02.015
38. Yu Y, Wang Z, Hou Z, et al (2019) 3D printing of hierarchical graphene lattice for advanced Na metal anodes. *ACS Appl Energy Mater* 2:3869–3877. https://doi.org/10.1021/acsaem.9b00540
39. Guo M, Dou H, Zhao W, et al (2020) Three dimensional frameworks of super ionic conductor for thermodynamically and dynamically favorable sodium metal anode. *Nano Energy* 70:104479. https://doi.org/10.1016/j.nanoen.2020.104479
40. Liu B, Lei D, Wang J, et al (2020) 3D uniform nitrogen-doped carbon skeleton for ultra-stable sodium metal anode. *Nano Res* 13:2136–2142. https://doi.org/10.1007/s12274-020-2820-y
41. He X, Jin S, Miao L, et al (2020) A 3D hydroxylated MXene/carbon nanotubes composite as a scaffold for dendrite-free sodium-metal electrodes. *Angew Chemie – Int Ed* 59:16705–16711. https://doi.org/10.1002/anie.202006783
42. Liu X, Lei X, Wang Y-G, Ding Y (2021) Prevention of Na corrosion and dendrite growth for long-life flexible Na-air batteries. *ACS Cent Sci* 7:335–344. https://doi.org/10.1021/acscentsci.0c01560
43. Chen K, Huang G, Ma JL, et al (2020) The stabilization effect of CO_2 in lithium-oxygen/CO_2 batteries. *Angew Chemie – Int Ed* 59:16661–16667. https://doi.org/10.1002/anie.202006303
44. Soni CB, Sungjemmenla J, Vineeth SK, Kumar V (2021) Challenges in regulating interfacial-chemistry of the sodium-metal anode for room-temperature sodium-sulfur batteries. *Energy Storage* 4:e264. https://doi.org/10.1002/est2.264
45. Tai Z, Liu Y, Yu Z, et al (2022) Non-collapsing 3D solid-electrolyte interphase for high-rate rechargeable sodium metal batteries. *Nano Energy* 94:106947. https://doi.org/10.1016/j.nanoen.2022.106947
46. Zhao Y, Adair KR, Sun X (2018) Recent developments and insights into the understanding of Na metal anodes for Na-metal batteries. *Energy Environ Sci* 11:2673–2695. https://doi.org/10.1039/C8EE01373J
47. Chen X, Shen X, Hou T-Z, et al (2020) Ion-solvent chemistry-inspired cation-additive strategy to stabilize electrolytes for sodium-metal batteries. *Chem* 6:2242–2256. https://doi.org/10.1016/j.chempr.2020.06.036
48. Mandl M, Becherer J, Kramer D, et al (2020) Sodium metal anodes: deposition and dissolution behaviour and SEI formation. *Electrochim Acta* 354:136698. https://doi.org/10.1016/j.electacta.2020.136698
49. Lutz L, Alves Dalla Corte D, Tang M, et al (2017) Role of electrolyte anions in the Na-O_2 battery: implications for NaO2 solvation and the stability of the sodium solid electrolyte interphase in glyme ethers. *Chem Mater* 29:6066–6075. https://doi.org/10.1021/acs.chemmater.7b01953
50. Zhao Y, Goncharova LV., Lushington A, et al (2017) Superior stable and long life sodium metal anodes achieved by atomic layer deposition. *Adv Mater* 29:1606663. https://doi.org/10.1002/adma.201606663
51. Zhao Y, Goncharova LV, Zhang Q, et al (2017) Inorganic-organic coating via molecular layer deposition enables long life sodium metal anode. *Nano Lett* 17:5653–5659. https://doi.org/10.1021/acs.nanolett.7b02464
52. Luo W, Lin C, Zhao O, et al (2017) Ultrathin surface coating enables the stable sodium metal anode. *Adv Energy Mater* 7:1601526. https://doi.org/10.1002/aenm.201601526
53. Wang H, Wang C, Matios E, Li W (2017) Critical role of ultrathin graphene films with tunable thickness in enabling highly stable sodium metal anodes. *Nano Lett* 17:6808–6815. https://doi.org/10.1021/acs.nanolett.7b03071
54. Tian H, Seh ZW, Yan K, et al (2017) Theoretical investigation of 2D layered materials as protective films for lithium and sodium metal anodes. *Adv Energy Mater* 7:1602528. https://doi.org/10.1002/aenm.201602528
55. Wu S, Qiao Y, Jiang K, et al (2018) Tailoring sodium anodes for stable sodium-oxygen batteries. *Adv Funct Mater* 28:1706374. https://doi.org/10.1002/adfm.201706374
56. Liang F, Qiu X, Zhang Q, et al (2018) A liquid anode for rechargeable sodium-air batteries with low voltage gap and high safety. *Nano Energy* 49:574–579. https://doi.org/10.1016/j.nanoen.2018.04.074
57. Lee Y, Lee J, Lee J, et al (2018) Fluoroethylene carbonate-based electrolyte with 1 M sodium bis(fluorosulfonyl)imide enables high-performance sodium metal electrodes. *ACS Appl Mater Interfaces* 10:15270–15280. https://doi.org/10.1021/acsami.8b02446
58. Zhu M, Zhang Y, Yu F, et al (2021) Stable sodium metal anode enabled by an interface protection layer rich in organic sulfide salt. *Nano Lett* 21:619–627. https://doi.org/10.1021/acs.nanolett.0c04158
59. Zhao S, Wang C, Du D, et al (2021) Bifunctional effects of cation additive on Na-O_2 batteries. *Angew Chemie Int Ed* 60:3205–3211. https://doi.org/10.1002/anie.202012787

60. Zhao Y, Liang J, Sun Q, et al (2019) In situ formation of highly controllable and stable Na_3PS_4 as a protective layer for Na metal anode. *J Mater Chem A* 7:4119–4125. https://doi.org/10.1039/C8TA10174D
61. Kumar V, Eng AYS, Wang Y, et al (2020) An artificial metal-alloy interphase for high-rate and long-life sodium-sulfur batteries. *Energy Storage Mater* 29:1–8. https://doi.org/10.1016/j.ensm.2020.03.027
62. Zheng X, Fu H, Hu C, et al (2019) Toward a stable sodium metal anode in carbonate electrolyte: a compact, inorganic alloy interface. *J Phys Chem Lett* 10:707–714. https://doi.org/10.1021/acs.jpclett.8b03536
63. Ma JL, Meng FL, Yu Y, et al (2019) Prevention of dendrite growth and volume expansion to give high-performance aprotic bimetallic Li-Na alloy-O_2 batteries. *Nat Chem* 11:64–70. https://doi.org/10.1038/s41557-018-0166-9
64. Amanchukwu CV, Chang HH, Gauthier M, et al (2016) One-electron mechanism in a gel-polymer electrolyte Li-O_2 battery. *Chem Mater* 28:7167–7177. https://doi.org/10.1021/acs.chemmater.6b03718
65. Ha TA, Fdz De Anastro A, Ortiz-Vitoriano N, et al (2019) High coulombic efficiency Na-O_2 batteries enabled by a bilayer ionogel/ionic liquid. *J Phys Chem Lett* 10:7050–7055. https://doi.org/10.1021/acs.jpclett.9b02947
66. Wang S, Chen Y, Jie Y, et al (2020) Stable sodium metal batteries via manipulation of electrolyte solvation structure. *Small Methods* 4:1900856. https://doi.org/10.1002/smtd.201900856
67. Tripathi AM, Su W-N, Hwang BJ (2018) In situ analytical techniques for battery interface analysis. *Chem Soc Rev* 47:736–851. https://doi.org/10.1039/C7CS00180K
68. Zhang C, Firestein KL, Fernando JFS, et al (2020) Recent progress of in situ transmission electron microscopy for energy materials. *Adv Mater* 32:1–19. https://doi.org/10.1002/adma.201904094
69. Wood KN, Kazyak E, Chadwick AF, et al (2016) Dendrites and pits: untangling the complex behavior of lithium metal anodes through operando video microscopy. *ACS Cent Sci* 2:790–801. https://doi.org/10.1021/acscentsci.6b00260
70. Ma B, Lee Y, Bai P (2021) Dynamic interfacial stability confirmed by microscopic optical operando experiments enables high-retention-rate anode-free na metal full cells. *Adv Sci* 2005006:1–11. https://doi.org/10.1002/advs.202005006
71. Yui Y, Hayashi M, Nakamura J (2016) In situ microscopic observation of sodium deposition/dissolution on sodium electrode. *Sci Rep* 6:1–8. https://doi.org/10.1038/srep22406
72. Zeng Z, Barai P, Lee SY, et al (2020) Electrode roughness dependent electrodeposition of sodium at the nanoscale. *Nano Energy* 72:104721. https://doi.org/10.1016/j.nanoen.2020.104721
73. Li X, Zhao L, Li P, et al (2017) In-situ electron microscopy observation of electrochemical sodium plating and stripping dynamics on carbon nanofiber current collectors. *Nano Energy* 42:122–128. https://doi.org/10.1016/j.nanoen.2017.10.050
74. Banis NM, Yadegari H, Sun Q, et al (2018) Revealing the charge/discharge mechanism of Na-O_2 cells by in situ soft X-ray absorption spectroscopy. *Energy Environ Sci* 11:2073–2077. https://doi.org/10.1039/C8EE00721G
75. Han M, Zhu C, Ma T, et al (2018) In situ atomic force microscopy study of nano-micro sodium deposition in ester-based electrolytes. *Chem Commun* 54:2381–2384. https://doi.org/10.1039/C7CC09751D
76. Sun H, Zhu G, Xu X, et al (2019) A safe and non-flammable sodium metal battery based on an ionic liquid electrolyte. *Nat Commun* 10:3302. https://doi.org/10.1038/s41467-019-11102-2
77. Zhang Z, Cui Y, Vila R, et al (2021) cryogenic electron microscopy for energy materials. *Acc Chem Res* 54:3505–3517. https://doi.org/10.1021/acs.accounts.1c00183
78. Kumar V, Wang Y, Eng AYS, et al (2020) A biphasic interphase design enabling high performance in room temperature sodium- sulfur batteries. *Cell Reports Phys Sci* 1:100044. https://doi.org/10.1016/j.xcrp.2020.100044

16 Advancements in Air Cathodes in Sodium-Air Batteries
Cathode Materials

Ajish Babu, S. K. Vineeth, Roshny Joy, Abhilash Pullanchiyodan, Nutan Gupta, and Prasanth Raghavan

16.1 INTRODUCTION

Depletion of fossil fuels and the use of non-renewable resources has been a major area of concern, particularly considering the demand for energy for the ever-growing population. The usage of renewable energy sources has gained extensive attention as part of the global switch towards the development of sustainable energy solutions and net zero goals. In this scenario, researchers have been focusing their efforts on the development of energy solutions, which can effectively store the energy [1,2]. Lithium-ion batteries are quite well known to trigger a revolution in battery development [3–5]. To meet the ever-increasing energy demands, extensive research into greener, alternative, and abundant materials is highly significant. For the past decade, due to its vast availability and cost-effectiveness, sodium has received considerable attention to be used as a battery material [6,7]. Among the sodium metal-based batteries, sodium air batteries (SABs) have significant importance considering their easy working methods, efficiency, energy storage capacity, and tailorable electrodes [8,9]. Researchers are still grappling with the concern of developing high-energy batteries to meet rising energy demands. Cathodes are vital and inevitable elements in metal air batteries to meet high specific capacity and energy requirements. Consequently, effective strategies to design cathode materials for high-performance batteries are pivotal.

In rechargeable metal air batteries, anode, electrolyte, and cathode are important for the electrochemical reaction. During the charging and discharging of the cell, an oxygen-evolution reaction (OER) and oxygen-reduction reaction (ORR) takes place. Catalyst helps in enhancing the efficiency of these reactions owing to their high electrical conductivity and ability to form different complexes [10,11]. Air cathodes, which consist of electrode material and a catalyst, are remarkably significant for the performance of SAB. During the electrochemical processes, the insoluble discharge products, mainly sodium superoxide and peroxide (NaO_2 and Na_2O_2), are obtained through the chemical reaction of Na and oxygen molecules. These compounds accumulate on the surface of the cathode, thereby acting as an obstacle by obstructing the voids for the diffusion process and thereby affecting the cell performance [12,13]. Besides, the air electrode plays a vital role in deciding the morphology and decomposition of discharged products concerning the functional group present on their periphery. Optimization of the cathode with appropriate surface area, porosity volume, and chemical stability to enhance the electrochemical reactions are the major approaches to enhance the performance of SABs [14]. Throughout the evaluation of several catalyst systems used in cathodes, the ORR and OER catalytic activities have a significant role in the performance of SABs [15,16]. To gain a dominant catalytic activity, lower overpotential,

DOI: 10.1201/9781003310174-16

and high discharge capacity, several factors are to be considered, which are listed as follows: (1) Related to the morphology of cathode electrodes, the need for highly porous structure for accumulating discharge products, easy diffusion of ions and electrons through the surface and inner layers, as well as tailoring the structure to improve reversibility and control the discharge products. (2) Increment in electrical property, adsorption of metal ions, and oxygen molecules through suitable heteroatom doping. (3) Enhancement in the surface area and structural modifications. The decomposition of electrolytes and corrosion of cathodes could hinder the ORR/OER activities as well as limit the electrochemical catalysis causing poor performance. Hence, the need for a cost-effective, reliable, robust catalyst along with the above-discussed factors is required and is significant while designing the cathode and catalyst materials for SAB. With this regard, this chapter discusses the importance of air cathodes in enhancing the electrochemical performance of SAB. Various types of cathode host materials, such as carbon materials, nanomaterials, doped carbon-based materials, and transition metal oxides, along with the developments in catalysts, are highlighted.

16.2 CATHODE ENGINEERING FOR SABS

The general construction of a SAB consists of metallic sodium as an anode, an electrolyte, and carbonaceous materials as a cathode, which has oxygen as the active material supplied by the environment. The reversible capacity is governed by the cathode material; hence, engineering the cathode has significance in designing high-performance SABs. The electrochemical performance of SABs mainly depends on the morphology and microstructure of air cathodes. The morphology, amount, and composition of discharged products are influenced by the design of the cathode [17]. For instance, introducing a functional group in cathode material can substantially change the behaviour of discharged products [18, 19]. A bifunctional catalyst material can accelerate the ORR and OER activity to enhance the electrochemical reactions [20]. Furthermore, cathode materials with a high specific area can accommodate an ample amount of discharge products, for providing a high specific discharge capacity for the cell [21]. In short, it is challenging to provide all the characteristics in a single cathode material. Nevertheless, tailoring the cathode materials with appropriate functional modifications for designing an efficient air electrode for SABs is possible.

16.2.1 Carbon Nanomaterials

Carbonaceous nanomaterials like graphene, CNT, and nanofibres have received great attention as an air electrode regarding the importance of their high specific surface area, high conductivity for electron transfer and well-interconnected porous and robust and flexible structure [22–24]. Liu and his colleagues [25] reported for the first time the usage of graphene nanosheets (GNS) as a cathode in ethylene glycol dimethyl ether (DME) electrolyte-based SAB. The fabricated cell recorded a high discharge capacity of 9,268 mAh g^{-1} under a current density of 200 mA g^{-1}, which is closest to the GNS-based cathode used for lithium air battery. Figure 16.1a demonstrates the high discharge capacity of GNS electrodes in $NaPF_6$/DME as compared to $NaClO_4$/DME electrolytes. Variations in dissolved sodium salt in both electrolytes are supposed to be a possible explanation for this shift. The ability to sustain the discharge capacity at that level had been seen only up to ten cycles mainly as a result of the discharged product agglomeration.

Porosities in the cathode play an important role in creating more active sites for the electrochemical process and storing abundant discharge products for improved performance of the battery. Enterria et al. [26] modified the microstructure of graphene aerogel with different sets of porosities of ~100 and 30 μm. They found that cells with low mean pores performed better with an efficiency of more than 95%. The 3D stacking and configuration of the graphene sheets, the sizeable volume

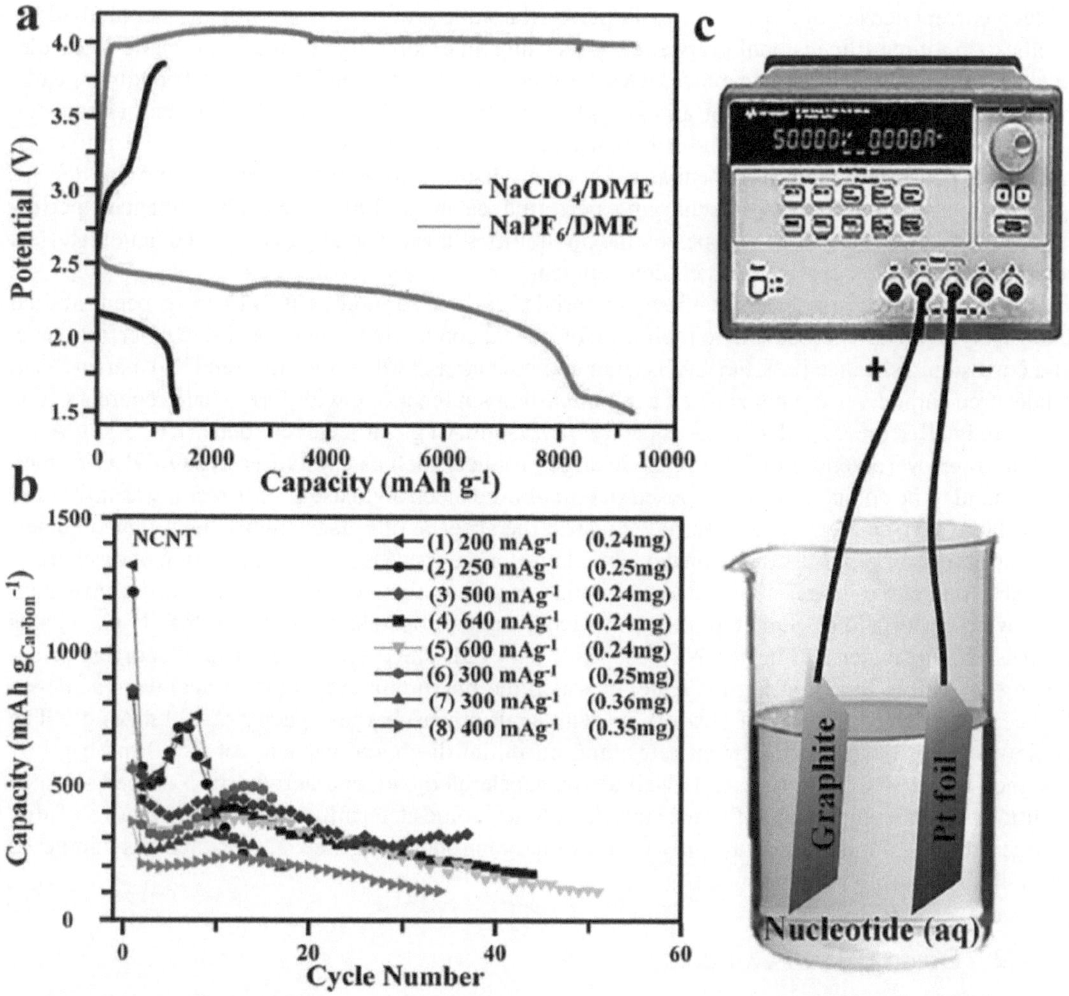

FIGURE 16.1 (a) Galvanostatic curves of GNS-based cathode for $NaClO_4$/DME and $NaPF_6$/DME electrolyte in SAB. Adapted and reproduced from Ref. [25], Copyright © 2013 The Royal Society of Chemistry. (b) Schematic representation of the electrochemical exfoliation of graphite using an aqueous solution of a given nucleotide (AMP, GMP, ATP, or FMN) as the electrolytic medium. Adapted and reproduced from Ref. [27], Copyright © 2019 American Chemical Society. (c) A curve showing the cyclability of N-CNT-based cathode at different current densities. Adapted and reproduced from Ref. [31], Copyright © 2015 Elsevier.

of voids ($0.22\,cm^3\,g^{-1}$), and the optimum size of the voids (5–25 nm) contributed to a good diffusion process and are attributed to a discharge capacity of 6.61 mAh cm^{-2}. Considering the increase of awareness of sustainable and green approaches for energy storage developments, Munuera et al. [27] developed an environmentally friendly method for graphene synthesis and its usage in SAB. RNA-related nucleotides like adenosine monophosphate, guanosine monophosphate, adenosine triphosphate, and flavin mononucleotide, were used as an exfoliating electrolyte and a colloidal stabilizing agent for preparing graphene in water. Figure 16.1b shows the schematic representation of the electrolytic cell setup having graphite and platinum foil as cathode and anode respectively, for exfoliating the graphene. A well-interconnected 3D aerogel cathode was designed from this dispersed graphene in water to obtain a highly porous structure allowing ample space for absorbing discharged products and thereby increasing the capacity of SAB. It maintained 95% stability up to 50 cycles under a current density of 0.5 mA cm^{-2}. Similarly, the same group came up with another method for synthesizing three-dimensional graphene sheets using an exfoliating agent as an organic molecule, nucleotide (adenosine monophosphate, AMP), for developing graphene-based eco-friendly air cathode [28]. Initially, the presence of an excessive amount of nucleotide on the surface of graphene sheets led to large overpotential and irregular electrochemical reactions. Controlling the amount of nucleotide resulted in a discharge capacity of 9.6 mAh cm^{-2}, under 0.2 mA cm^{-2} of current density with an outstanding charging-discharging performance up to 100 cycles. The enhanced performance of the obtained cell could open up the possibilities of eco-friendly synthesized nanomaterials to achieve the goal of sustainable cell development.

Apart from graphene-based cathode materials, carbon nanotubes (CNT) have been studied intensively as an air cathode due to their high electrical conductivity, electrochemical performance, and corrosion resistance [29]. For SABs, Jian and coworkers [30] communicated CNT paper-based binder-free cathode, using benzene as a major carbon source along with ferrocene precursors. The fabricated cell achieved a discharge capacity of 7,530 mAh g^{-1} at a current density of 500 mA g^{-1} with a lower overpotential of 200 mV. Discharge products such as crystalline $Na_2O_2{\cdot}2H_2O$ formed were found to be completely decomposed in ether-based electrolyte used. This result was attributed to the large surface area, lower electrode mass in the absence of binder, highly interconnected network structure for rapid electron transfer, abundant storage space for the accumulation of discharged products, and corrosion resistance due to the high graphitization degree of CNT. Similar investigations were performed by Sun et al. [31], in which nitrogen-doped self-stacking CNT (N-CNT) was used as cathode material. The N-CNT-based cell achieved. The cyclic life at different current densities for the cell is shown in Figure 16.1c, in which the maximum cycles (50 cycles) were achieved by keeping a current density of 600 mA g^{-1} with a cut-off discharge capacity of 100 mAh g^{-1}. The achieved better life at high current rates, and an initial discharge capacity of 7,500 mAh g^{-1}, is assigned to the N-CNT networks as well as the accelerated catalytic activity of N-doped sites. In addition, it was comprehended that catalytic activity could determine the microstructure of discharge products. Poor anomalous growth of discharge products on the CNT surface has hampered reversibility and cell performance.

16.2.2 Doped Carbon Cathodes

As the demand for high-performance batteries is increasing, modifications of existing systems with enhanced properties are significant to future advancements. Although carbon cathodes have many advantages over other materials as an air cathode, the low catalytic activity of pure carbon-based cathode materials acts as a barrier to the development of high-performance batteries. To overcome this challenge, efforts have been made to dope a heteroatom into a carbon-based cathode/catalyst for improving the intrinsic catalytical activity [32,33]. Accounting for the high catalytic activity of nitrogen, Li, and coworkers [34] researched nitrogen doped with GNS (N-GNS) based cathode and reported a superior electrochemical property compared to a GNS-based one. Cyclability of the first three cycles of N-GNS at a current density of 75 mA g^{-1} resulted in a quite restricted reduction of discharge plateaus compared to GNS-based cathode. Although a decrease in discharge capacity for both cathodes was confirmed along with the increase in current density, the N-GNS cathode showed higher discharge capacity than that of the undoped cathode. The difference in the morphology and particle size of the discharge product accumulated on the surfaces of N-GNS is assumed to be the reason for this enhanced discharged capacity. Similarly, Ma and Zhang [35] investigated nitrogen-doped (N-doped) carbon cathode using melamine and tannic acid as nitrogen and carbon sources respectively for effective electrochemical performance in SAB. The investigators primarily focused on the preparation methods of N-doped carbon cathode for analysing the performance. The techniques to create the N-doped carbon cathode are firstly from an N-containing precursor and the secondly via ammoxidation. They observed that the cathode prepared through the precursor method generates a significant amount of N content compared with that of the ammoxidation method, and also provides an increased surface area, with an initial discharge capacity of 6,905 mAh g^{-1}. Results indicate the first method as the preferable one for N-doped cathode designs for effective electrochemical performance. Owing to the extraordinarily high strength and mechanical characteristics of carbon fibre (CF), the same team developed a nitrogen-doped CF cathode coated with cobalt (Co-ECNCF) by the electrospinning method [36]. Doping of heteroatoms such as nitrogen and CoOx has provided abundant catalytically active sites and reduced overpotential. The highly porous interlinked framework of the electrode with a mean size of 130 nm is intended to enhance the easy transport of oxygen and electrons for better ORR/OER. The researchers discovered that the cell had achieved a discharge capacity of 6,102 mAh g^{-1} under a current density of 200 mAh g^{-1}, and a cyclic performance of 112 cycles while keeping a fixed discharge capacity of 550 mAh g^{-1}.

Motivations and curiosity developed from the excellent electrochemical activity of high electron-rich N-doped cathode materials prompted researchers to explore the potential catalytic activity of the counterpart element, electron-deficient heteroatom boron [37]. Owing to the significant electronegativity difference between oxygen and boron, adsorption of O_2 on boron is far easier to provide better O_2, resulting in better electrochemical properties [38]. Shu and coworkers [39] developed a mesoporous onion-like carbon structure doped with boron (B-OLC) to study catalytic behaviour in a non-aqueous battery system. B-OLC was primed by thermal treatment of nanodiamonds (5 nm as diameter) with boric acid followed by annealing at 1,500°C. To compare electrocatalytic activity, the investigators prepared undoped onion-like carbon (OLC) and Super P carbon black using similar methods. On electrochemical analysis, outstanding performance with a high initial discharge capacity of 10,200 mAh g^{-1} under 0.15 mA g^{-1} of current density with a stable cyclability over 125 cycles under a current density of under 0.3 mA g^{-1}, was achieved for B-OLC, which is very high compared to other two cathode structures as shown in Figure 16.2a. On further investigation of the discharge product behaviour, the decomposition of the discharge product ($Na_2O_2{\cdot}2H_2O$) while the charging process in B-OLC, was observed, which revealed a high reversibility of the cell. These enhanced electrochemical activities of B-OLC are attributed to the synergistic effect of pore size distribution, high surface area, unique mesoporous structure, and doping effect of B created abundant active sites for O_2 as well as provides easy electrolyte and oxygen transfer. When co-doped, the difference in electro negativity of N (3.04) and boron (2.05) can provide a synergistic effect, resulting in plentiful active sites and increased catalytic performance in carbon-based cathode materials [40]. Benti et al. [41] investigated the electrocatalytic behaviour and applicability of boron, nitrogen, and co-doped graphene-based air cathode through density functional theory. The schematic representation of differently doped structures prepared by the authors is shown in Figure 16.2b. The probable equivalent atomic sizes of the N, B, and C atoms and their strong possibility of accessing the graphene lattice for semiconductor behaviour are expected to boost the electrochemical properties of the catalysts. The investigated results suggest that, while comparing with normal graphene-based cathodes, the 1 and 2 boron-doped cathodes and three nitrogen (pyridinic nitrogen) doped cathodes showed relatively less overpotential (0.40 V), higher equilibrium potential (2.22 V), and considerable catalytic activity for ORR process.

A rational design for creating a cathodic structure that can accommodate maximum discharge products and provide abundant active sites is necessary to enhance the battery performance [42,43]. Sun et al. [44] prepared a hierarchical porous carbon sphere (PCS) by chemical vapour deposition (CVD) technique using MnO as a precursor for applicability as a cathode in SAB. The obtained MnO/C nanocomposite through the CVD technique was removed by the etching process to get the final three-dimensional (3D) PCS structure. The schematic representation of PCS preparation and the morphological images (SEM and TEM) of the obtained highly porous structure of both MnO/C spheres and fabricated PCS are shown in Figure 16.3. The cell with PCS cathode showed a high initial discharge capacity of 17,500 mAh g^{-1}, compared to the cell with a CB electrode with 7,790 mAh g^{-1} discharge capacity, measured at 500 mA g^{-1} of current density. With a cut-off discharge capacity of 500 and 1,000 mAh g^{-1}, the cell showed cyclic stability up to 400 and 150 cycles,

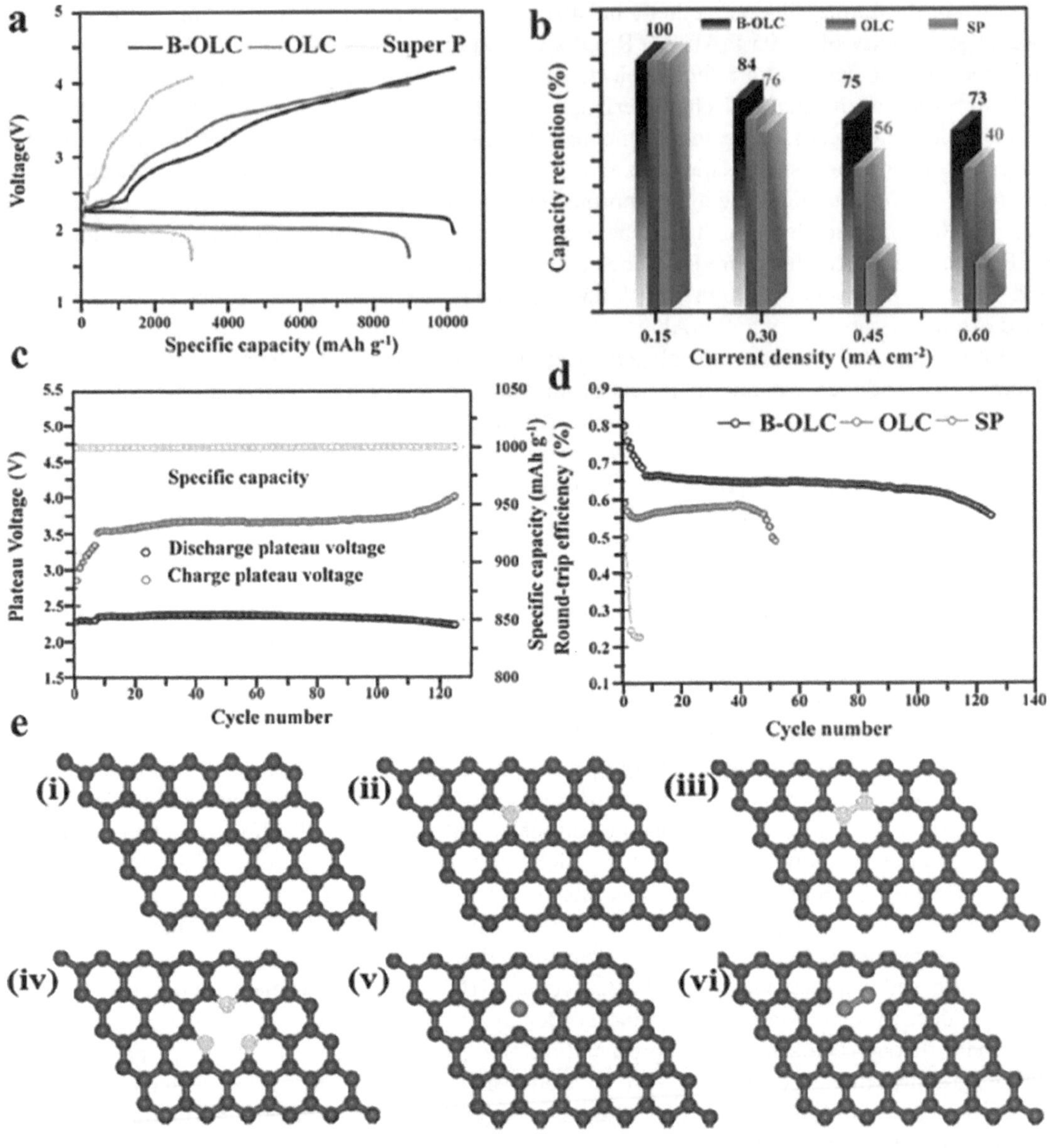

FIGURE 16.2 (a and 1) Discharge/charge curves of onion-like carbon structure doped with boron (B-OLC), OLC, and Super P cathodes in SAB at 0.15 mA cm^{-2} of current density; (2) graph showing capacity retention on different current densities of B-OLC, OLC and Super P based SAB; (3) plateau voltage vs the number of cycles for SAB with B-OLC at 0.3 mA cm^{-2} of current density; (4) round-trip efficiency vs the number of cycles curve for SAB with three kinds of cathode electrodes at a current density of 0.3 mA cm^{-2}. Adapted and reproduced from Ref. [39], Copyright © 2016 The Royal Society of Chemistry. (b) Schematic diagrams of the different structures prepared through doping of N and B, where the blue, orange, and brown spheres indicate N, B, and carbon atoms, respectively. (1) Bare graphene sheet, (2) 1-N-doped graphene sheet, (3) 2-N-doped graphene sheet, (4) 3-N-doped graphene sheet, (5) 1-B doped graphene sheet, (6) 2-B doped graphene sheet. Adapted and reproduced from Ref. [41], Copyright © 2020 The Royal Society of Chemistry.

respectively. PCS electrodes, having their unique porous structure, resisted the blockage of film-like discharge products (NaO_2) and hence paved the way for better oxygen diffusion in SAB with superior properties.

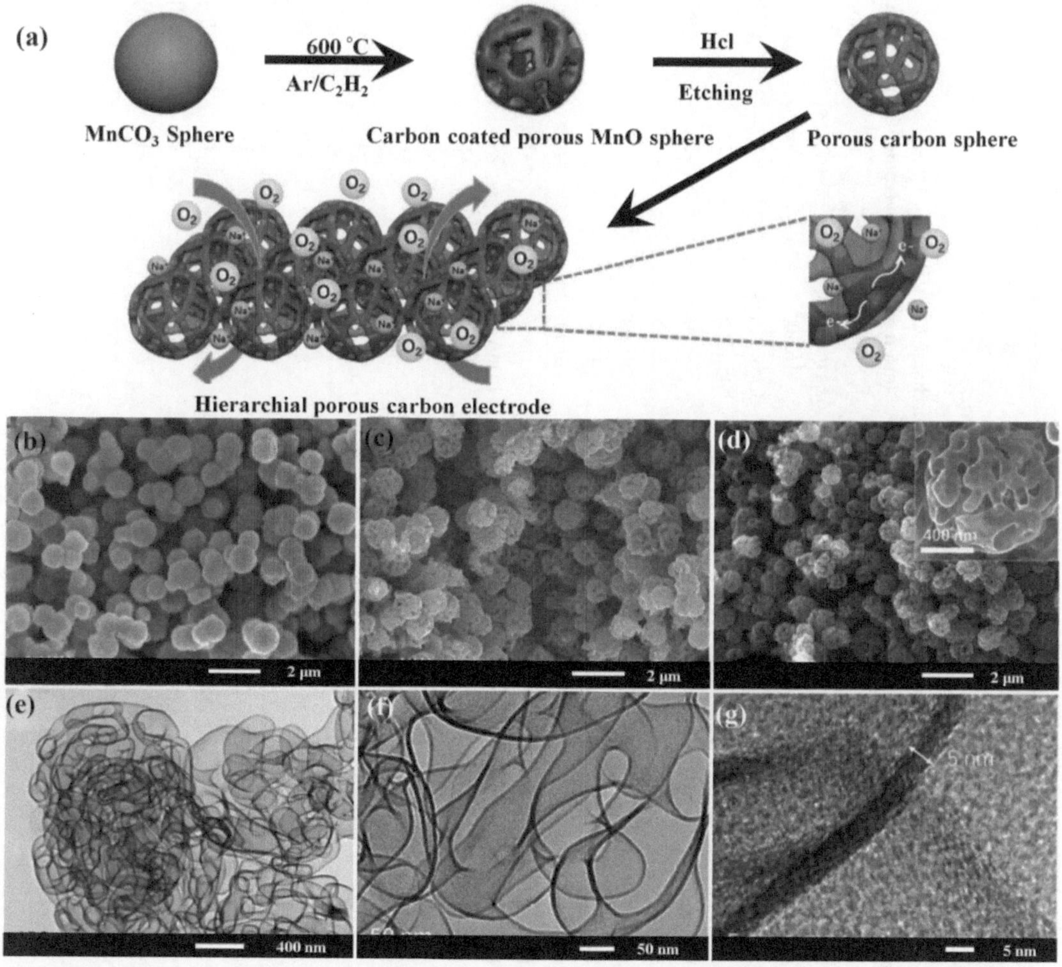

FIGURE 16.3 (a) Schematic representation preparation of PCS, (b–d) SEM images of (b) MnO/C spheres, (c) porous carbon sphere structure, (d) high-resolution image of PCS is shown inset, (e–g) TEM images of porous carbon sphere at different magnifications. Adapted and reproduced from Ref. [44], Copyright © 2017 Wiley-VCH.

Table 16.1 shows consolidated and comprehensive data on materials used for the air cathode, the electrolyte used, the discharge product detected, and their performance in terms of discharge capacity.

On accounting for the relevance of commercial air cathode for SAB, Ha and colleagues [68] studied commercially available carbon-based air cathode like H 23, Toray 090, BC 39, and MGL 190 in pyrrolidinium-based room temperature ionic liquid electrolyte battery. They found that hydrophobic behaviour and micropores in cathode materials play a significant role in cell performance. However, space to occupy the discharged insoluble products to avoid blockage of diffusion path, corrosion resistance of cathode material, good electrical conductivity, and better catalytic activity are some of the significant properties of cathode material to exhibit a high discharge or specific capacity with an enhanced cyclability on the charging–discharging process, which can be improved through tailoring the microstructure of these air cathode materials. Recently, Zheng and coworkers [69] developed a high level of nitrogen-doped carbon nanofibre (N-CNF) as cathode material for improving the SAB efficiency. The cathode system was fabricated through calcination of polypyrrole (PPy) nanofibre having a web-like structure with a size of about 60–80 nm.

TABLE 16.1
Recent Research for Air Cathodes in SAB [45]

	Air Cathode	Electrolyte	Discharge Product	Current Density ($mA\ g^{-1}$)	Discharge Capacity ($mA\ h\ g^{-1}$)	Overpotential (V)	Ref.
Carbon materials	N-doped carbon nanotubes	0.5M $NaCF_3SO_3$/DEGDME	NaO_2	150	N/A	~1.5	[46]
	Graphene aerogel	0.1 M $NaClO_4$/DME	NaO_2	100	6.61 $mA\ h\ cm^{-2}$	~0.3–0.4	[26]
	Porous carbon spheres	0.5 M $NaCF_3SO_3$/DEGDME	NaO_2	500	16,500	~0.4	[44]
	Reduced graphene oxide	1 M $NaCF_3SO_3$/TEGDME	NaO_2	0.05 $mA\ cm^{-2}$	~3.75 $mA\ h\ cm^{-2}$	~1.7	[47]
	Boron-doped reduced graphene oxide		NaO_2		~3.45 $mA\ h\ cm^{-2}$	~1.8	
	Carbon nanotubes	0.5 M $NaCF_3SO_3$/DEGDME	NaO_2, Na_2O_2	150	~660	~2.2	[48]
	Nitrogen-doped carbon	0.5 M $NaCF_3SO_3$/TEGDME	NaO_2, Na_2O_2	200	5,905	~1.0	[35]
	Boron-doped onion-like carbon	1 M $NaCF_3SO_3$/TEGDME	$Na_{2-x}O_2$	0.15 $mA\ cm^{-2}$	10,200	~1.3	[39]
	Ordered mesoporous carbon	0.5 M $NaCF_3SO_3$/PC	Na_2O_2, Na_2CO_3	100	7,987	~1.5	[49]
	Nitrogen-doped graphene aerogels	0.3 M $NaCF_3SO_3$/DEGDME	Na_2O_2	100	10,905	~1.0	[50]
	Nitrogen-doped carbon nanotubes	0.5 M $NaCF_3SO_3$/TEGDME	NaO_2, Na_2O_2	25	1,887	~3.0	[51]
	Nitrogen-doped carbon nanotubes	0.5 M $NaCF_3SO_3$/DEGDME	NaO_2, Na_2O_2	0.1 $mA\ cm^{-2}$	~11 $mA\ h\ cm^{-2}$	N/A	[42]
	Vertically aligned carbon nanotubes	0.5 M $NaCF_3SO_3$/TEGDME	NaO_2	67	~4,100	~0.2	[52]
	Carbon nanotube paper	0.5 M $NaCF_3SO_3$/DEGDME	Na_2O_2.2H2O	500	7,530	~0.2	[30]
	Nitrogen-doped graphene nanosheets	0.5 M $NaCF_3SO_3$/DEGDME	Na_2O_2	75	~8,600	N/A	[34]
	Graphene nanosheets	0.25 M $NaPF_6$/DME	NaO_2,	300	6,208	~1.7	[25]
	Diamond-like carbon thin films	1 M $NaPF_6$/EC : DMC (1:1)	Na_2O_2	0.1 C	1,884	~1.5	[53]
Transition metal-based materials	Co_3O_4 @ $MnCo_2O_{4.5}$	1 M $NaClO_4$/TEGDME	Na_2O_2, $Na_{2-x}O_2$	50	8,400	0.46	[54]
	RuO_2/Mn_2O_3/carbon nanofibre	0.5 M $NaCF_3SO_3$/TEGDME	NaO_2	0.05 $mA\ cm^{-2}$	9,352	~0.5	[55]
	High-spin Co_3O_4	0.5 M $NaCF_3SO_3$/TEGDME	NaO_2	100	N/A	~0.06	[56]
	CoB	0.5 M $NaCF_3SO_3$/TEGDME	NaO_2, Na_2O_2	100	11,482	~0.6	[57]
	C@$NiCo_2O_4$	1 M $NaClO_4$/TEGDME	Na_2O_2	50	6,500	~0.6	[58]
	Co-N-doped carbon fibres	0.5 M $NaCF_3SO_3$/TEGDME	Na_2O_2	200	6,102	~1.3	[36]
	Co_3O_4/C	1M $NaClO_4$/DME	Na_2O_2	0.05 $mA\ cm^{-2}$	2,000	~1.0	[59]

(Continued)

TABLE 16.1 (*Continued*)
Recent Research for Air Cathodes in SAB [45]

	Air Cathode	Electrolyte	Discharge Product	Current Density (mA g^{-1})	Discharge Capacity (mA h g^{-1})	Overpotential (V)	Ref.
	Co_3O_4@carbon nanotubes	0.5 M $NaCF_3SO_3$/DEGDME	NaO_2, Na_2O_2	150	~710	~2.1	[48]
	Co_3O_4/carbon nanotubes		NaO_2, Na_2O_2		~555	~2.1	
	$La_2Co_2O_7$	0.5 M $NaCF_3SO_3$/TEGDME	NaO_2, Na_2O_2	100	19,040	~1.0	[60]
	$CaMnO_3$	1 M $NaCF_3SO_3$/TEGDME	NaO_2, Na_2O_2	100	9,560	~1.6	[61]
	$NiCo_2O_4$	1 M $NaClO_4$/DME	Na_2O_2	202	1,762	0.96	[62]
Noble metal-based materials	RuO_2/C	0.5 M $NaCF_3SO_3$/TEGDME	NaO_2	100	6,157	~0.5	[63]
	Ru/carbon nanotubes	0.5 M $NaClO_4$/DEGDME	NaO_2, $Na_2O_2.2H_2O$	0.191 mA cm^{-2}	N/A	~1.0	[64]
	RuO_2/boron-doped reduced graphene oxide	1 M $NaCF_3SO_3$/TEGDME	$Na_{2-x}O_2$	0.05 mA cm^{-2}	~3.80 mA h cm^{-2}	~1.0	[47]
	Pd/ZnO/C	0.5 M $NaCF_3SO_3$/TEGDME	$Na_2O_2.2H_2O$	N/A	7.5 mA h cm^{-2}	~0.5	[65]
	Ag-reduced graphene oxide	1.0 M $NaPF_6$/TEGDME	NaO_2, Na_2O_2, Na_2O	0.1 mA cm^{-2}	120	~1.0	[66]
	Pt @ graphene nanosheets	1 M $NaClO_4$/PC	Na_2CO_3	0.1 mA cm^{-2}	7,574	~1.3	[67]

This highly interconnected structure of N-CNF boosts the electrochemical activity by shortening the oxygen pathways and enhancing the transport of the reactants. With NaO_2 as a major discharge product and excellent cyclic stability of 90 cycles, this cell showed a high initial discharge capacity of 8,554.7 mAh g^{-1} under a current density of 100 mA g^{-1}. In addition, the investigators found that the pyridinic and pyrrolic nitrogen doping can activate an outstanding ORR.

In general, cathodes used in SABs should have high porosity even to accumulate the discharge product and to show better catalytic activity. In practical application, most of the voids or pores in cathodes are not sufficiently utilized to achieve the actual cell performance. Considering this, Langsdorf and coworkers [70] developed a pulse discharging method by increasing the supply of oxygen atoms for effective utilization of voids and porosities in the cathode. The cell consisted of sodium triflate (NaOTf) electrolyte, and Ketjenblack as cathode. A cyclic current, within a small interval of time, was applied to the battery for the discharging process, called pulse discharging. Compared to the rapid reduction of oxygen molecules and formation of discharged products in the continuous charging process, an increase in O_2 concentration in electrolytes happens in each pulse discharging process. An increase in the ratio of resting to pulsing time of the current applied results in improving the cell capacity. The investigators concluded that the effect of pulse discharging highly relies upon the types of cathode (regarding 2D or 3D material and void space) and the ratio resting-pulsing time.

16.2.3 Catalytic Cathodes

One of the major constraints in SABs is the limitation of discharge capacity due to the sluggish ORR and OER reactions. Furthermore, the solid discharge products formed during the reactions obstruct the porous cathodes, leading to further reduction of cell capacity and performance. Noble metals and transition metal-based catalysts are usually used in metal air batteries due to their high catalytic activity. Generally, noble metals like platinum, ruthenium, silver, and gold are used for cathode materials in SABs. Zhang and colleagues [67] communicated the cell properties of platinum nanomaterials (5 nm size) imposed in graphene sheets as a cathode electrode with $NaClO_4$ salt in the PC electrolyte. The cathode was fabricated through the in situ growth of platinum nanoparticles on graphene nanosheets through hydrothermal technique. The authors compared the electrochemical properties of Pt@GNS with pristine GNS and observed a high initial discharge capacity of 7,574 mAh g^{-1} was obtained at a current density of 0.1 mA cm^{-2}, which was 40% more than the GNS-based electrode. This enhanced property of Pt@GNS is believed to be due to the effect of Pt, providing the factors like supply of abundant oxygen active sites, enhanced electrochemical activity of highly dispersed Pt nanoparticles in GNS, and uniform deposition of discharge products. Hashimoto et al. [71] performed a comparative analysis of SAB cells consisting of nano porous gold (NPG) cathode and a NASICON ceramic separator, in both alkyl carbonate-based non-aqueous electrolyte with $NaClO_4$ as sodium salt and an aqueous solution of NaOH. Superior properties in terms of the high discharge capacity, low formation of discharged products, and robustness towards the atmosphere and side reactions were observed in aqueous cells having gold-based cathode as compared to non-aqueous cells. The improved properties of aqueous cells than non-aqueous cells were concluded as a result of their lower resistance value and higher affinity for NPG electrodes. Kumar et al. [66] communicated a non-aqueous SAB based on a bifunctional silver-reduced graphene oxide (Ag-RGO) catalyst operated at room temperature. The cell obtained an 80% efficiency, with 566 mAh g^{-1} as the discharge capacity at a current density of 0.1 mA cm^{-2}. From the cell reactions, discharged products such as NaO_2, Na_2O_2, and Na_2O were evolved, and from morphological analysis (Figure 16.4a and b) it was observed that these evolved discharge products covered the electrode completely even after ten cycles. Similarly, Liu et al. [72] reported SAB using silver nanoparticles as a catalyst in fluorinated GNS (Ag@F-GNS) to reduce the voltage gap in F-GNS alone during the charging and discharging process. It was discovered that the addition of Ag to the F-GNS cathode facilitated the deposition of amorphous discharge product, NaF, on the Ag surface.

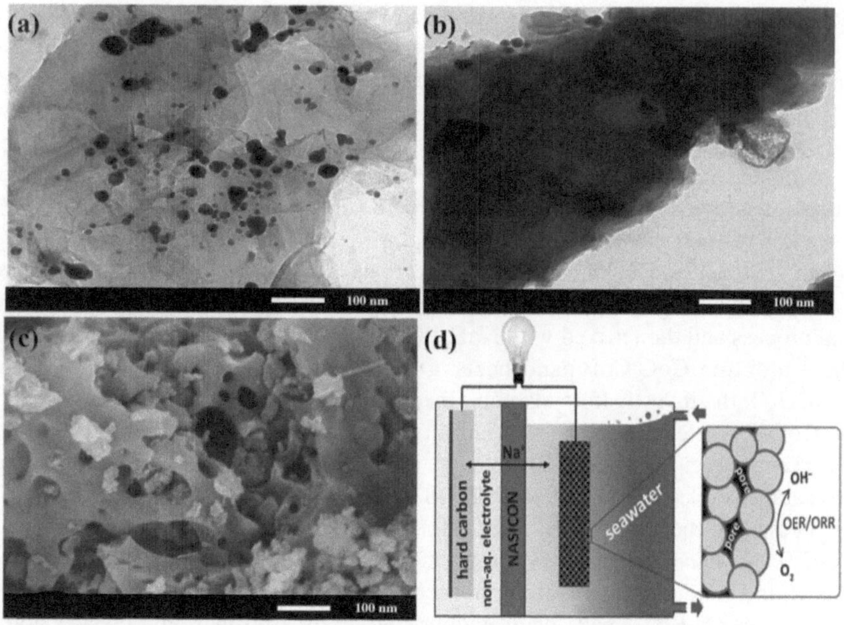

FIGURE 16.4 TEM images of Ag-RGO-based air electrodes (a) at starting of the cell operations and (b) after ten cycles. Adapted and reproduced from Ref. [66], Copyright © 2016 The Royal Society of Chemistry. (c) SEM image of meso and microporous N-doped graphitic carbon. Adapted and reproduced from Ref. [76], Copyright © 2017 The Electrochemical Society. (d) Schematic illustration of fabricated open-structured CMO cathode-based seawater battery and the possible ORR/OER reaction that occurs in the cathode. Adapted and reproduced from Ref. [78]. Copyright © 2016 American Chemical Society.

Furthermore, this effect reduced the polarization and the voltage gap by 480 mV without affecting the discharge capacity of the Ag@F-GNS cathode compared to the F-GNS cathode.

Discharge products that form during the cell reactions can play an important role in controlling the performance by controlling the charging potential in SABs. According to studies, SABs that produce Na_2O_2 as a major discharge product have higher theoretical capacity compared to those that form NaO_2 [73]. Using nano-sized ruthenium particles as a catalyst for CNT paper-based air electrodes, Kang et al. [64] prepared SABs that produce $Na_{2-x}O_2$ as a major discharge product for achieving high energy efficiency. On analysis, it was reported that Ru catalyst in Ru@CNT-based cells with a dual catalytic effect oxidized the discharged products by their strong affinity towards oxygen molecules to show a lower overpotential, and stable performance over 100 cycles compared with bare CNT-based electrode. Similarly, Guo et al. [74] investigated ruthenium nanoparticle embedded porous ketjen black (KB) carbon cathode fabricated through the reduction of ruthenium chloride in the presence of KB in ethylene glycol for Na-CO_2 battery. The ruthenium nanospheres of 5 nm mean size covered the KB substrate to obtain an enhanced catalytic activity. The composite electrode material delivered an initial specific capacity of 11,537 mAh g^{-1} with 94.1% of coulombic efficiency. Under a current density of 300 mA g^{-1} and 500 mA g^{-1}, the cell showed a stable cycle up to 90 and 40 cycles respectively.

Despite the fact that noble metals have excellent catalytic activity, their large-scale development is limited owing to their high cost. Besides that, using a single-component catalyst to meet the requirements of both ORR and OER reactions is challenging [75]. As a result, it is critical to design cathodes with effective catalysts that are bifunctionally active towards both ORR and OER and provide prolonged cycle life, to develop high-performance SABs. Attempts to develop metal-free, cheaper, carbon-based bifunctional catalyst for Na-air cells through a cost-effective and greener method was performed by Baskar and coworkers [76]. The catalyst was developed by treating the waste tree leaves derived from highly porous carbon with urea [$CO(NH_2)_2$]. Moreover, the authors studied the catalytic activity by doping the catalyst with nitrogen (N) and sulphur and observed a superior catalytic activity in N-doped graphitic carbon. The morphological analysis (Figure 16.4c) of N-doped graphitic carbon shows a highly porous structure which provides a higher catalytic activity. Furthermore, compared to undoped cathode and S-doped catalysts, the high electron donation tendency of N-doped catalysts leads to an increase in the rapid transport of ions and electrons along with an increase in active sites for catalysis, leading to achieve an efficiency of about 83% under a current density of 0.01 mA cm^{-2}. Similarly, for exploring the potential of bifunctional catalysts in SABs, Wang and colleagues [77] developed an effective CoO/CoP hetero-structured electrocatalyst having an O–P interface for SAB. Precursor of $Co(CO_3)_{0.5}(OH)0.11H_2O$ was synthesized through a hydrothermal process and then mixed with $NaH_2PO_3{\cdot}2H_2O$ for calcination in a nitrogen atmosphere to obtain these ultrathin CoO/CoP nanosheets. On varying the ratios of $Co(CO_3)_{0.5}(OH)0.11H_2O$ and $NaH_2PO_3{\cdot}2H_2O$, the investigators also obtained Co_3O_4 and CoP structures. On electrochemical analysis, the CoO/CoP electrocatalyst showed an exceptionally high initial discharge capacity of 12,654 mAh g^{-1} at a current density of 100 mA g^{-1}, and superior ORR and OER activity compared to the other two catalysts (Co_3O_4 and CoP with 7,048.7 and 12,176.2 mAh g^{-1}, respectively). The intercalation of P atoms into the CoO lattice and O atoms into the CoP lattice during phosphatization of $Co(CO_3)_{0.5}(OH)0.11H_2O$ nanosheets that caused the O–P interphase and are attributed to enhanced electron transfer and conductivity, resulting for better OER and ORR activity.

Pyrochlore oxides ($A_2B_2O_7$) have been widely studied as a bifunctional electrocatalyst due to their high charge transfer characteristics [79–81]. Considering the high catalytic activity and electrical conductivity of pyrochlore oxides, Kim and coworkers [82] investigated the electrochemical performance of single crystalline thallium rhodium oxide pyrochlore. Fabrication of the cathode was done by spraying the $Tl_2Rh_2O_7$ ink solution with a binder as polyvinylidene fluoride on carbon paper. $Tl_2Rh_2O_7$ nanoparticles developed using thallium (III) acetate and rhodium (III) nitrate solution showed an oxygen electrode activity of 0.82 V, which is quite superior to commercial Pt/C cathode systems. Through ex-situ X-ray spectroscopy, a good electrochemical performance was found for the cell having 0.208 V of low overpotential with a round-trip efficiency of 93.65%, which was quite high. As it is well known that the oxidation of A sites and B sites in a pyrochlore provides better catalytic activity, here similarly A-site Tl and B-site Rh ions were highly oxidized and created a rapid electron transfer pathway during OER and ORR for obtaining the better performance of the cell. Another research on thallium base pyrochlore structure was investigated by the same group. In this method the electrochemical performance of $Tl_2Ru_2O_7$ nanoparticles with dihydrogen phosphate ion ($H_2PO_4^-$) was introduced for boosting the electrochemical activity of aqueous Na-air cell. The lower electronegativity of $H_2PO_4^-$ improved the chemical and covalent character of Ru-O bonds and created a favourable oxidation behaviour for Tl and Ru ions during the charge/discharge process, resulting in an unreasonable OER and ORR activity than bare $Tl_2Ru_2O_7$ structure. Furthermore, the developed cell resulted in the lowest overpotential gap of 0.17 V with a high round-trip efficiency of up to 50 cycles in comparison with the commercially available Pt/C, Ir/C, and carbon paper-based cathodes [83].

Metal-organic frameworks (MOFs) are a class of crystalline materials with ultra-high porosity and high specific surface area [84]. Due to their excellent architectures with controllable pore size and structure, and high catalytic activity towards ORR, MOFs have been effectively utilized as an electrocatalyst for replacing noble metal catalysts [85,86]. Wu et al. [87] developed MOF-derived N-doped carbon nanotubes (MOF-NCNTs) as a cathode catalyst system for hybrid SAB having organic, aqueous electrolytes and NASICON as separators. The hollow and robust structure of CNT, the influence of doped nitrogen, and the highly porous structure assisted an easy oxygen and ion transfer, thus reported to be the ground for enhanced catalytic activity. The cell showed a lower voltage gap of 0.30 V along with a round-trip efficiency of 87% after 35 cycles when compared with commercially available Pt/C, RuO_2, NCNTs, MWNTs, and carbon paper cathode electrodes. Similarly, Abirami et al. [78] developed a MOF-derived nanoporous cobalt manganese oxide

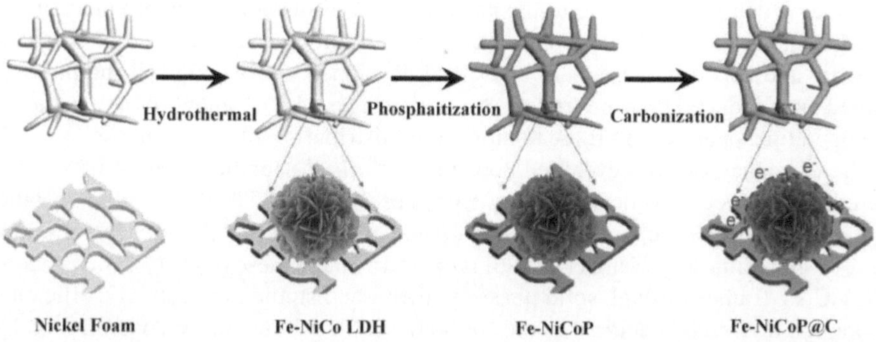

FIGURE 16.5 Schematic illustration of synthesis method of electrocatalyst for SABs. Adapted and reproduced from Ref. [93]. Copyright © 2020 Elsevier.

(CMO)-based cathode for rechargeable eco-friendly seawater hybrid SAB. The system consisted of nano cubed spinel-type CMO-based cathode, which was synthesized through the pyrolysis method, seawater catholyte, non-aqueous anolyte, sodium superionic conductor (NASICON) separator and Na anode as shown in Figure 16.4d. The prepared cell gained a discharge capacity of ~190 mAh g^{-1} with a lower overpotential of 0.53 V and a coulombic efficiency of more than 96% with stability up to 100 cycles. This enhanced performance of the cell is believed due to the high porosity of the cathode, abundant active sites, and high oxidation tendency of Co (+2.46) and Mn (+2.69) ions enabling the enhanced bifunctional catalytic activity. It is worth noting that further optimizing the porosities of MOF-derived catalysts could boost catalytic activity and improve battery performance. The fabrication of these complex structures with low yield and high cost, on the contrary, could stymie the future development of MOFs in batteries [88].

Transition metal phosphides have been widely used to replace noble metals as bifunctional electrocatalysts [89–91]. Among them, nickel cobalt phosphides (NiCoP) have gained great attraction as a catalyst due to the two transition metals that can possibly enhance the catalytic activities [92]. Kang et al. [93] synthesized iron (Fe) doped NiCoP as a cathode catalyst in HSAB. Batteries were fabricated based on three catalysts, NiCoP, Fe-NiCoP, and carbon-coated Fe-NiCoP (Fe-NiCoP@C) on nickel foam for comparative studies. Different routes like hydrothermal technique, phosphatization, and carbonization were adapted to synthesize the electrocatalyst to obtain a 3D microflower morphology as shown in Figure 16.5. On electrochemical performance analysis, Fe–NiCoP@C showed superior stability up to 500 cycles without any significant drop in potential with a round-trip efficiency of 93.26%. In addition, the Fe–NiCoP@C based cell achieved a power density of 621 mW g^{-1}, which is the highest among the other electrocatalyst (Fe-NiCoP (532 mW g^{-1}) and NiCoP (473 mW g^{-1}))-based cells. Fe acts as an OER and ORR active site, whereas carbon coating provides an additional way for rapid electron transport as well as stabilizes the interface between electrolyte and catalyst is attributed to the enhanced performance of Fe–NiCoP@C based SABs.

16.2.4 Transition Metal Oxide-Based Cathodes

Applicability of transition metal oxide gains quite an attractiveness due to their abundant availability, convenient processing methods, high stability, high intrinsic activity, and excellent ORR activities to act as a substitute for noble metal-based catalysts [94]. To create a carbon-free and binder-free air electrode, Liu et al. [62] attempted the first transition metal-based, $NiCo_2O_4$ nanosheets deposited in Ni foam substrate air cathode for non-aqueous SAB in 2014. The cell with 1 M $NaClO_4$ in DME as an electrolyte showed 1,762 mAh g^{-1} as the initial discharge capacity and a very low polarization of 0.96 V, under a current density of 20 mA g^{-1} with nanosheets of Na_2O_2 as the major discharge product. The cell showed a progressive increase of discharge and charge polarization after the initial cycle, which is mainly due to the accumulation of insoluble side products on the highly porous $NiCo_2O_4$ electrode. Organic solvents are used for traditional electrode preparation processes, which are toxic and harmful. Rosenberg et al. [95] developed a less toxic solvent and dry coating method called laser-induced forward transfer for preparing α-MnO_2 nanowires as a cathode catalyst in room-temperature operated SAB. A homogenous distribution of MnO_2 particles in electrodes enhanced the efficiency of the developed cell with an initial specific capacity of 1,215 Ah kg^{-1}. The prepared cell showed a cyclic stability of 55 cycles, keeping 85% of the initial specific charge. Furthermore, the electrode preparation method opens up a greener method for maintaining a long-term battery design. Similarly, Zeng et al. [96] prepared γ-manganese oxide (γ-MnO_2) nanoparticles embedded in CNT frames through solid phase synthesis and studied the cathodic efficiency of this setup. It obtained an initial specific capacity of 8,804 mAh g^{-1} and an overpotential of 140 mV at a current density of 200 mA g^{-1}. In addition, the cell showed stability up to 246 cycles with a cut-off of 1,000 mAh g^{-1} at 500 mA g^{-1} of current density. In comparison with different mass ratios of γ-MnO_2 and CNT, the optimum property was evolved while keeping it at a ratio of 1:1, due to the uniform distribution of γ-MnO_2 on the CNT surface. The observed disappearance of discharged product NaO_2 on the surface while charging revealed reversibility and promised a long cyclability.

To investigate the potential of cathode catalyst which comprises abundant alkaline earth metals and transition metal oxide, Hu and colleagues [61] developed a highly interconnected porous calcium manganese oxide ($CaMnO_3$) microspheres as catalyst in carbon cathode for ether electrolyte-based SAB. Due to the microporous structure, the catalyst provided abundant space for the accumulation of discharge products as well as widely available active sites for catalytic activities to achieve 9,560 mAh g^{-1} of discharge capacity under a current density of 100 mA g^{-1}, and with 95% of coulombic efficiency. In addition, a stable cyclability of up to 80 cycles with a capacity of 1,000 mAh g^{-1} at a current density of 200 mA g^{-1} was achieved.

Conventional cathode materials are prepared using conductive carbon materials and polymer binders such as polyvinylidene fluoride and polytetrafluoroethylene (PTFE). The use of these types of binders has been linked to a reduction in battery stability [97]. Furthermore, when considering the properties of ions, electrons, and oxygen transportation, flexible electrodes can outperform traditional rigid air cathode-based SABs in terms of electrochemical performance. Li and colleagues demonstrated flexible binder-free cathodes based on vertically grown Co_3O_4 nanowire on carbon textiles [98]. This structure, obtained through hydrothermal treatment followed by an annealing process, developed a low-resistance route for electron transport. Microscopic analysis of carbon textiles woven by CFs with a diameter of 10 μm, and different magnifications of Co_3O_4 nanowire with an average diameter of 40 nm hybrid structure are shown in Figure 16.6a–c. The superior catalytic activity of nano Co_3O_4 enhanced oxygen, ions, and electron transport for better electrochemical reactions, resulting in a higher discharge capacity of 4,687 mAh g^{-1}. The cell's flexibility, on the contrary, adds to its applicability in flexible electronic applications. Similarly, Liu and coworkers [58] studied the possibility of self-standing, binder-free, nanoneedle-structured $NiCo_2O_4$ fabricated on carbon paper cathode. Films like Na_2O_2 were obtained as the discharge products, where they occupied the enormous space between each needle-like methoxide structure. The schematic illustration of the process and fabricated $NiCo_2O_4$ structure is shown in Figure 16.6d. Provision of sufficient space for discharge products, excellent structural stability, and abundant active sites due to the urchin-like structure provided rapid electron transfer and oxygen diffusion path to reach an initial specific capacity of 6,500 mAh g^{-1} with a current density of 50 mA g^{-1} for the designed cell.

Another cobalt-based catalyst was developed and studied by Senthilkumar and coworkers [75]. They prepared a SAB based on bifunctional nanostructured cobalt phosphate ($Co_3(PO_4)_2$) as an electrocatalyst having grain morphology of about 100 nm thickness (Figure 16.6e). With the co-precipitation-synthesized catalyst, the SAB demonstrated good electrocatalytic activity with approximately 83% round-trip efficiency at a current density of 0.05 mA g^{-1}, compared to conventional platinum-carbon-based cathode. It is well established that the subsequent accumulation of discharged products blocks the electrode pathways that lead to cell failure. Conversely, the discharged

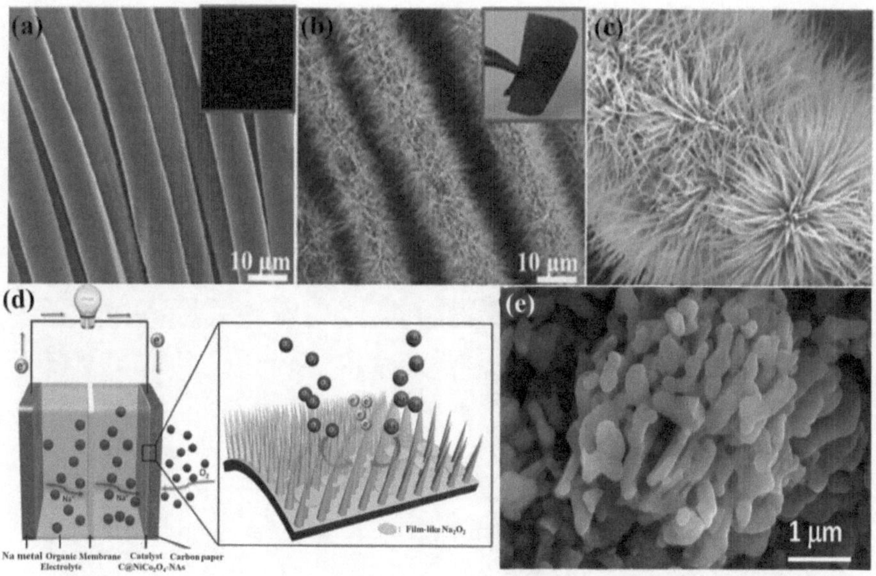

FIGURE 16.6 (a) SEM and optical images (inset) of carbon textiles. Photograph (inset) and SEM images of developed vertically grown Co_3O_4 NW/carbon textile cathode, (b) low magnification image, (c) high magnification image [98] (d) Schematic representation of electrochemical reaction inside SAB and an illustration of fabricated nanoneedle-like structure of $NiCo_2O_4$.[58] (e) HR-TEM image of grain-like structured $Co_3(PO_4)_2$. Adapted and reproduced from Ref. [75]. Copyright © 2016 Elsevier.

product dissolved and hence improved the cell performance. Unlike traditional fabrication systems like mechanical milling and grinding, atomic layer deposition (ALD) gives better catalyst-substrate adherence without disturbing the actual structure of the substrate. Sun and coworkers [48] embedded Co_3O_4 nanoparticles in CNT through the ALD method to achieve a better-dispersed catalyst–cathode system with high porosity and even distribution of Co_3O_4/CNT nanocomposite. The SEM images of fabricated highly porous and evenly distributed Co_3O_4/CNT nanocomposite are shown in Figure 16.7a and b. While comparing the electrochemical activities with mechanically ground CNT/Co_3O_4 nanocomposite and the pristine CNT, it was found that the cathode prepared through the ALD technique has shown a reduced overpotential, better reversibility, increased specific capacity, and high cyclability. Moreover, the Co_3O_4 catalyst played a significant role in decomposing the evolved discharge products like NaO_2, Na_2O_2, and Na_2CO_3 which enhanced the life of cells, which shows the promising superiority of ALD technique in the fabrication of metal oxide-based cathode electrodes for SAB. Porous cobalt boride (CoB) nanosheet-based cathode material was developed by Ma and coworkers [57] through argon-protected heat treatment with precursors such as cobalt chloride and sodium borohydride in a 1:8 ratio. With the high catalytic activity, electrochemical stability, and surface area of nanosheets, the cell has employed a high initial value of 11,482 mAh g^{-1} specific capacity under a current density of 100 mA g^{-1}, and a cyclic life of 74 cycles with a specific capacity of 2,000 mAh g^{-1}. High porosity and interconnected nanosheets provided enormous space for discharged particles and efficient catalytic activity, which could be attributed to this cell performance. Faktorovich-Simon and colleagues [99] made a comparison study of the electrochemical performance of different cathode catalysts like carbon, a spinel-type transition metal oxide $Ni_xCo_yO_z$ (independent NiO and Co_3O_4 phases), and titanium carbide (TiC), in which cathode based on $Ni_xCo_yO_z$ powder with a proportion of $NiCo_2O_4$, metallic nickel, metallic cobalt, and CoO in 45%, 30%, 15%, and 10%, respectively, gained a higher discharge capacity. On investigations, the highest dropping (~57%) of the peak area of TiC shows the inefficiency of that material for catalytic application in SAB. The metallic cobalt and nickel phases are assumed to be the reason behind the

FIGURE 16.7 SEM images of ALD fabricated Co_3O_4/CNT nanocomposite with (a) lower resolution and (b) high resolution [48]. (c and d) Morphology of $MnCo_2O_4$ microsphere obtained through SEM at different resolutions. Adapted and reproduced from Ref. [102], Copyright © 2016 Elsevier.

good catalytic activity of $Ni_xCo_yO_z$, while a poor adhesion of TiC powder on the electrode which could affect surface changes are believed to be the cause of such a degradation and poor efficacy of TiC based cathodes.

Spinel materials with their bi-metallic oxide structure have gained great attention in the energy storage field due to their higher electrical conductivity and electrochemical activity compared with single metallic oxide catalysts [100]. Kang et al. [101] studied the electrochemical performance of the dual-phase spinel $MnCo_2O_4$ nanoparticles with nitrogen-doped reduced graphene oxide hybrid (dp-$MnCo_2O_4$/N-rGO) catalyst system for hybrid SAB having NASICON solid separator and an organic electrolyte. On comparing the electrochemical property with commercially available Pt/C-based cathode, the dp-$MnCo_2O_4$/N-rGO electrocatalyst showed a superior property. Furthermore, the cell exhibited an overpotential of 0.39 V, which is relatively much lower than the Pt/C cathode. Highly interconnected 3D structures, dual-phase cobalt manganese spinel nanoparticles, the activity of introduced nitrogen dopants, high active surface area, and homogeneous distribution of oxide on graphene layers were believed as the potential cause for enhanced cell performance for prepared metal oxide-based cathodes. Shang and colleagues [102] reported the nanosized $MnCo_2O_4$ microsphere-based air cathode prepared through the hydrothermal method. The spherical-shaped $MnCo_2O_4$ with an average particle size of around 1.0–3.0 μm (Figure 16.7c and d), contributes to a highly porous structure and increase in the active sites for oxygen diffusion. The cell with the prepared cathode showed an initial discharge capacity of 7,709.4 mAh g^{-1} under a current density of 0.1 mA cm^{-2}, and a stable cyclability of 130 cycles by keeping a cut-off capacity of 1,000 mAh g^{-1}. The investigation hence showed the applicability of manganese-based catalysts with better electrocatalysis properties for the application in SAB.

Pyrochlore structured compounds ($A_2B_2O_6O'$), a combination of alkaline earth metal and transition metal oxides have been extensively applied in semiconductors, energy storage, electrical devices, catalysts, etc. The B_2O_6 framework of these compounds provides considerable flexibility and an easy electron conduction pathway. Li et al. [60] developed SAB with $La_2Sn_{2-x}Co_xO_7$ as a catalyst for the cathode. Substitution of A or B sites eventually changes the structure and the compound

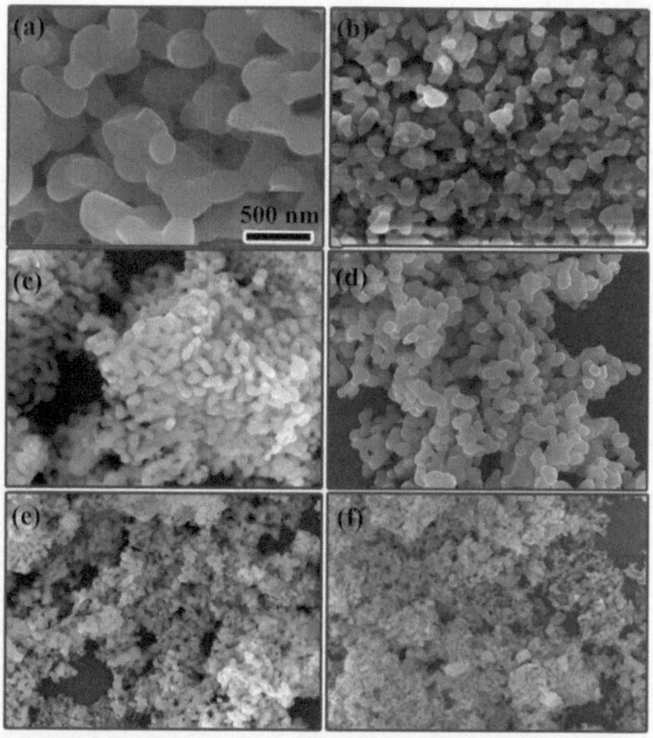

FIGURE 16.8 SEM images of $La_2Sn_{2-x}Co_xO_7$ by keeping different values for x. (a) $x=0$, (b) $x=0.2$, (c) $x=0.5$, (d) $x=1.0$, (e) $x=1.5$, (f) $x=2$. Adapted and reproduced from Ref. [60], Copyright © 2017 American Chemical Society.

leans towards a metallic character, which in turn creates a high catalytic activity. According to microscopic analysis, an increase of Co concentration in the compound gives rise to a reduction of particle size of $La_2Sn_{2-x}Co_xO_7$ and an increase in surface area and enhances catalytic activity without altering the morphology. The SEM images of prepared pyrochlore oxide at different 'x' values are shown in Figure 16.8, where a reduction of particle size as an increase in Co concentration is clearly seen. By keeping $x=2$ in the $La_2Sn_{2-x}Co_xO_7$ structure, the cell performed with an exceptionally high initial discharge capacity of 20,184.2 mAh g^{-1} reported to date for any SAB. The cell showed cyclic stability up to 177 cycles under current dentistry of 200 mAh g^{-1} by keeping a cut-off discharge capacity of 2,000 mAh g^{-1}. A considerable availability of active sites and vacancies for ORR and OER processes as well as the metallic property in these cells, could be the preferable reason for the exceptionally high performance of this system. Similar studies on pyrochlore oxides ($A_2B_2O_7$) have been reported by Kim et al. [103] on bismuth ruthenate pyrochlore oxide ($Bi_2Ru_2O_7$) nanoparticles as a bifunctional catalyst in cathode materials of hybrid SAB for the first time. The electron sharing tendency of bismuth cations (Bi^{3+}) in the delocalized orbital of $Bi_2Ru_2O_7$ created a fast charge transfer pathway during electrocatalysis. Moreover, the donated electrons by both Bi and Ru ions can easily diffuse through the surface and core of the layers to create a less resistive pathway for electron transfer. This mechanism in this investigation was assumed to be the reason for enhanced ORR and OER activities during the cycle to achieve a low overpotential of 0.211 V with a round-trip efficiency of 93.58% and high stability up to 50 cycles during operation.

According to various studies based on cathodes in SABs, many researchers believe that, rather than using an additional catalyst, catalytic activities of carbonaceous materials are satisfactory for acting as an air cathode in SABs [42,104]. In contrast, the use of noble metals and transition metal-based catalysts is also well proven to enhance electrode performance. Liu and coworkers [105] attempted an emerging technology in situ imaging of electrocatalysis activity in SAB using advanced environmental electron transfer microscopy to identify the chemistry during catalytic reactions. In this technique of atomic level characterization, the gold-coated manganese oxide (Au-MnO_2) nanowire processed by ion sputtering was used as an air cathode. The analysis showed that the nanowire volumetrically expanded 18 times after application of the negative potential compared to bare MnO_2-based electrodes. In addition, a bubble-like NaO_2 discharge product was formed and blended with Au particles, causing the volume to shrink over a period of time. Furthermore, in the reverse process, the NaO_2 discharge product was found to be decomposed to form Na_2O_2 and O_2. On the contrary, in analysis using bare MnO_2 cathode, no ORR was observed; instead, the nanowires swelled up to 217% due to Na^+ intercalation. The atomic level characterization opened a new insight into the need for a catalyst system, as well as a new method for comprehending real-time electrocatalysis in metal air batteries.

16.3 CONCLUSION

SABs have been significantly researched and incentivized as a promising future energy storage technology. SABs are assured to be a feasible alternative to other systems owing to their higher round-trip efficiency and stable charge-discharge plateaus. For meeting future energy demands, designing high-performance SABs with higher discharge capacity, cyclic stability, catalytic activity, and power density is significant. Despite the other components like the anode and electrolyte, the cathode holds a special significance in deciding the performance of SABs. Air electrodes mainly consist of carbonaceous materials, binders, and current collectors. To meet the future demand for energy, high-performance SABs are required, which cannot be satisfied using these elements. To improve the performance of SABs, various modifications, and inclusion of catalyst systems using nanomaterials, doped carbon species, noble metals, transition metals, and their oxides have been researched. Furthermore, structural modification and designs to accommodate the enormous amount of discharge products and to provide abundant active sites for the efficient electrochemical process have also been employed. Most of the above-discussed cathode systems provide a higher initial discharge capacity, and retaining their cyclic stability up to considerable counts remains difficult due to the eventual migration of oxides and superoxides. So control over the discharge products released during reactions and their properties is also significant. Furthermore, mass-scale development of air electrodes using the aforementioned materials must conquer various practical challenges such as availability of materials, precise cathode engineering, and economic and environmental concerns. In our view, designing an air electrode with materials having high porosity and catalytic activity, along with environmental considerations, is important for high-performance SABs to achieve the goal of sustainable energy storage.

REFERENCES

1. Choi N-S, Chen Z, Freunberger SA, et al (2012) Challenges facing lithium batteries and electrical double-layer capacitors. *Angew Chemie – Int Ed* 51:9994–10024. https://doi.org/10.1002/anie.201201429
2. Larcher D, Tarascon J-M (2015) Towards greener and more sustainable batteries for electrical energy storage. *Nat Chem* 7:19–29. https://doi.org/10.1038/nchem.2085
3. Ghazi ZA, Sun Z, Sun C, et al (2019) Key aspects of lithium metal anodes for lithium metal batteries. *Small* 15:1900687. https://doi.org/10.1002/smll.201900687
4. Jishnu NS, Vineeth SK, Das A, et al (2021) Electrospun PVdF and PVdF-co-HFP-based blend polymer electrolytes for lithium ion batteries. In: Balakrishnan NTM, Raghavan P (eds) *Electrospinning for Advanced Energy Storage Applications*, 1st ed. Springer, Singapore, pp. 201–234.
5. Scrosati B, Garche J (2010) Lithium batteries: status, prospects and future. *J Power Sources* 195:2419–2430. https://doi.org/10.1016/j.jpowsour.2009.11.048
6. Yadegari H, Sun X (2018) Recent advances on sodium-oxygen batteries: a chemical perspective. *Acc Chem Res* 51:1532–1540. https://doi.org/10.1021/acs.accounts.8b00139

7. Liu Q, Pan Z, Wang E, et al (2020) Aqueous metal-air batteries: fundamentals and applications. *Energy Storage Mater* 27:478–505. https://doi.org/10.1016/j.ensm.2019.12.011
8. Zheng Z, Wu C, Gu Q, et al (2021) Research progress and future perspectives on rechargeable Na-O_2 and Na-CO2 batteries. *Energy Environ Mater* 4:158–177. https://doi.org/10.1002/eem2.12139
9. Vineeth SK, Soni CB, Sun Y, et al (2022) Implications of Na-ion solvation on Na anode-electrolyte interphase. *Trends Chem* 4:48–59. https://doi.org/10.1016/j.trechm.2021.11.002
10. Yin W, Fu Z (2017) The potential of Na-air batteries. *ChemCatChem* 9:1545–1553. https://doi.org/10.1002/cctc.201600646
11. Cheng F, Chen J (2012) Metal-air batteries: from oxygen reduction electrochemistry to cathode catalysts. *Chem Soc Rev* 41:2172. https://doi.org/10.1039/c1cs15228a
12. Ha S, Kim J-K, Choi A, et al (2014) Sodium-metal halide and sodium-air batteries. *ChemPhysChem* 15:1971–1982. https://doi.org/10.1002/cphc.201402215
13. Lin X, Sun Q, Doyle Davis K, et al (2019) The application of carbon materials in nonaqueous Na-O_2 batteries. *Carbon Energy* 1:141–164. https://doi.org/10.1002/cey2.15
14. Chawla N, Safa M (2019) Sodium batteries: a review on sodium-sulfur and sodium-air batteries. *Electronics* 8:1201. https://doi.org/10.3390/electronics8101201
15. Liu S, Liu S, Luo J (2016) Carbon-based cathodes for sodium-air batteries. *New Carbon Mater* 31:264–270. https://doi.org/10.1016/S1872-5805(16)60012-4
16. Baek M-J, Choi J, Wi T-U, et al (2022) Strong interfacial energetics between catalysts and current collectors in aqueous sodium-air batteries. *J Mater Chem A* 10:4601–4610. https://doi.org/10.1039/D2TA00329E
17. Bender CL, Hartmann P, Vračar M, et al (2014) On the thermodynamics, the role of the carbon cathode, and the cycle life of the sodium superoxide (NaO_2) battery. *Adv Energy Mater* 4:1301863. https://doi.org/10.1002/aenm.201301863
18. Yadegari H, Franko CJ, Banis MN, et al (2017) How to control the discharge products in Na-O_2 cells: direct evidence toward the role of functional groups at the air electrode surface. *J Phys Chem Lett* 8:4794–4800. https://doi.org/10.1021/acs.jpclett.7b02227
19. Yadegari H, Li Y, Banis MN, et al (2014) On rechargeability and reaction kinetics of sodium-air batteries. *Energy Environ Sci* 7:3747–3757. https://doi.org/10.1039/C4EE01654H
20. Zhang Y-L, Goh K, Zhao L, et al (2020) Advanced non-noble materials in bifunctional catalysts for ORR and OER toward aqueous metal-air batteries. *Nanoscale* 12:21534–21559. https://doi.org/10.1039/D0NR05511E
21. Yadegari H, Sun Q, Sun X (2016) Sodium-oxygen batteries: a comparative review from chemical and electrochemical fundamentals to future perspective. *Adv Mater* 28:7065–7093. https://doi.org/10.1002/adma.201504373
22. El-Kady MF, Shao Y, Kaner RB (2016) Graphene for batteries, supercapacitors and beyond. *Nat Rev Mater* 1:16033. https://doi.org/10.1038/natrevmats.2016.33
23. Nomura A, Ito K, Kubo Y (2017) CNT sheet air electrode for the development of ultra-high cell capacity in lithium-air batteries. *Sci Rep* 7:45596. https://doi.org/10.1038/srep45596
24. Pendashteh A, Palma J, Anderson M, et al (2018) Doping of self-standing CNT fibers: promising flexible air-cathodes for high-energy-density structural Zn-air batteries. *ACS Appl Energy Mater* 1:2434–2439. https://doi.org/10.1021/acsaem.8b00583
25. Liu W, Sun Q, Yang Y, et al (2013) An enhanced electrochemical performance of a sodium-air battery with graphene nanosheets as air electrode catalysts. *Chem Commun* 49:1951. https://doi.org/10.1039/c3cc00085k
26. Enterría M, Botas C, Gómez-Urbano JL, et al (2018) Pathways towards high performance Na-O_2 batteries: tailoring graphene aerogel cathode porosity & nanostructure. *J Mater Chem A* 6:20778–20787. https://doi.org/10.1039/C8TA07273F
27. Munuera JM, Paredes JI, Enterría M, et al (2020) High performance Na-O_2 batteries and printed micro-supercapacitors based on water-processable, biomolecule-assisted anodic graphene. *ACS Appl Mater Interfaces* 12:494–506. https://doi.org/10.1021/acsami.9b15509
28. Enterría M, Gómez-Urbano JL, Munuera JM, et al (2021) Boosting the performance of graphene cathodes in Na-O2 batteries by exploiting the multifunctional character of small biomolecules. *Small* 17:2005034. https://doi.org/10.1002/smll.202005034
29. Ushijima K, Iwamura S, Mukai SR (2020) Simple and cost-effective method to increase the capacity of carbon nanotube sheet cathodes for lithium-air batteries. *ACS Appl Energy Mater* 3:6915–6921. https://doi.org/10.1021/acsaem.0c00991

30. Jian Z, Chen Y, Li F, et al (2014) High capacity Na-O_2 batteries with carbon nanotube paper as binder-free air cathode. *J Power Sources* 251:466–469. https://doi.org/10.1016/j.jpowsour.2013.11.091
31. Sun Q, Yadegari H, Banis MN, et al (2015) Self-stacked nitrogen-doped carbon nanotubes as long-life air electrode for sodium-air batteries: Elucidating the evolution of discharge product morphology. *Nano Energy* 12:698–708. https://doi.org/10.1016/j.nanoen.2015.01.018
32. Lee WJ, Maiti UN, Lee JM, et al (2014) Nitrogen-doped carbon nanotubes and graphene composite structures for energy and catalytic applications. *Chem Commun* 50:6818. https://doi.org/10.1039/c4cc00146j
33. Ryu J, Jung N, Lim D-H, et al (2014) P-modified and carbon shell coated Co nanoparticles for efficient alkaline oxygen reduction catalysis. *Chem Commun* 50:15940–15943. https://doi.org/10.1039/C4CC07143C
34. Li Y, Yadegari H, Li X, et al (2013) Superior catalytic activity of nitrogen-doped graphene cathodes for high energy capacity sodium-air batteries. *Chem Commun* 49:11731. https://doi.org/10.1039/c3cc46606j
35. Ma J, Zhang X (2016) Optimized nitrogen-doped carbon with a hierarchically porous structure as a highly efficient cathode for Na-O_2 batteries. *J Mater Chem A* 4:10008–10013. https://doi.org/10.1039/C6TA02793H
36. Ma J-L, Meng F-L, Xu D, Zhang X-B (2017) Co-embedded N-doped carbon fibers as highly efficient and binder-free cathode for Na-O_2 batteries. *Energy Storage Mater* 6:1–8. https://doi.org/10.1016/j.ensm.2016.09.002
37. Yang L, Jiang S, Zhao Y, et al (2011) Boron-doped carbon nanotubes as metal-free electrocatalysts for the oxygen reduction reaction. *Angew Chemie – Int Ed* 50:7132–7135. https://doi.org/10.1002/anie.201101287
38. Meng W, Wen L, Song Z, et al (2017) Metal-free boron-doped carbon microspheres as excellent cathode catalyst for rechargeable Li-O_2 battery. *J Solid State Electrochem* 21:665–671. https://doi.org/10.1007/s10008-016-3412-3
39. Shu C, Lin Y, Zhang B, et al (2016) Mesoporous boron-doped onion-like carbon as long-life oxygen electrode for sodium-oxygen batteries. *J Mater Chem A* 4:6610–6619. https://doi.org/10.1039/C6TA00901H
40. Chen W, Xu L, Tian Y, et al (2018) Boron and nitrogen co-doped graphene aerogels: facile preparation, tunable doping contents and bifunctional oxygen electrocatalysis. *Carbon* 137:458–466. https://doi.org/10.1016/j.carbon.2018.05.061
41. Benti NE, Tiruye GA, Mekonnen YS (2020) Boron and pyridinic nitrogen-doped graphene as potential catalysts for rechargeable non-aqueous sodium-air batteries. *RSC Adv* 10:21387–21398. https://doi.org/10.1039/D0RA03126G
42. Yadegari H, Banis MN, Xiao B, et al (2015) Three-dimensional nanostructured air electrode for sodium-oxygen batteries: a mechanism study toward the cyclability of the cell. *Chem Mater* 27:3040–3047. https://doi.org/10.1021/acs.chemmater.5b00435
43. Tan P, Shyy W, An L, et al (2014) A gradient porous cathode for non-aqueous lithium-air batteries leading to a high capacity. *Electrochem Commun* 46:111–114. https://doi.org/10.1016/j.elecom.2014.06.026
44. Sun B, Kretschmer K, Xie X, et al (2017) Hierarchical porous carbon spheres for high-performance Na-O_2 batteries. *Adv Mater* 29:1606816. https://doi.org/10.1002/adma.201606816
45. Zheng Z, Wu C, Gu Q, et al (2021) Research progress and future perspectives on rechargeable Na-O_2 and Na-CO_2 batteries. *Energy Environ Mater* 4:158–177. https://doi.org/10.1002/eem2.12139
46. Sun Q, Liu J, Xiao B, et al (2019) Visualizing the oxidation mechanism and morphological evolution of the cubic-shaped superoxide discharge product in Na-air batteries. *Adv Funct Mater* 29:1808332. https://doi.org/10.1002/adfm.201808332
47. Wu F, Xing Y, Lai J, et al (2017) Micrometer-sized RuO_2 catalysts contributing to formation of amorphous Na-deficient sodium peroxide in Na-O_2 batteries. *Adv Funct Mater* 27:1700632. https://doi.org/10.1002/adfm.201700632
48. Sun Q, Liu J, Li X, et al (2017) Atomic layer deposited non-noble metal oxide catalyst for sodium-air batteries: tuning the morphologies and compositions of discharge product. *Adv Funct Mater* 27:1606662. https://doi.org/10.1002/adfm.201606662
49. Kwak W-J, Chen Z, Yoon CS, et al (2015) Nanoconfinement of low-conductivity products in rechargeable sodium-air batteries. *Nano Energy* 12:123–130. https://doi.org/10.1016/j.nanoen.2014.11.057
50. Zhang S, Wen Z, Jin J, et al (2016) Controlling uniform deposition of discharge products at the nanoscale for rechargeable Na-O_2 batteries. *J Mater Chem A* 4:7238–7244. https://doi.org/10.1039/C6TA02336C
51. Sun Q, Yadegari H, Banis MN, et al (2015) Self-stacked nitrogen-doped carbon nanotubes as long-life air electrode for sodium-air batteries: elucidating the evolution of discharge product morphology. *Nano Energy* 12:698–708. https://doi.org/10.1016/j.nanoen.2015.01.018

52. Zhao N, Li C, Guo X (2014) Long-life Na-O_2 batteries with high energy efficiency enabled by electrochemically splitting NaO_2 at a low overpotential. *Phys Chem Chem Phys* 16:15646. https://doi.org/10.1039/c4cp01961j
53. Sun Q, Yang Y, Fu Z-W (2012) Electrochemical properties of room temperature sodium-air batteries with non-aqueous electrolyte. *Electrochem Commun* 16:22–25. https://doi.org/10.1016/j.elecom.2011.12.019
54. Liu Y, Chi X, Han Q, et al (2019) Metal-organic framework-derived hierarchical Co_3O_4@$MnCo_2O_{4.5}$ nanocubes with enhanced electrocatalytic activity for Na-O_2 batteries. *Nanoscale* 11:5285–5294. https://doi.org/10.1039/C8NR08818G
55. Tovini MF, Patil B, Koz C, et al (2018) Nanohybrid structured RuO_2/Mn_2O_3/CNF as a catalyst for Na-O_2 batteries. *Nanotechnology* 29:475401. https://doi.org/10.1088/1361-6528/aadfb7
56. Khajehbashi SMB, Xu L, Zhang G, et al (2018) High-performance Na-O_2 batteries enabled by oriented NaO_2 nanowires as discharge products. *Nano Lett* 18:3934–3942. https://doi.org/10.1021/acs.nanolett.8b01315
57. Ma J, Li N, Zhang Q, et al (2018) Synthesis of porous and metallic CoB nanosheets towards a highly efficient electrocatalyst for rechargeable Na-O_2 batteries. *Energy Environ Sci* 11:2833–2838. https://doi.org/10.1039/C8EE01472H
58. Liu Y, Chi X, Han Q, et al (2019) Vertically self-standing C@$NiCo_2O_4$ nanoneedle arrays as effective binder-free cathodes for rechargeable Na–O_2 batteries. *J Alloys Compd* 772:693–702. https://doi.org/10.1016/j.jallcom.2018.09.121
59. Yin W-W, Fu Z-W (2017) A highly efficient bifunctional heterogeneous catalyst for morphological control of discharged products in Na-air batteries. *Chem Commun* 53:1522–1525. https://doi.org/10.1039/C6CC09348E
60. Li N, Yin Y, Meng F, et al (2017) Enabling pyrochlore-type oxides as highly efficient electrocatalysts for high-capacity and stable Na-O_2 batteries: the synergy of electronic structure and morphology. *ACS Catal* 7:7688–7694. https://doi.org/10.1021/acscatal.7b02074
61. Hu Y, Han X, Zhao Q, et al (2015) Porous perovskite calcium-manganese oxide microspheres as an efficient catalyst for rechargeable sodium-oxygen batteries. *J Mater Chem A* 3:3320–3324. https://doi.org/10.1039/C4TA06287F
62. Liu W-M, Yin W-W, Ding F, et al (2014) $NiCo_2O_4$ nanosheets supported on Ni foam for rechargeable nonaqueous sodium-air batteries. *Electrochem Commun* 45:87–90. https://doi.org/10.1016/j.elecom.2014.05.021
63. Fathi Tovini M, Hong M, Park J, et al (2018) Instability of a noncrystalline NaO_2 film in Na-O_2 batteries: the controversial effect of the RuO_2 catalyst. *J Phys Chem C* 122:19678–19686. https://doi.org/10.1021/acs.jpcc.8b06024
64. Kang J-H, Kwak W-J, Aurbach D, Sun Y-K (2017) Sodium oxygen batteries: one step further with catalysis by ruthenium nanoparticles. *J Mater Chem A* 5:20678–20686. https://doi.org/10.1039/C7TA06584A
65. Ma L, Zhang D, Lei Y, et al (2018) High-capacity sodium peroxide based Na-O_2 batteries with low charge overpotential via a nanostructured catalytic cathode. *ACS Energy Lett* 3:276–277. https://doi.org/10.1021/acsenergylett.7b01143
66. Kumar S, Kishore B, Munichandraiah N (2016) Electrochemical studies of non-aqueous Na-O_2 cells employing Ag-RGO as the bifunctional catalyst. *RSC Adv* 6:63477–63479. https://doi.org/10.1039/C6RA13596J
67. Zhang S, Wen Z, Rui K, et al (2015) Graphene nanosheets loaded with Pt nanoparticles with enhanced electrochemical performance for sodium-oxygen batteries. *J Mater Chem A* 3:2568–2571. https://doi.org/10.1039/C4TA05427J
68. Ha TA, Pozo-Gonzalo C, Nairn K, et al (2020) An investigation of commercial carbon air cathode structure in ionic liquid based sodium oxygen batteries. *Sci Rep* 10:7123. https://doi.org/10.1038/s41598-020-63473-y
69. Zheng Z, Jiang J, Guo H, et al (2021) Tuning NaO_2 formation and decomposition routes with nitrogen-doped nanofibers for low overpotential Na-O_2 batteries. *Nano Energy* 81:105529. https://doi.org/10.1016/j.nanoen.2020.105529
70. Langsdorf D, Dahms T, Mohni V, et al (2020) Pulse discharging of sodium-oxygen batteries to enhance cathode utilization. *Energies* 13:5650. https://doi.org/10.3390/en13215650
71. Hashimoto T, Hayashi K (2015) Aqueous and nonaqueous sodium-air cells with nanoporous gold cathode. *Electrochim Acta* 182:809–814. https://doi.org/10.1016/j.electacta.2015.09.153
72. Liu W, Guo R, Wang Y, et al (2020) A low-overpotential sodium/fluorinated graphene battery based on silver nanoparticles as catalyst. *J Colloid Interface Sci* 565:70–76. https://doi.org/10.1016/j.jcis.2020.01.011

73. Landa-Medrano I, Li C, Ortiz-Vitoriano N, et al (2016) Sodium-oxygen battery: steps toward reality. *J Phys Chem Lett* 7:1161–1166. https://doi.org/10.1021/acs.jpclett.5b02845
74. Guo L, Li B, Thirumal V, Song J (2019) Advanced rechargeable Na-CO_2 batteries enabled by a ruthenium@porous carbon composite cathode with enhanced Na_2CO_3 reversibility. *Chem Commun* 55:7946–7949. https://doi.org/10.1039/C9CC02737H
75. Senthilkumar B, Khan Z, Park S, et al (2016) Exploration of cobalt phosphate as a potential catalyst for rechargeable aqueous sodium-air battery. *J Power Sources* 311:29–34. https://doi.org/10.1016/j.jpowsour.2016.02.022
76. Baskar S, Murugesan C, Barpanda P (2017) Highly porous N-doped graphitic carbon from bio-waste as bifunctional electrocatalyst for hybrid Na-air battery. *ECS Trans* 80:425–430. https://doi.org/10.1149/08010.0425ecst
77. Wang J, Gao R, Zheng L, et al (2018) CoO/CoP heterostructured nanosheets with an O-P interpenetrated interface as a bifunctional electrocatalyst for Na-O_2 battery. *ACS Catal* 8:8953–8960. https://doi.org/10.1021/acscatal.8b01023
78. Abirami M, Hwang SM, Yang J, et al (2016) A metal-organic framework derived porous cobalt manganese oxide bifunctional electrocatalyst for hybrid Na-air/seawater batteries. *ACS Appl Mater Interfaces* 8:32778–32787. https://doi.org/10.1021/acsami.6b10082
79. Park J, Risch M, Nam G, et al (2017) Single crystalline pyrochlore nanoparticles with metallic conduction as efficient bi-functional oxygen electrocatalysts for Zn-air batteries. *Energy Environ Sci* 10:129–136. https://doi.org/10.1039/C6EE03046G
80. Parrondo J, George M, Capuano C, et al (2015) Pyrochlore electrocatalysts for efficient alkaline water electrolysis. *J Mater Chem A* 3:10819–10828. https://doi.org/10.1039/C5TA01771H
81. Lee KT, Lee BW, Camaratta MA, Wachsman ED (2013) Enhanced oxygen reduction reaction with nano-scale pyrochlore bismuth ruthenate via cost-effective wet-chemical synthesis. *RSC Adv* 3:19866. https://doi.org/10.1039/c3ra44633f
82. Kim M, Ju H, Kim J (2018) Single crystalline thallium rhodium oxide pyrochlore for highly improved round trip efficiency of hybrid Na-air batteries. *Dalt Trans* 47:15217–15225. https://doi.org/10.1039/C8DT03217C
83. Kim M, Ju H, Kim J (2019) Dihydrogen phosphate ion functionalized nanocrystalline thallium ruthenium oxide pyrochlore as a bifunctional electrocatalyst for aqueous Na-air batteries. *Appl Catal B Environ* 245:29–39. https://doi.org/10.1016/j.apcatb.2018.12.031
84. Zhou H-C, Long JR, Yaghi OM (2012) Introduction to metal-organic frameworks. *Chem Rev* 112:673–674. https://doi.org/10.1021/cr300014x
85. Song Z, Liu W, Cheng N, et al (2017) Origin of the high oxygen reduction reaction of nitrogen and sulfur co-doped MOF-derived nanocarbon electrocatalysts. *Mater Horizons* 4:900–907. https://doi.org/10.1039/C7MH00244K
86. Ying J, Li J, Jiang G, et al (2018) Metal-organic frameworks derived platinum-cobalt bimetallic nanoparticles in nitrogen-doped hollow porous carbon capsules as a highly active and durable catalyst for oxygen reduction reaction. *Appl Catal B Environ* 225:496–503. https://doi.org/10.1016/j.apcatb.2017.11.077
87. Wu Y, Qiu X, Liang F, et al (2019) A metal-organic framework-derived bifunctional catalyst for hybrid sodium-air batteries. *Appl Catal B Environ* 241:407–414. https://doi.org/10.1016/j.apcatb.2018.09.063
88. Xu X, Hui KS, Dinh DA, et al (2019) Recent advances in hybrid sodium-air batteries. *Mater Horizons* 6:1306–1335. https://doi.org/10.1039/C8MH01375F
89. Song M, He Y, Zhang M, et al (2018) Controllable synthesis of Co_2P nanorods as high-efficiency bifunctional electrocatalyst for overall water splitting. *J Power Sources* 402:345–352. https://doi.org/10.1016/j.jpowsour.2018.09.042
90. Zhang X, Yu X, Zhang L, et al (2018) Molybdenum phosphide/carbon nanotube hybrids as pH-universal electrocatalysts for hydrogen evolution reaction. *Adv Funct Mater* 28:1706523. https://doi.org/10.1002/adfm.201706523
91. Jiang Y, Lu Y, Lin J, et al (2018) A hierarchical MoP nanoflake array supported on Ni foam: a bifunctional electrocatalyst for overall water splitting. *Small Methods* 2:1700369. https://doi.org/10.1002/smtd.201700369
92. Liu C, Zhang G, Yu L, et al (2018) Oxygen doping to optimize atomic hydrogen binding energy on NiCoP for highly efficient hydrogen evolution. *Small* 14:1800421. https://doi.org/10.1002/smll.201800421
93. Kang Y, Wang S, Zhu S, et al (2021) Iron-modulated nickel cobalt phosphide embedded in carbon to boost power density of hybrid sodium-air battery. *Appl Catal B Environ* 285:119786. https://doi.org/10.1016/j.apcatb.2020.119786

94. Xue Y, Sun S, Wang Q, et al (2018) Transition metal oxide-based oxygen reduction reaction electrocatalysts for energy conversion systems with aqueous electrolytes. *J Mater Chem A* 6:10595–10626. https://doi.org/10.1039/C7TA10569J
95. Rosenberg S, Hintennach A (2015) In situ formation of α-MnO_2 nanowires as catalyst for sodium-air batteries. *J Power Sources* 274:1043–1048. https://doi.org/10.1016/j.jpowsour.2014.10.187
96. Zeng X, Zhang X, Liu S, et al (2019) A highly efficient cathode catalyst γ-MnO_2@CNT composite for sodium-air batteries. *Sci China Chem* 62:727–731. https://doi.org/10.1007/s11426-018-9442-1
97. Cui Y, Wen Z, Liu Y (2011) A free-standing-type design for cathodes of rechargeable Li-O_2 batteries. *Energy Environ Sci* 4:4727. https://doi.org/10.1039/c1ee02365a
98. Li N, Xu D, Bao D, et al (2016) A binder-free, flexible cathode for rechargeable Na-O_2 batteries. *Chinese J Catal* 37:1172–1179. https://doi.org/10.1016/S1872-2067(15)61089-0
99. Faktorovich-Simon E, Natan A, Peled E, Golodnitsky D (2019) Comparison of the catalytic activity of carbon, spinel-based, and carbide materials in the Na-air battery. *Front Mater* 6:00249. https://doi.org/10.3389/fmats.2019.00249
100. Li Y, Han X, Yi T, et al (2019) Review and prospect of $NiCo_2O_4$-based composite materials for supercapacitor electrodes. *J Energy Chem* 31:54–78. https://doi.org/10.1016/j.jechem.2018.05.010
101. Kang Y, Zou D, Zhang J, et al (2017) Dual-phase spinel $MnCo_2O_4$ nanocrystals with nitrogen-doped reduced graphene oxide as potential catalyst for hybrid Na-air batteries. *Electrochim Acta* 244:222–229. https://doi.org/10.1016/j.electacta.2017.05.100
102. Shang M, Liu Y, Xia J, et al (2017) Synthesis and characterization of $MnCo_2O_4$ microspheres based air electrode for rechargeable sodium-air batteries. *Ceram Int* 43:3218–3223. https://doi.org/10.1016/j.ceramint.2016.11.147
103. Kim M, Ju H, Kim J (2019) Single crystalline $Bi_2Ru_2O_7$ pyrochlore oxide nanoparticles as efficient bifunctional oxygen electrocatalyst for hybrid Na-air batteries. *Chem Eng J* 358:11–19. https://doi.org/10.1016/j.cej.2018.09.204
104. Ortiz-Vitoriano N, Batcho TP, Kwabi DG, et al (2015) Rate-dependent nucleation and growth of NaO_2 in Na-O_2 batteries. *J Phys Chem Lett* 6:2636–2643. https://doi.org/10.1021/acs.jpclett.5b00919
105. Liu Q, Geng L, Yang T, et al (2019) In-situ imaging electrocatalysis in a Na-O_2 battery with Au-coated MnO_2 nanowires air cathode. *Energy Storage Mater* 19:48–55. https://doi.org/10.1016/j.ensm.2018.08.026

17 Innovative design strategies in Electrolyte Engineering and Separator Modifications for Sodium-Air Batteries

S. K. Vineeth, Ajish Babu, Roshny Joy, Abhilash Pullanchiyodan, Nutan Gupta, and Prasanth Raghavan

17.1 INTRODUCTION

The booming demand for energy requirement by the massively growing population has identified the requirement for sustainable energy sources. Solar, wind, geothermal, and tidal energy sources are being considered as candidates for sustainable energy sectors. The intermittent nature of these sources limits its attractivity. In this aspect, the employability of rechargeable batteries reshapes the human lifestyle [1]. Considering the developments and growth in rechargeable battery technology, the major portion is overruled by lithium-ion batteries (LIBs) [2]. Although LIB is a commercialized and established technology, the bottleneck challenge faced by LIBs is its limited availability of lithium (Li) resources [3]. Furthermore, the developments in LIB have almost reached saturation in the energy density of 200–250 Wh kg^{-1} [4]. To circumvent the situation to meet the demand for electrification and contribute to grid energy storage, sodium (Na) metal batteries have been proposed [5,6].

Na accounts for a high crustal abundance compared to Li, with more than 1,000 times, and is the fourth most abundant element on Earth [7,8]. Although the specific discharge capacity of Li is 3,800 mAh g^{-1}, and for Na it is 1,165 mAh g^{-1}, the cost of Na accounts approximately 4% of Li [9]. Coupled with the Na metal anode, a porous cathode with oxygen diffusion pathways and a sodium ion (Na^+)–conducting separator in an electrolyte makes up a sodium-air battery (SAB) [10]. Compared to its lithium counterpart, lithium-air battery, the high reversibility of electrochemical redox pair, oxygen-superoxide system ($O_{2\_}$ O_2^{2-}) in SAB, contributes to lower charging overpotential and high energy efficiency [11,12]. Furthermore, the reversibility of $O_{2\_}$ O_2^{2-} governs the oxygen reduction reaction (ORR) and oxygen evolution reaction (OER) mechanisms in the air cathode [13]. The sluggish nature of ORR and OER are mitigated by the usage of a bifunctional cathode catalyst [14,15].

As the developments of SAB are in infancy, to accelerate the progress, unveiling the physico-chemical aspects of SAB needs prime importance. Enhancing the material properties of anodes, cathodes, and electrolytes by engineering can undoubtedly push the electrochemical performance of SAB [16–18]. Achievement of better cyclability is mainly related to the ionic conductivity of an electrolyte and the presence of free solvent molecules. Parasitic side reactions, electrolyte decomposition, evaporation of electrolytes, etc. are the main limitations which need to be rectified in the engineering of electrolytes [19]. Optimizing the electrolyte composition is an effective approach to enhance the electrochemical properties of a SAB. Moreover, the electrode–electrolyte interphase also had a direct correlation to the performance of a cell [20].

 DOI: 10.1201/9781003310174-17

Morphology and the chemistry of discharge products in a cell depend on the selection of active salts and solvents in an electrolyte. Hence, it is unequivocally important to implement electrolyte engineering and stabilize the anode part. This chapter highlights the state of the art in the development of SAB with a focus on anode and electrolyte parts.

17.2 ELECTROLYTES FOR ENHANCED ELECTROCHEMICAL REACTIONS

In general, the role of an electrolyte is ionic conduction inside a cell. Redox cell reaction produces electrons and ions, as electronic transport is facilitated by the metallic connections in a cell, while ionic conduction is facilitated by the electrolytes. Ionic conduction and migration, viscosity of electrolyte, electrochemical stability window, etc. are the important parameters which governs the suitability of electrolyte. Solvation dynamics are dependent on the interactions between ions and solvent molecules [21]. Solvation–desolvation dynamics of electrolytes govern the reversibility in a sodium metal battery. The solvated state Na^+ in an electrolyte solution is complexed by solvent molecules. Depending on the type of electrolyte solvent (carbonates and ether), the sluggishness of solvation Na^+ varies [22]. As ionic mobility and reversibility of a cell are dependent on solvation–desolvation dynamics, theoretical calculations are used as an aid to evaluate the dynamics [23–25]. SABs can be classified based on the type of electrolytes used [26]. Aqueous SABs use sodium hydroxide (NaOH) solutions with a separator, usually NAtrium Superionic CONductors (NASICON). Na is oxidized to ions during the discharge process, and it diffuses from anolyte to catholyte. During charging, at the cathode, OER proceeds and ions are deposited on the metallic anode [27]. Non-aqueous electrolyte-based SAB consists of carbonate or ether-based solvents containing sodium perchlorate ($NaClO_4$), sodium hexafluorophosphate ($NaPF_6$), and sodium trifluoromethanesulphonate ($NaSO_3CF_3$, sodium triflate) as Na salts. The most explored carbonate solvents are ethylene carbonate (EC), dimethyl carbonate (DMC), and propylene carbonate (PC). In ether-based solvents, dimethyl ether (DME) and tetra ethylene glycol dimethyl ether (TEGDME) are the most reported systems. The composition of the discharged product depends on the solvent and salt used. EC/DMC solvent with $NaPF_6$ showed the presence of Na_2O_2, Na_2CO_3, and NaOCO-R as the discharge products [28]. Discharge products formed at the cathode in non-aqueous batteries mainly consist of Na_2O_2 and NaO_2, and these products negatively affect the cyclability and power density and increase the overpotential [29]. Moreover, non-aqueous SAB requires purified oxygen as an active material, which increases the cost. Apart from aqueous and non-aqueous electrolytes, ionic liquid (IL) electrolytes and solid-state electrolytes have been reported recently. Ideally, an electrolyte for SAB should have a wide electrochemical window, be chemically stable against superoxide anion (O_2^-), have high ionic conduction, have lower vapour pressure, and have high oxygen solubility [30]. Hence to develop electrolytes which have resistance to the attack of superoxide radicals and to suppress parasitic side reactions with a high electrochemical window, various approaches have been postulated. A detailed discussion based on various classes of electrolytes is listed in the following section.

17.2.1 AQUEOUS/SOLID OR MIXED ELECTROLYTES

Hayashi et al. [31] fabricated SAB with a mixed aqueous/aprotic electrolyte and a solid ceramic, high ionically conducting NASICON-based separator (Figure 17.1). The ceramic NASICON separator used has high stability towards the aqueous phase and Na anode and with ionic conduction of 2×10^{-3} S cm^{-1} measured at 50°C. Oxidation at the anode caused a 30% decrement in the theoretical discharge value. Moreover, the thickness of the NASICON separator affected the areal output power. It was also reported that by the optimization of the concentration of NaOH in electrolytes with a proper cathode, the cell performance can be engineered.

Liang et al. [32] reported the usage of mixed aprotic and aqueous electrolytes (NaOH) separated by a NASICON solid separator of high ionic conductivity for SAB primary cells, showing high discharge

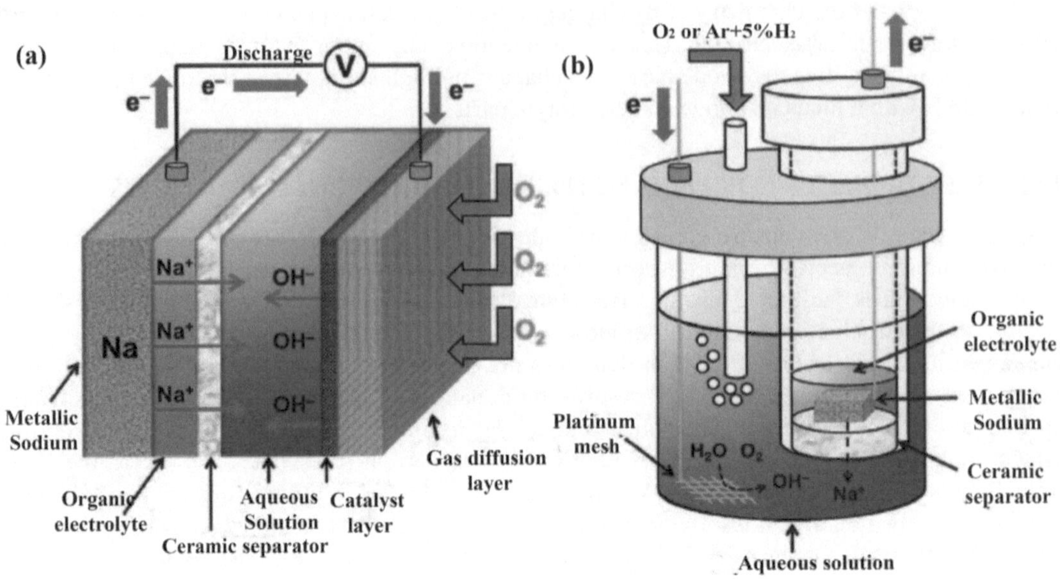

FIGURE 17.1 Schematic representation of cell architecture (a) components of SAB (b) a representation of the fabricated cell used for the study. Adapted and reproduced with permission from Ref. [31]. Copyright © 2013 The Electrochemical Society.

performance. The air cathode consists of activated carbon by trimanganese tetraoxide (Mn_3O_4) as a catalyst with polytetrafluoroethylene (PTFE) as a binder. Carbonate-based and ether-based electrolytes were employed with $NaSO_3CF_3$ and $NaClO_4$ was added to the electrolyte. At 25°C the cell with carbonate organic electrolyte and NASICON separator of 0.83 mm thickness showed a maximum discharge of 21 mW cm^{-2}. Moreover, electrochemical impedance spectroscopy (EIS) showed that the ohmic resistance of the NASICON separator contributed to the total cell resistance; hence engineering on the separator could further enhance the performance of SAB. Sahgong et al. [33] were the first to communicate the comparative performance of Vulcan XC72R and Pt/C coated and uncoated as air cathode catalysts for SAB. Polyvinylidene fluoride (PVDF) was used as a binder for preparing a slurry containing a catalyst, which was then deposited over the carbon paper air cathode. The electrolyte employed was TEGDME with 1 M of $NaCF_3SO_3$ as Na salt. Na metal attached to a nickel mesh as an anode was dipped in NaOH solution prior to its insertion in a pouch cell. The schematic representation of chemical reaction taking place on cell operation is shown in Figure 17.2. While analysing the cell chemical reactions, the OER, where the Na^+ through separator diffuses into anolyte, and the ORR, where oxidized Na^+ from Na metal passes to anolyte while discharged. Electrochemical measurements showed the rechargeability of cells with enhanced properties of Pt/C coated cathode showing a 2.85 V discharge voltage and 84.3% discharge efficiency.

In aqueous SAB, the binder should offer better adhesiveness and should be hydrophobic. Weak adhesion properties of binder lead to detachment of catalyst and corrosion of current collectors. Baek et al. [34] investigated the applicability of catechol-derivative hydrophilic binders for the electrocatalyst–current collector interface and reported that, in addition to enhanced adhesion, electrochemical properties showed improvement. Improved adhesion resulted in enhanced wettability at catalyst–current collector interface. Furthermore, better adhesion contributed to the prevention of catalyst and cathode separation, which was observed from morphological analysis and from theoretical analysis. Electrochemical performance showed a shift in the positive side. Compared to commercial PVDF binder, the cell with catechol group containing binder showed 15% enhancement in cycling, reduced overpotential value, improved ORR performance by four times, and improvement in energy efficiency and power.

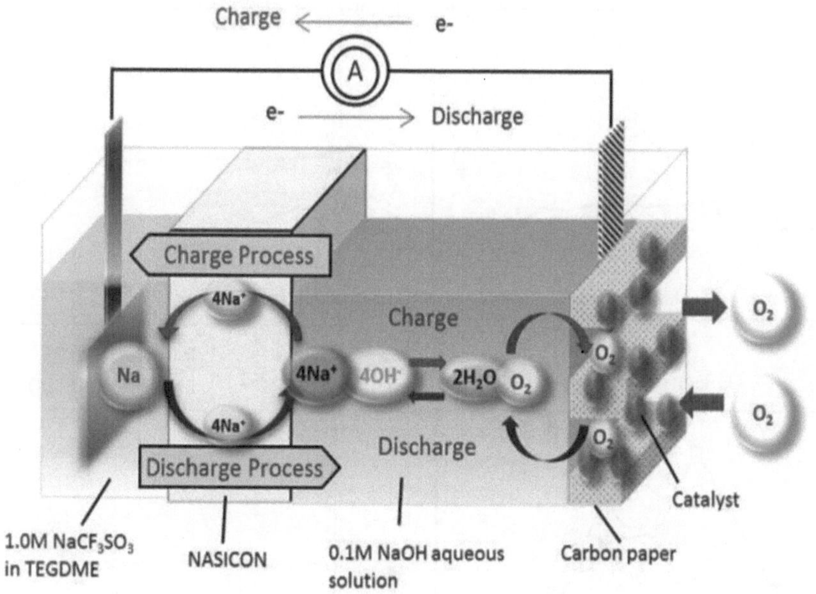

FIGURE 17.2 Schematic representation of aqueous sodium–air cell with the cell architecture, the process of charge–discharge, and various materials used. Adapted and reproduced with permission from Ref. [33]. Copyright © 2015 Elsevier.

17.2.2 Non-Aqueous Electrolytes

Morphology and composition of discharge products have a direct influence on the electrochemical performance of SABs. Das et al. [35] reported that by employing a mixture of O_2 and CO_2 as the feed for a primary SAB, an increment of 200%–300% hike in energy density can be achieved. Morphological analysis showed the presence of discharge products being filled at the air cathode, which have a chemical composition as Na_2O_2, Na_2CO_3, and sodium oxalate ($Na_2C_2O_4$). The cell was fabricated with non-aqueous electrolytes, namely, 1 M $NaClO_4$ in TEGDME/tetraglyme and an IL electrolyte with 0.75 M $NaCF_3SO_3$ in 1-ethyl-3-methylimidazolium trifluromethanesulphonate. Electrochemical analysis showed that IL electrolyte cells with 40% as CO_2 feed showed a capacity of 3,500 mAh g^{-1}, and cells with glyme, consisting of 63% CO_2 feed, showed a capacity of 2,882 mAh g^{-1} (Figure 17.3). These values are 2.6 and 2.1 times higher than that of a cell with 100% O_2 feed. The discharge profile of two identical tetraglyme-based cells cycled with 37% O_2 and 63% CO_2 feed showed similar profiles with a polarization of ~500 mV. However, from the alternating current (AC) impedance spectra, Na cells stored in O_2/CO_2 atmosphere showed less electrode–electrolyte interfacial stability on comparison to a cell stored in argon atmosphere (Figure 17.3).

17.2.3 Glyme-Based Electrolytes

Electrolyte influences the formation of stable SEI at the anode–electrolyte interface [20]. Stable SEI formation mitigates the issue of dendritic growth; thereby, the cycle life of the cell increases [20]. Seh et al. [36] developed an electrolyte containing $NaPF_6$ in glymes (mono-, di-, and tetraglymes), which enabled highly reversible and dendrite-hindered room temperature plating–stripping processes on Na metal anode. It was reported that when operated at a current density of 0.5 mA cm^{-2}, the cell showed 99.9% of Coulombic efficiency over 300 cycles. These results were believed to be due to the formation of an inorganic SEI layer, which majorly consisted of sodium fluoride (NaF) and Na_2O. The SEI layer further protected metallic Na from electrolyte solvent to be permeable and hence acted to be unfavourable for the dendrite growth. Lutz et al. [37] studied the discharge

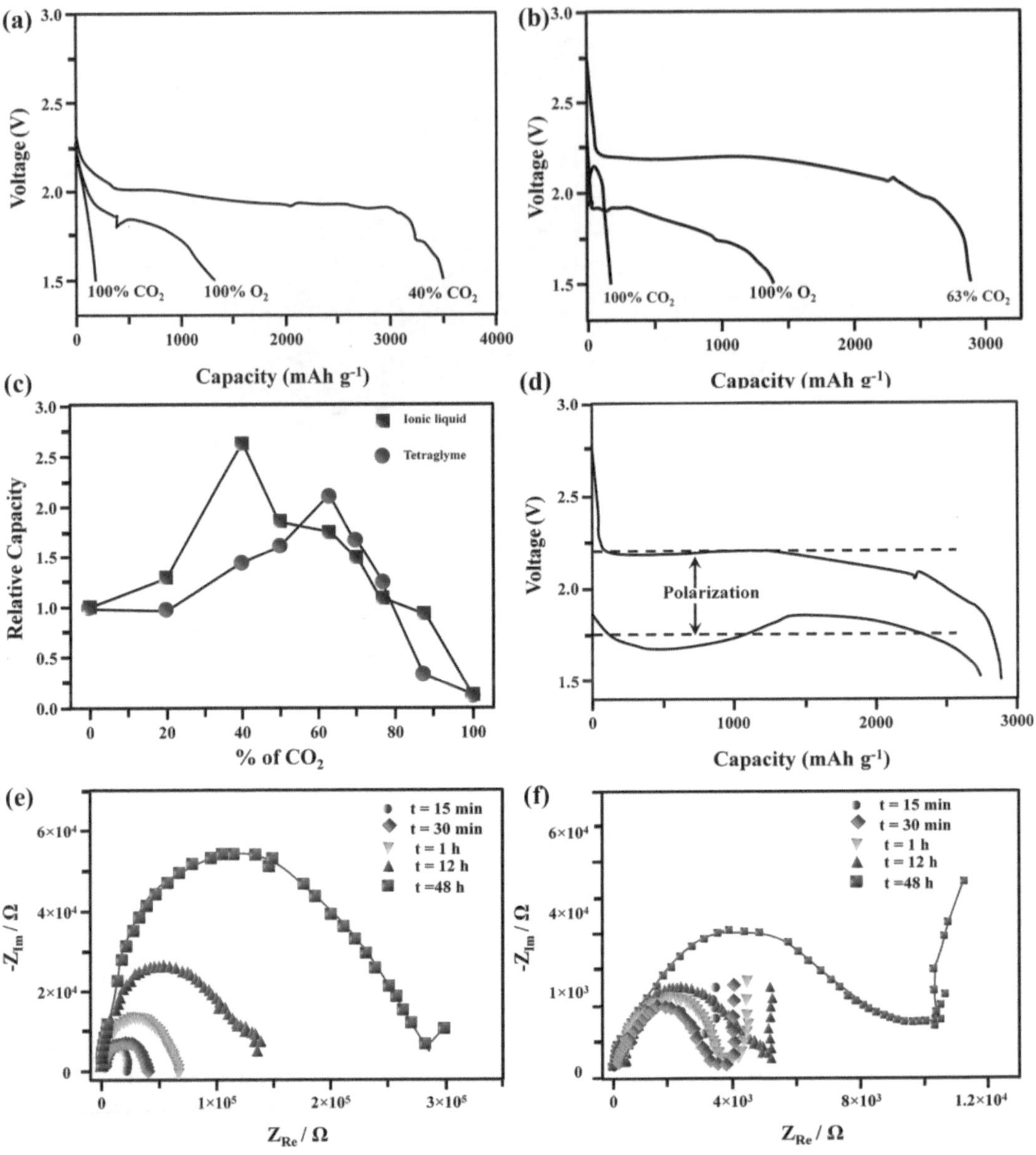

FIGURE 17.3 Discharge profiles of SAB with carbon dioxide (CO_2)/oxygen (O_2) mixture as feed. (a) Cell with ionic liquid electrolyte, (b) cell with tetraglyme electrolyte, (c) change in capacity with respect to the concentration of CO_2, (d) discharge profile for SAB with tetraglyme (feed consists of 37% O_2 and 63% CO_2), and (e and f) AC impedance spectra of SAB under argon atmosphere and CO_2/O_2 atmosphere, respectively. Adapted and reproduced with permission from Ref. [35]. Copyright © 2013 Elsevier.

capacity of SAB based on various glyme–ether solvents such as DME, DEGDME, and TEGDME and reported the significance of the low desolvation barrier needed for the growth of NaO_2. The work observed the chelation effects of Na^+ with the solvent. And it was observed that, as the chain length of ethers increased, a kinetic barrier was formed, which contributed to the death of premature cells by passivizing the electrode, and hence short-chain ethers promoted the cubic crystal formation and showed high capacity [37]. Li et al. [38] reported that, in glyme systems, chain length is one of the parameters, which regulates solvation properties and transportation of charges, thereby influencing battery performance. Ab initio molecular dynamics simulation results predicted that ionic

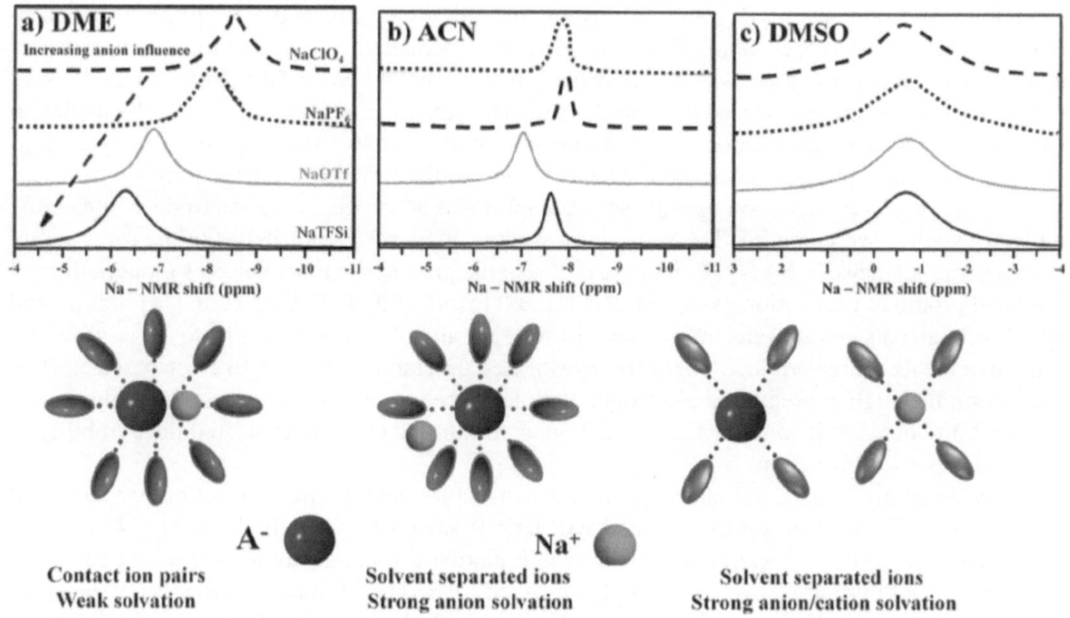

FIGURE 17.4 Visualization of the shift in the peaks of ^{23}Na-NMR of various solvents such as (a) DME,(b) ACN, and (c) DMSO. The blue line indicates ClO_4^-; the green colour corresponds to PF_6^-; the pink line denotes OTf^-; and orange corresponds to $TFSi^-$ anions. Adapted and reproduced with permission from Ref. [22]. Copyright © 2017 American Chemical Society.

conductivity is influenced by the ion association behaviour and hopping transport of the anionic species. Another work from the group investigated the influence of conductive electrolyte Na salt anion and related to the formation of NaO_2 in SAB [22]. By varying the Na salt and its solvation in different solvents were studied by sodium-nuclear magnetic spectroscopy (Na-NMR). Solvents used were DME, acetonitrile (ACN), dimethyl sulphoxide (DMSO), and tetrabutylammonium trifluoromethanesulphonate, and Na salts were NaOTf, NaTFSI, $NaPF_6$, and $NaClO_4$. From ^{23}Na-NMR, it was reported that Na^+ ion weakly solvates in DME, and in DMSO strong solvation of Na^+ is observed from the chemical shift and is shown in Figure 17.4. Moreover, the PF_6^- anion with DME formed SEI, which enhanced the cycling, and the opposite happened to the bis(trifluoromethane) sulfonimide anion ($TFSI^-$) anion. Overall, the work concluded that the selection of Na salt anion has a better role on the formation of SEI at the anode than the solvation of Na^+.

The chemical composition of discharge products also depends on the type of electrolytes used and influences the performance of SAB. Kim et al. [39] identified the reversible discharge products for carbonate-based and glyme-based SAB. It was reported that the major composition of reversible discharge products from carbonate-based SAB consisted of Na_2CO_3. On the contrary, for ether-based SABs, it was hydrated Na_2O_2. Both the cells delivered a capacity of 2,800 and 6,000 mAh g^{-1}. Zhao et al. [40] investigated the cell chemistry of non-aqueous SABs based on various electrolytes such as EC/PC, IL electrolytes such as *N*-methyl-*N*-propylpiperidinium bis-(trifluoromethansulphonyl) imide (PP13TFSI), and ether-based electrolyte, TEGDME containing $NaSO_3CF_3$ as Na salt. The air cathode consisted of vertically aligned carbon nanotubes, where the discharged products were deposited. It was observed that in a carbonate electrolyte (with p*Ka* of ~20.9), only 8.92% NaO_2 as discharge product could be detected, and the majority was Na_2CO_3 and sodium alkyl carboxylate. In PP13TFSI with p*Ka* ≈ 30, 38.5% consisted of NaO_2 with the remaining as other parasitic products. TEGDME with p*Ka* of ~46–52 showed better stability with 92.7% discharge product as NaO_2. The work showed the important relationship between p*Ka* of electrolyte for enhancing the performance and efficiency of SAB.

Faktorovich-Simon et al. [41] found that ageing of electrolytes affects the electrochemical properties of a SAB. The processes of ORR and OER were evaluated and related to the performance of SAB. The electrolyte used was polyethylene glycol dimethyl ether 500 (PEGDME500) with NaTFSI, where the O_2 was purged prior to the analysis. Ageing of solvent influences the oxidation overpotentials of reduction products. Preparation of polymer involves the usage of sodium alcoholate; however as the conversion was incomplete, residual sodium alcoholate in the polymer formed a stable complex with NaO_2. Upon ageing, NaO_2 complex gets adsorbed on the electrode, contributing to high oxidation overpotential. The stability of sodium superoxide contributes to the reversibility and energy efficiency of a SAB [42]. Storage of discharge products in ether solvents reduces the stability and promotes dissolution, which affects the cell reversibility [43]. Park et al. [44] highlighted the role of high concentrated electrolyte, with a lower amount of solvent, to mitigate the issue of dissolution of the discharge product. The work investigated the stability of NaO_2 in ether-based electrolyte and studied high concentrated electrolyte with low free solvent for enhancing the performance and durability of SAB. It was concluded that high concentrated electrolyte shifted the durability of the cell in a positive direction.

In any metal-air battery, the quantity, chemical structure, and nature of reaction products will indicate the mechanism inside the cell and will help in unveiling the cell chemistry [45]. In this aspect Liu et al. [46] used operando synchrotron radiation powder X-ray diffraction to study the characteristics of discharge products in a glyme-based SAB. The formation of the discharge product, NaO_2, showed an efficiency of 90% in SAB and is dependent on salts. Moreover, the Gutmann donor number of the electrolyte has a crucial role in the performance of SAB. During discharge it was observed that the cell with $NaPF_6$ salt showed lower capacity but increased in the domain size, and, when the salt was NaOTf, high capacity was shown. Qin et al. [47] investigated the degradation kinetics of NaO_2 by both experimental and theoretical studies in a diglyme-based SAB. The work studied the discharge, idling, and charge of different intervals to determine the rates of degradation and precipitation. For the analysis, the fabricated cell consisted of a Celgard separator in a diglyme electrolyte with 0.5 M $NaCF_3SO_3$. It was evidenced that, to reduce the degradation of NaO_2, the solvent should be selected in such a way that it should be with minimum solubility of NaO_2. The storage life of the cell showed enhancement when the electrolyte was circulated between the chambers.

With the background of stability of glymes with the metallic Na anode and with quasi-stability over O_2^- attack, glyme-based solvent had been a good candidate for Na-based batteries. Vitoriano et al. [48] studied the usage of glyme-based solvents and their effects on the performance of SAB by combining experimental and theoretical studies. The work correlated the chain length of various glymes, namely, DME, DEGDME, and TEGDME, having different chain lengths on the cell performance. For the electrochemical study, SAB was prepared using Na metal as the anode, the air cathode consisting of a gas diffusion layer (GDL), and Celgard H2010 as the separator, and glyme electrolyte with $NaClO_4$ as Na salt and purged with oxygen at 1 atm. From various analyses, viscosity played a crucial role in the transport of Na^+ and in DME. Compared to TEGDME, ion transport showed fastest in DME. TEGDME with the bulky and four chelation sites contributed to the sluggish transport of ions. However, the DEGDME system showed better cyclability, and considering transport and kinetics of reaction viscosity and stability, its candidature can be better in both high and low salt concentrations. The overall effect of various parameters is represented in Figure 17.5.

Ma et al. [49] investigated the effect of saturation of ether-based electrolyte with the performance of SAB. Na salt, NaOTf, of 0.5 M concentration was added to the TEGDME solvent. The work reported that in supersaturated electrolytes, dendritic formation was suppressed and the reason for that is the high Na^+ flux. Moreover, saturated electrolytes prevented the dissolution of NaO_2, contributed to the easy shuttling of O_2^- anion, and showed better cycling performance of the cell. To mitigate dendrite formation in Na anode, Ma et al. [50] investigated the effect of coating the bismuth layer over the anode and fabricated SAB. The hydrophilic coating over Na metal catalysed $NaSO_3CF_3$ in diglyme to penetrate the anode easily. SAB with bismuth-coated Na anode showed stability in discharge capability for 50 cycles with 1,000 mAh g^{-1} and a coulombic efficiency of 96.3%.

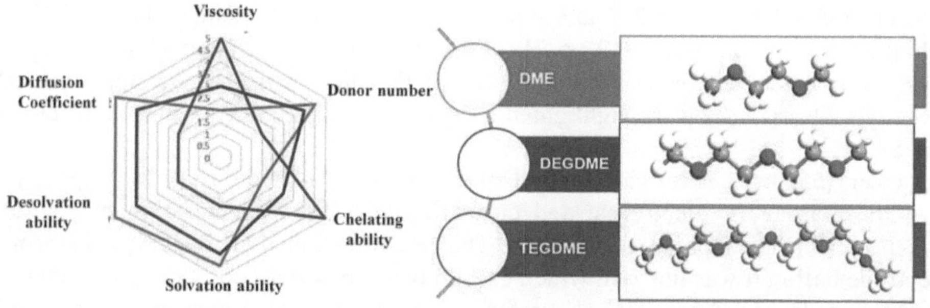

FIGURE 17.5 Schematic representation showing the parameters of glyme-based solvents. Adapted and reproduced with permission from Ref. [48]. Copyright © 2020 Elsevier.

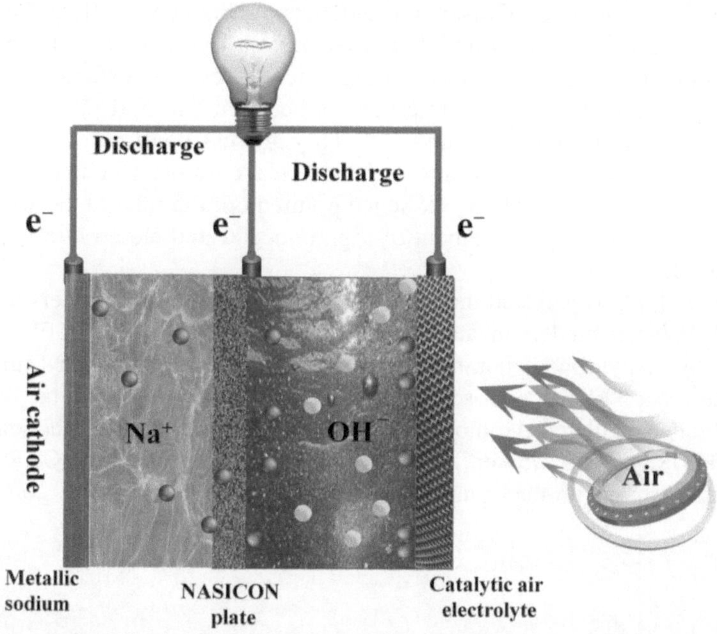

FIGURE 17.6 A schematic representation of hybrid SAB architecture, with anolyte as an aprotic electrolyte and catholyte as an aqueous electrolyte. Adapted and reproduced with permission from Ref. [54]. Copyright © 2016 Elsevier.

17.2.3.1 IL Electrolytes

Room-temperature IL electrolytes have a wide electrochemical window and high ionic conductivity with lower volatility, thereby ensuring smooth battery operation [51]. These electrolytes were found to be a good candidate for various metal-air batteries [52]. Inherently, IL electrolytes possess thermal stability with a wide liquid range, which allows operation at a wide temperature range. Coupled with this, the non-flammability of IL ensures enhanced safety compared to organic solvent base systems, allowing them as a potential candidate for large-scale industrially viable applications [53].

Kang et al. [54] were the first to report a hybrid SAB based on IL of bis(fluorosulphonyl) amide-based Na[FSA–C_2C_1im][FSA] as electrolyte and measured electrochemical properties at various temperatures: 25°C, 50°C, and 70°C. The hybrid cell fabricated with a solid electrolyte, consisting of NASICON that separated IL anolyte containing Na[FSA] salt and NaOH catholyte, is shown in Figure 17.6. The air cathode consisted of a GDL with Mn_3O_4/C as the catalyst. From the electrochemical performance analysis, the cell showed a theoretical capacity of 99%, which is approximately 835 mAh g^{-1} with 27.6 mW cm^{-2} as power density at a temperature of 50°C, which was the highest reported in SAB till 2016. However, the result of discharge–charge plot showed a flat charge over potential with superiority of OER than the ORR, which showed the requirement of a better catalyst. Overall, the work highlighted the potential applicability of aprotic IL electrolytes for hybrid SAB.

Zhang et al. [55] used *N*-butyl-*N*-methyl pyrrolidinium bis(trifluoromethylsulphonyl)imide [C4mpyr][TFSI] as an IL electrolyte and correlated with the concentration of Na salt with respect to the performance properties of SAB. The electrochemical analysis was performed on a three-electrode half-cell was made in which carbon paper-based air cathode, platinum coil wire as counter electrode, and Ag/AgTfO as a reference electrode. The study reported that with the increase in the concentration of NaTFSI, the discharge capacity enhanced and also a reduction in overpotential led to better cyclability in the SAB system. Morphological analysis performed over the cathode which correlated the effect of Na^+ ion with ORR concluded that the enhancement is related to the morphology of discharge products. At high concentrated Na salt (NaTFSI) containing electrolyte, a dense coverage of smaller particles with spherical morphologies was deposited. Moreover, at high concentrations of Na^+, the O_2^- forms coordination bonds with Na^+ instead of with $[C4mpyr]^+$. In addition, interactions between Na^+ and $[TFSI]^-$ ions contribute to the ion pair formation between Na^+ and O_2^-. This process contributes to the formation of deposition over the cathode rather than in the electrolyte. A simple schematic representation of the same is shown in Figure 17.7. The work hence pointed out the relation between Na^+ concentration with electrodes and the requirement of high concentrated electrolyte for enhancing the performance of SAB.

Recently, Ha et al. [56] investigated the effect of a series of polymeric binders for air cathode in an IL-based SAB. Polymer binders included in the study were PVDF and PTFE and an IL-based polymer binder, poly(diallyldimethylammonium) bis(trifluoromethanesulphonyl)imide (PDADMA TFSI). Solid-state ^{23}Na-NMR spectroscopy, Raman spectroscopy, and other characterization showed that the binder regulated the morphology and the composition of reaction products. The polymeric IL, PDADMA TFSI, worked as a better binder and enhanced the ORR with NaO_2 and Na_2O_2 as the main discharge products, and hence acted as a functional binder compared to PVDF

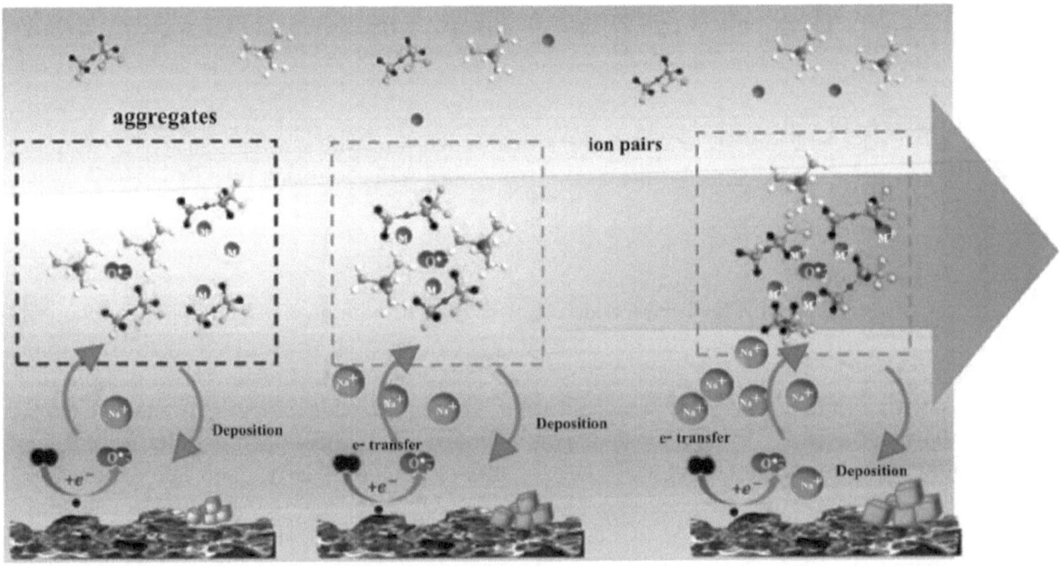

FIGURE 17.7 Schematic representation showing the relation between NaTFSI/[C4mpyr][TFSI] electrolyte concentration on ORR and the deposition. Adapted and reproduced with permission from Ref. [55]. Copyright © 2018 American Chemical Society.

and PTFE. PDADMA TFSI showed the lowest overpotential of 0.9 V due to the better interfaces between binder–electrode and electrolyte due to the effect of the IL environment, which contributed to enhancement in electrochemical performance.

17.3 MODIFICATION OF SEPARATOR

As safety is a major concern in batteries, separator functions limit the O_2 crossover and block dendrites, thereby enhancing the cycle life of SAB. Ansari et al. [57] developed a reinforced glass microfibre as a separator to minimize the O_2 crossover and suppress the dendrite formation in SAB. The separator structure was fabricated by the infusion of Al_2O_3 in the presence of an organic binder into a glass microfibre layer. The schematic of the preparation method and the scanning electron microscopy (SEM) images of the prepared separator are shown in Figure 17.8. The ability of the separator to hinder the oxygen crossover from the electrolyte to the anode was attributed to the long life of SAB. Moreover, the dried sodium triflate in DME-based electrolyte transform to form NaO_2 as a stable discharge product. Due to the effective performance of both the separator and the ultra-dried electrolyte, the cell showed stability of more than 400 cycles with a lower overpotential of less than 40 mV.

Trilayer separator based on titanium dioxide (TiO_2) nanoparticles to protect the Na anode in SAB was investigated by Yang et al. [58]. The reactivity of TiO_2 nanoparticles with Na dendrites inhibits their further growth to the cathodic surface, potentially mitigates the dendritic penetration, and hence alleviates cell failure. The trilayer separator consisted of TiO_2 sandwiched between polypropylene layers. Figure 17.9 is a schematic illustration of a possible dendrite formation mechanism on normal SAB and restriction of dendrite growth on trilayer separator-based SAB. It was observed from SEM analysis that the separator was filled with TiO_2 nanoparticles (Figure 17.9). SAB with TiO_2 nanoparticle-sandwiched separator showed an enhanced cycling from 82 to 137 cycles, at a discharge capacity of 1,000 mAh g^{-1} at 200 mA g^{-1} current density, compared to cells without sandwich separator.

FIGURE 17.8 (a) Schematic of the preparation method of RGMF; (b) SEM image of glass microfibre (GMF); and (c) SEM image of GMF after the infusion of the binder and oxide. Adapted and reproduced with permission from Ref. [57]. Copyright © 2018 Wiley-VCH.

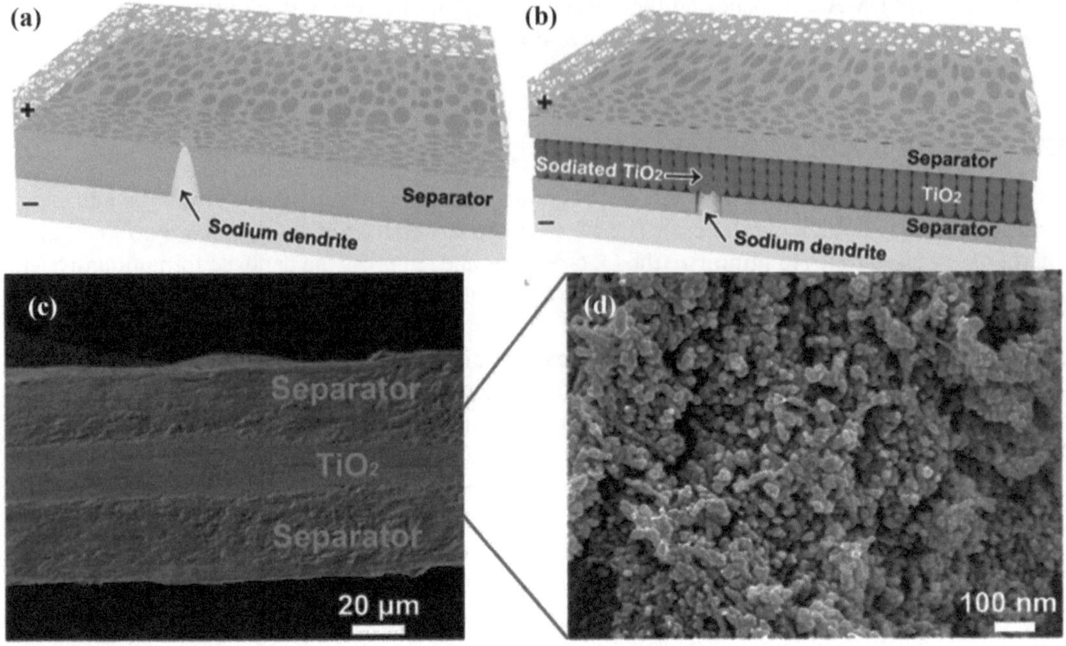

FIGURE 17.9 Schematic illustration of (a) dendrite growth to cathode causing cell failure in SAB, (b) inhibited dendrite growth by sodiated TiO_2-based separator in SAB, and (c and d) morphological analysis by scanning electron microscopy of titanium dioxide nanoparticle-sandwiched separator at various resolutions. Adapted and reproduced with permission from Ref. [58]. Copyright © The Royal Society of Chemistry.

Ma et al. [59] devised a low-cost method for inhibiting the growth of Na dendrites in SAB. The researchers created PVDF films with various structures, PTFE, and polyethylene oxide (PEO) films for comparative studies to better evaluate the impact of structure on dendrite suppression efficiency. When compared to other prepared interlayers, the fabricated electrospun fibrous-PVDF interlayer with a fibre diameter of 200–300 nm with a well-interlinked porous structure demonstrated superior properties. The fibrous configuration of PVDF acted as a barrier to the initiation of dendrites, and the strong C–F groups directed uniform Na^+ ion deposition by providing a higher affinity for Na than the C–O functional groups in PEO. Furthermore, when compared to PEO, PTFE, and other PVDF film-based SAB, the greater wettability of fibrous PVDF with electrolyte played a prominent part in hindering dendrite growth and achieving good electrochemical performance.

Sun et al. [60] developed a low-cost system using a cuprous oxide-modified nickel foam host for Na anode protection from dendrite growth in SAB. The host matrix was developed through a displacement reaction in an aqueous system. With a nanoflower morphology, the cuprous oxide reacted with Na^+ to minimize the nucleation energy during the plating process, to achieve a uniform Na deposition. Moreover, the 3D porous structure of Ni foam played a key role in moderating the deposition behaviour of Na. These synergistic effects of both cuprous oxide and Ni foam resulted in achieving a stable performance over 1,000 cycles at a current density of 1 mA cm^{-2}.

17.4 CONCLUSION

As the demand for sustainable energy storage solutions is the need of the hour, SABs have received immense research attention considering sustainability and cost-effective technology. Abundance of Na sources, environmental benignity, and high capacity of SABs make them a suitable candidate for industrial mass production technology. As metallic Na, being a reactive soft metal, is used as an anode for SABs, it has inherent limitations. Reaction with electrolytes, formation of dendrites, unstable SEI formation, and non-uniform Na deposition are the major drawbacks. Hence, various strategies have been postulated, which preferentially protect Na from corrosion and guide uniform deposition. Electrolytes also play an important role in the electrochemical performance of SABs. The composition of a discharged product depends on the solvent and salt used, which determines the reversibility and cycle life of the cell. Hence engineering electrode–electrolyte interface also catalyses the performance. To unveil the complexities of SAB, a complete understanding of each battery component is prime important. It is expected that by a multifaced approach to stabilizing anode, engineering electrolyte, and tailoring cathode, there can be a proliferating advancement in the SAB technology.

REFERENCES

1. Li M, Lu J, Chen Z, Amine K (2018) 30 years of lithium-ion batteries. *Adv Mater* 30:1800561. https://doi.org/10.1002/adma.201800561
2. Jishnu NS, Vineeth SK, Das A, et al (2021) Electrospun PVdF and PVdF-co-HFP-based blend polymer electrolytes for lithium ion batteries. In: Balakrishnan NTM, Prasanth R (eds) *Electrospinning for Advanced Energy Storage Applications*, 1st ed. Springer, Singapore, pp. 201–234
3. Delmas C (2018) Sodium and sodium-ion batteries: 50 years of research. *Adv Energy Mater* 8:1703137. https://doi.org/10.1002/aenm.201703137
4. Masias A, Marcicki J, Paxton WA (2021) Opportunities and challenges of lithium ion batteries in automotive applications. *ACS Energy Lett* 6:621–630. https://doi.org/10.1021/acsenergylett.0c02584
5. Sungjemmenla J, Soni CB, Vineeth SK, Kumar V (2022) Exploration of the unique structural chemistry of sulfur cathode for high-energy rechargeable beyond-Li batteries. *Adv Energy Sustain Res* 3:2100157. https://doi.org/10.1002/aesr.202100157
6. Sungjemmenla J, Soni CB, Vineeth SK, Kumar V (2021) Unveiling the physiochemical aspects of the matrix in improving sulfur-loading for room-temperature sodium-sulfur batteries. *Mater Adv* 2:4165–4189. https://doi.org/10.1039/D1MA00247C
7. Vineeth SK, Soni CB, Sanjaykumar C, Yamauchi Y, Han M, Kumar V (2023) A quasi-solid state polymer electrolyte for high-rate and long-life sodium-metal batteries. *J Energy Storage*, 73:108780. https://doi.org/10.1016/j.est.2023.108780
8. Hwang J-Y, Myung S-T, Sun Y-K (2017) Sodium-ion batteries: present and future. *Chem Soc Rev* 46:3529–3614. https://doi.org/10.1039/C6CS00776G
9. Abraham KM (2020) how comparable are sodium-ion batteries to lithium-ion counterparts? *ACS Energy Lett* 5:3544–3547. https://doi.org/10.1021/acsenergylett.0c02181
10. Zheng Z, Wu C, Gu Q, et al (2021) Research progress and future perspectives on rechargeable Na-O_2 and Na-CO_2 batteries. *Energy Environ Mater* 4:158–177. https://doi.org/10.1002/eem2.12139
11. Yadegari H, Sun X (2018) Recent advances on sodium-oxygen batteries: a chemical perspective. *Acc Chem Res* 51:1532–1540. https://doi.org/10.1021/acs.accounts.8b00139
12. Aurbach D, McCloskey BD, Nazar LF, Bruce PG (2016) Advances in understanding mechanisms underpinning lithium-air batteries. *Nat Energy* 1:16128. https://doi.org/10.1038/nenergy.2016.128
13. Enterría M, Reynaud M, Paredes JI, et al (2022) Driving the sodium-oxygen battery chemistry towards the efficient formation of discharge products: The importance of sodium superoxide quantification. *J Energy Chem* 68:709–720. https://doi.org/10.1016/j.jechem.2021.12.014
14. Yin W-W, Fu Z-W (2017) A highly efficient bifunctional heterogeneous catalyst for morphological control of discharged products in Na-air batteries. *Chem Commun* 53:1522–1525. https://doi.org/10.1039/C6CC09348E
15. Zhao S, Wang C, Du D, et al (2021) Bifunctional effects of cation additive on Na-O_2 batteries. *Angew Chemie – Int Ed* 60:3205–3211. https://doi.org/10.1002/anie.202012787
16. Wu S, Qiao Y, Jiang K, et al (2018) Tailoring sodium anodes for stable sodium-oxygen batteries. *Adv Funct Mater* 28:1706374. https://doi.org/10.1002/adfm.201706374
17. Vineeth SK, Tebyetekerwa M, Liu H, Soni CB, Zhao XS, Kumar V (2022) Progress in the development of solid-state electrolytes for reversible room-temperature sodium–sulfur batteries. *Mater Adv*, 3(16):6415–6440. https://doi.org/10.1039/D2MA00428C
18. Yang X, Peng C, Hou M, et al (2022) Rational design of electrolyte solvation structures for modulating $2e^-/4e^-$ transfer in sodium-air batteries. *Adv Funct Mater* 2201258:1–11. https://doi.org/10.1002/adfm.202201258

19. Soni CB, Sungjemmenla J, Vineeth SK, Kumar V (2022) Challenges in regulating interfacial-chemistry of the sodium-metal anode for room-temperature sodium-sulfur batteries. *Energy Storage* 4:e264. https://doi.org/10.1002/est2.264
20. Vineeth SK, Soni CB, Sun Y, et al (2022) Implications of Na-ion solvation on Na anode-electrolyte interphase. *Trends Chem* 4:48–59. https://doi.org/10.1016/j.trechm.2021.11.002
21. Vineeth SK, Kumar V (2023) Introduction: Sodium-sulfur battery technology. In *Room-Temperature Sodium-Sulfur Batteries* (pp. 1–16). CRC Press. https://doi.org/10.1201/9781003388067-1
22. Lutz L, Alves Dalla Corte D, Tang M, et al (2017) Role of electrolyte anions in the Na-O_2 battery: implications for NaO_2 solvation and the stability of the sodium solid electrolyte interphase in glyme ethers. *Chem Mater* 29:6066–6075. https://doi.org/10.1021/acs.chemmater.7b01953
23. Vineeth SK, Yamauchi Y, Kumar V (2023) Electrolytes for room-temperature sodium-sulfur batteries: A holistic approach to understand solvation. In *Room-Temperature Sodium-Sulfur Batteries* (pp. 79–114). CRC Press. https://doi.org/10.1201/9781003388067-4
24. Okoshi M, Chou C-P, Nakai H (2018) Theoretical analysis of carrier ion diffusion in superconcentrated electrolyte solutions for sodium-ion batteries. *J Phys Chem B* 122:2600–2609. https://doi.org/10.1021/acs.jpcb.7b10589
25. Eng AYS, Soni CB, Lum Y, et al (2022) Theory-guided experimental design in battery materials research. *Sci Adv* 8:eabm2422. https://doi.org/10.1126/sciadv.abm2422
26. Hashimoto T, Hayashi K (2015) Aqueous and nonaqueous sodium-air cells with nanoporous gold cathode. *Electrochim Acta* 182:809–814. https://doi.org/10.1016/j.electacta.2015.09.153
27. Yin W, Fu Z (2017) The potential of Na-air batteries. *ChemCatChem* 9:1545–1553. https://doi.org/10.1002/cctc.201600646
28. Sun Q, Yang Y, Fu ZW (2012) Electrochemical properties of room temperature sodium-air batteries with non-aqueous electrolyte. *Electrochem Commun* 16:22–25. https://doi.org/10.1016/j.elecom.2011.12.019
29. Das SK, Lau S, Archer LA (2014) Sodium-oxygen batteries: a new class of metal-air batteries. *J Mater Chem A* 2:12623–12629. https://doi.org/10.1039/c4ta02176b
30. Yadegari H, Sun X (2020) Sodium-oxygen batteries: recent developments and remaining challenges. *Trends Chem* 2:241–253. https://doi.org/10.1016/j.trechm.2019.12.003
31. Hayashi K, Shima K, Sugiyama F (2013) A mixed aqueous/aprotic sodium/air cell using a NASICON ceramic separator. *J Electrochem Soc* 160:A1467–A1472. https://doi.org/10.1149/2.067309jes
32. Liang F, Hayashi K (2015) A high-energy-density mixed-aprotic-aqueous sodium-air cell with a ceramic separator and a porous carbon electrode. *J Electrochem Soc* 162:A1215–A1219. https://doi.org/10.1149/2.0421507jes
33. Sahgong SH, Senthilkumar ST, Kim K, et al (2015) Rechargeable aqueous Na-air batteries: Highly improved voltage efficiency by use of catalysts. *Electrochem Commun* 61:53–56. https://doi.org/10.1016/j.elecom.2015.10.004
34. Baek MJ, Choi J, Wi TU, et al (2022) Strong interfacial energetics between catalysts and current collectors in aqueous sodium-air batteries. *J Mater Chem A* 10:4601–4610. https://doi.org/10.1039/d2ta00329e
35. Das SK, Xu S, Archer LA (2013) Carbon dioxide assist for non-aqueous sodium-oxygen batteries. *Electrochem Commun* 27:59–62. https://doi.org/10.1016/j.elecom.2012.10.036
36. Seh ZW, Sun J, Sun Y, Cui Y (2015) A highly reversible room-temperature sodium metal anode. *ACS Cent Sci* 1:449–455. https://doi.org/10.1021/acscentsci.5b00328
37. Lutz L, Yin W, Grimaud A, et al (2016) High Capacity Na-O_2 batteries: key parameters for solution-mediated discharge. *J Phys Chem C* 120:20068–20076. https://doi.org/10.1021/acs.jpcc.6b07659
38. Li K, Subasinghege Don V, Gupta CS, et al (2021) Effect of anion identity on ion association and dynamics of sodium ions in non-aqueous glyme based electrolytes - OTf vs TFSI. *J Chem Phys* 154:184505. https://doi.org/10.1063/5.0046073
39. Kim J, Lim H-D, Gwon H, Kang K (2013) Sodium-oxygen batteries with alkyl-carbonate and ether based electrolytes. *Phys Chem Chem Phys* 15:3623. https://doi.org/10.1039/c3cp43225d
40. Zhao N, Guo X (2015) Cell chemistry of sodium-oxygen batteries with various nonaqueous electrolytes. *J Phys Chem C* 119:25319–25326. https://doi.org/10.1021/acs.jpcc.5b09187
41. Faktorovich-Simon E, Natan A, Peled E, Golodnitsky D (2018) Oxygen redox processes in PEGDME-based electrolytes for the Na-air battery. *J Solid State Electrochem* 22:1015–1022. https://doi.org/10.1007/s10008-017-3843-5
42. Landa-Medrano I, Pinedo R, Bi X, et al (2016) New insights into the instability of discharge products in Na-O_2 batteries. *ACS Appl Mater Interfaces* 8:20120–20127. https://doi.org/10.1021/acsami.6b06577
43. Xiao N, Rooney RT, Gewirth AA, Wu Y (2018) The long-term stability of KO_2 in K-O_2 batteries. *Angew Chemie – Int Ed* 57:1227–1231. https://doi.org/10.1002/anie.201710454

44. Park H, Kim J, Lee MH, et al (2018) Highly durable and stable sodium superoxide in concentrated electrolytes for sodium-oxygen batteries. *Adv Energy Mater* 8:1801760. https://doi.org/10.1002/aenm.201801760
45. Choudhury S, Wan CTC, Al Sadat WI, et al (2017) Designer interphases for the lithium-oxygen electrochemical cell. *Sci Adv* 3:1–12. https://doi.org/10.1126/sciadv.1602809
46. Liu C, Rehnlund D, Brant WR, et al (2017) Growth of NaO_2 in highly efficient Na-O_2 batteries revealed by synchrotron in operando x-ray diffraction. *ACS Energy Lett* 2:2440–2444. https://doi.org/10.1021/acsenergylett.7b00768
47. Qin B, Chan K-Y, Li C-YV (2020) Studies of superoxide degradation kinetics and electrolyte management for a reversible NaO_2 battery. *ACS Sustain Chem Eng* 8:4317–4324. https://doi.org/10.1021/acssuschemeng.9b07500
48. Ortiz Vitoriano N, Ruiz de Larramendi I, Sacci RL, et al (2020) Goldilocks and the three glymes: how Na^+ solvation controls Na-O_2 battery cycling. *Energy Storage Mater* 29:235–245. https://doi.org/10.1016/j.ensm.2020.04.034
49. Ma JL, Zhang WC, Wang XD, et al (2021) Revealing the mechanism of saturated ether electrolyte for improving the long-cycling stability of Na-O_2 batteries. *Nano Energy* 84:105927. https://doi.org/10.1016/j.nanoen.2021.105927
50. Ma M, Lu Y, Yan Z, Chen J (2019) In situ synthesis of a bismuth layer on a sodium metal anode for fast interfacial transport in sodium-oxygen batteries. *Batter Supercaps* 2:663–667. https://doi.org/10.1002/batt.201900020
51. Vineeth SK, Sreeram P, Vlad A, Joy R, Raghavan P, Pullanchiyodan A (2023) Polymer blend nanocomposite electrolytes for advanced energy storage applications. In *Polymer Blend Nanocomposites for Energy Storage Applications* (pp. 203–238). Elsevier. https://doi.org/10.1016/B978-0-323-99549-8.00016-9
52. Soavi F, Monaco S, Mastragostino M (2013) Catalyst-free porous carbon cathode and ionic liquid for high efficiency, rechargeable Li/O_2 battery. *J Power Sources* 224:115–119. https://doi.org/10.1016/j.jpowsour.2012.09.095
53. Monti D, Jónsson E, Palacín MR, Johansson P (2014) Ionic liquid based electrolytes for sodium-ion batteries: Na^+ solvation and ionic conductivity. *J Power Sources* 245:630–636. https://doi.org/10.1016/j.jpowsour.2013.06.153
54. Kang Y, Liang F, Hayashi K (2016) Hybrid sodium-air cell with Na[FSA-C2C1im][FSA] ionic liquid electrolyte. *Electrochim Acta* 218:119–124. https://doi.org/10.1016/j.electacta.2016.09.113
55. Zhang Y, Ortiz-Vitoriano N, Acebedo B, et al (2018) Elucidating the impact of sodium salt concentration on the cathode-electrolyte interrrolidinium-based ionic liquid electrolyte in a half-cell has been explored in this work. Increasing the concentration of sodium salt in an ionic liquid electrolyte produc. *J Phys Chem C* 122:15276–15286. https://doi.org/10.1021/acs.jpcc.8b02004
56. Ha TA, Li H, Wang X, et al (2021) Functional binders based on polymeric ionic liquids for sodium oxygen batteries using ionic liquid electrolytes. *ACS Appl Energy Mater* 4:434–444. https://doi.org/10.1021/acsaem.0c02352
57. Ansari Y, Virwani K, Yahyazadeh S, et al (2018) A highly stable sodium-oxygen battery using a mechanically reinforced membrane. *Adv Energy Mater* 8:1802603. https://doi.org/10.1002/aenm.201802603
58. Yang H, Sun J, Wang H, et al (2018) A titanium dioxide nanoparticle sandwiched separator for Na-O_2 batteries with suppressed dendrites and extended cycle life. *Chem Commun* 54:4057–4060. https://doi.org/10.1039/C8CC00993G
59. Ma J-L, Yin Y-B, Liu T, et al (2018) Suppressing Sodium dendrites by multifunctional polyvinylidene fluoride (PVDF) interlayers with nonthrough pores and high flux/affinity of sodium ions toward long cycle life sodium oxygen-batteries. *Adv Funct Mater* 28:1703931. https://doi.org/10.1002/adfm.201703931
60. Sun J, Zhang M, Ju P, et al (2020) Long-life sodium metal anodes achieved by cuprous oxide-modified ni foam host. *Energy Technol* 8:1901250. https://doi.org/10.1002/ente.201901250

18 Rechargeable Zn-Air Battery *Current Challenges and Opportunities*

Kee Wah Leong, Yifei Wang, and Dennis Y.C. Leung

18.1 INTRODUCTION

To date, lithium-ion batteries are considered the most widely recognized battery technology in the energy and electronics industry. With a superior specific power of over 800 W kg^{-1} [1] and up to 5,000 stable discharge–charge cycles [2], Li-ion batteries enabled a mass deployment of electric vehicles with promising potential to replace conventional internal combustion engine vehicles. Li-ion batteries are still limited by a practical energy density of around 250 Wh kg^{-1} [3]. The expensive and toxic lithium metal also poses cost concerns and environmental threats for widespread utilization. The shortcomings of Li-ion batteries have motivated a rising research interest in alternative battery technologies, such as the metal-air battery.

Powered by an air-breathing cathode, metal-air batteries exhibit a higher theoretical energy density than Li-ion batteries due to weight reduction at the cathode. Table 18.1 compares the distinct theoretical properties of metal-air batteries with different earth-abundant metals (Zn, Fe, Al, Mg) as anodes. While Al and Mg have the highest energy densities and theoretical cell voltages, they are more prone to severe corrosion and self-discharge in aqueous electrolytes and are not rechargeable. On the contrary, Zn and Fe are more stable in aqueous electrolytes due to their lower activity, enabling a high rechargeability [4,5]. In particular, Zn-air batteries exhibit a remarkable volumetric energy density of 6,136 Wh L^{-1} [5], which has attracted the attention of many researchers.

Compared to the Li-ion battery, Zn-air batteries exhibit a lower theoretical specific capacity, but the practical specific capacity achieved is actually higher than many currently commercialized Li-ion batteries. Apart from the capacity, Table 18.2 shows that the currently achievable energy density of the Zn-air battery is even superior to the Li-ion battery. Besides, the Zn metal is more earth abundant and cheaper than the Li metal, making Zn-air batteries an attractive alternative to Li-ion batteries.

Despite the promise of Zn-air batteries, the currently achievable cycle life is still far from enabling practical operations, as shown in Table 18.2. Since the discovery of Zn-air batteries in

TABLE 18.1
Comparison of the Theoretical Energy Densities and Nominal Cell Voltage of Different Metal-Air Batteries [5]

	Specific Energy Density (Wh kg^{-1})	Volumetric Energy Density (Wh L^{-1})	Nominal Cell Voltage (V)
Zn-air	1,218	6,136	1.66
Fe-air	1,080	3,244	1.28
Al-air	5,779	10,437	2.71
Mg-air	5,238	9,619	3.09

 DOI: 10.1201/9781003310174-18

TABLE 18.2
Comparison of Battery Performance, Metal Abundance, and Price of Li-Ion and Zn-Air Batteries to Date [2, 3, 6–10]

	Theoretical/Currently Achievable Specific Energy Density (Wh kg^{-1})	Theoretical/Currently Achievable Specific Capacity (mAh g^{-1})	Currently Achievable Cycle Life	Metal abundance (Fraction of Earth's Crust)	Cost (USD $ kWh^{-1})
Li-ion	400/100–250	1,800/100–400	5,000 cycles	1.80×10^{-5}	250
Zn-air	1,218/350–500	819/700–800	150 cycles	7.60×10^{-5}	100

the nineteenth century, their commercialization has been limited to primary cells for low-power electronic devices, such as hearing aids [11]. The rechargeability had been largely hindered by irreversible Zn and inefficient electrochemical kinetics. However, in the recent decade, novel material advances and engineering techniques offered new perspectives that revived interest in Zn-air batteries, propelling exciting research progress.

18.2 BATTERY CONFIGURATION AND MECHANISM

A Zn-air battery is composed of the following components: a Zn anode, an aqueous alkaline electrolyte, an ion separator, and an air cathode, consisting of a catalyst layer and a gas diffusion layer. As illustrated by Figure 18.1, the oxygen reduction reaction (ORR) occurs at the air cathode when the battery undergoes discharge, where oxygen in the atmosphere reacts with the electrolyte to form OH^- ions. The OH^- would then combine with Zn^{2+} ions, the oxidation product at the Zn anode, and generate zincate ions ($Zn(OH)_4^{2-}$). After several repeated cycles, $Zn(OH)_4^{2-}$ would

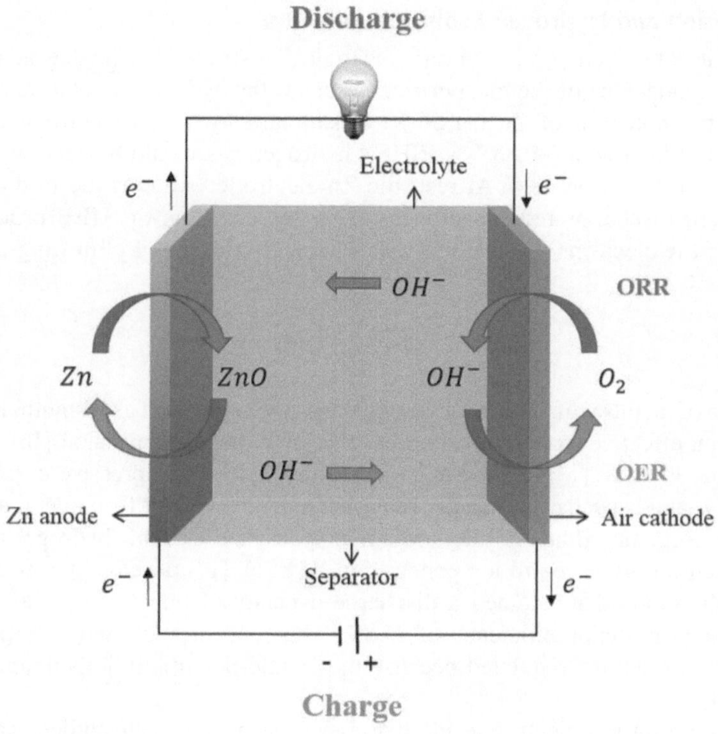

FIGURE 18.1 Schematic representation of a rechargeable Zn-air battery configuration.

eventually become supersaturated in the electrolyte and decompose into ZnO, as described in Equations 18.1–18.3. During battery charging, these reactions will be reversed. Zn will be regenerated at the anode through the backward reaction of Equation 18.1a and b, while oxygen will be produced at the cathode via the oxygen evolution reaction (OER), shown by the backward reaction of Equation 18.2.

$$\text{Anode: } Zn + 4OH^- \rightarrow Zn(OH)_4^{2-} + 2e^- \tag{18.1a}$$

$$Zn(OH)_4^{2-} \rightarrow ZnO + H_2O + 2OH^- \tag{18.1b}$$

$$\text{Cathode: } O_2 + 2H_2O + 4e^- \rightarrow 4OH^- \tag{18.2}$$

$$\text{Overall: } 2Zn + O_2 \rightarrow 2ZnO \tag{18.3}$$

The theoretical voltage of the rechargeable Zn-air battery is 1.66 V at standard conditions. The practical discharge voltages are often less than 1.2 V, while a recharge voltage of 2 V or higher is typically required to reverse the reactions [5,12]. The dominant cause of this is the inefficient kinetics of the redox reactions, while Zn anode deformation in the alkaline electrolyte further limits the reversibility of the battery. The following section will summarize the main obstacles in the development of Zn-air batteries and the current strategies used to address these issues.

18.3 CURRENT CHALLENGES AND OPPORTUNITIES

18.3.1 Zn Anode

18.3.1.1 Corrosion and Hydrogen Evolution Reaction

The rechargeability of current Zn-air batteries is mainly restricted by the degradation and deformation of the Zn anode. One of the most critical issues is the hydrogen evolution reaction (HER). Since the reduction potential of Zn (–1.26 V vs standard hydrogen electrode (SHE)) is more negative than that of hydrogen (–0.83 V vs SHE), hydrogen gas would be generated on the anode surface, as shown in Equation 18.4. At rest, the Zn electrode will corrode in the alkaline electrolyte, causing self-discharge and capacity loss. During charging, the HER process would also consume some of the electrons provided to the zinc electrode, thereby limiting the battery efficiency [13–15].

$$Zn + 2H_2O \rightarrow Zn(OH)_2 + H_2 \tag{18.4}$$

The incorporation of metals with a higher hydrogen overpotential, such as bismuth, lead, and nickel, was found to be an effective strategy to suppress the HER and Zn corrosion [16]. Using a Zn–Bi alloy as the anode, Jo et al. [13] created a Zn-air battery with an impressive corrosion inhibition efficiency of 91.5% and a discharge capacity retention of 99.5%. Apart from metal alloys, researchers have explored organic additives like sodium polyacrylate (PANa) hydrogel and polyaniline (PANI) due to their ability to retard ion penetration. Jo et al. [17] developed a Zn-air battery with a PANI-coated Zn anode that retained a discharge capacity of 565 mAh g^{-1} after 24 h storage, showing a corrosion inhibition efficiency of 75.8%. The concentration and incorporation method of additives must be precisely regulated because they could potentially impede mass transfer and increase costs [14].

Similar to anode additives, electrolyte additives have also been widely studied. Rather than metal compounds, surfactants with organic inhibitors have gained prevalence due to their low cost, safety, and eco-friendliness. When surfactants in the electrolytes are adsorbed on the surface of the metal anode, the superficial active area could be shielded from water and ionic access, thereby protecting Zn from corrosion [18]. The application of organic inhibitors such as polyoxyethylene (40) nonyl-phenyl (PNE) [18], ethylenediaminetetraacetic acid (EDTA) [19], and polyethylene glycols (PEGs) [20] have all demonstrated effectiveness in corrosion suppression.

18.3.1.2 Passivation

Passivation is the formation of an insulating layer of ZnO or $Zn(OH)_2$ on the anode surface, due to the saturation and precipitation of $Zn(OH)_4^{2-}$ ions after numerous repeated cycles. As a barrier to the anode, the passivating film would critically limit Zn utilization by restricting its oxidation and deposition and eventually block access to the Zn electrode completely [21].

Passivation could be effectively mitigated through the formation of a solid electrolyte interface on the metal anode. Using electrodeposition methods, Schmid et al. [22] prepared a silica-coated Zn anode that generated a layer of $Si(OH)_4$ gel at the electrode–electrolyte interface, which significantly reduced the severity of passivation, as shown in Figure 18.2a–c. The interface also obstructed water access and successfully suppressed the HER by 40%. Passivation could also be prevented by increasing the solubility of ZnO in the electrolyte. With the use of dimethyl sulfoxide (DMSO) as an electrolyte additive, Hosseini et al. [23] observed that the oxide surface film could be suspended and dissolved, thereby restricting surface passivation. A promising rechargeability of over 600 cycles was achieved, as illustrated in Figure 18.2d–e.

Another effective strategy is to directly modify the anode structure and optimize its porosity. With a porous structure, the surface area of the Zn electrode could be increased to improve activity, while the interconnected pores would benefit mass transfer and weaken the passivation effect.

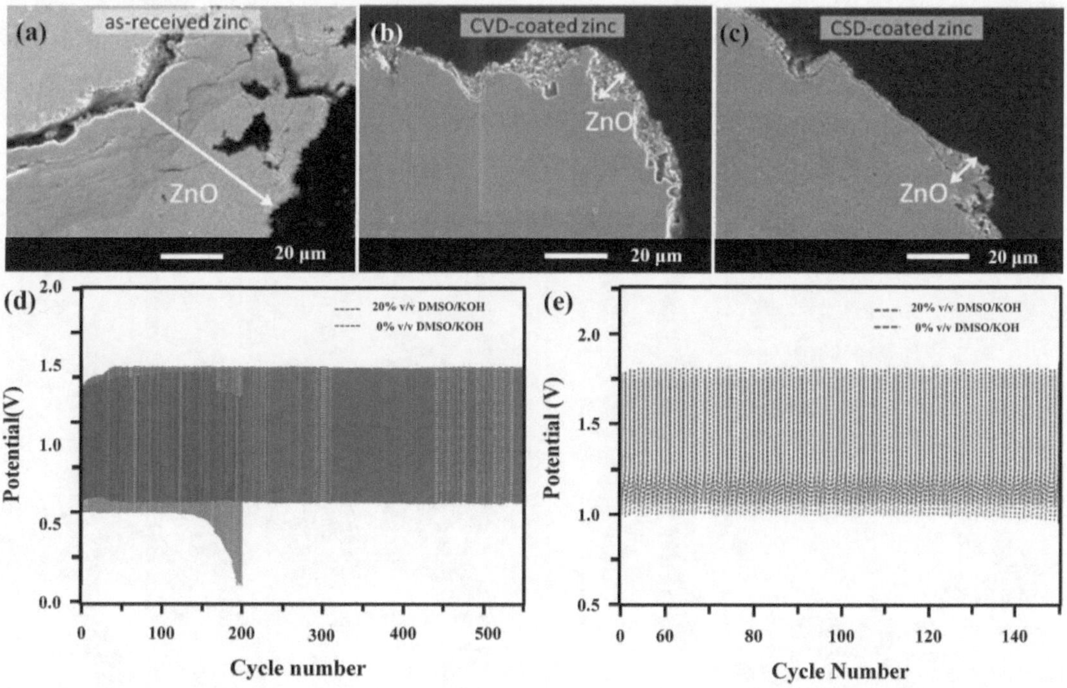

FIGURE 18.2 (a–c) SEM images showing the thickness of the passivation layer in (a) uncoated Zn, (b) chemical vapor deposition (CVD)-coated Zn, and (c) chemical solution deposition (CSD)-coated Zn after discharge. Adapted and reproduced from Ref. [22]. Reproduced with permission from Elsevier. (d–e) Galvanostatic discharge–charge profiles of the batteries with and without DMSO additives, discharged at 75 mA cm^{-2} and charged at 25 mA cm^{-2} for 5 mAh. (d) During the 1st–600th cycles and (e) during the 50th–150th cycles. Adapted and reproduced from Ref. [23]. Copyright © 2019 Springer Nature.

Using a porous Zn with a 3D network frame structure, Liu et al. [24] fabricated a Zn-air battery that achieved stable performance and a narrow discharge–charge voltage gap of 0.63 V at 5 mA cm^{-2} over 65 cycles.

18.3.1.3 Dendrite Formation

Dendrite formation on the Zn anode is a critical obstacle limiting the cycle life of the Zn-air battery. The precipitation of $Zn(OH)_4^{2-}$ ions leads to the uneven deposition of Zn onto the metal anode due to non-uniform concentration gradients during recharge. Without being fully absorbed, rigid and spiky crystalline structures are formed. These dendritic protrusions could severely impact the battery stability and anode reversibility and, even worse, pierce the separator and cause a short circuit [5,25].

To mitigate dendrite growth, metallic coatings have once again proven to be an effective strategy. For example, Aremu et al. [26] reported dendrite-free cycling with a Zn anode coated with bismuth oxide, potassium sulfide, and lead (II) oxide additives. With higher chemical stability, these metals can form a stable, conductive matrix to support a more uniform deposition of Zn during charging [16,27]. Structural modulation of the Zn anode is also a common solution here. Through the creation of 3D porous structures, Zn deposition during battery recharge could be effectively confined within the interior cavities, thereby diminishing irregular dendrite initiation on the exposed surfaces, as illustrated in Figure 18.3 [28].

Apart from anode design, researchers have explored the use of alternative electrolytes to inhibit dendrite growth. Aqueous alkaline electrolytes, such as KOH, are conventionally used in Zn-air batteries, yet they are considered a rather aggressive environment for a reversible Zn electrode. To prevent anode deformation, water-in-salt electrolytes have recently gained tremendous research interest. Water-in-salt electrolytes are highly concentrated metal salts dissolved in a small amount of water. For example, Wang et al. [29] developed a water-in-salt electrolyte with 1 m $Zn(TFSI)_2$ + 20 m LiTFSI, which successfully enabled dendrite-free battery operation for over 200 cycles, with a high

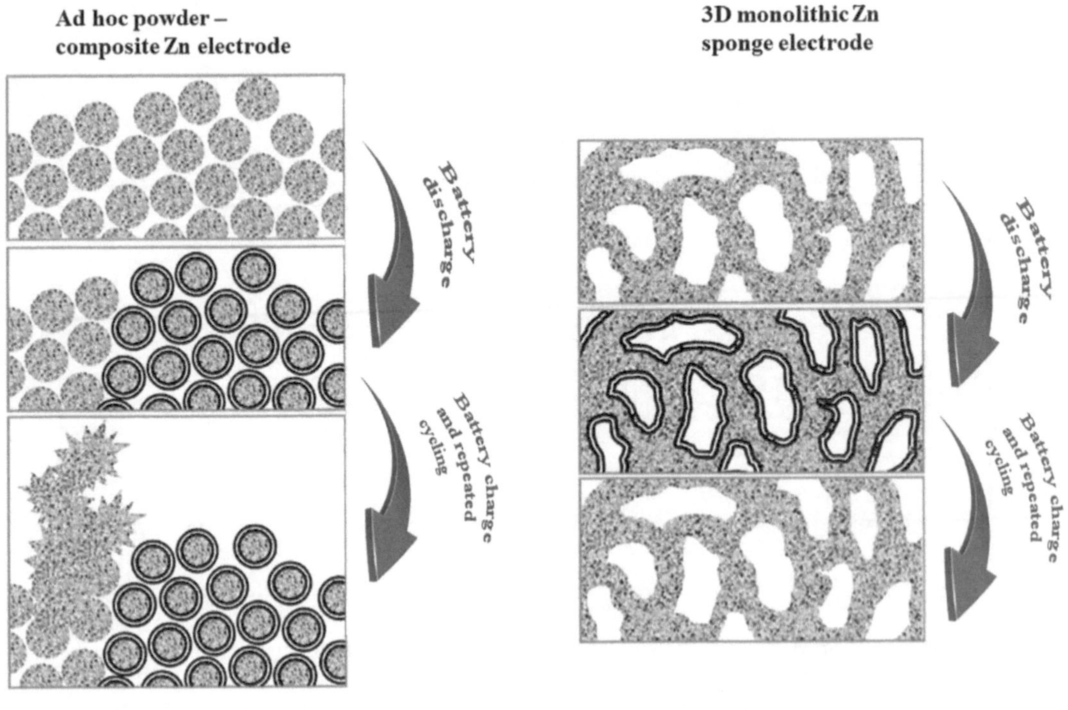

FIGURE 18.3 Schematic diagram of the dissolution–precipitation of ZnO during battery cycling in (a) conventional powder-based Zn anode and (b) the 3D Zn sponge anode. Adapted and reproduced from Ref. [28]. Copyright © 2014 Royal Society of Chemistry.

energy density of 300 Wh kg^{-1}. The quasi-neutral electrolyte could prevent dendrite formation caused by the saturation of $Zn(OH)_4^{2-}$ ions in conventional alkaline electrolytes, while the absence of free water molecules also mitigates other water-induced side reactions such as Zn corrosion.

Researchers have also experimented with ionic liquids—a non-aqueous molten salt with high chemical stability. Ingale et al. [30] developed a diethylmethylammonium trifluoromethanesulfonate (DEMATfO) ionic liquid that eliminated dendrite growth and exhibited stable operations over 700 h of cycling in a Zn-air battery, but the maximum power density was only 0.045 mW cm^{-2}. Ionic liquids generally have a lower ionic conductivity than the conventional aqueous electrolyte and are often more expensive. Therefore, further research efforts are required to enable the practical use of Zn-air batteries [31,32].

18.3.2 Electrolyte

18.3.2.1 Electrolyte Evaporation

During battery discharge, water is consumed at the cathode via the ORR. Even though the water is regenerated via the OER during recharge, water evaporation from the aqueous electrolyte is inevitable over time. Water evaporation increases the concentration of Zn^{2+} ions in the electrolyte and expedites the precipitation of zincates, which would further worsen dendrite formation and surface passivation [33,34]. As a result, water evaporation would significantly shorten the cycle life of the battery or even completely dry out and end the battery operations.

To address electrolyte evaporation, the use of non-aqueous ionic liquids, as described in Section 18.3.1.3, is a direct method due to the absence of water. Another popular strategy is the addition of polymer gels in the aqueous alkaline electrolyte, such as polyvinyl alcohol (PVA), polyacrylic acid (PAA), and polyacrylamide. These polymer gel materials have strong water retention abilities, and, because of their ion-confinement effect, they could slow down unwanted parasitic reactions at the Zn anode [35,36]. Yet, the increase in ionic resistance could simultaneously weaken the electrochemical performance of the battery. As a result, alkaline–gel hybrid electrolytes are often coupled with additives such as cellulose, chitosan particles, and potassium iodide (KI), to boost their ionic conductivity and stability [37]. For example, Song et al. [37] developed a PVA, PAA, and graphene oxide (GO) co-crosslinked, KI-containing, alkaline electrolyte (KI-PVAA-GO) with high water retention abilities, which enabled stable cyclic operations over 200 h with a high energy efficiency of 73% in a flexible Zn-air battery. Figure 18.4 displays its galvanostatic cycling profile and a demonstration of powering a smartwatch with the assembled solid-state Zn-air battery in different configurations.

18.3.2.2 Carbonate Formation

Another unfavorable side reaction in aqueous alkaline electrolytes is the formation of carbonate precipitates. Because of exposure to carbon dioxide in the ambient air, the electrolyte tends to generate insoluble carbonates, as shown in Equation 18.5. These by-products would lower the conductivity of the electrolyte and impede oxygen diffusion due to obstructions at the cathode, critically impacting the efficiency of cyclic operations [27,38].

$$CO_2 + 2OH^- \rightarrow CO_3^- + H_2O \tag{18.5}$$

A direct and common solution is to implement a flow battery system so that the electrolyte is continuously circulated in and out of the battery using external pumps, as illustrated in Figure 18.5. This effectively removes any carbonates and other by-products. In addition, excess zincate ions and precipitated oxides would be carried away, thereby eliminating dendrite growth and surface passivation [5]. In a Zn-air flow battery using a potassium persulfate electrolyte additive, Khezri et al. [39] reported that a narrow discharge–charge voltage gap of 0.6 V over 800 stable cycles was achieved. Yet, despite the electrochemical success of the flow battery, it is a rather unsustainable design. The additional energy consumption and bulky nature of the electrolyte circulation system are often considered impractical in real-life applications, especially for portable electronics.

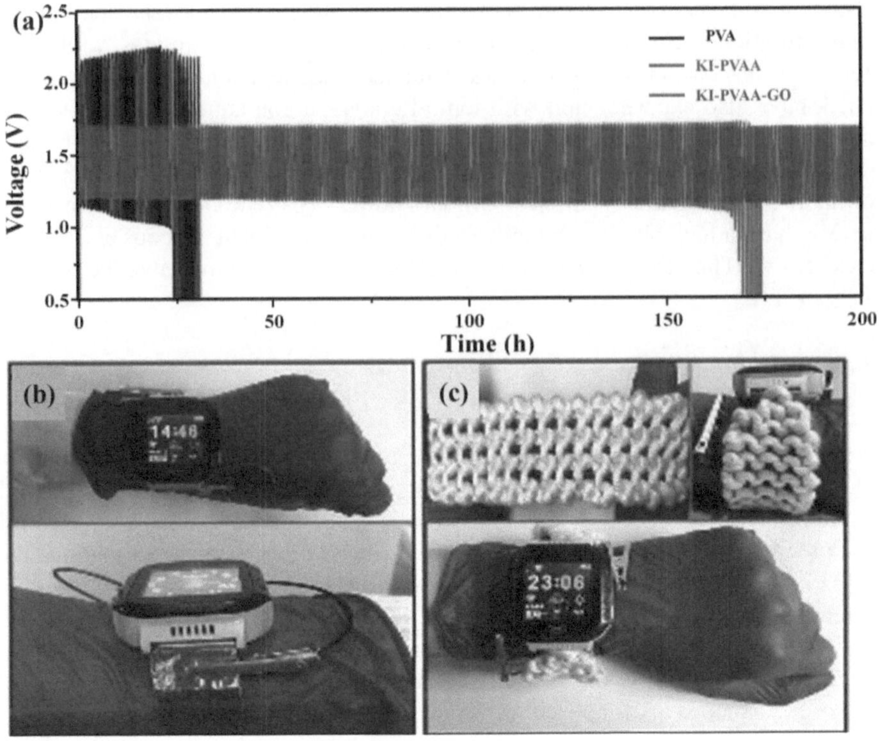

FIGURE 18.4 (a) Galvanostatic discharge–charge curves of sandwich-type Zn-air batteries using PVA, PVAA-GO, and KI-PVAA-GO as electrolytes at 2 mA cm^{-2}. (b and c) Demonstrations of a flexible Zn-air battery used to power a smartwatch in a (b) sandwich-type and (c) cable-type configuration. Adapted and reproduced from Ref. [37]. Copyright © 2020 John Wiley and Sons.

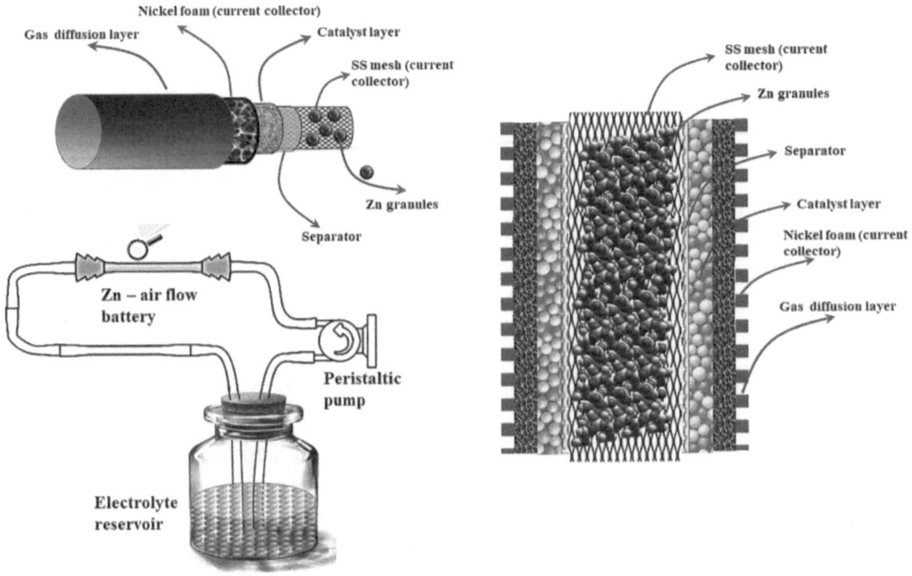

FIGURE 18.5 Schematic configuration and photographic image of a Zn-air flow battery. Adapted and reproduced from Ref. [23]. Copyright © 2019 Springer Nature.

18.3.3 Air Cathode

Last but not least, one of the most critical bottlenecks in the development of rechargeable Zn-air batteries is the sluggish oxygen reduction and evolution reactions at the air cathode [40]. Even if the anode irreversibility and electrolyte-related issues are completely solved, the high overpotentials of ORR and OER would still limit the energy efficiency, discharge–charge voltage performances, and power density of the battery. As a result, advanced bifunctional oxygen catalysts have been introduced to promote more efficient oxygen reaction kinetics. Current developments in oxygen catalysts can be classified into three main categories—transition metals, metal-free carbon catalysts, and metal–carbon composites.

18.3.3.1 Transition Metal Catalysts

Metal-based materials are promising oxygen catalysts in metal-air batteries because of their high electrical conductivity and electrochemical stabilities [41]. In particular, transition metal-based catalysts have become an attractive choice due to their low cost and material abundance. Moreover, transition metals (Ni, Co, Mn, Fe) and their alloys and oxides possess intrinsic electrochemical activity, which could accelerate the oxygen redox reactions significantly [40]. An example is the transition metal-based Co@$LiCoO_2$ nanofibers catalyst developed by Gui et al. [42], which exhibited rich oxygen vacancies that improved electrochemical activity and resulted in a Zn-air battery with superior cyclic performance over 200 h, outperforming the benchmark Pt/C-IrO_2 catalyst. Though less common, other transition metal derivatives have also attracted the attention of many researchers in the past few years, including transition metal nitrides [43], transition metal sulfides [44], and transition metal selenides [45]. Table 18.3 reports some of the latest Zn-air battery performances achieved by transition metal-based catalysts.

18.3.3.2 Carbon-Based Catalysts

Alternatively, carbon-based materials have been investigated as bifunctional oxygen catalysts due to their chemical stability, tunable morphology, and superior electronic conductivity. Without the use of metals, these catalysts are inexpensive and more environmentally friendly [51].

TABLE 18.3
Selection of Zn-Air Batteries Using Transition Metal-Based Oxygen Catalysts in Recent Literature

Cathode	OCV (V)	PPD (mW cm^{-2})	Discharge Capacity (mAh g^{-1})	Discharge-Charge Voltage Gap (V)	Cycling Stability	References
AlFeCoNi-Cr	1.55	125	800	0.86 (20 mA cm^{-2})	120 h	[46]
Co@ $LiCoO_2$ nanofibers	1.43	198	791	0.8 (10 mA cm^{-2})	200 h	[42]
Oxygen defect Co_3O_{4-x} nanosheets		122	791	0.5 (5 mA cm^{-2})	450 cycles, 150 h	[47]
Fe-doped Co_3O_4 nanosheets		268.6		0.75 (2 mAcm^{-2})	300 cycles, 100 h	[48]
$CoIn_2Se_4$ nanosheet	1.37	107	733	0.71 (10 mA cm^{-2})	400 cycles	[45]
S-doped $LaCoO_3$	1.47	92	747	0.95 (2 mA cm^{-2})	300 cycles, 100 h	[49]
$LaCo_{0.8}Ru_{0.2}O_{3-\delta}$	1.28	136	433	0.78 (5 mA cm^{-2})	1,440 cycles, 240 h	[50]

OCV, open circuit voltage; PPD, peak power density.

TABLE 18.4
Selection of Zn-Air Batteries Using Metal-Free Carbon Catalysts in Recent Literature

Cathode	OCV (V)	PPD (mW cm^{-2})	Discharge Capacity (mAh g^{-1})	Discharge–Charge Voltage Gap (V)	Cycling Stability	References
N, S-doped carbon	1.48	149	769	1.03 (5 mA cm^{-2})	600 cycles, 100 h	[54]
N, S-doped carbon		144		0.9 (10 mA cm^{-2})	88 h	[55]
N, F-doped carbon	1.33	130		0.852 (20 mA cm^{-2})	147 cycles, 49 h	[56]
3D honeycombed porous N, O-doped carbon		261		0.72 (10 mA cm^{-2})	1,165 cycles, 388 h	[57]
P, S-doped carbon nitride sponges	1.51	198	830	1.3 (2 mA cm^{-2})	200 cycles, 40 h	[58]

OCV, open circuit voltage; PPD, peak power density.

The conventional carbon materials used are carbon black and porous carbon, but advanced carbons such as graphene and carbon nanotubes (CNTs) often display higher catalytic activity due to their larger surface area and higher electronic conductivity [52]. Other graphene derivatives like GO and reduced graphene oxide (rGO) possess edge defects which further promote catalytic activity [53]. In fact, the physical and electronic structures of carbon-based catalysts are often modulated to enrich active sites. As shown in Table 18.4, the recent metal-free carbon catalysts reported in the literature are mostly heteroatom-doped. Doping introduces electron depletion to carbon atoms to increase catalytic activity. Besides, 3D porous carbon structures could create a larger surface area to further expose active sites.

18.3.3.3 Metal–Carbon Catalysts

Metal–carbon composites are becoming an increasingly popular bifunctional oxygen catalyst due to the combined advantages of both the metal and carbon. When incorporated with carbon supports, transition metals exhibit stronger durability and higher corrosion resistance. In addition, particle agglomeration can be prevented, enabling a more uniform dispersion of active sites, thereby enhancing catalytic performance [59,60]. Dong et al. [61] developed a metal–carbon catalyst using cobalt-doped Mn_3O_4 nanocrystals supported on graphene nanosheets (Co-Mn_3O_4/G), which demonstrated a narrow discharge–charge voltage gap of 0.91 V over 315 h at 10 mA cm^{-2} in a Zn-air battery.

Recently, metal-organic frameworks (MOFs) have gained massive interest in the preparation of metal–carbon composite as an oxygen catalyst. Composed of evenly arranged metal ions coordinated to organic molecules, MOFs exhibit high performance, structural tunability, and fabrication efficiency. Using a MOF-derived co-embedded in hollow N-doped carbon tubes (Co@hNCTs) as an oxygen catalyst, Zhou et al. [62] recently reported a narrow discharge–charge voltage gap of 0.85 V over 500 h at 5 mA cm^{-2}. Figure 18.6 illustrates the template-guided synthesis method of the Co@hNCT metal–carbon composite.

Table 18.5 summarizes selected Zn-air battery performances enabled by the development of metal–carbon composites, which are the dominant type of bifunctional oxygen catalysts in recent studies. Current research on catalyst materials has significantly improved the electrochemical reaction kinetics at the air cathode, resulting in lowered discharge–charge voltage gaps and longer cycle lives. Nonetheless, catalyst material designs so far have mainly originated from trial-and-error experimental methods. To achieve performance breakthroughs, more in-depth studies are required to enable the rational and efficient development of bifunctional oxygen catalysts [7].

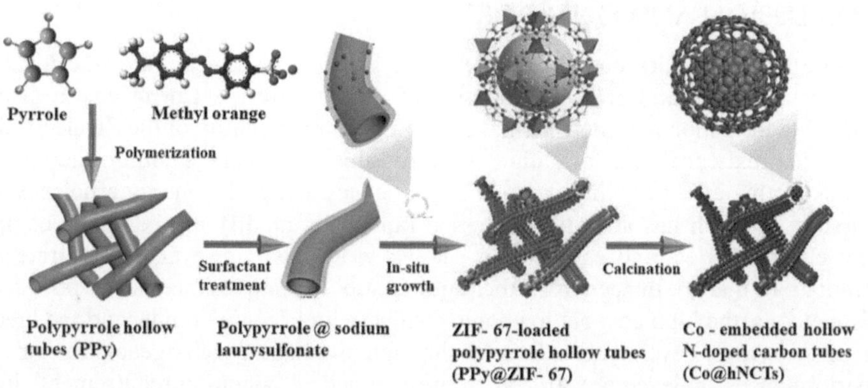

FIGURE 18.6 Schematic representation of the preparation of Co@hNCT. Adapted and reproduced from Ref. [62]. Copyright © 2020 Elsevier.

TABLE 18.5
Selection of Zn-Air Batteries Using Metal–Carbon Catalysts in Recent Literature

Cathode	OCV (V)	PPD (mW cm^{-2})	Discharge Capacity (mAh g^{-1})	Discharge–Charge Voltage gap (V)	Cycling Stability	References
Co-Mn_3O_4/G		115.2		0.91 (10 mA cm^{-2})	945 cycles, 315 h	[61]
Pt-SCFP/C	1.4	122	790.4	0.77 (5 mA cm^{-2})	240 cycles, 80 h	[63]
CoFe@ N,P-doped CHS	1.516	123.1		0.591 (10 mA cm^{-2})	2,000 cycles, 40 h	[64]
Co_9S_8/ N, S-doped carbon	1.408	60	738	0.93 (5 mA cm^{-2})	105 cycles, 35 h	[65]
M-MnO_2/ CNTs		342.5		0.72 (10 mA cm^{-2})	129 h	[66]
MnO_x@C		116		1.1 (10 mA cm^{-2})	215 cycles	[67]
Co_8FeS_8@ N,S-doped carbon	1.471	158.6		0.85 (2 mA cm^{-2})	300 cycles, 50 h	[68]
Co-N-C nanobrushes	1.47	246		0.8 (2 mA cm^{-2})	140 h	[69]
Co-embedded graphitic carbon nitride nanosheet	1.446	85.3	675.7	0.6 (10 mA cm^{-2})	1,000 cycles, 183 h	[70]
MOF-derived Co@N-CNT	1.45	149	746	0.85 (5 mA cm^{-2})	500 h	[62]
MOF-derived 2D hierarchical Fe-N-C	1.45	156	798		130 h	[71]
MOF-derived Co_9S_8/GN		186		0.52 (2 mA cm^{-2})	2,000 cycles, 147 h	[72]

Notes: OCV, open circuit voltage; PPD, peak power density; SCFP, $Sr(Co_{0.8}Fe_{0.2})_{0.95}P_{0.05}O_{3-\delta}$; CHS, carbon nanovesicles assembled hollow spheres; CNT, carbon nanotubes; GN, graphene nanosheet.

18.4 CONCLUSION AND OUTLOOK

Zn anode failure, electrolyte dry-out, and inefficient oxygen reaction kinetics are all critical barriers impeding the development and practical adoption of Zn-air batteries. Corrosion, surface passivation, and dendrite formation have detrimental effects on the reversibility of the Zn electrode, while electrolyte dry-out and carbonate formation severely limit the battery stability. Sluggish oxygen redox reactions further affect the discharge–charge voltages and cyclic energy efficiency. To mitigate these issues, research has already progressed rapidly by modifying the anode composition, exploring novel electrolytes, and engineering catalyst materials. However, these strategies have their own limitations that are in need of further optimization. Examples include the poor scalability of synthesis methods, the high cost but low ionic conductivity of ionic liquids, and the inefficiency of trial-and-error catalyst development. Despite the high promise of rechargeable Zn-air batteries, the latest performances achieved are still far from the practical operation requirements in electric vehicles and other portable electronics. Therefore, to realize their full potential, continuous research studies in the above areas are urgently needed to enable long-term cycle life and high energy efficiency in rechargeable Zn-air batteries.

REFERENCES

1. Iclodean, C., et al. Comparison of different battery types for electric vehicles. *IOP Conf Ser Mater Sci Eng*. 2017. 252: 012058
2. Tan, P., et al., Flexible Zn-and Li-air batteries: recent advances, challenges, and future perspectives. *Energy Environ Sci*. 2017. 10(10): 2056–2080.
3. Cano, Z.P., et al., Batteries and fuel cells for emerging electric vehicle markets. *Nat Energy*. 2018. 3(4): 279–289.
4. Emregül, K. and A.A. Aksüt, The behavior of aluminum in alkaline media. *Corros Sci*. 2000. 42(12): 2051–2067.
5. Fu, J., et al., Electrically rechargeable zinc-air batteries: progress, challenges, and perspectives. *Adv Mater*. 2017. 29(7): 1604685.
6. Li, Y. and J.J. Lu, Metal-air batteries: will they be the future electrochemical energy storage device of choice? *ACS Energy Lett*. 2017. 2(6): 1370–1377.
7. Fu, J., et al., Recent progress in electrically rechargeable zinc-air batteries. *Adv Mater*. 2019. 31(31): 1805230.
8. Nitta, N., et al., Li-ion battery materials: present and future. *Mater Today*. 2015. 18(5): 252–264.
9. Hosseini, S., et al., Discharge performance of zinc-air flow batteries under the effects of sodium dodecyl sulfate and pluronic F-127. *Sci Rep*. 2018. 8(1): 1–13.
10. Kee Wah Leong, Yifei Wang, Meng Ni, Wending Pan, Shijing Luo, Dennis YC Leung, Rechargeable Zn-air batteries: Recent trends and future perspectives. *Renew. Sustain. Energy Rev*. 2022. 154: 111771.
11. Zhang, J., et al., Zinc-air batteries: are they ready for prime time? *Chem Sci*. 2019. 10(39): 8924–8929.
12. Li, Y. and H. Dai, Recent advances in zinc-air batteries. *Chem Soc Rev*. 2014. 43(15): 5257–5275.
13. Zhang, X.G., *Corrosion and Electrochemistry of Zinc*. 1996: Springer Science & Business Media, Singapore.
14. Zhao, Z., et al., Challenges in zinc electrodes for alkaline zinc-air batteries: obstacles to commercialization. *ACS Energy Lett*. 2019. 4(9): 2259–2270.
15. Lee, S.-M., et al., Improvement in self-discharge of Zn anode by applying surface modification for Zn-air batteries with high energy density. *J Power Source*. 2013. 227: 177–184.
16. Yano, M., et al., Effect of additives in zinc alloy powder on suppressing hydrogen evolution. *J Power Source*. 1998. 74(1): 129–134.
17. Jo, Y.N., et al., Improving self-discharge and anti-corrosion performance of Zn-air batteries using conductive polymer-coated Zn active materials. *J Ind Eng Chem*. 2019. 76: 396–402.
18. Deyab, M.A., Application of nonionic surfactant as a corrosion inhibitor for zinc in alkaline battery solution. *J Power Source*. 2015. 292: 66–71.
19. Huang, M.C., et al., Improved electrochemical performance of Zn-air secondary batteries via novel organic additives. *J Chinese Chem Soc*. 2018. 65(10): 1239–1244.

20. Dobryszycki, J. and S. Biallozor, On some organic inhibitors of zinc corrosion in alkaline media. *Corros Sci*. 2001. 43(7): 1309–1319.
21. Kim, H.-I., et al., Influence of ZnO precipitation on the cycling stability of rechargeable Zn-air batteries. *J Appl Electrochem*. 2015. 45(4): 335–342.
22. Schmid, M. and M. Willert-Porada, Electrochemical behavior of zinc particles with silica based coatings as anode material for zinc air batteries with improved discharge capacity. *J Power Source*. 2017. 351: 115–122.
23. Hosseini, S., et al., The influence of dimethyl sulfoxide as electrolyte additive on anodic dissolution of alkaline zinc-air flow battery. *Sci Rep*. 2019. 9(1): 1–12.
24. Liu, P., et al., Porous zinc anode design for Zn-air chemistry. *Front Chem*. 2019. 7: 656.
25. Wang, R., D. Kirk, and G.X. Zhang, Effects of deposition conditions on the morphology of zinc deposits from alkaline zincate solutions. *J Electrochem Soc*. 2006. 153(5): C357.
26. Aremu, E.O., D.-J. Park, and K.-S. Ryu, The effects of anode additives towards suppressing dendrite growth and hydrogen gas evolution reaction in Zn-air secondary batteries. *Ionics*. 2019. 25(9): 4197–4207.
27. Xu, M., et al., Rechargeable Zn-air batteries: progress in electrolyte development and cell configuration advancement. *J Power Source*. 2015. 283: 358–371.
28. Parker, J.F., et al., Wiring zinc in three dimensions re-writes battery performance-dendrite-free cycling. *Energy Environ Sci*. 2014. 7(3): 1117–1124.
29. Wang, F., et al., Highly reversible zinc metal anode for aqueous batteries. *Nat Mater*. 2018. 17(6): 543–549.
30. Ingale, P., M. Sakthivel, and J.-F. Drillet, Test of diethylmethylammonium trifluoromethanesulfonate ionic liquid as electrolyte in electrically rechargeable Zn/Air battery. *J Electrochem Soc*. 2017. 164(8): H5224.
31. Alwast, D., et al., Ionic Liquid electrolytes for metal-air batteries: Interactions between O_2, Zn^{2+} and H_2O impurities. *J Electrochem Soc*. 2019. 167(7): 070505.
32. Kar, M., et al., Ionic liquid electrolytes as a platform for rechargeable metal-air batteries: a perspective. *Phys Chem Chem Phys*. 2014. 16(35): 18658–18674.
33. Li, Y., et al., The role of PTFE in cathode transition layer in aqueous electrolyte Li-air battery. *Electrochim Acta*. 2016. 191: 996–1000.
34. Xu, M., et al., Rechargeable Zn-air batteries: progress in electrolyte development and cell configuration advancement. *Journal Power Source*. 2015. 283: 358–371.
35. Ma, L., et al., Super-stretchable zinc-air batteries based on an alkaline-tolerant dual-network hydrogel electrolyte. *Adv Energy Mater*. 2019. 9(12): 1803046.
36. Tang, Y., et al., Ion-confinement effect enabled by gel electrolyte for highly reversible dendrite-free zinc metal anode. *Energy Storage Mater*. 2020. 27: 109–116.
37. Song, Z., et al., A rechargeable Zn-air battery with high energy efficiency and long life enabled by a highly water-retentive gel electrolyte with reaction modifier. *Adv Mater*. 2020. 32(22): 1908127.
38. Chen, P., et al., Recent progress in electrolytes for Zn-air batteries. *Front Chem*. 2020. 8: 00372.
39. Khezri, R., et al., Enhanced cycling performance of rechargeable zinc-air flow batteries using potassium persulfate as electrolyte additive. *Int J Mol Sci*. 2020. 21(19): 7303.
40. Tomboc, G.M., et al., Ideal design of air electrode-A step closer toward robust rechargeable Zn-air battery. *APL Mater*. 2020. 8(5): 050905.
41. Lang, X., Z. Hu, and C. Wang, Bifunctional air electrodes for flexible rechargeable Zn-air batteries. *Chinese Chem Lett*. 2020. 32: 999–1009.
42. Gui, L., et al., In situ exsolved Co nanoparticles coupled on $LiCoO_2$ nanofibers to induce oxygen electrocatalysis for rechargeable Zn-air batteries. *J Mater Chem A*. 2020. 8(38): 19946–19953.
43. Yu, X., et al., Recent advances on the modulation of electrocatalysts based on transition metal nitrides for the rechargeable Zn-air battery. *ACS Mater Lett*. 2020. 2(11): 1423–1434.
44. Sun, H., et al., Self-supported transition-metal-based electrocatalysts for hydrogen and oxygen evolution. *Adv Mater*. 2020. 32(3): 1806326.
45. Wang, J., et al., Developing indium-based ternary spinel selenides for efficient solid flexible Zn-air batteries and water splitting. *ACS Appl Mater Interfaces*. 2020. 12(7): 8115–8123.
46. Fang, G., et al., Multi-component nanoporous alloy/(oxy) hydroxide for bifunctional oxygen electrocatalysis and rechargeable Zn-air batteries. *Appl Catal B Environ*. 2020. 268: 118431.
47. Li, M., et al., Atomic layer Co_3O_{4-x} nanosheets as efficient and stable electrocatalyst for rechargeable zinc-air batteries. *J Catal*. 2020. 381: 395–401.

48. Liu, X.-Z., et al., Fe-doped Co_3O_4 polycrystalline nanosheets as a binder-free bifunctional cathode for robust and efficient zinc-air batteries. *Chem Commun*. 2020. 56(40): 5374–5377.
49. Ran, J., et al., Modulation of electronics of oxide perovskites by sulfur doping for electrocatalysis in rechargeable Zn-air batteries. *Chem Mater*. 2020. 32(8): 3439–3446.
50. Chandrappa, S.G., et al., The influence of ruthenium substitution in $LaCoO_3$ towards bi-functional electrocatalytic activity for rechargeable Zn-air batteries. *J Mater Chem A*. 2020. 8(39): 20612–20620.
51. Hu, C., et al., Carbon-based metal-free catalysts for energy storage and environmental remediation. *Adv Mater*. 2019. 31(13): 1806128.
52. Chen, D., H. Feng, and J. Li, Graphene oxide: preparation, functionalization, and electrochemical applications. *Chem Rev*. 2012. 112(11): 6027–6053.
53. Li, F., et al., Graphene oxide: a promising nanomaterial for energy and environmental applications. *Nano Energy*. 2015. 16: 488–515.
54. Lyu, D., et al., In situ molecular-level synthesis of N, S co-doped carbon as efficient metal-free oxygen redox electrocatalysts for rechargeable Zn-air batteries. *Appl Mater Today*. 2020. 20: 100737.
55. Cheng, Y., et al., Hierarchically porous metal-free carbon with record high mass activity for oxygen reduction and Zn-air batteries. *J Mater Chem A*. 2019. 7(16): 9831–9836.
56. Zhu, P., et al., An efficient metal-free bifunctional oxygen electrocatalyst of carbon co-doped with fluorine and nitrogen atoms for rechargeable Zn-air battery. *Int J Hydrog Energy*. 2020. 45(16): 9512–9521.
57. Xiao, X., et al., Robust template-activator cooperated pyrolysis enabling hierarchically porous honeycombed defective carbon as highly-efficient metal-free bifunctional electrocatalyst for Zn-air batteries. *Appl Catal B Environ*. 2020. 265: 118603.
58. Shinde, S.S., et al., Scalable 3-D carbon nitride sponge as an efficient metal-free bifunctional oxygen electrocatalyst for rechargeable Zn-air batteries. *ACS Nano*. 2017. 11(1): 347–357.
59. Tian, Y., et al., Rational design of sustainable transition metal-based bifunctional electrocatalysts for oxygen reduction and evolution reactions. *Sustain Mater Technol*. 2020. 25: e00204.
60. Zhu, Y.P., et al., Surface and interface engineering of noble-metal-free electrocatalysts for efficient energy conversion processes. *Acc Chem Res*. 2017. 50(4): 915–923.
61. Dong, M., et al., Cobalt-doped Mn_3O_4 nanocrystals embedded in graphene nanosheets as a high-performance bifunctional oxygen electrocatalyst for rechargeable Zn-air batteries. *Green Energy Environ*. 2020. 5: 499–505.
62. Zhou, Q., et al., Template-guided synthesis of Co nanoparticles embedded in hollow nitrogen doped carbon tubes as a highly efficient catalyst for rechargeable Zn-air batteries. *Nano Energy*. 2020. 71: 104592.
63. Wang, X., et al., High-performance platinum-perovskite composite bifunctional oxygen electrocatalyst for rechargeable Zn-air battery. *Adv Energy Mater*. 2020. 10(5): 1903271.
64. Niu, H.-J., et al., Assembled hollow spheres with CoFe alloyed nanocrystals encapsulated in N, P-doped carbon nanovesicles: an ultra-stable bifunctional oxygen catalyst for rechargeable Zn-air battery. *J Power Source*. 2020. 475: 228594.
65. Tang, K., et al., Inverse-opal-structured hybrids of N, S-codoped-carbon-confined Co_9S_8 nanoparticles as bifunctional oxygen electrocatalyst for on-chip all-solid-state rechargeable Zn-air batteries. *Appl Catal B Environ*. 2020. 260: 118209.
66. Xu, N., et al., Hierarchical bifunctional catalysts with tailored catalytic activity for high-energy rechargeable Zn-air batteries. *Appl Energy*. 2020. 279: 115876.
67. Marsudi, M.A., et al., Manganese oxide nanorods decorated table sugar derived carbon as efficient bifunctional catalyst in rechargeable Zn-air batteries. *Catalysts*. 2020. 10(1): 64.
68. Chen, L., et al., Co_8FeS_8/N, S-doped carbons derived from Fe-Co/S-bridged polyphthalocyanine: efficient dual-function air-electrode catalysts for rechargeable Zn-air batteries. *ACS Sustainable Chem Eng*. 2020. 8(35): 13147–13158.
69. Luo, H., et al., Self-catalyzed growth of Co-N-C nanobrushes for efficient rechargeable Zn-air batteries. *Small*. 2020. 16(20): 2001171.
70. Niu, W.-J., et al., A hybrid transition metal nanocrystal-embedded graphitic carbon nitride nanosheet system as a superior oxygen electrocatalyst for rechargeable Zn-air batteries. *Nanoscale*. 2020. 12(38): 19644–19654.
71. Yuan, R., et al., Two-dimensional hierarchical Fe-N-C electrocatalyst for Zn-air batteries with ultrahigh specific capacity. *ACS Mater Lett*. 2019. 2(1): 35–41.
72. Sun, X., et al., Exploiting a high-performance "double-carbon" structure Co_9S_8/GN bifunctional catalysts for rechargeable Zn-air batteries. *ACS Appl Mater Interfaces*. 2020. 12(34): 38202–38210.

19 Emerging Cost-Efficient Aluminium-Air Battery Technology-Anode Materials

S. K. Vineeth, Pritam V. Dhawale, Roshny Joy, Abhilash Pullanchiyodan, Jou-Hyeon Ahn, Nutan Gupta, and Prasanth Raghavan

19.1 INTRODUCTION

After the commercialization of lithium-ion batteries (LIBs) by Sony Corporation in 1991, a technological revolution occurred in energy storage for portable devices and electric vehicles (EVs) [1,2]. However, considering the availability of raw materials for LIB, the reactivity of Li metal, along with challenges in recycling, impedes further revolution, which led to the search for alternate systems [3,4]. Metal-air batteries (MABs) constitute a major class of systems, which have a high energy density, and are one of the competent for post-LIB technology era [5,6]. Among various MABs such as lithium-air battery (LAB), zinc-air, magnesium-air, sodium-air, and aluminium-air batteries (AABs), LAB is the one capable of providing the highest specific energy density followed by AAB [7]. MAB uses metals or metal alloys as the negative electrode active material, and the oxygen in the air reaches the electrochemical reaction interface through the gas diffusion and reacts with the anode to release the electrical energy [8]. Nevertheless, aluminium (Al) is the most stable and recyclable metal and is cost-efficient as well. AABs are more intriguing than lead/acid or nickel metal hydride in EVs in terms of range, cost, and lifecycle [9]. AABs would complement LIBs to provide a potential range extender, though stops for a water top-up would be needed every couple of 100 miles. Table 19.1 shows a comparative study of various metal-air battery technologies.

Likewise, AABs have been used in military applications where their high capacity and light weight are of huge advantage [10]. Al anode is exceptionally light, and as the active cathode material is oxygen from ambient air, the whole package can be much lighter than many other battery types [11]. The low atomic weight and trivalent state of the Al anode offer a high specific capacity (2.98 Ah gAl^{-1}) and high energy density (theoretically 8.1 Wh gAl^{-1}), which are greater than those of current LIBs [12]. Despite the great potential of AABs to be employed at commercial-scale implementation, several issues need to be addressed: (1) self-discharge: the parasitic corrosion reaction of Al and electrolyte are responsible for causing self-discharge and shortening the battery shelf life; (2) passivation of the anode by discharge products: formation of gel-like aluminium hydroxide and/or hydrated alumina passivates the anode surface and prevents further discharging; (3) complex air electrode design with an active catalyst, gas diffusion layer, and suitable current collector, to promote oxygen diffusion/reduction and prevent carbonate formation/deposition [13]. Thus far, AABs have just been suitable for non-rechargeable batteries.

Research developments are underway to develop mechanically replaceable Al anodes so that entire battery packs or the Al anode can be swapped promptly at refilling stations. Various researchers have investigated and implemented techniques to mitigate the corrosion of an Al anode by

DOI: 10.1201/9781003310174-19

TABLE 19.1
A Comparative Study of Aluminium-Air Battery with Various MAB Technology

Batteries	Overall Reaction	Voltage (V)	Theoretical Specific Capacity (Ah kg^{-1})	Theoretical Energy Density (Wh kg^{-1})
Aluminium-air	$4Al+3O_2+6H_2O \rightarrow 4Al(OH)_3$	2.71	1,030	2,791
Lithium-air	$2Li+O_2 \rightarrow Li_2O_2$	2.96	1,170	3,463
Sodium-air	$Na+O_2 \rightarrow NaO_2$	2.27	487	1,105
	$2Na+O_2 \rightarrow Na_2O_2$	2.33	687	1,600
Potassium-air	$K+O_2 \rightarrow KO_2$	2.48	377	935
Magnesium-air	$Mg+0.5O_2+H_2O \rightarrow Mg(OH)_2$	3.09	920	2,843
Zinc-air	$Zn+0.5O_2 \rightarrow ZnO$	1.65	658	1,085

adding additives to electrolytes, using gel and solid electrolytes, coating anodes with ceramics, taking advantage of non-aqueous electrolytes, alloying Al anode with magnesium (Mg), tin (Sn), gallium (Ga), indium (In), manganese (Mn), zinc (Zn), lead (Pb), antimony (Sb), and so on [14]. However, AAB technology suffers corrosion of anode material from parasitic hydrogen evolution during discharge; this has been a long-standing barrier to the commercialization of AABs. This chapter highlights the recent developments in anode engineering for AAB. Various strategies of anode engineering and tailoring have been overviewed in the following sections.

19.2 AABS: SOLUTION FOR SENSORS TO HYBRID EVS

Various MABs such as LAB, zinc-air, magnesium-air, AAB, iron-air, and calcium-air have gained attention due to their high theoretical capacity and energy density [15]. Future demands for Internet of things devices and power needs of the dramatically rising number of small-sized devices have stimulated explosive growth in the market for small, portable, and disposable electronics. To facilitate clean and green energy and meet portable power requirements at a minimal cost, fruitful research needs to be done for the miniaturization of the flow battery system for small devices.

In aluminium-air flow batteries (AAFBs), the passivation layer on the surface of the Al anode was well addressed using various electrolytes. In this context, Phusittananan et al. [16] reported a dual-electrolyte system by introducing anolyte as ethylene glycol/ethanol in different volume ratios such as 0%, 5%, 10%, 20%, and 30% in 3 M potassium hydroxide (KOH) for enhanced Al dissolution kinetics and mitigating anode corrosion. Their electrochemical properties have been investigated via the hydrogen evolution test (gasometric method), potentiodynamic polarization (Tafel polarization), and electrochemical impedance spectroscopy (EIS). Overall, results demonstrated that the dual-electrolyte AAFB proved to be an effective approach to control anodic corrosion and passivation. Dual-electrolyte AAFB performed sufficiently well and provided excellent discharge capacity.

In 2015, Fan et al. [17] communicated the relation between the crystallographic orientation of Al anode with its electrochemical performance for AAB. Polycrystalline Al, Al (001), (110), and (111) single crystals of Al were tested in 4 M NaOH and KOH as electrolytes. Al (001) showed better performance with a capacity density of 2,541.4 mAh g^{-1} in 4 M KOH and showed low corrosion. The work concluded that, for making better-performance alkaline AABs, crystallographic orientation needs to be optimized and designed.

Ren et al. [18] found that the incorporation of Mg to Al and alloying for the anode in alkaline (4 M KOH) AAB showed corrosion resistance with good performance. Among various anode materials such as pure Al, Al–Sn, Al–Mg, and Al–Mg–Sn alloys, Al–Mg showed the least hydrogen evolution with a better specific capacity of 2,723.7 mAh g^{-1}. It is due to the negative electrode potential of Mg compared to the Al matrix, which results in the formation of a galvanic couple and hence reduces the corrosion.

The inclusion of theoretical calculations and simulations correlates with experimental observations and hence offers deeper insights into the fundamental mechanism [19]. Liu et al. [20] used corrosion inhibitors of the type quaternary ammonium surfactants: cetyl trimethyl ammonium bromide (CTAB), dihexadecyl dimethyl ammonium bromide (DDAB), and dodecyl dimethyl benzyl ammonium bromide (DDBAB), for the protection of anode in an AAB with an alkaline electrolyte. The work found that the efficiency of inhibitors is in the order of DDBAB>CTAB>DDAB, mainly due to the difference in the structure and interaction with Al ions, and was studied from various analyses including hydrogen evolution and electrochemical impedance study. Theoretical calculations employing the molecular structure of surfactants, using the density functional theory, correlated with the structure and inhibition effects. The presence of surfactant contributed to enhancing the specific capacity along with an increase in the voltage, contributing to the increase in the performance of AAB.

Mori et al. [21] reported an AAB assembly with a ceramic Al ion conductor $Al_2(WO_4)_3$ to protect the Al anode from corrosion by contact with an alkaline sodium hydroxide (NaOH) electrolyte. The ceramic Al ion conductor was placed both over the anode and underneath the air cathode. The cell successfully worked for a week. Moreover, the formation of $Al(OH)_3$ and aluminium oxide (Al_2O_3), evidenced from X-ray diffraction (XRD) analysis, as a by-product showed the successful reaction product of the charge–discharge process. They worked to prevent the formation of by-products on the electrodes by adding ceramic fillers such as Al_2O_3 and aluminium tungsten oxide between the electrolyte and the electrode. Figure 19.1 shows the illustration of AAB with ceramic

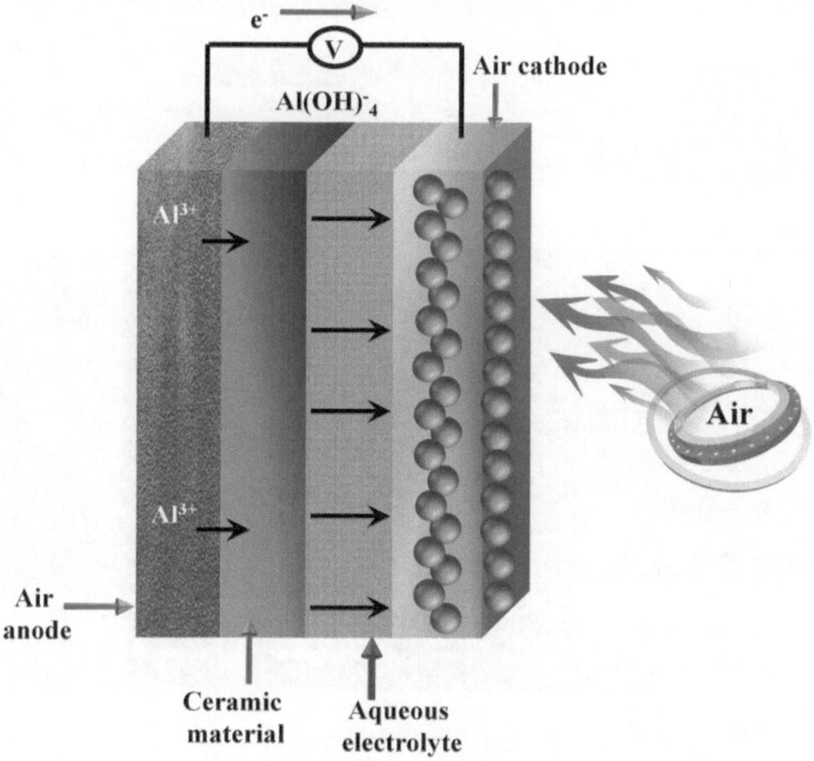

FIGURE 19.1 Representation of AAB with ceramic material placed between the electrolyte, the anode, and the air cathode. Adapted and reproduced with permission from Ref. [22]. Copyright © 2014 The Electrochemical Society.

material between electrolytes and electrodes. It was observed that the cell was rechargeable when the electrolyte medium was alkaline (NaOH and KOH) due to the ionic conductivity of sodium or potassium ion along with the $Al(OH)^{4-}$ ion. The work succeeded in preventing the evaporation of sodium chloride (NaCl) electrolyte by adding glycerin, which has a high boiling point. However, by the addition of glycerin, the cell capacity was reduced as glycerin acted against the electrochemical reaction in the electrolyte.

In another study, Mori et al. [23] reported that, in NaCl electrolyte-based AAB, using trite aluminium oxide to cover the Al anode, which is an Al ion conductor, can prevent the corrosion of the anode by the contact with electrolyte. Moreover, the accumulation of reaction by-products was suppressed as the aluminium oxide acted as a buffer. The work reported the electrolyte evaporation, and it needs to be further rectified. Hong et al. [24] were successful in developing a low-cost, durable, and efficient oxygen reduction reaction (ORR) electrocatalyst based on cerium-doped cupric oxide nanoplatelets (Cu–Ce–O) for alkaline AAB. Figure 19.2 shows the schematic illustration of various steps involved in its synthesis and characterizations with the morphological images obtained from scanning transmission electron microscope (STEM), transmission electron microscopy (TEM), and high-resolution TEM (HRTEM) of Cu–Ce–O oxide. The synergistic effects of copper and cerium contributed to enhanced electrochemical performance, and hence its electrocatalytic activity was comparable with commercial platinum nanoparticles supported on carbon (Pt/C) catalysts. Moreover, the air cathode prepared by coating carbon black (CB) over Cu–Ce–O (Cu–Ce–O/CB) showed a better discharge voltage plateau, and it outperformed commercial Pt/C.

The unique advantages of Al such as remarkably high theoretical specific energy density (8.1 kW h kg^{-1}), high theoretical voltage (2.7 V), abundance of raw materials (Al is the most abundant metal element in Earth's crust at 8.1 wt.%), low cost compared to zinc metal, and safe operation make it

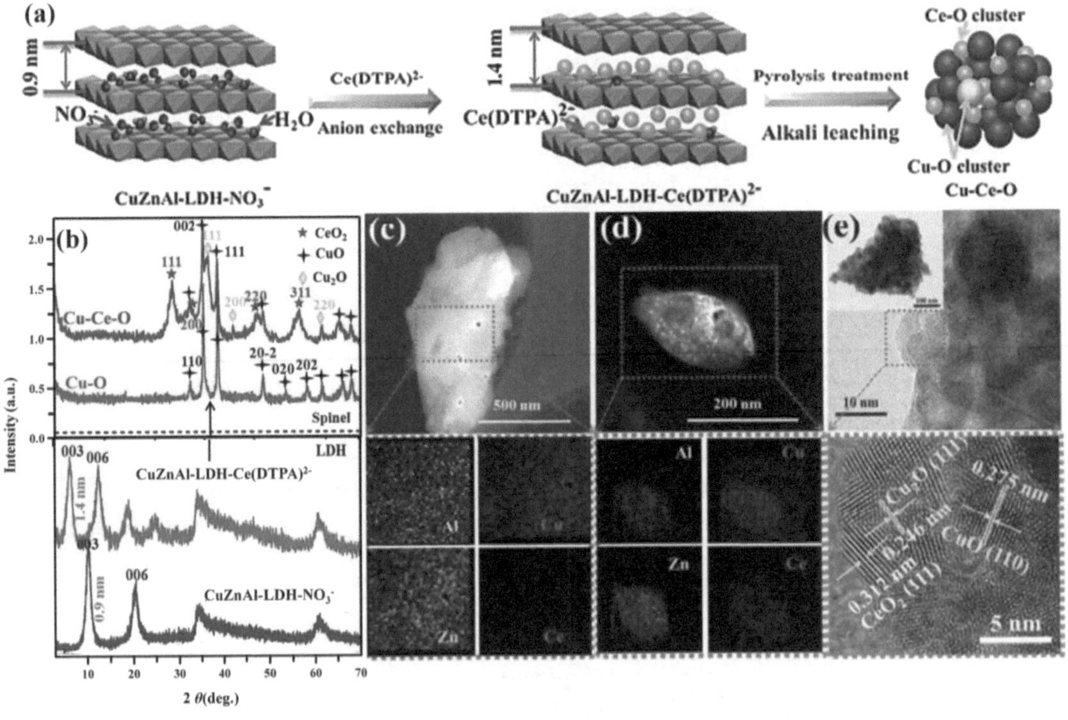

FIGURE 19.2 (a) Schematic representation of the synthesis of Cu–Ce–O oxide; (b) XRD of compounds formed during the reaction; (c) STEM and elemental mapping analysis of (d) STEM and elemental mapping analysis of (e) the morphology of Cu–Ce–O oxide from TEM and HRTEM. Adapted and reproduced with permission from Ref. [24]. Copyright © 2017 American Chemical Society.

a potential candidate for fabricating the MABs [25–27]. A common problem with all MABs is the evaporation of electrolytes with time, leading to the failure of the system. In AABs, conversion of Al into $Al(OH)_3$ and Al_2O_3 as by-products leads to the deterioration of the Al anode [28]. Hence, conversion of $Al(OH)_3$ and Al_2O_3 back to Al is very difficult, disabling the rechargeability of batteries. Continuous efforts have been put together by researchers to fabricate rechargeable AABs, which may have wide applications ranging from small-sized/miniature devices to hybrid EVs.

19.3 ANODE RESEARCH IN AABS

In AABs various anode compositions have been widely investigated and compared with the commercial ones. Ideally, an anode must have a high negative open circuit voltage (OCV) with high corrosion resistance, the least anodic polarization, and high anode efficiency. Various techniques have been introduced to enhance the efficiency of the anode. Alloying Al with other metals is one of the most explored techniques, thereby decreasing anodic overpotential loss and mitigating the self-corrosive nature of the Al anode [29]. For alloying, (1) the element should have solubility homogenization and uniform distribution with Al metal; (2) it must have a low melting point to form a solid solution; (3) the hydrogen overpotential must be high; (4) the element should be higher in electrochemical series with higher nobility; and (5) it should be compatible with the electrolyte and soluble in it [30]. Moreover, by alloying, the overpotential reduces and hence parasitic reactions are suppressed. Although, from the thermodynamic aspect, the Al anode is expected to deliver a potential of −2.35 V in alkaline and −1.66 V in saline electrolytes, the voltage drop is experienced in practical operation [31]. Lowering the potential is mainly contributed by the formation of oxide and hydroxide layers, causing a delay in the voltage [32]. In addition to this, parasitic reactions, which corrode Al, especially in aqueous electrolytes, lower the effective utilization of Al with hydrogen evolution, which impedes the coulombic efficiency [33]. Hence, various approaches have been employed to rectify the aforementioned issues. The following section highlights the development in anode engineering for AAB.

19.3.1 PRESSED, PRINTED, AND COATED AL ANODES

Various processing treatments of the Al anode have alleviated the self-corrosion in the anode, enhancing discharge performance and hence contributing to the overall improvement in the performance of AAB [34]. The process refines the microstructure and leads to better homogenization in the case of Al alloy anodes, enhancing their discharge properties and anodic efficiency. Fan et al. [35] fabricated AABs with different coarse-grain Al anodes using the equal channel angular pressing method producing ultrafine grained (UFG) Al and coarse-grained (CG) microstructure of Al, as an anode in alkaline electrolytes (NaOH and KOH) and NaCl electrolyte. Figure 19.3 shows the scanning electron microscopic (SEM) images of UFG and CG Al anodes after discharge using NaCl and NaOH electrolytes and observed the neutral condition provided by NaCl corroded equally to both UFG and CG Al. In alkaline conditions both UFG and CG Al performed well, with UFG Al having better corrosion resistance.

Yin et al. [36] investigated the applicability of Al alloy, Al–0.5Mg–0.1Sn–0.05Ga–0.05In (wt.%), as anode for alkaline and NaCl electrolyte-based AABs. Al alloys were as-cast and as-rolled for studying their effects on performance as anodes. It was reported that as-rolled alloy showed better discharge properties with high anode efficiency and voltage in both alkaline (4 M NaOH) and neutral (2 M NaCl) solutions. In an alkaline medium, rolled alloy showed a better corrosion inhibition rate but in a neutral (NaCl) solution higher rate of corrosion.

Xu et al. [37] found that usage of calcium-containing Al anode for alkaline AAB showed higher capacity density with better anode efficiency. The incorporation of calcium modified the microstructure with a finer grain size of the commercial Al alloy 1070. It was also reported that as the content of calcium increased, polarization resistance and discharge voltage increased with better

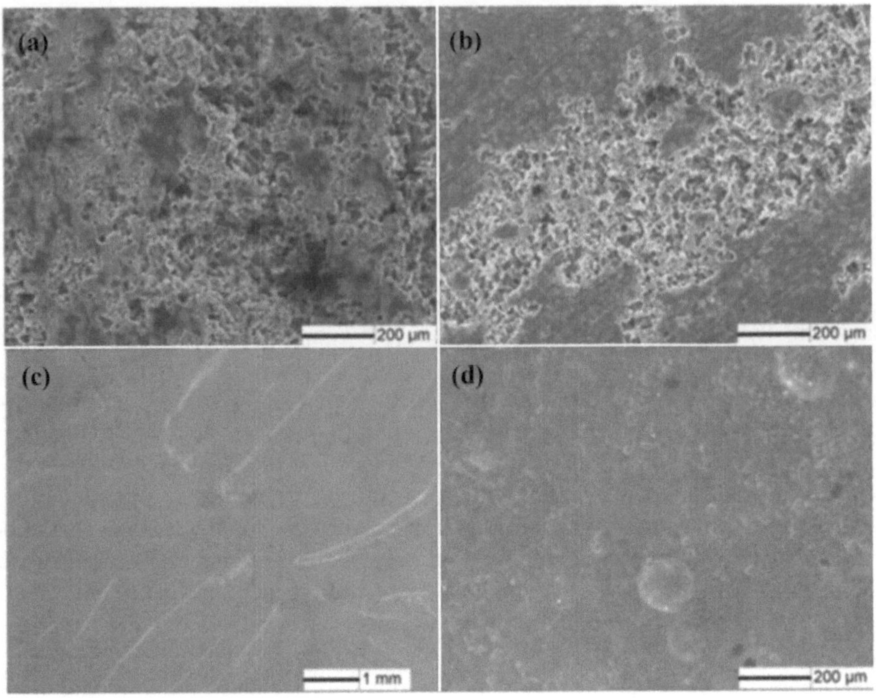

FIGURE 19.3 SEM images of various Al anodes after discharge using electrolytes: (a) CG Al anode with electrolyte as 2 M NaCl, (b) UFG Al anode with electrolyte as 2 M NaCl, (c) CG Al anode with electrolyte as 4 M NaOH, and (d) UFG Al anode with electrolyte as 4 M NaOH. Adapted and reproduced with permission from Ref. [35]. Copyright © 2019 Elsevier.

anode efficiency. Mutlu et al. [38] also worked on improving anode performance in AAB by depositing copper using chemical and electrochemical processes. It was found that by cost-effective electrochemical processes, active centres were created for the copper-deposited Al, and it sustained corrosion by forming protective films on contact with an alkaline electrolyte. Yu et al. [39] reported printed Al as an anode for gel KOH electrolyte AAB by laser sintering technique with viscous Al slurry. Moreover, later sintering was employed for removing the organic solvent, which was in the slurry, hence enhancing the conductivity and electrochemical performance of AAB. The air cathode was made from conductive carbon paper coated with platinum mixed with CB, which is hydrophobic in nature with a porous structure. Moreover, the gel electrolyte is made from polyvinyl alcohol and water solution with KOH, which favoured high ionic conductivity in the cell. Overall, the work successfully fabricated AAB that delivered an energy density of 227 mWh g^{-1} and a capacity of 239 mAh g^{-1} when operated at 0.95 V.

The effect of the electrochemical performance of Al anodes with different grain sizes was studied by Han et al. [40]. Anodes were prepared with different grain sizes by the addition of Al–Ti–B refiner in the casting process. It was evidenced from various tests that the smallest grain-sized Al anode of ~100 μm showed uniform microstructure and hence exhibited best performance with a high specific capacity and large working voltage. The work showed a cost-effective and simple procedure for enhancing the performance of AAB.

Wu et al. [41] investigated the performance of rolled and heat-treated Al–0.5Mg–0.1Sn–0.05Ga Al alloy with various grain sizes for the anode in alkaline AAB. Compared to a fine-grained structure, CG microstructure showed a lower rate of corrosion tested by the hydrogen evolution test. In 4 M NaOH solution, the CG microstructure (6,532 mm), produced by a combination of solidification and rolling process followed by annealing, showed anodic efficiency of 84.9%. Moreover, the specific capacity reached a value of 2,529.7 $AhKg^{-1}$ and a peak energy density of 3,743.9 $Whkg^{-1}$ at

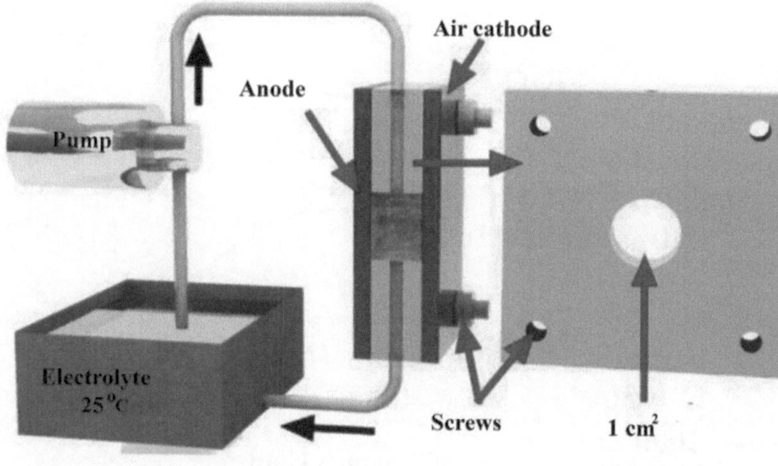

FIGURE 19.4 Schematic representation of AAB for testing. Adapted and reproduced with permission from Ref. [42]. Copyright © 2020 Elsevier.

40 mA cm^{-2}. Zheng et al. [42] studied the effect of friction stir processing (FSP) on low-cost commercial purity grade Al (CP-Al) as anode for alkaline AAB. It was reported that the self-corrosion nature of the anode alleviated upon FSP with improvements in the electrochemical activity and anodic efficiency. The average grain size reduced considerably from 500 to 3 micrometre with fragmentation and reduction in Al–Fe phase. AAB was fabricated with FSP Al anode, MnO_2 catalyst-loaded air cathode, and 4 M NaOH as electrolyte with a 1 cm^2 area of electrodes kept exposed to the electrolyte. Figure 19.4 is the schematic representation of the fabricated AAB during electrochemical analysis. Anode made by FSP delivered stable discharge performance and at a current density of 20 mA cm^{-2}; the energy density was recorded as 1,674 mWh g^{-1}.

19.3.2 Alloy Anodes: Commercial vs Fabricated

Nestoridi et al. [43] conducted a comparative study on the usage of pure Al as an anode and Al with Mg, Sn, and/or Ga as an alloy for NaCl-based electrolytes and its applicability to AABs. Evidence of corrosion was observed both visually and from SEM, when the anode was inserted in 2 M NaCl. It was seen that H_2 bubbles were formed on the anode, and subsequently a white precipitate was formed with surface irregularities, which is attributed to the formation of Al_2O_3 or $Al(OH)_3$. Alloys with Mg showed resistance to corrosion. Moreover, it was found that Sn initiated the corrosion and Ga helped prevent the passivation of the initiated corrosion. It was also found that corrosion depends not only on the composition of the alloy but also on the treatment of the alloy such as heat treatments and mechanical work for making the alloy and the distribution of minor elements.

Pino et al. [44] fabricated primary alkaline AABs with commercial Al alloys (Al2000 Unclad, Clad, and Al7000) as an anode in a KOH electrolyte. Among various alloys, unclad Al2000 alloy showed better performance with low self-corrosion, lower anodic polarization, specific capacity, and high electrochemical activity. Figure 19.5 shows a comparative polarization curve of various anodes, Al2000 Clad, Al2000, and Al7000 with NiOOH and air cathodes, in a 4 M KOH solution. The work further proposed adding additives to minimize the self-corrosion of the anode.

Smoljko et al. [45] investigated the applicability of using indium- and tin-containing alloys of Al as an anode with NaCl as an electrolyte for AABs. The work compared five various grades of Al, namely, pure Al (99.999%), technical-grade Al (99.8%), Al with 0.1% In, 0.2% tin (Sn), and a combination of 0.1% In and 0.2% Sn with Al as alloy for anode in AAB. As per the hydrogen evolution test, introducing In and Sn reduces electrode polarization with increasing anode corrosion.

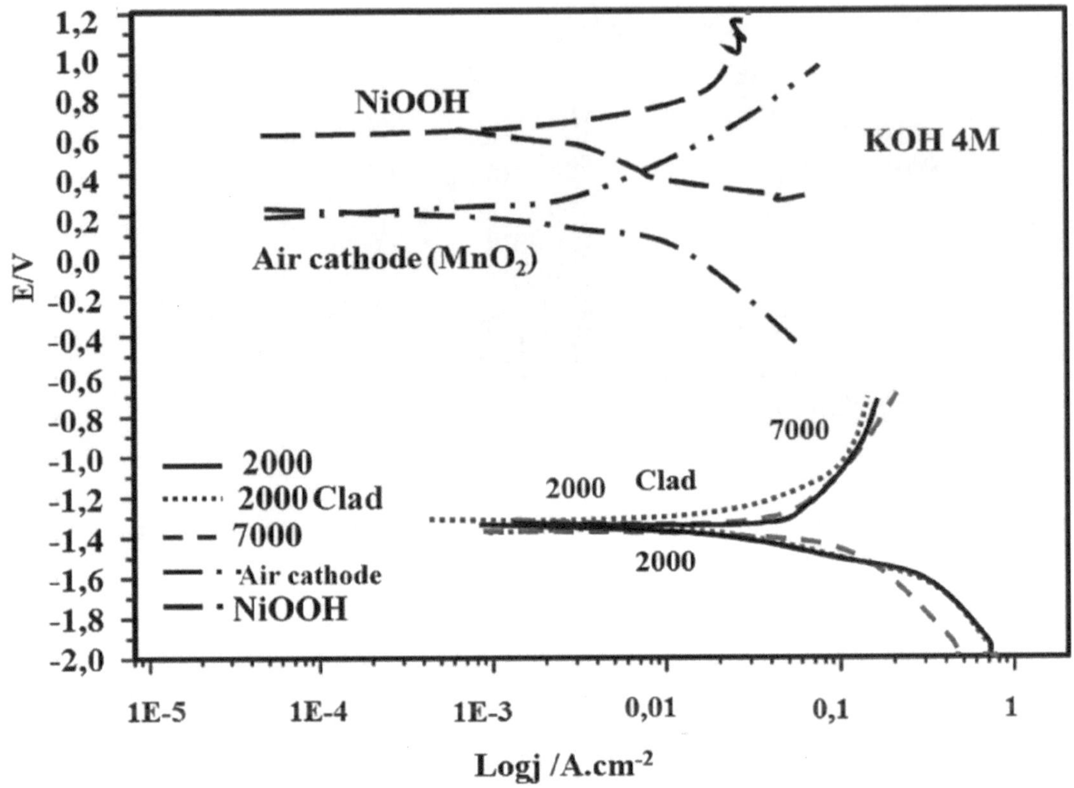

FIGURE 19.5 Comparative polarization curves of Al2000Clad, Al2000, and Al7000 anodes in a 4 M KOH solution and with NiOOH and air cathodes at a scan rate of $v = 1$ mV s^{-1} and 25°C. Adapted and reproduced with permission from Ref. [44]. Copyright © 2014 Elsevier.

However, Al–In alloy with 0.1% In showed lower corrosion with high anode efficiency. Ma et al. [46] investigated the performance of Al, Zn, Al–Mg–Ga–Sn, and Al–Mg–Ga–Sn–Mn as anodic material for AAB by testing its corrosion behaviour in a 2 M NaCl solution. It was seen that when Al–Mg–Ga–Sn–Mn showed better corrosion resistance with higher operating voltage, better anodic utilization of 85.3%, and better performance than others. EIS and SEM results of the anode after discharge showed that Al–Mg–Ga–Sn–Mn has the highest charge transfer resistance (R_t) corresponding to lower corrosion and has loose cracks in the surface. The incorporation of Mn in the alloy caused a reduction in corrosion.

Sun et al. [47] investigated the performance of an anode made by alloying Al with Ga, In, and Sn for application in AABs. The alloys, namely Al–0.1 wt.% Ga, Al–0.1 wt.% In, and Al–0.1 wt.% Sn, were tested in 1 M KOH solution and 2 M NaCl solutions. Electrochemical measurements were performed using a traditional three-electrode glass cell of 300 mL capacity and with platinum foil as the counter-electrode. It was seen that Al–0.1 wt.% In or Al–0.1 wt.% Sn alloys showed better corrosion resistance and performance with better anodic efficiency in 1 M KOH than pure Al, because of the deposition of In and Sn over the Al surface due to the high hydrogen evolution overpotential, hence protecting the anode from corrosion. In the 2 M NaCl solution, all the alloys showed better performance by activating the Al and showed a lower self-corrosion rate. Moreover, AAB with Al–0.1 wt.% In as anode at 10 mA cm^{-2} showed a power density of 3,267.6 Wh Kg^{-1} and Al–0.1 wt.% Sn as anode delivered a power density of 3,343.2 Wh Kg^{-1}.

Pino et al. [48] fabricated AABs based on various alloys of Al (Al2024 and Al7475, where Cu and Zn, respectively, were the dominant species) as anode material and compared them with a commercial Al alloy anode. Alkaline gel electrolytes containing zinc oxide (ZnO) and zinc chloride ($ZnCl_2$)

as corrosion inhibitors to Al were added as filler. The work studied clad and unclad Al alloys. The gel electrolytes were prepared using KOH as solvent containing ZnO and $ZnCl_2$ inhibitors, to which the gelling agents, acrylic acid and *N,N'*-methylene-bisacrylamide, were added. Polymerization initiator potassium dithionite ($K_2S_2O_4$) was added to initiate the polymerization reaction and performed till the product was a transparent gel. The Al alloys were (Al2024 clad and unclad (Al2C and Al2U), Al7475 clad and unclad (Al7C and Al7U), and Al1085 unclad (Al1U)) tested in the gel cell assembly vs two air cathodes. SEM and energy-dispersive X-ray analysis showed a reduction in corrosion with the usage of ZnO inhibitors. Among various Al alloys, AAB containing Al7U exhibited a maximum specific capacity of 426 mAh g^{-1}, which can be comparable with the performance of low-cost commercial Al alloys, showing the importance of the work.

Ma et al. [49] investigated the electrochemical behaviour of various Al alloys as anodes for alkaline and neutral NaCl electrolyte-based AABs. It was observed that in 2 M NaCl electrolytes, Al–0.5Mg–0.1Sn–0.02Ga –0.1Si showed a high cell potential of 1.13 V with anodic utilization of 78.1%. However, Al–0.5Mg–0.1Sn–0.02In–0.1Si alloy showed better results in 4 M NaOH solution with a cell potential of 1.47 V and 29.2% anodic utilization. In both alloys the incorporation of Si contributed to enhancement in discharge performance and reduced the rate of corrosion. EIS analysis and SEM images of anode after discharge are in line with the observation, showing the enhancement in performance when Si is added to the Al as alloy.

Ilyukhina et al. [50] conducted a comparative study on the performance of pure Al and its alloy with indium (Al–0.45In) as anode for alkaline AAB. Polarization characters varied as the reaction product layers were different in alloy and pure Al anode. The alloy showed 1.934 V as OCV and at 1.3 V; the cell delivered a maximum power density of 240 mW cm^{-2}. Moghadam et al. [51] found that the incorporation of urea and thiourea in a 5 M KOH electrolyte could resist the corrosion of Al–Mg–In–Mn alloy without compromising the discharge performance. Moreover, anodic utilization of Al alloy in thiourea showed the maximum value of 86.1%, followed by 85.4% in urea, and the least was shown in the blank electrolyte. The corrosion is prevented by the adhesion of urea and thiourea over the alloy, which was evidenced by hydrogen evolution test by electrochemical methods: Tafel polarization, EIS, and galvanostatic discharge, which confirmed its applicability as anode for AAB.

Liang et al. [52] found that Al alloys (Al–0.05Ga–0.05Sn–0.05Pb–*x*Mg) with varying content of Mg for the anode in AAB have a great influence on the performance due to the difference in the corrosion behaviour. The study was conducted by varying the Mg content in the alloy from 0.05 to 0.15 wt.%. It was found that the alloy with 0.1 wt.% showed proper grain size and active areas with active grain size from SEM and energy-dispersive spectroscopy and XRD. The same alloy showed lower H_2 evolution and mass loss with a negative polarization potential value of −1.54 V and a higher anode utilization of 98%.

Liu et al. [53] investigated the applicability of alloying pure Al with varying content of pure Sb for the anode application with corrosion protection and enhanced performance in AABs. The alloys used in the study were Al–0.02Sb, Al–0.06Sb, Al–0.18Sb, and Al–0.36Sb alloy. Sb with its ability to form intermetallic AlSb compound inhibits corrosion. Morphological analysis of various alloy anodes after discharge (Figure 19.6) reveals the presence of grain boundaries and AlSb precipitates from the morphological images. Pure Al showed deep pits and peeling units with grain boundaries. It was also found that oxidation of micro-nano AlSb provided a film over the cathode, which decreased the formation of passive film on the alloy anode. From various compositions, Al–0.06Sb0.053 wt.% Sb in the alloy showed the best results with anodic efficiency of 98.2% at a current density of 40 mA cm^{-2} and at a current density of 20 mA cm^{-2}; the cell delivered a high energy density of 3,781 Wh kg^{-1}.

Zhang et al. [54] investigated the performance of Al–Sb alloy as an anode for AAB. The alloy is prepared by heat treatment, and its discharge performance as an anode was evaluated by changing process conditions. It was observed that Al–0.02Sb alloy, which was treated at 550°C for 3 h, showed better anodic efficiency (92.6%) and 2,758 Ah kg^{-1} as discharge capacity with the least

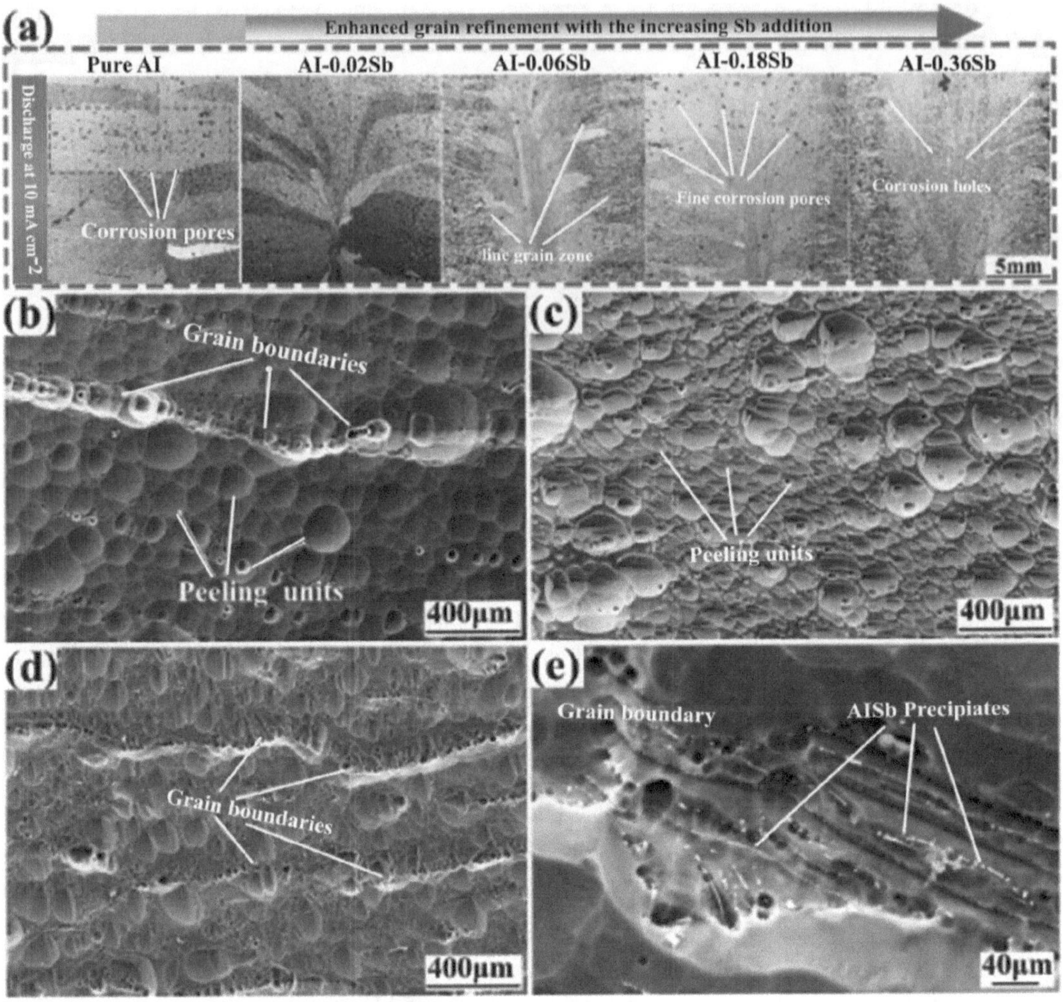

FIGURE 19.6 Morphological analysis, after discharge at 10 $mA \cdot cm^{-2}$ for 6h, of Al–Sb alloy anode. (a) Various macroscopical images with increasing content of Sb in Al–Sb alloys. (b–e) SEM images of (b) Pure Al as anode; (c) anode as Al–0.02Sb alloy; (d) and (e) Al–0.18Sb alloy anode. Adapted and reproduced with permission from Ref. [53]. Copyright © 2019 Elsevier

evolution of H_2. Moreover, the prolonged treatment contributed to grain growth, leading to the precipitation of alloy, hence contributing negatively towards the anodic performance.

Wu et al. [55] studied the electrochemical properties and corrosion behaviour of various Al alloys containing In, Ga, Bi, Mg, and Sn in a 4M NaOH solution. The work reported that, when indium was added to alloy (Al 0.5–Mg 0.1–Sn 0.05–In), anodic efficiency was enhanced along with the enhancement in discharge activity, while suppressing the H_2 evolution reaction. Morphological analysis by SEM recorded after the discharge showed the presence of smaller pores with smooth surfaces, hence supporting the better performance of Al 0.5–Mg 0.1–Sn 0.05–In as a suitable anode for AAB.

In other reports Wu et al. [56] demonstrated that by incorporation of Ga and indium into the Al and alloying, it can reduce the self-corrosion and improve the electrochemical properties of 2M NaCl solution-based AAB. Introduction of In generates discharge products (In and $In(OH)_3$)), which form passivation film over the anode; Ga incorporation dissolves the film and negatively contributes to corrosion. X-ray photoelectron spectroscopy (XPS) results reveal the discharge products present

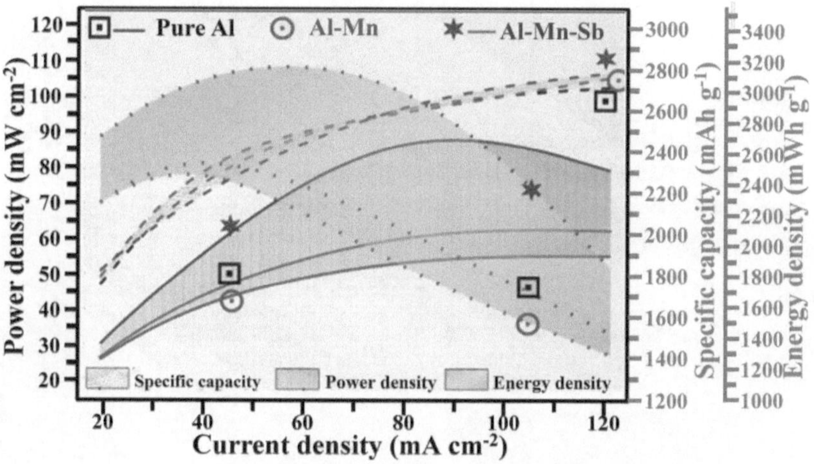

FIGURE 19.7 Discharge performance of various Al anodes in 4 M NaOH solution. Adapted and reproduced with permission from Ref. [57]. Copyright © 2020 Elsevier.

in the electrolyte when various alloys have been used. Moreover, the incorporation of both can contribute to enhanced discharge activity with higher anode efficiency.

Liu et al. [57] investigated Al–Mn–Sb alloy as an anode material for AABs delivering high power/energy. The work compared the performance of pure Al, two alloys of Al (Al–0.05Mn and Al–0.05Mn–0.05Sb) in a 4 M KOH solution, and the discharge performance, which is shown in Figure 19.7. From electrochemical characterizations, it was found that Al–0.05Mn–0.05Sb alloy exhibited a higher discharge voltage of 1.206 V and peak power density of 96 mW cm^{-2}. The Al–0.05Mn–0.05Sb alloy showed the least H_2 evolution rate of 0.32 mL $min^{-1}cm^{-2}$ and high corrosion resistance compared to pure Al and Al–0.05Mn alloy. It can be due to the protective nature of the micro-nano AlSb precipitate formed on the grain boundaries.

Hosseini et al. [58] reported that the functional groups of the additive (acetoxy groups, which are weak bases) showed inhibition to the undesirable self-corrosion of Al anode in AAB, fabricated using KOH as an electrolyte. The introduction of additives decelerated the evolution of H_2, hence contributing to the enhanced electrochemical performance of cells. Yang et al. [59] incorporated polyol compounds of butantetraol (BTT) and dithiothreitol (DTT) as corrosion inhibitors by incorporating in a 4 M NaOH electrolyte for the protection of AA5052 alloy as anode in AAB. It was observed from infrared studies and XPS that coordination bonds between polyols and Al ions confirmed the formation of complex films which protected the anode from further corrosion. The presence of hydrosulphonyl group in dithiothreitol contributed to the formation of a six-membered ring in an alkaline medium, which acts as a protective film. Figure 19.8 is the schematic illustration of protecting the AA5052 alloy anode with butantetraol and dithiothreitol containing 4 M NaOH as electrolytes. The work highlighted the need for an effective additive to suppress self-corrosion of the Al alloy anode, thereby enhancing the performance of AABs.

The effect of annealing Sn–Ga–Mg containing an alloy of Al and its applicability as anode for AAB was investigated by Han et al. [60]. It was seen that varying annealing temperature to Al–0.08Sn–0.08Ga–0.5Mg alloy anodes has direct effects on the electrochemical performance of AAB. When the annealing temperature was at 350°C for 10 h, AAB with 4 M NaOH as electrolyte showed a current density of 120 mA cm^{-2}, energy density is 3,379.15 mWh g^{-1}, and capacity density is 2,670.99 mWh g^{-1}.

Katsoufis et al. [61] conducted a study where commercial-grade Al alloy, Al-6061, containing mainly Mg and Si (1.44–4.15 wt. %), was investigated for their application as anode for AAB. Air cathode was fabricated using CB (Vulcan XC72)-coated carbon cloth with and without a few drops of diamminedinitritoplatinum(II) as platinum (Pt) catalyst and with aqueous KOH of

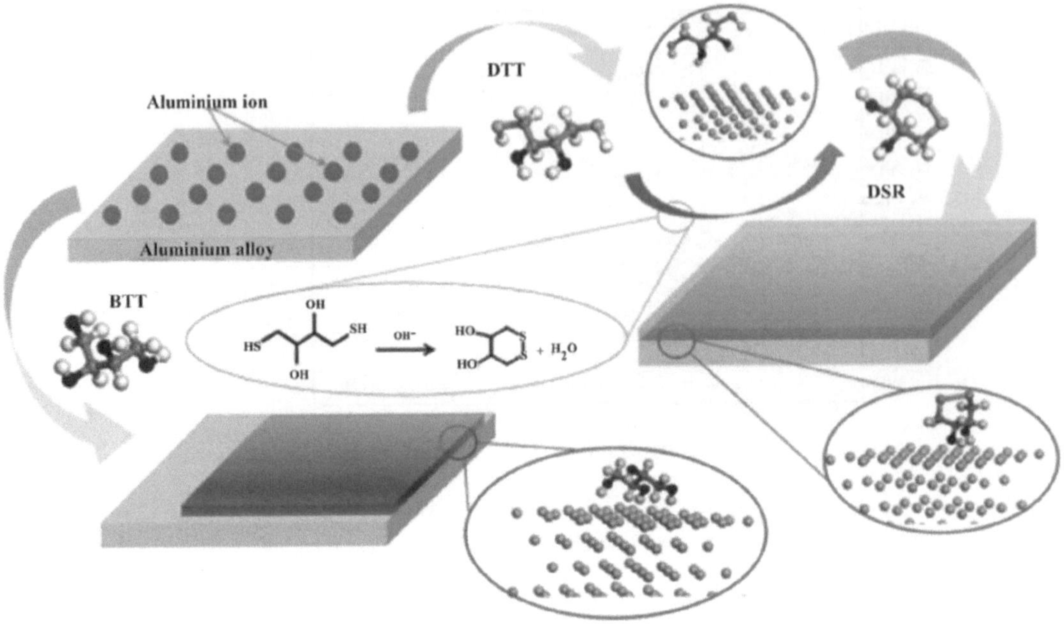

FIGURE 19.8 Schematic illustration showing the protection for AA5052 alloy anode with butantetraol and dithiothreitol containing 4 M NaOH as electrolyte. Adapted and reproduced with permission from Ref. [59]. Copyright © 2020 Elsevier.

various concentrations as electrolyte. It was found that the device worked both in the presence and in the absence of O_2. In the absence of O_2, hydrogen peroxide was supplied, which resulted in the generation of high OCV. A higher current was delivered when carbon cloth with carbon black-containing Pt was employed as the air cathode. Since, in the Al-6061 alloy, the grain boundaries were prominent with the presence of impurities, the H_2 evolution reaction due to corrosion was evident.

Recently, Wu et al. [62] found that the incorporation of Ca into Sn-containing Al alloy enhances the discharge properties with high anodic efficiency of in an alkaline AAB. The authors reported that self-corrosion was reduced by the precipitation of the CaSn and Al_4Ca phases. This was due to the formation of a protective film by the oxidation of CaSn forming a calcium hydroxide ($Ca(OH)_2$) film. The Al–0.1Sn–0.08Ca alloy showed an anodic efficiency of 83.75%. Moreover at 20 mA cm^{-2} AAB showed a discharge capacity of 2,495.75Ah kg^{-1} and 3,793.54 Wh kg^{-1} as peak energy density. The work reported that at high current density the protective $Ca(OH)_2$ film discharge behaviour was negatively affected. Hence this work showed the applicability of Al–0.1Sn–0.08Ca alloy anode for low current density AABs.

Ren et al. [63] conducted a study to investigate the effect of impurity elements in Al anode relating to its performance properties. It was evidenced from various analyses that the presence of impurity elements like Fe and Si in the Al anode can contribute negatively towards self-corrosion of the anode and lowering anodic efficiency. Moreover, the presence of Zn and Ga in Al activated electrochemical activity.

19.4 RECENT DEVELOPMENT IN AAB: FLEXIBLE, PAPER-BASED AABS

Early developments in the AABs were reported in the 1960s [64]. Following that many developments and innovations in AAB technology have been reported, such as the development of solid-state AABs, the introduction of printable AABs, alloy anodes, and additives to suppress H_2 evolution, are a few examples [9,65]. Portable and cost-effective, lightweight AABs have been developed in recent times [66]. Pan et al. [67] developed cotton-based AA, consisting of alkaline electrolytes, which showed a peak power density of 73 mW cm^{-2}. Wang et al. [68] reported various works on the development of paper-based AAB suitable for operating in mini-watt applications, where cellulose-based paper is used as electrolyte substrate. The incorporation of paper-based solid electrolytes eliminated leakage issues and ensured flexible battery packing [69]. Instead of a liquid electrolyte, the cell consists of porous cellulose paper, acting as a gel electrolyte, and showed OCV of 1.5 V with a peak power density of 3.8 mW cm^{-2}.

In another work, Wang et al. [68] developed a mini-watt application battery, which performed until the anode was exhausted. The advantage of such system is that it can be replaced, and the components of the cell can be reassembled as each of the components is independent. The incorporation of electrolyte by dropping to the cellulose paper is delivered by capillary action, and hence the electricity is generated based on the reactions represented in Figure 19.9. The possible side reaction is also mentioned in Figure 19.9, which accounts for a small proportion. The cell showed 1,273 mAh g^{-1} as its specific capacity and showed an OCV of 1.6 V, with a peak power density of 21 mW cm^{-2}.

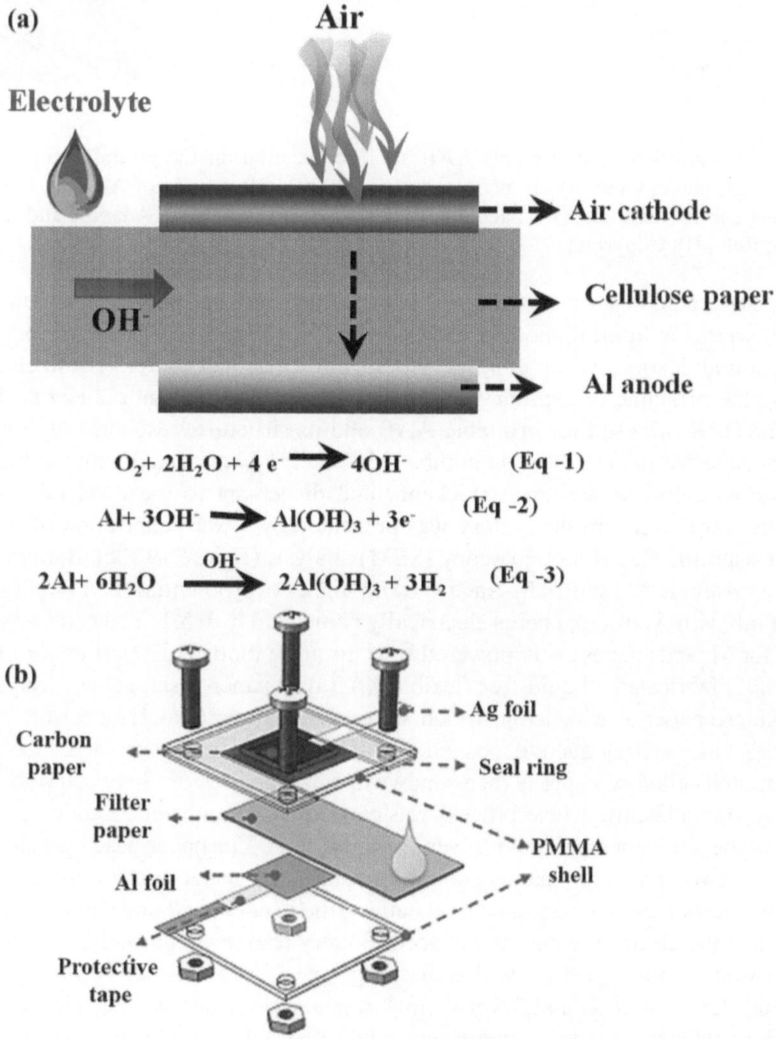

FIGURE 19.9 Prototype on paper-based AAB with (a) the working principle; and (b) components of the cell. Adapted and reproduced with permission from Ref. [70]. Copyright © 2017 Elsevier.

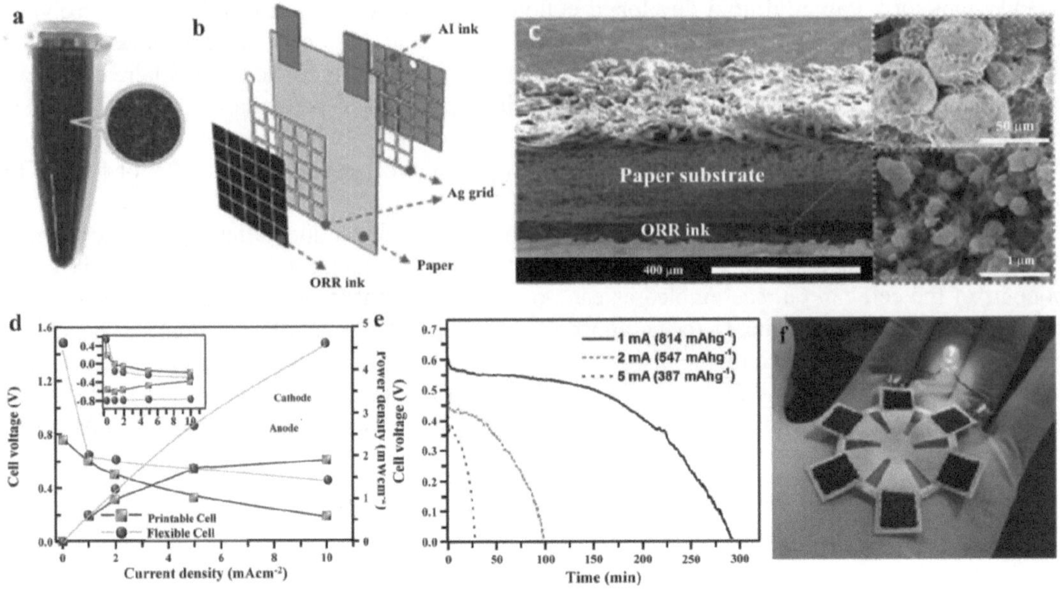

FIGURE 19.10 Overall view of a printable AAB. (a) The ink used for the printable cell; (b) structure of AAB; (c) surface morphology (SEM) of anode and cathode; (d) performance of AAB in 4M NaCl; (e) discharge at different current densities; (f) a pack of six cells connected to a LED. Adapted and reproduced with permission from Ref. [71]. Copyright © 2019 Elsevier.

Moreover, when operated in saline condition it operated for 5.6h at 1 mA cm^{-2}. Furthermore, when a five-pack cell series configuration delivered an OCV of 7.7V with 97% stacking efficiency [70].

In a short communication, Wang et al. [71] used cellulose paper to deliver the electrolyte automatically using the principle of capillary action and Ag grids as current collectors. Figure 19.10a and b shows the ORR ink used for printable AAB and its structure, respectively. For the preparation of ink, manganese dioxide/carbon nanotube (MnO_2/CNT) was mixed with the binder (Nafion), and carboxymethyl cellulose acting as thickener and dispersant to these Al microparticles was then added. The short circuit in the battery was prevented by lower penetration of ink, which was observed from scanning electron microscopy (SEM) analysis (Figure 19.10c). Battery performance of the printable system tested with 4M NaCl showed higher overpotential than the flexible type due to the usage of ink with Al microspheres electrically connected by CNT. The cell showed a continuous discharge for 6h and successfully powered light emitting diode (LED), shown in Figure 19.10f.

Wang et al. [72] fabricated a liquid-free flexible AAB on a paper-based gel electrolyte. Electrolyte consists of cellulose paper as a skeleton, which stores the alkaline (NaOH) electrolyte with sodium polyacrylate (SPA) as a gelling agent deposited over it, hence aiming for portable electronic applications. SPA-deposited cellulose paper is then sandwiched at high pressure between poly(methyl methacrylate) (PMMA) moulds; the whole process was carried out in ambient conditions. The schematic representation of the fabrication process is given in Figure 19.11. The paper-based gel electrolyte-based AAB consisted of two prototypes, namely, a mechanical-rechargeable battery and a flexible battery. Moreover, in the mechanical-rechargeable type battery, the battery shell and cathode can be reusable, and the anode and the electrolyte can be replaced. Battery testing performed by an electrochemical workstation showed promising results with a discharge-specific capacity of 900mA g^{-1} (98.2% pure Al foil as anode), 1.5V as OCV, and 3.5mW cm^{-2} as peak power density. Furthermore, battery performance in a four-stack series configuration showed 87.5% stacking efficiency with an OCV of 5.6V.

Katsoufis et al. [73] investigated paper-based thin film alkaline AAB. It consisted of a thin, porous paper soaked in 2M KOH electrolyte, a commercial-grade Al alloy plate, Al-6061, as an anode, and electrocatalyst on a carbon black deposited carbon cloth (CB/CC) as air cathode.

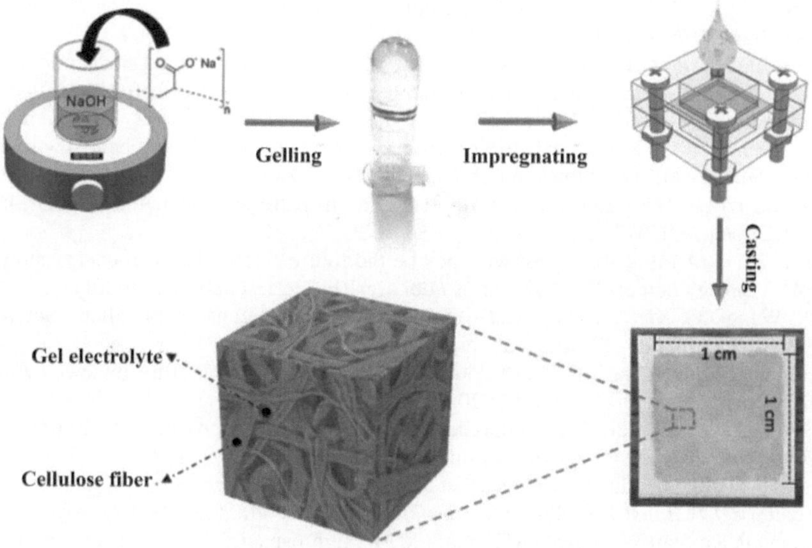

FIGURE 19.11 Flow chart and fabrication process represented schematically. Adapted and reproduced with permission from Ref. [72]. Copyright © 2019 Elsevier.

The incorporation of MnO_2-loaded ORR catalyst to the carbon black-containing carbon cloth (MnO_2-CB/CC) showed enhancement in the electrochemical performance of the cell. The cell with MnO_2-CB/CC showed OCV of 1.45 V with 28 mW cm^{-2} as the maximum power density, and, during the discharge, the maximum energy yield was 5.5 mWh cm^{-2}.

19.5 CONCLUSION

Coupled with the abundance and cost-effective characteristics of Al, and considering the booming demand for batteries for EVs and portable electronics, AAB has been identified as a potential candidate. Recent developments on solid-state AABs, utilizing cost-effective cellulose-based electrolytes coupled with Al foil anode for mini-watt, which are mechanically rechargeable, are promising techniques for the next generation of smart devices. To meet the commercialization level, there are many bottleneck challenges associated with AAB. In practice, AAB experiences voltage drops, which negatively affects its performance. Oxides and hydroxides formed at the anode contribute to the drop, which hampers the utilization of Al and contributes to the evolution of hydrogen. Alloying anode is one of the most explored methods to mitigate the aforementioned challenges. Microstructural engineering by grain size reduction and tailoring anode by heat treatment showed inhibition to corrosion. Additives can also alter the corrosion of the anode; functional groups of additives decelerated the evolution of H_2, thereby enhancing electrochemical performance. The inclusion of theoretical calculations and simulations can guide experimentalists to improve the momentum of research. Hence juxtaposing the approaches and developing newer techniques to rectify the challenges can upgrade the performance of AAB, thereby a step towards the realization of effective utilization of this battery technology.

REFERENCES

1. Jishnu NS, Vineeth SK, Das A, et al (2021) Electrospun PVdF and PVdF-co-HFP-based blend polymer electrolytes for lithium ion batteries. In: Balakrishnan NTM, Prasanth R (eds) *Electrospinning for Advanced Energy Storage Applications*, 1st ed. Springer, Singapore, pp. 201–234
2. Vineeth SK, Soni CB, Sanjaykumar C, Yamauchi Y, Han M, Kumar V (2023) A quasi-solid state polymer electrolyte for high-rate and long-life sodium-metal batteries. *J Energy Storage* 73:108780. https://doi.org/10.1016/j.est.2023.108780

3. Vineeth SK, Soni CB, Sun Y, et al (2022) Implications of Na-ion solvation on Na anode-electrolyte interphase. *Trends Chem* 4:48–59. https://doi.org/10.1016/j.trechm.2021.11.002
4. Sungjemmenla J, Soni CB, Vineeth SK, Kumar V (2021) Unveiling the physiochemical aspects of the matrix in improving sulfur-loading for room-temperature sodium-sulfur batteries. *Mater Adv* 2:4165–4189. https://doi.org/10.1039/D1MA00247C
5. Cheng F, Chen J (2012) Metal-air batteries: from oxygen reduction electrochemistry to cathode catalysts. *Chem Soc Rev* 41:2172. https://doi.org/10.1039/c1cs15228a
6. Olabi AG, Sayed ET, Wilberforce T, et al (2021) Metal-air batteries-a review. *Energies* 14:7373. https://doi.org/10.3390/en14217373
7. Li Y, Lu J (2017) Metal-air batteries: will they be the future electrochemical energy storage device of choice? *ACS Energy Lett* 2:1370–1377. https://doi.org/10.1021/acsenergylett.7b00119
8. Zhang X, Wang XG, Xie Z, Zhou Z (2016) Recent progress in rechargeable alkali metal-air batteries. *Green Energy Environ* 1:4–17. https://doi.org/10.1016/j.gee.2016.04.004
9. Goel P, Dobhal D, Sharma RC (2020) Aluminum-air batteries: a viability review. *J Energy Storage* 28:101287. https://doi.org/10.1016/j.est.2020.101287
10. Liu Y, Sun Q, Li W, et al (2017) A comprehensive review on recent progress in aluminum-air batteries. *Green Energy Environ* 2:246–277. https://doi.org/10.1016/j.gee.2017.06.006
11. Viliševskis A, Briedis U, Juchnevičienė Ž, et al (2020) Design improvement of flexible textile aluminium-air battery. *J Text Inst* 111:985–990. https://doi.org/10.1080/00405000.2019.1676521
12. Teabnamang P, Kao-ian W, Nguyen MT, et al (2020) High-capacity dual-electrolyte aluminum-air battery with circulating methanol anolyte. *Energies* 13:2275. https://doi.org/10.3390/en13092275
13. Mori R (2020) Recent developments for aluminum-air batteries. *Electrochem Energy Rev* 3:344–369. https://doi.org/10.1007/s41918-020-00065-4
14. Wu P, Wu S, Sun D, et al (2021) A review of Al alloy anodes for Al-air batteries in neutral and alkaline aqueous electrolytes. *Acta Metall Sin (English Lett* 34:309–320. https://doi.org/10.1007/s40195-020-01140-x
15. Lee J-S, Tai Kim S, Cao R, et al (2011) Metal-air batteries with high energy density: Li-air versus Zn-air. *Adv Energy Mater* 1:34–50. https://doi.org/10.1002/aenm.201000010
16. Phusittananan T, Kao-Ian W, Nguyen MT, et al (2020) Ethylene glycol/ethanol anolyte for high capacity alkaline aluminum-air battery with dual-electrolyte configuration. *Front Energy Res* 8:1–12. https://doi.org/10.3389/fenrg.2020.00189
17. Fan L, Lu H, Leng J, et al (2015) The effect of crystal orientation on the aluminum anodes of the aluminum-air batteries in alkaline electrolytes. *J Power Sources* 299:66–69. https://doi.org/10.1016/j.jpowsour.2015.08.095
18. Ren J, Ma J, Zhang J, et al (2019) Electrochemical performance of pure Al, Al-Sn, Al-Mg and Al-Mg-Sn anodes for Al-air batteries. *J Alloys Compd* 808:151708. https://doi.org/10.1016/j.jallcom.2019.151708
19. Eng AYS, Soni CB, Lum Y, et al (2022) Theory-guided experimental design in battery materials research. *Sci Adv* 8:eabm2422. https://doi.org/10.1126/sciadv.abm2422
20. Liu Y, Zhang H, Liu Y, et al (2019) Inhibitive effect of quaternary ammonium-type surfactants on the self-corrosion of the anode in alkaline aluminium-air battery. *J Power Sources* 434:226723. https://doi.org/10.1016/j.jpowsour.2019.226723
21. Mori R (2013) A new structured aluminium-air secondary battery with a ceramic aluminium ion conductor. *RSC Adv* 3:11547–11551. https://doi.org/10.1039/c3ra42211a
22. Mori R (2015) Addition of ceramic barriers to aluminum-air batteries to suppress by-product formation on electrodes. *J Electrochem Soc* 162:A288–A294. https://doi.org/10.1149/2.0241503jes
23. Mori R (2014) A novel aluminium-air rechargeable battery with Al_2O_3 as the buffer to suppress byproduct accumulation directly onto an aluminium anode and air cathode. *RSC Adv* 4:30346–30351. https://doi.org/10.1039/c4ra02165g
24. Hong Q, Lu H, Wang J (2017) CuO nanoplatelets with highly dispersed ce-doping derived from intercalated layered double hydroxides for synergistically enhanced oxygen reduction reaction in Al-air batteries. *ACS Sustain Chem Eng* 5:9169–9175. https://doi.org/10.1021/acssuschemeng.7b02076
25. Farsak M, Kardaş G (2018) 2.12 Electrolytic materials. In: *Comprehensive Energy Systems*. Elsevier, Amsterdam, The Netherlands. pp. 329–367
26. Yuan D, Zhao J, Manalastas W, et al (2020) Emerging rechargeable aqueous aluminum ion battery: status, challenges, and outlooks. *Nano Mater Sci* 2:248–263. https://doi.org/10.1016/j.nanoms.2019.11.001
27. Korolev IV, Bakirtzis AI, Burdyukov DA, Kravchenko MV (2021) Prospects for the use of aluminum-air batteries in solar energy in Greece. *IOP Conf Ser Earth Environ Sci* 808:012011. https://doi.org/10.1088/1755-1315/808/1/012011

28. Mori R (2019) All solid state rechargeable aluminum-air battery with deep eutectic solvent based electrolyte and suppression of byproducts formation. *RSC Adv* 9:22220–22226. https://doi.org/10.1039/c9ra04567h
29. Wu S, Zhang Q, Ma J, et al (2020) Interfacial design of Al electrode for efficient aluminum-air batteries: issues and advances. *Mater Today Energy* 18:100499. https://doi.org/10.1016/j.mtener.2020.100499
30. Macdonald DD, English C (1990) Development of anodes for aluminium/air batteries - solution phase inhibition of corrosion. *J Appl Electrochem* 20:405–417. https://doi.org/10.1007/BF01076049
31. Gelman D, Shvartsev B, Ein-Eli Y (2014) Aluminum-air battery based on an ionic liquid electrolyte. *J Mater Chem A* 2:20237–20242. https://doi.org/10.1039/c4ta04721d
32. Mokhtar M, Talib MZM, Majlan EH, et al (2015) Recent developments in materials for aluminum-air batteries: a review. *J Ind Eng Chem* 32:1–20. https://doi.org/10.1016/j.jiec.2015.08.004
33. Li Q, Bjerrum NJ (2002) Aluminum as anode for energy storage and conversion: a review. *J Power Sources* 110:1–10. https://doi.org/10.1016/S0378-7753(01)01014-X
34. Palanisamy S, Rajendhran N, Srinivasan S, et al (2021) A novel nano-YSZ-Al alloy anode for Al-air battery. *J Appl Electrochem* 51:345–356. https://doi.org/10.1007/s10800-020-01493-2
35. Fan L, Lu H, Leng J (2015) Performance of fine structured aluminum anodes in neutral and alkaline electrolytes for Al-air batteries. *Electrochim Acta* 165:22–28. https://doi.org/10.1016/j.electacta.2015.03.002
36. Yin X, Yu K, Zhang T, et al (2017) Influence of rolling processing on discharge performance of Al-0.5Mg-0.1Sn-0.05Ga-0.05In alloy as anode for al-air battery. *Int J Electrochem Sci* 12:4150–4163. https://doi.org/10.20964/2017.05.40
37. Xu T, Hu Z, Yao C (2019) The effects of Ca addition on corrosion and discharge performance of commercial pure aluminum alloy 1070 as anode for aluminum-air battery. *Int J Electrochem Sci* 14:2606–2620. https://doi.org/10.20964/2019.03.53
38. Mutlu RN, Yazıcı B (2019) Copper-deposited aluminum anode for aluminum-air battery. *J Solid State Electrochem* 23:529–541
39. Yu Y, Chen M, Wang S, et al (2018) Laser sintering of Al nanoparticles for Al-air batteries. *J Laser Appl* 30:032605. https://doi.org/10.2351/1.5040651
40. Han Y, Ren J, Fu C, et al (2020) Electrochemical performance of aluminum anodes with different grain sizes for Al-air batteries. *J Electrochem Soc* 167:040514. https://doi.org/10.1149/1945-7111/ab754a
41. Wu Z, Zhang H, Zou J, et al (2020) Enhancement of the discharge performance of Al-0.5Mg-0.1Sn-0.05Ga (wt.%) anode for Al-air battery by directional solidification technique and subsequent rolling process. *J Alloys Compd* 827:154272. https://doi.org/10.1016/j.jallcom.2020.154272
42. Zheng X, Zhang T, Yang H, et al (2020) Friction stir processing induced elctrochemical performance improvement of commercial Al for Al-air battery. *Electrochim Acta* 354:136635. https://doi.org/10.1016/j.electacta.2020.136635
43. Nestoridi M, Pletcher D, Wood RJK, et al (2008) The study of aluminium anodes for high power density Al/air batteries with brine electrolytes. *J Power Sources* 178:445–455. https://doi.org/10.1016/j.jpowsour.2007.11.108
44. Pino M, Cuadrado C, Chacón J, et al (2014) The electrochemical characteristics of commercial aluminium alloy electrodes for Al/air batteries. *J Appl Electrochem* 44:1371–1380. https://doi.org/10.1007/s10800-014-0751-6
45. Smoljko I, Gudić S, Kuzmanić N, Kliškić M (2012) Electrochemical properties of aluminium anodes for Al/air batteries with aqueous sodium chloride electrolyte. *J Appl Electrochem* 42:969–977. https://doi.org/10.1007/s10800-012-0465-6
46. Ma J, Wen J, Gao J, Li Q (2014) Performance of Al-0.5 Mg-0.02 Ga-0.1 Sn-0.5 Mn as anode for Al-air battery in NaCl solutions. *J Power Sources* 253:419–423. https://doi.org/10.1016/j.jpowsour.2013.12.053
47. Sun Z, Lu H, Fan L, et al (2015) Performance of Al-air batteries based on Al-Ga, Al-In and Al-Sn alloy electrodes. *J Electrochem Soc* 162:A2116–A2122. https://doi.org/10.1149/2.0681510jes
48. Pino M, Chacón J, Fatás E, Ocón P (2015) Performance of commercial aluminium alloys as anodes in gelled electrolyte aluminium-air batteries. *J Power Sources* 299:195–201. https://doi.org/10.1016/j.jpowsour.2015.08.088
49. Ma J, Wen J, Ren F, et al (2016) Electrochemical performance of Al–Mg–Sn based alloys as anode for Al-air battery. *J Electrochem Soc* 163:A1759–A1764. https://doi.org/10.1149/2.1241608jes
50. Ilyukhina AV, Zhuk AZ, Kleymenov BV, et al (2016) The influence of temperature and composition on the operation of Al anodes for aluminum-air batteries. *Fuel Cells* 16:384–394. https://doi.org/10.1002/fuce.201500032
51. Moghadam Z, Shabani-Nooshabadi M, Behpour M (2017) Electrochemical performance of aluminium alloy in strong alkaline media by urea and thiourea as inhibitor for aluminium-air batteries. *J Mol Liq* 242:971–978. https://doi.org/10.1016/j.molliq.2017.07.119

52. Liang R, Su Y, Sui XL, et al (2019) Effect of Mg content on discharge behavior of Al-0.05Ga-0.05Sn-0.05Pb-*x*Mg alloy anode for aluminum-air battery. *J Solid State Electrochem* 23:53–62. https://doi.org/10.1007/s10008-018-4093-x
53. Liu X, Zhang P, Xue J (2019) The role of micro-naoscale AlSb precipitates in improving the discharge performance of Al-Sb alloy anodes for Al-air batteries. *J Power Sources* 425:186–194. https://doi.org/10.1016/j.jpowsour.2019.04.012
54. Zhang P, Liu X, Xue J, Wang Z (2019) Evaluating the discharge performance of heat-treated Al-Sb Alloys for Al-air batteries. *J Mater Eng Perform* 28:5476–5484. https://doi.org/10.1007/s11665-019-04287-6
55. Wu Z, Zhang H, Guo C, et al (2019) Effects of indium, gallium, or bismuth additions on the discharge behavior of Al-Mg-Sn-based alloy for Al-air battery anodes in NaOH electrolytes. *J Solid State Electrochem* 23:2483–2491. https://doi.org/10.1007/s10008-019-04341-2
56. Wu Z, Zhang H, Qin K, et al (2020) The role of gallium and indium in improving the electrochemical characteristics of Al-Mg-Sn-based alloy for Al-air battery anodes in 2 M NaCl solution. *J Mater Sci* 55:11545–11560. https://doi.org/10.1007/s10853-020-04755-8
57. Liu X, Zhang P, Xue J, et al (2021) High energy efficiency of Al-based anodes for Al-air battery by simultaneous addition of Mn and Sb. *Chem Eng J* 417:128006. https://doi.org/10.1016/j.cej.2020.128006
58. Hosseini S, Xu TH, Masoudi Soltani S, et al (2022) The efficient acetoxy-group-based additives in protecting of anode in the rechargeable aluminium-air batteries. *Int J Hydrogen Energy* 47:501–516. https://doi.org/10.1016/j.ijhydene.2021.10.030
59. Yang H, Li X, Wang Y, et al (2020) Excellent performance of aluminium anode based on dithiothreitol additives for alkaline aluminium/air batteries. *J Power Sources* 452:227785. https://doi.org/10.1016/j.jpowsour.2020.227785
60. Han Z, Xu Y, Zhou S, et al (2020) The effect of annealing on electrochemical performances of an Al-Sn-Ga-Mg alloy as an anode for Al-air batteries in alkaline electrolytes. *J Electrochem Soc* 167:100541. https://doi.org/10.1149/1945-7111/ab9b97
61. Katsoufis P, Mylona V, Politis C, et al (2020) Study of some basic operation conditions of an Al-air battery using technical grade commercial aluminum. *J Power Sources* 450:227624. https://doi.org/10.1016/j.jpowsour.2019.227624
62. Wu Z, Zhang H, Tang S, et al (2021) Effect of calcium on the electrochemical behaviors and discharge performance of Al-Sn alloy as anodes for Al-air batteries. *Electrochim Acta* 370:137833. https://doi.org/10.1016/j.electacta.2021.137833
63. Ren J, Fu C, Dong Q, et al (2021) Evaluation of impurities in aluminum anodes for Al-air batteries. *ACS Sustain Chem Eng* 9:2300–2308. https://doi.org/10.1021/acssuschemeng.0c08415
64. Zaromb S (1962) The use and behavior of aluminum anodes in alkaline primary batteries. *J Electrochem Soc* 109:1125. https://doi.org/10.1149/1.2425257
65. Mokhtar M, Majlan EH, Ahmad A, et al (2018) Effect of ZnO Filler on PVA-alkaline solid polymer electrolyte for aluminum-air battery applications. *J Electrochem Soc* 165:A2483–A2492. https://doi.org/10.1149/2.0381811jes
66. Shen LL, Zhang GR, Biesalski M, Etzold BJM (2019) Paper-based microfluidic aluminum-air batteries: toward next-generation miniaturized power supply. *Lab Chip* 19:3438–3447. https://doi.org/10.1039/c9lc00574a
67. Pan W, Wang Y, Kwok HYH, Leung DYC (2019) A low-cost portable cotton-based aluminum-air battery with high specific energy. *Energy Procedia* 158:179–185. https://doi.org/10.1016/j.egypro.2019.01.067
68. Wang Y, Kwok HYH, Pan W, et al (2019) Combining Al-air battery with paper-making industry, a novel type of flexible primary battery technology. *Electrochim Acta* 319:947–957. https://doi.org/10.1016/j.electacta.2019.07.049
69. Wang Y, Pan W, Kwok H, et al (2019) Low-cost Al-air batteries with paper-based solid electrolyte. *Energy Procedia* 158:522–527. https://doi.org/10.1016/j.egypro.2019.01.146
70. Wang Y, Kwok HYH, Pan W, et al (2019) Parametric study and optimization of a low-cost paper-based Al-air battery with corrosion inhibition ability. *Appl Energy* 251:113342. https://doi.org/10.1016/j.apenergy.2019.113342
71. Wang Y, Kwok H, Pan W, et al (2019) Innovative paper-based Al-air batteries as a low-cost and green energy technology for the miniwatt market. *J Power Sources* 414:278–282. https://doi.org/10.1016/j.jpowsour.2019.01.018
72. Wang Y, Pan W, Kwok HYH, et al (2019) Liquid-free Al-air batteries with paper-based gel electrolyte: a green energy technology for portable electronics. *J Power Sources* 437:226896. https://doi.org/10.1016/j.jpowsour.2019.226896
73. Katsoufis P, Katsaiti M, Mourelas C, et al (2020) Study of a thin film aluminum-air battery. *Energies* 13:1447. https://doi.org/10.3390/en13061447

20 Emerging Cost-Efficient Aluminum-Air Battery Technology
Cathode Materials

S. K. Vineeth, Pritam V. Dhawale, Roshny Joy, Abhilash Pullanchiyodan, Jou-Hyeon Ahn, Nutan Gupta, and Prasanth Raghavan

20.1 INTRODUCTION

Traditional fossil-fuel-based energy is confronted by many challenges, and it cannot feed the demand of the proliferating population. As renewable energy sources are intermittent, efficient energy storage solutions must be explored. Commercialization of lithium-ion batteries (LIB) by Sony Corporation in 1991 fuelled the development of portable devices and electric vehicles (EVs), which completely depend upon energy storage devices [1]. Although the past few decades have witnessed remarkable development in LIB, shortage of raw materials, cost-consideration, and environmental impact, especially due to the mining of raw materials, paved the way for alternate systems than LIB [2,3]. Hence a stream of post-LIB battery technology was evolved, where the aim is to use materials that are abundant, cost-effective, and have high energy density [4–6]. Metal-air batteries (MAB) are a class of battery technology, where the electrocatalyst contains a cathode, utilizing oxygen from the ambient environment, metal or metal alloys as anode, and an electrolyte (aqueous or non-aqueous) [7]. The general working principle involves the usage of oxygen from the ambient environment, oxidization of metal anode, and formation of solid metal oxides, which get recycled [8].

In the MAB technology, the first primary zinc-air batteries were introduced in 1878 by Maiche, and commercial products were introduced in 1932 [9]. A great development occurred in 1960, with the introduction of aqueous MABs. Over the past decade, various MABs have been intensively investigated due to their high energy density/capacity and their relatively low costs. There were many challenges and limitations for researchers to scale them up to the commercialization level [10]. Aluminum (Al) is the most abundant metal in Earth's crust (8.1 wt. %) and is much easier and more cost-efficient to recycle than the components used in LIBs [11]. Ideally, the waste product in an aluminum-air battery (AAB) is aluminum hydroxide, which can be recycled to produce more Al, but, in general, the creation of rechargeable AABs is challenging, because the electrolyte reaction works by corroding the Al. AABs produce electricity from the reaction of oxygen in the air with Al and exhibit a high theoretical energy density of 8,100 Wh kg^{-1} [12]. An EV with an Al-based battery has the potential for up to eight times the range of a LIB with a significantly lower total weight. Furthermore, it is possible to mechanically recharge the battery with new Al anodes; hence, it opens up an eco-friendly technology with greater availability, coupled with a cost-effective way to store electricity [13].

As the air cathode in AAB has a profound effect on electrochemical properties, various developments have been reported in cathode engineering. The air cathode consists of a gas diffusion layer (GDL), electrocatalyst, and a metallic current collector [14]. In the complex cathode research design,

DOI: 10.1201/9781003310174-20

high-performance electrocatalysts can trigger various reactions such as oxygen reduction reaction (ORR), hydrogen evolution reaction (HER) occurring at air cathode, and the oxygen evolution reaction (OER) [15].

For the realization of high-performance AABs, efforts should be underway for regulating the ORR/OER kinetics and structural engineering of the air cathode. Therefore, this chapter highlights cathode engineering for AABs with a focus on materials, methods, and their effect on the performance of the system. As the electrocatalyst is an integral part of an air cathode, discussions relating to the development in tailoring catalysts are also discussed in detail. As AABs are promising candidates in MABs, continuous development in exploring newer materials, engineering the mechanism, and tailoring the architecture, there can be an acceleration in the commercialization and employment of AABs in the future mobility through EV technology.

20.2 CATHODE ENGINEERING FOR ALUMINUM AIR BATTERIES

The performance of AAB is largely dependent on the materials used for the anode and the cathode. Improper selection of cathode catalyst leads to a negative effect on the efficiency and performance of the cells. Challenges associated with cathode are the clogging of pores by the insoluble products such as carbonates in the alkaline electrolyte, electrode resistance, overvoltage, lower surface area for catalytic adsorption, and cost considerations linked with cathode [16]. Hence, bifunctional catalysts which facilitate both ORR and OER were reported [17]. For the ORR, a hydrophobic catalyst over a porous GDL is required, whereas, for OER, a hydrophilic catalyst containing a porous cathode sintered on a metallic substrate is the requirement [18]. The selection criteria of cathode and electrocatalyst are based on the polarization curves, cyclic voltammogram (CV), and Koutecky–Levich (K–L) plots, which unveil the electron transfer number and ORR pathways [19,20].

Another important parameter of a catalyst is its ability to endure ORR and OER conditions, for efficient working and longer operation. In most cases these purposes are satisfied by noble metal-based commercial catalysts such as Pt/C and Ru/IrO_2 [21–23]. The term 'electrocatalyst' for AAB technology has been explored for synergistically enhanced kinetics for both ORR and OER. Although these catalysts are highly efficient, the cost associated with it and the poor performance in alkaline AAB are the main bottlenecks for limiting its wide cost-effective commercialization. Hence, there has been a lot of development in the synthesis of alternative transition metal, carbon-based nanomaterials, such as carbon nanotubes (CNTs), and metal-organic framework (MOF)-based precursors as cost-effective and efficient electrocatalysts [24]. For the fabrication of high-performance AABs, the intrinsic sluggish rate for the ORR needs to be rectified. Per the reported electrocatalysts, the ORR electrocatalysts can be classified into metal-based and carbon-based materials.

20.2.1 Nanoparticles as Electrocatalysts

Nanoparticles are a class of materials with dimension in the nanorange; these particles have a larger surface area compared to micro-sized particles [25–27]. Nanoparticles as electrocatalysts for metal-air batteries have been shown to accelerate the sluggish kinetics of ORR and OER [28]. Liu et al. [29] prepared electrocatalysts based on $LiMn_2O_4$ (LMO) nanoparticles, modified nitrogen-doped reduced graphene oxide nanosheets by hydrothermal method, for AAB. For comparison, catalysts with LMO-modified undoped reduced graphene oxide (LMO/rGO) nanosheets and pure nitrogen-doped reduced graphene oxide (rGO) nanosheets were also evaluated. It was evidenced from various analyses such as CV, steady-state polarization curves, and EIS in 0.1 M KOH electrolyte that the LMO supported N-rGO and showed faster charge transfer, positive onset potential, and half-wave potential with high cathodic peak current. Figure 20.1 shows the results of CV by using various electrode catalysts in O_2- and N_2-saturated 0.1 M KOH electrolyte with a scan rate of 50 mV s^{-1} and at a constant catalyst loading of 0.238 mg cm^{-2}.

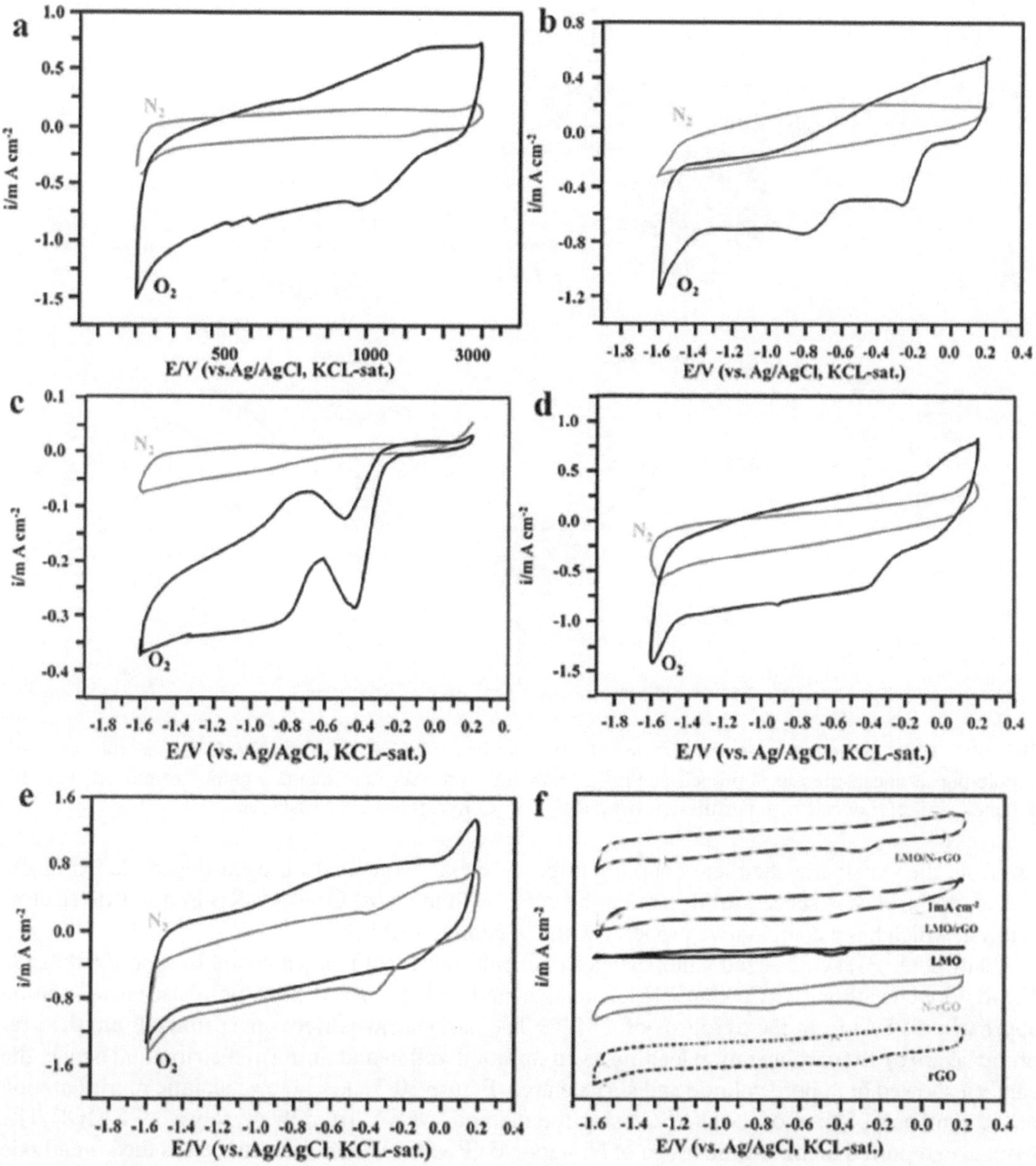

FIGURE 20.1 Results of cyclic voltammograms by using various electrode catalysts in O_2 and N_2 saturated 0.1 M KOH electrolyte with a scan rate of 50 mV/s and at a constant catalyst loading of 0.238 mg cm^{-2} (a) rGO, (b) N-rGO, (c) LMO, (d) LMO-rGO, and (e) LMO/N-rGO. CV curves of (f) overlaid graph with all the electrocatalysts in O_2-saturated 0.1 M KOH aqueous solution. Adapted and reproduced with permission from Ref. [29]. Copyright © 2015 Elsevier.

Liu et al. [30] found that a composite of cobalt oxide (Co_3O_4) and nitrogen-doped Ketjen black (Co_3O_4/N-KB) prepared by hydrothermal method worked as a high-performance catalyst for ORR for alkaline AAB. The anode material consists of polished Al, and the air cathode consists of three layers: a catalytic layer, Nickel foam as the current collector, and a GDL, prepared by the hot pressure method. Composite showed better ORR current density with more positive half-wave potential than Co_3O_4 and N-KB separately and also compared to commercial Pt/C. The main reason for the enhanced catalytic activity, which almost equaled the performance of commercial Pt/C, is due to the creation of an active

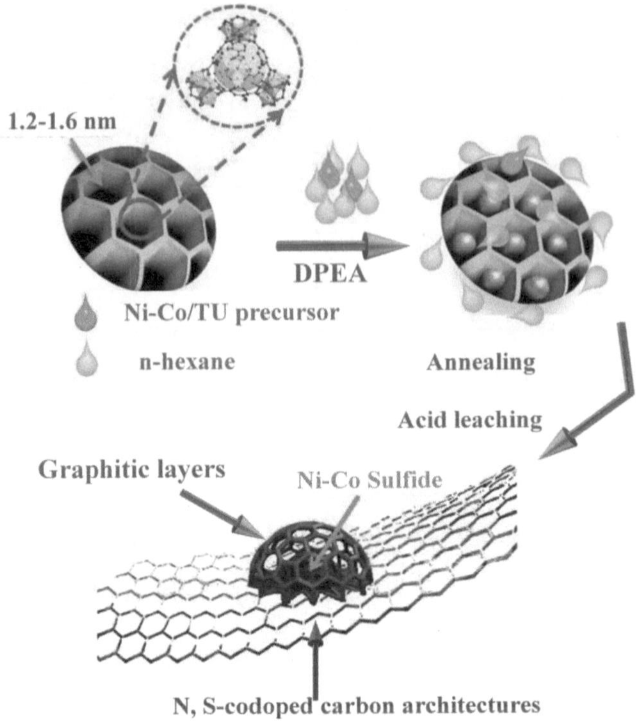

FIGURE 20.2 Schematic representation of the synthesis of Ni–Co–S@G/NSC using the method, double-phase encapsulation approach (DPEA), from the pyrolysis of a metal-organic framework (MOF). Adapted and reproduced with permission from Ref. [31]. Copyright © 2017 Elsevier.

site from the synergetic chemical coupling effect of cobalt oxide and nitrogen-doped Ketjen black. Hence, the work was successful in fabricating AAB with low-cost Co_3O_4/N-KB as an ORR electrocatalyst, which has a comparative property with the commercial Pt/C.

Wang et al. [31] synthesized nanoparticles of cobalt-nickel-sulfide, which are immobilized inside N,S-co-doped carbon (Ni-Co-S@G/NSC) using a method known as the double-phase encapsulation approach (DPEA) from the pyrolysis of a MOF. The nanoparticles were encapsulated into the graphitic layers by the carbonization leading to an immobilization and uniform distribution; hence, the catalyst showed high pore volume and surface area. Figure 20.2 shows the schematic representation of the synthesis of Ni–Co–S@G/NSC using the method, DPEA, from the pyrolysis of a MOF. The work also reported the first-time usage of Ni–Co–S@G/NSC ORR electrocatalyst as the air cathode material for AAB. The performance of the prepared catalyst was comparable to the commercial Pt/C catalyst. The catalyst showed electrochemical stability with a high-capacity density and power density of 2,947 mAh g^{-1} and 261.3 mW cm^{-2}, respectively. The work showed a cost-effective efficient way for the synthesis of high-performance low-cost ORR electrocatalyst having properties comparable to commercial Pt/C electrocatalyst.

To enhance the electrocatalytic activity in AAB, Ryu et al. [32] reported the use of silver-based nanoparticles as a highly efficient ORR electrocatalyst. Silver atoms were incorporated into manganese oxide having a nanoplate-like morphology through a metal seed-mediated growth method, which produced silver manganate nanoplates (SMNp) having dislocations on its surface. Hence, the irregularities in the surface contribute to the high catalytic activity for ORR. The work further investigated the performance of SMNp electrocatalysts by alkaline aluminum-air flow batteries. Figure 20.3 shows the working and design; it consists of an alkaline electrolyte tank as an anode and Al metal- and catalyst-loaded air electrode as the cathode. The system delivered stable output and a high energy

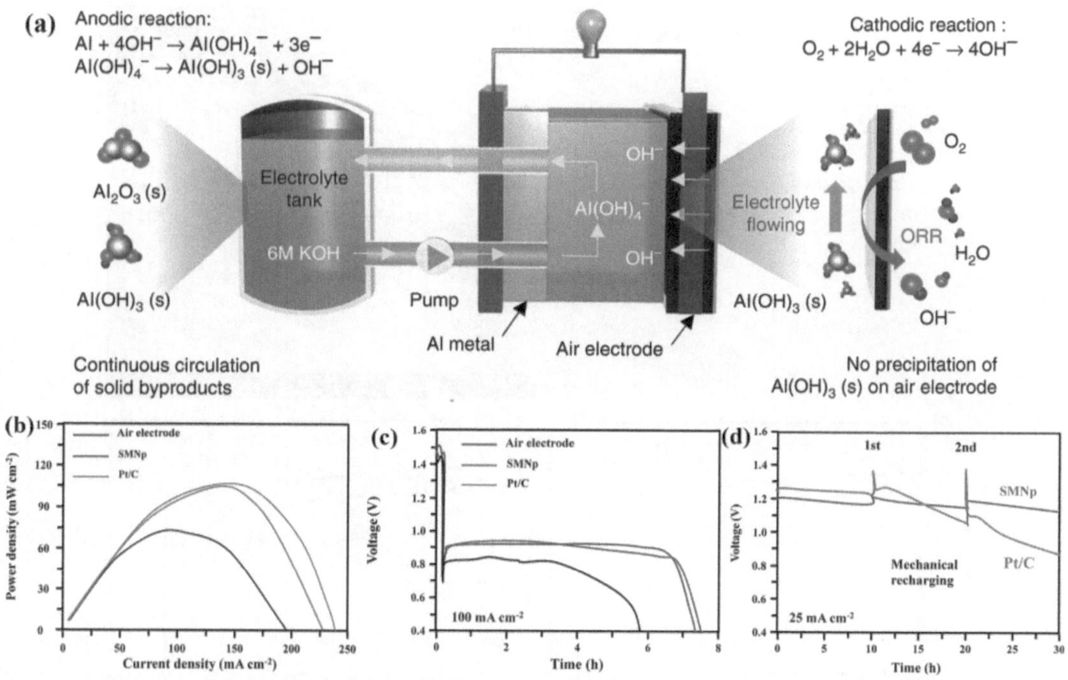

FIGURE 20.3 Schematic representation of working of an aluminum-air flow battery with SMNp, its working, and electrochemical performance analysis. Adapted and reproduced with permission from Ref. [32]. Copyright © 2012 Nature.

density of ~2,507 Wh kg_{Al}^{-1} as gravimetric energy density and ~6,890 Wh L_{Al}^{-1} of volumetric energy density, showing its potential applicability for the next-generation energy storage and generation.

Liu et al. [33] reported the ORR electrocatalytic activity of a hybrid MOF based on carbon and reduced graphene oxide (rGO) as a cathode catalyst for alkaline AAB. The hybrid nanocage electrocatalyst, $ZnO/ZnCo_2O_4$/C@rGO, was prepared from a template of Co-based MOFs, known as zeolitic imidazolate framework (ZIF-67) and graphene oxide by hydrothermal and pyrolytic methods. The nanocage mesoporous structure of the ORR electrocatalyst provided high surface area and active sites, which prevented the aggregation of transition metal oxides. Further evaluation of the performance of $ZnO/ZnCo_2O_4$/C@rGO and $ZnO/ZnCo_2O_4$/C ORR electrocatalysts were investigated by using them as cathode catalyst embedded in carbon cathode with Al foil as an anode, 2 M KOH as an electrolyte, and carbon-based air cathode with catalyst and the whole as an Al air coin cell. Figure 20.4 shows the components of the coin cell and a comparative study of the properties of $ZnO/ZnCo_2O_4$/C@rGO and $ZnO/ZnCo_2O_4$/C ORR electrocatalysts. The results showed that, compared to $ZnO/ZnCo_2O_4$/C cell, the $ZnO/ZnCo_2O_4$/C@rGO with the presence of nanosheets of rGO and its nanocage structure showed better discharge voltage, specific capacity, power density, and open circuit voltage (OCV). The synergistic interaction of the nanocage structure of ZnO/ $ZnCo_2O_4$/C and the nanosheet morphology of rGO contributed to a positive shift in the performance of the hybrid electrocatalyst.

The trifunctional electrocatalytic activity of copper–cobalt intermetallic sulfide ($CuCo_2S_4$) having a morphology of hollow nanoneedle arrays grown on Ni foam (NF) through a facile hydrothermal method was observed and reported by Shao et al. [34]. Commercial NF was reacted with $Cu(NO_3)_2{\cdot}6H_2O$ and $Co(NO_3)_2{\cdot}6H_2O$ forming a precursor, CuCo-layered double hydroxides (LDH)/NF, which were then kept at 350°C in a tube furnace followed by sulfurization forming $CuCo_2S_4$/NF hollow nanoneedle arrays. The hollow nanoarray structure was evidenced by SEM, TEM, and HRTEM morphological

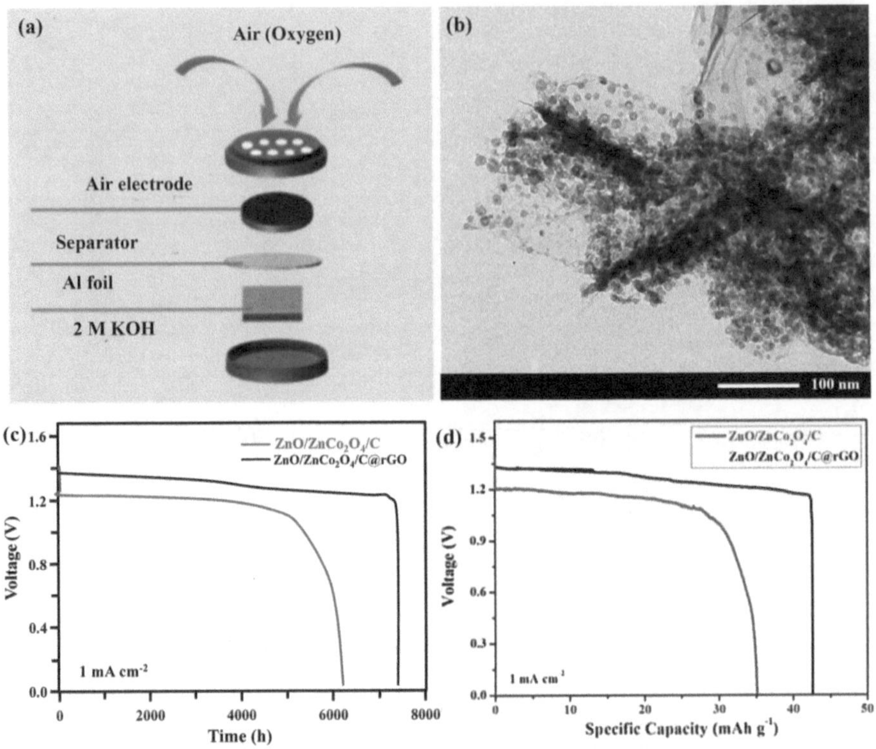

FIGURE 20.4 Components of an aluminum-air coin cell, (b) and (c) a comparative study upon using ZnO/ $ZnCo_2O_4$/C and ZnO/$ZnCo_2O_4$/C@rGO hybrid as ORR electrocatalysts, showing the (b) discharge curves at ambient conditions with 1 mA cm^{-2} current density, and (c) specific discharge capacity. Adapted and reproduced with permission from Ref. [33]. Copyright © 2017 American Chemical Society.

analysis, and X-ray diffraction and X-ray photoelectron spectroscopy analysis confirmed the products after the synthesis. Electrochemical analysis was measured in an alkaline condition provided by KOH. The results showed the trifunctional activity by the enhanced performance for water splitting at a current density of 10 mA cm^{-2}; the AAB with $CuCo_2S_4$/NF cathode catalyst showed synergistic activity with an overpotential of 61 mV for HER and for OER; the value was recorded as 189 mV. Moreover, the catalyst-loaded air cathode showed an OCV of 1.81 V and a power density of 160 mW cm^{-2}. Figure 20.5 shows the performance properties of $CuCo_2S_4$/NF electrocatalyst in alkaline AAB and its stability curve. Also, the photographs corresponding to the two-electrode configuration during water splitting connected to alkaline AAB containing $CuCo_2S_4$/NF as a catalyst are also shown. The high surface area of nanoarray contributed to the enhancement of electrocatalytic activities, which made the catalyst a multi-functional, cost-efficient, and effective catalyst for energy storage application in an AAB.

Zhu et al. [35] embedded Co nanoparticles in CNT arrays as a flexible catalyst and demonstrated its electrocatalytic activity for the application as an air cathode catalyst in AAB and Zn-air batteries. Over the flexible carbon cloth, using the atomic layer deposition process, a zinc oxide (ZnO) seed layer was deposited followed by a hydrothermal reaction with hexamethylenetetramine and $Zn(NO_3)_2 \cdot 6H_2O$. The ZnO-grown cloth is reacted with 2-methylimidazole (2-MIM) and mixed with $Co(NO_3)_2 \cdot 6H_2O$ at various concentrations: 0, 0.02, and 0.2 mmol, producing carbon cloth with ZnO@Zn/Co-ZIF, ZnO@Zn/Co-ZIF-1, and ZnO@Zn/Co-ZIF-2 core-shell nanoarrays, respectively. In the final step, the hollow nitrogen-doped CNT array embedded with cobalt nanoparticle (HCA-Co) electrocatalyst was prepared by template solution reaction and then annealing for 2 h at 800°C, which made inside ZnO to evaporate, and Zn/Co-ZIF converts to nitrogen-doped CNT

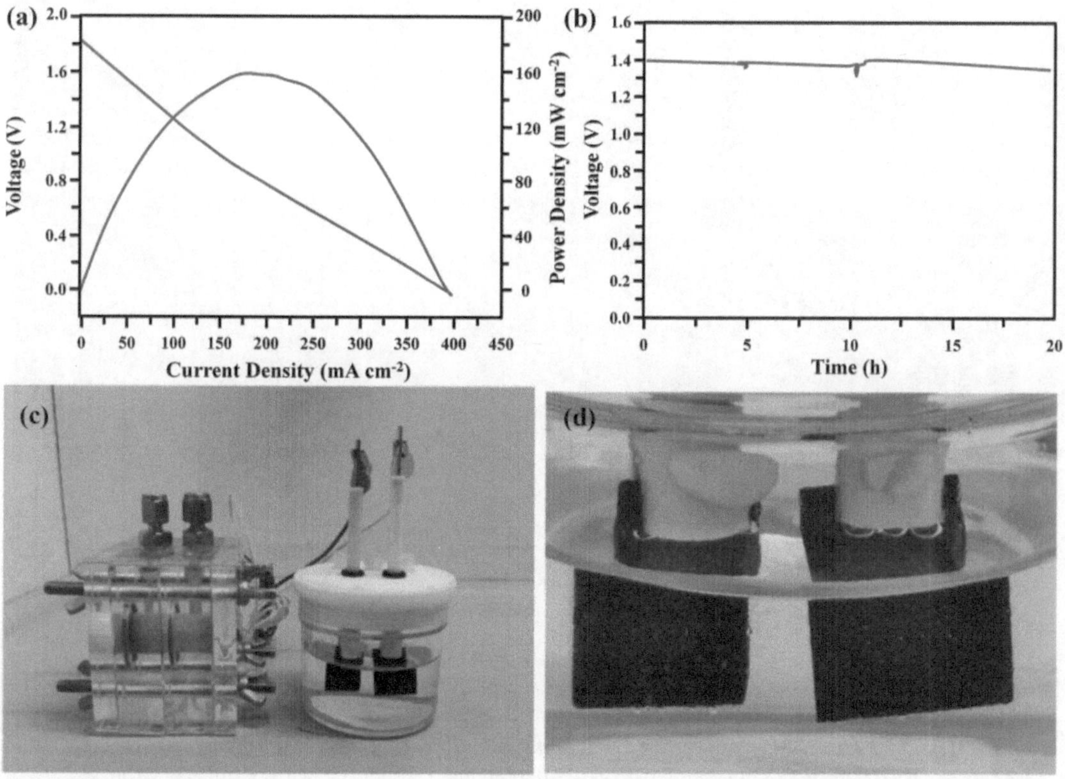

FIGURE 20.5 (a) Discharge curve and the power density curves of alkaline AAB equipped with the electrocatalyst, $CuCo_2S_4$/NF with alkaline electrolyte (4 M KOH solution) with a peristaltic pump, (b) stability curve measured at 10 mA cm^{-2} current density (c and d) photographs corresponding to the two-electrode configuration during water splitting connected to alkaline AAB with $CuCo_2S_4$/NF as catalyst. Adapted and reproduced with permission from Ref. [34]. Copyright © 2020 Elsevier.

embedded Co nanoparticles. Samples were named as HCA-Co, HCA-Co-1, and HCA-Co-2 based on the precursor from which they are annealed ZnO@Zn/Co-ZIF, ZnO@Zn/Co-ZIF-1, and ZnO@Zn/Co-ZIF-2 respectively. Various morphological analyses such as SEM and TEM showed the uniform nanorod structure of ZnO over carbon fibers and core-shell structure, and the morphology turned to a hollow structure after the annealing process while embedding Co nanoparticles. XRD analysis showed that HCA-Co-1 is relatively smaller than HCA-Co-2. Figure 20.6 shows the morphological images of the prepared electrocatalyst. Moreover, AAB with HCA-Co-1 air cathode catalyst showed 1.966 V as an open circuit potential with better flexibility than Pt/C.

Guo et al. [36] reported the performance of a three-dimensional boron-doped carbon microsphere (B-CM) as an effective electrocatalyst for AAB. At a current density of 50 mA g^{-1}, AAB with mesoporous B-CM catalyst showed a discharge capacity of 300 mAh g^{-1} and after 30 cycles of charging and discharge the capacity of 584 mAh g^{-1}. From XRD and Raman spectroscopy analysis, it was evidenced that boron atoms were in a boron carbide state with defects in the microsphere of carbon. The high electrocatalytic activity is linked to the high surface area, which enhanced the active sites contributing to enhanced ORR.

The applicability of iron group metal sulfide (MS) immobilized inside a porous N,S-enriched carbon ((MS@G/NSC) as an electrocatalyst for alkaline AAB was investigated by Hong et al. [37]. The electrocatalyst was prepared from a nitrogen-rich MOF (NH_2-MIL-101(Al)) by DPEA and then pyrolysis and washing with 10 wt.% hydrofluoric acid (HF). The DPEA process and the pyrolysis immobilize MS nanoparticles to get encapsulated in the porous graphitic layers as the matrix.

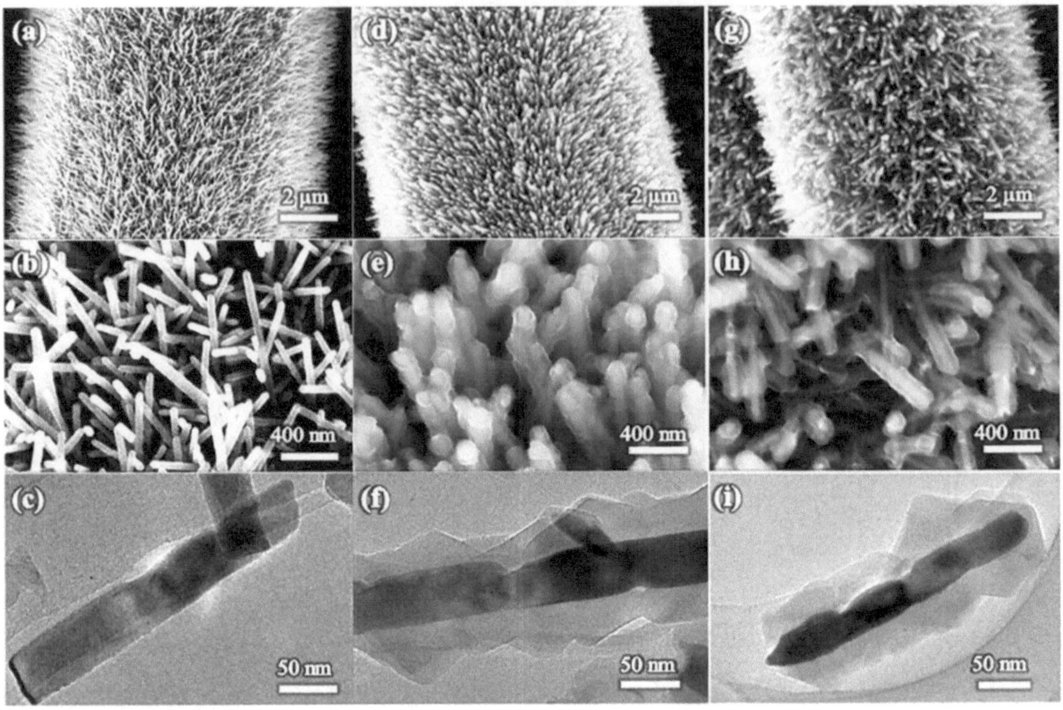

FIGURE 20.6 SEM and TEM images of electrocatalysts during the synthesis process (a–c) ZnO, (d–f) ZnO@Zn/Co-ZIF-1, and (g–i) ZnO@Zn/Co-ZIF-2. Adapted and reproduced with permission from Ref. [35]. Copyright © 2020 Elsevier.

Figure 20.7 schematically represents the synthesis process. It was observed that when iron sulfide (FeS) was immobilized inside the N,S-enriched graphite layer, the activity was enhanced with low Tafel slope and positive onset potential. In addition, the work reported that the prepared catalyst outperformed commercial Pt/C. AAB prepared by FeS@G/NSC as an air cathode catalyst delivered a peak power density of 160.1 mW cm^{-2} with stability, and at a current density of 50 mA cm^{-2}, the discharge voltage plateau was recorded near to 1.47 V.

Binary FeCo-nitrogen-doped carbon (FeCo-NC) material-loaded Fe/Co species and in situ generated interconnected CNT composite were found to be efficient ORR electrocatalysts for AAB by Guo et al. [38]. AAB with the synthesized electrocatalyst delivered 188 mW cm^{-2} as specific power; at 10.0 mA cm^{-2} current density, the cell showed a discharging voltage of 1.70 V with 1.88 V as OCV. The work successfully developed a low-cost non-noble metal-based electrocatalyst having comparable properties with Pt/C for the manufacture of cost-effective AAB.

Mori et al. [39] studied the performance of rechargeable AAB with 1-ethyl 3-methylimidazolium chloride as an ionic liquid electrolyte. The investigation used titanium carbonitride (TiN), titanium carbide (TiC), and titanium diboride (TiB_2) as air cathodes. The work concluded that the accumulated $Al(OH)_3$ and Al_2O_3 were suppressed when the air cathodes used were TiN and TiC. The work was the first to report the suppression of the byproduct accumulation of electrodes, thereby giving a stable rechargeable AAB. Following the work, to reduce the byproduct accumulation in electrodes, Mori et al. [40] investigated various air cathode materials such as carbon-based, MOF non-oxide ceramic materials for AAB. The work reported that, when using TiC as a cathode, the byproducts such as $Al(OH)_3$ and Al_2O_3 were suppressed, hence providing the opportunity for developing rechargeable AABs.

The usage of aluminum terephthalate, a MOF for the air cathode, and the ionic liquid, 1-ethyl 3-methylimidazolium chloride, as an electrolyte showed positive results for the development of rechargeable AABs [41]. Stable CV profiles were shown without the presence of $Al(OH)_3$ and Al_2O_3

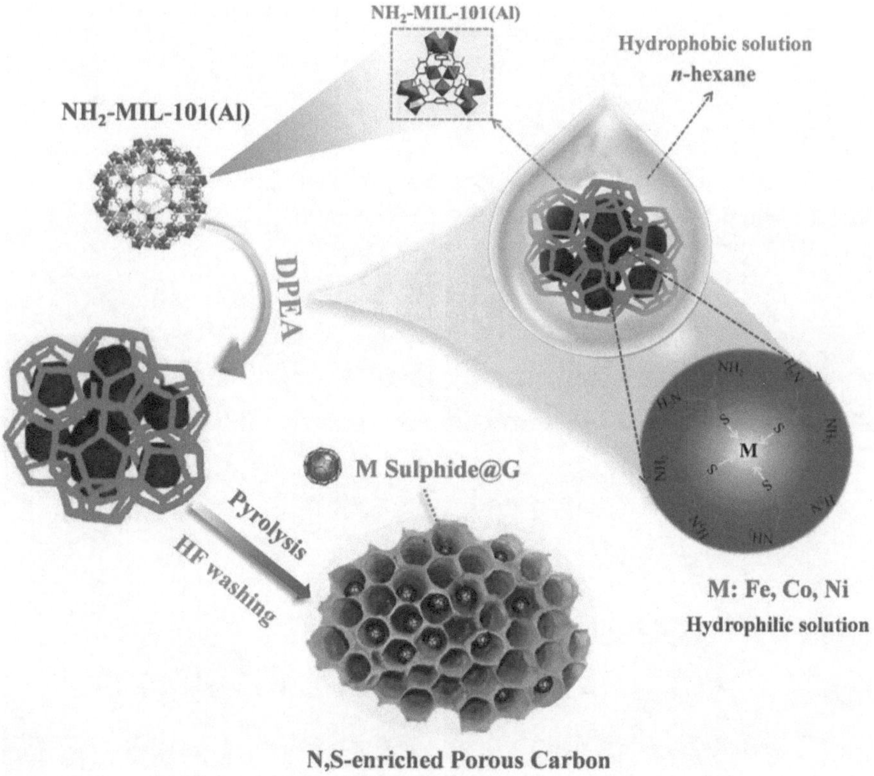

FIGURE 20.7 Schematic representation of the synthesis procedure for the MS@G/NSC electrocatalyst by DPEA process followed by pyrolysis in a nitrogen atmosphere and washing with 10 wt.% hydrofluoric acid (HF). Adapted and reproduced with permission from Ref. [37]. Copyright © 2019 Elsevier.

in the Al anode and with a low interfacial cell impedance considering long-term electrochemical reactions. The work was the first reported study where MOF was used as an active material for rechargeable AAB.

Li et al. [42] developed an ORR electrocatalyst by bimetallic co-doping with copper, iron, and nitrogen with carbon material (Cu–Fe–N–C) for AAB. The synthesis steps included copper and iron chlorides ($FeCl_2 \cdot 4H_2O$ and $CuCl \cdot 4H_2O$) calcination with 1,10-phenanthroline and melamine. Compared to iron and nitrogen co-doped carbon materials (Fe–N–C), the Cu–Fe–N–C showed better properties with an increase of half-wave potential by 25 mV. Furthermore, from density functional theory, it was shown that, by introducing Cu, the adsorption of O_2 improved, thus contributing to ORR.

20.2.2 Nitrogen-Doped Carbon-Based Catalysts

Nitrogen, being an efficient dopant having a comparable size with carbon and possessing the ability to form delocalized conjugation by the lone pair electrons, thereby functioning as an electrocatalyst in many applications. Considering the same, nitrogen-doped carbon nanomaterials has been explored as an effective ORR electrocatalysts in AAB applications [43]. The formation of delocalized conjugation with carbon contributed to the formation of more active sites due to the spin density and distribution of charge. Furthermore, the surface area of carbon-based nanoparticles is much higher, which also contributes to enhancing the electrocatalytic properties.

Liu et al. [44] developed an ORR catalyst based on porous nanoneedle-shaped $CoMn_2O_4$ (CMO) on nitrogen-doped CNT and three-dimension (3D) graphene (NCNTs/3D graphene) for AAB. Figure 20.8 shows the morphological images of catalysts. The nano electrocatalyst, prepared by

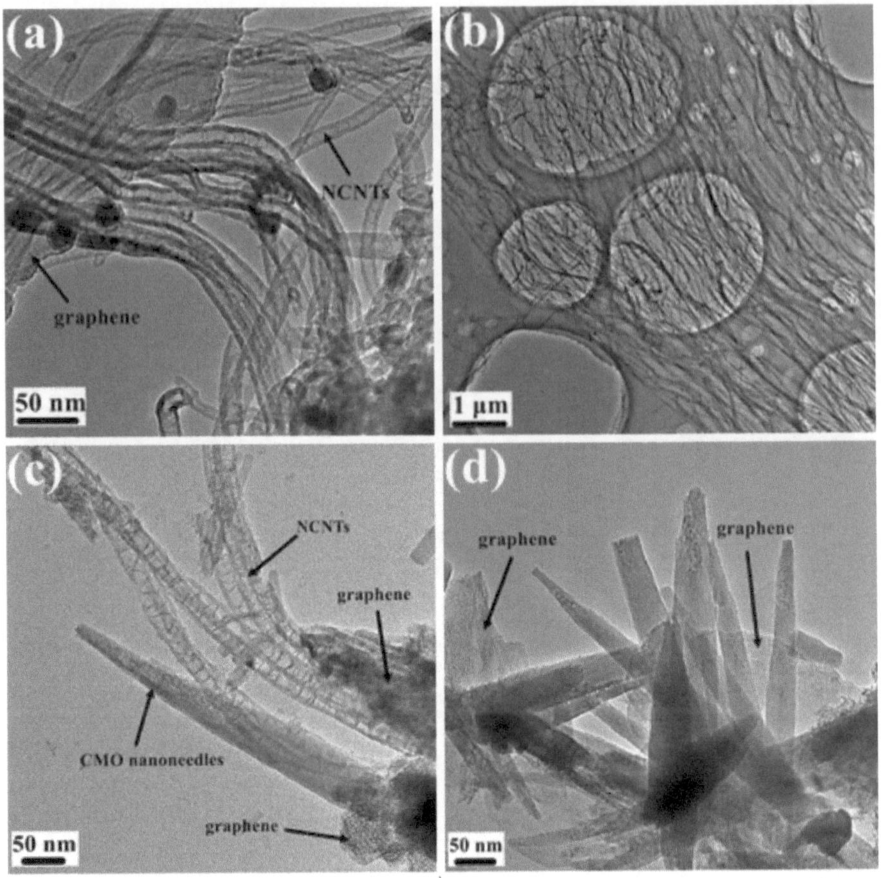

FIGURE 20.8 Morphological images (TEM) of electrocatalysts (a and b) NCNTs/3D graphene at a resolution of 50 and 1,000 nm in different regions, respectively. (c and d) CMO/NCNTs/3D graphene at a resolution of 50 nm in different regions. Adapted and reproduced with permission from Ref. [44]. Copyright © 2019 American Chemical Society.

in situ chemical vapor deposition (CVD) and hydrothermal process with nickel foam as substrate, when tested in alkaline conditions showed enhancement in the performance compared to commercial 20 wt.% Pt/C electrocatalyst by measurements such as linear sweep voltammograms and CV. Formation of active sites of C–N with 3D interconnected and porous morphology with a synergistic effect of NCNTs/3D graphene and nanoneedles of CMO resulted in the enhancement in properties. Coin cell of Al air system with cathode catalyst as CMO/NCNTs/3D graphene composite showed OCV of 1.55 V with 312.8 mAh g^{-1} as specific capacity measured at 1 mA cm^{-2} current density.

Another investigation on cathode polymer-binder-free efficient ORR based on ultrathin-nanosheet clusters of Co_3O_4 with nitrogen-doped CNT and 3D graphene (Co_3O_4/NCNTs/3D graphene) for AAB was conducted by Liu et al. [45]. Co_3O_4/NCNTs/3D graphene prepared using the CVD technique and thermal treatments over NF outperformed the commercial 20 wt.% Pt/C electrocatalyst. Furthermore, Al air coin cell with Co_3O_4/NCNTs/3D graphene as ORR showed 482.80 mAh g^{-1} as specific capacity at a current density of 1 mA cm^{-2} with 4.88 mW cm^{-2} as maximum power density and OCV of 1.52 V, showing the potential applicability for further development.

Although the polymer inherently has binding properties, which hold the materials in the cathode, a decrement in the electrical conductivity of the cathode with possibilities for side reactions hampers the overall performance of the cell [46–48]. In the pursuit of exploring polymer-binder-free cathode, Liu et al. [49] developed an efficient ORR catalyst based on Co@CoO_x/nitrogen-doped CNT with helical morphology (Co@CoO_x/HNCNT) for AAB. The catalyst prepared over carbon fiber paper by hydrothermal and in situ single-step CVD showed enhanced performance in alkaline media. It was shown that the optimized growth time is 1 h, and the optimized catalyst, Co@CoO_x/HNCNT-1 h, used as a binder-free cathode for Al air coin cell showed 3.86 mW cm^{-2} as the maximum power density with an OCV of 1.48 V, and at a discharge current density of 1 mA cm^{-2}, the specific capacity was recorded as 367.31 mAh g^{-1}.

Wang et al. [50] developed a hierarchically nitrogen-doped carbon-based porous 3D ORR electrocatalyst for AAB from chitosan by calcination and gas-foaming assisted strategy method with defects in the structure. AAB fabricated using the porous ultrathin catalyst showed high and stable electrocatalytic properties with an energy density of 2,453.4 Wh kg^{-1} and 401 mW cm^{-2} as power density at a current density of 200 mA cm^{-2} with OCV of 1.85 V. It is due to the structural uniqueness of the catalyst with a high surface area and porous structure that the oxygen/electrolyte diffusion happens in more active sites, thereby contributing to performance enhancement.

20.3 FUTURE ADVANCEMENTS IN AABS

A long-range, cost-efficient, recyclable AAB provides energy density much greater than that of LIB of comparable weight/volume and can provide large breakthroughs in EV technology because of the large theoretical energy density. The batteries are constructed from plastic, Al, and a non-toxic, safe electrolyte. A discharged Al air cell would simply be replaced with a recycled one, swapping an empty power source for a new, fully charged one at a refueling machine serving multiple 'swap stations'. While rechargeable LIBs can cost too much to replace, tremendous research in this direction is still currently in progress.

Recently, paper material has been integrated into electrochemical devices due to its advantages over other platforms. Paper is ubiquitous, universally available, and an inexpensive alternative to other materials including glass and silicon. Also, cellulose fibers make up the majority of various types of paper and allow the passive flow of fluids through capillary action, which consequently removes the need for external pumps. Paper-like energy storage devices are becoming more attractive due to their easy integration with electronic devices. Moreover, paper's porosity makes it beneficial for the manipulation of electrons and ion transport throughout the structure. Current and continuous research in the area of AABs have potential to significantly advance the design of AABs, creating next-generation high-efficiency, low-cost, and eco-friendly energy sources.

Understanding the experimental observations and unveiling the complexities of electrochemical mechanisms, theory-guided experimental observations are gaining acceptance in battery technology [51]. Various theoretical calculations, mathematical modeling, and simulations are being promoted by researchers to relate the structure–property relationship of battery materials, cycle life, state of health, and remaining useful life [51]. Real-time monitoring of cells can rectify the ambiguity in the electrochemical mechanism; one such technique is operando characterizations [52].

Many innovations have been reported in anode modification, electrolyte engineering, and the designing of cells. The usage of solid electrolytes, alloy anodes, ionic liquid electrolytes, electrolyte additives, and corrosion inhibitors has also enhanced the performance of AAB [53,54]. One such recent work by Xia et al. [55], where the applicability of Co-MnO_2/C as an electrocatalyst for AAB with $AlCl_3$ and ionic liquid based on urea as an electrolyte was reported. Intercalation of Co with MnO_2 contributed to a positive shift in the surface area. AAB with a catalyst having 40 mole percent of Co to Mn showed better performance with good reversible charge and discharge of 2 V during a cycle of 30, showing its potential applicability for further developing rechargeable AAB.

20.4 CONCLUSION

Search for clean energy alternatives and renewable energy sources, including solar energy, wind, tidal, and geothermal energy, have been explored as a result of fossil fuel depletion and increasing environmental considerations. The realization of the intermittence of these sources resulted in the development of energy storage devices. As the ambiguity on the sustainability of LIB technology was predicted considering the demand for EVs, the demand for high energy density batteries accelerated. AABs with a high theoretical energy density of 8,100 Wh kg^{-1} and abundant availability of aluminum attracted researchers. Advances in AAB are in their infancy and are associated with several challenges. Material chemistry related to air cathode and electrocatalyst plays a determinant role in the performance of cells. Covering up the pores of air cathode by the insoluble products (mainly carbonates), in alkaline electrolyte systems, leads to overpotential, and lowering the active surface area for the catalytic adsorption decelerates the electrochemical performance. Electron transfer resistance is another factor which impedes the performance. Designing GDL and tailoring the properties by binding the catalyst directly to the current collector metal mesh can minimize the resistance. The development of an efficient bifunctional catalyst can enhance the kinetics of ORR and OER. Various metals, metal oxides, and its derivatives in the nano-dimension have been employed as effective bifunctional catalysts. The development in AABs has identified the challenges and established approaches to rectify the same. Multi-faced approaches, along with the collaboration of materials design, experimental innovations, and theoretical calculations, function as a potential step for enhancing the research momentum. As AABs have been identified as a potential source for powering future mobility, unveiling the properties of cathode and electrolyte can further accelerate the development.

REFERENCES

1. Jishnu NS, Vineeth SK, Das A, et al (2021) Electrospun PVdF and PVdF-co-HFP-based blend polymer electrolytes for lithium ion batteries. In: Balakrishnan NTM, Prasanth R (eds) *Electrospinning for Advanced Energy Storage Applications*, 1st ed. Springer, Singapore, pp. 201–234
2. Vineeth SK, Soni CB, Sun Y, et al (2022) Implications of Na-ion solvation on Na anode-electrolyte interphase. *Trends Chem* 4:48–59. https://doi.org/10.1016/j.trechm.2021.11.002
3. Vineeth SK, Soni CB, Sanjaykumar C, Yamauchi Y, Han M, Kumar V (2023) A quasi-solid state polymer electrolyte for high-rate and long-life sodium-metal batteries. *J Energy Storage* 73:108780. https://doi.org/10.1016/j.est.2023.108780
4. Walter M, Kovalenko MV, Kravchyk KV (2020) Challenges and benefits of post-lithium-ion batteries. *New J Chem* 44:1677–1683. https://doi.org/10.1039/c9nj05682c
5. Sungjemmenla J, Soni CB, Vineeth SK, Kumar V (2022) Exploration of the unique structural chemistry of sulfur cathode for high-energy rechargeable beyond-Li batteries. *Adv Energy Sustain Res* 3:2100157. https://doi.org/10.1002/aesr.202100157
6. Sungjemmenla J, Soni CB, Vineeth SK, Kumar V (2021) Unveiling the physiochemical aspects of the matrix in improving sulfur-loading for room-temperature sodium-sulfur batteries. *Mater Adv* 2:4165–4189. https://doi.org/10.1039/D1MA00247C
7. Olabi AG, Sayed ET, Wilberforce T, et al (2021) Metal-air batteries-a review. *Energies* 14:7373. https://doi.org/10.3390/en14217373
8. Cheng F, Chen J (2012) Metal-air batteries: from oxygen reduction electrochemistry to cathode catalysts. *Chem Soc Rev* 41:2172. https://doi.org/10.1039/c1cs15228a
9. Zhang J, Zhou Q, Tang Y, et al (2019) Zinc-air batteries: are they ready for prime time? *Chem Sci* 10:8924–8929. https://doi.org/10.1039/C9SC04221K
10. Li Y, Lu J (2017) Metal-air batteries: will they be the future electrochemical energy storage device of choice? *ACS Energy Lett* 2:1370–1377. https://doi.org/10.1021/acsenergylett.7b00119
11. Shen LL, Zhang GR, Biesalski M, Etzold BJM (2019) Paper-based microfluidic aluminum-air batteries: toward next-generation miniaturized power supply. *Lab Chip* 19:3438–3447. https://doi.org/10.1039/c9lc00574a
12. Liu Y, Sun Q, Li W, et al (2017) A comprehensive review on recent progress in aluminum-air batteries. *Green Energy Environ* 2:246–277. https://doi.org/10.1016/j.gee.2017.06.006

13. Wang Y, Kwok H, Pan W, et al (2019) Innovative paper-based Al-air batteries as a low-cost and green energy technology for the miniwatt market. *J Power Sources* 414:278–282. https://doi.org/10.1016/j.jpowsour.2019.01.018
14. Lee J-S, Tai Kim S, Cao R, et al (2011) Metal-air batteries with high energy density: Li-air versus Zn-air. *Adv Energy Mater* 1:34–50. https://doi.org/10.1002/aenm.201000010
15. Gu P, Xu Y, Zhao Y, et al (2017) Electrocatalysis of rechargeable non-lithium metal-air batteries. *Adv Mater Interfaces* 4:1700589. https://doi.org/10.1002/admi.201700589
16. Mokhtar M, Talib MZM, Majlan EH, et al (2015) Recent developments in materials for aluminum-air batteries: a review. *J Ind Eng Chem* 32:1–20. https://doi.org/10.1016/j.jiec.2015.08.004
17. Yan S, Xue Y, Li S, et al (2019) Enhanced bifunctional catalytic activity of manganese oxide/perovskite hierarchical core-shell materials by adjusting the interface for metal-air batteries. *ACS Appl Mater Interfaces* 11:25870–25881. https://doi.org/10.1021/acsami.9b06141
18. Jörissen L (2006) Bifunctional oxygen/air electrodes. *J Power Sources* 155:23–32. https://doi.org/10.1016/j.jpowsour.2005.07.038
19. Loh A, Trudgeon DP, Li X, et al (2022) Selection of oxygen reduction catalysts for secondary tri-electrode zinc-air batteries. *Sci Rep* 12:6696. https://doi.org/10.1038/s41598-022-10671-5
20. Tomboc GM, Yu P, Kwon T, et al (2020) Ideal design of air electrode-A step closer toward robust rechargeable Zn-air battery. *APL Mater* 8:050905. https://doi.org/10.1063/5.0005137
21. Chang Z, Xu J, Zhang X (2017) Recent progress in electrocatalyst for Li-O_2 batteries. *Adv Energy Mater* 7:1700875. https://doi.org/10.1002/aenm.201700875
22. Vineeth SK, Tebyetekerwa M, Liu H, Soni CB, Zhao XS, Kumar V (2022) Progress in the development of solid-state electrolytes for reversible room-temperature sodium–sulfur batteries. *Mat Adv*, 3(16):6415–6440. https://doi.org/10.1039/D2MA00428C
23. Liu J, Jiao M, Lu L, et al (2017) High performance platinum single atom electrocatalyst for oxygen reduction reaction. *Nat Commun* 8:1–9. https://doi.org/10.1038/ncomms15938
24. Wang Z-L, Xu D, Xu J-J, Zhang X-B (2014) Oxygen electrocatalysts in metal-air batteries: from aqueous to nonaqueous electrolytes. *Chem Soc Rev* 43:7746–7786. https://doi.org/10.1039/C3CS60248F
25. Vineeth SK, Gadhave RV, Gadekar PT (2019) Nanocellulose applications in wood adhesives-review. *Open J Polym Chem* 09:63–75. https://doi.org/10.4236/ojpchem.2019.94006
26. Singh HK, Patil T, Vineeth SK, et al (2020) Isolation of microcrystalline cellulose from corn stover with emphasis on its constituents: corn cover and corn cob. *Mater Today Proc* 27:589–594. https://doi.org/10.1016/j.matpr.2019.12.065
27. Vineeth SK, Gadhave RV, Gadekar PT (2020) Glyoxal cross-linked polyvinyl alcohol- microcrystalline cellulose blend as a wood adhesive with enhanced mechanical, thermal and performance properties. *Mater Int* 2:0277–0285.
28. Wang YJ, Fang B, Zhang D, et al (2018) *A Review of Carbon-Composited Materials as Air-Electrode Bifunctional Electrocatalysts for Metal-Air Batteries*. Springer, Singapore.
29. Liu Y, Li J, Li W, et al (2015) Spinel $LiMn_2O_4$ nanoparticles dispersed on nitrogen-doped reduced graphene oxide nanosheets as an efficient electrocatalyst for aluminium-air battery. *Int J Hydrogen Energy* 40:9225–9234. https://doi.org/10.1016/j.ijhydene.2015.05.153
30. Liu K, Zhou Z, Wang H, et al (2016) N-Doped carbon supported Co_3O_4 nanoparticles as an advanced electrocatalyst for the oxygen reduction reaction in Al-air batteries. *RSC Adv* 6:55552–55559. https://doi.org/10.1039/c6ra10486j
31. Wang J, Lu H, Hong Q, et al (2017) Porous N,S-codoped carbon architectures with bimetallic sulphide nanoparticles encapsulated in graphitic layers: highly active and robust electrocatalysts for the oxygen reduction reaction in Al-air batteries. *Chem Eng J* 330:1342–1350. https://doi.org/10.1016/j.cej.2017.08.072
32. Ryu J, Jang H, Park J, et al (2018) Seed-mediated atomic-scale reconstruction of silver manganate nanoplates for oxygen reduction towards high-energy aluminum-air flow batteries. *Nat Commun* 9:1–10. https://doi.org/10.1038/s41467-018-06211-3
33. Liu Y, Jiang H, Hao J, et al (2017) Metal-organic framework-derived reduced graphene oxide-supported ZnO/$ZnCo_2O_4$/C hollow nanocages as cathode catalysts for aluminum-O_2 batteries. *ACS Appl Mater Interfaces* 9:31841–31852. https://doi.org/10.1021/acsami.7b08647
34. Shao L, Liang ZX, Chen H, et al (2020) $CuCo_2S_4$ hollow nanoneedle arrays supported on Ni foam as efficient trifunctional electrocatalysts for overall water splitting and Al-air batteries. *J Alloys Compd* 845:155392. https://doi.org/10.1016/j.jallcom.2020.155392

35. Zhu C, Ma Y, Zang W, et al (2019) Conformal dispersed cobalt nanoparticles in hollow carbon nanotube arrays for flexible Zn-air and Al-air batteries. *Chem Eng J* 369:988–995. https://doi.org/10.1016/j.cej.2019.03.147
36. Guo T, Qin X, Gao X, et al (2019) Boron-doped carbon microspheres as the catalyst for rechargeable Al-air batteries. *Fullerenes Nanotub Carbon Nanostructures* 27:299–304. https://doi.org/10.1080/1536383X.2018.1562445
37. Hong Q, Lu H, Cao Y (2019) Improved oxygen reduction activity and stability on N, S-enriched hierarchical carbon architectures with decorating core-shell iron group metal sulphides nanoparticles for Al-air batteries. *Carbon* 145:53–60. https://doi.org/10.1016/j.carbon.2019.01.002
38. Guo M, Zhang X, Yang T, et al (2020) Binary FeCo-N-doped carbon/carbon nanotube composites for efficient oxygen reduction and high-performance aluminum-air battery. *J Power Sources* 456:227933. https://doi.org/10.1016/j.jpowsour.2020.227933
39. Mori R (2017) Suppression of byproduct accumulation in rechargeable aluminum-air batteries using non-oxide ceramic materials as air cathode materials. *Sustain Energy Fuels* 1:1082–1089. https://doi.org/10.1039/c7se00087a
40. Mori R (2017) Rechargeable aluminum-air battery using various air-cathode materials and suppression of byproduct formation on both anode and air cathode. *ECS Meet Abstr* MA2017-02:565. https://doi.org/10.1149/ma2017-02/5/565
41. Mori R (2017) Electrochemical properties of a rechargeable aluminum-air battery with a metal-organic framework as air cathode material. *RSC Adv* 7:6389–6395. https://doi.org/10.1039/c6ra25164a
42. Li J, Chen J, Wan H, et al (2019) Boosting oxygen reduction activity of Fe-N-C by partial copper substitution to iron in Al-air batteries. *Appl Catal B Environ* 242:209–217. https://doi.org/10.1016/j.apcatb.2018.09.044
43. Liu S, Cao Z, Meng Y, et al (2021) Aerophilic Co-embedded N-doped carbon nanotube arrays as highly efficient cathodes for aluminum-air batteries. *ACS Appl Mater Interfaces* 13:26853–26860. https://doi.org/10.1021/acsami.1c00837
44. Liu Y, Zhan F, Wang B, et al (2019) Three-dimensional composite catalysts for Al-O_2 batteries composed of CoMn2O4 nanoneedles supported on nitrogen-doped carbon nanotubes/graphene. *ACS Appl Mater Interfaces* 11:24. https://doi.org/10.1021/acsami.9b04861
45. Liu Y, Yang L, Xie B, et al (2020) Ultrathin Co_3O_4 nanosheet clusters anchored on nitrogen doped carbon nanotubes/3D graphene as binder-free cathodes for Al-air battery. *Chem Eng J* 381:122681. https://doi.org/10.1016/j.cej.2019.122681
46. Gadhave RV, Vineeth SK, Mahanwar PA, Gadekar PT (2021) Effect of addition of boric acid on thermo-mechanical properties of microcrystalline cellulose/polyvinyl alcohol blend and applicability as wood adhesive. *J Adhes Sci Technol* 35:1072–1086. https://doi.org/10.1080/01694243.2020.1832775
47. Vineeth SK, Gadhave RV, Gadekar PT (2022) Investigation of crosslinking ability of sodium metabisulphite with polyvinyl alcohol-corn starch blend and its applicability as wood adhesive. *Indian Chem Eng* 64:197–207. https://doi.org/10.1080/00194506.2021.1887769
48. Lopez J, Mackanic DG, Cui Y, Bao Z (2019) Designing polymers for advanced battery chemistries. *Nat Rev Mater* 4:312–330. https://doi.org/10.1038/s41578-019-0103-6
49. Liu Y, Wang B, Sun Q, et al (2020) Controllable synthesis of Co@CoO_x/helical nitrogen-doped carbon nanotubes toward oxygen reduction reaction as binder-free cathodes for Al-air batteries. *ACS Appl Mater Interfaces* 12:16512–16520. https://doi.org/10.1021/acsami.0c01603
50. Wang Y, Hao J, Yu J, et al (2020) Hierarchically porous N-doped carbon derived from biomass as oxygen reduction electrocatalyst for high-performance Al-air battery. *J Energy Chem* 45:119–125. https://doi.org/10.1016/j.jechem.2019.10.005
51. Eng AYS, Soni CB, Lum Y, et al (2022) Theory-guided experimental design in battery materials research. *Sci Adv* 8:eabm2422. https://doi.org/10.1126/sciadv.abm2422
52. Soni CB, Sungjemmenla J, Vineeth SK, Kumar V (2022) Challenges in regulating interfacial-chemistry of the sodium-metal anode for room-temperature sodium-sulfur batteries. *Energy Storage* 4:e264. https://doi.org/10.1002/est2.264
53. Ryu J, Park M, Cho J (2019) Advanced technologies for high-energy aluminum-air batteries. *Adv Mater* 31:1804784. https://doi.org/10.1002/adma.201804784
54. Chawla N (2019) Recent advances in air-battery chemistries. *Mater Today Chem* 12:324–331. https://doi.org/10.1016/j.mtchem.2019.03.006
55. Xia Z, Zhu Y, Zhang W, et al (2020) Cobalt ion intercalated MnO_2/C as air cathode catalyst for rechargeable aluminum-air battery. *J Alloys Compd* 824:153950. https://doi.org/10.1016/j.jallcom.2020.153950

21 Emerging Cost-Efficient Aluminium-Air Battery Technology

Electrolyte Materials

S. K. Vineeth, Pritam V. Dhawale, Roshny Joy, Abhilash Pullanchiyodan, Jou-Hyeon Ahn, Nutan Gupta, and Prasanth Raghavan

21.1 INTRODUCTION

From the realization that the dependency on fossil-fuel-based energy sources is not reliable for a longer duration, there has been appreciable development in the extraction of energy from renewable energy sources. These sources are intermittent, which catalysed the growth of energy storage technology. Batteries constitute a major class of energy storage devices that work via electrochemical reactions [1]. After the successful commercialization of lithium-ion batteries (LIB) by Sony Corporation in 1991, revolutionary developments are visible in battery technology [2]. The development in LIBs has reached an energy density of ~100~200 Wh kg^{-1}, which is insufficient to satisfy the requirement of electric vehicles (EVs) [3,4]. Hence, to meet the requirement of EV technology and systems which are cost-efficient and reliable, researchers and technologists opened the area of post-LIB technology. One such development is the focus of research on high-energy-density batteries, where metal-air batteries (MABs) have prominent candidature [5]. MABs have the advantage of both high energy density with safety in operation [6]. Moreover, the employment of a cathode, which utilizes oxygen from ambient air, lowers the weight and is hence a cost-effective technology for the post-LIB era [7,8]. In MAB technology, aluminium (Al)-based MAB, aluminium air battery (AAB), was considered a potential candidate for powering EVs, mainly due to the high theoretical energy density of 8,100 Wh kg^{-1}, with abundance and cost-effective considerations [9]. In addition, the added advantage of Al includes the high theoretical electrochemical equivalent value of 2.98 A h g^{-1} [10,11].

Although AAB technology has huge potential, the developments in AABs are still in infancy; hence challenges faced by the technology are massive. Self-corrosive nature of Al (in open circuit and discharge condition), in alkaline electrolyte, and by-product formation, namely aluminium oxide (Al_2O_3) and aluminium hydroxide ($Al(OH)_3$), impede the electrochemical performance of the system [12]. Hydrogen evolution reaction (HER) occurs mainly due to the parasitic reactions of Al with water molecules, lowering the potential of AAB [12,13]. Furthermore, intercalation of trivalent Al ion (Al^{3+}) to the air cathode and reversibility issues with the hydrated alumina formation results in the loss of charge on storage affecting the shelf life of the system [12]. As electrolytes in a cell provide internal connection, engineering electrolyte plays a crucial role in enhancing performance [14].

In AAB, electrolyte regulates HER and electrochemical reactions at the anode. In addition to alloying of anode, microstructure engineering through tailoring grain boundaries by heat treatment and mechanical tailoring of the electrolyte have also shown enhancement in electrochemical performance. Incorporating electrolyte additives and corrosion inhibitors mitigates the HER and

DOI: 10.1201/9781003310174-21

enhances anodic utilization. Moreover, the employment of solid-state electrolytes, hydrogels, ionic liquids (IL), electrolytes, etc. also shows promising properties. The main aim of this chapter is to highlight various research, developments, and innovations reported in electrolyte engineering for AABs. A detailed analysis of various strategies employed in electrolyte tailoring has been discussed. As AABs are promising technology for the post-LIB era, it is believed that by continuous development in materials engineering and through judicious selection of design aspects, AABs can be upgraded as an advanced system for future energy storage requirements.

21.2 DEVELOPMENTS IN ELECTROLYTES FOR AAB

Electrolyte in electrochemical energy storage devices acts as an internal connection, where the ions are transported between the electrodes [15]. Tremendous research in the area of electrolytes for AABs has contributed to continuous enhancement in the electrochemical performance of AABs. The self-corrosive nature of the Al anode and the formation of the oxide and hydroxide layer over the anode mitigate the electrochemical performance [16]. Apart from anode modifications, engineering electrolytes by various approaches have been shown to rectify the aforementioned issues. Incorporating electrolyte additives, varying the electrolyte systems, addition of corrosion inhibitors, etc. are some of the most explored approaches [17]. Corrosion inhibitors function by reacting with Al^{3+} present in the anode, thereby forming a protective layer over it, which further prevents corrosion [18]. Moreover, the usage of non-aqueous electrolytes can contribute to mitigating the rate of self-corrosion; however, it negatively results in decreasing the energy density of the cell. The following section highlights developments in electrolytes for AAB.

21.2.1 Hydrogels Electrolytes

To upgrade AAB, being useful for flexible electronics and wearable devices, hydrogel-based electrolytes are employed. Polymeric materials are employed for the fabrication of host materials for hydrogel electrolytes. As polymeric materials have flexibility and mechanical and thermal properties, these are preferred for hydrogel fabrication [19–22]. Palma et al. [23] observed high ionic conductivity on an alkaline (1 M and 8 M KOH) AAB based on polysaccharides (Xanthan and κ-carrageenan) and observed formation of elastic gel in the alkaline medium provided by KOH. The elastic gels of three different grades were prepared: GEL01 contains xanthan in 1 M KOH, GEL02 constitutes xanthan in 8 M KOH, and GEL03 represents κ-carrageenan in 8 M KOH at three different magnifications. X-ray diffraction (XRD) showed the high amorphous nature of gels. Electrochemical impedance spectroscopy (EIS) technique showed the ionic conductivity in the range of 10^{-2} S cm^{-1} with high energy density of 33 mWh cm^{-2}. Scanning electron microscopic (SEM) images in Figure 21.1 show the granular morphology of GEL03 κ-carrageenan gel, which contributed positively towards enhancing the performance of the cell, by transferring the reaction product through the electrolyte.

Palma et al. [24] also investigated the applicability of xanthan hydrogels in acidic (HCl), alkaline (KOH), and water medium for primary AABs. It was reported that when xanthan was in acidic media (5 M HCl), anode corrosion was lower and the cell at high current density showed high cell capacity above 70 mAh cm^{-2}. In an alkaline medium of KOH, the electrolyte showed good discharge at low current density with high ionic conductivity.

Moreover, in continuation to this study, it was observed that when the pH of the xanthan hydrogel electrolyte was acidic, anodic efficiency was the highest with high discharge performance [25]. However, from the cyclic voltammetry test, the cell showed poor reachability due to the higher reactivity associated with high hydrogen content in the acidic condition. In alkaline conditions the cell showed a rechargeable nature when the content of xanthan was in high amounts but poor anodic efficiency and electrolytic decomposition [25]. The result confirmed that at acidic conditions, xanthan acts as a self-corrosion inhibitor.

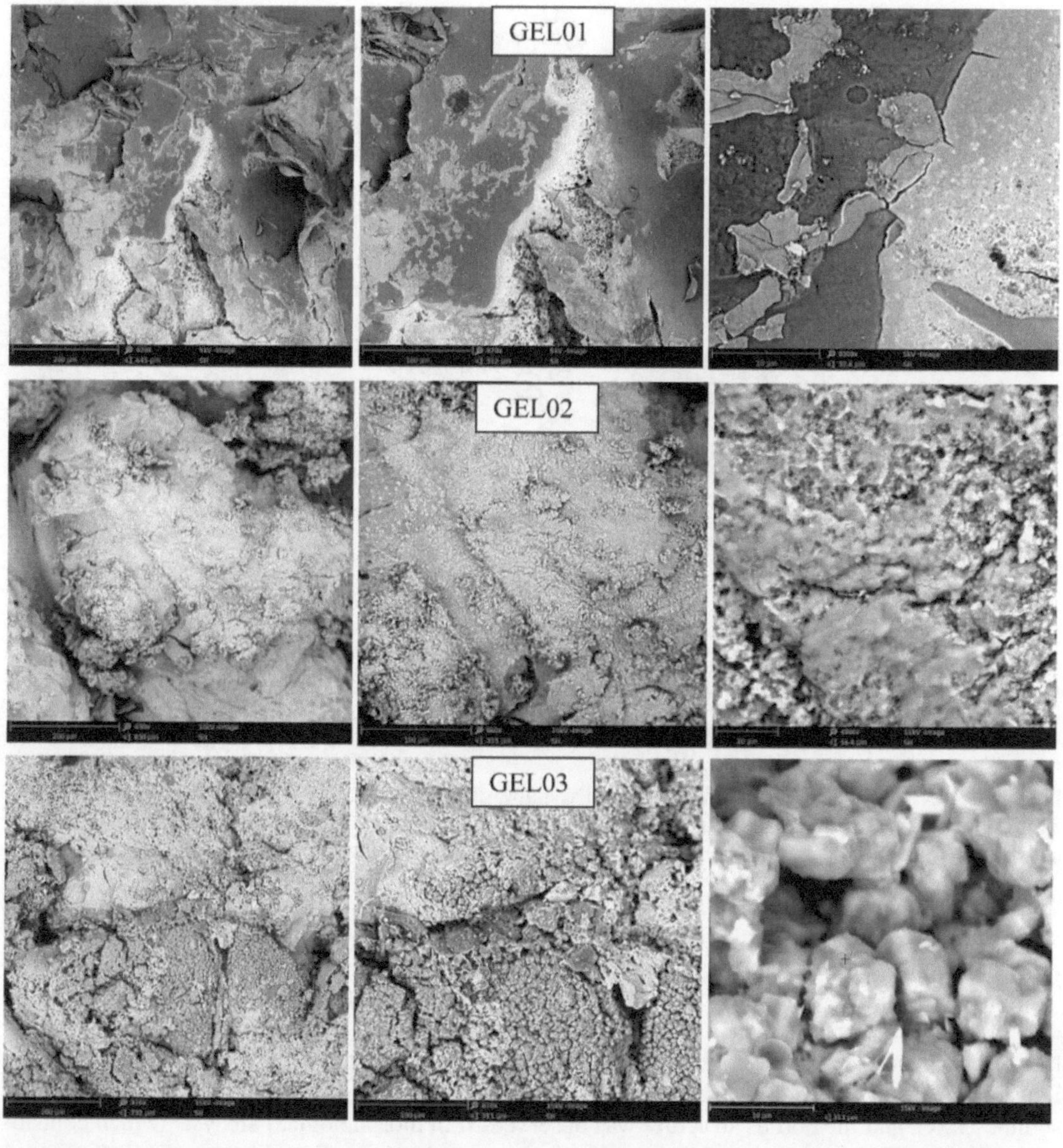

FIGURE 21.1 SEM images of various hydrogels after discharge cycles, where GEL01 represents xanthan in 1 M KOH, GEL02 represents xanthan in 8 M KOH, and GEL03 contains κ-carrageenan in 8 M KOH at three different magnifications. Adapted and reproduced with permission from Ref. [23]. Copyright © 2017 Elsevier.

Liu et al. [26] reported recyclable Al-air coin cells using metal oxide-chitosan hydrogel, namely SnO_2-chitosan, SiO_2-chitosan, and ZnO-chitosan, membranes as separators. Out of the various membranes SiO_2-chitosan membrane showed better performance with higher power density and longer and improved discharge performance. It was also seen that the generation of SiO_3^{2-} on the reaction between the alkaline electrolyte, 4 M KOH, and SiO_2 prevented the corrosion in the aluminium anode. Moreover, the work demonstrated a successful eco-friendly recycling process after discharge, where acetic acid was used for dissolving the chitosan membrane and removing $Al(OH)_3$ formed over the surface of anode and filtered using a filter paper and funnels. Finally, for neutralizing acetic acid and precipitating aluminium ions, sodium bicarbonate was used; the detailed process

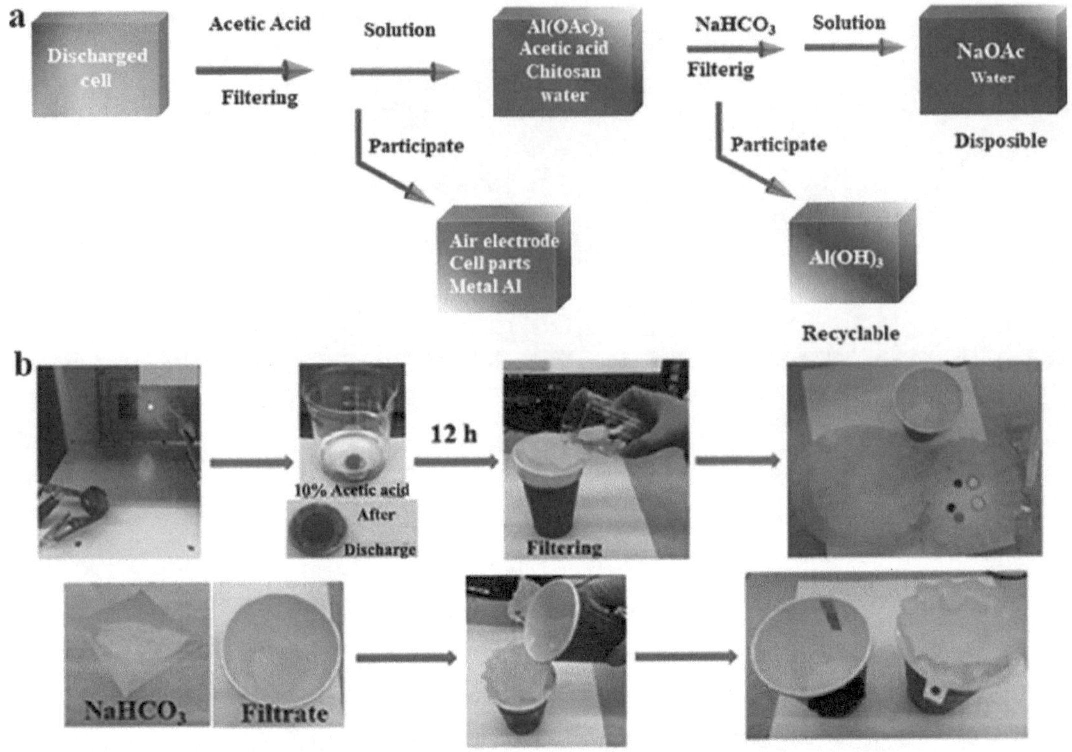

FIGURE 21.2 (a) Process of recycling presented as a flow diagram and (b) stepwise image representing recycling of a discharged Al-air coin cell using acetic acid, sodium bicarbonate ($NaHCO_3$), and filter paper and funnel. Adapted and reproduced with permission from Ref. [26]. Copyright © 2018 American Chemical Society.

is shown in Figure 21.2. Overall, the work opened up the possibilities for the further development of recyclable and portable coin cells based on the Al-air system.

Sun et al. [27] fabricated flexible AAB with sodium polyacrylate (SPA)–sodium carboxymethyl cellulose (CMC) as a polymer hydrogel electrolyte that are reusable electrolytes. Preparation of the hydrogel involves free radical polymerization of SPA with varying amounts of sodium CMC. The Fourier transform infrared spectroscopy (FTIR) result showed the successful synthesis of SPA–sodium CMC, and the porous structure was observed from morphological analysis, which contributed to high electrolyte uptake while ensuring ionic conductivity. Five percent Na-CMC loaded polymer composite showed the highest electrolyte uptake with high ionic conductivity of 0.324 S cm^{-1} due to the decrease in pore size and introduction of more –OH and –COO– groups. Moreover, due to the physical and covalent cross-linked structure of SPA 5% CMC, it showed enhancement in mechanical properties compared to pure SPA. AAB with SPA 5% CMC electrolyte showed power density, which is twice that of the pure SPA electrolyte with a discharge voltage of 1.86 V at a current density of 1 mA cm^{-2}. Two AABs connected in series showed an open circuit voltage (OCV) of over 3.9 V, which could successfully charge a mobile phone; the same worked when attached to the straps of the smartwatch, shown in Figure 21.3, showing its applicability in flexible and wearable electronics.

Sun et al. [28] successfully fabricated a high-performance wearable AAB with agarose gel electrolyte (AGE). AAB was fabricated with air cathode using a nickel foam with nitrogen-doped grapheme, and introduction to AGE ensured good mechanical properties like bendability, flexibility, prevention of electrolyte leakage due to its high toughness, formation of stable anode–electrolyte interface (solid–quasi-solid interface), and inhibition of self-corrosion of the anode by the formation

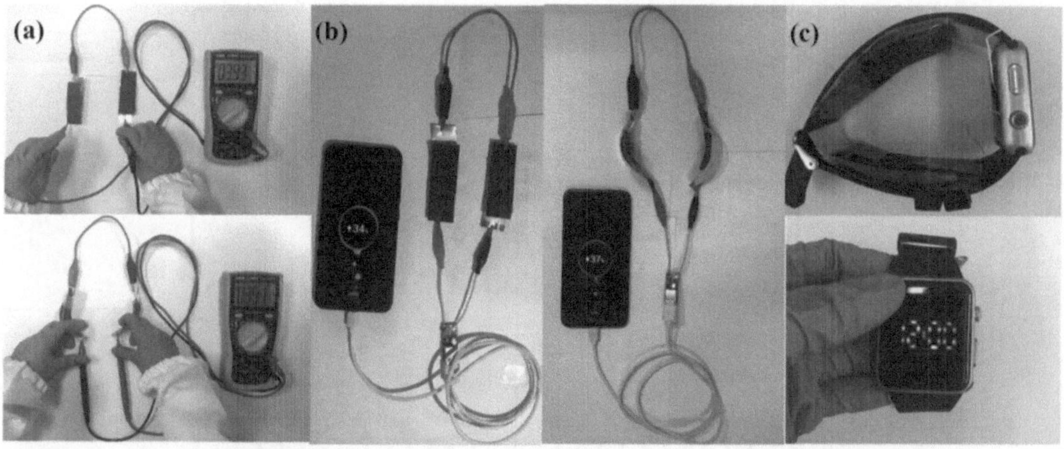

FIGURE 21.3 Applicability of the AAB in flexible and wearable electronics: (a) measurement of OCV, (b) AAB-connected mobile charger, and (c) flexible and bendable AAB for smartwatch. Adapted and reproduced with permission from Ref. [27]. Copyright © 2020 The Electrochemical Society.

of a passivation layer, hence favouring the fabrication of wearable AAB. It was seen that a decrement of 402% in the self-corrosion occurred when AGE was used. Moreover, the performance properties of AAB were enhanced. AAB with a 10 cm^2 soft package system exhibited a high specific capacity of 2,148.5 mAh g^{-1} with a discharge time of 20.1 h. The authors were successful in making a LED array and a smartphone charger pack with the so-prepared AAB. Figure 21.4 shows the conceptualization of wearable AGE-based AAB.

21.2.2 Solid, Acid, and Semi-Solid Electrolytes

Migliardini et al. [29] were the first to report a study on solid and strong acids as electrolytes for AAB. The work made use of the inherent gelling nature of xanthan gum by incorporating HCl as an electrolyte. For the fabrication of AAB, the air cathode was based on carbon cloth with Pt catalyst supported on active carbon and with polyvinylidene fluoride as a binder. From weight loss measurements, EIS and potentiodynamic polarization curves, it was evidenced that the xanthan–HCl hydrogels imparted corrosion inhibition properties to the anode with good electrochemical performance. AAB with 1.9 10^{-2} S cm^{-1} as the highest ionic conductivity and 96% inhibition efficiency was obtained when xanthan gum/HCl electrolyte with 1.40 g mL^{-1} of xanthan gum and 24% HCl was taken. The cell showed a cell capacity of 40 mAh cm^{-2} with a maximum anodic efficiency of 83% with low dendrite formation.

Hibino et al. [30] communicated the usage of $Sn_{0.92}Sb_{0.08}P_2O_7$ as a solid-state electrolyte for AAB. The capability of $Sn_{0.92}Sb_{0.08}P_2O_7$ for hydroxide ion exchange by charge compensation through the partial substitution of Sn^{4+} with Sb^{5+} makes them suitable for electrolyte application. Moreover, the compound is anhydrous; hence, without H_2 generation reaction electrodeposition can be made possible. It was observed that the cell showed poor rate capability due to the inability of reversible reduction of aluminate species to aluminium. Hence the work concluded that further improvement in the anode is needed for good cyclability and performance of AAB.

From a study by Mokhtar et al. [31], it was seen that incorporating ZnO as filler in polyvinyl alcohol (PVA)/alkaline solid polymer electrolyte for AAB could enhance electrochemical performance. The addition of 4 wt.% ZnO to PVA in 2 M KOH electrolyte at room temperature contributed to improvement in the ionic conductivity with a value of 1.61×10^{-2} S cm^{-1}, measured by impedance measurement. The semi-crystalline nature of PVA was altered when ZnO was added to PVA, as evidenced by XRD results. The electrochemical properties were studied by fabricating an aluminium

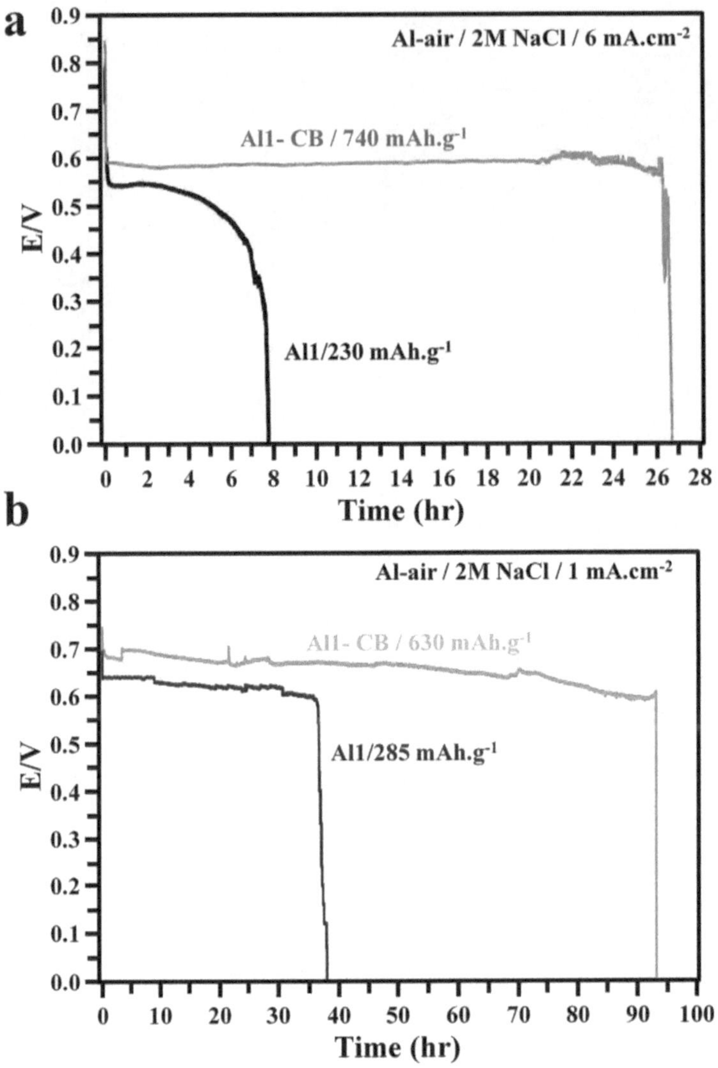

FIGURE 21.4 (a) Schematic illustration and concept of a wearable agarose gel electrolyte-based AAB having its various layers; and (b–e) discharge mechanism OH transportation and formation of the passivation layer. Adapted and reproduced with permission from Ref. [28]. Copyright © 2021 Elsevier.

air coin cell with the air cathode as carbon-containing nickel mesh with MnO_2 as an electrocatalyst and an aluminium plate of 99.5% purity as anode with KOH-soaked PVA membrane containing ZnO filler as a solid electrolyte. Measurement of the electrochemical stability window for the electrolyte by a linear sweep voltammetry test showed that enhancement on adding ZnO filler, PVA/4 wt.% ZnO/2 M KOH, showed a cell potential value of ±2.15 V. Overall ZnO-incorporated PVA as a polymer electrolyte showed promising results for its further development in AABs.

Mori et al. [32] fabricated AAB based on a mixture of aluminium chloride ($AlCl_3$), urea, CMC, and glycerin as solid electrolytes in the ratio of 3:2:1:1, respectively. Air cathode was made in the shape of a pellet using titanium nitride (TiN) and polyvinylidene difluoride (PVDF) at a molar ratio of 1:0.3. It was observed from the cyclic voltammetry measured at a rate of 10 mV s^{-1} and from the charging and discharge cycles that the prepared AAB exhibited stable electrochemical reactions. Moreover, it was found that a thin layer was formed over the air cathode, but it did not contribute to hindering electrochemical reactions.

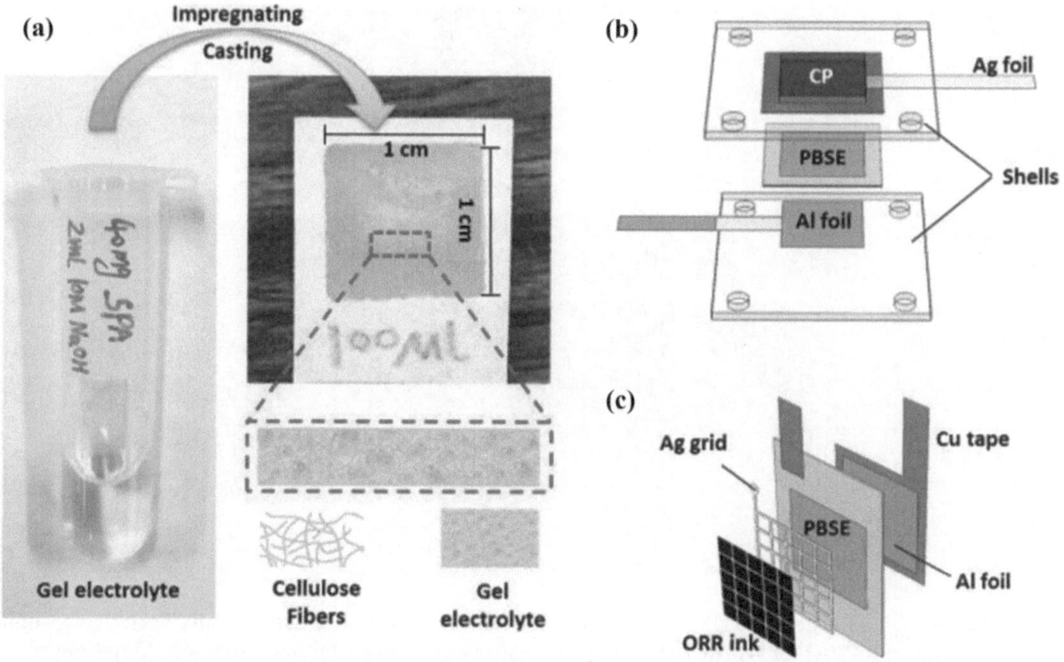

FIGURE 21.5 Schematic representation of (a) fabrication process of PBSE and (b–c) rechargeable AAB components. Adapted and reproduced with permission from Ref. [33]. Copyright © 2019 Elsevier.

Wang et al. [33] fabricated paper-based solid electrolyte (PBSE) for a mechanical rechargeable AAB operating in miniwatt range applications. The gel electrolyte was prepared by dissolving SPA and NaOH and after one day casting the gel over filter paper (cellulose paper) and then sandwiching it between plates of poly(methyl methacrylate), followed by the evaporation of water to form a solid state. Figure 21.5 is the schematic representation of the fabrication process of PBSE and the components of AAB. The AAB showed an OCV corresponding to 1.5 V and Al specific capacity of 900.8 mAh g^{-1} with 3.8 mW cm^{-2} as a peak power density. Hence, the flexible PBSE-AAB could deliver a good performance and can be suited for miniwatt range applications.

Gaele et al. [34] investigated to study various electrolytes for AAB. An aluminium air cell was fabricated with the membrane of PVA treated with HCl (PVA–HCl) as anolyte and PVA with xanthan gum, followed by mixing with saturated KCl (XaKCl), was employed as catholyte and ultrapure aluminium as an anode. Figure 21.6 shows the flexible PVA-based membranes: PVA–HCl and XaKCl. Platinum-containing carbon cloth was used as the air cathode catalyst with polyvinylidene fluoride as a binder. XaKCl was placed to improve the cathode–electrolyte interface, which made the system a dual electrolyte (DE) type configuration. The electrolyte displayed an ionic conductivity of 2×10^{-3} S cm^{-1}. By the DE technique, the discharge performance of the cell showed enhancement with a possibility of recharging the cell.

21.2.3 Alkaline Electrolyte

Alkaline electrolytes belong to the category of aqueous electrolytes, where the bifurcation of alkaline, neutral, and acidic solutions is based on the pH of the system. Alkaline electrolytes are one of the most described systems among aqueous electrolytes, mainly due to the higher achievable power and theoretical cell potential. NaOH and KOH are the common alkaline electrolyte media reported. Alkaline media catalyses the corrosion rate of Al, leading to the evolution of hydrogen at more positive potentials which impedes the rechargeability [35,36]. Moreover, the formation of a passivation

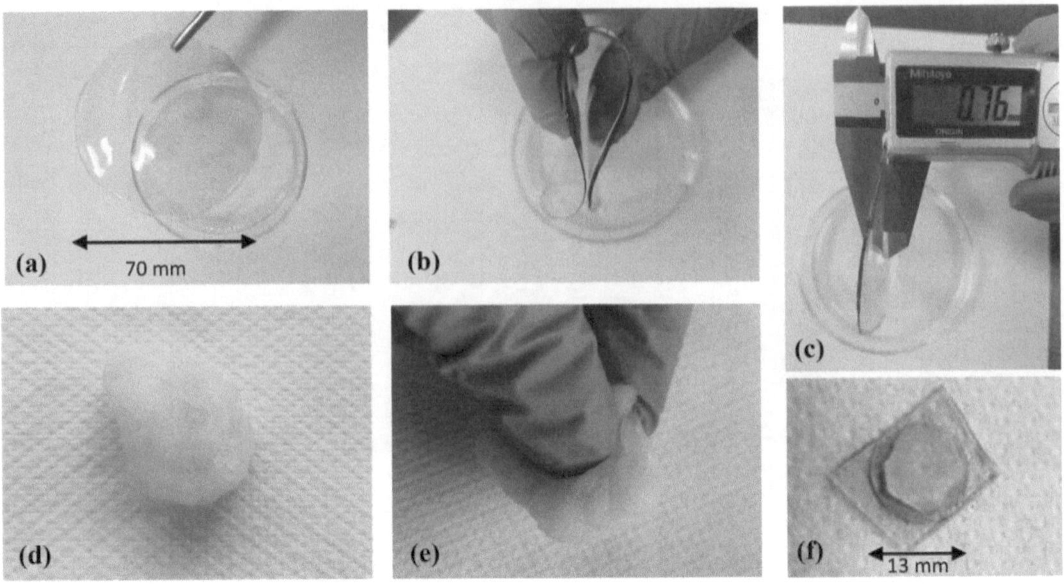

FIGURE 21.6 (a–c) Flexible and slight yellowish membrane of PVA–HCl and (d–f) XaKCl gel, (f) a 13-mm diameter anode with PVAHCl | XaKCl. Adapted and reproduced with permission from Ref. [34]. Copyright © 2021 Springer Nature.

layer consisting of oxide and hydroxide leads to high overpotentials for anodic dissolution. Alkaline electrolyte experiences carbonate precipitation mainly by the side reaction of carbon dioxide and hydroxide. Electrode pores are clogged by the insoluble carbonates [9].

$$\text{Side reaction}: CO_2 + OH^- \rightarrow CO_3^{2-} + H_2O \tag{21.1}$$

Various approaches have been evaluated to mitigate the aforementioned challenges, of which electrolyte additives and corrosion inhibitors are the most reported along with alloy Al anode, in alkaline electrolyte systems [37]. Wang et al. [38] incorporated L-cysteine and cerium nitrate as electrolyte additives for an alkaline AAB and found a synergistic corrosion inhibition in the Al alloy anode (AA5052) by the formation of a protective film over the alloy, which further prevented the corrosion. The rare earth complex compounds were mixed with the 4 M NaOH electrolyte. Figure 21.7 shows the morphological image of Al alloy with and without additives immersed in 4 M NaOH electrolyte, showing that the combination of additives forms a protective film over the alloy (Figure 21.7d). Moreover, it was found that by the combination of L-cysteine and cerium nitrate, polarization resistance was enhanced, and 20 mM L-cysteine and 2 mM cerium nitrate mixture showed 76% inhibition efficiency. Infrared spectra of the 20 mM L-cysteine and 2 mM cerium nitrate mixture confirm the new bond formation, which helped in retarding the corrosion of the alloy in the alkaline electrolyte. Hence, the work concluded that by adding organic rare earth compound mixtures to the electrolyte, a decrease in the corrosion of the Al alloy anode can be attained.

Wang et al. [39] conducted another study where among the various dicarboxylic acids such as succinic acid (SUA), adipic acid (ADA), and sebacic acid (SEA) as electrolyte additives in the alkaline ethylene glycol solution for AA5052 AAB, ADA showed best corrosion inhibition towards the Al alloy anode. Figure 21.8 shows the SEM images of Al alloy immersed in 4 M NaOH-ethylene glycol electrolyte for 1 h with and without dicarboxylic acid additives. It can be seen in Figure 21.9a that in 1 mM ADA, a uniform smooth film formed over the Al alloy. In addition, from the infrared spectra, it was observed that the formation of a complex film of ADA-Al^{3+} (due to the observed bond interaction between carboxylic acid groups of ADA and Al^{3+}) contributed to better corrosion

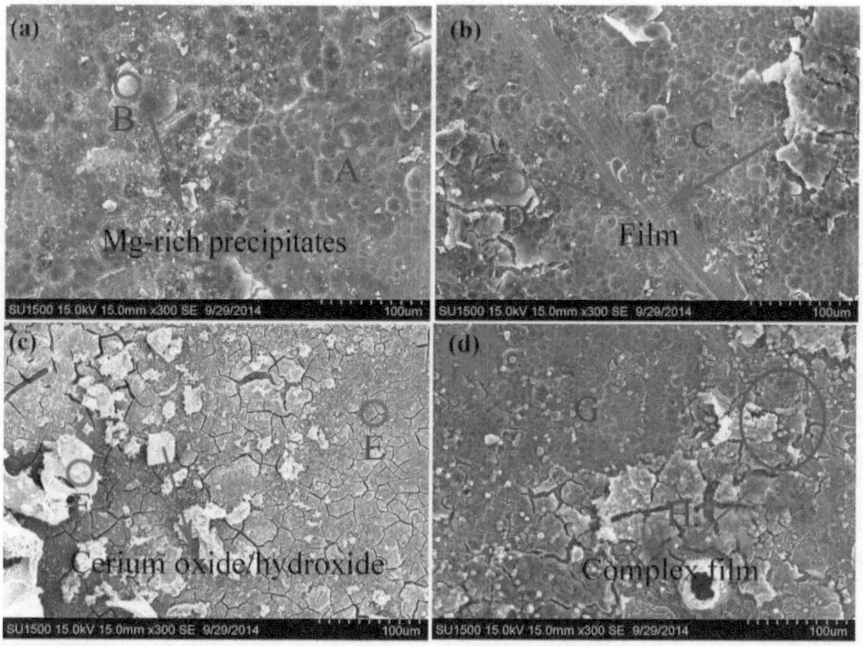

FIGURE 21.7 SEM images of Al alloy immersed in 4 M NaOH electrolyte for 1 h with and without organic rare earth additives: (a) without additive; (b) sample with 20 mM L-cysteine; (c) sample with 2 mM cerium nitrate; (d) sample with 20 mM L-cysteine and 2 mM cerium nitrate. Adapted and reproduced with permission from Ref. [38]. Copyright © 2015 Elsevier.

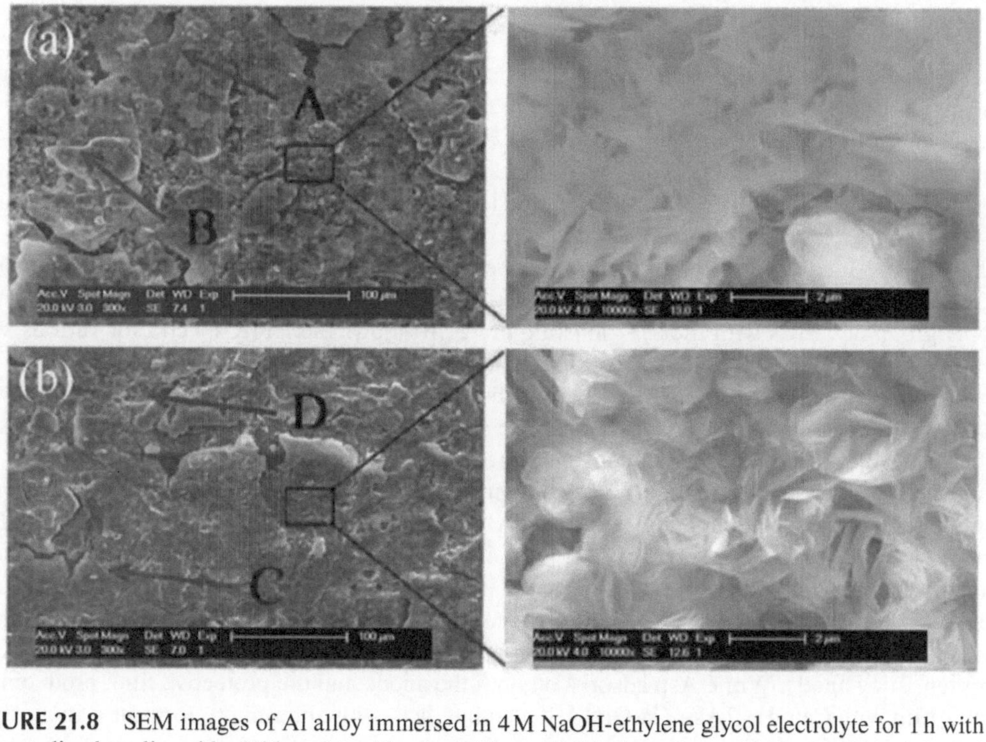

FIGURE 21.8 SEM images of Al alloy immersed in 4 M NaOH-ethylene glycol electrolyte for 1 h with and without dicarboxylic acid additives (a) without additive; and (b) electrolyte with 1 mM SUA. Adapted and reproduced with permission from Ref. [39]. Copyright © 2015 Elsevier.

FIGURE 21.9 SEM images of Al alloy immersed in 4 M NaOH-ethylene glycol electrolyte for 1 h with and without dicarboxylic acid additives: (a) electrolyte with 1 mM ADA; and (b) electrolyte with 1 mM SEA. Adapted and reproduced with permission from Ref. [39]. Copyright © 2015 Elsevier.

protection. Also, from the quantum energy calculations, ADA with two carboxylic acid end groups has a smaller energy gap, confirming its ability to form complex cyclic structures with Al^{3+}.

Liu et al. [40] worked on preventing corrosion of the AA5052 alloy anode for an alkaline AAB by incorporating electrolyte additives. It was evidenced from the hydrogen evolution test that incorporating 10 g L^{-1} CMC and 8 g L^{-1} ZnO was found to have a synergistic effect on retarding corrosion of AA5052 alloy anode in 4 M NaOH.

The FTIR results taken from a film formed over the surface of AA5052 alloy after immersing in 4 M NaOH containing ZnO and CMC showed the coordination bond developed between the –COOH group of CMCs with the Zn^{2+} ions. Figure 21.10 shows the FTIR spectra for various systems. It can be observed that the peak of CMC molecules at 1,620 cm^{-1} shifted to 1,555 cm^{-1} with a narrow band at 3,600–2,800 cm^{-1}, indicating the formation of the complex coordination bond. Moreover, CMC with a cyclic polymeric structure adsorbs on the surface of Al and interacts easily with the Zn^{2+} ions, protecting the anode from corrosion.

Kang et al. [41] performed a study where calcium oxide (CaO) and L-aspartic acid (Asp) as corrosion inhibitors were incorporated into alkaline AAB based on AA5052 aluminium alloy anode. The inhibitors were added as electrolyte additives in 4 M NaOH. Maximum inhibition with enhancement in discharge performance was shown by the incorporation of 10 mM CaO and 4 mM Asp into the electrolyte. Per the theoretical and experimental studies, the formation of calcium hydroxide by the reaction of CaO with the alkaline electrolyte produces a protective film over the AA5052 aluminium alloy anode. While Asp adsorbs on both the anode and the protective film, producing a complex film with $Ca(OH)_2$ (Asp-$Ca(OH)_2$), it also interacts with aluminium, forming a monolayer

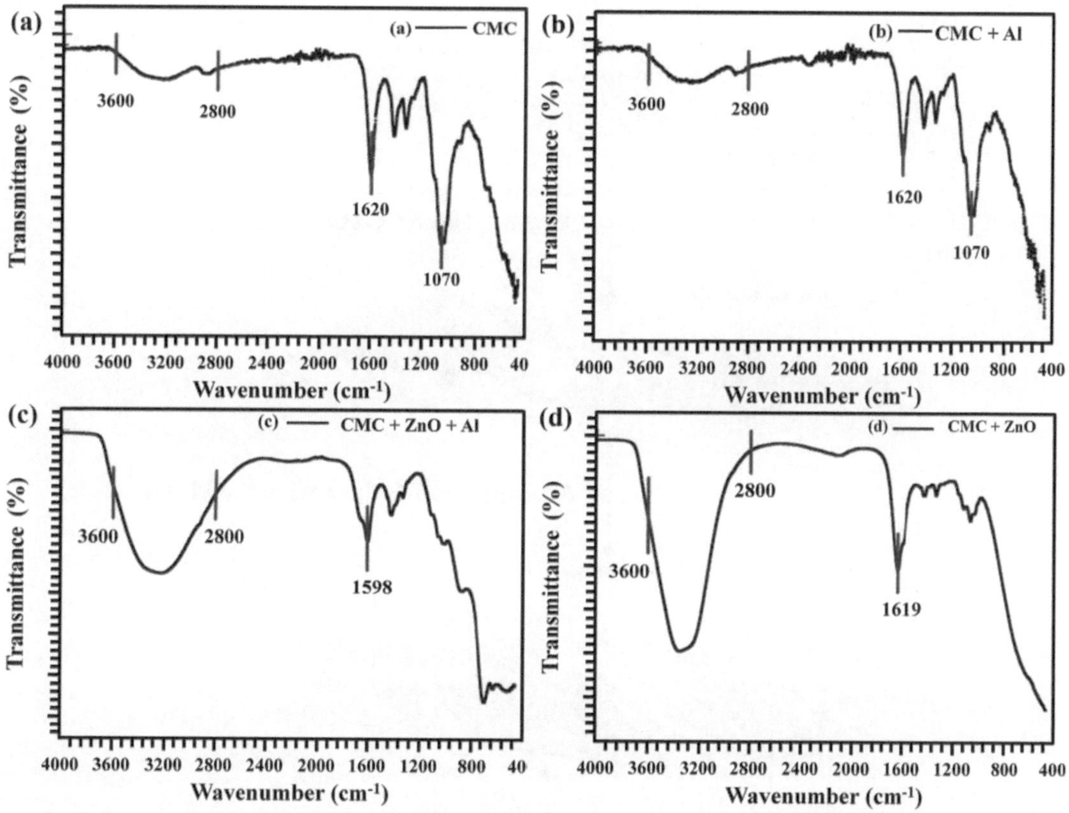

FIGURE 21.10 FTIR spectra of various electrolyte additives obtained after contact of A5052 alloy in 1 h of immersing in 4 M NaOH: (a) pure CMC; (b)10 g L^{-1} CMC; (c) 8 g/L ZnO and 10 g L^{-1} CMC; and (d) spectra of the film stripped from the surface of AA5052 alloy immersed in 4 M NaOH containing 8 g L^{-1} ZnO and 10 g L^{-1} CMC. Adapted and reproduced with permission from Ref. [40]. Copyright © 2016 Elsevier.

composite film (Al-Asp), which further protects the anode by preventing dissolution. Figure 21.11 shows the mechanism of corrosion inhibition by the additives.

Theoretical guidance unveils the complexities of electrochemical mechanisms in batteries; hence, reliance on correlating theory and experimental observation can provide better insights [42]. Therefore, with these results, Zhou et al. [43] studied the effects of HER in AAB using experimental and mathematical modelling of cell performance. The HER was measured in various concentrations of KOH. Figure 21.12 represents the working of AAB and the reactions occurring. The reaction of the anode with OH^- produces Al $(OH)_4^-$ and is transported towards the air cathode. In the cathode, the reaction between water and oxygen produces OH^-, which is then transported to the anode. In addition to this, the parasitic reactions at the anode side contribute to the evolution of H_2. The study showed that parasitic side reactions involving the evolution of H_2 and bubble shielding on the electrode contribute mainly to the kinetic loss in the case of aqueous AAB, which can be controlled by the concentration of OH^- and the electrolyte velocity. Moreover, to minimize the parasitic side reactions, high current operation of the cell is recommended.

Wu et al. [44] reported that a hybrid corrosion inhibitor consisting of alkyl polyglucoside (APG) and potassium stannate (K_2SnO_3) could protect the aluminium anode and deliver an inhibition efficiency of 94.14% in alkaline AAB. The cell fabricated with hybrid inhibitors showed a cell voltage of 1.3 V with 2.5 times more discharge mass-specific capacity compared to one without additives.

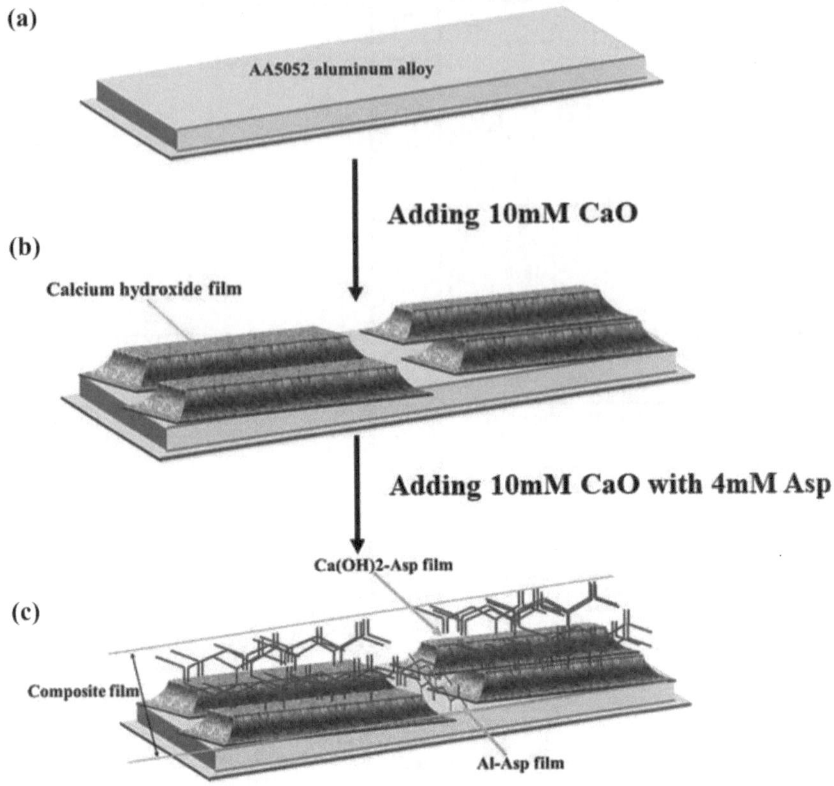

FIGURE 21.11 Corrosion inhibition mechanism by the incorporation of calcium oxide and L-aspartic. (a) AA5052 aluminium alloy surface; (b) formation of $Ca(OH)_2$ film; (c) formation of complex films, Asp-$Ca(OH)_2$ and Al-Asp. Adapted and reproduced with permission from Ref. [41]. Copyright © 2019 Elsevier.

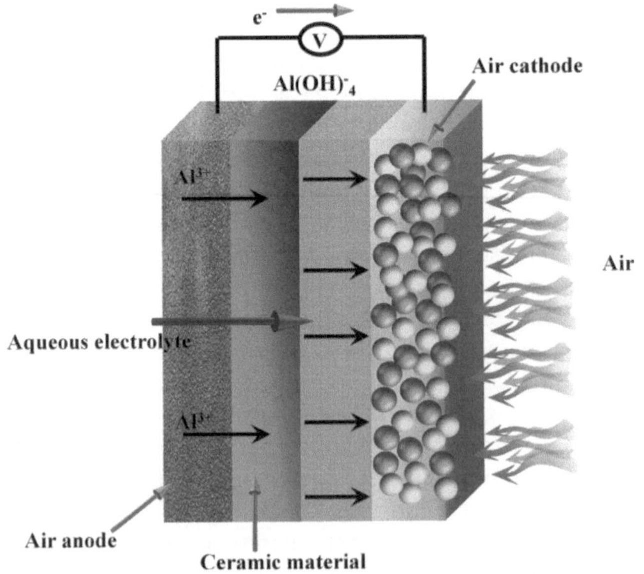

FIGURE 21.12 Illustration of AAB with a cell height of 17 cm and 0.2 cm as cell gap. The H_2 evolution reaction at the aluminium anode is represented in red colour. Air cathode is based on steam-activated carbon with 5% cobalt tetamethoxyphenly porphyrin. Adapted and reproduced with permission from Ref. [43]. Copyright © 2019 Elsevier.

APG reduced the stannate and hence enhanced the interface by forming a protective layer over the aluminium anode. The protective layer further inhibited the corrosion by preventing self-discharge. Hence, the work opened up further research in exploring synergistic hybrid corrosion inhibitors for anode protection in AABs.

To mitigate the issue of self-corrosion in the anode for AAB, Yang et al. [45] incorporated zinc sulphate ($ZnSO_4$) and sodium alginate (SA) as alkaline electrolyte (4 M NaOH) additives. From the hydrogen evolution test and EIS, it was evidenced that the cathodic inhibitor, $ZnSO_4$, contributes positively to decrease the rate of corrosion by enhancing discharge performance and preventing cathodic reactions. Moreover, compared to the corrosion inhibition efficiency of $ZnSO_4$ and SA, the hybrid $ZnSO_4$/SA additive outperformed. By adding $ZnSO_4$ to 4 M NaOH electrolyte, a protective zinc layer was formed, which suppressed the HER but contributed negatively to the electrochemical activity. When the hybrid $ZnSO_4$/SA additive was added to 4 M NaOH, due to the stabilization of the zinc layer formed by the action of adsorbed SA, it contributed to the enhanced corrosion inhibition efficiency with an enhancement in discharge capacity and voltage. Figure 21.13 shows the proposed mechanism on the effect of $ZnSO_4$ and SA additives on decreasing the rate of corrosion, thereby enhancing the discharge performance of AAB. Compared to the alkaline blank electrolyte, the electrolyte with 10 mM of $ZnSO_4$ in 1 g L^{-1} SA showed an enhanced discharge capacity of 64.6%. The work underlined the need for a better selection of corrosion inhibitors for reducing self-corrosion while enhancing discharge performance in AAB.

21.2.4 Ionic Liquids

The usage of room-temperature IL as an electrolyte for AAB has seen an enhancement in electrochemical performance. The advantages of using IL as an electrolyte in AAB include a larger electrochemical window, lower vapour pressure both at high temperature and normal temperature, non-flammability and safety, good ionic conductivity, and thermal and chemical stability [46]. As ILs have shown resistance to corrosion, decrement in energy density is the main challenge faced by the system [47]. Gelman et al. [47] were the first to report a non-aqueous AAB with more than 70% of theoretical Al capacity, based on 1-ethyl-3-methylimidazolium oligo-fluorohydrogenate (EMIm(HF)2.3F) as an IL electrolyte. The work reported the novel AAB with high energy densities of 2,300 Wh Kg^{-1} and 6,200 Wh L^{-1} with a low corrosion rate of Al anode with IL. Moreover, the electrolyte exhibited high environmental stability with changes in humidity and exposure to the atmosphere.

Following the work, a detailed investigation of the performance of EMIm(HF)2.3F hydrophilic IL for non-aqueous primary AAB was performed by Gelman et al. [46]. Electrolyte interaction with aluminium anode resulted in the formation of a protective layer (Al–O–F), which alleviates corrosion and contributes positively towards activation. It was observed that by the Al–O–F interaction, EMIm(HF)2.3F could dissolve the aluminium oxide layer formed during the electrochemical reaction in the cell. Moreover, the work pointed out that the dip in discharge is mainly due to the poor wetting of the air cathode before discharge; hence, wetting the cathode for more than 16 h could rectify the problem. The work also pointed out the need for optimizing the cathode; the capacity of the cell could also be affected by the functionality of the cathode.

Revel et al. [48] successfully fabricated non-aqueous AAB using aluminium chloride/1-ethyl-3-methylimidazolium chloride ($AlCl_3$/EMImCl) as an IL electrolyte, which resulted in a high current density and low self-discharge rate. The drawback was on the air cathode, which generated overpotential than the potential stability of electrolyte.

Deyab et al. [49] used (1-allyl-3-methylimidazolium bis(trifluoromethylsulphonyl)imide) (IL) as an additive to arrest self-corrosion in an alkaline AAB. The incorporation of 1.5 mM of IL to 4 M NaOH electrolyte showed the best result with a capacity density of 2,554 mAh g^{-1}. Hydrogen evolution inhibition efficiency calculated from hydrogen evolution rate measurement showed that 1.5 mM IL incorporated with 4 M NaOH electrolyte showed a value of 94.28%, showing its ability

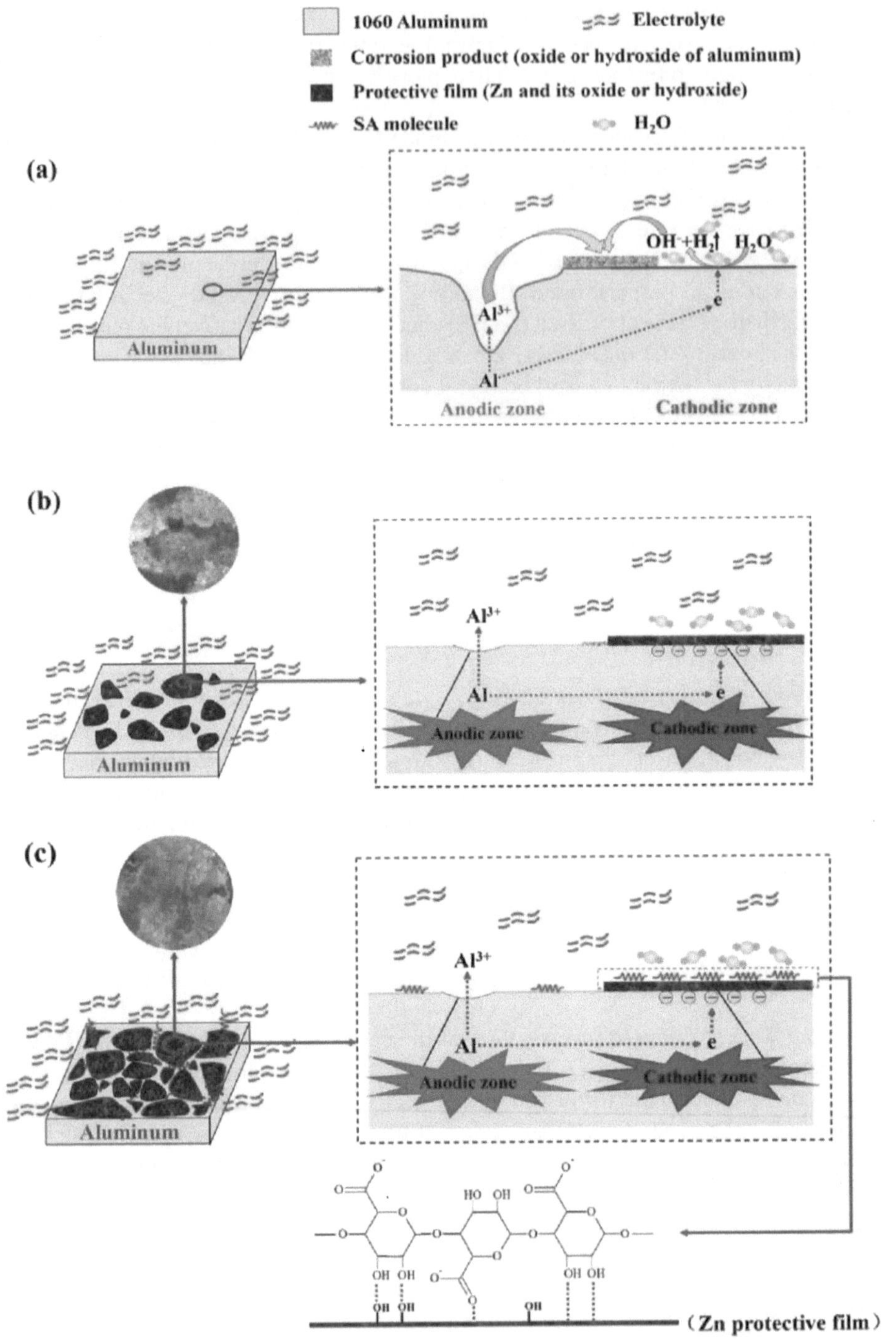

FIGURE 21.13 Proposed mechanism on the effect of $ZnSO_4$ and SA additives contributing to decrease the rate of corrosion, hence enhancing the discharge performance of AAB:. (a) without any additives; (b) $ZnSO_4$ as additive; and (c) $ZnSO_4$/SA as hybrid additive. Adapted and reproduced with permission from Ref. [45]. Copyright © 2021 Elsevier.

to resist corrosion. Moreover, corrosion inhibition efficiency from polarization measurements and electrostatic force microscopy measurements supports the observation that IL incorporated mitigated parasitic reaction and hence inhibited corrosion.

Levy et al. [50] used tetrabutylammonium dihydrogen trifluoride ($TBAH_2F_3$) as an IL electrolyte for AAB, which resulted in a high cell capacity, which is approximately 70% of the theoretical value. The presence of fluoro-hydrogenated ions ($H_2F_3^-$) was evidenced from the FTIR analysis, which was responsible for the activation of the Al surface leading to high anodic currents. The formation of $Al_xO_yF_z$ on the surface of aluminium anode acted as a protective layer against corrosion. For the enhancement of ionic conductivity, propylene carbonate was incorporated, making it a hybrid electrolyte, which led to a decrease in viscosity; hence, enhancement was shown in ionic conductivity. By EIS measurements it was evidenced that for ensuring high discharge potentials, proper wetting of the cathode by electrolyte is needed. The work concluded that desirable electrochemical properties can be achieved by the proper selection of hybrid electrolyte systems with IL.

Welch et al. [51] reported the usage of non-aqueous IL [EmIm]Cl with $AlCl_3$ as an electrolyte for AAB, to prevent parasitic side reactions and inert oxide film formation over the electrodes. The study showed that by the optimum chloride ion in the electrolyte, the surface tension decreases and hence the oxide film formation can be controlled. The electrochemical behaviour of AAB was also studied using a physics-based modelling method. Moreover, the work showed that the oxygen solubility in electrolyte and its diffusion to porous air cathode is a factor which has a definite role in the performance of the cell.

Xia et al. [52] reported the usage of IL based on urea as an electrolyte for AAB, where the electrocatalyst consisted of Co-MnO_2/C with $AlCl_3$. Enhancement in the surface area was achieved by the intercalation of Co with MnO_2. Cells with 40 mole% of Co to Mn showed reversible charge and discharge of 2 V during 30 cycles.

21.2.5 Saltwater Electrolytes

In 2008, Nestoridi et al. [53] reported a detailed study on the effect of the addition of Mg, Sn, and Ga to Al anode in 2 M brine solution and correlated microstructure on the electrochemical performance of AAB. In addition to the composition of the anode, mechanical and heat treatment showed an influence on the electrochemical performance. Mori et al. [54] communicated $Al_2(WO_4)_3$ Al^{3+} conductor suitable for AAB with saltwater as an electrolyte. The ceramic ion conductor prevented self-corrosion of Al anode, and the cell worked for 1 month. The electrochemical properties were enhanced due to the introduction of conductive carbon and $Al_2(WO_4)_3$ as the air cathode material. By the usage of NaCl as an electrolyte, instead of $Al(OH)_3$ and Al_2O_3, the products detected were sodium salts of aluminium, $Na_{11}WO_3$, etc.

Pino et al. [55] reported AAB with carbon-treated commercial Al alloy and Al1085 and Al7475 as anode for AABs with NaCl as an electrolyte, and the cell showed enhanced discharge by three times compared to the untreated alloy anode AAB. Of the various carbonaceous materials used for the treatment, namely, carbon black (CB), grapheme, and pyrolytic graphite, superior properties were shown by CB. The electrolyte used in the study consists of a 2 M NaCl and a commercial E4A air cathode containing MnO_2 as the catalyst. Figure 21.14 shows a comparative study on the discharge plots of AAB with and without CB-treated Al alloy at various current densities. Moreover, a 3D-printed 4s1p AAB was fabricated using CB-treated Al alloy, and it achieved a discharge plateau of 2 V performed at 5 mA cm^{-2} current density for 18 h and a specific capacity of 1,200 mAh g^{-1}.

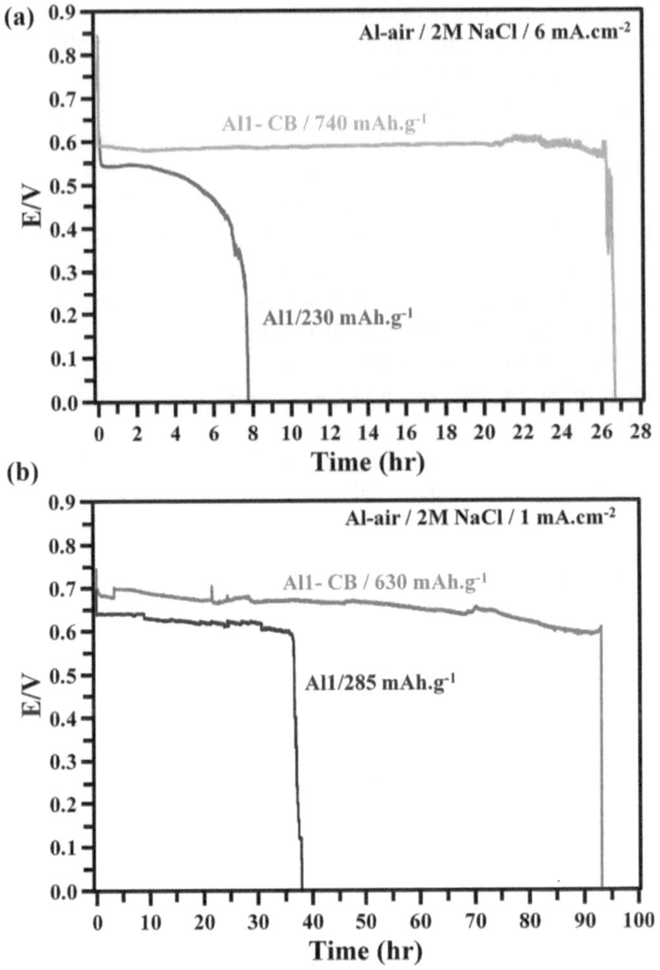

FIGURE 21.14 Comparative graph of discharge plots of AAB with and without CB-treated Al alloy at various current densities: (a) 6 mA cm^{-2} and (b) 1 mA cm^{-2}. Adapted and reproduced with permission from Ref. [55]. Copyright © 2016 Elsevier.

21.3 CONCLUSION

Among batteries for the post-LIB era, AABs have proven their candidature as they provide high theoretical energy density, high electrochemical equivalence, and a larger abundance of materials employed in the fabrication of AAB. Considering commercialization, developments in AAB are impeded by its inherent challenges. Self-corrosive nature of Al, even in open circuit and discharge conditions, formation of oxide and hydroxide barriers, HER, clogging of electrode pores by insoluble carbonates, susceptibility towards self-discharge, etc. are the main challenges. Alloy anodes and morphological modification of the anode structure have enhanced the properties. In addition to anode modifications, the electrolyte engineering approach has shown promising results. Employing electrolyte additives and corrosion inhibitors has been explored. Furthermore, developments of non-aqueous electrolytes such as ILs and solid-state and semi-solid-state electrolytes showed the inhibition of corrosion and rectified the issues of electrolyte leakage and self-discharge. Application-based developments in AAB have focussed on solid, semi-solid, and hydrogel-based electrolytes, thereby achieving the realization of wearable devices. Therefore, by judicious selection of electrolytes and additives with modification of the anode by alloying and engineering the air cathode, high-energy-density AABs can be developed.

REFERENCES

1. Sungjemmenla J, Soni CB, Vineeth SK, Kumar V (2021) Unveiling the physiochemical aspects of the matrix in improving sulfur-loading for room-temperature sodium-sulfur batteries. *Mater Adv* 2:4165–4189. https://doi.org/10.1039/D1MA00247C
2. Jishnu NS, Vineeth SK, Das A, et al (2021) Electrospun PVdF and PVdF-co-HFP-based blend polymer electrolytes for lithium ion batteries. In: Balakrishnan NTM, Prasanth R (eds) *Electrospinning for Advanced Energy Storage Applications*, 1st ed. Springer, Singapore, pp. 201–234
3. Deng J, Bae C, Denlinger A, Miller T (2020) Electric vehicles batteries: requirements and challenges. *Joule* 4:511–515. https://doi.org/10.1016/j.joule.2020.01.013
4. Miao Y, Hynan P, von Jouanne A, Yokochi A (2019) Current Li-ion battery technologies in electric vehicles and opportunities for advancements. *Energies* 12:1074. https://doi.org/10.3390/en12061074
5. Duffner F, Kronemeyer N, Tübke J, et al (2021) Post-lithium-ion battery cell production and its compatibility with lithium-ion cell production infrastructure. *Nat Energy* 6:123–134. https://doi.org/10.1038/s41560-020-00748-8
6. Li Y, Lu J (2017) Metal-air batteries: will they be the future electrochemical energy storage device of choice? *ACS Energy Lett* 2:1370–1377. https://doi.org/10.1021/acsenergylett.7b00119
7. Vineeth SK, Tebyetekerwa M, Liu H, Soni CB, Zhao XS, Kumar V (2022) Progress in the development of solid-state electrolytes for reversible room-temperature sodium–sulfur batteries. *Mat Adv* 3(16):6415–6440. https://doi.org/10.1039/D2MA00428C
8. Chawla N (2019) Recent advances in air-battery chemistries. *Mater Today Chem* 12:324–331. https://doi.org/10.1016/j.mtchem.2019.03.006
9. Liu Y, Sun Q, Li W, et al (2017) A comprehensive review on recent progress in aluminum-air batteries. *Green Energy Environ* 2:246–277. https://doi.org/10.1016/j.gee.2017.06.006
10. Elia GA, Kravchyk KV, Kovalenko MV, et al (2021) An overview and prospective on Al and Al-ion battery technologies. *J Power Sources* 481:228870. https://doi.org/10.1016/j.jpowsour.2020.228870
11. Wang Q, Miao H, Xue Y, et al (2017) Performances of an Al-0.15 Bi-0.15 Pb-0.035 Ga alloy as an anode for Al-air batteries in neutral and alkaline electrolytes. *RSC Adv* 7:25838–25847. https://doi.org/10.1039/c7ra02918g
12. Goel P, Dobhal D, Sharma RC (2020) Aluminum-air batteries: a viability review. *J Energy Storage* 28:101287. https://doi.org/10.1016/j.est.2020.101287
13. Mokhtar M, Talib MZM, Majlan EH, et al (2015) Recent developments in materials for aluminum-air batteries: a review. *J Ind Eng Chem* 32:1–20. https://doi.org/10.1016/j.jiec.2015.08.004
14. Vineeth SK, Soni CB, Sun Y, et al (2022) Implications of Na-ion solvation on Na anode-electrolyte interphase. *Trends Chem* 4:48–59. https://doi.org/10.1016/j.trechm.2021.11.002
15. Prasanth R, Shubha N, Hng HH, Srinivasan M (2014) Effect of poly(ethylene oxide) on ionic conductivity and electrochemical properties of poly(vinylidenefluoride) based polymer gel electrolytes prepared by electrospinning for lithium ion batteries. *J Power Sources* 245:283–291. https://doi.org/10.1016/j.jpowsour.2013.05.178
16. Mori R (2020) Recent developments for aluminum-air batteries. *Electrochem Energy Rev* 3:344–369. https://doi.org/10.1007/s41918-020-00065-4
17. Vineeth SK, Soni CB, Sanjaykumar C, Yamauchi Y, Han M, Kumar V (2023) A quasi-solid state polymer electrolyte for high-rate and long-life sodium-metal batteries. *J Energy Storage* 73:108780. https://doi.org/10.1016/j.est.2023.108780
18. Egan DR, Ponce De León C, Wood RJK, et al (2013) Developments in electrode materials and electrolytes for aluminium-air batteries. *J Power Sources* 236:293–310. https://doi.org/10.1016/j.jpowsour.2013.01.141
19. Gadhave RV, Vineeth SK, Dhawale PV, Gadekar PT (2020) Effect of boric acid on poly vinyl alcohol-tannin blend and its application as water-based wood adhesive. *Des Monomers Polym* 23:188–196. https://doi.org/10.1080/15685551.2020.1826124
20. Gadhave RV, Vineeth SK, Mahanwar PA, Gadekar PT (2021) Effect of addition of boric acid on thermo-mechanical properties of microcrystalline cellulose/polyvinyl alcohol blend and applicability as wood adhesive. *J Adhes Sci Technol* 35:1072–1086. https://doi.org/10.1080/01694243.2020.1832775
21. Wu F, Zhang K, Liu Y, et al (2020) Polymer electrolytes and interfaces toward solid-state batteries: recent advances and prospects. *Energy Storage Mater* 33:26–54. https://doi.org/10.1016/j.ensm.2020.08.002

22. Singh HK, Patil T, Vineeth SK, et al (2020) Isolation of microcrystalline cellulose from corn stover with emphasis on its constituents: corn cover and corn cob. *Mater Today Proc* 27:589–594. https://doi.org/10.1016/j.matpr.2019.12.065
23. Di Palma TM, Migliardini F, Caputo D, Corbo P (2017) Xanthan and κ-carrageenan based alkaline hydrogels as electrolytes for Al/air batteries. *Carbohydr Polym* 157:122–127. https://doi.org/10.1016/j.carbpol.2016.09.076
24. Di Palma TM, Migliardini F, Gaele MF, Corbo P (2019) Physically cross-linked xanthan hydrogels as solid electrolytes for Al/air batteries. *Ionics (Kiel)* 25:4209–4217. https://doi.org/10.1007/s11581-019-02965-y
25. Di Palma TM, Migliardini F, Gaele MF, Corbo P (2021) Aluminum-air batteries with solid hydrogel electrolytes: effect of pH upon cell performance. *Anal Lett* 54:28–39. https://doi.org/10.1080/00032719.2019.1708923
26. Liu Y, Sun Q, Yang X, et al (2018) High-performance and recyclable Al-air coin cells based on eco-friendly chitosan hydrogel membranes. *ACS Appl Mater Interfaces* 10:19730–19738. https://doi.org/10.1021/acsami.8b04974
27. Sun Y, Fu C, Ren J, et al (2020) Highly conductive and reusable electrolyte based on sodium polyacrylate composite for flexible Al-air batteries. *J Electrochem Soc* 167:080502. https://doi.org/10.1149/1945-7111/ab8802
28. Sun P, Chen J, Huang Y, et al (2021) High-strength agarose gel electrolyte enables long-endurance wearable Al-air batteries with greatly suppressed self-corrosion. *Energy Storage Mater* 34:427–435. https://doi.org/10.1016/j.ensm.2020.10.009
29. Migliardini F, Di Palma TM, Gaele MF, Corbo P (2018) Solid and acid electrolytes for Al-air batteries based on xanthan-HCl hydrogels. *J Solid State Electrochem* 22:2901–2916. https://doi.org/10.1007/s10008-018-4003-2
30. Hibino T, Kobayashi K, Nagao M (2013) An all-solid-state rechargeable aluminum-air battery with a hydroxide ion-conducting Sb(v)-doped SnP_2O_7 electrolyte. *J Mater Chem A* 1:14844–14848. https://doi.org/10.1039/c3ta12707a
31. Mokhtar M, Majlan EH, Ahmad A, et al (2018) Effect of ZnO filler on PVA-alkaline solid polymer electrolyte for aluminum-air battery applications. *J Electrochem Soc* 165:A2483–A2492. https://doi.org/10.1149/2.0381811jes
32. Mori R (2019) All solid state rechargeable aluminum-air battery with deep eutectic solvent based electrolyte and suppression of byproducts formation. *RSC Adv* 9:22220–22226. https://doi.org/10.1039/c9ra04567h
33. Wang Y, Pan W, Kwok H, et al (2019) Low-cost Al-air batteries with paper-based solid electrolyte. *Energy Procedia* 158:522–527. https://doi.org/10.1016/j.egypro.2019.01.146
34. Gaele MF, Migliardini F, Di Palma TM (2021) Dual solid electrolytes for aluminium-air batteries based on polyvinyl alcohol acidic membranes and neutral hydrogels. *J Solid State Electrochem* 25:1207–1216. https://doi.org/10.1007/s10008-021-04900-6
35. Deltombe E, Pourbaix M (1958) The electrochemical behavior of aluminum-potential pH diagram of the system Al-H_2O at 25C. *Corrosion* 14:16–20. https://doi.org/10.5006/0010-9312-14.11.16
36. Cho YJ, Park IJ, Lee HJ, Kim JG (2015) Aluminum anode for aluminum-air battery - part I: influence of aluminum purity. *J Power Sources* 277:370–378. https://doi.org/10.1016/j.jpowsour.2014.12.026
37. Buckingham R, Asset T, Atanassov P (2021) Aluminum-air batteries: a review of alloys, electrolytes and design. *J Power Sources* 498:229762. https://doi.org/10.1016/j.jpowsour.2021.229762
38. Wang D, Li H, Liu J, et al (2015) Evaluation of AA5052 alloy anode in alkaline electrolyte with organic rare-earth complex additives for aluminium-air batteries. *J Power Sources* 293:484–491. https://doi.org/10.1016/j.jpowsour.2015.05.104
39. Wang DP, Zhang DQ, Lee KY, Gao LX (2015) Performance of AA5052 alloy anode in alkaline ethylene glycol electrolyte with dicarboxylic acids additives for aluminium-air batteries. *J Power Sources* 297:464–471. https://doi.org/10.1016/j.jpowsour.2015.08.033
40. Liu J, Wang D, Zhang D, et al (2016) Synergistic effects of carboxymethyl cellulose and ZnO as alkaline electrolyte additives for aluminium anodes with a view towards Al-air batteries. *J Power Sources* 335:1–11. https://doi.org/10.1016/j.jpowsour.2016.09.060
41. Kang QX, Wang Y, Zhang XY (2019) Experimental and theoretical investigation on calcium oxide and L-aspartic as an effective hybrid inhibitor for aluminum-air batteries. *J Alloys Compd* 774:1069–1080. https://doi.org/10.1016/j.jallcom.2018.09.391
42. Eng AYS, Soni CB, Lum Y, et al (2022) Theory-guided experimental design in battery materials research. *Sci Adv* 8:eabm2422. https://doi.org/10.1126/sciadv.abm2422

43. Zhou C, Bhonge K, Cho KT (2020) Analysis of the effect of hydrogen-evolving side reaction in the aqueous aluminum-air battery. *Electrochim Acta* 330:135290. https://doi.org/10.1016/j.electacta.2019.135290
44. Wu S, Zhang Q, Sun D, et al (2020) Understanding the synergistic effect of alkyl polyglucoside and potassium stannate as advanced hybrid corrosion inhibitor for alkaline aluminum-air battery. *Chem Eng J* 383:123162. https://doi.org/10.1016/j.cej.2019.123162
45. Yang L, Wu Y, Chen S, et al (2021) A promising hybrid additive for enhancing the performance of alkaline aluminum-air batteries. *Mater Chem Phys* 257:123787. https://doi.org/10.1016/j.matchemphys.2020.123787
46. Gelman D, Shvartsev B, Wallwater I, et al (2017) An aluminum - ionic liquid interface sustaining a durable Al-air battery. *J Power Sources* 364:110–120. https://doi.org/10.1016/j.jpowsour.2017.08.014
47. Gelman D, Shvartsev B, Ein-Eli Y (2014) Aluminum-air battery based on an ionic liquid electrolyte. *J Mater Chem A* 2:20237–20242. https://doi.org/10.1039/c4ta04721d
48. Revel R, Audichon T, Gonzalez S (2014) Non-aqueous aluminium-air battery based on ionic liquid electrolyte. *J Power Sources* 272:415–421. https://doi.org/10.1016/j.jpowsour.2014.08.056
49. Deyab MA (2017) 1-Allyl-3-methylimidazolium bis(trifluoromethylsulfonyl)imide as an effective organic additive in aluminum-air battery. *Electrochim Acta* 244:178–183. https://doi.org/10.1016/j.electacta.2017.05.116
50. Levy NR, Lifshits S, Yohanan E, Ein-Eli Y (2020) Hybrid ionic liquid propylene carbonate-based electrolytes for aluminum-air batteries. *ACS Appl Energy Mater* 3:2585–2592. https://doi.org/10.1021/acsaem.9b02288
51. Welch C, Mohammad AK, Hosmane NS, et al (2020) Effect of aluminum oxide on the performance of ionic liquid-based aluminum-air battery. *Energies* 13:2014. https://doi.org/10.3390/en13082014
52. Xia Z, Zhu Y, Zhang W, et al (2020) Cobalt ion intercalated MnO_2/C as air cathode catalyst for rechargeable aluminum-air battery. *J Alloys Compd* 824:153950. https://doi.org/10.1016/j.jallcom.2020.153950
53. Nestoridi M, Pletcher D, Wood RJK, et al (2008) The study of aluminium anodes for high power density Al/air batteries with brine electrolytes. *J Power Sources* 178:445–455. https://doi.org/10.1016/j.jpowsour.2007.11.108
54. Mori R (2014) A novel aluminium-air secondary battery with long-term stability. *RSC Adv* 4:1982–1987. https://doi.org/10.1039/c3ra44659j
55. Pino M, Herranz D, Chacón J, et al (2016) Carbon treated commercial aluminium alloys as anodes for aluminium-air batteries in sodium chloride electrolyte. *J Power Sources* 326:296–302. https://doi.org/10.1016/j.jpowsour.2016.06.118

Index